# Advances in Intelligent and Soft Computing 110

**Editor-in-Chief: J. Kacprzyk**

## Advances in Intelligent and Soft Computing

**Editor-in-Chief**

Prof. Janusz Kacprzyk
Systems Research Institute
Polish Academy of Sciences
ul. Newelska 6
01-447 Warsaw
Poland
E-mail: kacprzyk@ibspan.waw.pl

---

Further volumes of this series can be found on our homepage: springer.com

Vol. 96. A. Gaspar-Cunha, R. Takahashi, G. Schaefer, and L. Costa (Eds.)
*Soft Computing in Industrial Applications, 2011*
ISBN 978-3-642-20504-0

Vol. 97. W. Zamojski, J. Kacprzyk, J. Mazurkiewicz, J. Sugier, and T. Walkowiak (Eds.)
*Dependable Computer Systems, 2011*
ISBN 978-3-642-21392-2

Vol. 98. Z.S. Hippe, J.L. Kulikowski, and T. Mroczek (Eds.)
*Human – Computer Systems Interaction: Backgrounds and Applications 2, 2011*
ISBN 978-3-642-23186-5

Vol. 99. Z.S. Hippe, J.L. Kulikowski, and T. Mroczek (Eds.)
*Human – Computer Systems Interaction: Backgrounds and Applications 2, 2011*
ISBN 978-3-642-23171-1

Vol. 100. S. Li, X. Wang, Y. Okazaki, J. Kawabe, T. Murofushi, and Li Guan (Eds.)
*Nonlinear Mathematics for Uncertainty and its Applications, 2011*
ISBN 978-3-642-22832-2

Vol. 101. D. Dicheva, Z. Markov, and E. Stefanova (Eds.)
*Third International Conference on Software, Services and Semantic Technologies S3T 2011, 2011*
ISBN 978-3-642-23162-9

Vol. 102. R.S. Choraś (Ed.)
*Image Processing and Communications Challenges 3, 2011*
ISBN 978-3-642-23153-7

Vol. 103. T. Czachórski, S. Kozielski, and U. Stańczyk (Eds.)
*Man-Machine Interactions 2, 2011*
ISBN 978-3-642-23168-1

Vol. 104. D. Jin and S. Lin (Eds.)
*Advances in Computer Science, Intelligent System and Environment, 2011*
ISBN 978-3-642-23776-8

Vol. 105. D. Jin and S. Lin (Eds.)
*Advances in Computer Science, Intelligent System and Environment, 2011*
ISBN 978-3-642-23755-3

Vol. 106. D. Jin and S. Lin (Eds.)
*Advances in Computer Science, Intelligent System and Environment, 2011*
ISBN 978-3-642-23752-2

Vol. 107. P. Melo-Pinto, P. Couto, C. Serôdio, J. Fodor, and B. De Baets (Eds.)
*Eurofuse 2011, 2011*
ISBN 978-3-642-24000-3

Vol. 108. Y. Wang (Ed.)
*Education and Educational Technology, 2011*
ISBN 978-3-642-24774-3

Vol. 109. Y. Wang (Ed.)
*Education Management, Education Theory and Education Application, 2011*
ISBN 978-3-642-24771-2

Vol. 110. L. Jiang (Ed.)
*Proceedings of the 2011 International Conference on Informatics, Cybernetics, and Computer Engineering (ICCE2011) November 19–20, 2011, Melbourne, Australia, 2011*
ISBN 978-3-642-25184-9

Liangzhong Jiang (Ed.)

# Proceedings of the 2011 International Conference on Informatics, Cybernetics, and Computer Engineering (ICCE2011) November 19–20, 2011, Melbourne, Australia

## Volume 1: Intelligent Control and Network Communication

Springer

**Editor**

Liangzhong Jiang
South China Normal University
Yifeng Dasha Building C 1707
Huang Pu Da Dao Zhong
Cui Yuan Jie 8
Tian He Qu
Guangzhou, 510000
China
E-mail: jiangliangzhong@126.com

ISBN 978-3-642-25184-9 e-ISBN 978-3-642-25185-6

DOI 10.1007/978-3-642-25185-6

Advances in Intelligent and Soft Computing ISSN 1867-5662

Library of Congress Control Number: 2011940773



*Typeset* by Scientific Publishing Services Pvt. Ltd., Chennai, India

Printed on acid-free paper

5 4 3 2 1 0

springer.com

# Preface

2011 International Conference on Informatics, Cybernetics, and Computer Engineering (ICCE2011) will be held in November 19-20, 2011, Melbourne, Australia.

Informatics is the science of information, the practice of information processing, and the engineering of information systems. Informatics studies the structure, algorithms, behavior, and interactions of natural and artificial systems that store, process, access and communicate information. It also develops its own conceptual and theoretical foundations and utilizes foundations developed in other fields. Since the advent of computers, individuals and organizations increasingly process information digitally. This has led to the study of informatics that has computational, cognitive and social aspects, including study of the social impact of information technologies. Loosely, it can be thought of as "studying how to design a system that delivers the right information, to the right person in the right place and time, in the right way," and is intimately tied to workflow discussions and standards.

Cybernetics is the interdisciplinary study of the structure of regulatory systems. Cybernetics is closely related to information theory, control theory and systems theory, at least in its first-order form. (Second-order cybernetics has crucial methodological and epistemological implications that are fundamental to the field as a whole.) Both in its origins and in its evolution in the second half of the 20th century, cybernetics is equally applicable to physical and social (that is, language-based) systems.

Contemporary cybernetics began as an interdisciplinary study connecting the fields of control systems, electrical network theory, mechanical engineering, logic modeling, evolutionary biology, neuroscience, anthropology, and psychology in the 1940s, often attributed to the Macy Conferences.

Other fields of study which have influenced or been influenced by cybernetics include game theory, system theory (a mathematical counterpart to cybernetics), perceptual control theory, sociology, psychology (especially neuropsychology, behavioral psychology, cognitive psychology), philosophy, and architecture and organizational theory.

Computer engineering, also called computer systems engineering, is a discipline that integrates several fields of electrical engineering and computer science required to develop computer systems. Computer engineers usually have training in electronic engineering, software design, and hardware-software integration instead of only software engineering or electronic engineering. Computer engineers are involved in many hardware and software aspects of computing, from the design of individual microprocessors, personal computers, and supercomputers, to circuit design. This field of engineering not only focuses on how computer systems themselves work, but also how they integrate into the larger picture.

Usual tasks involving computer engineers include writing software and firmware for embedded microcontrollers, designing VLSI chips, designing analog sensors, designing mixed signal circuit boards, and designing operating systems. Computer engineers are also suited for robotics research, which relies heavily on using digital systems to control and monitor electrical systems like motors, communications, and sensors.

International Conference on Informatics, Cybernetics, and Computer Engineering is the premier forum for the presentation of new advances and research results in the fields of Informatics, Cybernetics, and Computer Engineering. The conference will bring together leading researchers, engineers and scientists in the domain of interest from around the world. All submitted papers are written in English. Each paper is reviewed by 2-3 reviewers. Finally, there are 270 papers included in the Proceedings. Therefore, I think you will get lots of useful information from the Proceedings.

Special thanks goes to editors, staff of association and every participants of the conference. It's you make the conference a success. We look forward to meeting you next year.

# ICCE 2011 Organization Committee

## Honorary Chairs

| | |
|---|---|
| Wei Lee | Melbourne ACM Chapter, Australia |
| Junwu Zhu | University of Western Sydney, Australia |

## Program Chairs

| | |
|---|---|
| Qihai Zhou | Shanghai Jiaotong University, China |
| Alireza Yazdizadeh | International Science and Engineering Center, Hong Kong |

## Publication Chair

| | |
|---|---|
| Liangzhong Jiang | South China Normal University,China |

## International Committee

| | |
|---|---|
| Wei Li | Asia Pacific Human-Computer Interaction Research Center, Hong Kong |
| Xiaoming Liu | Nankai University, China |
| Xiaoxiao Yu | Wuhan University, China |
| Chi Zhang | Nanchang University, China |
| Bo Zhang | Beijing University, China |
| Lei Zhang | Tianjin Institute of Urban Construction, China |
| Mark Zhou | Hong Kong Education Society, Hong Kong |
| Liangming Zhang | Hong Kong Education Society, Hong Kong |
| Alireza Yazdizadeh | International Science and Engineering Center, Hong Kong |
| Wenjin Hu | Jiangxi Computer Study Institute, China |
| Qun Zeng | Nanchang University, China |
| Alireza Yazdizadeh | International Science and Engineering Center, Hong Kong |
| Junwu Zhou | University of Western Sydney, Australia |

# Table of Contents

## Intelligent Control and Network Communication

# A Trust Management Model Based on Reputation for Virtual Community

Shuqin Zhao, Yucui Guo, and Shang Lv

Beijing University of Posts and Telecommunications,
Beijing, P.R. China
{iamzhaoshuqin,yucuig}@163.com, shangat@sohu.com

**Abstract.** To improve the accuracy of trust value among members in virtual community(VC), a trust model based on reputation was proposed. Superior to the existing approaches, the activity degree and the length of the trust chain were taken into account and corresponding punishment mechanism was introduced, and a novel computational method was put forward. In addition, the model can motivate members to participate in activities. Simulations show that this model is effective and feasible to virtual communities.

**Keywords:** virtual community, trust management, reputation.

## 1 Introduction

With the development of information technology, virtual community becomes a promising component of the socialized network. Iresearch Company estimated that there were already over 50 million virtual communities in China by the end of 2007, and more than 50 percent people always participated in virtual communities[1]. Undoubtedly, virtual communities will play a much more important role in social life.

A virtual community is comprised of members who share a common interest, interact repeatedly, generate communal resources and carry on transactions and so on. Most information is produced and also consumed by the community members. The definition of the virtual community differs based on the context of varied purposes. On the whole, VC shows several characters as follow:

(1) Members with various backgrounds are anonymous and may be distinct in capability or behavior.

(2) The internal organization of VC is loose and coreless. Its management and organization depend on the community members themselves.

(3) There is little even no limitation for members to join in or leave, so the construction of a virtual community is unstable.

These attributes expose some security vulnerabilities including the problem of finding trustworthy communication partners, some members provide ineffective information, a member group may conspire to attack others and so on. So trust management is a necessary choice to identify trusted entities and isolated un-trusted ones. An effective trust management mechanism is crucial to the development of VC.

There is a large amount of research that focused on virtual community [2-9]. The update method of direct trust value and reputation is introduced according to the transaction results in [8]. The CCR(Cross-Community Reputation) model by sharing and combining reputation data from different communities is proposed in [9]. Most existing models don't give the method of calculating trust value.

A trust evaluation model based on the reputation and recommendation is put forward to improve the accuracy of trust value in this paper, which plays a critical role in identifying an appropriate member to service the requesting member. Some definitions and parameters are introduced to describe recommendation; different from other models, community activity degree of members and the length of recommendation chain are taken into account during the computing of trust value. And the validity of the model is proved by simulation experiments.

## 2 Some Definitions

Trust is generally divided into two parts: direct trust ($DT$) and recommendation trust ($RT$), and $DT$ is determined by the direct experience between two members, $RT$ is obtained from others' recommendation. In our model, all the members come from the same community. In order to describe the model, some definitions are presented first.
Known relationship: if member $i$ knows $j$, they are known relationship, noted as $Know(i,j)$.
Acquaintance relationship: if $i$ knows $j$ and $j$ knows $i$, and they have direct transaction experience, they are acquaintance relationship, noted as $Acq(i,j)$, i.e. $Acq(i,j) = Know(i,j) \Lambda Know(j,i)$.
Acquaintance relationship set: It is constituted of all $i$ 's acquaintances, noted as $Acq(i)$.
Recommendation relationship: The ordered pair $r < c_m, c_n >$ expresses that member $c_m$ gives some recommendation for $c_n$
Recommendation chain:
$e =< i, c_1, c_2, \ldots\ldots, c_n, j >$ is a recommendation chain from $i$ to $j$ if $\forall k = 1,2,\ldots\ldots,n, \exists Acq(c_k, c_{k+1})$ , $Acq(i, c_1)$ and $Acq(c_n, j)$ , and $DT(c_k, c_{k+1}) \le DT(c_{k+1}, c_{k+2})$, where $c_k\ (k = 1,2,\ldots\ldots, n)$ are recommenders.

For every member, four parameters are noted as:

$DT(i,j)$—$i's$ direct trust value about $j$

$RT(i,j)$—$i's$ recommendation trust value about $j$

$R_i$—$i's$ community reputation value

$A_i$—$i's$ community active degree value

and all $\in [0,1]$.

Community reputation value $R_i$ is derived from the feedback of other members who have direct transactions with $i$ during the period $\Delta t = [t_{current} - \Delta t, t_{current}]$:

$$R_i = \frac{\sum_{k \in Acq(i)} F_{ki} \times CR_k}{\sum_{k \in Acq(i)} CR_k} \tag{1}$$

$F_{ki}$ is the feedback from member $k$:

$F_{ki} = \frac{sRating_{ki}}{sRating_{ki} + fRating_{ki}}$ where $sRating_{ki}$ and $fRating_{ki}$ are the total rating of successful interactions and failed interactions of transaction $k$ with $i$ respectively, and $CR_k \in [0,1]$ is the credibility of member $k$.

Community active degree value $A_i$ [8] represents the active status of member $i$ in the community and it is a parameter related to the time,

$$A_i = e^{-A_f \times A_t} \tag{2}$$

$A_f$ is a decay factor and $0 < A_f < 1$. $A_t = t - t_{last}$, in which $t$ represents the current time and $t_{last}$ represents the time that the last direct transaction $i$ with others. If $i$ just supplies recommendation information, $A_f$ will decrease, which makes $A_i$ fade slower than before. If $i$ has a new transaction with others, $A_t$ will be reset to 0.

## 3 Model Design

Generally, direct trust plays a more important role than recommendation trust, but most members, such as a new member, may have only a little or even no interaction history with others. So it is necessary to consider recommendation trust.

### 3.1 Trust Value Computing

Let $T(i,j)$ denote $i's$ trust value about $j$, we have

$$T(i,j) = (aDT(i,j) + bRT(i,j)) \times A_j \tag{3}$$

in formula (3), $a+b=1$, and parameter $a$ symbolizes $i$ 's confidence in its own experience with $j$. Its value is related to the direct trust value about $j$:

$$a = \ln(1 + \frac{e-1}{DTH} \times \min(DT(i,j), DTH))$$

Where $DTH$ is a threshold, If $DT(i,j) \geq DTH$ , we have $a = 1$ , which means that $i$ has enough interaction experience with $j$ to evaluate its trust value about $j$ .

### 3.2 Recommendation Trust Value Computing

If all members in $Acq(i)$ don't have the resource $i$ needs, it must enquire to others until find certain member (may be more than one) prossessing it. A number of recommendation chains may be found according to some certain rules. Therefore we need to combine these chains to get the synthetic trust value to decide whether believe it (or who is believable).

*1) A single recommendation chain*

With regards to a single recommendation chain $e = < i, c_1, c_2, \ldots\ldots, c_n, j >$ , some measures are introduced to solve this problem in existing models, but most of them only consider the members' reputation. In our model, we consider both the reputation and the length of the trust chain:

$$RT(i,j,e_l) = \omega_i \times DT(i,c_1) + \sum_{k=1}^{n-1} \omega_k \times DT(c_k, c_{k+1}) + \omega_n \times DT(c_n, j) \tag{4}$$

where $RT(i,j,e_l)$ represents the integrated recommendation trust value of $e_l$ . $\omega_i$ and $\omega_k$ $(k = 1,2,\ldots\ldots,n)$ are weighting factors, computed as follow:

$$\omega_i = \frac{R_i^{\beta} A_i^{\gamma}}{\sum_{s=0}^{n} \left(\frac{1}{s+1}\right)^{\alpha} R_s^{\beta} A_s^{\gamma}}$$

and $$\omega_k = \frac{\left(\frac{1}{k+1}\right)^{\alpha} R_k^{\beta} A_k^{\gamma}}{\sum_{s=0}^{n} \left(\frac{1}{s+1}\right)^{\alpha} R_s^{\beta} A_s^{\gamma}} \qquad k = 1,2,\ldots\ldots,n$$

In which $\alpha, \beta, \gamma$ are parameters that control the relative importance of the length versus reputation and active degree value. Here, we extend the algorithm put forward in [10] which defines the transition probability from one town to another in AS(Ant System) to three parameters.

*2) Trust Chains Fusion*

The synthetic $RT$ of all recommendation chains is computed as below:

$$RT(i,j) = \sum_{l=1}^{m} \omega_l \times RT(i,j,e_l) \tag{5}$$

where $\omega_l = \dfrac{e^{RT(i,j,e_l)-RTH}}{\sum_{s=1}^{m} e^{RT(i,j,e_s)-RTH}}$, $RTH$ is the threshold and $m$ is the number of recommendation chains.

### 3.3 Trust Feedback Mechanism

In order to identify the malicious members, the feedback mechanism is introduced and it is divided into two parts: service effectiveness and recommendation accuracy. When a transaction is completed, the trustee's reputation will be updated according to the formula (1), and the trustor update the direct trust value about it. If a member provides incredible recommendation to others, its reputation value will be reduced to the half of before, or else added by 0.1.

$$R_{ij}^{new} = \begin{cases} R_{ij}^{old}/2 \\ R_{ij}^{old} + 0.1 \end{cases} \tag{6}$$

## 4 Simulation

Suppose that there are 1000 members in a virtual community and 95 percent members have 1000 transactions with others on average, that is to say, even a new member can communicate with others via recommendation.

Figure1 shows that the trust value accuracy estimated with our model is much higher than the model in reference [11] and declines much slower than that obviously. Even there are 60 percent malicious members in the community, the trust value accuracy is still 66 percent, which proves that this model can effectively resist malicious attacks.

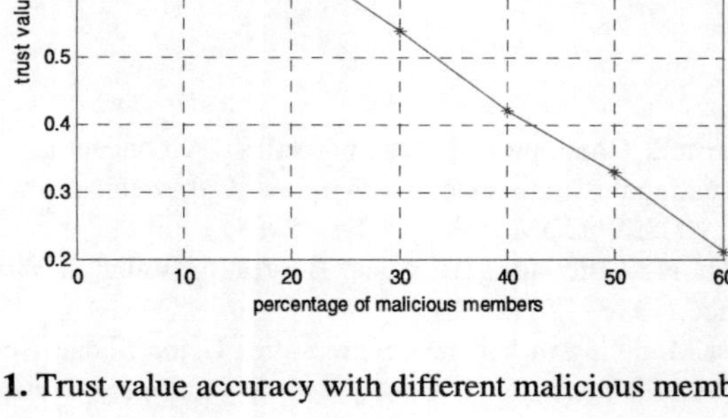


**Fig. 1.** Trust value accuracy with different malicious members rate

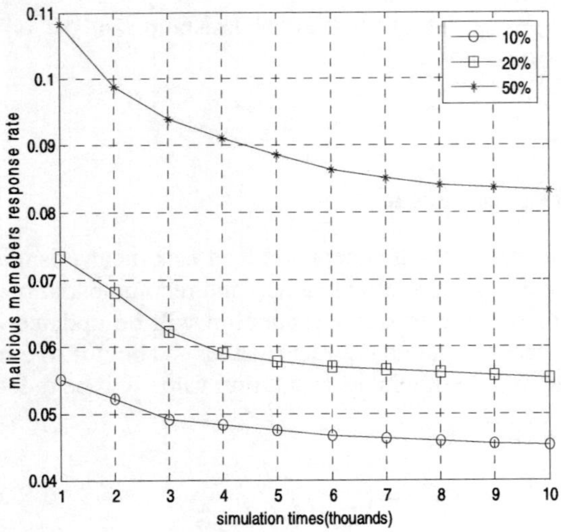


**Fig. 2.** Malicious members response rate with different malicious members rate

Simulation results in Figure2 shows that the malicious member response rate increases with the percent of malicious members rise. However, with the increasing of simulation times, the response rate has obviously declined, which means that the feed back mechanism and the punishment mechanism have a positive effective.

## 5 Conclusion

Trust and reputation are important sources of information that we gathering about each other in our daily life. They seem even more important in the online world, due to the nature of virtual community. This paper presented a model based on reputation and considered the activity degree and the length of the trust chain which can get satisfactory trust value accuracy. The model can work well in worse network environment having more malicious members, and the feedback mechanism and punishment mechanism plays a very good role in this environment. The simulation has proved the validity of the model.

## References

1. Si, W., Jian, J.: Virtual Community based marketing: A conceptual model for tourism industry. In: International Conference on Wireless Communications, Networking and Mobile Computing 2008, WiCOM 2008, pp. 1–5 (2008)
2. Porter, C.E., Donthu, N.: Cultivating Trust and Harvesting Value in Virtual Communities. Management Science 1, 113–128 (2008)
3. Kolaczek, G.: Trust Modeling in Virtual Communities Using Social Network Metrics. In: Proceedings of 2008 3rd International Conference on Intelligent System and Knowledge Engineering, pp. 1421–1426 (2008)

4. Boella, G., van der Torre, L.: Security Policies for sharing Knowledge in virtual communities. IEEE Transactions on Systems, Man, and Cybernetics-Part A: Systems and Humans 5, 439–450 (2006)
5. Wu, J.-J., Chen, Y.-H., Chung Trust, Y.-S.: factors influencing virtual community members: A study of transaction communities. Journal of Business Research 3, 1–8 (2009)
6. Deng, G., Zhou, H., Zhang, P.: Study of incentive mechanism and internet behavior of virtual communities. Web Society 8, 213–218 (2009)
7. Boella, G., van der Torre, L.: Security Policies for Sharing Knowledge in Virtual Communities. IEEE Transactions on Systems, Man, and Cybernetics-Part A: Systems and Humans 36(3), 439–450 (2006)
8. Kui, M., Yue, W., Xu, Z., et al.: A Trust Management Model for Virtual Community. In: Proceedings of the Fifth International Conference on Computer and Information Technology (CIT 2005), pp. 741–745 (2005)
9. Grinshpoun, T., Gal-Oz, N., Gudes, A.M.E.: CCR: A Model for Sharing Reputation Knowledge Across Virtual Communities. In: 2009 IEEE/WIC/ACM International Conference on Web Intelligence and Intelligent Agent Technology – Workshops, pp. 34–41 (2009)
10. Dorigo, M., Maniezzo, V., Colorni, A.: The Ant System: Optimization by a colony of cooperating agents. IEEE Transactions on Systems, Man, and Cybernetics-Part B: Cybernetics 1, 29–41 (1996)
11. Zuo, Y., Hu, W.-C., O'Keefe, T.: Trust Computing for Social Networking. In: 2009 Sixth International Conference on Information Technology: New Generations, pp. 1534–1539 (2009)

# Fault Location of Aircraft Cable Based on One-Terminal Electrical Quantities

Xiaolin Liu[1,2], Zhenzhen Zhang[1], Tao Jing[1], and Xudong Shi[1]

[1] College of Aeronautical Automation Civil Aviation University of China
TianJin, China
[2] College of Electrical Engineering Yanshan University
Qinhuangdao, China
missxiaolin1997@163.com

**Abstract.** An airplane cable fault location method is proposed to solve the problem of being difficult to locate the fault. The differential equation mathematical model of aircraft cable fault location based on the one-terminal electrical quantities is created. The BP neural network is selected as the control algorithm which can identify the unknown parameters of the model. The experiment result shows that the mathematical model is rational and effective.

**Keywords:** aircraft cable, fault location, mathematical model, BP neural network.

## 1 Introduction

Aircraft cable is the important part in modern aircraft which serves as the transmission medium of dynamic electric source, control signal and data information for control system, the aerial electronic system and so on. Its performance affects flight security directly. How to inspect the fault location of the cable rapidly and exactly is an important task for aircraft crew to maintenance the airplane, which is pointed out in [1]-[3]. At present, the method based on the two-terminal electrical quantities [4]-[6] is researched to locate the fault of the cable. However, because many kinds of cables in the airplane are sorted and deposited in the gaps of airplane, it is very difficult to measure the two-terminal electrical quantities of the cable, which is described in [7]. Therefore, in the case of an actual electric transmission line, the aircraft cable fault location system based on the one-terminal electrical quantities should be designed, which can be applied to engineering practice.

## 2 Mathematical Model Researches

The mathematical model reflects the logical and quantitative relation of the input, internal state and output of the system, which can supply the gist for the computer to calculate and dispose. Thus, in order to locate the aircraft cable fault accurately, it is very important to create the mathematical model of the system parameters.

In the case of the distributed parameter model of the cable, the characteristic can be reflected by four parameters, which are resistance, inductance, conductance and capacitance. The active power loss of the insulating medium and the electrical leakage is light, because the length of the aircraft cable is no more than 100 meters. Therefore, the parameters of the capacitance and conductance can be ignored and the equivalent circuit of the fault cable condition system is introduced in this paper, as shown in Fig. 1. According to the superposition theorem of the circuit, the Fig. 1 can be constructed by Fig. 2(a) and Fig. 2(b).

In Fig. 1 and Fig. 2, the ends of the cable are lettered $M$ and $N$. The value of $h$ is the length of the cable. The point of $P$ is assumed to be the fault position and the value of $x$ is the distance from the point of $P$ to the point of $M$. The value of $R$ and $L$ are the resistance and the inductance per unit length of the cable respectively. The value of $r$ and $R_P$ are the load and the transition resistance respectively. The value of $e_M$ is the power source. The value of $u_M$, $i_M$ and $i_N$ are the voltage and current instantaneous value of the ends of the fault cable condition system. Similarly, $u'_M$ and $i'_M$ are the voltage and current value of the normal cable condition system, $u''_M$, $i''_M$ and $i''_N$ are the voltage and current value of the pure-fault cable condition system.

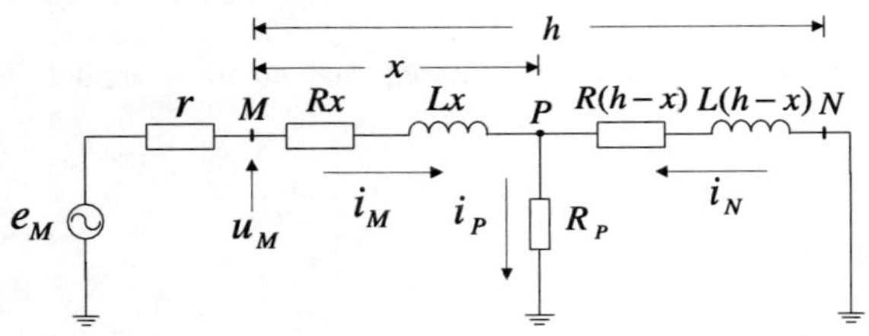


**Fig. 1.** Equivalent circuit of fault cable condition system

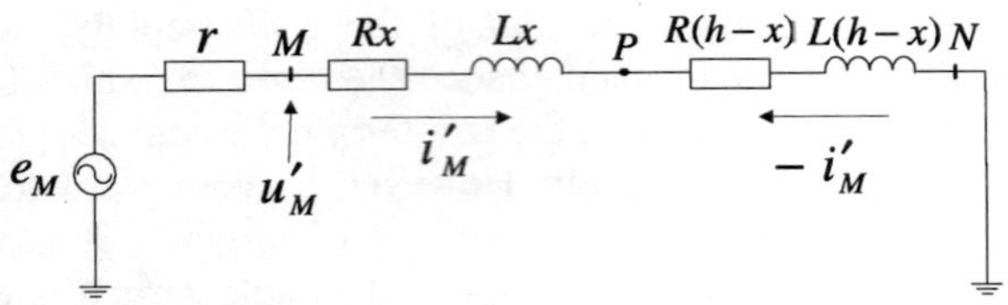


**Fig. 2(a).** Equivalent circuit of normal cable condition system

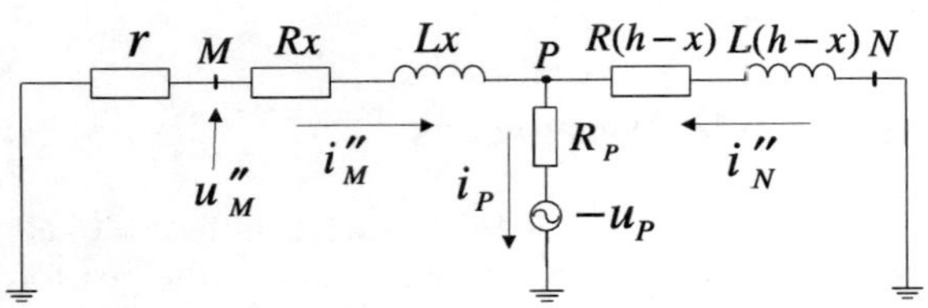


**Fig. 2(b).** Equivalent circuit of pure-fault cable condition system

**Fig. 2.** Decomposition of equivalent circuit of fault cable condition system

According to the Kirchhoff's voltage and current law and Fig. 2(b), (1) can be deduced.

$$ri''_M + Rxi''_M + Lx\frac{di''_M}{dt} = L(h-x)\frac{di''_N}{dt} + R(h-x)i''_N \tag{1}$$

Moreover, according to the superposition theorem of the circuit, (2) can be deduced.

$$\begin{cases} i_M = i'_M + i''_M \\ i_N = -i'_M + i''_N \end{cases} \tag{2}$$

According to (1) and (2), the current value of $i''_M$ and $i''_N$ can be eliminated, as shown in (3).

$$r(i_M - i'_M) + Rx(i_M - i'_M) + Lx\frac{d(i_M - i'_M)}{dt}$$

$$= L(h-x)\frac{d(i_N + i'_M)}{dt} + R(h-x)(i_N + i'_M) \tag{3}$$

First, in (3), the current value of $i_N$ is the measurement parameter of the point of $N$, which should be eliminated. According to Fig. 1, (4) can be deduced.

$$u_M = Rxi_M + Lx\frac{di_M}{dt} + R_p(i_M + i_N) \tag{4}$$

According to (4), the current value of $i_N$ and $\frac{di_N}{dt}$ can be deduced, as shown in (5).

$$\begin{cases} i_N = \dfrac{u_M - Rxi_M - Lx\dfrac{di_M}{dt}}{R_P} - i_M \\ \dfrac{di_N}{dt} = \dfrac{\dfrac{du_M}{dt} - Rx\dfrac{di_M}{dt} - Lx\dfrac{d^2 i_M}{dt^2}}{R_P} - \dfrac{di_M}{dt} \end{cases} \tag{5}$$

Second, in the course of the actual aircraft maintenance, it is very difficult to find a normal cable, the length of which is equal to the same type of the fault cable. Therefore, the current value of $i'_M$ in (3) can't be obtained. In order to eliminate $i'_M$, the equivalent circuit of the normal condition system of the random length cable should be considered.

According to the theory of the distributed parameter circuit, the Fig. 2(a) is equivalent to Fig. 3(a). The equivalent circuit of normal cable condition system, the length of which is $l$, is shown in Fig. 3(b), where the value of $i'_{Ml}$ is the current instantaneous value.

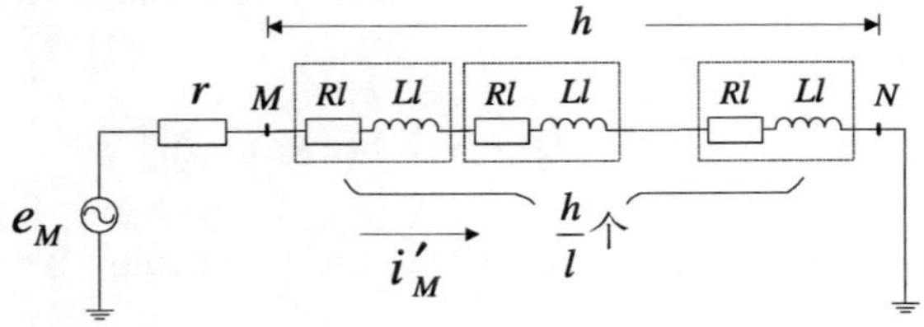


**Fig. 3(a).** Equivalent circuit of normal $h$ length cable condition system

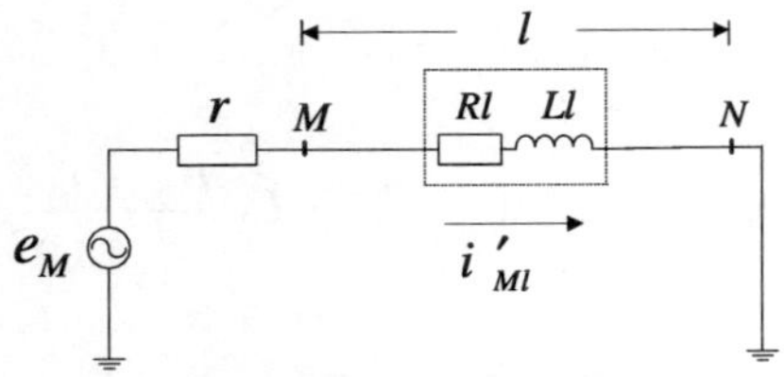


**Fig. 3(b).** Equivalent circuit of normal random length cable condition system

**Fig. 3.** Equivalent circuit of normal cable condition system.

According to Fig. 3(a), (6) can be deduced, where the value of $\omega$ is the angular frequency of the system.

$$e_M = i'_M \left[ r + \frac{h}{l}\sqrt{(Rl)^2 + (\omega Ll)^2} \right] \tag{6}$$

In like manner, according to Fig. 3(b), (7) can be deduced.

$$e_M = i'_{Ml} \left[ r + \sqrt{(Rl)^2 + (\omega Ll)^2} \right] \tag{7}$$

According to (6) and (7), (8) can be constructed.

$$i'_{Ml} = \frac{r + \sqrt{(Rl)^2 + (\omega Ll)^2}}{r + \frac{h}{l}\sqrt{(Rl)^2 + (\omega Ll)^2}} i'_{Ml} \tag{8}$$

Lastly, according to (3), (5) and (8), the differential equation mathematical model of aircraft cable fault location based on the one-terminal electrical quantities can be denoted by (9).

$$(Ru_M + R^2 h i_M + 2RhL\frac{di_M}{dt} + L^2 h\frac{d^2 i_M}{dt^2} + L\frac{du_M}{dt})x$$

$$+(-R^2 i_M - 2RL\frac{di_M}{dt} - L^2\frac{d^2 i_M}{dt^2})x^2$$

$$+\left[ (i_M - \frac{r + \sqrt{(Rl)^2 + (\omega Ll)^2}}{r + \frac{h}{l}\sqrt{(Rl)^2 + (\omega Ll)^2}} i'_{Ml})(r - Rh) \right.$$

$$+Lh\left(\frac{di_M}{dt}-\frac{r+\sqrt{(Rl)^2+(\omega Ll)^2}}{r+\frac{h}{l}\sqrt{(Rl)^2+(\omega Ll)^2}}\frac{di'_{Ml}}{dt}\right)\Bigg] R_P$$

$$= Rhu_M + Lh\frac{du_M}{dt} \tag{9}$$

In (9), the value of $R$, $L$, $\omega$, $r$, $R_P$, $h$, $l$ and $x$ are model parameters, among which $R$, $L$, $\omega$, $r$, $h$ and $l$ are known , but $x$ and $R_P$ are unknown. The instantaneous value of $i_M$, $i'_{Ml}$ and $u_M$ are measurement parameters, which can be obtained by the measurement equipments of the fault or normal cable condition system. The value of the difference equations are substituted for the instantaneous value of $\frac{di_M}{dt}$, $\frac{d^2 i_M}{dt^2}$, $\frac{di'_{Ml}}{dt}$ and $\frac{du_M}{dt}$ of the differential equations, as shown in (10), where the value of $T_S$ is the sampling period and the value of $k$ is the sampling time. Therefore, (9) can be simplified, as shown in (11), where the value of $a_k$, $b_k$, $c_k$, $d_k$ satisfies (12).

$$\begin{cases} \dfrac{di_M(k)}{dt}=\dfrac{i_M(k+1)-i_M(k-1)}{2T_S} \\ \dfrac{d^2 i_M(k)}{dt^2}=\dfrac{i_M(k+1)+i_M(k-1)-2i_M(k)}{{T_S}^2} \\ \dfrac{di'_{Ml}(k)}{dt}=\dfrac{i'_{Ml}(k+1)-i'_{Ml}(k-1)}{2T_S} \\ \dfrac{du_M(k)}{dt}=\dfrac{u_M(k+1)-u_M(k-1)}{2T_S} \end{cases} \tag{10}$$

$$a_k x + b_k x^2 + c_k R_p = d_k \,(k=1,2,\cdots,n) \tag{11}$$

$$\begin{cases} a_k = Ru_M(k)+R^2 h i_M(k)+2RhL\dfrac{di_M(k)}{dt}+L^2 h\dfrac{d^2 i_M(k)}{dt^2}+L\dfrac{du_M(k)}{dt} \\ b_k = -R^2 i_M(k)-2RL\dfrac{di_M(k)}{dt}-L^2\dfrac{d^2 i_M(k)}{dt^2} \\ c_k = \left(i_M(k)-\dfrac{r+\sqrt{(Rl)^2+(\omega Ll)^2}}{r+\frac{h}{l}\sqrt{(Rl)^2+(\omega Ll)^2}}i'_{Ml}(k)\right)(r-Rh)+Lh\left(\dfrac{di_M(k)}{dt}-\dfrac{r+\sqrt{(Rl)^2+(\omega Ll)^2}}{r+\frac{h}{l}\sqrt{(Rl)^2+(\omega Ll)^2}}\dfrac{di'_{Ml}(k)}{dt}\right) \\ d_k = Rhu_M(k)+Lh\dfrac{du_M(k)}{dt} \end{cases} \tag{12}$$

## 3 Model Parameters Identification

The neural network has the characteristic of concurrent processing information and the ability of self-study in response to the control request in order to adapt to the complex circumstance. Moreover, the neural network can control nonlinear object effectually, which is described in [8]. Consequently, in order to identify the unknown parameters $x$ and $R_P$ of the differential equation mathematical model of aircraft cable fault location, the BP (Back Propagation) neural network is selected to process the data collected on the spot.

The control process can be divided into two parts. The first part is to train the neural network. The initial value of the network is set and the network is trained by using of training samples. Thus, the structure of the network can be initialized. The second part is to test the neural network. After the weight value of the convergent network is adopted, the testing samples are input and the learning effect of the network is tested.

In the BP neural network, there are one input layer, one implication layer and one output layer. The input layer has four neural cells, which are $a_k$, $b_k$, $c_k$ and $d_k$. The output layer has two neural cells, which are the distance $x$ and the transition resistance $R_P$. After the study time and error limiting value of the network are set, the Simulink toolbox of MATLAB is applied to train the network. The training samples of the network are made up of a great deal of actual measured datum.

Thus, on the premise that the $n$ groups of data by the measurement equipments of the system are obtained, the unknown parameters $x$ and $R_P$ of the mathematical model can be identified. Therefore, the aircraft cable fault can be located under the control of algorithm, which can be used in the computer.

## 4 Matlab Simulation

In order to validate the differential equation mathematical model of aircraft cable fault location, the aircraft cable is used in the MATLAB simulation experiment, the type of which is 55A0812-20-4 described in [10]. The parameters of the cable are given as follows that $R = 0.03015\Omega / m$ and $L = 0.2695\mu H / m$. The parameters of the system are given as follows that $h =$ 100$m$, $e_M = 115\angle 0°$ and $r = 10\Omega$. The simulation models are shown in Fig. 4 and Fig. 5. According to the result of the experiment, the relative error is less than 0.2%, which satisfies the needs of the actual engineering and illuminates that the system is rational and effective.

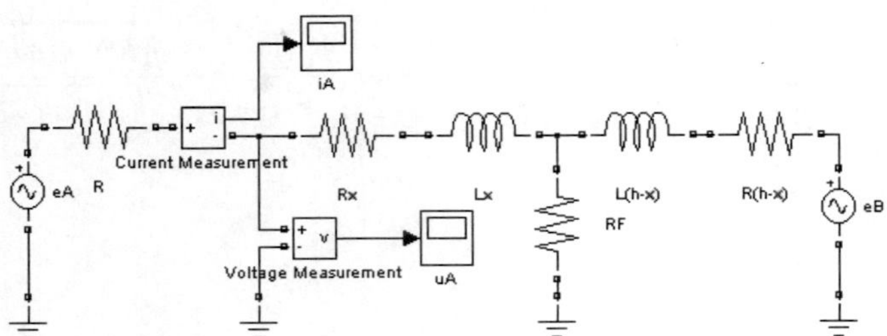


**Fig. 4.** Simulation system model of fault condition

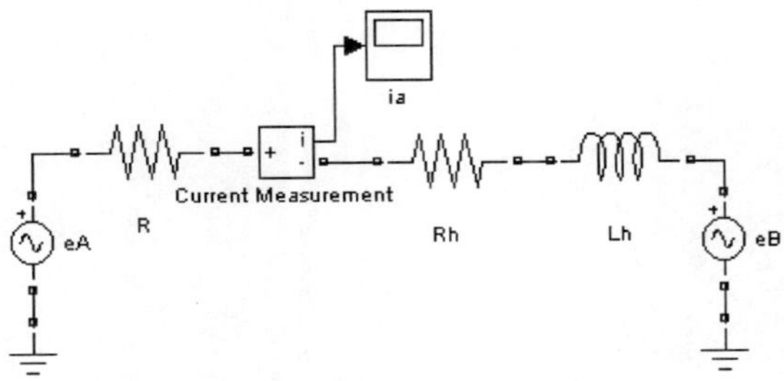


**Fig. 5.** Simulation system model of normal condition

## 5 Conclusions

First, the differential equation mathematical model of aircraft cable fault location based on the one-terminal electrical quantities is created according to the theory of the distributed parameter circuit.

Then, the BP neural network is selected as the control algorithm, which can identify the unknown parameters of the mathematical model and supply the gist for computer to calculate and dispose.

**Acknowledgment.** It is a project supported by the union project of the National Natural Science Foundation of China and Civil Aviation Administrator of China (60776809) and the science and technology projects of Civil Aviation University of China (ZXH2009D020).

## References

1. Yang, Y., Wang, Y.: Portable intelligent measuring system design for cable of airplane. Aeronautical Manufacturing Technology (2), 88–90 (2005)
2. Luo, X.B., Zhao, Z.P.: Integrated system of automatic fault diagnosis of aerial cable. Industrial Control Computer 16(3), 5–7 (2003)
3. Mao, P., Sun, Y.M., Zhang, Z.N.: Study of fault location for high voltage over-head transmission line using neural networks model system with redundant neuron. Proceeding of the CSEE 20(7), 28–33 (2000)
4. Shao, W.Q., Jiama, L.H., Song, G.B.: Permanent fault identification method based on capacitance paremeter. Journal of Xi'an Jiaotong University 42(6), 708–712 (2008)
5. Song, S.P., Song, G.B., Jiao, Z.B.: Novel time-domain fault location algorithm by two-terminal currents. Journal of Xi'an Jiaotong University 43(4), 101–105 (2009)
6. Suonan, J.L., Zhang, Y.N., Jun, Q.: Time domain fault location method based on transmission line parameter identification using two terminals data. Power System Technology 30(8), 65–70 (2006)
7. Kang, F., Wang, L.W.: Design and implementation of airplane-cable fault location used by civil aviation. Journal of Civil Aviation University of China 20(6), 6–10 (2002)
8. Xiong, X.Y., Lin, J.H.: A cable fault location method based on power balance theory. Relay 30(12), 21–23 (2002)

# Application Research on Computer Rank Examination Basing on Data Mining

Xue Xiao-feng, Xue Xiang-hong, and Ye Fei-yue

School of Computer Engineering
Jiangsu Teachers University of Technology
Jiangsu Changzhou
{xxf,xxh1226,yfy}@jstu.edu.cn

**Abstract.** Analyzed the grade of C language of computer rank examination of Jiangsu province and questionnaire data by using of data mining technology. Design and Realize the data mining process of computer rank examination performance based classification of decision tree. Identify the hidden curriculum information to provide guidance for the C language.

**Keywords:** Data Mining, Computer Rank Examination, Classification, Decision Tree.

## 1 Introduction

Computer rank examination of Jiangsu province is the province's colleges and universities to strengthen non-computer professional teaching basic computer courses, improve teaching quality, by the Jiangsu Provincial Department of Education established "Jiangsu Province Universities Computer Rank Examination Steering Committee" and "Computer Rank Examination center" are responsible for the implementation of colleges and universities in Jiangsu Province's computer rank examination system. Computer Rank Examination of Jiangsu province, began in the fall of 1993. Computer Rank Examination of Jiangsu Province, has been widely recognized by the province's colleges, universities and employers, due to item difficulty and other reasons, the pass rate is generally low, especially the two language type of examination. In order to better guide examinations and courses teaching the use of data mining techniques in this paper the classification of excavation, told me the school level data were mining test scores obtained in the course of study the implicit information.

## 2 Classification

### 2.1 The Basic Concept

Classification (Classification) in data mining is a very important task, is currently the most widely used commercially. The purpose of classification is to identify a group to describe the typical characteristics of the data model or function to be able to identify the attribution or categories of the unknown data,. Classification model can be learned to obtain through data mining classification algorithms from a set of training sample data (the type of attribution is known) to learn to obtain [1]. Categories can be used to predict the forecast aims to re-use historical data records to automatically derive the promotion of a given description of the data so that it can predict the future data.

### 2.2 The Classification Method and Steps

Commonly used in data classification methods: decision tree classification, Bayesian classification, neural networks. And K-best approach of classification, case-based reasoning, genetic algorithms, rough sets and fuzzy sets method [2].

Data classification is generally realized in two steps: Firstly, create a classification model, classification model rules for each data set describes the basic characteristics of certain types of objects; secondly, the model has been created using the categories of the unknown tuple return into one or several classes. Classification model is summed up from a finite group that contains data set, in which the collection of data used to create the model is called the training set, a single tuple of the training set is called the training samples.

### 2.3 Decision Tree Classification

Decision tree is a kind of tree used to represent the people in order to make a decision-making in a series to determine the process tree. The basic idea of decision tree method is to use the training set data to automatically construct decision tree, and then to determine any instance on the basis of this decision tree. At present, the decision tree method is mainly used for classification, it provides a kind of method rules, which shows "under what conditions, what would be the value". The decision tree with an intuitive, clear, easily understood characteristics of the user, while the decision tree is generated with easy rules, algorithms, advantages such as scalability [3].

The use of decision tree to classify, that is, the selection of an attribute partition in internal node, each branch is split as part of a leaf node represents a distribution. Decision tree generation algorithm is divided into two steps: the generation of the tree, the data began in root node, recursively broken down: tree pruning, removing some might be noise or exceptional data. Using the decision tree is to partition the unknown data, that is used in accordance with the division of property on the tree layer by layer down until you encounter a leaf node.

### 2.4 ID3 Algorithm Description

ID3 algorithm is the world's most influential and the first decision tree method. This method first check all the fields in the database to identify with the greatest information gain Gain (A) of the field as a decision tree node, then according to the different values of the field to establish a branch of the tree for each subset of the branch and create duplicate nodes and the lower tree branches, until the results of a subset of the same class. Algorithm is described as follows [4]:

Algorithm: Generate_decision_tree from a given training set generated a decision tree.

Input: Training samples samples, indicated by the discrete attribute values; the candidate attribute set attribute_list.

Output: a decision tree.

Method:

a) Create a node N;

b) if samples are in the same class C then return N as a leaf node to class C as a marker;

c) if attribute_ list is empty, then return N as a leaf node, marked as samples of the most common category;

d) Choose attribute_ list has the highest information gain attributes test_ attribute;

e) Mark node N as test_ attribute;

f) While the known values of $a_i$ in each of the test_ attribute;

g) from node N grown test _attribute = a condition of branch;

h) Let test_attribute = yes samples in the collection of samples;

i) if $s_i$ is empty, then add a leaf, marked as samples of the most common type; else with a Generate_ decision_ tree (, attribute__list-test_attribute) return nodes.

### 2.5 Generates Classification Rules

Classification rules can be generated by the decision tree. We can use this form IF-THEN to represent rules. When generating rules, each leaf node create a rule, each division become a condition of the rule, the category of leaf nodes is THEN content.

## 3 Application of Decision Tree Classification

### 3.1 Data Preprocessing

After surveys and studies of grade test scores, process the data as following, obtain data Table 1:

a) "Computer sound foundation" into "yes" and "No" categories;

b) the "timely completion of operations" into "yes" and "No" categories;

c) "Evaluation of teacher school" into "excellent", "qualified" and "unsatisfactory" categories;

d) "weekly hours on the plane" into "" 2 "," 2-3 "and" "3" categories;

e) "the pass of the grade test" into "yes" and "No" categories;

**Table 1.** Training Set Data of Computer Rank Examination

| A good foundation and No | Timely completion of work | Class evaluation of teachers | The number of hours per week on the machine | Pass or not |
|---|---|---|---|---|
| not | not | excellent | <2 | not |
| yes | not | excellent | <2 | not |
| not | not | excellent | 2-3 | yes |
| not | not | pass | >3 | yes |
| not | yes | not pass | >3 | yes |
| yes | yes | not pass | >3 | not |
| yes | yes | not pass | 2-3 | yes |
| not | not | pass | <2 | not |
| not | yes | not pass | <2 | yes |
| not | yes | pass | >3 | yes |
| yes | yes | pass | <2 | yes |
| yes | not | pass | 2-3 | yes |
| not | yes | excellent | 2-3 | yes |
| yes | not | pass | >3 | not |

### 3.2 Decision Tree Construction

The table above is the training set of the result of data clean-up test scores of students in grades information. In accordance with the pass grade or not to construct a decision tree model. According to principles of ID3 algorithm, using the following steps to construct :

Step 1: Calculate the expectations of the information required for the sample classification results

When a sample is divided into two classes, class C1 is set up as the pass grade class, C2 is did not pass grade class, then S1 = 9, S2 = 5, a total of S = 14, calculated for a given sample classification performance expectations required information:

$$I(S_1, S_2) = I(9,5) = -\frac{9}{14}\log_2\frac{9}{14} - \frac{5}{14}\log_2\frac{5}{14} = 0.94$$

Step 2: Calculate the information gain of each attribute

a) First calculate properties of "a good foundation or no" information gain

For "a good foundation or no" = "Yes", $S_{11}$=3, $S_{21}$=3,

$$I(S_{11}, S_{21}) = -\frac{3}{6}\log_2\frac{3}{6} - \frac{3}{6}\log_2\frac{3}{6} = 1$$

For "a good foundation or no" = "No", $S_{12}$=6, $S_{22}$=2,

$$I(S_{12}, S_{22}) = -\frac{6}{8}\log_2\frac{6}{8} - \frac{2}{8}\log_2\frac{2}{8} = 0.81$$

Calculated according to "a good foundation or no" division of the sample given the expectations of the information required is:

$$\text{E (“a good foundation or no”)} = \frac{6}{14} * I(S_{11}, S_{21}) + \frac{8}{14} * I(S_{12}, S_{22}) = 0.89$$

Then, this division of information gain is:

Gain("a good foundation or no") $=I(S_1, S_2)$ - E ("a good foundation or no") =0.94-0.89=0.05

b) Use the same method to calculate the "timely completion of work", "school teacher evaluation", "number of hours per week on the machine" information gain 0.15,0.03,0.25, respectively.

Step 3: Determine the test properties

As the property, "the number of hours per week on the machine" information gain the highest it has been selected for testing attributes, for the establishment of the first node, and the sample is divided into three parts. Then calculate each sub-tree recursively in accordance with the above method, came to the conclusion of the decision tree shown in Figure 1.

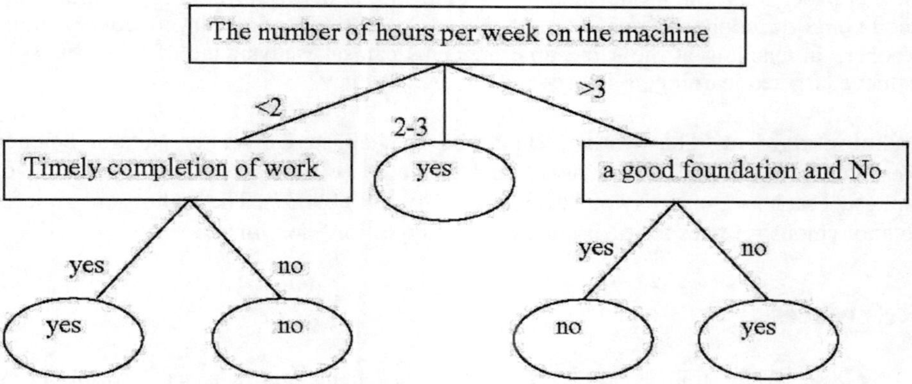


**Fig. 1.** Decision Tree

### 3.3 Extraction of Classification Rules

In this case, the rules can be extracted out of the following categories:

IF "a week-on the number of" <2 AND "timely completion of operations" = "yes" THEN pass rate = 14.3%

IF "a week-on the number of" = "2-3" THEN pass rate = 28.6%

IF "a week-on the number of"> 3 AND "a good foundation and No" = "No" THEN pass rate = 21.4%

IF "a week-on the number of" <2 AND "timely completion of operation" = "No" THEN did not pass rate = 21.4%

IF "a week-on the number of"> 3 AND "a good foundation and No" = "yes" THEN did not pass rate = 14.3% From the extracted rules, available to the following conclusions:

a) The timely completion of work of students, 86% of students have passed grade examinations;

b) The number of weekly on the machine "= 2 students, 78% of students have passed grade examinations; 3) the students of a sound foundation, 50% did not pass grade exams.

c) In view of these conclusions, teachers, in the follow-up teaching, should give full attention to the management of students on timely layout and marking student assignments, encouraging students to use their leisure time, strengthen the training on the machines. At the same time, we can see that the students grade level of the computer-based has little impact on exam passed or not, if able to do homework on time, adhere to practice on the machines,they can entirely pass grade exams.

## 4 Conclusion

In this paper, data mining technology is used in classifying in mining for grade test scores and questionnaire survey data, the use of derived rules can effectively guide teachers in teaching at the same time, students can take advantage of these rules, to achieve targeted learning, to improve learning efficiency.

**Acknowledgment.** This work is supported by Jiangsu University Natural Science Research Program(No.08KJD520006) and theYouth scientific research foundation of Jiangsu Teachers University of Technology (No. KYY09037). The authors are grateful to anonymous referees for their corrections, suggestions and improvements.

## References

1. Su, X.-N., Yang, J.-L., et al.: Data warehouse and data mining. Tsinghua University Press, Beijing (2006)
2. Xia, H.-S.: Data Warehouse and Data Mining Technology. Science Press, Beijing (2004)
3. Fang, Z.-G., Yu, L.-H.: Data Mining in Computer Rank Examination of exploration. Hebei Vocational and Technical College 8(2), 50–51 (2008)
4. Chen, W., Cheng, H.-J.: ID3 algorithm is constructed student test scores Top-analysis of the decision tree. Computer Knowledge and Technology 5(3), 744–746 (2009)

# Popular Misunderstandings and Improvement Strategies for Network Marketing

Ling Feng and Anbao Wang*

School of Computer and Information,
Shanghai Second Polytechnic University
Shanghai 201209 China
Ling2037QQ@163.com, abwang@it.sspu.cn

**Abstract.** with the development of information technology, network marketing is active in the business of enterprises. However some enterprises have failed to grasp the network marketing because of the network marketing misunderstandings. This paper listed five misunderstandings of network marketing and gave the brief description of the causes. At the same time, it also put forward five improvement strategies to help some SMEs (Small and medium-sized enterprises) out of the failure of network marketing.

**Keywords:** network marketing, SMEs, misunderstandings, improvements.

## 1 Introduction

By the end of 2000, the volume of global e-commerce transaction reached nearly 200 billion dollars, especially in the Europe; E-commerce permeated each industry field. E-commerce has fundamentally changed the internal operations, external cooperation and exchange mechanism. According to an e-commerce investigation report, the "The Economist" magazine once pointed that the e-commerce of United States is developing rapidly, the volume of e-commerce has reached 20 billion dollars, by 2004, the figure could rise to 184 billion dollars and by 2010, online transactions may account for 15% to 20% of U.S. retail market [1]. With the development of information technology, the internet digital revolution also swept half of China. According to the 24th Chinese Internet Development Statistics report shows that at the end of June 2009, the scale of Chinese netizens has reached 338 million, compared with the end of 2008, the number increased 13.4%. In addition, the rate of Internet penetration has reached 25.5% and the scale of netizens continues to expand, Internet penetration steadily increasing [2]. Such a great size and potential market has built a huge platform for Chinese e-commerce development, and the network marketing as the important implementation means of e-commerce also has been concerned by many rese[1]archers in this digital revolution.

* Corresponding Author.

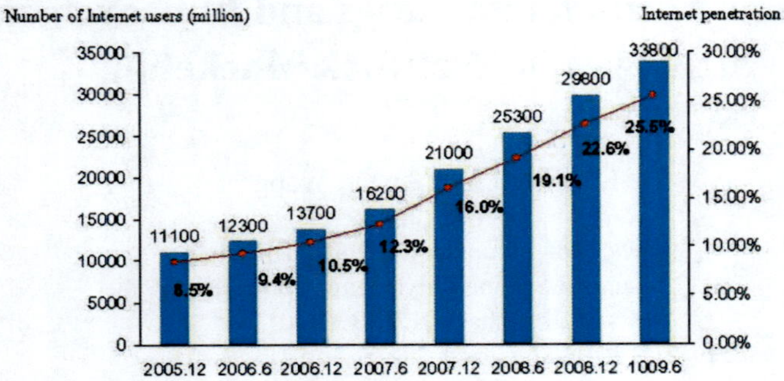


**Fig. 1.** The report of the 24th Chinese internet development statistics

## 2 Related Work

In recent years, with the emergence of e-commerce, network marketing, which party crossed the ocean into China since 1996, has been from the initial "diorama" to the marketing "barrel of a gun." During this period, through the method of combining theory and practice, numerous experts and scholars interpreted the definition of network marketing, pointed out the opportunities and challenges that network marketing may bring for the Chinese enterprises. and summarized successful factors from some domestic and international network marketing cases. There is no doubt that the prevalence of network marketing led SMEs into spring. However, the "spring" is not perfect yet. Some companies seized the opportunity to earn a lot, but the others made fruitless efforts. So the paper is mainly about analyzing and exploring the reasons for the companies which have a loss on network marketing, and put forward some improvement strategies on the basis of previous studies.

## 3 The Concept and Misunderstandings of Network Maketing

With the development of electronic commerce, the majority of people described the network marketing as a new marketing method, which departs from the traditional theoretical foundation and makes the network marketing become the meretricious "face project", so in order to truly understand the network marketing, enterprises still need to B2B (Back to Basic) [3]. Philip Kotler once said that marketing is a social process, through which the individuals and organizations can get what they need by production, supply and exchanging goods and services with others [4]. And network marketing makes use of the Internet tools and resources to promote the agreement of trade, and then obtains the materials they need. And with the improvement of the application of network marketing, its characteristic of low-cost, high-efficient and wide-cover is becoming more and more obviously, attracting numerous enterprises. However, due to the concept limitations, some of the people make mistakes in the awareness of network marketing and the actual operation, which is commonly known as "network marketing popular misunderstandings."

### 3.1 The First Misunderstanding: The Enterprise Taking Action Quickly Will Surpass the One Doing Slowly

Businessmen always emphasize a word — "fast", which refers to seize the market before others. This is just a good marketing policy, but some Internet companies have been inspired by this and then put forward a new idea that the survival of the fastest. Unfortunately, this view has not withstood the test in historical changes. In fact, in the virtual market, the first-mover, of course, have the advantages of competition, but the efforts of the latecomer also can not be ignored. [3] As the first Chinese Internet services provider, the company of Yinweihai was hailed as the leader of the Chinese ISP industry, however, the ambitions of the company had been broken years because of its narrow profit model, immature marketing idea and imperfect management system and then the company became the fragments in Chinese Internet history. At the same time, neither did people notice that in Shenzhen, a chat software marked by a small penguin (OICQ later renamed QQ) was quietly spreading until it ranked the first of the casual chat software, swepting China. As a result [4], the pre-emptive of the traditional meaning is not an absolute criterion. In the final analysis, whether the quick ones will surpass the slow ones or the quick ones surpass the quick ones, depends on the specific problems.

### 3.2 The Second Misunderstanding: Network Marketing Will Take into Effect Immediately

The view that network marketing can get instant results comes from some SMEs, they focus their attention on the sales and ignore the influence of brands, especially in the recession period. In the survival period, this view is understandable, but once enterprises enter into the development period, brand construction becomes particularly important, because network marketing is the cost-effective way of brand construction.[6] Brand, what we commonly known as "recognition" is a kind of intangible assets, so the brand construction is the promotion and maintenance of intangible assets. Compared with the most of the tangible assets, the construction of intangible assets can not help consuming more time and efforts. Only being based on integrity and regarding product quality and product features as the core, can enterprises foster consumers' credit awareness and enhance brand image. The higher the credibility is, the better reputation becomes, and the performance effects of network marketing will naturally appear, thus, network marketing is a protracted war, which transform the cumulative effect of brands into market sales. Those who consider network marketing as the "catalyst" will lose in the bubble of network marketing in the end.

### 3.3 The Third Misunderstanding: Pricing War Will Be Arose by Network Marketing

Many people believed that compared with traditional market, virtual market is a more complete market. Because of the decreasing search costs, the price competition among suppliers will lead to the increasingly narrowed profit margins. In fact, the price dispersion of virtual market is even greater than traditional market's.

However, the highest price might not be less popular while the lowest price might also not attract thousands of people and more popular because compared with the traditional price factors, in the virtual market, consumers pay more attention to the business' integrity and information security.

Entry cost of Virtual market is low, thus the business' qualification is inevitably mixed, and customers are difficult to distinguish the authenticity of business-to-goods only by a diagram. Therefore, the businesses which have good reputation and high brand credit can naturally win consumers' favor and get a higher premium. In addition, it's not what people have expected that if consumers want to change suppliers, they just need to click the mouse, but there are objective conversion costs; they have to transmit their payment information on an unfamiliar path, in which hidden risks. [3]

### 3.4 The Fourth Misunderstanding: Viral Marketing (Marketing through Word of Mouth) Is the Same as Virus of Marketing (Effect-Focused Marketing)

Viral marketing was originally a popular method of network marketing, which was designed to meet some demand as the starting point. The method uses the principle of public praise to sell products to those who have the same requirements and diffuses by layers. Because its mode of transmission is similar to the spread of the virus on the Internet, it gets the name. In fact, it is a useful marketing method, which is high speed, low-cost and highly targeted. However, faced with the network marketing, some enterprises either be eager for instant success and quick profits or don't know what course to take, as the result they made the mistakes when they applied the principle of network marketing, and formed a malignant marketing that is similar to the real virus — "virus" marketing. Compared with the former, this marketing method is more focused on the efficiency of transmission rather than the product itself. "Virus" marketing is pervasive and sends junk mails and information by groups in different kinds of websites and Blogs, attempting to replace the marketing quantity with the number of transmission. Perhaps, this kind of marketing method may increase the views of the enterprises' Web site in the short term and achieve a certain effect of network marketing, but on the long-term interests, the method will be harmful to the enterprises' image, reducing consumers' confidence and the abnormal idea of blind pursuing high returns and low-paid even will make the enterprises loss the opportunities of network marketing.

### 3.5 The Fifth Misunderstanding: SMEs (Small and Medium Sized Companies) Are Incapable of Carrying out Network Marketing

According to the survey of shandong commission shows that in Shandong, the application of internet mainly focus on the level of information exchange, 16% of enterprises use E-commerce functions, 10% of enterprises use the third-party e-commerce platform, but only the 13% of enterprises really applied e-commerce to the business activities. It proved that the majority of SMEs just regarded the network as the platform for gathering and exchanging information [6]. Some bosses considered that because of the relatively small scale and limited financial resources, SMEs often lack the necessary infrastructures and high-end network marketing professionals, which keeps the majority of SMEs away from the network marketing. However, with the steady development of Chinese e-commerce, the problem of lacking facilities and talents is being eased gradually, a number of educational institutions have offered the training of network marketing and the construction of logistics and some other hardware facilities has also been in large-scale. What the most important is that as the stepping-stone of the new internet economic era, although we can not subjective consider that network marketing will be able to bring huge profits for the enterprises, without doubt it represents business opportunity, and the opportunity is often associated with the wonderful. Moreover, the interaction of network marketing contributes to the achievement of enterprises' goals, the 24-hours service of network marketing helps enterprises establish the corporate brand image and increase loyalty of customers; and finally its low-cost even can strengthen the competitive advantages for enterprises.

# 4 Five Application Points of Network Marketing

A rolling stone gathers no moss, if an enterprise wants to maintain smooth operation; good marketing channel plays an important role. The application of network marketing not only can contribute to the promotion of the traditional marketing channel, improving product sales and shortening the time of financial turnaround but also can reduce the cost of marketing communications through the network platform, realizing the interactions between consumers and businesses. So operating network marketing correctly is important to enterprises' success.

## 4.1 Holistic Thinking and Strategy Formulation

Network marketing is the main tool of competing for virtual market, but it does not mean that network marketing is "fast food culture", which can be grasped immediately. Like playing the chess, need to think twice before acting, If enterprises wants to depend on network marketing to rise, it also need holistic thinking and strategy formulation. Thinking determines the outlet, what the enterprises need to think firstly is whether the business opportunity has the development prospect. Secondly, the network marketing is a line composed by two points--online sales and logistics operation. So it should be comprehensive and thoughtful when enterprises make plan for the network marketing, especially the logistics, was regarded as the e-commerce's "bottleneck" by many enterprises. However, if there is no corresponding facilities' support, but we are still expecting the network marketing to create the sales myth, which is impossible and unrealistic. Thirdly, it is the marketing strategy. A novel marketing strategy can attract customers and increase customers' loyalty in the first time.

Amazon's rise depends on its unique and efficient logistics distribution. Facing the crisis of e-commerce, Amazon developed logistics as promotions. Furthermore Amazon sacrificed the logistics cost to dominate the market and carried on the "new, simple, solid, speed" marketing strategy to attract customers expanding the market share. It is visible that even the network marketing also need to devise strategies, can it gain the victory.

### 4.2 Pay Attention to Brand Construction

Network marketing is divided into two camps, brand marketing and effective marketing. And now, the majority of SMEs are at the core of effective marketing. However, it is believed that the second-rate enterprises make the Battle of Prices and the first-class enterprises make the Battle of Brands. So SMEs must raise great attention to brand construction, if they want to make use of network marketing to develop.

*1) Ensure the product quality*
Ensuring the product quality is the basis for brand construction. Even the best marketing strategy and the lowest purchase cost can not turn the tide for the enterprise if consumers aren't satisfied with product quality.

*2) Provide consumers with more detailed information[7]*
Enterprises can provide consumers with more detailed company information, such as the marketing concept, composition structure, to enhance corporate professional image and increase consumers' confidence.

*3) Make full use of the existing brand awareness and influence[7]*
Internet Brands are the extension and development of the traditional brands. When the enterprises plan to create internet brand, they could make full use of the awareness and influence of their existing brand, and then register it as domain name, such as the Crest Web site: http://www.crest. Com

*4) Focus on the customers' feedback*
It is a respect that enterprises put focus on the consumers' feedback.. Through this method, enterprises can both win customers' favor and establish the good brand image.

*5) Perseverance*
Constant dripping wears away a stone, for the same reason, brand construction also needs time accumulation. Some famous brands which surround us in daily life, such as General Motors, Nike, KFC all have their own historical stories, thus, brand construction should be persistent for decades.

### 4.3 Applying the "Three More" Policy

The policy of "three more" refers to "read more classics", "think more cases" and "participate in more practices". As the saying goes, "watering the roots of a tree in order to let it grow stronger" Any studies of knowledge and skills can not be away from the basic theory. Reviewing the list of five network marketing mistakes, it is easy to find that some of the mistakes were caused by some misunderstandings which were taken out of context. Thus SMEs should read more classics of the network marketing to increase knowledge and enhance their theoretical training

Secondly, SMEs should think more cases. Learning without thinking, people will be fooled by appearance of knowledge. It is not conductive to the absorption of knowledge. Therefore, SMEs should strive to achieve "think twice" when they study various types of network marketing cases—Think of the marketing strategy, think of the key to success and think of the improvement program. By the way, enterprises can broaden their horizons and understand the complexity of the real problem of marketing. Thirdly, take part in more practices. If the condition permits, SMEs should participate in the actual practice of network marketing as much as possible, in order to enhance their physical operating capacity, accumulate marketing experience and eliminate the phenomenon of the armchair strategist.[3]

## 4.4 The Application of Integrated Marketing

Integrated marketing is a series of marketing work which contains brand planning, implementation and supervision in order to establish, maintain and spread brand and strengthen customers' relationship. Integrated marketing converge each independent marketing work into a whole to generate synergistic effect, which is integration. It is similar to the operation of machine, which doesn't depend on a gear or a lever, only because of the close coordination of parts, without the cooperation, even the most efficient machine is useless.

The application of integrated marketing is also conducive to the realization of the concept of goal-oriented. Consumers are at the core of the integrated marketing, thus, the integration is used to strengthen the communication with the stakeholders through the market and transfer the focus of marketing activities and communication activities to the "the concept of goal-oriented " as far as possible, forming a circle which is at the center of target. In addition, the integrated marketing also has many other effects such as maximize the effect of communication; reduce the transaction costs and etc.

## 4.5 Varied Network Marketing Ways

Network Marketing is "synchronized swimming ", which need a pairs of slender legs and a set of nice gestures to "seduce" the audience. Through the network "combination punches," Barack Obama won the crown of president in one fell swoop [8]. In Mr Obama's campaign website, users can click and watch Obama's video, purchase products with Obama' s mark online and download his speech as ring for the mobile phone directly. The multifarious network marketing makes Obama swim in the oceans of the Internet, pulling users' mouse like performing the "synchronized swimming".

Network means to Obama just like TV to Roosevelt, which is a two-way breakthrough in thinking and Technology. Since the initial product sale to recent concept sale, and then to today's political image sale, the area of network marketing has been constantly expanding. Heinz promoted his product sales by guiding consumer's concept and Obama regarded internet as the propaganda position, displaying his image to win votes. Network marketing now is transferring from real to virtual which tends to be more abstract and complicated. However its effect is acting in an opposite way, rendering the results gradually. Various ideas of network marketing make direct sale not the only shortcuts to increase sales. It may give us an unexpected effect if we save the nation through twisted means occasionally.

## 5 Conclusion

The characteristic of internet is low cost, convenient, quick, and open-source and brings the sunshine for the SMEs, especially in the financial crisis. However, because of the limitations of concept, quite a few SMEs produced some ambiguous of network marketing, and then formed various misunderstandings. It is noteworthy that in the process of seeking solutions, SMEs can't only learn experience from failure. Only regarding theories as dot and five measures as radius, can enterprises draw a round for themselves.

## References

1. Li, Q.: Overview of foreign e-commerce development. MESSAGE Haiding (5) (2001)
2. The 24th Internet Report III: Basic Rescourses Overview [EB/OL], 07–16 (2009), http://tech.qq.com/
3. Liu, X.: Network marketing introduction, pp. 33–34. Tsinghua University Press (2005)
4. Kotler, P.: China top brands (1), 32–35 (2001)
5. Hu, Y.: From the yin haiwei to Sims blog [OL/EN] (June 2008), http://bbs.city.tianya.cn/
6. Liu, Y.: Several major network marketing misunderstandings of the financialk crisis [OL] (March 2009), http://www.efu.com/
7. Jiang, X.: Cyber- Marketing. Tsinghua University Press (2003)
8. Shanghai shopping district [OL/EN] (January 2009), http://bbs.china-china.biz/

# Research of RFID Service Frame Based on OSGi

He Fugui and Zhang Mei

Beijing Vocational College of Labour and Social Security
Beijing, China
hefugui@163.com, zh_mei2005@sina.com

**Abstract.** The paper researches RFID service frame based on OSGi, OSGi is an oriented-service model, RFID provides a new type of service. The paper presents related research, and proposes a new RFID service frame based on OSGi, the frame may be used in mobile device. The paper presents the structure, and describes realized method for each part. Because mobile devices are popular the frame will has a widely use in future.

**Keywords:** RFID, OSGi, Service Frame, Reader, bundle.

## 1 Introduction

OSGi(Open Service Gateway Initiative) is an oriented-service model in essential, it manages some software components: bundles, these bundles run on OSGi and may be regarded as application or library, these bundles are some services in fact and can be install dynamically. OSGi[1] is initially designed for home gateway. Because OSGi is an open structure, more and more devices to use services are using OSGi. RFID(Radio Frequency Identification) is a new technique for automatic identification and is widely used in logistics, RFID service is provided by RFID middleware, RFID service may fuse with other service by RFID middleware to run on OSGi, thus system may provide better service.

RFID is a network system, it own large scale, and its maintenance is difficult. OSGi is dynamic, it may be easily configured by CM(Configuration Admin Service), and OSGi supports dynamic managing for services, these services can be automatically installed, run, uninstalled. OSGi may link distributed RFID system by RFID middleware.

## 2 Analyses of Related Research

Reference[2] explain that middleware in next generation is dynamically configurable and can satisfy special application and supports cross-platform.

Reference[3] presents a light RFID system structure based OSGi, see Figure 1, software modules of RFID are divided into different service layer, such as device layer, event layer, communication layer and data agent layer. The communication layer use SOAP(Single Object Accesses Protocol) protocol.

Reference[4] presents RFID service frame based on plug-in for mobile device, see Figure 2, it is consist of seven parts, each part is consist of a clan of plug-in, the plug-in can dynamically replace, each plug-in adopt independent interface that separate with realization.

The RFID service frame[5] researched by Shanghai Jiao Tong University see Figure 3, the structure orients service, and can provide configurable ability for developing RFID middleware, and support different RFID application.

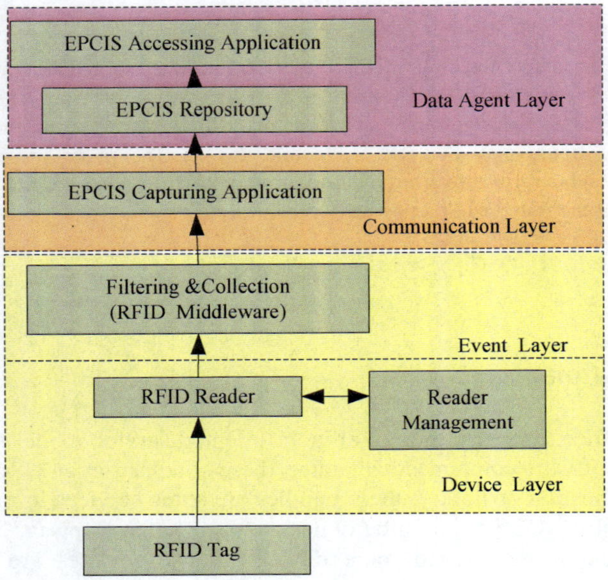


**Fig. 1.** A light RFID system structure based on OSGi

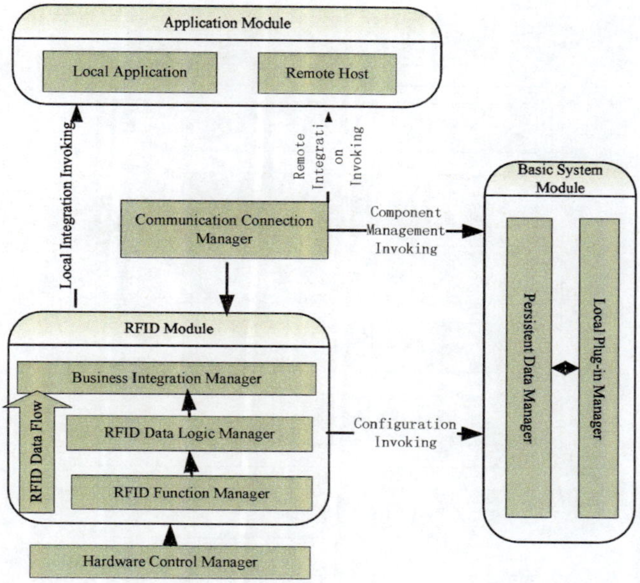


**Fig. 2.** RFID service frame on mobile device

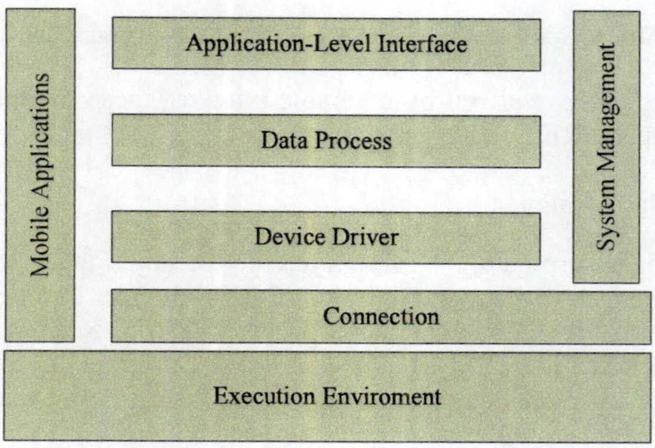


**Fig. 3.** Structure of mobile RFID service

## 3 Configurable RFID Service Frame Based on OSGi

In future, RFID will have an extensive application, for example in the product manufacturing, supply chain and family service etc. RFID will be a network type application because of RFID characteristics; in fact, RFID application is a kind of service. OSGi is a kind of oriented-service frame, with developing of RFID and OSGi, the RFID service frame based on OSGi will be popular.

Based on analysis of existing research, the paper provides a new RFID service frame based on OSGi: Configurable RFID service frame based on OSGi, see Figure 4.

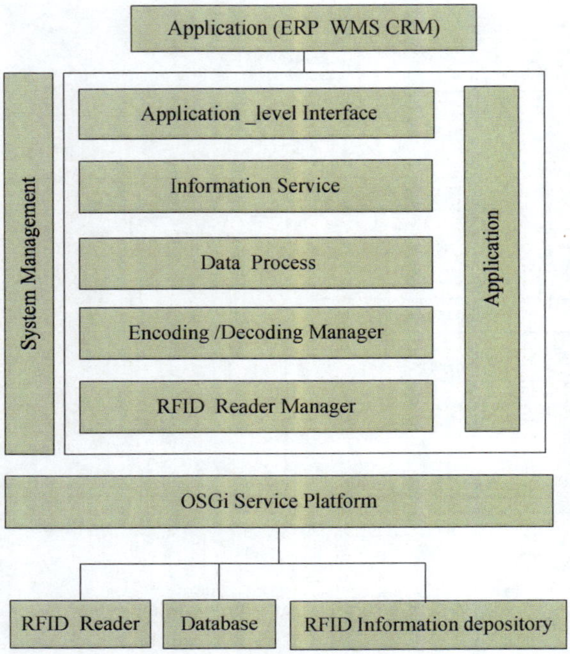


**Fig. 4.** Configurable RFID service frame based on OSGi

Each RFID function is realized by assemble bundles, these bundles run on OSGi, these bundles complete RFID function altogether.

### 3.1 RFID Reader Manager

RFID Reader Manager provides a unified interface for upper layer, and also provides some filter for primary-level events, the frame of RFID Reader Manager, See Figure 5, it include three parts: RFID logic Reader, Device Profile, Reader. Different RFID Reader provide a unified logical by the Device Profile, user may configure the Device Profile through the interface that system provides. Future RFID reader will support UnPNP and Jini technique, OSGi has a good support to them.

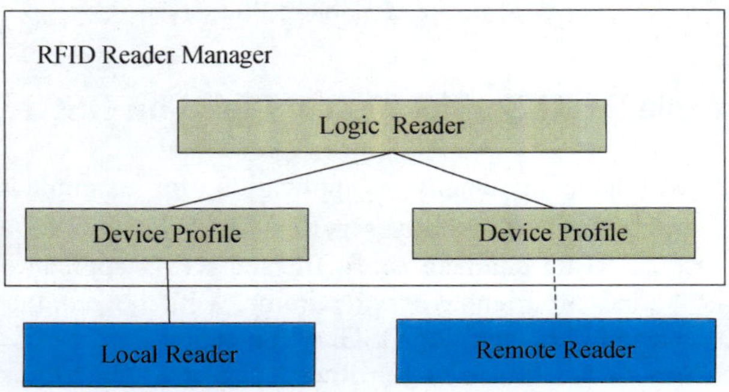


**Fig. 5.** Frame of RFID Reader manage

### 3.2 Encoding/Decoding Manager

The standard of RFID is different in different region, at present, a unified RFID' standard do not exist. RFID encoding/decoding plug-in is like adapter, user may use different RFID code. Microsoft RFID infrastructure [6] also use the type of plug-in. in order to use the plug-in in OSGi, the plug-in is realized by OSGi bundle because OSGi bundles can dynamic load and unload, data queue save the data from Reader. The frame sees Figure 6.

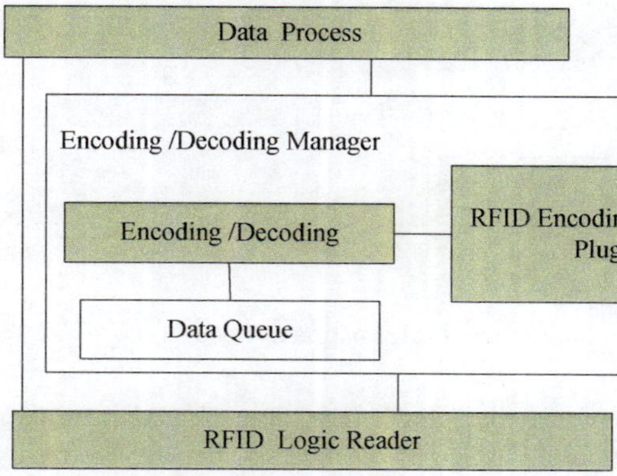


**Fig. 6.** Frame of Encoding/decoding manager

### 3.3 Data Process

The data to obtain from RFID Reader need to filter and aggregation, at present some better Reader contain data aggregation and filtration.

Reference[7] provide a event-handled frame based on XML, XML describe the rules that dada filter, user may modify these rules.

OSGi itself contains XML interpreter, and has a good support to XML. User may subscribe or register event in RFID Reader manager. The structure is Flexible because data filtration and aggregation can be dynamically replaced, the frame see Figure 7.

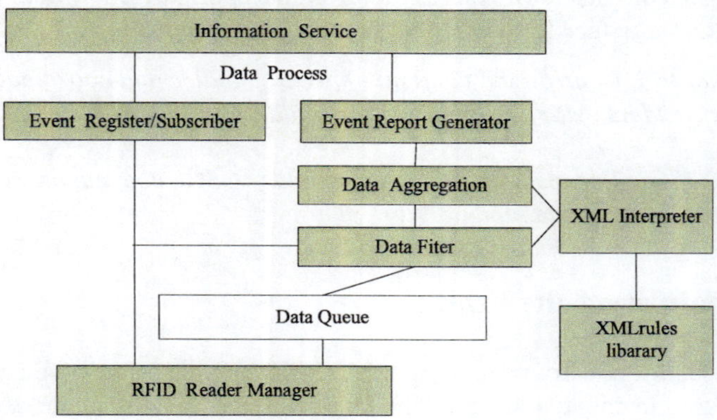


**Fig. 7.** Frame of Data Process

### 3.4 Information Service

Information Service realizes advance dada process, and complete protocol conversion, see Figure 8.

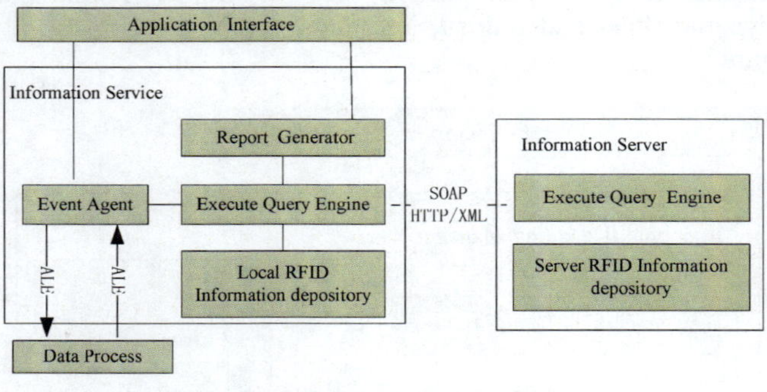


**Fig. 8.** Frame of Information Service

Each RFID label marks an object, the object has some related information, such as product name, date of manufacture, place of production, period of validity, price, and catalogue, etc. if the object is a product. Information Service mainly completes to process these data.

The part mainly include event agent and EPC(Electronic Product Code) information manage

*1) Event agent. According to different event agent will turn to different handling, and provide ALE(Application Level Events) interface.*

2) *EPC information manages. Provide EPC- related search, Query engine may query local RFID information depository, remote query will request a XML format Documentation.*

### 3.5 System Manager

System manager completes to manage, control and configure for whole system; and provide operating interface for user. Main function is as follows.

*1) System control. Control and configure special bundles and coordination them.*

*2) Manage Readers. Manage and monitor Readers to link to system, display their state.*

*3) Operate OSGi. Encapsulate OSGi commands, OSGi commands may complete bundles to view, load, unload, update, start, stop*

## 4 The Features of the Frame

The paper presents a new frame: Configurable RFID service frame based on OSGi, the frame can be used in client, not server; the frame may be used in mobile device.

Compare with existing RFID service frame based on OSGi, the frame possess following characteristics.

- In device manager, the device profile is used; system may operate different RFID Reader, such as remote Reader.
- Encoding/decoding plug-in is used, the system possess compatibility and can use different RFID protocol and satisfy future using. Encoding/decoding plug-in may be dynamically updated.
- When process data, the rules adopt XML language, thus user may configure these rules, the Flexibility increase.
- Information service may inquire local RFID information library and remote RFID information library, the remote library also may is not the library to run on OSGi, such as SUN RFID server.
- The whole system emphasize the configuration , and make full use of the OSGi' feature, and is suitable for future development

## 5 Conclusion

In future, oriented-service applications are popular, with the widely use of mobile device, the services to run on mobile device will be better used. RFID is a new type technique for automatic identification, and OSGi is a configurable service frame. The paper presents the Configurable RFID service frame based on OSGi, and describes realization for each part. At present, practical application of the frame is further researching.

## References

1. OSGi technical whitepaper, `http://www.osgi.org/documents`
2. Khedo, K.K.: Requirements for Next Generation Middleware Implementations. In: International Multi-Conference on Computing in the Global Information Technology, ICCGI 2006, pp. 53–53 (August 2006)
3. Sun, K.-H., Hou, T.-W., Wu, Z.-Y.: Pluggable RFID components and a Lightweight RFID System based on OSGi. In: International Conference on Consumer Electronics 2007, ICCE 2007. Digest of Technical Papers, January 10-14, pp. 1–2 (2007)
4. Liu, J., Wang, D., Sheng, H.: RFID Service Framework with Plug-in Technology on Devices. In: 2007 1st Annual RFID Eurasia, September 5-6, pp. 1–5 (2007)
5. Jie, W., Dong, W., Huanye, S.: Design an OSGi Extension Service for Mobile RFID Applications. In: IEEE International Conference on e-Business Engineering, ICEBE 2007, October 24-26, pp. 323–326 (2007)
6. Microsoft RFID infrastructure, `http://www.microsoft.com/china/CTC/Newsletter/200605/RFID.htm`
7. Xu, J., Cheng, W., Liu, W., Xu, W.: XML Based RFID Event Management Framework. In: 2006 IEEE Region 10 Conference, November 14-17, pp. 1–4 (2006)

# Analysis of the Radiation Mode of the Leaky Coaxial Cable with Vertical Slots

Qian Jiansheng[1], Yang Chen[1], Hu Qianqian[1], and Zhou Liujun[2]

[1] School of Information and Electrical Engineering
China University of Mining and Technology (CUMT)
XuZhou, China
qianjsh@cumt.edu.cn, {xiaoyuer_2005,huqian851122}@126.com
[2] Xuzhou institute of Architectural Technology
XuZhou, China
ljzhouxz@163.com

**Abstract.** Starting from the Floquet theorem, the radiation mode of leaky coaxial cable is studied detailedly in this article, the existing condition of the radiation mode is given, the work frequency band of the radiation direction is analyzed and the radiation direction mode under each radiation mode is discussed. To manifest the leaky cable's each radiation mode intuitively, some concrete parameter is designed in the paper. By using the matlab software simulation, the field pattern and radiation direction pattern are emulated in this paper.

**Keywords:** leaky coaxial cable, radiation mode, radiation direction, frequency band.

## 1 Introduction

A radiating cable functions both as a transmission line and as an antenna simultaneously. Through a series of slots or gaps on the outer conductor. The leaky coaxial cable (also called LCX for short) can partly leak out the electromagnetic wave energies to the space, and the field intensity will be weakened evenly along the route without any fluctuation.

The main merits of leaky coaxial cable are as follows:The band width;Even signal coverage area;Wide range of application;Low cost.

In view of the advantages of the cable, it is used widely in subway, tunnels, underground shopping malls, mine fields and other closed space[1].

There are three main performance indicators of LCX[2]: the using frequency band, the coupling loss and the transmission loss. This paper will focused on the discussing of LCX's radiation pattern.

## 2 The Field Distribution of LCX

The aim in notching on outer conductor of LCX is to signal radiation energy from the cable out,so that we can gain the ends that transmitting and receiving radio waves.

Figure 1 shows the LCX with slot cycle P. The slotted outlet location point will produce surface waves along the cable which spread positively or reversedly and affect each other.

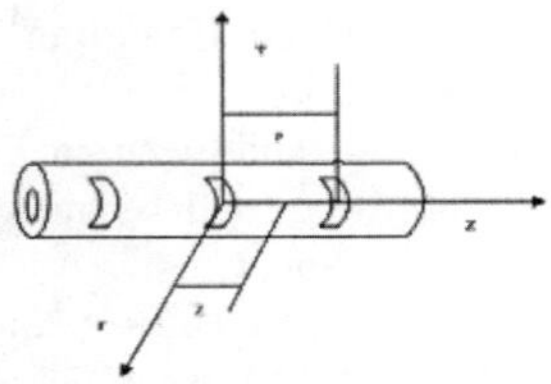

**Fig. 1.** LCX slotted structure and cylindrical coordinates

The following will show the magnetic field strength around LCX with cylindrical coordinate:

$$E(r,\varphi,z) = M(\eta,r,\varphi)Z(z)e^{j\beta_0 z}e^{-\alpha z} \tag{1}$$

$\beta_0$: Propagation constant of guided wave in LCX, which can be obtained from $k_0\sqrt{\varepsilon_r}$;

$\alpha$ : The attenuation constant of guided wave;

$\varepsilon_r$: The relative dielectric constant of medium in LCX.

$M(\eta,r,\varphi)$ in the formula above is a function related to $\eta$ and $r,\kappa$, it can be expressed as:

$$M(\eta,r,\varphi) = \sum_{n=-\infty}^{+\infty} F_n[J_n(\eta_r),H_n^{(2)}(\eta_r)]G_n(\varphi) \tag{2}$$

$F_n$ in formula (2) is a operator which is related to specific boundary conditions and the radial component of electric field. $J_n(\eta_r)$ is the first class Bessel function and $H_n^{(2)}(\eta_r)$ is the second Hankel function; $G_n(\varphi)$ is a function about $\varphi$, n is an integer.

According to Floquet theorem, in structure that has infinite length and periods, the field of the corresponding points of each cycle, is only different in a complex of constant coefficients, so Z (z) in formula (1) is a function whose period is P.Its Fourier Expand Series is [3]:

$$Z(z) = \sum_{M=-\infty}^{+\infty} Z_m e^{-j\frac{2m\pi}{P}z} \tag{3}$$

"m" is an integer, "Zm" is a Fourier coefficients.Bring (2) into (1):

$$E(r,\varphi,z) = e^{-\alpha z}\sum_{m=-\infty}^{+\infty} Z_m M(\eta_m,r,\varphi)e^{-j\beta_m z} \tag{4}$$

of which:

$$\beta_m = \beta_0 + \frac{2\pi m}{P} \tag{5}$$

The formula named (4) is just the expressions of electric field strengths around the cable.

## 3 Radiation Patterns and Radiation Angle of LCX

### 3.1 Radiation Patterns of LCX

If $\eta_m^{\ 2} \leq 0$, formula (2) will turn into:

$$M(\eta_m, r, \varphi) = \sum_{n=-\infty}^{+\infty} F_n{}'[I_n(\eta_{mr}{}'), K_n(\eta_{mr}{}')]G_n(\varphi), \eta_m{}' = \sqrt{-\eta_m^2} \tag{6}$$

$I_n(\eta_{mr}{}')$ and $K_n(\eta_{mr}{}')$ are modified Bessel function.

The field fall in radial direction without radiation. Only when $\eta_m^{\ 2} \geq 0$,there is radiation in order to ensure the radial.

The longitudinal wave propagation constant and the radial propagation constants satisfy the following relationship:

$$\beta_m^{\ 2} + \eta_m^{\ 2} = k_0^{\ 2} \tag{7}$$

From the discussion above we can see:

$\beta_0 = k_0\sqrt{\varepsilon_r}$, $k_0 = 2\pi f / c$

Bring it into (3-2):

$$\eta_m^2 = (2\pi f/c)^2 - (2\pi f\sqrt{\varepsilon_r}/c + \frac{2\pi m}{P})^2 > 0 \tag{8}$$

We can obtain:

$$-mf_1 < f < -mf_2 \tag{9}$$

In which:

$$f_1 = \frac{c}{P(\sqrt{\varepsilon_r}+1)}, \ f_2 = \frac{c}{P(\sqrt{\varepsilon_r}-1)}$$

"c"is the propagation velocity of electromagnetic waves in free space, whose unit is "m / s".

By (9) we can see:

LCX radiate radiation wave to space, taking central axis of LCX as apex angle and taking cone-angle as constant phase front [4].When "$f < f_1$", there was only the surface wave; when $f > f_1$,the minus one sub-module began to radiate; when $f = 2f_1$,the minus two sub-module started radiating, it meant that when $f > 2f_1$, high-order modes began to appear and interfere with each other. Therefore, $f_1$ to $2f_1$ is the single-mode radiation area, which have a high practical value.

### 3.2 The Radiation Angle of Radiation Mode of LCX

$\sin\varphi_m = \beta_m / k_0$ showed the relationship between the radiation direction and the propagation constant of various models. In which:

$\beta_m = \beta_0 + \frac{2\pi m}{P}$, $\beta_0 = k_0\sqrt{\varepsilon_r}$, $k_0 = 2\pi f / c$

Bring them into $\sin\varphi_m = \beta_m / k_0$, we can obtain:

$$\sin\varphi_m = \sqrt{\varepsilon_r} - \lambda / p \tag{10}$$

Which means:

$$\varphi_m = \sin^{-1}(\sqrt{\varepsilon_r} - \lambda / p) \tag{11}$$

When m =-1, $f = f_1 = \frac{c}{P(\sqrt{\varepsilon_r}+1)}$, $\varphi_{-1} = -90^0$ ,that meant that there is only the surface wave; if m <-1, there's only surface wave, too; when m>-1, the-1one sub-module began to radiate; with the increasing of "f", the angle of radiation named $\varphi_{-1}$ would increase gradually,when the "f" =f2, $\varphi_{-1} = 90^0$, the radiation wave became into the surface-wave again; increase"f" continuously , when "$f > 2f_1$", -2 times ,-3 times, -4 times ... ... high-order mode began to appear.

When "m =- 1"or "m <-1", only the surface waves exist. But because of its weak anti-interference ability, short transmission distance in space and the bad overall usefulness, surface waves is not desirable. Radiation wave is a good choice for transmitting signals. Radiation waves of high-order mode would seriously interfere with the -1 times mode and they also interference each other, which led to large fluctuations of electromagnetic field[5].

The following examples simulated with MATLAB reflected the analysis above intuitively.

The parameters of LCX were set as follows:
Length: L = 70m; Slot cycle: P = 0.333m;
Relative dielectric constant: = 1.25;
The speed of light: c = c=$3\times10^8$ m/s;

According to $f_1 = \frac{c}{P(\sqrt{\varepsilon_r}+1)}$, we can figure out: $f_1$ = 425.35MHz. So the LCX's single-mode radiation band is 425.35MHz—850.70MHz. Figure 2 showed the radiation patterns in various frequency.

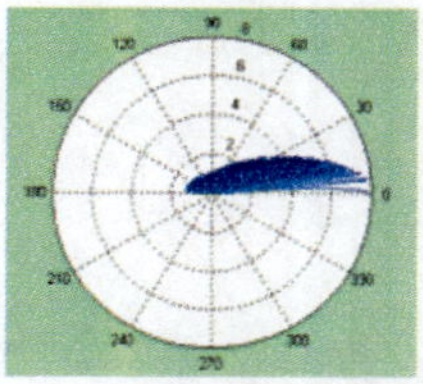

(a)f=300MHz

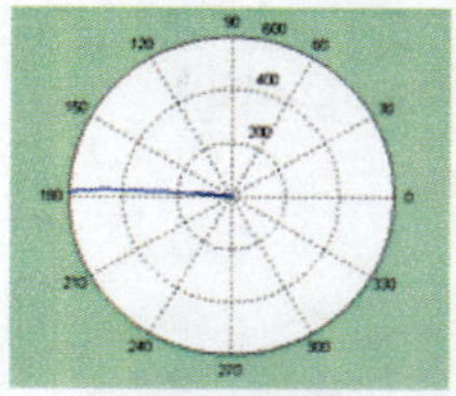

(b)f=425.35MHz

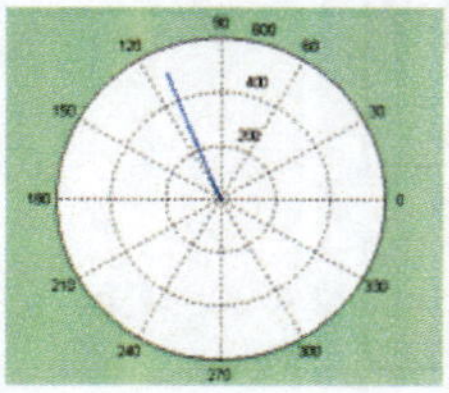

(c) f=600MHz

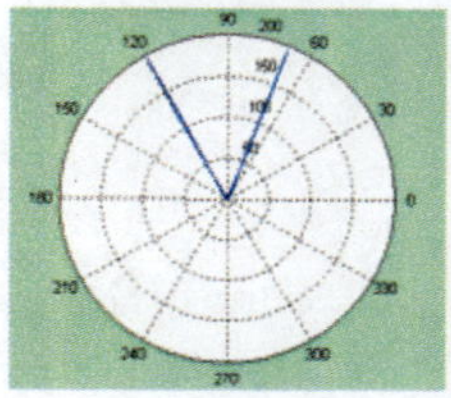

(d)f=1000MHz

**Fig. 2.** Radiation patterns of LCX

As shown in figure (a): $f$ = 200MHz, there was only surface waves and radiation directions were very scattered; In figure (b), $f = f_1$=425.35MHz, $\varphi_{-1} = -90^0$, which was consistent with analysis; In figure (c) $f$ = 600MHz, there was only one direction of radiation ,LCX worked in single-mode region. The radiation was stability; In figure (d) $f$ = 1000MHz.There were two radiation directions. It meant that a variety of models had begun to emerge.

## 4 Three Kinds of Band in Different Radiation Patterns

We analysed the conditions for LCX to produce radiation wave, and showed the radiation directions of each mode with simulation diagram. In order to discuss the three radiation patterns deeply, we will analysed the bands of each mode correspondingly. According to interference that came from the outside space, the radiation area can be divided into the following three-band [6]:

(1)Surface wave mode: $f < f_1$;

(2)Single-mode radiation mode: $f_1 \le< f \le 2 f_1$;

(3)Multi-mode radiation mode: $f > 2 f_1$.

In order to describe the three working bands more intuitively, we still use a simulation to analyze the problem. Set the parameters as follows:

Length: L = 1000m; Slot cycle: P = 0.333m;

Relative dielectric constant: = 1.25;

The speed of light: c = c=$3 \times 10^8$ m/s;

The distance between antenna and LCX: d = 2m.

So its single-mode radiation band is: 425.35-850.70MHz. The simulation results were shown in figure 3:

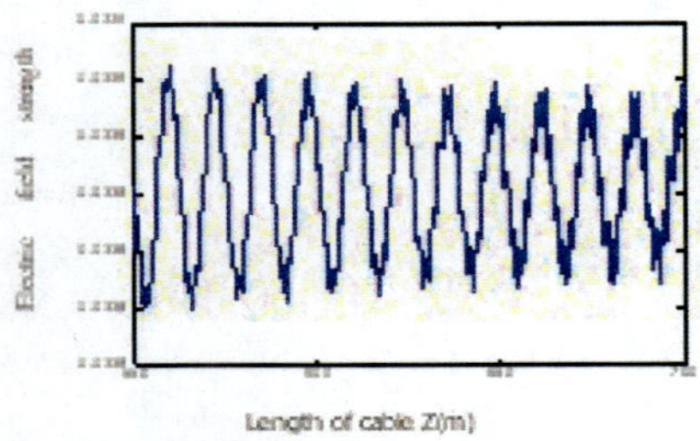


(a)Electric field strength distribution along the cable at f=200MHz

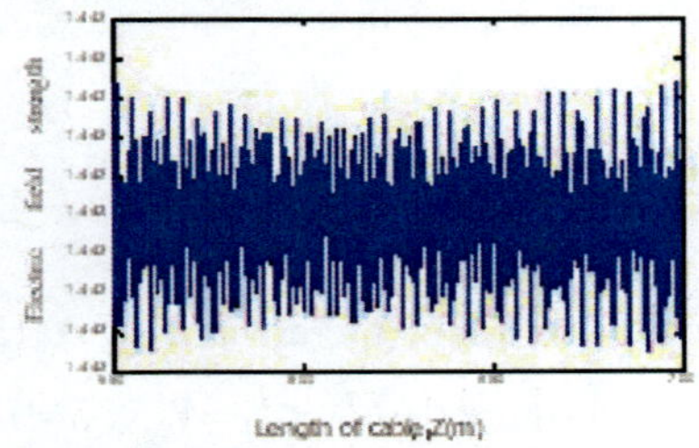


(b)Electric field strength distribution along the cable at f=600MHz

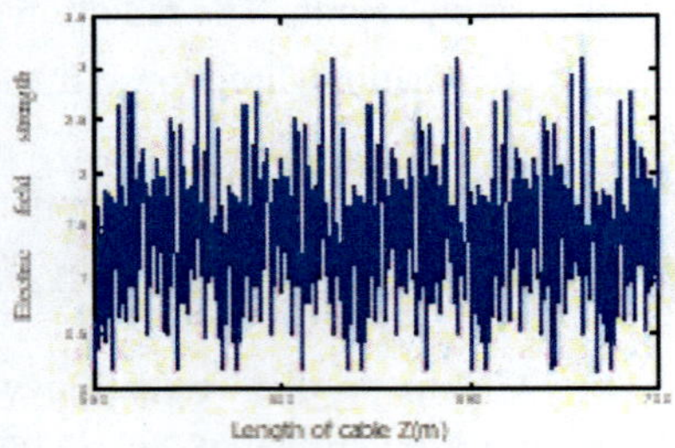


(c)Electric field strength distribution along the cable at f=1000MHz

**Fig. 3.** The electric field strength distribution of the three working bands

In figure (a), f = 200MHz, LCX work in the surface wave region. The maximum value of field strength is only 0.0339V / m at this time. We above ever mentioned that with the increase of the distance between antenna and LCX ,the electrical field strength of surface wave field would decreased rapidly ,the simulation results confirm it well; In figure (b), f = 600MHz. LCX worked in single-mode radiation area. The figure showed that there was only a little fluctuation in field strength, less than 0.001V / m or so. It shown that when in the single-mode radiation area, the field distribution is even and steady; When f = 1000MHz, LCX worked in multi-mode radiation region. The field strength is unstable and uneven,because of mutual interference between high-order mode.

In figure 4, we supposed that it was 500 meters away from the LCX. Set f= 200MHz, 600MHz 1000MHz respectively. The simulation diagram showed the process of distance between viewpoint and LCX. By comparing of the three curve, when in the surface wave region (f = 200MHz), the electrical field strength of surface wave field would decreased rapidly, with the increase of the distance between viewpoint and LCX.

Through the analysis above, it can be known that the higher single-mode radiation area is very useful. In order to improve its utilization efficiency, we must find ways to curb high order harmonic generation. The more, the better. Then the band will be wider, which means a lot to the practical application of LCX. At present, the measures to curb high order harmonic generation is mainly opening a series of new slots in the outer conductor. These slots must have the same size and shape as the original slot. Adjust the position between the old and the new slot and then high order harmonic generation can be suppressed out [7].

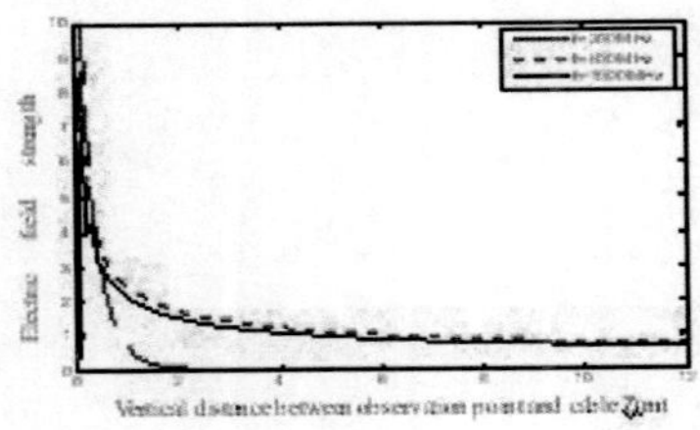

**Fig. 4.** Distribution of radiation field of different frequencies

## 5 Conclusion

In this paper, we discussed the of the radiation pattern of the vertical slot of the leaky coaxial cable theoretically. We known from the discussion that LCX can produce stable, homogeneous electromagnetic field to transmit signals in the single-mode radiation area. So it is hoped that LCX can only work in single-mode radiation patterns. With the rapid development of mobile communication, a wider frequency band is needed to deliver signal, so the study for ultra-wideband of LCX is of great significance.

## References

1. Shuai, F.: Study of leaky coaxial cables of single-mode radiation and ultra-wideband. Harbin Institute of Technology, Harbin (2007)
2. Li, Q.-W., Chun, Y., Liu, X.: Design for radiation leakage coaxial cable-type. Optical Fiber and Cable and Its Application Technology (4), 30–32 (2004)
3. Wang, J.-H., Jian, S.-S.: Analysis of leakage coaxial cable's radiation pattern and high-order mode suppression. Communication Journal 21(12), 18–22 (2000)
4. Kishimoto, T., et al.: LCX Communication System. In: IEICE17-18P, Japan (1982)
5. Sakabe, I.: Design of a VHF-UHFsuper-broadband leaky coaxial cable (LCX). Sumitomo Electr. Tech. Rev. (39), 57–61 (1995)
6. Sakabe, et al.: VHF-UHF ultra-wideband leaky coaxial cable. Optical Fiber and Cable and Its Application Technology (6), 41–45 (1995)
7. Wang, J.H., Mei, K.K.: Design of leaky coaxial cable with periodic slots. Radio Science 37(5), 2-1–2-10 (2002)

# Decision-Making Modeling of Military Engineering Repair Operations Based on Organizational Design

Fengshan Wang, Wanhong Zhu, and Houqing Lu

Engineering Institute of Corps of Engineerings
People's Liberation Army University of Science and Technology
Nanjing, China
wfs919@tom.com

**Abstract.** A decision-making modeling method was proposed towards military engineering repair operations based on Organizational Design Theory. According to material, energy, and information shape features, "material flow" and "information flow" was designed, which described organizational process on wartime repair operations, and then repair decision-making organizational design method was elaborated under the resource restraints and optimization. In this method, decision-making optimized model was established on wartime military engineering repair operations, which successfully achieved conformity and optimization on resource, task, entity, and other elements.

**Keywords:** organizational design, repair operations, military engineering, state, decision-making modeling, matter flow.

## 1 Introduction

Complexity is the essential property in military engineering wartime repair system. Wartime repair is not the linear relationship between the whole and the parts, but the open complex system with material, energy and information features, which showed a wide range of non-linear mechanism, and had random, fuzzy and other uncertain signs.

How to implement the integration and optimization in wartime repair system, is how to organize or design the repair strategy, which was the core issue in military engineering repair decision-making modeling. Dong-sheng Yang, and et al applied organization contingency theory into the complex battle space, and presented analysis and design method in the battlefield forces organization [1], which was successfully applied into the operational process design strategy and optimization [2,3], such as the strategic artillery operation issues [4], and provided a positive reference for the difficult problems in complex military engineering repair operations.

Organization and self-organization principle was the central content of Complex System Theory [5]. Organizational design theory as an important study tool of complex system theory, was applied into wartime engineering repair decision-making optimized structure, which aimed at the full integration of multi-dimensional repair forces, units and operations.

# 2 Organizational Process in Wartime Military Engineering Repair Structure

## 2.1 Organizational Structure on Wartime Engineering Repair Operations

Wartime military engineering repair operations ("$Org$") showed as one open system, including battlefield environment, organizational element, structure, process and organizational output. Under the organizational process of repair operations, when repair mission ("$Mis$") was translated into the outside repair task ("$Tas$"), association basis was provided among organizational elements, namely balancing the loads and task collaboration.

According to the requirement of repair mission, the organizational implementation could be seen as input and output conversion, which was linked by repair tasks. This conversion was "action flow", which existed within the organization or between the organization and the environment, including matter and information flow in the repair operations.

"Flow" design, meant to study out "Operation programs and plans", which finalized repair mission [1].

Set "$f_{Mat}$" as "matter flow" and "$f_{Inf}$" as "information flow" in organizational process about military engineering repair operation. Then, its conversion process could be described as the following relationship:

$$Org( f_{Mat}, f_{Inf} ) = f( Mis \rightarrow Tas ) \quad (1)$$

Organizational design aimed to obtain the best efficiency ("$ef$") in carrying out repair mission through material flow and information flow, described as following relationship:

$$Org'( f_{Mat}, f_{Inf} ) = \max\{ ef_{( f_{Mat}, f_{Inf} )}( Mis \rightarrow Tas )\} \quad (2)$$

"Matter flow" reflected organizational process strategy. In essence, it was how to design the best repair operations strategy and complete the complex repair tasks, under the limited repair resources. This design included how to generate interactive task sequence, and how to determine the mission requirements on resource, function and capacity.

## 2.2 Concepts and Symbols in Wartime Engineering Repair Organizational Strategy

Set function parameter as "$fun$", which was the functional measure about repair resources. And "$fun$" was the basis of relating organizational repair resource and mission. Set "$fun_1$, $fun_2$, ..., $fun_k$" as the description and sorts about all the repair resource functions within the repair organization, and set "$\overline{fun_1}$, $\overline{fun_2}$, ..., $\overline{fun_k}$" as the repair functional measure. Then, set "$F_{sys} = (\overline{fun_1}, \overline{fun_2}, ..., \overline{fun_k})$" as the organizational resource capacity state, in which "$k$" denoted the functional type number within the repair resource organization.

Define repair resource type as "$mal_i$", including such as repair force, equipment, material and other aspects, which was the carrier in organizing repair functional capacity. And set the constraint measure of repair resource platform as "$M_{sys} = (\overline{mal_1}, \overline{mal_2}, ..., \overline{mal_n})$".

Repair mission could be decomposed into various different sub-tasks in accordance with certain principle. Repair sub-tasks set were defined as "$Tas = \{tas_1, tas_2, ..., tas_m\}$", and resource capacity need was defined as "$W = \{wor_1, wor_2, ..., wor_k\}$", in which "$wor_j$ ($1 \le j \le k$)" meant the "$j$" type function to achieve the repair goal.

Define the sequence relation among the sub-tasks as "$Gr_{Tas}$", which described the reliant connections qualitatively, such as priority relation, basic information flow, in-out relation, and others. Set "$Gr_{Tas} = (V_{Tas}, E_{Tas})$", in which "$V_{Tas}$"denoted the task node, "$tas_i \in V_{Tas}$ ($1 \le i \le m$)", and "$E_{Tas}$" denoted the directional link among the sub-tasks, "$L_{ij} =< Tas_i, Tas_j > \in E_{Tas}$".

Repair task space was composed by "$Tas$" and "$Gr_{Tas}$", denoted as "$Roo_{Tas}$ =( $Tas, Gr_{Tas}$ )". "$Roo_{Tas}$"contains all possible organizational task process, which achieved the repair goal. Namely, if "$Gr'_{Tas}(roo_{Tas}) \rightarrow Mis$" appeared in certain "$roo_{Tas}$", then "$roo_{Tas} \in Roo_{Tas}$" existed surely.

### 2.3 Organizational Process Description on Wartime Repair Operations

"Matter Flow" design in military engineering repair organization could be decompounded into "$f_{Mat}(A)$" and "$f_{Mat}(B)$". "$f_{Mat}(A)$" meant the generation of repair task space, and "$f_{Mat}(B)$" meant the functional mapping process from task space to resource space. Military engineering repair organizational process could be expressed as the following formula.

$$\begin{cases} f_{Mat}(A):(Tas;Gr_{Tas}) \xrightarrow{\max\{ef_{Mat}(Gr_{Tas})\}} f_{Mat}(A) \\ f_{Mat}(B):(Roo_{Tas};P) \xrightarrow{\max\{ef_{Inf}(Gr_{Tas})\}} f_{Mat}(B) \end{cases} \quad (3)$$

Where in "(3)", "$P$" meant the organizational resources space. "$Tas$" was to achieve the repair efficiency ("$ef_{Mat}$"). Moreover, and the mapping from task space to resource space, was to establish the operational efficiency ("$ef_{Inf}(Gr_{Tas})$") of repair resources.

## 3 Decision-Making Organizational Design on Wartime Military Engineering Repair Operations

"$Roo_{Tas}$" was limited by organizational repair resource status. And, "$fun$" basically carried on the relationship between repair task space and organizational resource status. The corresponding relation must be erected between the capacity requirement of task space ("$W(Roo_{Tas})$") and resource function. This relationship could be expressed as the following description in "(4)".

$$W(roo_{Tas}) \xleftrightarrow{fun} F_{sys} = (\overline{fun_1}, \overline{fun_2}, ..., \overline{fun_k}) \tag{4}$$

In the repair task space ("$Roo_{Tas}$"), the capacity need of task process ("$Gr'_{Tas}$") must be limited to organizational resource capacity state ("$F_{sys}$"), as shown in "(5)".

$$\begin{cases} tas_1(wor_i) + tas_2(wor_i) + ... + tas_m(wor_i) \le \overline{fun_i} \\ W(roo_{Tas}) = [\sum_{i=1}^{|V_{Tas}|} tas_i(wor_1), ..., \sum_{i=1}^{|V_{Tas}|} tas_i(wor_k)] \le F_{sys} \end{cases} \tag{5}$$

"$f_{Mat}(A)$" expressed the optimal process under the organizational repair resource capacity constraints. Hereby, repair task set must carry out the goal, the requirements from task to resources must correspond with repair resource capacity one by one through the functional requirements, and the repair capacity could not exceed the repair resource capacity.

Therefore, "$f_{Mat}(A)$" was designed as the following mathematical description.

$$st\begin{cases} \max \quad ef_{Mat}(Gr'_{Tas}) \\ W(roo_{Tas}) \xleftrightarrow{fun} F_{sys} = (\overline{fun_1}, \overline{fun_2}, ..., \overline{fun_k}) \\ Gr'_{Tas}(roo_{Tas}) \rightarrow Mis \\ W(roo_{Tas}) = [\sum_{i=1}^{|V_{Tas}|} tas_i(wor_1), ..., \sum_{i=1}^{|V_{Tas}|} tas_i(wor_k)] \le F_{sys} \end{cases} \tag{6}$$

Under the functional mapping process from "$Tas$" to "$M_{sys}$", "$f_{Mat}(B)$" organized repair resource distributing strategy, which aimed at the best operational program for "$f_{Mat}(A)$". "$f_{Mat}(B)$" mainly considered on the repair efficiency as "$ef_{Inf}$", whose maximal goal could be interpreted as minimizing the spent time.

Therefore, "$f_{Mat}(B)$" was designed as the following mathematical description.

$$\begin{aligned} f_{Mat}(B): & \max\{ef_{Inf}(roo_{Tas} \xleftrightarrow{F_{sys}} M_{sys})\} \\ & = \min\{Time(roo_{Tas} \xleftrightarrow{F_{sys}} M_{sys})\} \end{aligned} \tag{7}$$

Where in "(7)", "$ef_{Inf}(roo_{Tas} \xleftrightarrow{F_{sys}} M_{sys})$" expressed the mapping efficiency function, and "$Time(roo_{Tas} \xleftrightarrow{F_{sys}} M_{sys})$" expressed the mapping time function from "$Tas$" to "$M_{sys}$".

# 4 Decision-Making Optimized Model on Wartime Military Engineering Repair Operations

## 4.1 Organizational Description on Military Engineering State Evolution

At the " $t_k$ " moment, engineering status was denoted as " $Sta(t_k)$ ". Engineering status conversion (" $Tr_{sta}$ ") occurred under the joint action of organizational repair operations and damage attack events, whose state evolution constructed the complex military engineering damage and repair situation.

Repair strategy was denoted as " $Tra(Dec)$ ". Under the organizational repair process strategy (" $Dec$ ") in optimal coordination design of complex repair system, " $Tra(Dec)$ " described the evolution process of " $Sta$ " and " $M_{sys}$ ". Then, " $Tra(Dec)$ " could be described with the status association equation about repair operation, damage case, and repair efficiency [6], as shown in "(8)".

$$\begin{cases} Sts(t_k+1) = \alpha(Sta(t_k), roo_{tas}(t_k), Dam(t_k)) \\ M_{sys}(t_k+1) = \beta(M_{sys}(t_k), roo_{tas}(t_k), Dam(t_k)) \\ roo_{task}(t_k) = \vartheta(Sta(t_k), M_{sys}(t_k)) \\ Dam(t_k) = \varpi(Sta(t_k), M_{sys}(t_k)) \end{cases} \tag{8}$$

Where in "(8)", " $Dam(t_k)$ " denoted the damage case at the time of " $t_k$ " moment. " $\alpha$ " denoted the engineering status conversion, and " $\beta$ " denoted the repair resource conversion, which was related with repair operation and damage events. " $\vartheta$ " and " $\varpi$ " respectively denoted the function relation of repair operations and damage case, which both associated with engineering status and resource status.

## 4.2 Strategy on Repair Organizational Process

Under the constraint condition of repair resources, the evolution of " $Sta$ " was limited to " $F_{sys}$ ", and met the restriction of " $M_{sys}$ ".

In the allowed condition of repair resource, repair capacity, and time resource, " $Dec$ " wanted to carry out the conversion about " $Sta(t_{Begin}) \rightarrow Sta(t_{End})$ " at the time of " $t_{Begin} \rightarrow t_{End}$ ". Set " $\Theta$ "as the feasible repair strategy set, namely " $Dec \in \Theta$ ". Then, the repair strategy process could be mathematical described, as shown in "(9)".

$$\Theta = \{ Dec \mid Tra( Sta( t_{Begin} ), Dec, Sta( t_{End} ))\}$$
$$s.t. \quad \begin{cases} F( Dec ) \le F_{sys} \\ Time( Dec ) \le Time( Org ) \le t_{End} - t_{Begin} \end{cases} \tag{9}$$

Where in "(9)", "$Tra(Sta(t_{Begin}), Dec, Sta(t_{End}))$" denoted the engineering status conversion process from the begin state ("$Sta(t_{Begin})$") to the end state ("$Sta(t_{End})$") under the organizational repair process strategy ("$Dec$"). "$Time(Dec)$" denoted the spent time resource. "$F(Dec)$" denoted the spent repair resource, which was consistent with the limited organizational resource capacity status, reflected in "(5)".

### 4.3 Wartime Military Engineering Repair Decision-Making Optimized Model

In essence, the organizational repair process strategy carried on optimal repair resource distribution, and generated the best repair operation in accordance with repair goal, which aimed at the best strategy in order to solve the resource conflict and time limitation [7].

Set "$h(mal_j \to fun_i)$" as the unit number of a certain repair resource ("$mal_j$") in carrying out a certain repair function ("$fun_i$"). Then, the repair resource must be limited in the resource constraints, as shown in "(10)".

$$\sum_{i=1}^{k} h(mal_j \to fun_i) \le \overline{mal_j} \tag{10}$$

Set "$v(mal_j \to fun_i)$" as the function measure of a certain repair resource ("$mal_j$") in carrying out a certain repair function ("$fun_i$"). Then, the role of this repair resource functions must satisfy this condition, as shown in "(11)".

$$\sum_{j=1}^{n} [v(mal_j \to fun_i) \bullet h(mal_j \to fun_i)] \ge [fun_i] \tag{11}$$

Where in "(11)", "$[fun_i]$" indicated the demand amount of "$fun_i$". Set "$pri_j$" represent the unit cost of "$mal_j$", and "$[pri]$" as the overall repair cost estimate. Then, the cost constraint could be set as the following expression.

$$\sum_{j=1}^{n} [pri_j \bullet \sum_{i=1}^{k} h(mal_j \to fun_i)] \le [pri] \tag{12}$$

Thus, as mathematical description of "(9)" under the repair process strategy was mapped into the comprehensive repair field, comprehensive repair process strategy was generated as the following mathematical description.

$$\begin{cases} \sum_{i=1}^{k} h(mal_j \to fun_i) \le \overline{mal_j} \\ \{[fun_i] \le \sum_{j=1}^{n} [v(mal_j \to fun_i) \bullet h(mal_j \to fun_i)] \le \overline{fun_i} \\ F(Dec) = \{ \sum_{j=1}^{n} [v(mal_j \to fun_i) \bullet h(mal_j \to fun_i)] \} \le F_{sys} \\ \sum_{j=1}^{n} [pri_j \bullet \sum_{i=1}^{k} h(mal_j \to fun_i)] \le [pri] \\ Time(Dec) \le Time(Org) \le t_{End} - t_{Begin} \\ i = 1,2,...,k;\ j = 1,2,...,n \end{cases} \tag{13}$$

Where, "(13)" reflected the resource and time constraints in the repair decision-making structure. Hereby, a comprehensive repair set as "$\Theta$" could be obtained through the reasoning and coordinating process.

Repair decision-making aimed at the maximal survival efficiency and reasonable cost-effective ratio. Under the resource-oriented constraints, the optimized model was established on military engineering comprehensive repair process strategy, which was expressed as "$E$".

$$E = \frac{\sum_{j=1}^{n} [v(mal_j \to fun_i) \bullet h(mal_j \to fun_i)]}{\sum_{j=1}^{n} [pri_j \bullet \sum_{i=1}^{k} h(mal_j \to fun_i)]} \tag{14}$$

Set "$E$"as objective function, then military engineering integrated repair model was constructed, in order to achieve the best comprehensive repair strategy as "$Dec^*$", namely:

$$Dec^* = \underset{Dec \in \Theta}{arg\,max} \quad P_e\{ E \mid Tra^*(Dec) \} \tag{15}$$

Where, "(13), (14), (15)" described the design of decision-making optimized model on military engineering repair operations, which was represented on comprehensive and integrated repair-oriented organizational theory in the field of complex repair system, carried on the global optimization and coordination.

## 5 Conclusion

Model design mastered the repair law thoroughly, carried on complex repair system rule, and carried forward the repair tactical development Then, this model must coordinate or optimize by means of heuristic dynamic algorithms, so as to obtain the feasible comprehensive repair strategy, and to build a comprehensive repair strategy set.

Some academic teams have carried out in-depth study of military organizational operations design, and achieved fruitful results [1]. This theory successfully solved the integration and optimization issues in complex military engineering repair decision-making system. However, due to complexity and dynamic points in military engineering damage case, and repair structure in the constant evolutionary progress, this design process was exceptionally complex. Therefore, the decision-making modeling method was only a kind of organizational design idea, which need test the effectiveness of organizational repair strategy with simulation and other means, and would be the next focus work.

## References

1. Yang, D.-S., Xiu, B.-X., Peng, X., Liu, Z., Zhang, W.: Analyzing and Designing the Military Organization in the Battlefields of Information Age. Journal of National University of Defense Technology 27, 116–122 (2005)
2. Bu, X.-J., Yang, D.-S., Sha, J.-C., Zhang, W.-M.: The Process Design Strategy and its Optimized Model for Operational Organization. Fire Control and Command Control 31, 8–12 (2006)
3. Peng, X.-H., Yang, D.-S., Liu, Z., Huang, J.-C.: Effect based Action Strstegy of Joint Operation. Fire Control and Command Control 32, 12–15 (2007)
4. Xie, L.-Y., Chen, L., Xueqin: The studt on Effect-based Artillery Operational organization strategy. Ship Electronic Engineering 27, 4–18 (2007)
5. Wang, Z.-Y., Ma, Y.-P., Li, K.: Study on Application of Compound System Theory in Joint Operations. Journal of System Simulation 15, 1675–1677 (2003)
6. Zhong, L., Zhang, W., Yang, D.-S., Huang, J.: Operations Plan System Technology. Defense Industry Press, Beijing (2007)
7. Yang, D.-S., Zhang, W.-M., Liu, Z., Zhu, C.: Research on methodology of designing C2 organizations. Journal of Systems Engineering 20, 606–612 (2005)
8. Liu, X.-D., Song, B.-F.: Theories and Methods for Decision Making in Complex Engineering System Conceptual Design: A Survey. Systems Engineering Theory and Practice 24, 72–77 (2004)

# Wartime Repair Decision-Making Framework for Military Engineering on Complex System Theory

Fengshan Wang, Wanhong Zhu, and Houqing Lu

Engineering Institute of Corps of Engineerings
People's Liberation Army University of Science and Technology
Nanjing, China
wfs919@tom.com

**Abstract.** To achieve conformity and optimization on resource, task, entity, and other elements, wartime repair decision-making framework was proposed towards military engineering based on complex system theory. On the basis of conceptual description about wartime military engineering damage and repair system, system characteristics were posted in repair structure, and then system laws in military engineering repair structure were constructed with complex system science. According to material, energy, and information shape features, wartime engineering repair operations were abstracted, decision-making framework was presented in the complex operation system. It provided certain reference on predominating repair laws thoroughly, and aimed to gain best operational efficiency.

**Keywords:** wartime repair, decision-making framework, system laws, complex system theory, military engineering.

## 1 Introduction

System counterwork is the typical feature in informational warfare operations, which outstandingly emphasizes the best "combining vertical" and "horizontal" among the operational organization of resources, platforms, information and decision-making, that is, "Stress the intensive combat forces, Pursue optimal operation effectiveness" [1].

Wartime engineering repair or maintenance is one important and necessary measure in the whole combat system, which must adapt to the war's "systems integration" features, and give full play to the system effectiveness of military engineering.

Complexity is the essential property in military engineering wartime repair system. Wartime repair is not the linear relationship between the whole and the parts, but the open complex system with material, energy and information features, which showed a wide range of non-linear mechanism, and had random, fuzzy and other uncertain signs.

How to implement the integration and optimization in wartime repair system, is how to organize or design the repair strategy, which was the core issue in military engineering repair decision-making modeling. Dongsheng Yang et al, applied organization contingency theory into the complex battle space, and presented analysis and design method in the battlefield forces organization [2], which was successfully applied into the strategic artillery operation issues [1], and provided a positive reference for the difficult problems in complex military engineering repair system.

Design about complex wartime engineering repair decision-making framework, namely the repair elements as independent nodes in complex wartime repair system, explored wartime repair laws from system conception and complexity angles. In accordance with organization and self-organization principle, let the wartime repair mission or tasks as the goal, carry on its own observation, judgment, join, transform, coordination, choice and action, adapting to the uncertain engineering damage and complexity with the changing and evolving of organizational structure and behavior.

# 2 Complex Wartime Repair Description on Military Engineering

## 2.1 Conceptual Description on Wartime Military Engineering Damage and Repair System

Conceptual design is the first and most creative phase in the complex engineering system design process [3]. In the abstract description of wartime engineering repair structure, primary principle is to satisfy the functional requirements. At the point of system view [4], the issue showed the input, output, and relationship among the state variables. When system input was converted into output, it showed a specific function.

Around the repair mission and function constraints, the related composition and phenomenon was abstracted from repair mission space, which expressed as entity, action, interaction and other recognizable elements. Engineering repair structure feature could be drawn from dynamic, static, functional levels, as shown in Fig.1.

Any stage or part in repair decision-making design must surround this function theme, which went throughout the design process. Function drove model structure design, which carried on study about essential operation characteristics from the static and dynamic perspective dimension.

Within the repair system, "Static dimension" expressed entities, and their static relationship and structure, including ammunition entity, military engineering entity, repair resource entity, whose static structure was established with inheritance, derivation, combination, aggregation, correlation, and other methods.

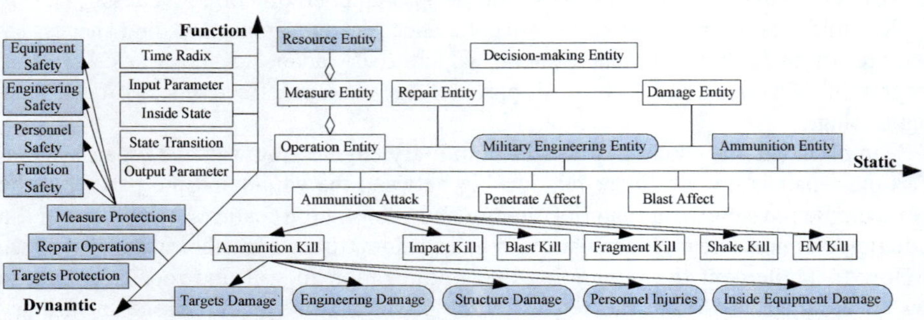


**Fig. 1.** Three dimension description about wartime military engineering damage and repair system

Wartime repair operations, actions, and other information was contained into "Dynamic dimension", which expressed the dynamic damage operation against military engineering, as well as the dynamic characteristics in the battlefield repair actions. For example, air attacks from ammunitions against military engineering, included ammunition penetration and explosion effects. It carried out the impact, explosive blast, fragmentation, shock and vibration, and other anti-personnel kill, which resulted in engineering damage, structure damage, personnel injuries, inside equipment damage and other target damage.

## 2.2 System Characteristics in Engineering Repair Structure

System integration is the essential requirement in the development of wartime military engineering repair system. This system represented complex mutual relationship, interaction, and penetration structure, and its elements, structure, environment, with the overall shape, structure, boundary, characteristics, action, function, as well as holistic space occupancy, time outspread and other features [5]. Therefore, it was a typical problem with complex system conception and characteristics.

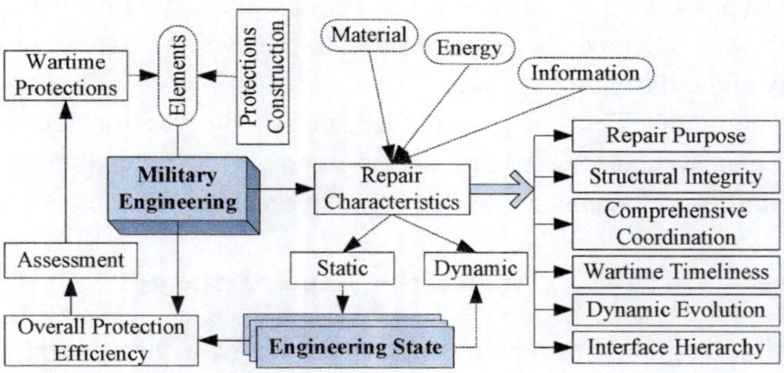


**Fig. 2.** System characteristics in engineering repair structure

Elements of complex repair structure weren't only independent of each other, but also interrelated, and mutual coupled, which showed a specific overall structure in accordance with the coordinated system, as shown in Fig.2, with some particular static and dynamic characteristic.

Fig.2 showed that the overall structure of complex repair system carried on coordination between peacetime and wartime, along with material, energy and information transfer and conversion, and leading to engineering status updates, with a non-linear, uncertainty, infinite-dimensional, distributed and multi-level features. State described the static characteristics of military engineering at a particular moment. Wartime repair actions promoted the different states transfer, reflecting the dynamic characteristics of military engineering wartime repair operations.

The overall objective of complex systems is to pursue the overall optimization [4]. As complex system theory was mapped into the military engineering repair fields, which called for global coordination and optimization design, so that all the elements could be in a well coordination condition, and complete repair mission together.

Global coordinative optimization of complex military engineering wartime repair system showed the following characteristics.

- Repair Purpose: The coordination of complex wartime repair system took on a clear objective, namely to achieve the overall efficiency.
- Structural Integrity: Multi-measures in the complex wartime repair system released the overall effect through the interaction and relationship, and the overall effect exceeded the sum effect of the various element measures.
- Comprehensive Coordination: This complex system carried on the unity among function, structure, organization, management, internal and external coordination.
- Wartime Timeliness: Peacetime and wartime measure, built protection against the organization and coordination of the time constraint.
- Dynamic Evolution: The coordination of complex system was not a static relationship, but reflected the integration of offensive and defensive system by a certain way and orderly movement.
- Interface Hierarchy: The coordination existed in the external environment and complex repair system, existed among the various repair subsystems, and existed among the internal elements and hierarchy relations.

## 2.3 System Laws in Military Engineering Repair Structure

Complex wartime engineering repair system goes on with the general system laws, including the entity, attribute, operation, and other elements. Interaction characterized the mutual relationship among the wartime engineering repair system, depicting the material flow, energy flow, information flow in the transfer and exchange process, which reflected the overall functions.

Elements, structure, environment jointly determine the function of systems [4]. Functional study on two levels, on the one hand with internal system function, namely the mutual interaction and effect among the system elements, reflecting the output capability of internal elements; on the other hand with outside system function, namely the interaction and effect between the system elements and the environment, reflecting the external ability of the whole system. As function was mapped into the real engineering repair world, it called for internal resource optimization and allocation, and called for external engineering protection capability.

Give structure, function, feature and other universal laws with complex systems science theory [5]. Then, the conception system and methodology was constructed, as shown in Fig. 3.

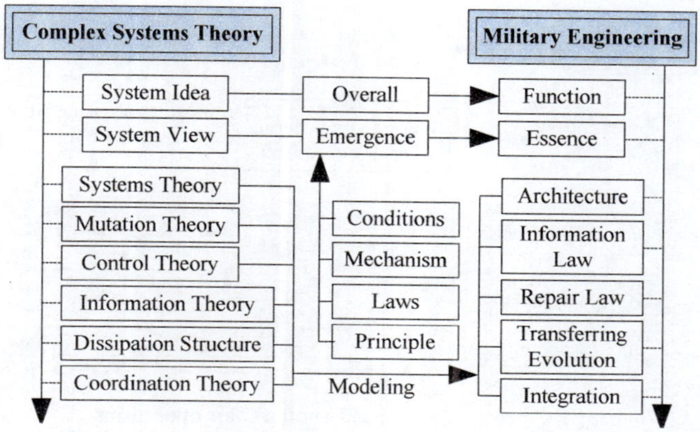


**Fig. 3.** Description about military engineering repair laws

Salient feature of system idea is to emphasize integrity, whose external performance was system function [4]. Engineering repair was "non-additive as a whole", but the emergent property. So, it must grasp the real problem as "system function", namely master "substantial repair" to support the emergent property, to protect and support combat functions.

Grasp the whole emergence of systems science, whose task was to study complex phenomena and issues of wartime military engineering repair structure at the point of the emergence view, to reveal the conditions, mechanisms, rules, principles about the emergence characteristics. With the application of information, control, mutation, coordination, dissipation structure theory in system science contents, complex systems modeling was instructed and carried on the complex wartime military engineering repair levels, including system architecture, information, repair technology, as well as system, information, and control laws in the transferring evolution by improved war technology.

# 3 Decision-Making Framework Design about Complex Wartime Repair System

## 3.1 Abstract Description on Wartime Repair Operations

Under the operational course of wartime military engineering repair system, entities represented mutual dependence, support, and restraint, whose structure, status, characteristic, operation and function evolved along over time and space process [8].

Repair operations were the driver engine in describing military engineering repair system and structure, which carried out engineering status variance as power source. Operation process is the overall orderly behavior in carrying on wartime engineering repair mission, namely the implementation flow in decision-making framework.

Abstract wartime military engineering repair process, as shown in Fig.4, describing ordered structure about engineering repair decision-making system, whose core element was repair activities.

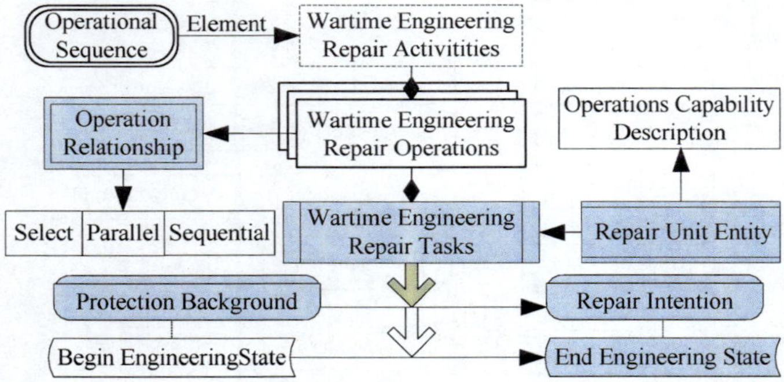


**Fig. 4.** Abstract description about repair operations

As shown in Fig.4, abstract description about repair operations included the mission and environment description, the operation process and relationship description, as well as the main operational capability description. Repair task was constituted by a series of actions, whose relations included sequential, parallel, select, etc. Core characterization was to achieve military engineering from the protective background to repair intention through repair mission, in which protective background depicted the begin state, and repair intention portrayed the required state at the operation end.

Repair decision-making process went along with the whole repair course, in which units carried on operations. Therefore, engineering state transferred into another required state.

### 3.2 Decision-Making Framework on Wartime Engineering Repair System

The overall coordination and optimization of military engineering was key mechanism to achieve the overall goal of complex wartime engineering repair system, namely how to realize the transition from disorderly to orderly. Otherwise, the synergy of the internal parameters was the key effect in the global coordination and optimization mechanism of complex military engineering repair system, which mostly depended on the organization process within the complex system.

With three-stage organization design theory proposed by Levchuk [6,7], organization theory was applied into wartime military engineering repair system, establish optimization and allocation from repair tasks to resources in accordance with the constraints between repair resource and task requirements. Set the clustering from repair resources to decision-making entity with the various decision-making ability; Set the level and collaborative relationship of decision-making entities under the efficient organizational collaborative need.

Specific description about military engineering repair decision-making framework was shown in Fig.5. In short, according to the repair goal, it resulted in repair task sequence, and planned repair resource requirements, so that repair missions were mapped with repair resources. And then repair resources were clustered, and collaborative relation was defined among the decision-making entities.

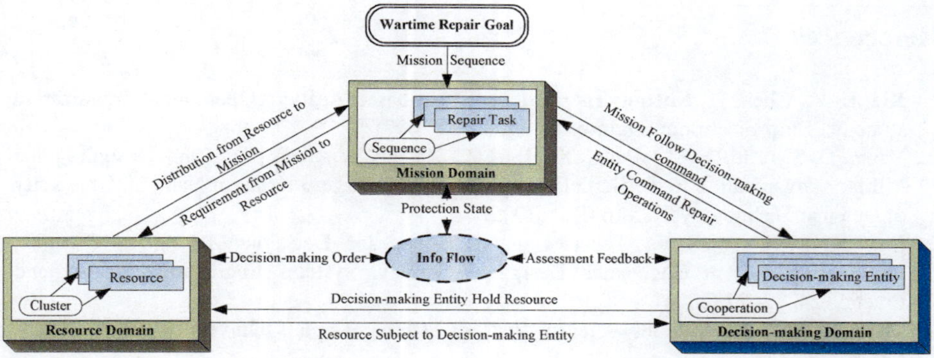


**Fig. 5.** Decision-making framework organizational design about wartime military engineering repair structure

As shown in Fig.5, repair decision-making framework included three domains and six kinds of relationships. Among them, domains included mission domain, resource domain, decision-making domain; relationships included cooperation among decision-making entities, sequence among repair tasks, cluster among resources, as well as relationships between decision-making entities and repair resources, relationships between decision-making entities and repair tasks, and relationships between repair tasks and resource [2].

Engineering repair decision-making framework meant the complete organizational process, in which repair entities carried out the mission under the specific environment, physical protection, including the implementation of repair mission, action programs of operational project, and the resource allocation and deployment in the repair process, as well as the command and control optimization.

## 4 Conclusion

Related with decision-making entities, repair resources, and missions, framework brought on the overall efficiency, and made decision-making activities take on new operation type. Therefore, it could obtain the best cost-effective goal, and get the most good repair strategy in the task & resource planning, and the objectives & strength matching.

Decision-making framework carried on coordination, organization, and optimization in complex wartime engineering repair system, which got the optimal system state solution from the system mission input, and whose design highlighted the integration of resources, processes, goals, and other elements. It was the program planning in protection-building and wartime protection, which was an effective tool to support the plan generation and evaluation.

## References

1. Xie, L.-Y., Chen, L., Xueqin: The studt on Effect-based Artillery Operational organization strategy. 'Ship Electronic Engineering 27, 4–18 (2007)
2. Yang, D.-S., Xiu, B.-X., Peng, X., Liu, Z., Zhang, W.: Analyzing and Designing the Military Organization in the Battlefields of Information Age. Journal of National University of Defense Technology 27, 116–122 (2005)
3. Liu, X.-D., Song, B.-F.: Theories and Methods for Decision Making in Complex Engineering System Conceptual Design: A Survey. Systems Engineering Theory and Practice 24, 72–77 (2004)
4. Miao, D.-S.: Elite Systems Science, 2nd edn. China People's University Press, Beijing (2006)
5. Zhu, W.-H., Wang, F.-S.: Research on the Protection of Important Economic Targets Based on System Science. China Safety Science Journal 17, 23–27 (2007)
6. Levchuk, G.M., Levchuk, Y.N., et al.: Normative Design of Organizations-Part I: Mission Planning. IEEE Transactions on Systems, Man,and Cybernetics 32(3), 346–359 (2002)
7. Levchuk, G.M., Levchuk, Y.N., et al.: Normative Design of Organizations-Part II: Organizational Structure. IEEE Transactions on Systems, Man,and Cybernetics 32(3), 360–375 (2002)
8. Liu, Z., Zhang, W., Yang, D.-S., Huang, J.: Operations Plan System Technology. Defense Industry Press, Beijing (2007)

# A Study on the Customer Profitability Analysis Using Activity-Based Costing

Zhao Wei

School of Economics & Management
Shenyang Ligong University
Shenyang, China
zweitech@126.com

**Abstract.** Activity-Based Costing (ABC) which has become an important aspect of manufacturing or service organizations can be defined as a methodology that measures the cost and performance of activities, resources and cost objects. It can be considered as an alternative paradigm to traditional cost-based accounting systems. Customer Profitability Analysis (CPA) is an important path to a more successful business operation, offering new insights into enhancing company performance. A well-designed and implemented ABC system is a powerful aid to management evaluation and decision-making, thereby improving organizational performance. This paper was aimed at unravelling the practical complexities of implementing ABC for CPA purposes.

**Keywords:** Activity-Based Costing (ABC), Customer Profitability Analysis (CPA), Activities; Cost.

## 1 Introduction

Over the past decade there has been increasing importance placed on customer profitability analysis (CPA) in practice. Companies have been extremely focused on their customers, trying to enhance customer service, paying more attention to customer's needs and measuring customer's satisfaction. CPA is an important and powerful tool.It allows companies to calculate and understand profitability at the individual customer and product level and to analyse underlying revenue and cost drivers.It will also helps you to understand which customers and which products are most profitable and therefore critical to your company. This new transparency enables you to re-evaluate your customer and product portfolio,and take strategic decisions over which products or customers should be retained, developed,or even neglected.

Activity-based costing (ABC) provides more meaningful information about the drivers of costs, the activities performed in a firm,and the relationship between costs and products, customers, markets, and segments. In addition to supplying more detailed and better cost and profitability information,an ABC analysis enables managers to evaluate processes from an activity viewpoint,leading to identification of non value-adding activities and process inefficiencies. A well-designed and implemented ABC system is a powerful aid to management evaluation and decision-making, thereby improving organizational performance.

## 2 Brief Review of Activity-Based Costing (ABC)

### 2.1 What Is ABC

ABC has been called one of the most important management innovations the last hundred years. In contrast to traditional cost-accounting systems, ABC systems first accumulate overhead costs for each organizational activity, and then assign the costs of the activities to the products, services, or customers (cost objects) causing that activity. As one might expect, the most critical aspect of ABC is activity analysis.

Activity analysis is the processes of identifying appropriate output measures of activities and resources (cost drivers) and their effects on the costs of making a product or providing a service. Significantly, as discussed in the next section, activity analysis provides the foundation for remedying the distortions inherent in traditional cost-accounting systems.

ABC was developed in the manufacturing sector of the United States during the 1970s and 1980s. It is a practice in which activities are identified and all related costs of performing them are calculated, providing actual costs chargeable. The focus of activity based costing is activities. Thus identifying activities is a logical first step in designing an activity based costing. An activity is an event, task or unit of work with a specified purpose. For example; designing products, setting up machines, operating machines and distributing products.

### 2.2 Difference between ABC and Traditional Cost Accounting

It is a well-known fact that the traditional costing systems utilise a single, volume-based cost driver. This is the reason why the traditional product costing system distorts the cost of products. In most cases this type of costing system assigns the overhead costs to products on the basis of their relative usage of direct labour. For this reason traditional cost systems often report inaccurate product costs.

What is really the difference between ABC and traditional cost accounting methods? Despite the enormous difference in performance, there is three major differences shown as Fig. 1 below:

In traditional cost accounting it is assumed that cost objects consume resources whereas in ABC it is assumed that cost objects consume activities.

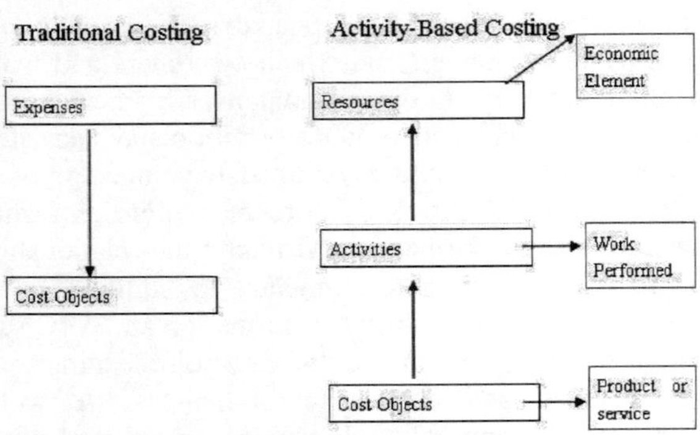


**Fig. 1.** Database contexts

Traditional cost accounting mostly utilizes volume related allocation bases while ABC uses drivers at various levels.

Traditional cost accounting is structure-oriented whereas ABC is process-oriented.

The direction of the arrows in Fig. 1 are different because ABC brings detailed information from the processes up to assess costs and manage capacity on many levels whereas traditional cost accounting methods simply allocate costs, or capacity to be correct, down onto the cost objects without considering any 'cause and effect' relations.

### 2.3 Benefits of ABC

An ABC analysis provides management with a wealth of financial and operational information. The benefits of ABC include the following:

- Costs are associated with activities that create those costs.
- Profitability can be calculated from multiple perspectives, such as product line, customer, or market.
- It provides information about "hidden" losers and winners, i.e. which product lines/customers/markets have lower profit margins than was originally thought and which give better profit margins.
- It provides cost rates for organizational activities that are helpful for benchmarking and making process decisions.
- It aligns with business process reengineering work by helping managers to put a price tag on non value-added activities, such as waste or rework.

## 3 Brief Review of Customer Profitability Analysis(CPA)

### 3.1 What Is CPA

The more recent growth of customer profitability analysis heralded a wave of interest in alternative costing methodologies. Initially, activity-based costs theorising was conducted in the context of product costing (e.g., Cooper and Kaplan, 1988)[1]. Such theorising illustrated how activities might serve as an appropriate unit of analysis in a costing exercise and highlighted the benefits that may be delivered from considering a variety of cost objects. It is in this context that the potential of differential cost allocation to customers (i.e., treating customers as cost objects) has commanded renewed interest.

Customer profitability is generally defined as the matching, assignment and/or allocation of company revenues and costs to its customers in order to understand relative profitability. Based on our experience, less than one third of all companies claim to know the profitability of their customers. And many of these firms have used arbitrary methods to calculate customer profitability that may cause them to make bad business decisions. In some cases, those customers thought to be the most profitable turn out to not even be profitable at all. For example, at one large manufacturer costs deemed to be customer related were allocated using a percentage of revenue dollars. For a high revenue customer, this method could allocate more costs to them than they truly consume. And conversely, a low sales volume customer may appea to be more profitable than they really are. If this manufacturer acted on this misinformation, it might deteriorate rather than improve its overall profitability.

Before you begin to format your paper, first write and save the content as a separate text file. Keep your text and graphic files separate until after the text has been formatted and styled. Do not use hard tabs, and limit use of hard returns to only one return at the end of a paragraph. Do not add any kind of pagination anywhere in the paper. Do not number text heads-the template will do that for you.

### 3.2 Benefits of CPA

CPA is an important path to a more successful business operation, offering new insights into enhancing company performance. In the current economic climate, these insights are more essential than ever before.CPA gives benefits across the whole company: sales initiatives start to focus on profitable customers, service improvements meet cost objectives, and process improvements reflect customer needs. A true understanding of customer and product profitability can help an organisation to:

- Evaluate and optimise the product/customer portfolio by identifying the most profitable customer-product mix.
- Personalise customer relationships, and target customer acquisition and retention programmes around customer profitability.
- Harmonise and link management information and performance management tools and establish an integrated performance management framework for finance, supply chain, sales and marketing, with selected key performance indicators reflecting the key strategic and operational performance drivers.
- Re-evaluate existing product prices and rebates, to stimulate customer behaviour through incentives and focussed customer development activities' without neglecting cost objectives.
- Identify and analyse key customer-and product-specific resource drivers and processes supporting strategic and operational cost management (e.g. activity based costing, cost-to-serve)[2].

## 4 Steps for Implementation of CPA Using ABC

To conduct an ABC analysis, organizational data are needed about costs incurred, work performed, and the cost objectives (for example products, customers, markets) of the analysis. An ABC analysis can be done for one department, for an entire manufacturing operation, or for the whole organization. However, it is often an advantage to start with a smaller-sized project (a single department or a plant) as the learning curve is steep.

### 4.1 Gather Cost Data

Segment the data into cost pools, where by each cost pool represents a related set of costs. Successful implementation of CPA using ABC will only be possible if you have the necessary resources and commitment as determined. If you are unwilling or unable to commit the necessary resources, then you'd better staying off the ABC and CPA highway. For example, a maintenance cost pool might consist of maintenance labor costs, supervision, tools and equipment used for maintenance, and training costs for maintenance workers[3].

### 4.2 Define the Activities of the Business

Using the following framework:

- Facility-level activities: Those related to overall operations (for example management, human resources, security, legal). Facility activities sustain the general processes in a facility. Facility level costs include items normally described as general overheads. This may include financing and other management cost not attributable to any of the previous levels of activities.
- Cost-object level activities: Activities that support a product (for example design, testing, engineering), or a customer (for example order processing, shipping, technical support), or a market (for example advertising, sales support). Activity drivers for this type of activity can be number of machine set-ups to produce a batch of products or engineering change notices.
- Batch-level activities: Activities related to a batch, equally and at the same time(for example set-up, material handling, inspection). Batch activities are performed on batches of products rather than individual product units. Activity drivers for this type of activity can be number of movements or number of runs[4].
- Unit-level activities: Activities performed for each unit of activity(for example direct labor). Once the activities of work are determined, assign the resource costs to the activities. unit level activity is an activity performed on units of products. Activity drivers for this type of activity can be direct labour hours or machine hours.

### 4.3 Assign the Activity Costs to the Cost Objectives

According to the cost object's use of each activity. Determining the allocation basis requires process knowledge and transactional data; often statistical analysis can be used to verify relationships between transactions performed and costs incurred.

Activity based costing requires that attention be given to a different product costing approach. The primary contribution of the activity based costing system is the recognition that cost drivers may encompass more than one facet of an organisation. Utilizing an activity based costing approach correctly revealed that low-volume products having significant transaction costs were unprofitable[5].

The successful selection of activity drives will result in dividing the costs into value-added and non-value-added components, with cost drivers separated into volume-related and transaction-related categories. This will result in approaching cost drivers as a way to manage and control expenses.

When selecting an activity driver, care should be taken as not all cost drivers are volume related and sometimes the drivers associated with transactions have a greater cost than the cost drivers associated with volume.

Activity drivers merge the requirements that cost objects places on activities. The importance of selecting activity drivers accurately cannot be over emphasised. This will impact on the accuracy of the costing of cost objects.

## 5 Conclusion

With organizational competitiveness, sustainability and cost management having higher visibility in enterprises, profitability analysis is something that no company can take for granted. Knowing customer cost and customer profitability is critical for a company today. Knowing your total costs for particular processes and activities allows you to focus on reducing and controlling them. Knowing costs for a specific customer allows you to reduce, change or charge for activities/services provided to them.

Many firms have benefited from designing and implementing ABC allocation systems. Using ABC tools has helped these organizations to understand customer profitability more clearly, and has provided meaningful information about processes and costs associated with delivering goods and services. A well-designed and implemented ABC system is a powerful aid to management evaluation and decision-making, thereby improving organizational performance.

## References

1. Cooper, R., Kaplan, R.: How cost accounting dissorts product costs. Management Accounting 39(10), 20–28 (1988)
2. Kaplan, R., Cooper, R.: Cost & Effect: Using Integrated Cost System to Drive Profitability and Performance. Harvard Business School Press, Boston (1998)
3. Liaoxiaoli, Dengjie: Discussion on Activity Based Costing. Foreign Investment in China (5) (2008)
4. Daly, J.L.: Pricig for profitability: Activity-Based Pricing for Competitive Advantage. Wiley, New York (2001)
5. Panfei, Guo, X.: Research on Activity Budget. Accounting Research (11) (2004)
6. Sharman, P.: A practice look at activity-based costing, `http://www.focusmanagement.com`
7. Ansari, S., Swenson, D., Bell, J., Kim, W.: Best Practice in Target Costing. Management Accounting Quarterly 4(2), 12–17 (2003)
8. Bailey, J.: Implementation of ABC Systems by UK Companies. Management Accounting, 30–32 (February 1991)
9. Bauer, H., Fischer, M.: Product Life Cycle Patterns for Pharmaceuticals and their Impact on R&D Profitability of Late Mover Products. International Business Review 9, 703–725 (2000)
10. Jiang, W.: Cost Management of SCM. In: Proceedings of the 4th International Conference on Material Handling & Logistics Systems (2002)

# Study and Design on the Model of Personalized Recommendation System in E-Commerce Based on Multi-Agent

Yun Xue, Xiaoping Shen, and Jianbin Chen

E-Commerce Department
Business College of Beijing Union University
Beijing, China
{yun.xue,xiaopin.shen,jianbin.chen}@bcbuu.edu.cn

**Abstract.** With the development of the Internet, e-commerce has become an important form for current consumer purchases or browsing commodity. It was a problem on how to provide personalized recommendation e-commerce service for users under growing and changes in the amount of commodity information. According to the Agent and Web data mining technology, this paper presents a base on Agent of e-commerce personalized recommendation system structure that is composed of customers, personalized recommendation system and e-commerce sites. The personalized recommendation system is divided into 3 layers, including the presentation layer, the business logic layer and the data service layer. The function of each part and multi-Agent cooperative work flow of the personalized recommendation system are designed. The system could meet intelligent personalized recommendation and personalized demand for consumers under the e-commerce environment, which has certain practical and promoted value in the future.

**Keywords:** Agent, Web Data Mining, Recommendation System, E-commerce, Personalization.

## 1 Introduction

As e-commerce is in widespread use, it will evolve a kind of world-wide, far-flung, dynamic commodity information repository gradually and become one of the main methods of buying or browsing commodity by world-wide consumer. However, along with the development of e-commerce, many problems appear, such as, the problem of "information overload" caused by the online commodity information's geometrical progression increase, and the problem of "information disorientation" during the process of finding commodity they needed in the great commodities. At present, one kind of search engine can be recommended to users is traditional, such as Yahoo and Google, and the other kind is the existing system, such as recommended systems based on collaborative filtering and content. Although they are used in different areas, there are some limitations. For examples, on the one hand, the coverage rate of search engine is limited, so recall ratio is low. And its inquiry is based on key words, not be connected with users' individual requirements, so precision ratio is not good. On the other hand, the existing recommendation system is lacking in individuation, intelligent, instantaneity, initiative and so on. So how to supply individual service to users from great increasing and changing commodity information is becoming a problem of e-commerce to solve urgently. The key technologies are the Agent and the web data mining technology. If they are used to e-commerce recommendation system, it can meet users' requirement actively, instantly and intelligently in order to realize e-commerce's personalization.

## 2 Related Notion

### 2.1 Personalized Service and Recommendation

The research of personalized information service originated in 1995, when Robert Armstrong form CARNEGIE MELLON University presented personalized guided system at the meeting of American Artificial Intelligence Association[1]. At present, personalized service is the research hot point all over the world information service, and the traditional general information service is transferred to it gradually. Personalized service aims at certain points regarding users' interest, identity and meeting their specific demands[2]. The commerce recommendation system is a kind of effective measure to meet personalization demands of users' in e-commerce, which provides customers with some personalized information. According to plenty of customers' historical shopping information, it can help customers decided which commodity to buy or forecast their future choices and recommend commodity to customers imitating salesman in order to make customers feel personalized service absolutely[3].

### 2.2 Agent Technology and Multi-Agent System

Agent is a term applying widely which generally means an entity which has perception, problem solving capability and communication ability with outside world. Agent technology is one of the latest development technologies of the artificial intelligence technology. The definition is: Agent is a computer system under certain circumstances, which can do autonomy to its circumstance to meet its goal of design. The basic characteristics are as follows: autonomy, communication ability, cooperative skills, pro-activeness, viability, perception, intelligence, sustainability and so on[4]. Multi-agent system is made up of multiple Agent systems. It can do reasonable arrangement to the targets and resource of multiple Agent systems which have different targets to coordinate their behaviors in order to achieve their each target farthest. Each Agent is also an autonomy computational entity with target, knowledge and capability. Multiple Agent systems form the solving web by cooperating. By this way, it not only improves each Agent's basic ability, but also solves the complicated questions which single Agent cannot solve. Adopting Multi-Agent system can overcome the limitations existed in e-commerce recommendation system. Now it becomes many researchers' hot point which touches aspects of information retrieval, e-commerce, social simulation, virtual environment and so on[5,6].

### 2.3 Web Data Mining

Web data mining makes data mining apply to web, and discovers potential and useful patterns or information from vast quantities of web documents and service. According to the different behavior of data mining, it can be divided into three types[7]: web content mining, web structure mining and web usage mining. Web usage Mining is a behavior pattern through mining log files and related data of corresponding sites to find viewers and customers. The methods of data mining consist of path analysis, association rules, sequence pattern, classification rules, statistical analysis, dependency modeling, cluster analysis and so on[8]. Data mining technology based on web is the key technology of realizing personalized e-commerce which is made possible by collecting users' all kinds of information through web data mining, analyzing users' data, and creating access mode, demand model and hobby mode.

## 3 The Design of Model

### 3.1 System Framework Model

Based on the agent technology, web data mining technology and the popularity of e-commerce system at present, to be able to absorb the design concept of web services which fit web application and have loosely-coupled property more effectively, we are currently busy with building a personalized e-commerce recommendation system model[9,10]. As shown in Figure 1, this system model consists of three parts: customers, personalized recommendation system and several e-commerce websites. From which, the personalized recommendation system can be divided into 3 layers: presentation layer, business logic layer and data service layer.

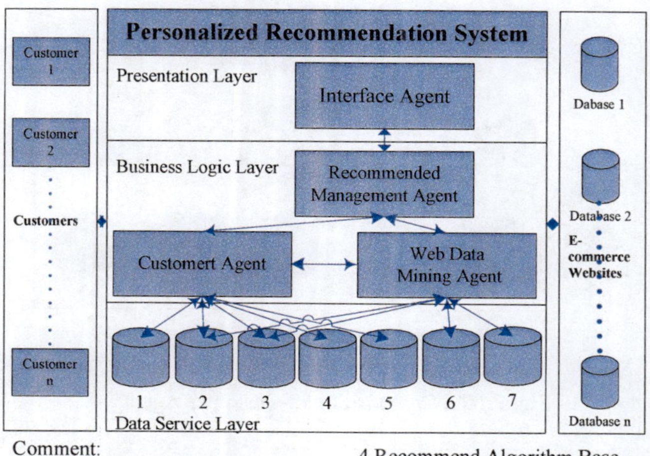


Comment:

1.Temporary Commodity Database
2.Customers Interest Knowledge Base
3.Customers Business Database
4.Recommend Algorithm Base
5Appraisal Database
6.Web Logs Database
7.Rule Base

**Fig. 1.** E-commerce personalized recommendation system model based on Multi-Agent

The main features of this model:

- Because of the reactive, initiative and autonomy of Agent technology, the system can react fast under the constantly changing circumstance and make fact timely.
- The sociality of Agent technology ensures cooperative ability between system interiors and system and users or external systems.
- Agent technology can reduce server's working pressure and increase efficiency.
- Web data mining technology provides customer purchasing or visiting behavior analysis and it serves as basis for this recommendation system[11].
- Recommending different commodity information to different customers according to customer business database and interest knowledge base.
- Achieving the target of "information find customer" using Agent's automation. Before customers' application, the multi-Agent system carried out automatically during some time. By this way, if the commodity data is updated, the system could feedback the new information to customers.
- Customer Agent and web data mining Agent have self-learning ability and data mining can realize the dynamic update information resource during the system's free time[12].

### 3.2 Customer Agent

Customer Agent is a multi-Agent system, which consists of search Agent, appraisal Agent and recommendation Agent. Their sub-roles are shown in Figure 2.

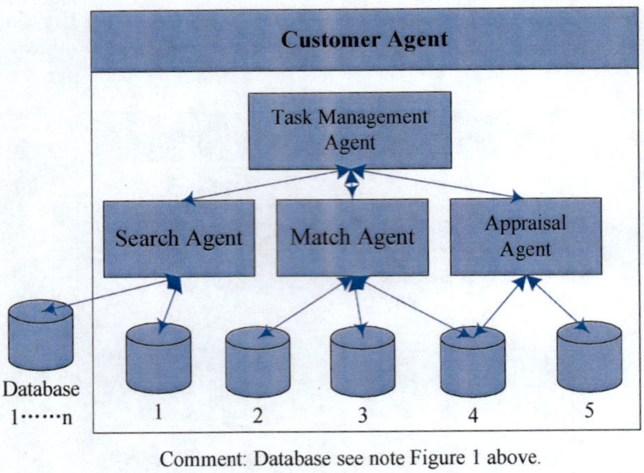


**Fig. 2.** The Composition of Customer Agent

*1) Search Agent*

Search Agent accepts task management Agent's request to research related commodity information from a large number of commodity database and store candidate commodity information to the temporary commodity database.

*2) Match Agent*

According to different property of commodity, Match Agent chooses corresponding recommendation algorithms to create initial commodity list using temporary commodity database, customers interest knowledge base and customers business database. Match Agent uses recommendation algorithms stored in recommend algorithm base to match, and the base has algorithms based on collaborative filtering, content, client browser, knowledge and clustering rule recommendation. Match Agent can be used flexibly with different commodity and recommendation request to make system strong initiative, flexibility and adaptability. Then transfer it to Appraisal Agent to make appraisal.

*3) Appraisal Agent*

Appraisal Agent checks the recommended result produced by Match Agent according to some appraisal models from appraisal database and ranks the candidate commodity according to the appraisal from high to low. Then amend the recommendation initial list according to appraisal feedback results. After a number of loops, the final recommended result will be produced. At the same time, the recommended rules and results can be stored in appraisal database for next time use and their own learning-enhancing. Appraisal Agent provide important basis in producing effective recommended result.

## 3.3 Web Data Mining Agent

Web data mining Agent consists of log analysis Agent, customers interest Agent and data mining Agent, which is also a multi-Agent system. Their subtasks are shown in Figure 3.

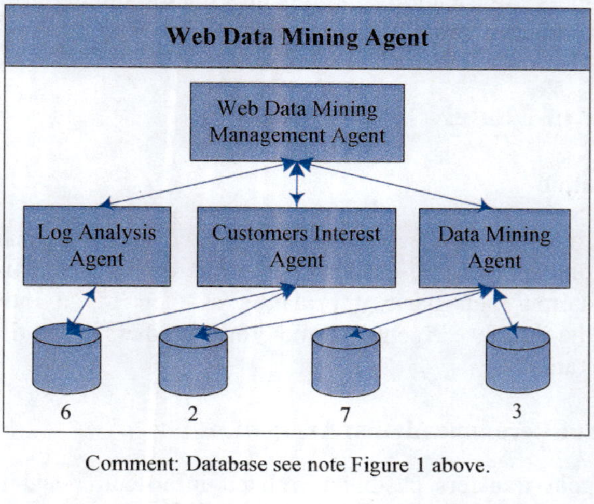


Comment: Database see note Figure 1 above.

**Fig. 3.** The Composition of Web Data Mining Agent

*4) Web log analysis Agent*

Web log analysis Agent mainly dynamically analyzes the web server log documents about customers in which customers' browsing and visiting information are recorded during their purchasing and browsing process. It makes up of much record, such as Server log (User's IP, Sever Name, Requested File's URL, Time Browsing Process Costs, Cookie ID and so on), Error log(Connection lost, Authorization failed, Time-out and so on), Cookie(User's Status, Access Paths) and so on.

*5) Customers interest Agent*

On one hand, the customers interest Agent dynamically follows the trail of customers' interest, for example, visit behavior, frequency, content and so on. Then mining this information, finding potential and unknown information, in order to acquire the customers' interest, extract their features, analyze and update personal interest information in customers interest knowledge base, in order to make the recommended result of system could reflect the user's latest interest. On the other hand, it statically gets the related information which consists of registration information and interest information supported by customers to sever. The static information makes the Agent get customers' initial interest model, and adjusts it according to customers' behavior to provide reference for recommended algorithm.

*6) Data mining Agent*

On one hand, Data mining Agent detects the changes of commodity information and customers' click-stream. If there is something changed, it will do corresponding analysis and update rules in the rule base to ensure providing latest rules. On the other hand, there is plenty of transaction data during their purchase process on e-commerce website, which records their purchase information, such as product name, price, quantity and suppliers' information. And this information mining is very important to analyze customers' interest and loyalty to suppliers.

# 4 Workflow of Model

## 4.1 Authentication

After logging in, the interface Agent of personalized recommendation system will authenticate customers' identity, either initial users or registered users. The initial visitors should submit their basic information as their initial interest model. To registered users, the interface Agent will monitor customers' operation and browsing behaviors momentarily.

## 4.2 Allocation of Recommendation Assignment

The interface Agent transfers customer- related information and recommendation request to recommendation management Agent who will allocate the recommendation task to customer Agent and web data mining Agent. They will do corresponding process according to customers' identity information and complete recommendation task with other Agents.

### 4.3 Recommendation Results' Calculation

Search Agent takes responsibility of collecting commodity records supplied on the database by each related e-commerce websites and returns to candidate commodity list; Web data mining Agent collects customers' interest information further and get their potential information according to their purchase behavior and interest they had. Match Agent calculates the recommendation result according to candidate commodity list and corresponding information provided by the system.

### 4.4 Amendment of Recommendation Result

Match Agent and Appraisal Agent make several appraisal, check candidate commodity list and present final recommendation result to customers by interface Agent.

### 4.5 Assessment and Feedback

The customers choose commodity according to the commodity list made by recommendation system and give their assessment and feedback to the result. And the system will save the recommendation rules, result and the customers' real purchase information and feedback information into appraisal database and customers business database which can be used to the system self-learning and next recommendation.

## 5 Conclusion

With the rapid development of e-commerce based on Internet, the personalized recommendation system of e-commerce should be paid more attention in theory and practice. This paper introduced the study and design of a personalized recommendation system of e-commerce based on Multi-Agent which adopts Agent technology and web data mining technology. And it introduced the functional component and workflow of this personalized recommendation system in details. By the effective integration between Agent technology and web data mining technology, the personalized recommendation system of e-commerce will be more intelligent, independent and flexible and solve some problems existed in present personalized recommendation system. It is worth of further promotion. At present, the work of this model's research and design has been accomplished. However, the further research of how to apply the Agent technology better and how to get better recommend algorithm to this system should be done in the future.

## References

1. Yang, L., Mao, Y.: Personalization: Customize Your Network Service. Software Engineer 7 (2003)
2. James, C., Charles, L.: Personalized Information Environment: An Architecture for Customizable Access to Distributed Digital Libraries. Library Magazine (6) (1999)

3. Yan, W., Ling, X.: E-government theory and practice, pp. 2–17. Tsinghua university publisher, Beijing (2004)
4. Jennings, N.R., Wooldrige, M.J.: Agent Technology. Foundations, Appliations, and Markwets. Springer, Heidelberg (1998)
5. Yan, Y., Kuphal, T., Bode, J.: Application of Multiagent Systems in Project Management. Int. J. Production Economics (68) (2000)
6. Wooldridge, M., Chunyi, S., et al.(translates): Multi-Agent system introduction. Electronics industry publisher, Beijing (2003)
7. Scafer, J.B., Konstan, J.A., Riedll, J.: E-commerce Recommendation Applications. Data Mining and Knowledge Discovery 5(1/2), 115–153 (2001)
8. Zhao, H., Zhao, Y.: Web data mining application in electronic commerce. China Financial Computer (2004)
9. Zhou, B., Chen, K.: A Web Mining Model Based on Muti-agent Echnology and Its Application. Journal of Hubei University of Technology 02 (2009)
10. Lei, L.: Application of Agent in E-Commerce. Science Mosaic 03 (2007)
11. Yu, H., Huang, X., Hu, X., Wan, C.: Knowledge Management in E-commerce: A Data Mining Perspective. In: International Conference on Management of e-Commerce and e-Government, pp. 152–155 (2009)
12. Yu, C., Ying, X.: Application of Data Mining Technology in E-Commerce. International Forum on Computer Science-Technology and Applications 1, 291–293 (2009)

# Nonlinear Regression for Analog Data Based on BP Neural Network

Yin-Zhen Jiang and Yi-Huai Wang

Soochow University
Institute of Computer Science and Technology
Suzhou, China
{jiangyinzhen,yihuaiw}@suda.edu.cn

**Abstract.** In order to solve the nonlinear regression problem in analog data collection system and achieve nonlinear analog correction by programming, a design method based on artificial neural network (ANN) is presented. After analyzing the lack of other regression methods, Error back-propagation (BP) algorithm is selected. The software on PC is designed to verify the feasibility and effectiveness of the BP algorithm, which is compared to least squares (LS) algorithm.

**Keywords:** Embedded system, Analog data collection, BP neural networks, Nonlinear regression.

## 1 Introduction

In the field of embedded monitoring and control, collecting analog data such as temperature, humidity and pressure is an important work. The relation between digital data converted by MCU and actual analog data is nonlinear commonly because of the sensor or data collection circuit, which consists of filter circuit, commutated circuit, nonlinear emendation circuit and so on. Actual physical data is identified by MCU through sensor, convertor, amplifier and AD convertor module[1], thus the physical data (y) is converted to the digital data (x), which can be accepted by MCU. During the process of collecting analog data, nonlinear of any circuit can lead to nonlinear between y and x, expressed by y=f(x). In order to solve nonlinear question, the following methods are used commonly.

1. For small nonlinear system, uses linear approximation.
2. If the number of A/D converter bits is low (such as 4-bit or 8-bit), uses point-by-point regression.
3. If the number of A/D converter bits is high (such as more than 8-bit), uses sub-linear regression.
4. Uses polynomial regression such as least squares algorithm.
5. Comprehensive use of many regression methods such as sub-linear and polynomial regression.

The only one purpose of above methods is to eliminate the deviation caused by nonlinear problem as far as possible, reduce the cost of hardware design and make x truly represent y using software.

Artificial neural network (ANN)[2] is a highly nonlinear dynamic system. It has been widely used in predicting the behavior of complex nonlinear systems. For the nonlinear regression problem in A/D conversion of data collection system, ANN should have its role. The study found that ANN method can eliminate the deviation and learn the function relationship between y and x efficiently in contrast to other regression methods if the process of analog-to-digital is regarded as a “Black Box”.

In the following section, theoretical basis, model selection and feasibility of the application are described, as well as the software design method of nonlinear regression using ANN in actual measurement and control system is discussed.

## 2 Problems and Enlightenment in Traditional Nonlinear Regression

When the analog data collection process is regarded as a black box, if the mapping (f) between the output (y) and input (x), that is, y = f (x), is found, as long as f is a monotonically increasing and continuous function, the basic requirements of the data collection system are meet. Generally, it is easy to guarantee that f is a monotonically increasing and continuous, but the concrete expression of f is difficult to determine. In traditional methods, the least squares principle (Legendre 1806) is commonly used, with the thinking of "make the smallest sum of error squares", to find the specific forms of expression y=f(x) from a large number of random experimental data, such as $(X_1, Y_1)$, $(X_2, Y_2)$, $(X_3, Y_3)$ ..., $(X_k, Y_k)$, but sometimes the relationship between x and y can not be expressed with any known mathematical formula.

In view of the successful implementation of artificial neural networks in the field of nonlinear application, for data acquisition system regarded as a black box, if the hardware measures ensure (f: x→y) is a monotonically increasing and continuous, and the number of learning samples meet certain conditions, the artificial neural network will be trained using teacher's demonstration and the parameters of the network structure will be determined. This network structure represents a "Black Box" which expresses (f: x→y), shown in Figure 1. However, the artificial neural network is not looking for a specific mapping mathematical expression, but learning and training the samples through a specific network (model). After the network training is completed, the network structure represents the map. Although this process can not come to a concise mathematical formula, it represents a more complex mapping. By this network structure (network layers, unit number of each layer, the connection weights and threshold values, etc. have been finalized), when x (Digital value) inputs, y (Analog value) will be generated as an output, it is the recall process. The network whose parameters have been set down becomes an expert to solve the special problems. For the physical data of each path, the network parameters are different. If these parameters are recorded, the purpose of eliminating non-linear signal conversion can be achieved according to the corresponding network algorithms.

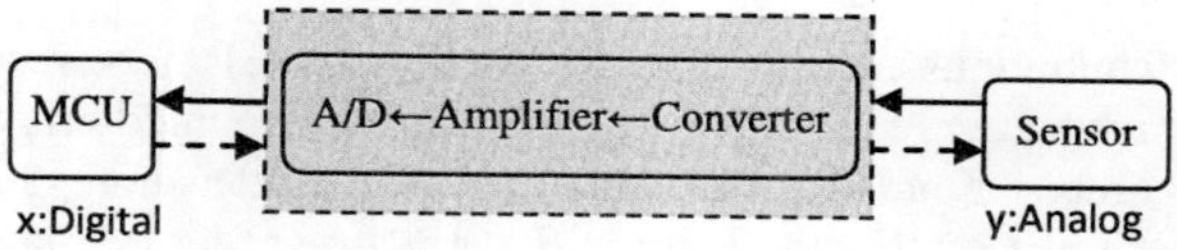


**Fig. 1.** A Black Box Expresses {f : x→y}

## 3 Basic Theory

The most important feature of ANN is the ability to learn from the environment, and to distribute the learning outcomes in the network structure. Learning mode is divided into three types according to by the information provided by environment: non-supervised learning, reinforcement learning and supervised learning. For the particular issue of analog non-linear regression, it belongs to a supervised learning area.

ANN is commonly interlinked by many perceptrons. It is called feed-forward network model if the neurons input/output are single directional. Feed-forward network model is one of the most widely used ANN type, and is also adopted in this study.

Theoretical studies have shown that, for the feed-forward networks in the ANN models, if the activation function is continuous (eg sigmoid function), the network output can approximate a continuous function with any accuracy degree. This conclusion can be described as the following theorem [3].

Theorem If $\varphi(\bullet)$ is a bounded and a monotone continuous function, Id represents a d-dimensional unit hypercube. C(Id) expresses the set of continuous function defined on Id, for any function $f \in C(Id)$ , and $\varepsilon>0$, an integer M and a set of real constant $\alpha_i$, $\theta_i$ and $w_{ij}$ are existed, where i=1,2,…,M, j=1,2,…,d makes the network output

$$F(x_1,\cdots,x_d)=\sum_{i=1}^{M}\alpha_i\varphi(\sum_{j=1}^{d}w_{ij}x_j-\theta_i) \tag{1}$$

close in upon $f(\bullet)$, just as

$$\left|F(x_1,\cdots,x_d)-f(x_1,\cdots,x_d)\right|<\varepsilon,\forall(x_1,\cdots,x_d)\in R^1 \tag{2}$$

Theorem states that the feed-forward network containing only one hidden layer is a universal function approximator, and a hidden layer is sufficient (though not necessarily the best). Therefore, it can be selected to solve the problem of non-linear regression.

## 4 Selection of ANN Model

Error back-propagation algorithm [4], called BP algorithm for short, is one of the main training algorithms for multi-layer feed-forward neural network. In view of the widely use and mature theory, BP algorithm is selected for nonlinear regression in this paper. As mentioned above, the feed-forward network with one hidden layer only can approximate any continuous function. So the three-layer BP network is used in the next. three layers respectively are input layer, hidden layer (middle layer) and output layer. As for the number of hidden layer units, cross-validation strategy can be used. In order to facilitate discussion, Figure 2 shows the structure of a three-layer BP network containing only one hidden layer. The following indicates the structure and algorithms mainly from the point of programming.

If m is the number of learning model, n is the dimension of input vector, p is the number of middle layer unit and q is the number of output layer unit, physical quantities in Figure 2 will be described as follows:

The vectors of Input pattern: a(k,i), k=1,2,…,m is the learning model number, i=1,2,…,n is the input layer unit number.

The connection weights between i unit of input layer and j unit of middle layer: w(i,j), i=1,2,…,n is the input layer unit number, j=1,2,…,p is the middle layer unit number.

The connection weights between j unit of middle layer and t unit of output layer: v(j,t), j=1,2,…,p is the middle layer unit number, t=1,2,…,q is the output layer unit number.

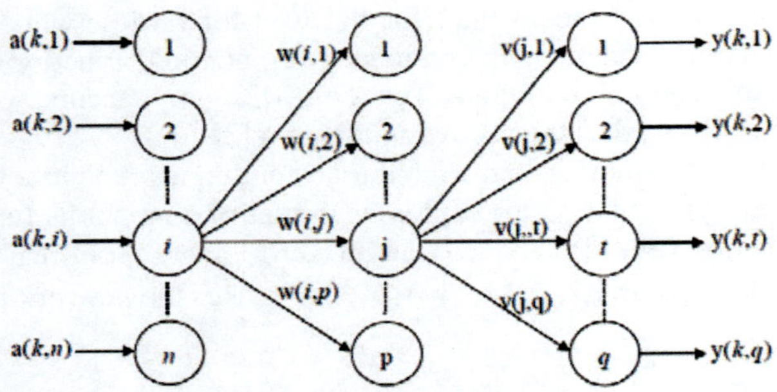


**Fig. 2.** Structure of 3-Layer BP ANN

In addition to them above, there are physical quantities such as the output array of learning model y(k,t), the input s(k,j), the output b(k,j), the output threshold o(j) and the generalization error e(k,j) of each middle layer unit, the input l(k,t), the output c(k,t), the output threshold r(t) and the generalization error d(k,t) of each output layer unit. The range of subscript variable is same as above. For clarity, Figure 2 shows a group only.

The learning process of BP algorithm is available in many literatures [5-7]. The following gives a brief description of the recall process.

If the learning is successful, the connection weights and threshold values of the network structure are determined. The tasks could be submitted to the network and obtain the output from the network. If the input submitting to the network is A={$A_i$, i=1,2,…,n}, here n is the number of dimensions for the input. From the A, each cell connection weights of the input layer and the middle layer and each cell threshold values of the middle layer, each cell input of the middle layer can be calculated by the following expression:

$$S_j = \sum_{i=1}^{n} W_{ij} A_i - O_j, j = 1,2,\cdots,p \tag{3}$$

First, calculated each cell output of the middle layer through Sigmoid function $b_j$=f($s_j$), then, through each cell output of the middle layer, each cell connection weights between the middle layer and output layer, and threshold values of the output layer, each cell input of the output layer can be calculated by the following expression:

$$l_t = \sum_{j=1}^{p} V_{jt} b_j - r_t, t = 1,2,\cdots,q \tag{4}$$

Each cell output of the output layer via Sigmoid function is $Y_t$=f($l_t$) ,t = 1,2, ..., q. It is the result of recall.

For the problem of analog non-linear regression, it belongs to a case of single input and single-output. It means i = 1 and t = 1 because of n = 1 and q = 1. Single-input A1 is the AD value, denoted by x, Y1 is the real analog, denoted by y, while $W_{ij}$ and $V_{jt}$ are denoted by $W_j$ and $V_j$ for short.The result of recall output is:

$$y = f(x) = \frac{1}{1+e^{-(\sum_{j=1}^{p} \frac{1}{1+e^{-(x * w_j - o_j)}} V_j - r_1)}} \tag{5}$$

In fact, formula (5) is the usual elementary functions, where the parameters $W_j$, $O_j$, $V_j$ and r are the connection weights of neural network and obtained by learning the neural network, p is the number of middle layer units. Practice shows that the value of P can be 3 for the problem of analog non-linear regression. So in Figure 2, an input unit, three hidden units and one output unit are necessary.

## 5 Programming for Comparison

Corresponding software is developed to solve the problem using BP neural network, shown in Figure 3. For a set of specific sample points, after the setting of study times and the number of Hidden Layer Units, click “OK” button then “Study” button in Figure 3, the textboxes will display the current global error, connection weights ω, the threshold value θ, the connection weights υ, as well as the threshold value γ dynamically, there ω, θ, υ and γ correspond to w, o, v and r in the above-mentioned formula respectively. After the click of “Draw” button, two curves are painted in the picture box, the blue one is for least-squares (LS) [8], and the red one is for BP neural network. At the same time, the textboxes show coefficients and residual standard deviation of LS and BP for comparison. The longer the study time is, the smaller the residual standard deviation is and the closer the correlation coefficient of BP neural network is to 1. Figure 3 shows the example of 12-bit AD converter, the abscissa stands for AD value up to 4095, and the vertical axis represents the actual analog value. It can be seen that a larger error occurs in least-squares regression method, and for BP neural networks, the error is also large after learning 2 seconds, but up to 141 seconds, BP is superior to LS obviously.

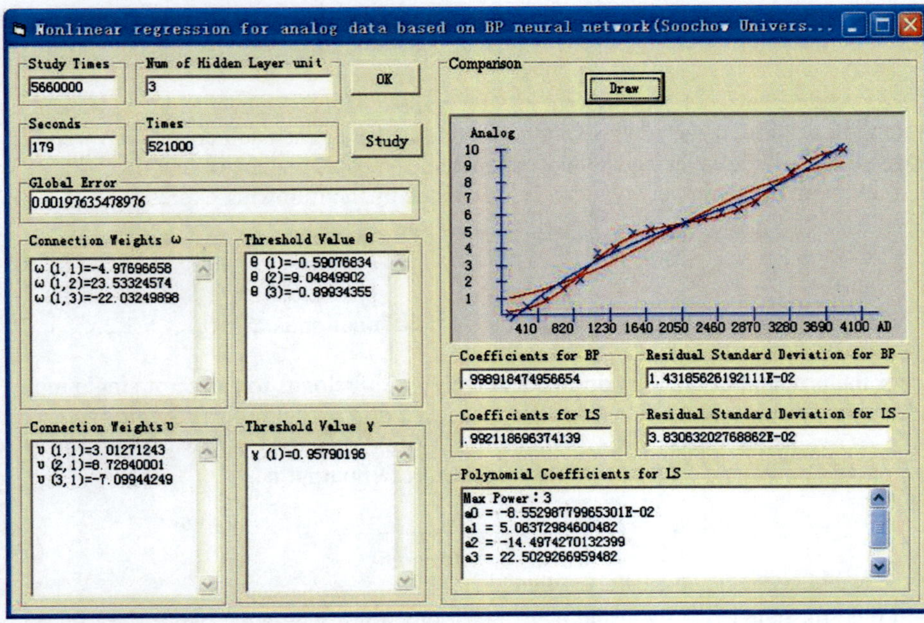


**Fig. 3.** Software Interface

**Table 1.** Research Data about COD and TN

| | 1 | 2 | 3 | 4 | 5 | 6 | 7 | 8 | 9 | 10 | 11 | 12 | 13 | 14 |
|---|---|---|---|---|---|---|---|---|---|---|---|---|---|---|
| AD Value | 900 | 1100 | 1300 | 1500 | 1700 | 1900 | 2100 | 2300 | 2500 | 2700 | 2900 | 3100 | 3300 | 3500 |
| COD(mg/L) | 3.20 | 4.20 | 4.50 | 5.30 | 4.90 | 5.10 | 5.33 | 5.42 | 5.51 | 5.56 | 6.10 | 7.10 | 8.70 | 9.60 |
| TN(mg/L) | 0.87 | 1.03 | 1.09 | 1.22 | 1.32 | 1.42 | 1.66 | 2.61 | 2.92 | 3.20 | 3.40 | 3.65 | 3.75 | 3.95 |

Table1 shows the research data about Chemical Oxygen Demand (COD) and Total nitrogen (TN) from Taihu water quality. Figure 4 AND Figure 5 show the comparison of two methods: LS and BP.

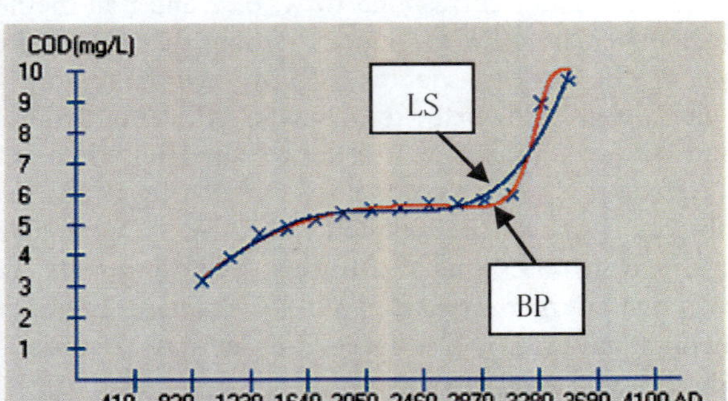


**Fig. 4.** Regression Curve for COD

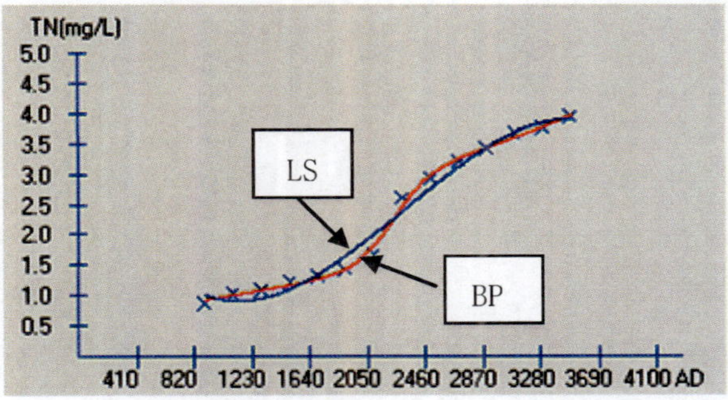


**Fig. 5.** Regression Curve for TN

## 6 Conclusions

This study successfully demonstrates the application of BP neural network to solve the problem of nonlinear regression. The feasibility and superiority of BP method is better than LS method. In the embedded systems for analog data acquisition, the nonlinear relationship between actual physical data (y) and digital data (x) received by MCU is ubiquitous. So this study is meaningful for reducing the hardware linearization costs.

**Acknowledgment.** This research was fully supported by National Natural Science Foundation Grant 60871086 of the People's Republic of China.

## References

1. Minkina, W.: Nonlinear models of temperature sensor dynamics. Sensors & Actuators A 30, 209–214 (1992)
2. Khanna, T.: Foundation of Neural Networks. Addison Wesley, New York (1990)
3. Wang, Y., Wang, L.: The Nonlinear Regression Based on BP Artificial Neural Network. Computer Engineering and Applications 12, 79–82 (2004)
4. Yu, X.A.: Backpropagation learning framework for feedforward neural networks. In: Proceedings - IEEE International Symposium on Circuits and Systems, pp. 700–702 (2001)
5. Kathirvalavakumar, T., Thangavel, P.: A modified backpropagation training algorithm for feedforward neural networks. Neural Processing Letters 23(2), 111–119 (2006)
6. Zhi, H., Wu, H., Ren: A new adaptive backpropagation algorithm based on Lyapunov stability theory for neural networks Man. IEEE Transactions on Neural Networks 17(6), 1580–1591 (2006)
7. Orlando, J.D., Hagan: Backpropagation algorithms for a broad class of dynamic networks. IEEE Transactions on Neural Networks 18(1), 14–27 (2007)
8. Tzu, C.R., Lin, C.-W., Chiang: Partial least-squares algorithm for weights initialization of backpropagation network. Neurocomputing 50(1), 237–247 (2003)

# A New Type of Control Algorithm of High-Power Active Power Filter

Fen Gong[1], Xiangyang Xia[1], Shiwu Luo[2], Feng Zhou[1], and Linju Xu[1]

[1] College of Electrical Engineering, Changsha University of Science and Technology, Changsha 410077, China
[2] Chongqing Power Education Training Centre, Chongqing 400053, China
gongfen86814@126.com

**Abstract.** In view of active power filter's control is difficult to keep better unity in the dynamic performance and control precision, it puts forward a improved quasi-sliding mode variable structure control algorithm based on repetitive prediction control in HAPFSIC. For the control reference signal is a cycle volume, combined with rapid response of sliding mode variable structure control algorithm and the advantages of a good real-time non-steady-state error of repeated prediction control algorithm, it imports the control volume of repetitive prediction control algorithm as the equivalent control of improved discrete quasi-sliding mode variable structure controller, which makes it has three change state of switching on both sides as well as the within band. The simulation shows that: this new type of compound control method that overcomes the sliding mode variable structure control's discrepancy regulation, a larger burr of current switch, and also avoids the shortcomings of repeated control's longer time to reach steady-state, achieves organic combination of two control methods.

**Keywords:** Active Power Filter (APF), repetitive prediction control, improved quasi-sliding mode variable structure control, cycle volume, and equivalent control.

## 1 Introduction

Compared with passive filter, APF is a dynamic and flexible mean of harmonic treatment, so it becomes the first choice and development direction for harmonic control, which requires the control to APF has a good real-time, dynamic performance and control precision [1]. However, it mostly depend on the control method that is becoming the research emphasis of active filtering technology at present. With in-depth study, a series of new control methods have been proposed, such as repetitive control, single-cycle control, variable structure control, etc [2-4]. Although they have their own advantages, the study was mainly targeted at APF systems of low-voltage and small-capacity, there is not so much research on APF control algorithms which is suitable for the conditions of middle-high voltage and large-capacity at domestic, according to the pre-deep theory discussion, it can't get satisfactory results from completely copying APF control algorithm. Therefore, in the engineering applications of HAPFSIC, it is a very important aspect to know how to control it effectively, but also is major problems that other types of APF have to face.

## 2 Modeling Hapfsic

Presently, it has proposed many methods of modeling APF[5], but it is too cumbersome, and the output modeling method is difficult to analysis the effect on HAPFSIC's internal parameters to the characteristics of device. In this article, by introducing the concept of switching function, it establishes a transient mathematical model's unified expression of HAPFSIC, and improves conditions for accurate analysis of the course of their work. At first, in the assumption that IGBT is the ideal switching components used in inverter, that is, ignoring the dead-time, inverter's on-off process and so on,. HAPFSIC's single-phase equivalent circuit is shown in Fig 1. In that figure, $Z_{C2}$ and $Z_L$ separately mean the value of capacitive reactance and inductive reactance of fundamental series resonant branch, $Z_{C1}$ is injection capacitive reactance, $Z_{Sa}$ is the value of power grid slip resistance, $U_{Sa}$ is grid voltage, $U_1$ is the voltage of fundamental series resonant slip, $Z_C$ is the equivalent impedance of inverter loss, output filter and coupling transformer converting to power grid side, $I_C$ is the outflow coupling transformer's current, $U_C^{'}$ is the value of inverter's output voltage $U_C$ converting to power grid side of coupling transformer, $I_{La}$ is the load harmonic source current, $U_2$ is the voltage of fundamental series resonant injection branch.

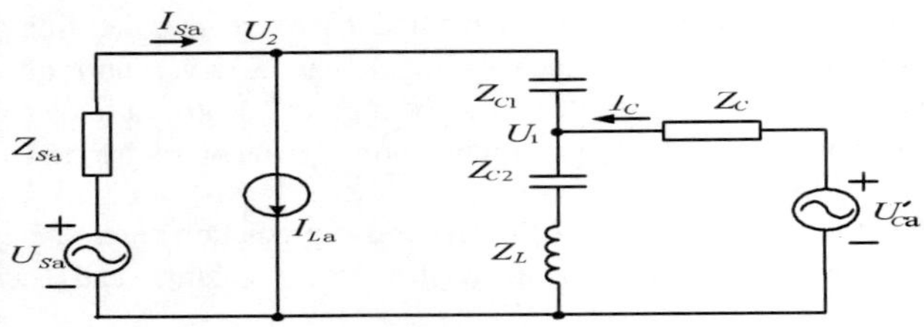


**Fig. 1.** A-phase equivalent circuit of HAPFSIC

Corresponding to the on-off state of each single-phase bridge arm, the switching function is defined as follows:

$$S_i = \begin{cases} 1 \\ -1 \end{cases} \qquad i = a, b, c$$

When $S_i = 1$, the on-tube of bridge arm is on, and the under-tube is off; When $S_i = -1$ ,the result is adverse. So we can get the switching circuit model as shown in Fig 2:

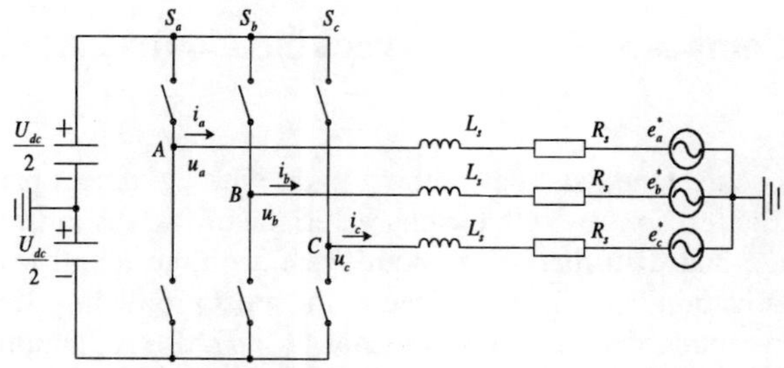


**Fig. 2.** The simplified HAPFSIC equivalent switching circuit

According to Kirchhoff law, the DC side's capacitance is C, so the DC-side's current should meet:

$$\begin{cases} i_{dc} = C\dfrac{dU_{dc}}{dt} \\ i_{dc} = S_a i_a + S_b i_b + S_c i_c \end{cases} \tag{1}$$

$$C\frac{dU_{dc}^2}{dt} = 4u_a i_a + 4u_b i_b + 4u_c i_c \tag{2}$$

Alliance with (1) and (2),we can get the mathematical model of HAPFSIC expressed with switching function:

$$DX' - AX = U \tag{3}$$

In above formula:

$$X = \left[ i_a \quad i_b \quad i_c \quad \frac{dU^2_{dc}}{dt} \right]^T, U = \left[ u_a - e_a^* \quad u_b - e_b^* \quad u_c - e_c^* \quad 0 \right]^T$$

$$A = \begin{bmatrix} -R_s & & & \\ & -R_s & & \\ & & -R_s & \\ 4u_a & 4u_b & 4u_c & 0 \end{bmatrix}, D = diag(L_S \quad L_S \quad L_S \quad C)$$

In order to analyze conveniently, only considering the harmonic domain, and applying to the concept of complex impedance in Fig2.2, It comes to HAPFSIC's three-phase system model as shown in formula (3). In condition of stable voltage in DC side, by reasonably controlling the output value of voltage source inverter can effectively reduce the current harmonic, also is related to the on-off state of each single-phase bridge arm; due to APF is characteristic of non-continuity action, as well as a natural variable structure system. So the sliding mode variable structure control has a very distinct advantage in application of APF.

## 3 A Novel Composite Current Tracking Control Algorithm of APF

Sliding mode can be designed and has nothing to do with the object parameters, which makes variable structure control with the characteristic of fast response, insensitivity to parameters' change and disturbance, without system online identification as well as simple physics realization, so it is considered as the most promising control method[6]. To carry out sliding mode variable structure control in HAPFSIC, mainly includes two aspects: (1) To select switch function, or determine the switching surface $S(k)=0$; (2) To get control rate. Considering HAPFSIC application, frequency of power harmonic to be filtered should be lower, they are usually below 1KHz, compared to switching frequency of inverter applied in HAPFSIC which is 10KHz～20KHz,so in a short control period, the output value of current reference $i_C^*(t)$ and voltage $u_1(t)$ of fundamental wave series resonant branch approximate to constants.

HAPFSIC's control target is to make inverter's output current $i_C(t)$ completely tracking the reference volume $i_C^*(t)$, so we can get the switching function of HAPFSIC discrete sliding mode variable structure as follows[7]:

$$S(k)=\Delta i_C(k)=i_C^*(k)-i_C(k) \tag{4}$$

According to Fig2.1,the difference equation of A-phase system is:

$$i_C(k+1)=\frac{L_C-R_C}{L_C}i_C(k)+\frac{1}{L_C}\dot{u}_{Ca}(k)-\frac{1}{L_C}u_1(k) \tag{5}$$

Putting (5) into (4),and $i_C^*(k)$ is a constant, so:

$$S(k+1)-S(k)=\frac{R_C}{L_C}i_C(k)-\frac{1}{L_C}\dot{u}_{Ca}(k)+\frac{1}{L_C}u_1(k) \tag{6}$$

According to the theory of discrete sliding mode variable structure control, under the condition of making system stable, it can achieve controlling using following formula, as shown in index close to the rate.

$$\frac{dS(k)}{T}=-\varepsilon\, sgn(S(k))-qS(k) \tag{7}$$

In addition, $\varepsilon>0$, $q>0$, T stands for sampling period(that is a control cycle), sgn(•) is symbol function.

Combined (6)and (7),we can get:

$$\dot{u}_{Ca}(k)=u_1(k)+R_Ci_C(k)+\varepsilon TL_C\,\mathrm{sgn}(S(k))+qTL_CS(k) \tag{8}$$

Due to $\varepsilon$ and $q$ are set parameters, we can make:

$$\varepsilon T=\frac{\varepsilon_0}{L_C},\quad qT=\frac{R_C}{L_C} \tag{9}$$

Putting (9) into(8), so:

$$u_{Ca}^{'}(k)=u_1(k)+R_C i_C^{*}(k)+\varepsilon_0\,\mathrm{sgn}(S(k))=u_0(k)+\varepsilon_0\,\mathrm{sgn}(S(k)) \tag{10}$$

Therefore:

$$u_{Ca}^{'}(k)=\begin{cases}u_0(k)+\varepsilon_0 & S(k)>0\\ u_0(k)-\varepsilon_0 & S(k)<0\end{cases} \tag{11}$$

In above formula, $\varepsilon_0>0$ is proportional to system reaching speed $\varepsilon$, also means reaching speed that is different on both sides of switching surface. When $S(k)>0$, $\varepsilon_0$ is $\varepsilon^+$, the value of $u_{Ca}^{'}(k)$ is $u_{Ca}^{+}(k)$; when $S(k)<0$, $\varepsilon_0$ is $\varepsilon^-$, the value of $u_{Ca}^{'}(k)$ is $u_{Ca}^{-}(k)$. Because $\varepsilon_0$ is a parameter that can be set, and expressed as arbitrary positive number, which means $\varepsilon^+$ and $\varepsilon^-$ are more than 0, can guarantee system in stable condition. So it is more simple to get control rate.

In actual sliding mode variable structure system, due to the time lag, discrete system and so on, there must be buffeting, the sliding mode of buffeting discrete systems is a "quasi-sliding mode". Therefore, discrete sliding mode variable structure control can only guarantee the current tracking error is less than $\Delta$ (not 0), but they can not track reference input signal with non-static, so the actual sliding mode variable structure control rate is a kind of discrepancy regulating rate. In addition, a majority of the load (such as rectifiers, computer and other electrical equipment) change slowly in their current (as opposed to power grid fundamental cycle), and turns on periodicity, in steady state, the load current and the harmonic current components are relatively stable, it could be considered that the load current is changing in cycle; and repetitive control algorithm uses the cyclical nature of load disturbance, "memorizes" the location of the disturbance occurred, amends step by step and improves the output waveform, which can get good waveform in steady-state, at the same time, for the characteristic of a cycle delay, so it can get no difference and real-time control.

In despite of repetitive control[8] can reduce the static errors, and completely compensate the system's phase deviation, it can't decrease the settling time of system to load disturbance within a fundamental cycle, so its dynamic performance is poor. In this paper, it adopts a repetitive prediction control method that can be seen as repetitive correction to the plat calculation algorithm. The so-called plat calculation, which means that considering the sampling value of harmonic current as forecast value of the next cycle, to provide for the use of advanced control algorithms in next cycle, the basic relationship is:

$$\hat{i}_c(k)=i_c(k) \tag{12}$$

$i_c(k)$ is the sampling value of harmonic current at time of k, and $\hat{i}_c(k)$ means the predictive value of harmonic current in next cycle also at same time.

When the load's harmonic current waveform arises repeatedly in cycle, the error between predictive value and the actual current value would also appear repeatedly as a fixed waveform. If we use repetitive correction algorithm, which will clear up the repeated error and is made up of the cycle integral links, the head compensation links and the proportion links. The principle of repetitive correction algorithm is: detecting the prediction error e at any time, adding the compensation value into sampling volume of harmonic current before next cycle, making the prediction results be closer to the actual value.

The purpose of predictive algorithm is to give the instruction current forecast volume of next time advancing a bat, the forecast error

$$e(k) = i_c(k) - \hat{i}_c(k-1) \tag{13}$$

From above formula, $e(k)$ is the error which is calculated at time of k, actually, it is the prediction error at time of k-1; $\hat{i}_c(k-1)$ is the prediction value accounted at the moment of k-1,expecting that it is close to sample value $i_c(k)$ of harmonic current after correction at time of k.

It shows that the calculation of prediction error has a lag of a bat, so the compensation volume which is proportional to will work in advance in the next cycle. The repetitive predictive controller is shown as Fig 3:

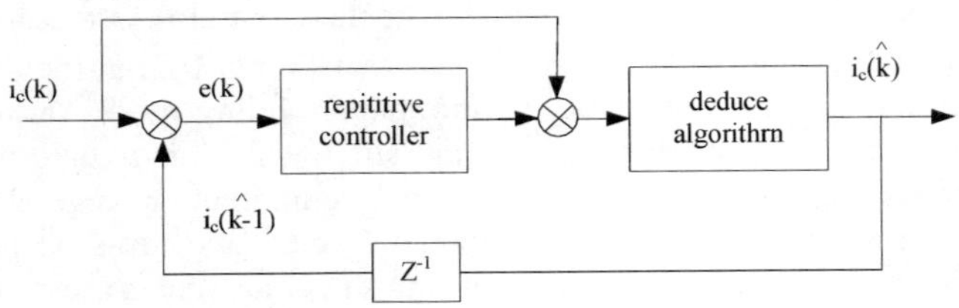


**Fig. 3.** The repetitive prediction controller

Improved discrete quasi-sliding mode variable structure controller and repetitive prediction controller are paralleled in forward channel of control system. The repetitive prediction controller inhibits tracking error between control object and original instruction, it detects the repetitive tracking errors of original system, correction value is added to the original error in advance to decrease the error. However, the improved discrete variable structure controller can induce the mutation of tracking error and regulate it immediately. The quality of output wave can not be well ensured, but the output voltage will not change suddenly, after a period, repetitive prediction controller regulate it to decrease the tracking error, then its control effect getting weaker till the system reach a new stable operation state. The methods not only overcome shortage in differential regulation of improved discrete variable structure controller, difficult in filtering of switching harmonic and so on, but also avoid time delay of repetitive control. The new compound controller is shown in Fig 4.

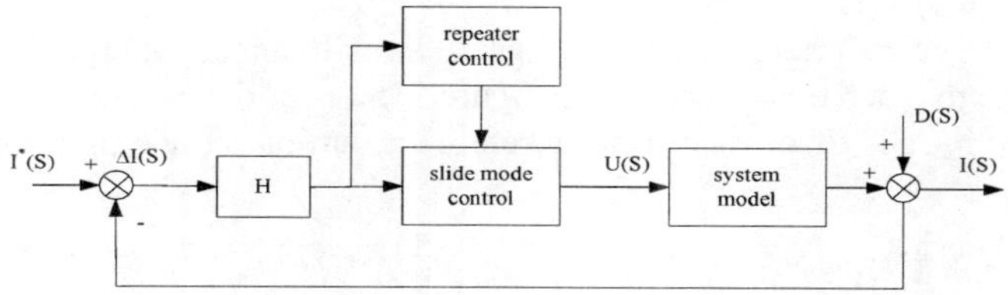


**Fig. 4.** The realization chart of compound controller

However, the compound control rule is:

$$s_a(k)=\begin{cases} 0 & e_a(k)>H \\ 1 & e_a(k)<-H \\ p(v_{eqa}(k)) & |e_a(k)|\le H \end{cases} \tag{14}$$

It can get $s_b(k)$ and $s_c(k)$, $p(v_{eq}(k))$ is corresponding to switching state of $v_{eq}(k)$ that is determined by repetitive prediction algorithm. From the above control rule, we can see that the greater the ring width H is, the much slower system response is; when H is equivalent to the maximum error, the new composite control is repetitive control; the smaller the H is, the greater the switching frequency of power devices is, even it may exceed the limits.

## 4 Simulation Analysis and Verification

In order to fully verify the comprehensive advantages of new type of composite control algorithm in terms of rapid speed, switching harmonic content, control error and so on, it makes a comparative study which includes sliding mode variable structure control, repetitive control and new type of composite control algorithm. In course of simulation, three-phase power supplies line voltage with 10KV, frequency is 50 HZ; the value of 4 times harmonic current is 50 A; in fundamental resonant circuit, the capacitance is $690.2\mu F$, and the inductance is $15.5mH$. In active part, we use three-phase bridge rectifier circuit, DC side voltage of inverter is 600V, output filter inductance is $0.5mH$, capacitance is $120\mu F$. The ring width of sliding mode variable structure control and repetitive control is 2A. The mode of control switch is used the method of triangle-wave modulation, the range of triangle-wave is -100~+100, frequency is 10 KHZ. In the delay cycle link of repetitive control system, the value of N is 12800/50=256, the filter Q is 1. Compensator $s(k)$ makes the low-frequency gain correction of control object to be 1, the system uses second-order low-pass filter. Anyway, passive filter is put into power grid at first, active power filter is at time of 0.2 seconds.

(1) Comparison of rapidity

In Fig 5, three control algorithms can get the better control effect at 0s~1s, it explains that waveform of grid current is very close to sine wave after putting single injection active power filter. Both sliding mode variable structure control and the new composite control make the current tracking error fast get to a small range, but repetitive controller need more time to reach this effect. In this figure, i_Sa is grid current; i_Ca is inverter's output current; i_Ca is Error waveform of reference current's tracking error.

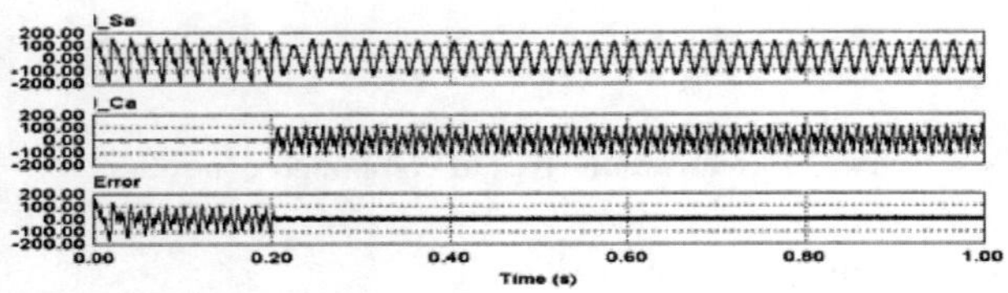


(a) The waveform under the condition of sliding mode variable structure control

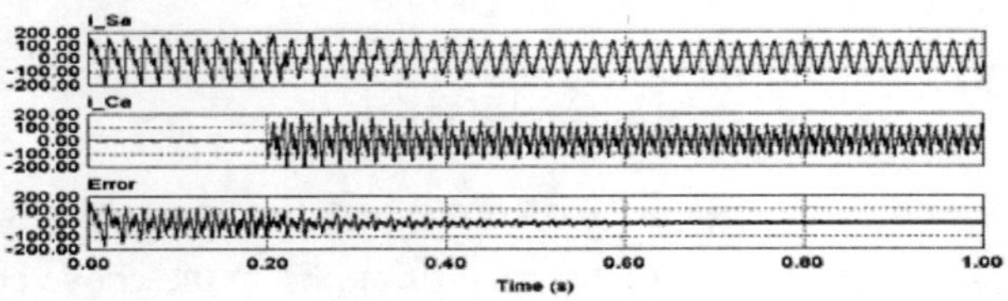


(b) The waveform under the condition of repetitive control

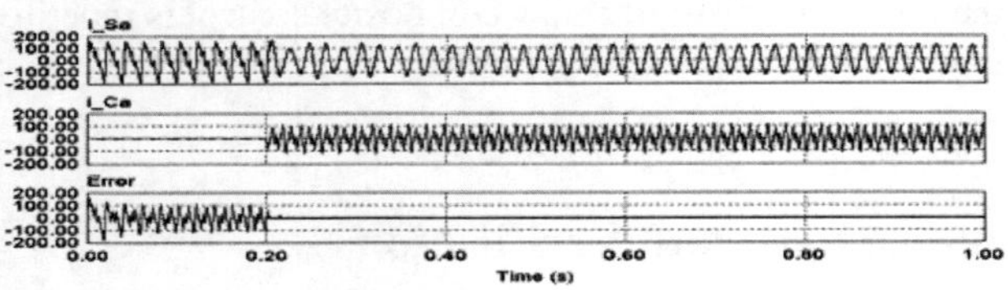


(c) The waveform under the condition of the new composite control

**Fig. 5.** The parallel to current waveform

(2) Comparison of control error

If the harmonic source are parallel with many current sources, such as the 4 times is 24A,11 times is 11A,13 times is 9A. We can get the value of reference current's tracking errors in different control method as shown in Table1.

**Table 1** The value of reference current's tracking errors in different control method (unit: A)

| Harmonic number | | Fundamental | 4 times | 11 times | 13 times |
|---|---|---|---|---|---|
| sliding mode variable structure control | current error | 0.13 | 0.70 | 0.57 | 0.53 |
| repetitive control | current error | 0..39 | 0.29 | 0.15 | 0.17 |
| new composite control | current error | 0.02 | 0.23 | 0.20 | 0.19 |

From the table, the effect of new composite control is best, and tracking error of every harmonic current is about 0.2A, fundamental error is only 0.02A;but the tracking error of sliding mode variable structure control is 3 times more, and fundamental error is 6.5 times more than the former. In repetitive control, both contents of low-frequency harmonic of grid current and tracking error of middle-low frequency harmonic in *Error* are better , but the part of high-frequency harmonics are less, which is due to the repetitive control do not make the system reach steady-state temporarily.

## 5 Conclusion

The new composite control method of HAPFSIC which combines with the fast response speed of quasi-sliding mode variable structure and no real-time steady-state error of repetitive predictive control algorithm, takes into account the unity of dynamic performance and control precision. The simulation shows: when current tracking error is beyond the setting ring width, the system will work under the rate of sliding mode variable structure control, and current tracking error reduces rapidly. It makes system to have good dynamic performance, can reach realization with no real-time and steady-state error of current tracking control, as well as makes the system have a good steady-state performance.

## References

1. Wang, Z., Yang, J.: Harmonic suppression and reactive power compensation. Mechanical Industry Press, Beijing (1998)
2. Ting, Q., Lv, Y.Z., Jin, H.: Based on one cycle control the active filter control strategy bicyclo. CSEE 23(3), 34–37 (2003)
3. Liu, F., Zou, Y.P., Li, H.: Repetitive control based on voltage source inverter output current waveform control method. CSEE 25(19), 58–63 (2005)
4. Tu, C.M., Luo, A. (Instructor): New resonant impedance-type hybrid active power filter RITHAF study. Doctoral Thesis, Central South University (2003)
5. Luo, A., Fu, Q., Phil, M.: High-capacity Hybrid Power Filter for Harmonic Suppression and Reactive Power Compensation in the Power Substation. CSEE 24(9), 115–123 (2004)
6. Singh, B., Al-Haddad, K., Chandra, A.: A New Control Approach to Three-phase Active Filter for Harmonics and Reactive Power Compensation. IEEE Trans. On PS 13(1), 133–138 (1998)
7. Fu, Q., Luo, An.(Instructor): High-power harmonic active control strategy for management of applied research and engineering. Doctoral Thesis, Central South University (2004)
8. Doulai, P., Ledwich, G.: Co-ordinated Three-phase Hysteresis-control Scheme for Active Power Filtering. IEEE Proceedings of Electric Power Applications 139(5), 457–464 (1993)

# High-Capacity Hybrid Active Power Filter for the Power Substation

Fen Gong[1], Xiangyang Xia[1], Shiwu Luo[2], and Feng Zhou[1]

[1] College of Electrical Engineering, Changsha University of Science and Technology, Changsha 410077, China
[2] Chongqing Power Education Training Centre, Chongqing 400053, China
gongfen86814@126.com

**Abstract.** Non-linear loads, such as diode, thyristor converters and arc furnaces are typical sources of harmonic currents. A capacitor clamped voltage source inverter for high-capacity hybrid active power filter (HHAPF) to compensate harmonics and reactive power. If the utility voltage is distorted, APF allows similar level of distortion in the compensated source current. The HHAPF which is composed of series resonators injection APF and passive filter (PF) by parallel hybrid can make static compensation of reactive power by capacitor and it can eliminate harmonics by the co-function of active filter and passive filter. The serial resonant LC network resonance occurs in the fundamental frequency, its fundamental current impedance is approximate 0, the fundamental current will all flow to this network, but not to flow into the coupling transformer and the inversion. Practice shows that the HHAPF has good filter effect and certain capacity of reactive power compensation.

**Keywords:** High-capacity, active power filter, passive filter, reactive power, fundamental current.

## 1 Introduction

The wide use of nonlinear loads such as power electronic devices in the power distribution systems for DC supply or inverter based applications causes significant power quality degradation in power distribution networks in terms of current/voltage harmonics, power factor and resonance problems.

Harmonics can be limited either by improving the design of nonlinear loads (e.g., increasing the number of pulses, enlarging the ac choke inductance, considering active front-ends, etc) or by adding filters to the power system. Although passive filters have been traditionally used for this task, the ongoing sustained developments in power electronics have made the use of active power filter. It is largely due to the advantages of APF's over passive filter. This paper presents a shunt APF based on a three phase voltage source inverter (VSI). The operational principle of the APF is to inject a harmonic current in parallel with the load.

## 2 Main Circuit of HHAPF and Its Principle

In this system,PF can suppress some special order harmonic, such as fifth and seventh order harmonics, and compensate reactive power, and active power filter can improve the characteristic of suppressing harmonic current and degrade resonance of passive filters, the power capacity of a power converter in hybrid active power filter is about one-tenth of nonlinear loads. From the configuration of HHAPF in Fig 1, we can know that HHAPF is made up of the series resonant active power filter with injection and PF by paralleling them. The static compensation of the reactive power is made by capacitance while active and passive component work together to suppress harmonics. Passive component of PF makes high-capacity reactive power compensation, and the impedance of fundamental frequency will be lower. Because in FSRC resonant with fundamental frequency, its impedance is nearly 0 which equals to the short circuit channel of fundamental current, so all the fundamental current flow through PF will flow into FSRC, and will not flow into coupling transformer and inverter; To PF, FSRC only withstand small part of fundamental voltage, a large part is taken by PF, therefore, reactive power compensation is made by PF, the HHAPF will not increase the capacity of the active component.

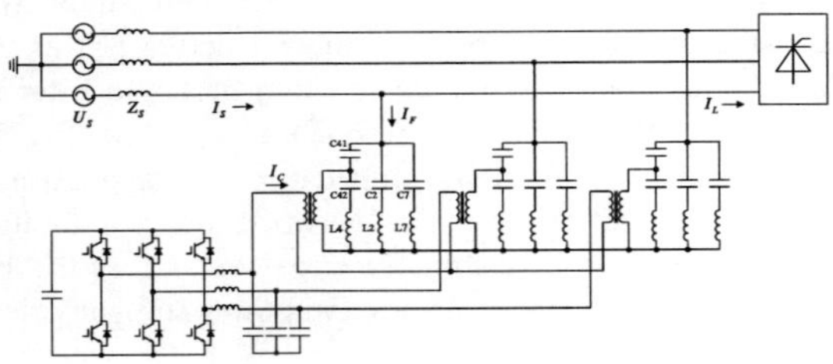

**Fig. 1.** The high-power hybrid active power filter

The single-phase equivalent circuit of the proposed HHAPF is shown in Fig 2. C41, C42, and L4 are selected to oscillate at fourth harmonic frequency in order to suppress fourth harmonic and compensate reactive power of the main feeder. For C2 and L2 , and C7 and L7 are selected to oscillate at second and seventh harmonic frequency respectively to suppress second and seventh harmonic and compensate reactive power.

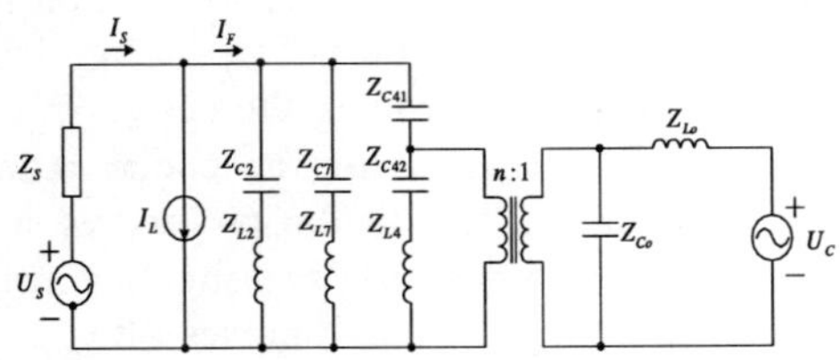

**Fig. 2.** The single-phase equivalent circuit of HHAPF

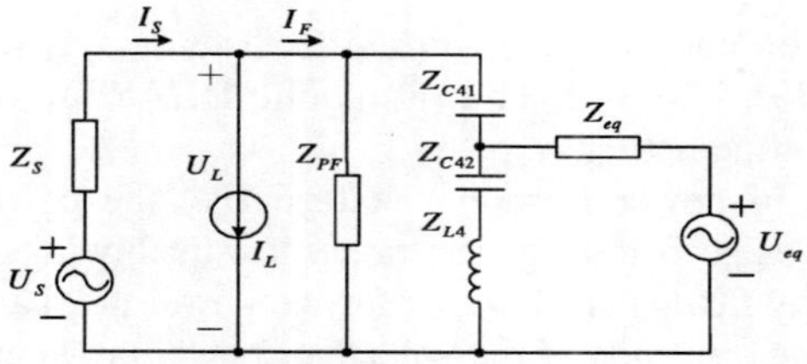


**Fig. 3.** The equivalent circuit of HHAPF

$$Z_{PF}=(Z_{C2}+Z_{L2})//(Z_{C7}+Z_{L7})=\frac{(Z_{C2}+Z_{L2})(Z_{C7}+Z_{L7})}{Z_{C2}+Z_{L2}+Z_{C7}+Z_{L7}} \tag{1}$$

$$Z_{eq}=n^2 Z_{Co}//n^2 Z_{Lo}=\frac{n^4 Z_{Co} Z_{Lo}}{n^2 Z_{Co}+n^2 Z_{Lo}}=\frac{n^2 Z_{Co} Z_{Lo}}{Z_{Co}+Z_{Lo}} \tag{2}$$

$$U_{eq}=nU_C\frac{n^2 Z_{Co}}{n^2 Z_{Co}+n^2 Z_{Lo}}=\frac{nZ_{Co}}{Z_{Co}+Z_{Lo}}U_C \tag{3}$$

$$\begin{aligned} Z'_{eq} &=((Z_{C42}+Z_{LA})//Z_{eq}+Z_{C41})//Z_{PF} \\ &=\frac{Z_{PF}(Z_{eq}(Z_{C41}+Z_{C42}+Z_{LA})+Z_{C41}(Z_{C42}+Z_{LA}))}{Z_{eq}(Z_{C41}+Z_{C42}+Z_{LA}+Z_{PF})+(Z_{C42}+Z_{LA})(Z_{C41}+Z_{PF})} \end{aligned} \tag{4}$$

$$\begin{aligned} U'_{eq} &=\frac{\dfrac{(Z_{C42}+Z_{LA})(Z_{C41}+Z_{PF})}{Z_{C41}+Z_{C42}+Z_{LA}+Z_{PF}}}{\dfrac{(Z_{C42}+Z_{LA})(Z_{C41}+Z_{PF})}{Z_{C41}+Z_{C42}+Z_{LA}+Z_{PF}}+Z_{eq}}\bullet\frac{Z_{PF}}{Z_{C41}+Z_{PF}}U_{eq} \\ &=\frac{Z_{PF}(Z_{C42}+Z_{LA})}{Z_{eq}(Z_{C41}+Z_{C42}+Z_{LA}+Z_{PF})+(Z_{C42}+Z_{LA})(Z_{C41}+Z_{PF})}U_{eq} \end{aligned} \tag{5}$$

$$\begin{cases} U_L = U_S - Z_S I_S \\ I_S = I_L + I_F \\ U_L = Z'_{eq} I_F + U'_{eq} \end{cases} \tag{6}$$

Form equation (6):

$$I_S=\frac{Z_{eq}I_L+U_S-U'_{eq}}{Z_S+Z_{eq}}=\frac{Z_{eq}I_L+U_S-K\dfrac{nZ_{Co}}{Z_{Co}+Z_{Lo}}U_C}{Z_S+Z_{eq}} \tag{7}$$

$$K=\frac{Z_{PF}(Z_{C42}+Z_{LA})}{Z_{eq}(Z_{C41}+Z_{C42}+Z_{LA}+Z_{PF})+(Z_{C42}+Z_{LA})(Z_{C41}+Z_{PF})}$$

From equation (7), we can see that the harmonic suppression is made by the output voltage UC controlling inverter to alter Is. We can also find that the performance of high-capacity hybrid active power filter is influenced by the following factors:

(1) The fluctuating of load harmonic current IL. The floating directly influence the harmonic current of electric power, Thus, it require HHAPF to have quick, dynamic response to adapt to its fluctuating.

(2) The fluctuating of power harmonic voltage Us. The power harmonic voltage comes from two ways, one is that the generator has low quality in electrical power, when it provides high-capacity fundamental waves for the power net, harmonic power is also produced, then cause the distortion of the voltage of power provider. On the other hand, harmonic load in other spots of the net will also cause the fluctuating of the harmonic voltage in power system. To high-capacity hybrid active power filter and harmonic load in other compensation spot, their effect is equivalent to adding a harmonic voltage source in fundamental wave. The existence of this extra harmonic voltage source, especially the harmonic load changes in the spot cause the fluctuating of harmonic voltage hichl result in the changes of compensation current in the branch of PF and it will greatly influence the harmonic compensation remaining current of the HHAPF.

(3) The fluctuating of impedance IS of the power system. Because of the changes of the power system surrounding, i.e. the temperature alteration and wearing of the circuit will lead to changes of the impedance of the power system.

(4) The fluctuating of the net frequency. The fluctuating of the net frequency will cause changes of parameter K and the harmonic impedance ZPF .The changes of the parameter will possibly affect the performance of HHAPF.

## 3 Simulation Results

To verify the characteristic of the proposed HHAPF, Simulation is done according to Fig1.The major parameters used in the simulation system are specification of the PF and APF, according to the power substation. Fig 4 shows current waveforms for different types of power filters. Simulated results illustrate the effectiveness of the proposed active power filter. The THD of the main feeder current is reduced from 30.5%. In the case of no harmonic suppression (Fig 4(a) and Fig 5(a)), to the 15.2% in the case of only passive power filter (Fig 4(b)), to only6.1% for passive filter and active power filter (Fig 4(c)and Fig 5(b)).

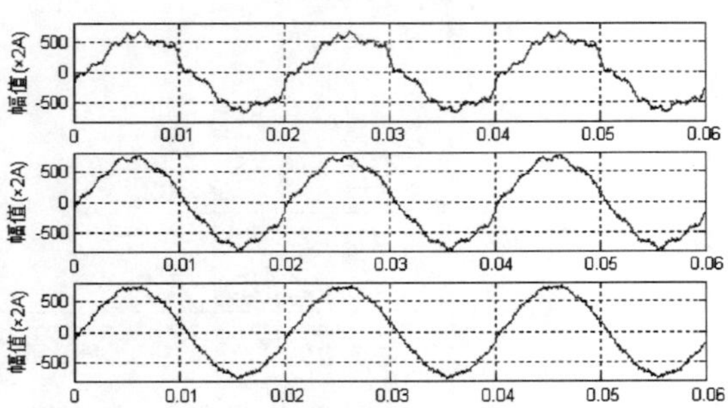


**Fig. 4.** Current waveforms

(a) current waveform without filter
(b) current waveform only with passive power filter
(c) current waveform with passive and active power filter

(a)The spectrum of source current without filter

(b) The spectrum of source current with passive and active power filter

**Fig. 5.** The spectrum of source current

## 4 Conclusions

A high-capacity hybrid power filter with passive power filter and active power filter of injecting harmonic current has been presented and analysed. It was proved that the proposed hybrid power filter can compensate high-capacity power factor by reactive power filter and inject high-capacity harmonic current of nonlinear loads, to improve power factor and suppress the harmonic of the main feeder simultaneously. Both simulation and industrial application results show that the proposed system is effective in compensation of power factor and harmonic suppression.

## References


1. Rastogi, M., Mohan, N., Edris, A.A.: Hybrid Active Filtering of Harmonic Currents in Power Systems. IEEE Trans. on PD 10(4), 1994–(2000)
2. Singh, B., Haddad, K.A., Chandra, A.: A review of active filter for power quality improvement. IEEE Trans. on IE-46, 950, 960–971 (1990)
3. Moran, R.L., Dixon, J., Espinoza, J.: A simple control scheme for active power filter. IEEE Proc.-Gener. Transm. Distrib 14(4), 485–490 (2002)
4. Lina, W., Qing, F., An, L.: An.Damping of Harmonic in Industrial Power System. Automation of Electric Power Systems 25(20) (2001)
5. Balbo, N., et al.: Simplified Hybrid Active Filters for Harmonic Compensation in Low Voltage Industrial Application. In: Proceedings of IEEE ICHPS VI, Bologna (1994)
6. Bhattacharya, S., Cheng, P.T., Divan, D.M.: Hybrid solutions for improving passive filter performance in high power applications. IEEE Trans. on IS-33, 930, 17–21 (1997)
7. Lina, W., Qing, F., An, L.: Damping of Harmonic in Industrial Power System. Automation of Electric Power Systems 25(20), 41–44 (2001)
8. Bhattacharya, S., Cheng, P.T., Divan, D.M.: Hybrid solutions for improving passive filter performance in highpower applications. IEEE Trans.on IS-33, 930, 17–21 (1997)

9. Singh, B., Haddad, K.A., Chandra, A.: A review of active filter for power quality improvement. IEEE Trans. on IE-46, 950, 960–971 (1990)
10. Moran, R.L., Dixon, J., Espinoza, J.: A simple control scheme for active power filter. IEEE Proc.-Gener. Transm. Distrib. 14(4), 485–490 (2002)
11. Al-Zamil, A.M., Torrey, D.A.: A passive series, active shunt filter for high power application. IEEE Trans. on Power Electronic 200116910, 101–109 (2001)
12. Tan, T., Luo, A., Tang, X., Tu, C.: Development of High-capacity Hybrid Power Filter. Proceedings of the CSEE 24(3), 41–45 (2004)
13. A Project Support by Chongqing Power Technology Program (2010)

# Research on the Tender Evaluation of Scientific and Technical Projects Based on Fuzzy Comprehension Assessment

Zhang Min, Yin Zhongfeng, and Tang Zhaojie

Changde Army Agency Office
General Armament Department
Changde, China
mmjingwei@163.com

**Abstract.** In this paper, a complete method of evaluating tenders which is based on fuzzy mathematics theory was mainly put forward solving the existing problems in the bidding tender assessment. Then we applied this method to evaluating the tender of the scientific and technical projects. The results have proved that this method could make the evaluation of tenders more comprehensive, rational, and scientific.

**Keywords:** fuzzy comprehensive assessment, bidding and tender, tender assessment, scientific and technical projects.

## 1 Introduction

In recently years, the research on tender evaluation of scientific and technical projects was developed rapidly. But in practice, the real tender evaluation in scientific and technical projects does not exist due to several problems. It is also difficult to establish the effective judge model in cases of tender evaluation of scientific and technical projects. In order to solve these problems, we put forward using the fuzzy mathematics theory to set up an effective method in this paper and the results have proved that this method could make the evaluation of tenders more comprehensive, rational, and scientific.

## 2 Fuzzy Comprehensive Assessment Model

Presumed *U, V* are two finitude gathers, $U=\{u_1,u_2,\cdots,u_n\}$, $V=\{v_1,v_2,\cdots,v_m\}$. $\tilde{A}$ and $\tilde{B}$ are the fuzzy subset of $U$ and V, write as: $\tilde{A}=(a_1,a_2,\cdots,a_n)$, $0\le a_i\le 1$, $\tilde{B}=(b_1,b_2,\cdots,b_m)$, $0\le b_i\le 1$.

$\tilde{R}$ is a fuzzy relation of U and V[1],

$$\tilde{R}=\begin{bmatrix} r_{11} & r_{12} & \cdots & r_{1m} \\ r_{21} & r_{22} & \cdots & r_{2m} \\ \cdots & \cdots & \cdots & \cdots \\ r_{n1} & r_{n2} & \cdots & r_{nm} \end{bmatrix} \tag{1}$$

and one transformation is determined by the $\widetilde{R}$. If $\widetilde{A}$ of U is given, then $\widetilde{B}$ of V can be got, the fuzzy transformation is written as:

$$\widetilde{B} = \widetilde{A} \circ \widetilde{R} \tag{2}$$

## 3 Application in Tender Evaluation of Scientific and Technical Projects

The main procedure of fuzzy comprehensive assessment is construct the factors set choice set and weigh set.

### 3.1 Establish the Factors Set

The factors set is composed of the variety factor of the judged objects, namely $U = \{u_1, u_2, \cdots, u_n\}$. In the tender evaluation of scientific and technical projects, Qualification, Technology, Money, Management, Risk and Ability are the main factors. Money and Ability are the quantitative factors and the others are qualitative factors.

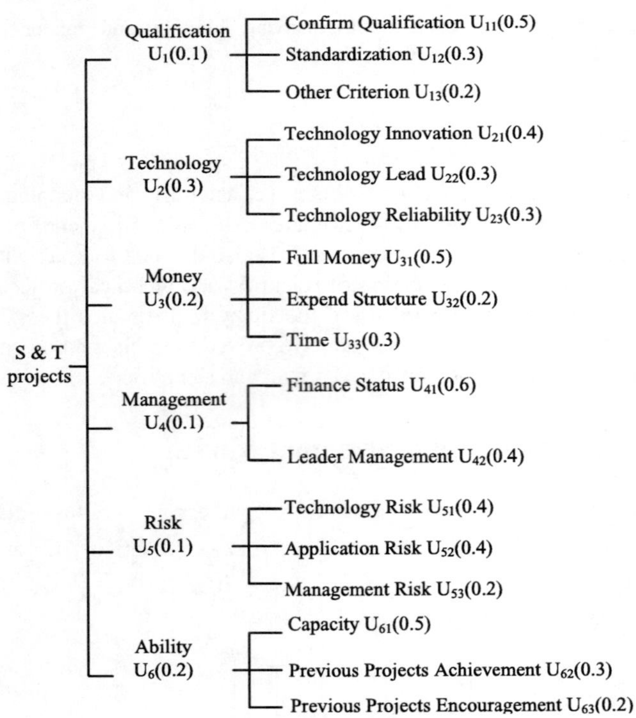


**Fig. 1.** Tender evaluation factors sketch figure

In this paper, we use a real tender evaluation sample to introduce the establishment steps of tender evaluation model. There are four research organizations attempting to obtain this scientific and technical project. All original data of the organizations were given in Table 1.

**Table 1.** Original data of the organizations

| Factors | A | B | C | D |
|---|---|---|---|---|
| Confirm Qualification | wonderful | very good | wonderful | wonderful |
| Standardization | good | wonderful | wonderful | very good |
| Other Criterion | very good | wonderful | very good | wonderful |
| Technology Innovation | very good | wonderful | wonderful | good |
| Technology Lead | wonderful | good | very good | wonderful |
| Technology Reliability | good | wonderful | very good | good |
| Full Money (Million Yuan) | 500 | 400 | 520 | 600 |
| Expend Structure (%) | 50 | 44 | 60 | 76 |
| Time (year) | 4.0 | 3.0 | 4.0 | 4.5 |
| Finance Status | very good | good | wonderful | wonderful |
| Leader Management | very good | wonderful | very good | wonderful |
| Technology Risk | wonderful | very good | wonderful | good |
| Application Risk | common | wonderful | very good | wonderful |
| Management Risk | wonderful | very good | wonderful | wonderful |
| Capacity | 6 | 8 | 8 | 7 |
| Previous Projects Achievement (%) | 94 | 97 | 100 | 98 |
| Previous Projects Encouragement | 6 | 3 | 5 | 5 |

### 3.2 Establish the Choice Set

The choice set is composed of the general results which are gained by the judge, namely $V = \{v_1, v_2, v_3, v_4\}$.

### 3.3 Establish the Weigh Set

In the factors set, the importance of different factors is not the same. In order to reflect the important degree of each factor, the weigh of each factor $u_i (i = 1,2,\cdots,n)$ must be given as $a_i(i=1,2,\cdots,n)$, and $\underset{\sim}{A}=(a_1,a_2,\cdots,a_n)$ is called the weigh set. Generally, $a_i(i=1,2,\cdots,n)$ are nonnegative and the sum of $a_i(i=1,2,\cdots,n)$ one:

$$\sum_{i=1}^{n} a_i = 1 \quad a_i \geq 0 \quad (i=1,2,\cdots,n) \tag{3}$$

which can be regard as important subjection of each factor $u_i(i=1,2,\cdots,n)$.

### 3.4 Determination of Subjective Function

There are four types of judge factor in scientific and technical projects, such as cost model, benefit model, middle model and interzone model[2]. Figure 2 describe the subjective function of the four types of judge model.

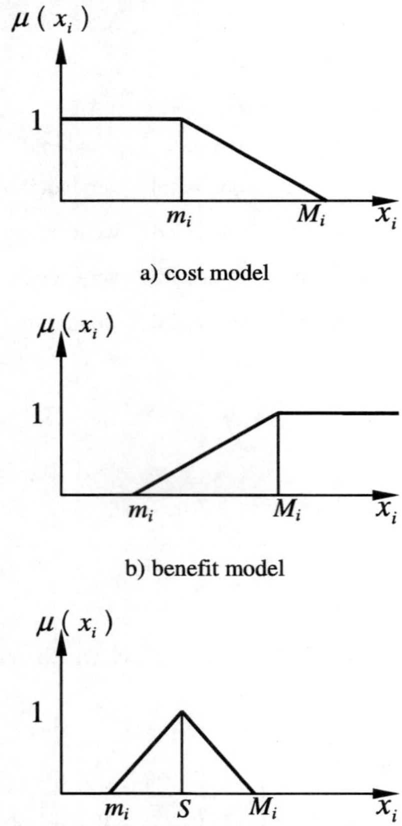


a) cost model

b) benefit model

c) middle model

**Fig. 2.** Subjective function sketch figure

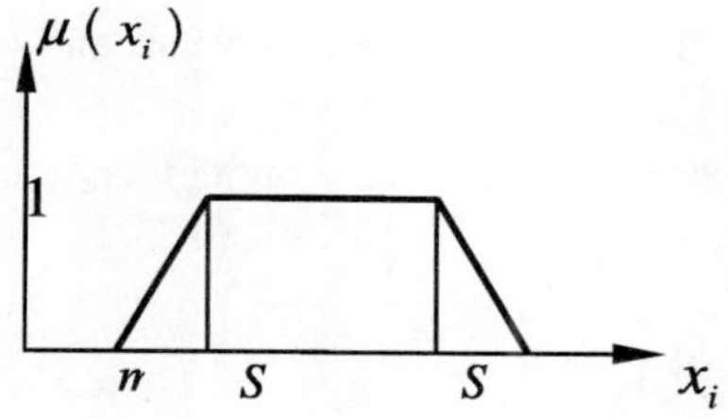


d) interzone model

**Fig. 2.** (*continued*)

In the tender evaluation of scientific and technical projects, Full Money $U_{31}$ belongs to the cost model. The function expression of cost model is

$$\mu_i = \begin{cases} 1 & x_i \le m_i \\ \dfrac{M_i - x_i}{M_i - m_i} & m_i < x_i \le M_i \\ 0 & x_i > M_i \end{cases} \quad (4)$$

In real application, $m_i$ is 500 million yuan, and $M_i$ is 700 million yuan. Therefore, the subjection of $U_{31}$ is {1.00, 1.00, 0.90, 0.50}.

Capacity, Previous Projects Achievement and Previous Projects Encouragement belong to the benefit model. The function expression of benefit model is

$$\mu_i = \begin{cases} 1 & x_i > M_i \\ \dfrac{x_i - m_i}{M_i - m_i} & m_i < x_i \le M_i \\ 0 & x_i \le m_i \end{cases} \quad (5)$$

In real application, $m_i$ is 4 persons and $M_i$ is 8 persons. Therefore the subjection of $U_{61}$ is {0.50, 1.00, 1.00, 0.75}.

Time belongs to middle model and the function expression of middle model is

$$\mu_i = \begin{cases} \dfrac{2\times(x_i - m_i)}{M_i - m_i} & m_i < x_i \le S \\ 0 & \text{others} \\ \dfrac{2\times(M_i - x_i)}{M_i - m_i} & S < x_i \le M_i \end{cases} \quad (6)$$

In real application, $m_i$ is 3 years, $S$ is 4 years and $M_i$ is 5 years. Therefore the subjection of $U_{33}$ is {1.00, 0.00, 1.00, 0.50}.

Expend Structure belongs to interzone model and the function expression of interzone model is

$$\mu_i = \begin{cases} 1 - \dfrac{S_1 - x_i}{\max\{S_1 - m_i \ , \ M_i - S_2\}} & x_i \le S_1 \\ 1 & S_1 < x_i \le S_2 \\ 1 - \dfrac{x_i - S_2}{\max\{S_1 - m_i \ , \ M_i - S_2\}} & x_i > S_2 \end{cases} \tag{7}$$

In real application, $m_i$ is 35%, $S_1$ is 50%, $S_2$ is 60% and $M_i$ is 75%. Therefore the subjection of $U_{32}$ is {1.00, 0.60, 1.00, 0.00}.

Ascertainment of qualitative factors need evaluate directly[3]. In real applications, the choice set (wonderful, very good, good, common, bad, very bad, worst) will be set as (1.00, 0.85, 0.75, 0.50, 0.35, 0.25, 0.00), and the subjection of factors will be given according to the conclusion of experts.

**Table 2.** Subjection of all factors

| Factors | A | B | C | D |
|---|---|---|---|---|
| Confirm Qualification | 1.00 | 0.85 | 1.00 | 1.00 |
| Standardization | 0.75 | 1.00 | 1.00 | 0.85 |
| Other Criterion | 0.85 | 1.00 | 0.85 | 1.00 |
| Technology Innovation | 0.85 | 1.00 | 1.00 | 0.75 |
| Technology Lead | 1.00 | 0.75 | 0.85 | 1.00 |
| Technology Reliability | 0.75 | 1.00 | 0.85 | 0.75 |
| Full Money (Million Yuan) | 1.00 | 1.00 | 0.90 | 0.50 |
| Expend Structure (%) | 1.00 | 0.60 | 1.00 | 0.00 |
| Time (year) | 1.00 | 0.00 | 1.00 | 0.50 |
| Finance Status | 0.85 | 0.75 | 1.00 | 1.00 |
| Leader Management | 0.85 | 1.00 | 0.85 | 1.00 |
| Technology Risk | 1.00 | 0.85 | 1.00 | 0.75 |
| Application Risk | 0.50 | 1.00 | 0.85 | 1.00 |
| Management Risk | 1.00 | 0.85 | 1.00 | 1.00 |
| Capacity | 0.50 | 1.00 | 1.00 | 0.75 |
| Previous Projects Achievement (%) | 0.94 | 0.97 | 1.00 | 0.98 |
| Previous Projects Encouragement | 1.00 | 0.50 | 1.00 | 1.00 |

### 3.5 Fuzzy Comprehensive Assessment

After the subjection of factors is determined, the model will be used in the tender evaluation of scientific and technical projects. The first stage judge is got by $\tilde{B}_i = \tilde{A}_i \circ \tilde{R}_i$, and the judge matrix can be got by the following expression,

$$\tilde{B}_1 = \tilde{A}_1 \circ \tilde{R}_1 =$$

$$\begin{bmatrix} 0.5 & 0.3 & 0.2 \end{bmatrix} \circ \begin{bmatrix} 1.00 & 0.85 & 1.00 & 1.00 \\ 0.75 & 1.00 & 1.00 & 0.85 \\ 0.85 & 1.00 & 0.85 & 1.00 \end{bmatrix}$$

$= [(0.5\times1.00+0.3\times0.75+0.2\times0.85),$
$(0.5\times0.85+0.3\times1.00+0.2\times1.00),$
$(0.5\times1.00+0.3\times1.00+0.2\times0.85),$
$(0.5\times1.00+0.3\times0.85+0.2\times1.00)]$
$= [0.895, 0.925, 0.970, 0.955]$.

Just as the above calculate, the others results will be obtained.

$\tilde{B}_2 = \tilde{A}_2 \circ \tilde{R}_2 = [0.865,\ 0.925,\ 0.850,\ 0.825]$,

$\tilde{B}_3 = \tilde{A}_3 \circ \tilde{R}_3 = [1.000,\ 0.620,\ 0.950,\ 0.400]$,

$\tilde{B}_4 = \tilde{A}_4 \circ \tilde{R}_4 = [0.850,\ 0.850,\ 0.940,\ 0.900]$,

$\tilde{B}_5 = \tilde{A}_5 \circ \tilde{R}_5 = [0.800,\ 0.910,\ 0.940,\ 0.900]$,

$\tilde{B}_6 = \tilde{A}_6 \circ \tilde{R}_6 = [0.732,\ 0.891,\ 1.000,\ 0.869]$.

The second stage can be got by synthesizing

$$\tilde{B} = \tilde{A} \circ \tilde{R} =$$

$$\begin{bmatrix} 0.1 & 0.3 & 0.2 & 0.1 & 0.1 & 0.2 \end{bmatrix} \circ \begin{bmatrix} 0.895 & 0.925 & 0.970 & 0.955 \\ 0.865 & 0.925 & 0.850 & 0.825 \\ 1.000 & 0.620 & 0.950 & 0.400 \\ 0.850 & 0.850 & 0.940 & 0.900 \\ 0.800 & 0.910 & 0.940 & 0.900 \\ 0.732 & 0.891 & 1.000 & 0.869 \end{bmatrix}$$

$= [0.8544, 0.8482, 0.9300, 0.7768]$.

According to the above result, C unit will be the best unit in the sample.

## 4 Conclusions

During the implement of fuzzy comprehensive assessment, the subjective factors will be weaken and the objective factors will be strengthen. The tender assessment model based on fuzzy comprehensive assessment will be used perfectly in the tender evaluation of scientific and technical projects.

## References

1. Han, L.: Application Fuzzy Mathematics, Beijing (1998)
2. Bao, Y.: Research on bidding and tender of scientific and technical projects. Huazhong Technical University (2002)
3. Tian, S.: Research on tender evaluation on architecture engineering (1998)

# Study of Resource Allocation Efficiency in Diversified Railway Logistics under Administrative Monopoly's Influence*

Lipeng Feng[1,2] and Jiwei Qin[3]

[1] Department of Economy and Management, Chongqing University of Arts and Sciences, Yongchuan, Chongqing, China
michelleflp@163.com
[2] College of Logistics, Southwest Jiaotong University, Chengdu, Sichuan, China
[3] College of Economy and Trade; Chongqing City Vocational College
Yongchuan, Chongqing, China
qinjiwei@sohu.com

**Abstract.** The paper analyzed whole resource allocation efficiency in diversified railway logistics from the broad aspect by constructing welfare loss formula. The result shows: since the major reform from 2003 in diversified railway, the trend of overall resource allocation efficiency in diversified railway logistics continues to increase, in which warehousing industry has optimal resource allocation efficiency, and the worst is transport supporting industry.

**Keywords:** Diversified railway logistics, Administrative Monopoly, resource allocation.

## 1 Introduction

Diversified railway industry is the product of railway reform and development in China. It is a special enterprise groups in the transition of Chinese railway and is funded by the railway. It is mainly engaged in railway transport-related accessories or other business areas outside the railway transport. Its emergence was in the early 80's, and development and growth was in the 90's. Since 21st century, it has developed many industries[1], such as transport industry (all transport businesses except railway transport), transport supporting industry (it includes freight forwarders, extension services, handling transportation, etc.), warehousing industry, wholesale and retail industry, accommodation and catering industry, real estate industry, leasing and business services industry, residents and other services industry, culture, sports and entertainment industry, agriculture, forestry, animal husbandry and fishery industry etc. Due to historical reasons, ministry of railway has a lot of tangible and intangible assets idle. It is the urgent problem for diversified railway department: How to make use of railways

* The project was supported by the ministry of railway.(2007F025); The introduction of talent project on ChongQing university of arts and science.

[1] Provided by the Ministry of Railways.

comprehensive advantages and all kinds of resources to improve the utilization of railway assets. Especially to diversified railway logistics (transport industry, transport supporting industry and warehousing industry), improve the efficiency of resource allocation is particularly important for it has higher investment in fixed assets.

Resource allocation efficiency has two meanings: first is on broad, macro-level. It is the whole efficiency of the allocation of social resources. It is achieved through the entire social and economic institutional arrangements; second is on narrow, micro-level. It is the efficiency of resource use, generally refers to the production efficiency of production units. It is achieved through internally production management and improved production technology. For diversified railway logistics industry, which is administrative monopoly, the former reflects the changes of whole resource allocation industry efficiency caused by administrative monopoly. So the paper studies it from broad aspect first.

## 2 The Allocation Efficiency of Whole Social Resources

We know perfectly competitive model meets Pareto optimal on equilibrium condition. At this point all resources achieve maximum efficiency, and social welfare are also to maximize. It can be said the resources allocation achieves perfect position. So resource allocation efficiency when market is perfectly competitive equilibrium becomes a benchmark to measure the state of other markets. We can come to resource allocation efficiency under different market states by comparing resources allocation on different market situation and on perfectly competitive equilibrium situation. These market conditions, of course, include a variety of monopoly. We can get the efficiency of macro-allocation of resources on a monopoly industry by social welfare loss for monopoly compared to social welfare on perfectly competitive market equilibrium. The greater the proportion of social welfare losses caused by Monopoly to GDP, the lower the performance of resource allocation on a monopoly industry.

Harberger established a model to measure social welfare loss for monopoly according to Marshall's theory of consumer surplus. Figure 1 shows the theory.

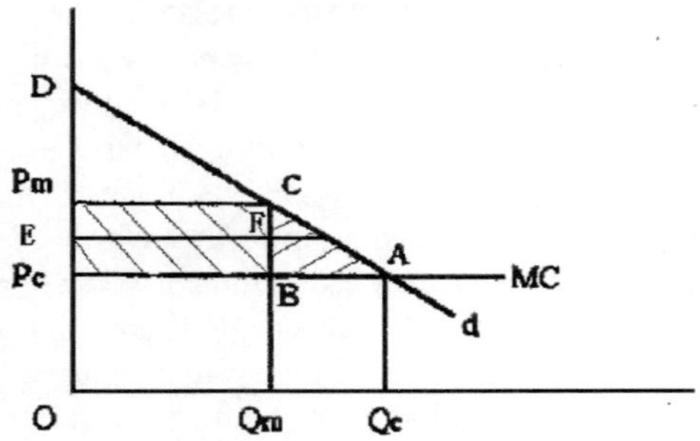


**Fig. 1.** Harberger triangles and Tulloch quadrilateral

Where, the vertical axis represents the price; the horizontal axis represents output; d is the demand curve.

We assume that all cost curves remain unchanged. So marginal cost MC is equal to Pc, where MC is supply curve on market. When the market is perfectly competitive, equilibrium price is Pc, equilibrium output is Qc and consumer surplus is the area of ΔADPc. When the market is complete monopoly, monopoly equilibrium price is Pm, equilibrium output is Qm and consumer surplus decreased to ΔCDPm. Harberger thinks the quadrilateral BCPmPc is that consumer surplus caused by the price rising from Pc to Pm transforms into monopoly profits. It is only a transfer of income. The total wealth of society does not change, and does not constitute a welfare loss. When the market is dominated by perfect competition into the complete monopolization, equilibrium output Qc down to Qm. The opportunity cost of Output (Qc-Qm) is the area of quadrilateral BAQcQm. The total welfare of (Qc-Qm) is area of trapezoidal ACQmQc. The deadweight welfare loss（DWL） or social welfare loss of monopoly is the area of ΔABC. It means the consequences that consumers are willing to purchase products at competitive prices Pc, but consumers turn to other alternatives because the price rises to Pm for monopoly. As Harberger first presents it is net loss in social welfare, economists look on a net loss of social welfare for monopoly as HarbergerΔ ABC. Its calculation formula is:

$$DWL = \frac{r^2 \eta p_m q_m}{2} \tag{1}$$

Where, r is sales profit margin. Harberger takes the average rate of return on capital as estimate of sales profit margins since the data are not readily available; η is demand Price elasticity. Harberger assumesη= 1.

Cowling and Mueller hold: In order to maximize profits, monopolists maintain production near the level which sales profit rate is equal to the reciprocal of price elasticity of demand. That is (Pm − Pc) / Pm = 1/η. Put it into (1), we get:

$$DWL = \frac{(p_m - p_c)q_m}{2} \tag{2}$$

Gordon Tullock analyzes quadrilateral BCPmPc. The social welfare loss of monopoly is ΔABC + quadrilateral BCPmPc. Quadrilateral BCPmPc is Tullock Quadrangle or Tullock Rectangle. The estimated formula of Tullock Rectangle (TR) from figure 1 is:

$$TR = (p_m - p_c)q_m \tag{3}$$

Cowling and Mueller constructed the formula (SCM) estimated welfare loss caused by monopoly based on Tullock Rectangle.

$$SCM = \frac{3(\pi + A)}{2} - T \tag{4}$$

Where, $\pi$ is economic profit. $\pi = (p_m - p_c)q_m$

A is advertising costs and T is taxes.

According to Liu and Jiang recommendations for improvements, the paper established welfare loss formula based on Cowling and Mueller, which is caused by administrative monopolies.

$$\text{SCM} = \text{Harberger triangles} + \text{Tulloch quadrilateral} - \text{Corporate Income Tax} \qquad (5)$$

Where, Tulloch quadrilateral is profit before tax in monopoly sectors. Harberger triangles are equal to half the monopoly profits by Cowling and Mueller. So

$$\text{SCM} = \frac{3}{2} \times \text{monopoly profits} \qquad (6)$$

## 3 Empirical Analysis

We get the loss of social welfare in diversified railway logistics accounting for the proportion of national income[2] from (6). After calculation, the related data is as in table 1.

**Table 1.** Profit in Diversified Railway and Nation Income from 2003 to 2007

| Profit (million) | 2003 | 2004 | 2005 | 2006 | 2007 |
|---|---|---|---|---|---|
| Transport industry | 26889 | 16615 | 23396 | 23732 | 26588 |
| transport supporting industry | 69643 | 54564 | 81744 | 91951 | 87656.5 |
| warehousing industry | 3983 | 6300 | 1154 | 1779 | 2670.89 |
| national income( Billion) | 135174 | 159587 | 184089 | 213132 | 251483 |

In order to facilitate observation, Figure 2 shows the trend based on table 1 and formula (6).

As can be seen from figure2:

1) Since 2003, indicator evaluated monopoly strength in diversified railway system, which is the loss of social welfare accounting for the proportion of national income, continues to decline. It reaches the minimum in 2004. It means: since the reform of "leapfrog development" from 2003, ministry of railway fully implemented the separation of major and secondary industry. A large number of related social services, such as primary and secondary schools, kindergartens, vocational schools and hospitals, transferred to the local government. The administrative monopoly power in diversified railway logistics has a falling trend with the reform in diversified railway logistics systems. It shows the reform in diversified railway logistics from 2003 is successful.

---

[2] Profit in diversified railway logistics is provided by Ministry of Railways; National Bureau of Statistics provides national income.

2) The indicator gradually rebounds from 2004. The main reason is: the reform related to operational aspects in diversified railway logistics has made some progress only in enterprise grouped, financial split and workers separation. Other separation reform has met with great resistance, and railway executives also quietly change the target. So the goal scheduled to be completed separation before year-end turns to further re-regulate and become bigger and stronger in diversified railway. The reform has a temporary lull in. The loss of social welfare caused by monopoly in diversified railway continues to increase. It notes the removal of the monopoly is still incomplete. The degree of monopoly on the market is still very serious. The strength of administrative monopoly further increased.

3) With the implementation of the railway leapfrog development, ministry of railways decided to withdraw the whole sub-head office in March 18, 2005, in order to meet new railway reform and development needs. Railway administration group directly implements the station and segment system. Diversified business development center has been subsequently withdrawn, which is administration for the original railway sub-head office diversified business center. All railway groups implement large-scale restructuring to multiple companies and affiliated companies according to the principle of specialization, in order to meet the requirements of "the separation of major". These multiple companies were managed by transport Stations Company and the former railway branch. To adapt to changes in management system, the railway administration group set up a regional multi-group, and it is a wholly-owned subsidiary in railway diversified business center. The measure makes administrative monopoly strength decline after 2005.

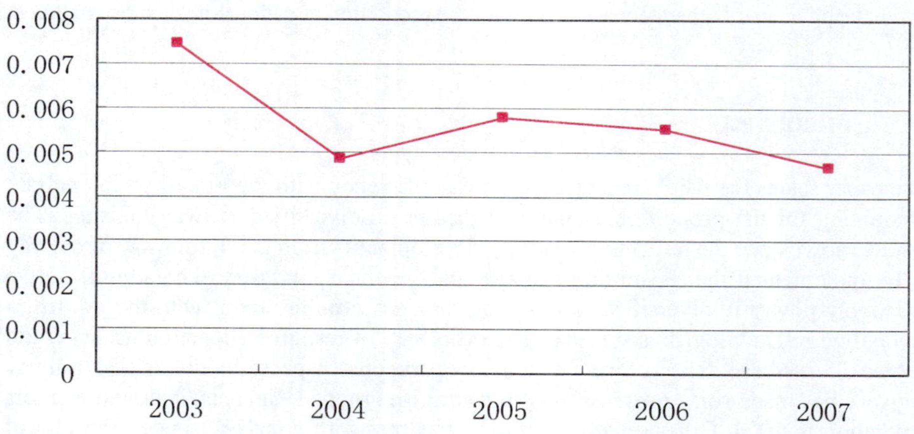


**Fig. 2.** The loss of social welfare in diversified railway logistics accounting for the proportion of national income form 2003 to 2007

For a more detailed analysis of the diversified railway logistics industry, the paper further analyzes transport industry, transport supporting industry and warehousingindustry in diversified railway logistics. Calculating the loss of social welfare accounting for the proportion of national income, we get figure 3.

From figure 3, we know transport supporting industry has the biggest social welfare losses in diversified railway logistics. It means warehousing industry has optimal resource allocation efficiency, with transport industry, transport supporting industry following, respectively. That is, transport supporting industry do more harm to people than other logistics industries. Ministry of railway should strengthen reform and measures on freight forwarders, extension services, handling transportation, etc.

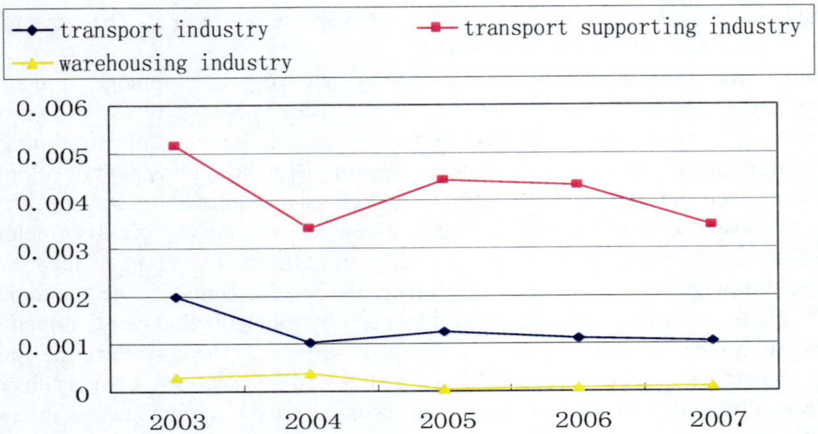


**Fig. 3.** the loss of social welfare in transport industry, transport supporting industry and warehousing industry separately accounting for the proportion of national income from 2003 to 2007

## 4 Conclusions

The paper scales the whole resource allocation efficiency with the loss of social welfare accounting for the proportion of national income in diversified railway logistics. The results show: since the reform of "leapfrog development" from 2003, ministry of railway fully implemented the separation of major and secondary industry. The administrative monopoly power in diversified railway logistics has a falling trend with the reform in diversified railway logistics systems. The efficiency of resource allocation has a certain degree of increase. The reform related to operational aspects in diversified railway logistics has made some progress only in enterprise grouped, financial split and workers separation in 2004. Other separation reform has met with great resistance. The loss of social welfare caused by monopoly in diversified railway continues to increase. The efficiency of resource allocation decreased. Ministry of Railways decided to withdraw the whole sub-head office in 2005 and directly implements the station and segment system. So since 2005, administrative monopoly strength continues to decline. The efficiency of resource allocation has increased. To transport industry, transport supporting industry and warehousing industry in diversified railway logistics, warehousing industry has optimal resource allocation efficiency, with transport industry, transport supporting industry following, respectively.

## References


1. He, W.: Industrial configuration coincidence degree method and its application on Resource allocation efficiency. Securities Market Herald (10) (2004)
2. Gao: Western Economics, p. 398. China Renmin University Press, Beijing (2000)
3. Samuelson: Economics, p. 129, 149. Huaxia Publishing House, Beijing (2002)
4. Harberger, A.C.: Monopoly and Resource Allocation. American Economic Review 44(2), 77–87 (1954)
5. Viscusi, W.K., Vernon, J.M., Harrington Jr., J.E.: Anti-monopoly and control of economic, p. 50. Machinery Industry Press (2004)
6. Cowling, K., Mueller, D.C.: The social cost of monopoly power. Economic Journal 88(12), 724–748 (1978)
7. Tullock, G.: An Economic Analysis of the rent-seeking activities, p. 7. Southwestern University of Finance Press (1999)
8. Cowling, K., Mueller, D.C.: The social cost of monopoly power. Economic Journal 88(12), 724–748 (1978)
9. Liu, Jiang: The cost estimation of administrative monopoly system in China's industrial. Jianghai Journal (1) (2003)
10. Feng: An empirical study of global scope economies model in diversified logistics. In: 2008 China International Conference on Logistics and Transportation, vol. 1, pp. 493–496 (2008)
11. Lin: Development Road Thoughts on diversified railway Enterprise. Railway Transportation and Economy 5, 23–26 (2005)
12. Yu: Speed up the reform of distribution system on diversified railway. Railway Economy Research 12, 38–42 (2006)
13. Xiao: A wide range thinking on investors in diversified railway. China Railway Science 12, 15–19 (2006)
14. Wang: Discussion on Investment and Tax Planning Banker with diversified railway. China Soft Science 4, 45–49 (2004)
15. Zhou, Xia: Union Risk Analysis and Prevention on diversified railway based on resources. Railway Transportation and Economy 6, 30–34 (2004)
16. Chames, A., Cooper, W.W., Rhodes, E.: Measuring the Efficiency of Decision Making Units. European Journal of Operational Research 2, 429–444 (1978)
17. Banker, R., Charnes, A., Cooper: Some Models for Estimating Technical And Scale Inefficiencies in Data Envelopment Analysis. Management Science 30, 1078–1092 (1984)
18. Chames, A., Cooper, W.W., Rhodes, E.: Measuring the efficiency decision making unites. Europeans Journal of Operational Research 12, 429–444 (1978)
19. Bookbinder, J.H., Qu, W.W.: Comparing the Performance of Major American Railroads. Transportation Reasearch Forum 33, 70–85 (1993)
20. Oum, T.H., Yu, C.: A Comparative Study of the OECD Countries's Railways. Journal of Transport Economics and Policy 3, 121–138 (1994)

# A New Condition Monitoring and Fault Diagnosis Method of Engine Based on Spectrometric Oil Analysis

Gao Jingwei, Hu Niaoqin, Jiang Lehua, and Fu Jianyi

College of Basic Education for Commanding Officers
National University of Defense Technology
Changsha, China
mmjingwei@163.com

**Abstract.** According to statistics, wear fault is about sixty percent to eighty percent of all the machinery faults. Spectrometric oil analysis is an important condition monitoring technique for machinery maintenance and fault diagnosis. Now, there are two existing mathematics analysis models based on spectrometric oil analysis, namely concentration model and gradient model. However, the above two models have respective disadvantages in condition monitoring and fault diagnosis of the engine. Then in this paper a new mathematics model, proportional model, was put forward monitoring wear condition and diagnosing wear faults of the engine. Proportional model use the relationship and correlation among the elements in the lubricating oil to detect wear condition and occurring faults in the engine. The steps of establishment of proportional model were described firstly. Then we used the experiments data to verify the feasibility of proportional model and gave limit values of proportional model. In order to validate the feasibility of proportional model, proportional model was applied to monitor wear condition and diagnose wear faults of an engine. The results from this paper have proved that the method based on proportional model is applicable in condition monitoring and fault diagnosis of the engine.

**Keywords:** condition monitoring, fault diagnosis; proportional model, spectrometric oil analysis.

## 1 Introduction

According to statistics, wear fault is about sixty percent to eighty percent of machinery faults[1]. Therefore, monitoring wear condition and diagnosing wear faults of the engine is very important in practice.

Wear procedure is complicated and always influenced with outside factors. Then it is difficult to measure wear degree of the engine and find wear faults in earlier period. Wear occurs whenever two or more surfaces interact, and can generate a lot of particles which can reflect wear condition and wear faults of the engine. Therefore, wear information in the lubricating oil will give important evidence in condition monitoring and faults diagnosis of the engine[2,3].

Spectrometric oil analysis technique is one of oil analysis methods that this technique applies spectrometer to analyze the concentration of metal elements, non-metal elements and additive elements in the lubricating oil, which can be used to monitor wear condition and diagnose wear faults of the engine[4]. In recent years, spectrometric oil analysis technique is used in industry more and more widely[5,6]. Nowadays, the existing data analysis model based on spectrometric oil analysis technique includes concentration model and gradient model[7]. But the above two models have respective disadvantages in practice. Then in this paper, we put forward the concept of proportional model first and overcome the disadvantages of the above two models.

## 2 Methodology

Proportional model uses the proportional relation among concentration of the elements to monitor wear condition of the elements. Proportional model can reflect the change of the relation among the elements and can evaluate wear condition of the elements. Therefore, proportion model develops a new approach in condition monitoring and fault diagnosis of the engine[8].

In practice, the lubricating oil of the engine will be sampling standard every 5 hours. Then we use spectrometer to get concentration of the elements in the lubricating oil.

We set concentration of the elements as a sequence of number, write as

$$X = \{x_1 \quad x_2 \quad x_3 \quad \cdots \quad x_n\} \tag{1}$$

where $x_i$ is the respective concentration of elements. In general, the order of the elements is Fe, Al, Cu, Pb, Cr, Mn, Mg, Na, Ba, B and Si.

Then concentration of all elements will divide each other and we will get the proportional value set Y,

$$Y=\{y_{12} \quad y_{13} \quad \cdots \quad y_{1n} \quad y_{23} \quad y_{24} \quad \cdots \qquad \cdots \quad y_{(n-1)n}\} \tag{2}$$

where $y_{ij}$ is the divide value between No. $i$ element and No. $j$ element.

## 3 Limit of Proportional Model

We know that proportional value of metal elements in the lubricating oil is consistent with running time. Therefore, we also use $3\sigma$ method to calculate limit value of proportional model. That means limit value will include warning limit and abnormal limit. Because proportional model shows the proportional value between one element and the other element, limit value should include two parts, namely upper limit value and lower limit value.

### 3.1 Upper Limit Value of Proportional Model

Here we take proportional value of Fe/Cr as an example. Firstly, we calculate proportional value of Fe/Cr in normal wear condition and set up a series of Fe/Cr proportional value. Then calculate the mean of Fe/Cr proportional value and obtain the mean G is 27.9253. Secondly, calculate the variance of Fe/Cr proportional value and obtain the variance $\sigma$ is 5.3001. Finally, we will obtain the warning upper limit

$$G_{wu}=G+2\sigma=38.5237 \quad (3)$$

and the abnormal upper limit

$$G_{fu}=G+3\sigma=43.8238. \quad (4)$$

### 3.2 Lower Limit Value of Proportional Model

Just as the above method, we can obtain the warning lower limit

$$G_{wl}=G-2\sigma=17.3251 \quad (5)$$

and the abnormal lower limit

$$G_{fl}=G-3\sigma=12.0250. \quad (6)$$

Fig. 1 gives the four limit value of proportional model.

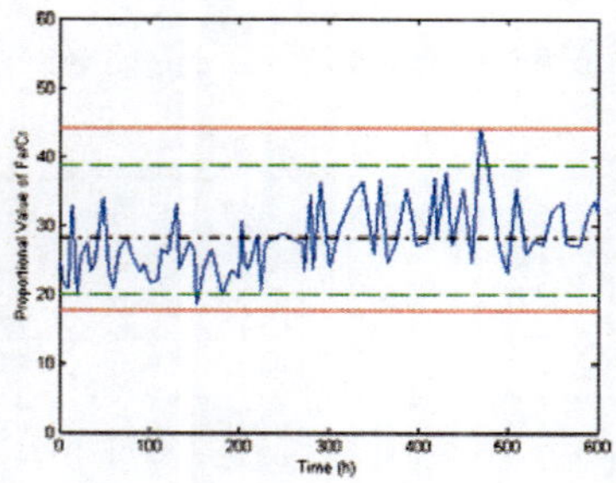


**Fig. 1.** Limit value of proportional model

In Fig. 1, the upper solid line is the abnormal upper limit line while the lower solid line is the abnormal lower limit line. And the upper dashed line is the warning upper limit line while the lower dashed line is the warning lower limit line. The dash-dot line is the normal line of proportional value.

After using the upper calculating method, limit value of this engine was obtained and was given in the following table 1.

**Table 1.** Limit value of one engine based on proportional model

| Proportional Relationship | Abnormal Upper Limit | Warning Upper Limit | Warning Lower Limit | Abnormal Lower Limit |
|---|---|---|---|---|
| Fe/Al | 17.38 | 14.52 | 4.66 | 3.87 |
| Fe/Cu | 4.62 | 3.95 | 1.30 | 1.06 |
| Fe/Pb | 12.97 | 11.19 | 4.75 | 4.06 |
| Al/Cu | 0.90 | 0.72 | 0.23 | 0.20 |
| Al/Pb | 1.47 | 1.28 | 0.63 | 0.55 |
| Cu/Pb | 7.22 | 5.88 | 1.49 | 1.21 |

After we get the limit value of proportional model, we can use proportional model to monitor wear condition and diagnose wear faults of the engine.

## 4 Presentation of Data

In this paper, proportional model was applied to monitor one engine's wear condition and diagnose wear faults. In this experiment, the engine was running for about 850 hours and the lubricating oil was sampled every 5 hours. Then we will use spectroscopy to measure the concentration of the elements in the lubricating oil. The following table 2 gives concentration of the main elements in the lubricating oil of one engine during 115 hours and 170 hours. From the table, it is difficult to analyze the spectrum data and detect wear faults of the engine.

**Table 2.** Concentration of the main elements in one engine during 115 hours and 170 hours (ppm)

| Time | Fe | Al | Cu | Pb | Cr | Mn | Mg | Na | Ba | B | Si |
|---|---|---|---|---|---|---|---|---|---|---|---|
| 115 | 25.52 | 2.48 | 9.29 | 3.11 | 0.96 | 0.49 | 9.73 | 6.59 | 7.42 | 0.23 | 9.53 |
| 120 | 25.61 | 2.74 | 10.15 | 3.06 | 0.99 | 0.49 | 9.73 | 6.40 | 6.66 | 0.21 | 9.69 |
| 125 | 27.79 | 2.59 | 11.00 | 3.12 | 0.84 | 0.50 | 9.39 | 6.50 | 7.66 | 0.25 | 9.21 |
| 130 | 30.38 | 2.97 | 12.35 | 3.59 | 1.28 | 0.54 | 9.32 | 6.62 | 7.37 | 0.23 | 8.47 |
| 135 | 32.82 | 3.14 | 13.33 | 3.55 | 1.19 | 0.60 | 9.23 | 6.82 | 7.43 | 0.35 | 10.57 |
| 140 | 34.66 | 3.34 | 15.54 | 3.66 | 1.31 | 0.57 | 9.14 | 7.03 | 7.78 | 0.25 | 9.54 |
| 145 | 10.80 | 1.37 | 6.36 | 1.85 | 0.46 | 0.28 | 9.56 | 6.99 | 7.25 | 0.22 | 8.06 |
| 150 | 16.87 | 1.57 | 7.66 | 1.61 | 0.54 | 0.28 | 8.64 | 6.95 | 6.99 | 0.23 | 8.56 |
| 155 | 13.59 | 1.63 | 8.74 | 1.85 | 0.64 | 0.31 | 9.81 | 7.54 | 7.87 | 0.25 | 9.30 |
| 160 | 19.35 | 2.04 | 8.25 | 2.11 | 0.77 | 0.36 | 10.93 | 8.50 | 8.65 | 0.26 | 8.49 |
| 165 | 19.21 | 1.76 | 8.27 | 2.38 | 0.86 | 0.38 | 10.73 | 9.58 | 9.65 | 0.28 | 8.40 |
| 170 | 24.46 | 1.96 | 9.67 | 2.47 | 0.99 | 0.40 | 11.40 | 10.01 | 10.23 | 0.30 | 9.42 |

## 5 Discussion

### 5.1 Concentration Model

Using concentration model limit, we can find that concentration of all elements is below the warning line. Here, we provide the relation between concentration of Na and Ba and running time of the engine.

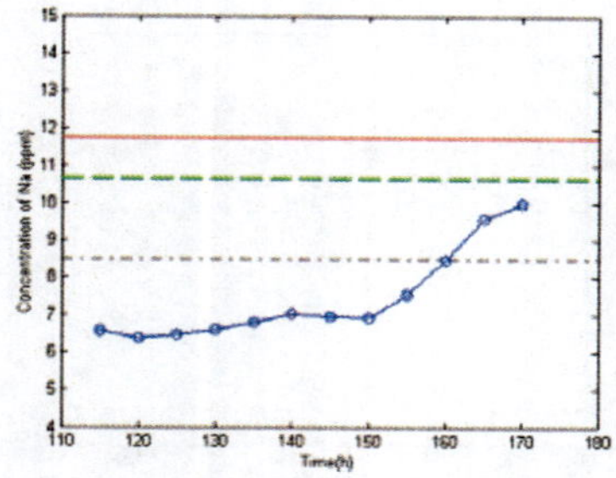


**Fig. 2.** Relation between concentration of Na and running time

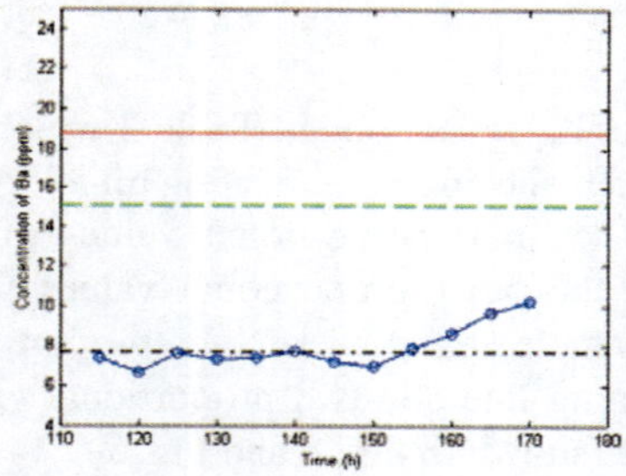


**Fig. 3.** Relation between concentration of Ba and running time

## 5.2 Gradient Model

Using limit of gradient model, we also can find that gradient of all elements were all below the warning line. Here, we give the relation between gradient of Na and Ba and running time of the engine.

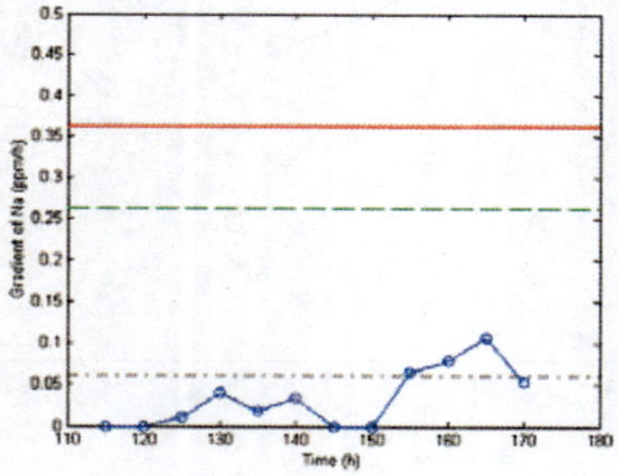


**Fig. 4.** Relation between gradient of Na and running time

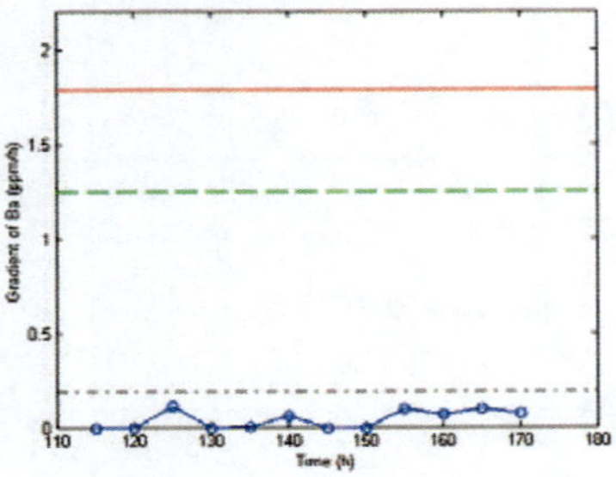


**Fig. 5.** Relation between gradient of Ba and running time

## 5.3 Proportional Model

Using proportional model limit, we can obtain Table 3 which gives proportional wear condition among the main elements in the lubricating oil.

From Table 3, we can see that proportional values of Na and Ba were both changing irregularly while the other proportional values were all in normal wear condition. Take Na as example, there were two proportional values (Fe/Na and Mn/Na) exceeding the warning line and two proportional values (Al/Na and Pb/Na) exceeding the abnormal line (shown in Fig. 6 and Fig. 7).

**Table 3.** Proportional wear condition among the main elements in the lubricating oil

| | Fe | Al | Cu | Pb | Cr | Mn | Mg | Na | Ba | B | Si |
|---|---|---|---|---|---|---|---|---|---|---|---|
| Fe | | | | | | | ☆ | ☆ | ☆ | | |
| Al | | | | | | | ☆ | ★ | ☆ | | |
| Cu | | | | | | | | | | | |
| Pb | | | | | | | | ★ | ★ | ☆ | |
| Cr | | | | | | | | | | | |
| Mn | | | | | | | | ☆ | | | |
| Mg | ☆ | ☆ | | | | | | | | | |
| Na | ☆ | ★ | | ★ | | ☆ | | | | | |
| Ba | ☆ | ☆ | | ★ | | | | | | | |
| B | | | | ☆ | | | | | | | |
| Si | | | | | | | | | | | |

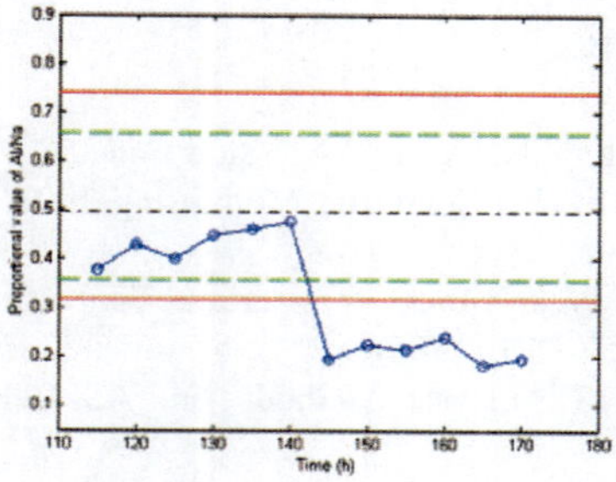


**Fig. 6.** Proportional value of Al/Na

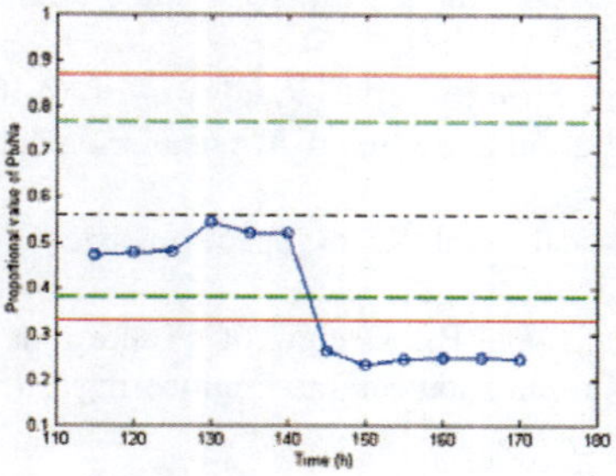


**Fig. 7.** Proportional value of Pb/Na

Then we believe that additive elements were almost in abnormal condition and we ought to exam the reason of this abnormal information.

At this moment, we can draw a conclusion that there was water leak fault occurring in the engine and we need to exam the engine immediately.

# 6 Conclusions

Through using spectrometric oil analysis technique, we can obtain a lot of useful wear information of the lubricating oil. From the experiment, we can find that proportional value among the elements in the lubricating oil will not be influenced with the oil-changing procedure and can reflect wear condition of the engine. Therefore, in this paper we use proportional model to monitor wear condition and diagnose wear faults of the engine. As a result, this engine condition monitoring and fault diagnosis method based on spectrometric oil analysis technology and proportional model can find the abnormal wear information and detect wear damage of the engine. The results have proved that this method is available and applicable in practice and puts forward a new method to monitor wear condition of the engine.

## References


1. Jiang, S., Dong, J.X., Jiang, Y.H., Yan, Y.N.: Study on the Maths Model of Monitoring Properties of Lubricating Oil. In: Condition Monitoring 1997, Xian, China, pp. 177–180 (1997)
2. Billatos, S.B.: A Statistic Wear Model for Certain Tool Materials with Application to Machining. Wear 112(1), 257–271 (1986)
3. Zhang, J.L.: Laboratory Oil Analysis Methods. In: ASIATRIB 1998, Beijing, China, pp. 413–419 (1998)
4. Gao, J., Zhang, P., Zhang, Y., Ren, G.: Study on Wearing Characteristics and Diagnosis based on Oil Spectrum Analysis. Transactions of CSICE 22(6), 571–576 (2004)
5. Lukas, M., Anderson, D.P., Yurko, R.J.: New Development and Functional Enhancements in RDE Used Oil Analysis Spectrometers. In: 1998 International Oil Analysis Conference, pp. 1–7 (1999)
6. Dahmani, R., Gupta, N.: Spectroscopic Analysis of Automotive Engine Oil. In: Instrumentation for Air Pollution and Global Atmospheric Monitoring, Newton, USA, pp. 179–183 (2001)
7. Ma, L.: Multivariate Statistical Analysis of Spectrometric Oil Test Data. In: IST 1993, China (1993)
8. Gao, J., Zhang, P., Ren, G., Li, B.: Design of Proportional Model for Oil Spectrum Analysis. Chinese Internal Combustion Engine Engineering 25(5), 34–37 (2004)

# Object Tracking by Mean Shift Dealing with Scale Increasing

Hai-ling Bi[1], Bao-feng Yuan[2], and Yu Fu[2]

[1] Equipment Support Department Military Transportation Institute
Tianjin, China
bhlryf@126.com

[2] Signal and Information Processing Laboratory
Chongqing Communication Institute Chongqing, China
yuanbaofeng330@sina.com

**Abstract.** In the process of object tracking, the scale of object changes along with the movement of object. But the opposite location of corner point only changes Rendering affine transform. At first, corner points were Distilled in different periods of time using an advanced Harris algorithm, and then the object Skeleton was reformed using these corners. Meanwhile, the displacement coefficient, the zoom coefficient, the rolling coefficient and the misplace coefficient were ascertained. At last, the tracking window was updated on line. The Mean Shift algorithm is used in this paper to adjust window. The experiment indicates that this algorithm can cope with the situation of scale changing, especial on the condition of Scale increasing.

**Keywords:** video object tracking, Harris algorithm, affine transform.

## 1 Introduction

With the continuous development of computer hardware and software, real-time video object tracking has been widely used in precision strike military targets, intelligent traffic monitoring, bank monitoring in all areas of national defense and social life.

Mean Shift algorithm is a highly efficient matching tracking algorithm[1]. The algorithm has less computation without the need for parameter estimation, and has robustness of the target block, deformation and rotation, so and it is subject to the extensive attention of many scholars.

In the framework of the algorithm. the size of the window determines the number of samples involved in the calculation. With the diversification of target location, the scale will be constantly changing. On the condition of the scale of the tracking window changing along whit the scale of the object, Algorithms can play a positive role. Comaniciu, however, in his classic article, only went for understatement of the adaptive window, Just do a change in the use of Bhattacharyya similarity coefficient to determine the optimal window. However, due to the existence of small-scale problem, namely, Bhattacharyya similarity coefficient easy to reach the optimal value of small-scale. So, getting a big target scale cases, the target window does not subsequently become larger, but become smaller, which will inevitably lead to instability in target tracking and even cause the goal to lose.

The adaptive window has been a problem in the Framework of Mean Shift algorithm. A large number of researchers have raised the issue of improved method. Collins introduces a Kernel in the literature[2], at first, location the positioning, then Then find the best goal of the nuclear scale on the basis of positioning by means of scale. However, because of its use of the Epanichnikov nuclear, its derivative is a constant. Therefore, this method is equivalent to positioning in space and taking a measure based on the average, thus improving effect on the target scale problems is not obvious. Peng Ning Song[3], etc.proposed a centroid matching adaptive window algorithm based on backward tracking according to Mean Shift algorithm becoming the target of the small features of a good adaptation, but there are Disadvantages of the computational complexity and not easy to achieving. ZUO Jun-Yi[4] etc. improve the algorithm by modifying the conditions and the selection of optimal window adaptive filter parameters. but there are also the computational complexity of problems. Qian Hui-min[5] etc. solve this problem by a Features that the amount of information through different scales show different characteristics . However, this method is more sensitive to image noise and object Deformation.

A large number of experiments show that Target changes in the location of the video can be described approximatively by affine model[6]. In addition, the corner point is a simple and practical Target Feature Representation. In this algorithm framework, In order to avoid illumination of the corner point extraction, First of all, extracts the HSV color model of the H (hue) component, extracts corner point using an improved Harris corner detection algorithm in different frames; Then, connects each corner point of each frame, Reconstruct of target skeleton; next,determines the displacement factor, scaling factor, misalignment factor through An analysis of the target skeleton; Finally, fixs the tracking windows by the affine coefficients obtained. experimental results show that the algorithm proposed above has a good adaptability.

## 2 An Improved Harris Corner Detection Algorithm

For a local window in the image, Translates the window for a small distance in all directions (usually eight field direction), If the gray-scale changes large at a threshold value, the center of the window should be called the Corner point. Classic Harris corner detection algorithm[7] has advantage of Simple calculation and high robustness. The specific process is as follows:

$$E_{u,v}(x,y)=\sum_{u,v}\omega_{u,v}(f(x+u,y+v)-f(x,y)) \tag{1}$$

Processes equation (1) using Taylor formula:

$$E_{u,v}(x,y)=\sum_{u,v}\omega_{u,v}(u\frac{\partial f}{\partial x}+v\frac{\partial f}{\partial y}+o(u^2+v^2))^2$$
$$\approx\sum_{u,v}\omega_{u,v}(u\frac{\partial f}{\partial x}+v\frac{\partial f}{\partial y})^2$$

$\omega_{u,v}$ is the value at position $(u, v)$ in the Gaussian function:

$$\omega_{u,v} = (1 / \sqrt{2\pi\sigma})e^{(-\frac{1}{2}(u^2+v^2)/2\sigma^2)} \tag{2}$$

$\frac{\partial f}{\partial x}$ , $\frac{\partial f}{\partial y}$ are the first-order differentials in direction $x$ and $y$,

$I_x$ denotes $\frac{\partial f}{\partial x}$, $I_y$ denotes $\frac{\partial f}{\partial y}$ ,so equation (1) can be expressed in a matrix form:

$$E_{u,v}(x, y) = [u, v] M^T [u, v] \tag{3}$$

$M = \begin{bmatrix} \langle I_x^2 \rangle & \langle I_x I_y \rangle \\ \langle I_x I_y \rangle & \langle I_y^2 \rangle \end{bmatrix}$, $\langle\rangle$ denotes Gaussian smoothing operator.

Seeking the two eigenvalues of matrix A, if both are higher than the threshold T, then regards the Center point of the window as the corner point.

Classic Harris corner detection algorithm has advantage of simpleness in principle and easement in realization. However, the algorithm is vulnerable to the impact of light, meanwhile, Single-threshold processing method has Disadvantage of Approximations. A number of edge points may be misjudged as a corner point. In an effort to avoid the effect of light, uses HSV color model of the H (hue) component instead of the original algorithm of gray feature In the classical framework of the Harris corner detection algorithm. In the threshold processing and corner point selection, uses the dual-threshold method and a self-defined pseudo-corner removing the principle, Definitemethods are as follows:

(1) First of all, set a higher threshold T1, set pixer which larger than the T1 as the corner point. the number set to N (N<4), if the N≥4, find the average gray value in the area of 8 dierctions through equation(4); and then sort the order, select 4 corner points; next, Skip to step (5). (In the simulation experiment, the 4 corner points are the four most representativ)

(2) Then, set up a threshold T2 which slightly smaller than the T1 (In this article, T2 = 0.85 × T1), sets the pixels as the pre-selected corner points which are greater than T2.

(3) Screens in the pre-selected corner points, do a comparison of pre-selected corner points and corner points which get in the step (1), For each pre-selected corner point, do a comparison of each corner point in the step (1) through the X coordinates and Y coordinates. If the distance of X coordinates or Y coordinates is less than a constant C, the point is the Final corner point.

(4) In the remaining primaries corner, find the average gray value in the area of 8 dierctions, and then sort the order, set the pre-(4-N) corner points as the true corner points.

(5) End

$$avg(x_{i,j}) = \sum_{i=0}^{7} x_{i,j} \Big/ 8 \tag{4}$$

Setting a threshold T3, if $avg(x_{i,j}) > T3$, Then determine the point as the corner point, or as the Pseudo Corner, $x_{i,j}$ is the pixel around the Pixel under test in the 8 directions For a vehicle image, uses the corner point detection algorithm, experimental results shown in Figure 1:

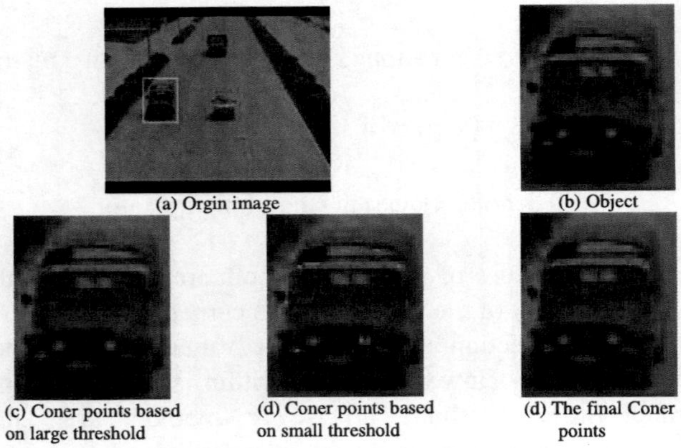


**Fig. 1.** The corner points graph of vehicle

Figure 1(a) is the original image; Figure 1(b) is the object image; Figure 1(c) is the image which is processed using large threshold; pink points are the corner points which are high degree; Figure 1(d) is the image which is processed using small threshold, light blue points are the per-corner points which are processed by small threshold; Do a comparison between Chart 1(c) and Figure 1(d), remove the pseudo-corner points, get Figure 1(e), pink points sre the final corner points.

## 3 Fix the Adaptive Tracking Window Using Affine Transformation

Affine transform is a linear transformation. Its most important feature is to maintain the point-collinearity (or coplanarity) as well as to maintain the parallel straight. translation transformation, scaling transformation, shear transformation, rotation transformation are typical affine transformations. Translation Transform is the coordinates moving from $(x,y)$ to $(x+\Delta x,y+\Delta x)$; The rule of Scaling transformation is that the abscissa or Ordinate of a point zooms in or out to the original Sx-fold; With the popular analogy, Shear transformation is similar to the nature of instability of Quadrilateral, such as fencing, can be Deformated from a rectangle to a diamond-shaped. Rotation transformation is the goal object rotates theta radians around origin point. Four kinds of typical affine transformation can be used a 3×3 matrix to represent, Its last line behavior (0,0,1). the original coordinates (x,y) transformed into the new coordinate (x',y') by The transformation matrix, The original coordinates and the new coordinates are considered three-dimensional column vectors whose last line behavior(1,1,1), the new left-column vector can be obtained through the original column vector left Multipling by transformation matrix, such as the equation (5):

$$\begin{bmatrix} x' \\ y' \\ 1 \end{bmatrix} = \begin{bmatrix} m_{00} & m_{01} & m_{02} \\ m_{10} & m_{11} & m_{12} \\ 0 & 0 & 1 \end{bmatrix} \begin{bmatrix} x \\ y \\ 1 \end{bmatrix} = \begin{bmatrix} m_{00} * x + m_{01} * y + m_{02} \\ m_{10} * x + m_{11} * y + m_{12} \\ 1 \end{bmatrix} \quad (5)$$

Four transformation matrixs of kinds of typical affine transformation are denoted such as equation (3~6):

$$\text{Translation Transform: } \begin{bmatrix} 1 & 0 & \Delta x \\ 0 & 1 & \Delta y \\ 0 & 0 & 1 \end{bmatrix} \quad (6)$$

$\Delta x, \Delta y$ is Translation distance

$$\text{caling transformation: } \begin{bmatrix} Sx & 0 & 0 \\ 0 & Sy & 0 \\ 0 & 0 & 1 \end{bmatrix} \quad (7)$$

$Sx, Sy$ is Scaling Coefficient

$$\text{Shear transformation: } \begin{bmatrix} 1 & shx & 0 \\ shy & 1 & 0 \\ 0 & 0 & 1 \end{bmatrix} \quad (8)$$

$shx, shy$ is Shear Coefficient

$$\text{Rotation transformation: } \begin{bmatrix} \cos(theta) & -\sin(theta) & 0 \\ \sin(theta) & \cos(theta) & 0 \\ 0 & 0 & 1 \end{bmatrix}. \quad (9)$$

$theta$ is Rotation Radian

At different times, Relative to the camera, the distance between targets is different, the size scale is different, as well as dislocation, rotation and other morphological changes. By a large number of studies, sSuch changes can be approximate descripted using the above-mentioned four kinds of typical affine transformation. Figure 2 presents the four kinds of affine transformation.

**Fig. 2.** The four kinds movement of an object

Through the introduction of section 2, Corner point is a good representation of the target feature. Extractes target corner points, and then connect each corner to a skeleton. Target position change in the video sequence transformation is a kind of affine transformation, so the skeleton position change can also be approximated considered as affine transform. Extractes Corner points in the video sequence of different periods using Harris corner detection algorithm presented in this paper. Then, connects the corner points, form a skeleton, the final Step is do affine analysis for this skeleton, so the displacement Coefficient, scaling Coefficient, shear coefficient, spin Coefficient can be obtained, and the tracking window can be updated on-line. As shown in Figure 3.

**Fig. 3.** The graph of corner point on the condition of affine transformation

By computer simulation, the location coordinates of corner points in each frame image can be obtained, displacement coefficient, scaling coefficient, shear coefficient, spin coefficient can be obtained using four transformation matrixs of affine transformations.

## 4 Simulation Results

This simulation platform is VC + +6.0, the computer is configured to P42.8G, 256MB of memory, the object Size in using video is Gradually increasing. In figure 4, the result is obtained by classial Mean shift object tracking algorithm. In figure 5, the result is obtained by the algorithm proposed in this paper.

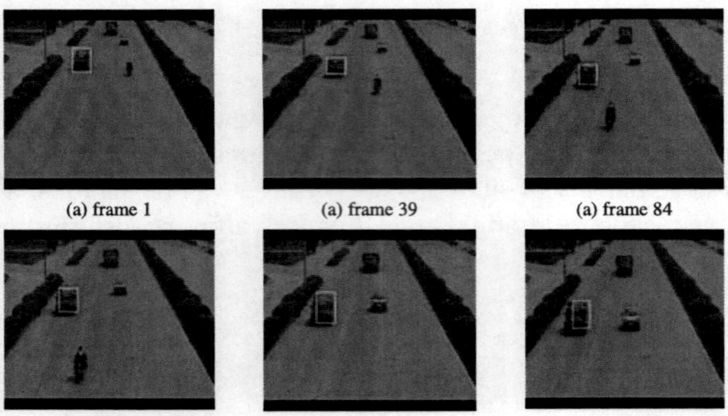

(a) frame 1 (a) frame 39 (a) frame 84

(a) frame 106 (a) frame 134 (a) frame 162

**Fig. 4.** The result of classial Mean shift algorithm

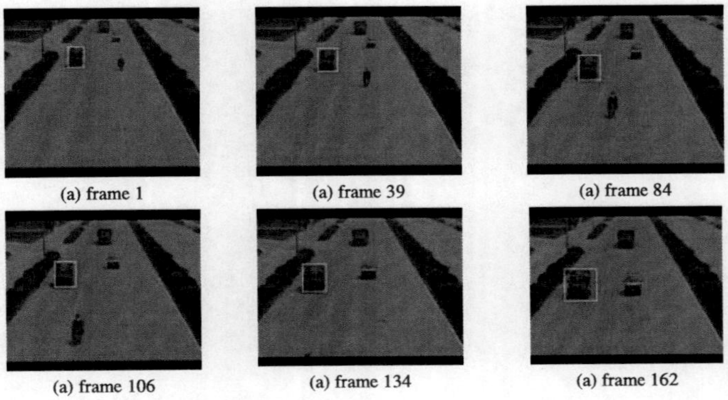


**Fig. 5.** The result of method proposed in this paper

As the target size increases, with the goal of movement, in the figure 5, the target window will not only not with the increasing of scale, but become smaller, meanwhile, the location of the tracking window will deviated. which also confirms the problem that classic Mean shift algorithm is easy to wander in the small scale; On the contrary, the simulation results of Figure 6 can be well adapted to the changes in the target window, which confirmed the feasibility of the proposed algorithm.

## 5 Summary

On the condition of target size increasing, an improved Harris corner detection algorithm is introduced to the classic Mean shift object tracking algorithm. The tracking accuracy is improved. But at the same time, the algorithm time complexity is inevitably increased. So, the next step is improving the algorithm to reduce the time complexity.

## References

1. Comaniciu, D., Ramesh, V., Meer, P.: Kernel-based object tracking. IEEE Transactions on Pattern Analysis and Machine Intelligence 25(5), 564–577 (2003)
2. Collins, R.T.: Mean-Shift blob tracking through scale space. In: Danielle, M. (ed.) IEEE Int'l Conf. on Computer Vision and Pattern Recognition, vol. 2, pp. 234–240. Victor Graphics, Baltimore (2003)
3. Peng, N.-S., Yang, J., Liu, Z., et al.: Automatic selection of kernel-bandwidth for Mean-Shift object tracking. Journal of Software 16(9), 1542–1550 (2005)
4. Zuo, J.-Y., Liang, Y., Zhou, C.-H., Pan, Q., Zhang, H.-C.: Researches on Scale Adaptation Strategy in Mean Shift Tracking Algorithm. Journal of Image and Graphics 13(9), 1750–1757 (2008)
5. Qian, H.-M., Mao, Y.-B., Wang, Z.-Q.: Mean Shift Tracking with Self-updating Tracking Window. Journal of Image and Graphics 12(2), 245–249 (2007)
6. You, C.-Y.: Analytic Geometry. Peking University Press, Beijing (2004)
7. Harris, C.G., Stephens, M.J.: A Combined Corner and Edge Detector. In: Proceedings of the 4th Alvey Vision Conference, Manchester, England (1988)

# Wireless Alarm System Basing on FPGA and SOPC

Yuemin Wan[1], Xiaoping Wang[1], Lei Xing[2], and Peng Zhang[1]

[1] Shijiazhuang TieDao University Department of Electrical and Electronics Engineering China, Shijiazhuang
{wanyuemin802,wxp1501,zhangpeng802}@163.com
[2] Beijing Railway Bureau BeiJing PDL Infrastructure Maintenance China, BeiJing
lengxueguao@163.com

**Abstract.** The sending system designed basing on FPGA (Field Programmable Gate Array) is great reliable, and includes the coding control circuit, modulation circuit, power amplifier circuit and the launching of circuit. The receiving system designed to SOPC (System on a Programmable Chip) as the core is good flexible and includes the power amplifier circuit, the demodulated circuit, the alarm circuit as well as the decoded circuit. Wireless alarm system which is small, easily installed, long in transmission distance, strong in anti-interference, low in false alarm rate, reliable as well as it is scalable and easy to modify, As time goes on, new features can be also added in if needed.

**Keywords:** wireless alarm, FPGA, SOPC, Nios II .

## 1 Introduction

Wireless alarm system is easily installed, good imperceptibility, strong in anti-interference, which can be used in the occasions of long-distance transmission and complex terrain, besides scalable, easily modified.

## 2 Sending System Basing on FPGA

Wireless alarm system is divided into two parts which is the sending system and the receiving system, one sending device is placed in the defensive zone. The sending system circuit is shown in figure 1, including detector interface circuit, crystal circuit, deployment status indication circuit, signal modulation circuit, power amplification circuit, sending circuit and coding control circuit. The EP2C5Q240C8 of Altera is selected as a FPGA, in which seven modules is designed with VHDL language in order to achieve coding and control circuit. In order to check the function of the system conveniently, when simulation, register module, divider module, interface module and control module are F5、F2、F3、F4、f6、o、Q2、R, the simulation wave of the Quartus II is shown in figure 2.

How it works: the power of the sending system is close, in the FPGA, the delay module achieves a 2-minute delay, the output K becomes high while the delay time is up and LED is on, the system is in the deployment status. The input D of the interface module becomes low while no alarm, the output Q2 becomes low through control module. Cyclic shift register module is in the working mode of parallel input address code because of the low of the D. Then, external transistor T1 is cut off, sending circuit do not send signal and lie in low-power state because there is no power. The input D of the interface module becomes high when alarm, in the rising edge of the clock pulse, the output Q2 becomes high because of trigger control module, external transistor T1 is conduct, the normally open point of the relay pulls in, sending circuit starts working through connecting to the power, at the same time, cyclic shift register module shifts to the working mode of serial shift, serial output address code with the baud rate of 256bps. When the address code is 0, the output frequency is 2048Hz, or the output frequency is 1024Hz. The output signal F controls the oscillation circuit constituted by the varactor to achieve FSK modulation, after frequency doubling and power amplifier, electromagnetic is sent to the air with the sending period of 0.4s. The first half-cycle of the first cycle sends, the second half-cycle stops sending, so that the sending system of other zone has the opportunity to send, after the second cycle it will repeat. The sending carrier frequency of the system is 30.225Hz with a frequency offset which is less then 4 KHz.

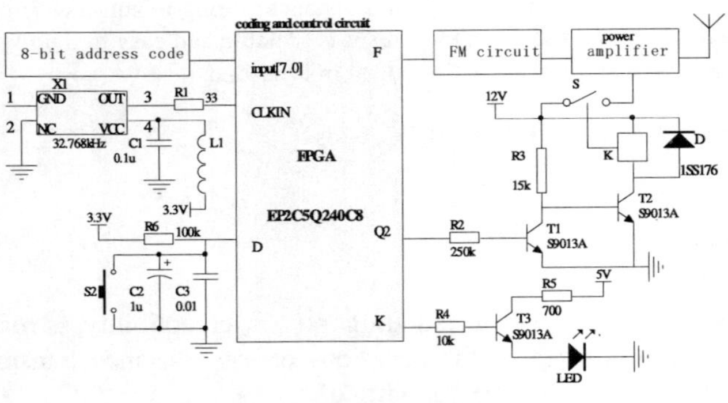


**Fig. 1.** Sending system overall circuit

## 3 Styling Receiving System Basing on SOPC

### 3.1 Principle of the Receiving System

A wireless receiver is installed in the monitoring center and responsible for the alarm monitoring of the all zone. The receiving system circuit is shown in figure 3. After receiving the alarm signal, the receiving system circuit amplifies frequency selectively through low-pass filter, high-frequency small-signal amplifier, IF amplification and mixing amplifier, then it is sent to the FSK demodulation circuit. Demodulation circuit adopts private integrated chip XR2211 to demodulate address code signal in the defensive zone. FSK demodulation signal decodes, checks, distinguish, store and display to drive alarm circuit by the SOPC. The alarm circuit adopts integrated chip NE556 to achieve alarm whistle.

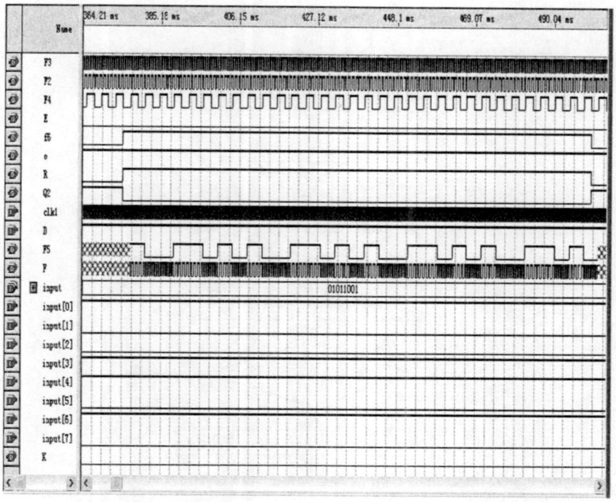

**Fig. 2.** Simulation wave of coding and control circuit

## 3.2 SOPC Design

The design of SOPC system mainly includes two parts which are hardware structures based on the Quartus II /SOPC Builder and software design based on the Nios II IDE[1]. These components are added to the hardware structures such as component CPU, component JTAG UART, component internal RAM, component external FLASH, component EPCS Serial Flash Controller, component ID, component PIO. Base address and interrupt are assigned automatically, meanwhile ext_flash is set up as the zone of the reset address, onchip_sram is set up as the memory address zone of exception address's running of software in external memory cell. Finally Nios II system is generated.

After the generation, Nios II IDE will produce custom-built HAL system library and match hardware's setup, as the interface of hardware and software, HAL's API is compatible with ANSIC, C language function is used to program.

SOPC system receives the signal of address encoded, then carry out odd parity, the correct signal is written in data buffer, the continuous address encoded is recognize three times, if all the results are equal, SOPC regards the result valid without interference, and triggers the alarm circuit to alarm ,LED will show the defensive zone's address encoded, so that operator on duty can take appropriate measures timely, receive process is programmed with C language, main flowchart is shown in figure 4.

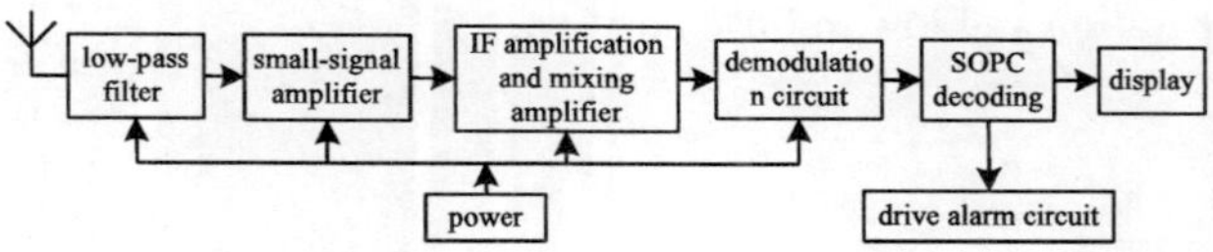


**Fig. 3.** Receiving system block diagram

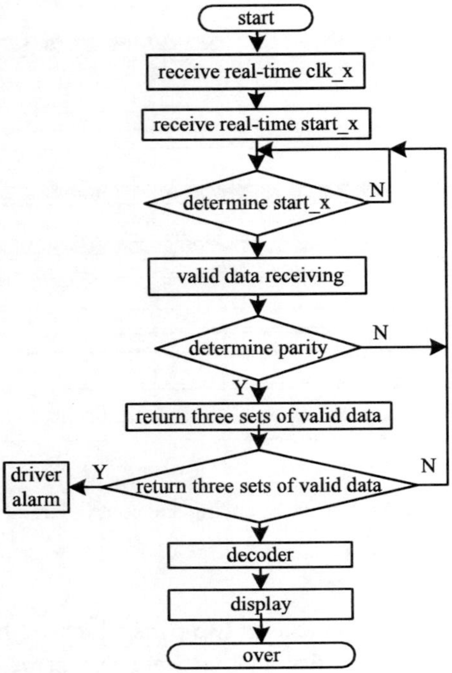


**Fig. 4.** Receiving data flow

## 4 Anti-interference Measures

The anti-jamming capability of sending system is increased immensely after two FSK modulation. In the receiving system, high frequency amplification, intermediate frequency amplification and SOPC system are put in two metal shield box separately, preventing external disturbance and internal high frequency disturbance[2], Because the sending and the receiving system are the hybrid system of analog and digital circuit, analog and digital circuit are designed and connected to ground separately.

## 5 Conclusions

In this text, wireless alarm system based on FPGA and SOPC is flexible for users to assemble and deploy different detector, such as human infrared detector, gas detector and smoke detector, the system take appropriate measures of anti-jamming, it is reliability in operation and low in false positive.

## References

1. ALTERA CORPORATION, "Nios II Hardware Development Tutorial" (unpublished)
2. Lee, J.S.: Speckle Analysis and smoothing of Synthetic Aperture Radar Images. Analysis and Smoothing of Synthetic Aperture Radar Images, 24–32 (November 2003)

# The Opportunities, Challenges and Tendency of Reverse Logistics

Ma Fen

Faculty of Management and Economics
Kunming University of Science and Technology
Kunming, China
martha_m65@163.com

**Abstract.** With the governments' emphasis on environmental issues, product resources shortage, waste materials recycling and re-using; the reverse logistics is changing its status. It is no longer "been looked upon as a necessary evil" [1]; today, more and more companies are viewing reverse logistics as a strategic activity to put it into organizational strategic consideration. Based on above reasons, this paper tried to discover the opportunities, challenges and tendency of reverse logistics. Firstly, it introduced the concept and drivers of reverse logistics; next, analyzed the opportunities and challenges of development reverse logistics; and then, displayed the practice of reverse logistics in the USA; at last, it discussed the tendency of reverse logistics management. This paper also implied that developing the third-party reverse logistics is the best solution for reverse management.

**Keywords:** reverse logistics, return management, reverse supply chain, E-commerce, environment protection.

## 1 Introduction

Economy globalization is compressing enterprises' profit margins; consequently, more and more enterprises are paying their attentions to logistics management, which known as "the third profit source". With the governments' emphasis on environmental issues, product resources shortage, waste materials recycling and re-using; the reverse logistics, which "has been looked upon as a necessary evil" [1] is changing its status. A few years ago, "Reverse logistics, the process of returning finished products for replacement, repair, or credit, often is viewed as a necessary but costly part of maintaining an efficient supply chain [2]." Today, more and more companies are seeing "reverse logistics as a strategic activity - one that can enhance supply chain competitiveness over the long term."[3]

## 2 The Feature and Drivers of Reverse Logistics

Globalization of trade and rapid development of E-commerce cause a large number of return products. Consequently, enterprises have to put the reverse logistics into their management consideration. According to Market Research Company Gartner's estimation that improperly handled returns can erode 30-35% of potential profits [4]. That is a tremendous number you can't ignore. Therefore, analyzing, studying and managing the reverse logistics is becoming an essential assignment for organizations.

## 2.1 The Concept of Reverse Logistics

There are a variety of names have been applied to reverse logistics, including "aftermarket logistics", "aftermarket supply chain" and "reverse supply chain". The Reverse Logistics Executive Council defines reverse logistics as "a specialized segment of logistics focusing on the movement and management of products and resources after the sale and after delivery to the customer [1]."

In other words, "reverse logistics is part of a broader supply chain management process called returns management [5]."

The statements above can be detailed as:

- "Returns management includes all activities related to returns flow, reverser logistics, effective gatekeeping, and even returns avoidance [5]."
- "Reverse logistics encompasses the traditional logistics activities of transportation and inventory management, but its focus shifts to getting product back from customers rather than moving product to customers [5] ."

## 2.2 The Feature of Reverse Logistics

The main tasks of reverse logistics are collection, transportation, reprocess or disposing of the return products and recyclable packages. In addition to the features of forward logistics, reverse logistics has also the following features:

- **Diversity of Objects.** The objects of reverse logistics involve not only economic goals, but also social indicators, such as environmental protection, social responsibilities, and other factors.
- **Complexity of System.** Because the return products and recyclable packages coming from the consumers or end-market are unpredictable in time, quantity and quality; under this situation, the complexity of reverse logistics system is increased.
- **Convergence of System.** Unlike the divergence of forward logistics, the reverse logistics system appears convergence. In this system, the return products flow from individual consumers or retailers gather to manufacturers.
- **Discordance within System.** Usually, the usable return products from customers or retailers do not match the demands of new customers; so it is difficult to create re-sales.
- **Information Technology-Intensive.** Steve Manning, vice president at Milpitas believes that reverse logistics is "an information technology-intensive business".[1]

- **Low input Costs.** The main return items are waste products or recyclable packages. Most of returns (from consumers or the retailer's) are transported to the suppliers and then reprocessed or re-treated in order to resale. So the input of reverse system is very low.

## 2.3 The Drivers of Reverse Logistics

The drivers forcing organization to develop reverse logistics can be classified in to six catalogues:

- **Returning.** Return products is increasing rapidly with globalization of economy and emergence of E-commerce. Reverse Logistics Association discovered that the value of return products is about 600 million U.S. dollars after Christmas in 1999; and was one billion U.S. dollars in 2000. Modern Materials Handling Magazine estimated that there were 11 billion U.S. dollars returns products in 2002 [6]. Just as Raj pointed out: "there has been significant growth in direct online selling, which can produce return rates of 40 percent for some catalog and online retailers [5]."
- **Recalling.** Product recall policy had implemented in American auto industry in 1960s, and spread rapidly into mobile phones, home appliances, commodities, and so on. China's General Administration for Quality Supervision, Inspection and Quarantine had also released *Administrative Regulations on Recalls of Defective Automobile Products* in March 12, 2004.
- **Laws, Regulations, Environmental Factors.** With the governments' efforts in environmental protection, a series of laws and regulations are conducting to prevent the pollutions caused by waste products and packages. Developing reverse logistics also involves the social responsibilities and ethics issues of enterprises.
- **Product Life Cycle.** Product life cycle is becoming shorter and shorter especially in the computer industry. Consumers are trending to cavil at goods along with rapid product development and product upgrade. More and more return products, package, and waste material will be needed to transport back to manufactories.
- **Diversification of Distribution Channels.** The emergence of direct sales increases return rates. The products return rate is 5% to 10% in general retailers, but this percentage could be as high as 35% by catalog sales and internet [7]. The management complexity and marketing costs both increase fast because of direct sales focus on global scale.
- **Power Shift of Supply Chain.** Tough competition and unbalance between supply and demand are expanding buyers' power. The right of consumers and retailers are fully protected by return warranty.

## 3 The Development of Reverse Logistics in the USA

The United States is one of the earliest countries to practice reverse logistics management. Most of manufacturers receive return producers from retailers to manufacturers directly. Absolutely, it is inefficiency and costly approach. Statistics shows that the total costs for handling returns is about 4% of sales in the USA each year [6].

In 2002, the percentage of return products values is about 6.3% of the total value in retail business in the USA [6]. A large number of returned products not only occupy a huge amount of expense, but also increase marketing costs, and waste marketing efforts. Indeed, poor return products management is a costly social expenditure.

Increasing return goods drives organizations looking for more effective ways to improve reverse logistics management.

Road Runner Sports is a multi-channel retailer of running shoes; the Company primarily sells through its catalogs but also offers a website and three retail stores on the West Coast. It has to deal with about 500-700 returns per day, or roughly 140,000-150,000 per year out of 1 million packages sold [6].

Road Runner gains competition edges by developing reverse logistics management to improve customers' services. Success reverse logistics management makes "Road Runner's return rate is 12%, well under the industry average of 15-20% returns" [6].

Best Buy Co. Inc., a $27 billion specialty retailer, had best-in-class reverse logistics capabilities several years ago. "It turned to third-party logistics provider Genco Distribution System to integrate a centralized returns management process that includes Genco's reverse logistics software." The result is that "Best Buy's in-house processing costs have gone down by about 50%" [6].

## 4 The Opportunities of Developing Reverse Logistics

In recent years, well-known enterprises put reverse logistics management into organizational strategic planning. The reverse logistics management is no longer being costs, but new profits and social value creators by cutting down costs. For example, Eastman Kodak develops recyclable camera; Hewlett-Packard and Epson produce refillable printer, which can reusable; and as well as Sears uses biodegradable packages to eliminate returning packages. These proactive activities not only reduce the wastes entering supply chain, but also lower operational costs. All of these reveal the strategic value of developing reverse logistics.

In summary, well-planning and well-working reverse logistics system could contribute to followings:

### 4.1 Improving Financial Performance

Organization can improve resources utilization, increase sales, cutting down product and marketing costs, and reduce total inventories through effective reverse logistics management.

### 4.2 Increasing Customer Value and Enhancing the Strategic Competitive Advantage

Customer value is the determinant of survival and development of enterprises. Excellent reverse logistics management is guarantee of improving customer satisfaction and maintaining customer loyalty.

### 4.3 Strengthening Information Integration and Advancing Product Quality Management System

It is a tough work to managing the information and product quality in reverse logistics. That means organizations have to strengthening information integration and developing quality management approach in order to manage reverse logistics well.

### 4.4 Promoting Enterprise's Brand

Environmental issues are society highlight today, and it involves organization's ethics and societal responsibilities. Efficient reverse logistics management is a visible indicator, which reflects the organization's philosophies of "green produce", "green marketing" and "green logistics".

### 4.5 Developing Employees' Sense of Social Responsibility

Nowadays, with the development science and technology, the social relationship is tighter than before, and the social conflict is also severer than ever. The reverse logistics practice is a significant means that awakes employees' sense of social responsibilities by experiencing reverse logistics activities.

## 5 The Challenges of Developing Reverse Logistics

The development of reverse logistics brings out a lot of opportunities, but also accompanies various challenges:

### 5.1 Inadequate Systems and Insufficient Resources

The big barriers of developing reverse logistics are inadequate systems and insufficient resources [4]. Along with customers becoming more demanding, shortening product life cycles and issuance of new regulations; organizations are more aware of requiring an advanced management system [4]. The requirements on reverse logistics system are especially high, and existing systems and resources are not match with these requirements. The gap between requirements and available resources should be filled in as soon as possible.

### 5.2 Take-Time and Costliness

"Managing the flow of goods through the forward-facing supply chain is hard enough, but keeping the reverse supply chain running smoothly presents a number of unique challenges [1]." An outsourced logistics and supply chain services provider-Fidelitone suggested that "the reverse supply chain is frequently much more labor intensive than the forward supply chain" [1]. It takes a lot organization's time and efforts to deal with reverse logistics process.

## 6 The Tendency of Reverse Logistics Development

With the expansion of reverse logistics demands, specialized reverse logistics companies – third-party reverse logistics companies (3PRLs)-are budding. Many America companies deliver their reverse logistics management to 3PRLs; this tendency is nurturing a new industry. 3PRLs can provide scientific and integrated logistics service to better match customers' requirements.

Genco, one of the leading 3PRLs companies, possesses 104 return product processing centers and processes about 4,000,000 products per year. Genco has become largest reverse logistics company in the USA, and has contracted with over 1,500 business partners [8].

Delivering reverse logistics to 3PRLs, undoubtedly, is a super-excellence solution for many organizations. Though, developing 3PRLs is a way adapted with social requirements.

The characteristics of 3PRLs can be described below:

### 6.1 Large-Scale Effect

Normally, the 3PRLs obtains scale effect by providing specialized reverse logistics services to several organizations simultaneously, in order to gain large-scale effect. For instance, Genco can serve Wal-Mart, Kmart and other retailers in same area at the same time. By the aid of large-scale effect, the 3PRLs create profits by cutting down costs effectively.

### 6.2 Intensification, Efficiency, and Detailed Labor Division

The 3PRLs can be divided into many branches, such as: logistics broker-agents, return transport firms, return warehouses, reprocessing firms, re-sales firms, disposing firms, and so on. Each firm is only responsible for some particular logistics processes, and specialized operation contributes to improve the total logistics' benefits and efficiency.

### 6.3 Maximization of Return Products Values

Many organizations utilize advanced management technique to maximize values of return products. Using management information systems; for example, Genco manages customers' information including return condition clauses of contracts, returns collection, storing and transporting, settlement, statistics, and other information. Genco employs more than 70 engineers to develop its reverse logistics software-"R-log", which is the most popular software used by many reverse logistics companies. It creates a huge profits and social benefits for Genco [8].

## 7 Conclusions

Just as Diane and David mentioned that "Reverse logistics has often been viewed as the unwanted step-child of supply chain management. It has been seen as a necessary cost of business, a regulatory compliance issue, or a 'green' initiative. But more companies are now seeing reverse logistics as a strategic activity - one that can enhance supply chain competitiveness over the long term [3]."

Reverse logistics management in the USA sprouted from 1990s. There are a number of third-party reverse logistics companies now. They provide reverse logistics service to various organizations. These organizations can free themselves from complicated reverse logistics processes to concentrate on developing core competitiveness. Practice of the USA proves that an effective reverse logistics should not be expenditure; but a powerful method which could create efficiency, enhance customer satisfaction and strengthen competitiveness.

Reverse logistics management in China is just in its infancy, and lacks professional reverse logistics providers. However, we have full confidence that huge market prospects of reverse logistics is attracting companies to explore this field. A new revenue model will be created from effective reverse logistics management in China; and further, a new business industry will be hatched.

## References

1. Andrew, K.R.: Meeting the Reverse Logistics Challenge. Supply & Demand Chain Executive 6, 44–47 (2005)
2. Anthony, C.: Smoothing the Reverse Flow. Frontline Solutions 6, 34–36 (2005)
3. Diane, A.M., David, J.C.: The Hidden Value in Reverse Logistics 9(5), 34–41 (2005)
4. Anonymous, "Technology-Reverse Logistics: Returning to profit". Cabinet Maker, 32 (April 8, 2005)
5. Raj, P.: Reverse Logistics: The four Rs of integrated returns. Materials Management and Distribution 50, 24–25 (2005)
6. Dave, B.: Moving forward in reverse. Logistics Today 46, 1–2 (2005)
7. Anthony, C.: Smoothing the Reverse Flow. Frontline Solutions 6, 34–36 (2005)
8. Anonymous, "60 seconds with...", Modern Materials Handling (Warehousing Management Edition), Boston, vol. 60, p. 70 (October 2005)

# Research on Optimization of Equipment Maintenance Plan Based on Entropy and TOPSIS

Fujiang Zhou[1,2], Xiaodo Zhu[1], and Hongwen Chen[3]

[1] Maintenance Engineering Institute Ordnance Engineering College Shijiazhuang, China
[2] Watercraft Engineering Department, Zhenjiang Watercraft College Zhenjiang, China
zfjiang0326@126.com, x.zhu@263.com
[3] Watercraft Command Department, Zhenjiang Watercraft College Zhenjiang, China
bbb@hit.edu.cn

**Abstract.** In order to ensure the decision-making of equipment maintenance plan more scientific and accomplish equipment support mission, this paper proposes the index system of the plan, applies the entropy to figure out the weight of the assessment index and optimizes equipment maintenance plan based on analyzing the influencing factors and optimizating purpose. The entropy and TOPSIS method is applied in the optimization of equipment maintenance plan. The result indicates that the method is more logical and has certain practical in the decision-making of the equipment maintenance plan.

**Keywords:** Entropy, TOPSIS, Equipment Maintenance, Optimization.

## 1 Introduction

Equipment maintenance plan is the primary explanation of equipment maintenance support missions and the programming of equipment maintenance support as a whole [1]. It includes maintenance styles, principle, levels, repair strategies, and so on. It is important to choose an economic and efficient maintenance plan under information-based condition of battle. Because a good plan can save maintenance resources and rapidly renew equipment campaigning capabilities. The optimization of equipment maintenance plan is the base of the decision-making plan, so research on optimizing plan has very high martial value in the military maintenance management and equipment Life Cycle Costs' management [2]. TOPSIS (Technique for 0rder Preference by Similarity to Ideal Solution) is a new method. There are some virtues, such as simple calculation, agile application, and so on. The paper applies the entropy and TOPSIS method in the optimization of equipment maintenance plan. The method can help people to rapidly choose an equipment maintenance plan.

## 2 Analysissing and Establishing Index System of Equipment Maintenance Plan

### 2.1 Influencing Factors

The factors that affect optimization of equipment maintenance plan include maintenance military efficiency, economy efficiency, maintenance avail, flexibility and security. The content of maintenance military efficiency includes military rapidity, quality, support degree, equipment capability renewing, and so on. The content of maintenance economy efficiency illuminates that expense is saved in the maintenance activities. The content of maintenance avail includes social resource economy and protecting circumstance efficiency. The content of maintenance efficiency mainly includes the flexibility of maintenance support group in the support area coverage. The content of maintenance security mainly illuminates that the group have an ability of accomplishing maintenance missions under the condition of enemy's destruction.

### 2.2 Optimizating Purpose

Many influencing factors are considered when equipment maintenance plan are drawn, analyzed, optimized. It is very important to make decision of the equipment maintenance plan. In the designing plan's stage, many maintenance plans can be compared and an optimization plan can be chosen. During being to put in practice, the designing plan can accord with actual circs and satisfy designing demand whether or not. In the carrying out plan's stage, the plan can be evaluated and adjusted in time. Research on decision-making and optimization of the equipment maintenance plan can improve equipment command decision-making level and make maintenance assignments to be accomplished.

### 2.3 Establishing the Index System of Equipment Maintenance Plan

While equipment maintenance being carried on, in the first instance, a task must be reasonably distinguished, maintenance resource must be scientific collocated, so as to complete that, maintenance support is provided in time and furthest brings into play itself efficiency[3,4]. Secondly equipment maintenance resources' flexibility and security must be considered. Because there is equipment support chaperonage and synchronization with campaign, maintenance support is hold in by circumstance and technique level, many factors affect the optimization of equipment maintenance plan. On the base of analyzing maintenance principia, factors and demand, the assessment index system of equipment maintenance plan is established.

*1) Maintenance military efficiency(C1):* It indicates military rapidity, quality, support degree, equipment capability renewing, quantity of repairing equipment, and so on.

*2) Maintenance economy efficiency(C2):* It indicates cost of repairing equipment, expenditure of maintenance spare part, maintenance man-hour, equipment transportation cost, maintaining cost for the future.

*3) Avail after Maintenance(C3):* It indicates wastage(oil and electricity) rate and drainage standard rate after equipment is repaired.

*4) Maintenance flexibility(C4):* It reflects flexibility of maintenance support group in the support area coverage and adaptive demand of support circumstance.

*5) Maintenance security(C5):* It reflects support groups' ability of accomplishing maintenance missions under the condition of enemy destroy.

In the middle of index system of equipment maintenance plan, C1 and C1 and C3 all reflect to accomplish equipment maintenance support missions' degree, C4 and C5 reflect to equipment maintenance's flexibility, security and adaptability.

# 3 Calculating Process of Equipment Maintenance Plan Optimization

## 3.1 Figuring out the Weight of Assessment Index

The concept of entropy comes from energetic. It usually refers to the degree of one system incertitude[5]. It's difficult to figure out the weight of assessment index. The tradition solving method is achieved by experts and AHP. The subjective setting makes the result warp to a certain extent. In the paper the entropy theory is applied to figure out the weight of assessment and make the result more scientific and reasonable.

There are $n$ equipment maintenance plans and $m$ assessment index. Primal matrix is

$R=(x_{ij})_{m\times n}(i=1,2,\cdots,m;j=1,2,\cdots,n)$, The assessment index is benefit attribute and cost attribute. Benefit attribute is $x'_{ij}=x_{ij}/\max x_{ij}$; Cost attribute is $x'_{ij}=x_{ij}/\min x_{ij}$. News matrix can be described: $x'_{ij}=(x'_{ij})_{m\times n}\ (i=1,2,\cdots,m;j=1,2,\cdots,n)$;The assessment index proportion is

$P_{ij}=(1+x'_{ij})/\sum_{j=1}^{n}(1+x'_{ij})$; The assessment index entropy is

$$H_i=-(\sum_{j=1}^{n}p_{ij}\ln p_{ij})/\ln(n); \tag{1}$$

The weight of assessment index is

$$w_i=(1-H_i)/\sum_{i=1}^{n}(1-H_i),\sum_{i=1}^{n}w_i=1; \tag{2}$$

## 3.2 Optimizating Process

TOPSIS is one of multi-targets decision-making methods. It includes the Ideal Perfect Value(IPV) and the Minus ideal Perfect Value (MIPV) of the decision problem to get the projects compositor[6]. TOPSIS is used in the paper to choose a best equipment maintenance plan.

If plan muster is $M=(M_1,M_2,\cdots,M_m)$ and assessment criteria muster is $C=(C_1,C_2,\cdots,C_n)$, index value is $x_{ij}=(i=1,2,\cdots,m;j=1,2,\cdots,n)$, [7]

Decision-making matrix $X$ is :

$$X=\begin{bmatrix} & C_1 & C_2 & \cdots & C_n \\ M_1 & x_{11} & x_{12} & \cdots & x_{1n} \\ M_2 & x_{21} & x_{22} & \cdots & x_{2n} \\ \vdots & \vdots & \vdots & \vdots & \vdots \\ M_m & x_{m1} & x_{m2} & \cdots & x_{mn} \end{bmatrix}$$

Standardization matrix is established by disposing $X$. Benefit attribute is

$$U(i,j) = \frac{x(i,j) - x_{\min}(j)}{x_{\max}(j) - x_{\min}(j)} \tag{3}$$

Cost attribute is $U(i,j) = \frac{x_{\min}(j) - x(i,j)}{x_{\max}(j) - x_{\min}(j)}$ (4)

$U(i,j)$ is target character value. After Standardization matrix multiplies the weight of assessment criteria, addition- weight decision-making matrix is $R = (r_{ij})_{m \times n}$,

$r_{ij} = w_i U_{ij}, i = (1,2,\cdots,m; j = 1,2,\cdots,n)$.The IPV $s^+$ and the MIPV $s^-$ can be confirmed.

$s_j^+ = \max_{1 \le i \le m} \{r_{ij}\}, (j = 1,2,\cdots,3)$, $c_j$ is benefit attribute.

Or
$$s_j^+ = \min_{1 \le i \le m} \{r_{ij}\}, (j = 1,2,\cdots,3) \tag{5}$$

$s_j^- = \min_{1 \le i \le m} \{r_{ij}\}, (j = 1,2,\cdots,n), c_j$ is cost attribute,

$$s_j^- = \min_{1 \le i \le m} \{r_{ij}\}, (j = 1,2,\cdots,n) \tag{6}$$

The distance between the mul-targets attribute's value with the IPV and MIPV can be calculated:

$$Sd_i^+ = \sqrt{\sum_{i=1}^{n} (S_j^+ - r_{ij})^2}, (i = 1,2,\cdots,m; j = 1,2,\cdots,n) \tag{7}$$

The competitive judging exponent will be getting:

$$\xi_i = \frac{Sd_i^-}{Sd_i^- + Sd_i^+} (i = 1,2,\cdots,m) \tag{8}$$

The best equipment maintenance plan can be assured by the value of $\xi$.

# 4 Application

There are four different equipment maintenance plans(A1、A2、A3、A4) which are established based on campaign condition and maintenance task. The assessment criteria value is arranged as [0 ,1]. The assessment index of equipment maintenance plan is described: C= ( C1,C2,C3,C4,C5),C1=Maintenance military efficiency;

C2=Maintenance economy efficiency; C3=Avail after Maintenance; C4=Maintenance flexibility; C5= Maintenance security. The experts and questionnaires can give different scores about the importance of each item. The primitive data is shown in Table 1. C1 and C4 and C5 are assessment index kind of benefit attribute, C2 and C3 are assessment index kind of cost attribute.

**Table 1.** Primitive Data

| PLAN | C1 | C2 | C3 | C4 | C5 |
|---|---|---|---|---|---|
| A1 | 0.96 | 0.86 | 0.75 | 0.80 | 0.65 |
| A2 | 0.75 | 0.75 | 0.70 | 0.70 | 0.85 |
| A3 | 0.90 | 0.80 | 0.80 | 0.85 | 0.70 |
| A4 | 0.85 | 0.82 | 0.65 | 0.75 | 0.75 |

Based on information Entropy and TOPSIS method, the judging matrix can be the (3) and (4) above:

$$Y = \begin{bmatrix} 0.5470 & 0.5319 & 0.5157 & 0.5148 & 0.4385 \\ 0.4318 & 0.4638 & 0.4813 & 0.4504 & 0.5734 \\ 0.5182 & 0.4948 & 0.5501 & 0.5470 & 0.4722 \\ 0.4894 & 0.5071 & 0.4469 & 0.4826 & 0.5060 \end{bmatrix}$$

The judging matrix can be unitary disposed:

$$Y' = \begin{bmatrix} 0.2774 & 0.2663 & 0.2586 & 0.2581 & 0.2203 \\ 0.2168 & 0.2322 & 0.2414 & 0.2259 & 0.2881 \\ 0.2601 & 0.2477 & 0.2759 & 0.2742 & 0.2373 \\ 0.2457 & 0.2539 & 0.2241 & 0.2419 & 0.2543 \end{bmatrix}$$

The weight of five assessment index can be figured out by (1) and (2):

$$w = [0.1731 \quad 0.1927 \quad 0.1750 \quad 0.2517 \quad 0.2075]$$

The addition- weight matrix can be figured out:

$$Z = \begin{bmatrix} 0.048 & 0.0513 & 0.0453 & 0.0650 & 0.0457 \\ 0.0375 & 0.0447 & 0.0422 & 0.0569 & 0.0598 \\ 0.045 & 0.0477 & 0.0483 & 0.0690 & 0.0492 \\ 0.0425 & 0.0489 & 0.0392 & 0.0609 & 0.0528 \end{bmatrix}$$

The distance between the targets attributes and the IPV and MIPV is calculated as follows:

$$sd^{+} = (0.0860 \quad 0.0830 \quad 0.0935 \quad 0.0944)$$

$$sd^{-} = (0.1035 \quad 0.0972 \quad 0.1001 \quad 0.1006)$$

The compositive judging exponent is got:

$$\xi = (0.5462 \quad 0.5394 \quad 0.5170 \quad 0.5159)$$

So we can get the optimization of equipment maintenance plan by it. $A1 > A2 > A3 > A4$.

## 5 Results

The optimization of equipment maintenance plan is a complicated and blurry task. In the paper the assessment index system of plan is proposed on the base of analyzing the optimization's content and purpose, TOPSIS model is applied in the optimization of equipment maintenance plan, the entropy is applied to figure out the weight of the assessment index. From the sample, it can get the reasonable result by calculating the basic data on the base of information entropy and TOPSIS method, the improved TOPSIS avoids decomposing of multi-process in the fuzzy evaluation. By the example, it can be applied in the optimization in the other fields of industry, so it has the actual meaning to solve above optimization problems for equipment maintenance support command.

## References

1. Gang, M., Kang, J., et al.: Military Equipment Maintenance Engineering Theory, pp. 40–46. China National Defence Press (2005)
2. Chen, X.: Modern Maintenance Theory, pp. 78–96. China National Defence Press (2002)
3. Song, T.: Equipment Indemnificatory Engineering, pp. 105–110. China National Defence Press (2005)
4. Song, J., et al.: Equipment Maintenance informationization Engineering, pp. 85–87. China National Defence Press (2005)
5. Mondal, T.K., Samanta, S.K.: Toploy of interval-valued intuitionistic fuzzy sets. Fuzzy Sets and Systems 19(3), 483–494 (2001)
6. Fu, Q.: Research OF about TOPSIS Method. Journal of Xiang Science and Technology College 25(1), 190–193 (2008)
7. Liu, Z., et al.: Research of Precept Evaluating Method Based on Entropy and TOPSIS. Journal of Mathematic Practice and Cognition 35(10), 114–119 (2005)

# Face Detection and Tracking Based on Weighted Histogram and Shape Constraints

Qin Ling and Tong Xian Hui

College of Electronics and Information Engineering
Nanjing University of Technology, NJUT
Nanjing, China
ql@njut.edu.cn

**Abstract.** In this article, a robust Two-dimensional face tracking is proposed which is based on histogram matching and shape constrain. First, an adaptive weighted histogram matching is used to describe the face of the regional information, make model for a more reasonable description of the regional characteristics, and increased robustness. Not only to avoid a model of skin color, but also increase the robustness described. Moreover, rotating or point of view in order to deal with the impact of changes or point of view, in the process of tracking the use of tracking the results of the current frame modified adaptive Histogram Matching, use shape constraints to the target position more accurately. The experimental results have shown that the proposed method not only attains high face detection rate, but also for block, illumination and face Pose variations are adaptability and real-time.

**Keywords:** Face detection, Weighting Histogram, Modal Matching, Shape constraints.

## 1 Introduction

Face detection and tracking relates to many realistic issues, the purpose and method varies based on the application. Two dimensional based tracking is enough for normal video conference and vision monitoring. In the two dimensional face tacking, the color of the skin usually has been usually used, but it's not robust enough when illumination changes or the backorder is also has a skin like color. Based upon this, this paper raised a new method of face detecting and tracking, it firstly uses the color histogram which carries the space information to perform the initial localization. This avoids the infinite searching strategy which is usually used when matching the histogram, it also significantly decrease the calculation as well as increase the robust. Then it uses the shape constraints module in a relatively small scope, searches the base position and scale of the human face. At last, it uses the λ2 - Test Measure Criterion to decide if an adjustment to the matching module is needed. The testing results demonstrate that this method has high performance in robustness and real-timing which can detect the human face quickly and effectively resolve the disturbance met during the real-time tracking.

# 2 Candidate Regions Detecting and Abstracting

## 2.1 Weighted Histogram Description

During the practical application, there can be complicated backgrounds in the video which might lead to the noise in the detection. In the face region description histogram, assuming that the entire color space being divided into ***m*** sub-regions, calculated the frequency of each pixel color vector falling into the according sub-regions and finally got the color histogram that consisted of m columns. Meanwhile, considering the affects to the establish of color histogram caused by the object side as well as the location of the pixel point in the target area, the algorithm added additional weighted resolution to each pixel point and that ends up to the more accurate color model of the target object. Since the image sensor output images of RGB type in which the luminance information was not stored separately, in order to make the tracking algorithm being sensitive to the illumination, RGB color space was translated to HIS space and the luminance component ***I*** was not taken into consideration. By doing this, the color histogram can be made by 8*8*1 columns. When building the color distribution model for an elliptic region with semi-major axis of ***w*** and semi-minor axis of ***h***, different pixel makes different attribution to the color histogram. For those pixel points that are far away from the central area, they are assigned with smaller weight number. Especially for those are near the edge of the region, the corresponding weight numbers are quite small and they have little or no even effect on the color histogram. The weight number function is:

$$k(r)=\begin{cases}1-r^2, r<1\\0, r\geq 1\end{cases} \tag{1}$$

***r*** - the distance between certain pixel point and the centre.

Using $p_y=\{p_y^{(u)}\}, u=1,2,\ldots,m$ to indicate the color distribution for the central point at location y, then

$$p_y^{(u)}=f\sum_{i=1}^{N}k\left(\frac{\| y-x_i \|}{\alpha}\right)\delta[h(x_i)-u] \tag{2}$$

*N* – the amount of the pixel points in the target area;
$X_i$ – certain point in the target area;
*H(xi)* – the according column in the color histogram for point $X_i$;
$\delta$– the Dirac function;
Parameter $\alpha=\sqrt{w^2+h^2}$ indicating the size of the target area;

Normalized factor $$f=\frac{1}{\sum_{i=1}^{N}k\left(\frac{\| y-x_i \|}{\alpha}\right)} \tag{3}$$

makes $\sum_{u=1}^{m}p_y^{(u)}=1$

As per definition (2.2), when figuring the color distribution level, each pixel point is assigned with a weight number based on the distance between the centre and itself. The smaller the distance is, the bigger weight number it gets, and vice versa. Through the sample matching and the necessary data corrosion, Fig.b shows the binary image for the original Fig.a.

While there are obstructions or complicated backgrounds, this also increased the robustness for the object description, because the further the pixel located, the unreliable it is.

**Fig. a.** Original Image

**Fig. b.** Binary Image

In addition, during the tracking processing, the object rotating and viewpoint changing are universal. As a result of this, simply using a fixed histogram sample is not proper. For example, in image 1.1 and 1.3, the histograms for target object are totally different. To resolve this, we refresh the required histogram for the next frame based on the tracking result of the current frame. See function (4):

$$p_N = \alpha \times p_T + (1-\alpha) \times p_R \tag{4}$$

$P_T$ – the current frame histogram sample;
$P_R$ – the current frame tracking results;
$P_N$ – the next frame histogram sample;
$a$ – the scale factor, range from 0 to 1.

There are amount of methods of histogram similarity measurement. This paper selected the statistics based measurement - $\lambda^2$ testing model and measurement. The reason is that it is proved to be a good histogram matching method with high performance of robust by Schiele:

$$\lambda^2(p_1, p_2) = \sum_{n-1}^{m} \frac{(p_1(u) - p_2(u))^2}{p_1(u) + p_2(u)}$$

$p_1$, $p_2$ – the two histograms that are waiting for comparison.

## 2.2 Shape Constrain Model

If the histogram of the moving object does not change in the movement, the histogram matching method can probably achieve the desired outcomes. However, in aspect of the face tracking issue especially when rotating and viewpoint switching existing in image 2, simply using histogram matching method does not meet the required robustness. Even we have applied formula 4 to refresh histogram sample, but this refreshing processing can still cause the accumulation of the sample error which will mislead the histogram to stray from the object after a period of time as long as there's no additional constrain at each frame. Apart from this, only using histogram matching method cannot efficiently follow the size changing of the target object in the tracking process. As a result, after the histogram matching, shape constrain is applied in this paper to locate the object more accurately, including the size changing of the object. The shape constrain model is to treat the face as an ellipse and maximize the normalized total gradient of the points at the ellipse edge, which is to say:

Using S = (*y, h*) to describe the ellipse, y is the centre and h is the radius. The goal of the shape constrain is to search for the best ellipse status S* within the area. This S* will make the normalized total gradient at ellipse edge to be maximum:

$$S^* = \arg\max_{\delta \in S} \left\{ \frac{1}{N} \sum_{i=1}^{N_h} |g_i| \right\} \tag{5}$$

$N_h$ – the amount of the pixel points at the ellipse edge.
$G_i$ – the gradient of the pixel point i at the ellipse.
$S$ – the local searching space:

$$S = \{s : |y - y_0| \le \Delta y, |h - h_0| \le \Delta h\}$$

# 3 Algorithm Steps and Result Analyze

## 3.1 Algorithm Steps

The parameter settings in the experiment are as follows: set scale factor in the update of the histogram template as $\alpha$=0.5; when using shape constrain, set the partial searching scope as $\Delta y$=2、$\Delta h$=1. After we get a match of the histogram, we combine it with the shape constrain module. Thus, the whole tracking algorithm steps can be summarized as following:

*a)* Initialize the histogram template of the current frame $p_t$, according to formula (2.2), calculate $p_{y0}$ of the current frame, and evaluate $\lambda^{2(p_{y0}, p_{t},)}$.

*b)* Save $\lambda^{2(p_{y0}, p_{t},)}$, according to formula (2.2) get the value $p_{y1}$ from next frame. When $\lambda^{2(p_{y1}, p_{t},)} > \lambda^{2(p_{y0}, p_{t},)}$, adjust $y_1$ to be $1/2(y_0+y_1)$.

*c)* According to the shape module defined in formula (5), in the partial searching scope ($h_0 \pm \Delta h, y_0 \pm \Delta y$), search for the best ellipse and assign to $h_0$ and $y_0$.

*d)* Use formula (2.4) and make the judgment that if a modification to the human face histogram template $p_t$ is needed, also getting prepared for the detecting of the next frame.

In the experiment, we manually initialize the template of the first frame, the setting of the parameters, such as $\alpha$, $\Delta y$, $\Delta h$ are all predetermined. They remain unchanged during the tracking session. effect varies in different environments, you need to test to find the best setting.

## 3.2 Result Analyze

In the experiment, we used the frames got from two separate real-time videos, the target in the first set of the frames is a female with long hair, the purpose is to detect her face when she moves in a relatively expanse area with obstacles. The second serial of frames is used to detect the human face when the target is swaying, turning around and the sudden change of the illumination.

To demonstrate the effectiveness of the algorithm, we compared it with the normal histogram matching algorithm. We used VC++6.0 to program the algorithm. We collected real-time video serials by using Haikang image collecting card, tested in different illumination conditions, backgrounds and positions, and with a process rate at 24-26 frame/s which achieves the real-time requests. Detecting results are demonstrated in Fig.1 and Fig.2, the analyze results in CHART I.

Figure.1-1

Figure.1-2

Figure.1-3

Figure.1-4

Figure.2-1

Figure.2-2

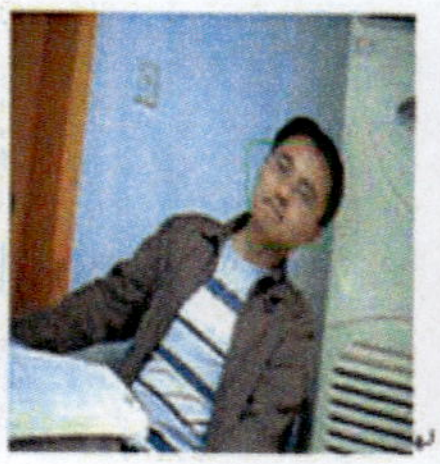
Figure.2-3

Figure.2-4

(**Notes:** To enhance the display, use green rectangle to highlight the human face)

**Table 1.** The Compare of the Human Face Detecting Algorithms

| Experiment Method | Image qty | Image Size | Hit Rate/% | Avg Time/ms |
|---|---|---|---|---|
| Usual Color Histogram Matching Algorithm | 150 | 326×288 | 82.7% | 56 |
| Algorithm based on Weighted Histogram and Shape constraints | 150 | 326×288 | 95.3% | 49 |

From the above testing results one could see, the human face detecting algorithm discussed in this paper has higher process speed while maintain the hit rate, it could fulfill the real-time application needs.

## 4 Conclusion

The arithmetic based on histogram matching and shape constrains which is discussed in this paper, narrows down the searching scope, enhances the detecting speed, and decrease the false detecting rate. It has decent robustness, and can achieve speedy human face localization, thus it fits for the real-time monitoring system. In the experiments, different kinds of testing serials will need different parameters, they need to be modified according to the real environment. In the further, one could take the tracking result of several former frames into consideration, dynamically predict and modify these parameters during the tracking, further perfect this arithmetic and enhance its detecting capability.

## References

1. Zheng, M.-E., Guan, Y.-P.: Face detection and tracking based on skin color and skin color moment. Computer Engineering and Design 29(4), 879–881 (2008)
2. Ai, H.-Z., Liang, L.-H., Xu, G.-Y.: Face Detection Based on Skin Color and Template. Journal of Software 12(12), 1784–1792 (2001)
3. Lu, C., Zhang, C., Wen, F.: Regional feature based fast human face detection. Journal of Tsinghua University (Science and Technology) 39(1), 101–105 (1999)
4. Gan, J.-Y., Chen, Y.-H., Gao, J.-H.: A Real-time Face Tracking Algorithm Based on Weighted Histogram and Mean-shift. Computer Simulation 25(11), 208–210 (2008)
5. Yang, M.H., Ahuja, N., Kriegman, D.: Detecing Faces in Images: A Survey. IEEE Trans. on PAMI 24(1), 34–58 (2002)

# Chirplet-Fractal Characteristics of Leakage Current on HV Insulators

Ali Hui[1] and Hui Lin[2]

[1] School of Electrical and Control Engineering
Xi'an University of Science and Technology Xi'an, China
huiali1975@yahoo.com.cn
[2] Member, IEEE, School of Automation
Northwest Polytechnical University Xi'an, China
linhui@nwpu.edu.cn

**Abstract.** With the increase of voltage grade in power transmission system and the aggravation of industrial pollution, flashover of contaminated insulators is getting more and more serious in high-voltage (HV) and ultra-high-voltage (UHV) systems, this threatens the security of power transmission severely. The key to improve the security of power system is to explore methods of monitoring states of insulator's surface by some eigenvalues. A new idea, using chirplet as the ruler to define fractal dimension, was proposed in this paper. Chirplet-fractal dimension is defined as the sum of residues of decomposed signals in formula 8. The fractal characteristics of leakage current (LC) are gotten from this dimension. The results show in figure 11, which the chirplet-fractal dimension efficiently describes the information of arcs discharge in LC, and it is also a good eigenvalue for flashover discrimination and risk prediction.

**Keywords:** dimensions, flashover, fractals, chirplet transforms, leakage current, high-voltage, insulator contamination, least squares methods, pattern recognition, risk prediction.

## 1 Introduction

As high-voltage (HV) and ultra-high-voltage (UHV) transmission line have been used more and more widely, flashovers from contaminated insulators became the primary factor threatening the reliability of HV transmission system, and the harm from pollution flashover has surpassed the thunder impact by far[1]. In case the flashover occurred, power cut accident would be caused generally in large scale, and the system cannot run afresh successfully in a short time[2], thereby great society influence and economical loss would be caused. At present, the most universal action for preventing pollution flashover is regular cleaning, using antipollution material, and adjusting creepage distance. However these methods have obvious defects such as resources waste, blindness, and so on, which are resulted from the lacking of accurate understanding of insulator surface condition. The key of improving the security of power system is exploring methods of monitoring states of insulator's surface by some eigenvalues. Thus, to choose an efficient eigenvalue is important.

At present, leakage current (LC) is the best discrimination standard with practical significance, because it generally reflects the voltage, the climate and the contamination. LC is a token of descending degree of insulation level, and is easier to supervise than other values, such as surface electric field, voltage distribution, infrared image[3]. Online supervision provides LC of natural contamination for flashover risk prediction. But supervised LC includes disturbs and noise signals because of complex local environment and atrocious weather, thus, it would lead to low veracity while LC is taken directly as the eigenvalue to identify different stage of flashover and predict risk.

In this paper, a new idea is proposed to use fractal to describe chirplet transform results of signals. It introjects chirplet transforms and fractal geometry, namely chirplet-fractal dimension. This dimension describes the change rule of the sum of residues of decomposed signals. The fractal characteristics of LC are gotten from chirplet-fractal dimension, and LC is obtained from artificial pollution test. From comparing different test data, the results show that the information of arcs discharge in LC can be reflected completely and effectively by chirplet-fractal dimension.

## 2 Fractal Dimension

### 2.1 Box Dimension

Fractal was firstly brought forward by Polish scientist Mandelbrot [4], and is used to study set topology, then applied to natural science and engineering later. In nature, the shape of objects and diverse figures can be divided into two kinds: one possesses characteristic scale that can be approached by line, face, and volume, and so on. Another doesn't possess character scale, whose shape is accidental and cannot be measured but has comparability between part and collectivity. This self-similar frame without characteristic scales called fractal. Fractal is used to research irregular complicated phenomenon existing widely in nature, and to describe the common structure of a great kinds of irregular sets and functions that cannot be described by traditional Euclidean geometry and calculus method.

In fractal theory, Fractal dimension is a significant parameter to depict fractal phenomenon, and it is the further development of traditional dimensions, using fraction to describe the degree and complicacy of the fractal collection filling space. Since the birth of fractal dimension, more than ten kinds of different fractal dimensions have been defined in allusion to different objects, for example, box dimension, Hausdorff dimension, similarity dimension, information dimension, and correlation dimension [5][6]. Among these dimensions, box dimension is the easiest and most widely used. As shown in Fig.1, the box dimension of a set S contained in n-dimension is defined as follows: for any $\delta>0$, let $N(\delta)$ be the minimum number of n-dimension cubes of side-length $\delta$ needed to cover S. The box dimension DB of S is[7]:

$$D_B = \lim_{\delta \to 0} \left[ \ln N(\delta) / \ln(\delta) \right] \tag{1}$$

**Fig. 1.** Definition of box dimension

The ideal fractal has infinite details. However, fractal phenomenon in nature generally doesn't have fractal character on condition of finite scale, so box dimension cannot be calculated by (1). The approximate method is usually used to compute it in practice, viz. take ruler, with certain length, $\Delta$ as the minimum (or maximum) side-length of meshes, then magnify (or minify) them to $k\Delta, k \in z^+, k < N_0$ step by step, note $N_{k\Delta}$ as the number of meshes that cover $S$ in side-length of $k\Delta$, and $N_{k\Delta}$ is shown as:

$$N_{k\Delta} = \sum_{j=1}^{N_0/k} ceil\{\max\{x[(j-1)k+1:jk]\} - \min\{x[(j-1)k+1:jk]\}/k\Delta\} \qquad (2)$$

where *ceil*(*y*) is taking integer upward, namely $ceil(m+\delta) = m+1$.

The famous Richardson curve[8] is gotten through plotting the curve of $\ln k\Delta \sim \ln N_{k\Delta}$, as shown in Fig.2. This figure can be divided into three regions A, B and C in terms of different slopes of the curve. $k\Delta$ is very small in region A, and fractals in nature generally doesn't have scale-free self-similarity. At the same time, $k\Delta$ is too large to reflect details of the curve in region C. So region B with preferable linearity is commonly regarded as the scale-free region. Suppose the start point and the end point of this region are $k_1$ and $k_2$, respectively, then $\ln k\Delta \sim \ln N_{k\Delta}$ satisfies linear regress model,

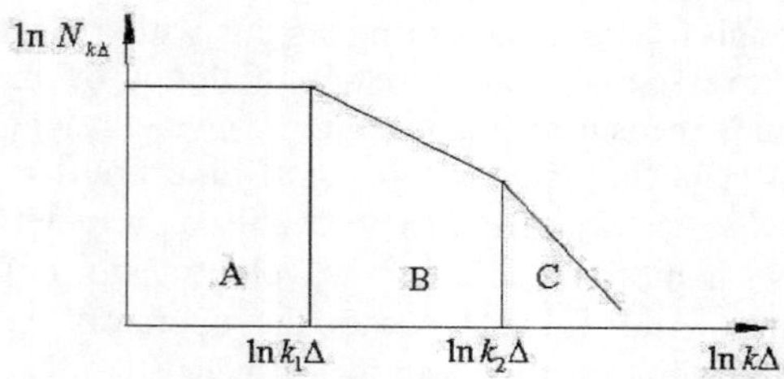


**Fig. 2.** Richardson curve

$$\ln N_{k\Delta} = -D_B \ln k\Delta + b, k_1 \le k \le k_2 \qquad (3)$$

where $D_B$ is the slope of the curve in region B and is defined as box dimension.

## 2.2 Chirplet-Fractal Dimension

According to the definition of wavelet transforms, radix function package, changed by translating and dilating mother chirplet ψ(t), is

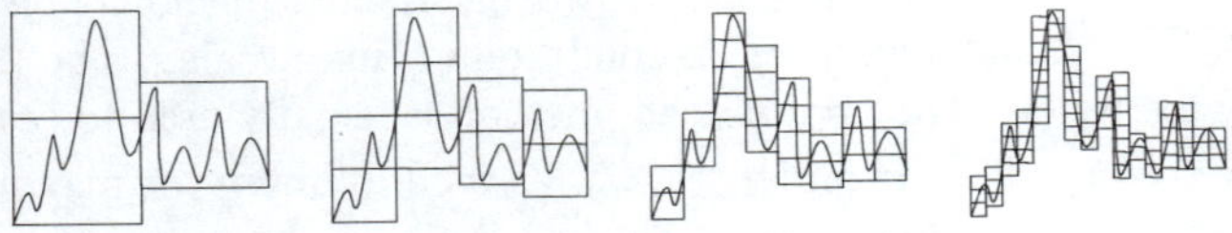

(a) $\delta=1$ (b) $\delta=1/2$ (c) $\delta=1/4$

$$\psi_{a,b}(t)=|a|^{-1/2}\psi\left[(t-b)/a\right] \qquad b\in R, a\in R^{+}, a\neq 0 \tag{4}$$

Any measurable function $f(t)\in L^2(R)$ can be constructed through $\{\psi_{a,b}(t)\}$ with different point of view and different time-frequency resolution. A fractal set $F$ with self-similarity, studied in fractal theory, can be formed similarly by function $\beta(t)$, with the compactly support set, namely:

$$\beta(t)=r^{H}\beta(rt) \qquad r,H>0 \tag{5}$$

where r is the self-similar affined operator, H is a parameter related with dimension.

Comparing operator $r$ with operator $a$, we can draw the following conclusion: the principle of wavelet transform from low frequency to high frequency is consistent with the thought of recognizing essence of things from collectivity to part, and from macroscopy to microcosm. Figure 3 shows the four-level wavelet decomposition results of time sequence signals decomposed by Db4 orthogonal wavelet. The upper four figures are Db4 orthogonal wavelets of different scales, and the following four figures are high frequency coefficients of wavelet decomposition. Fractal and wavelet transform have similar process by comparing Fig.1 with Fig.3, since the process of recognizing things through fractal is measuring signals with a length ruler of different scales, and studying the essential characteristics of things by measurement results, while wavelet transform is measuring signals with a wavelet ruler of different scales.

The chirplet transform is the extension of short-time Fourier transforms(STFTs) and wavelet transforms. It has time-frequency locality as wavelet transforms, and its time-frequency windows is more flexible than wavelet transform. Figure 4 shows five different transformations. And Table 1 shows the operators corresponding to the coordinate aces of the chirplet transform parameter space[9].

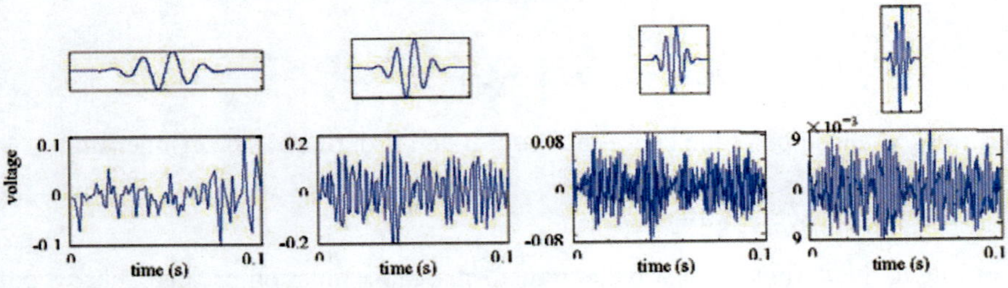


(a) $a=1$ (b) $a=1/2$ (c) $a=1/4$ (d) $a=1/8$

**Fig. 3.** Wavelet decomposition of time sequence signals

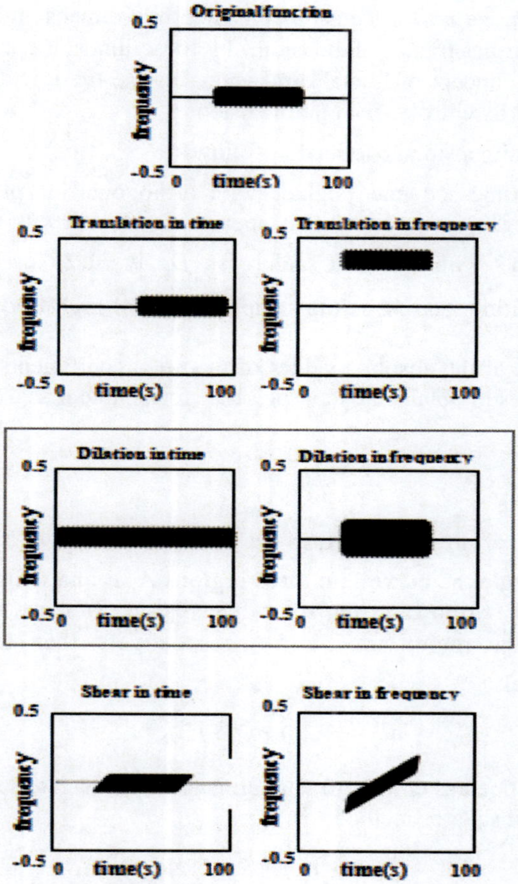


**Fig. 4.** Five affine transformations of TF plane

**Table 1.** Five Operators Corresponding to Chirplet Transform

| Description | 1-parameter notation | Composite notation | Time domain $g(t)$ |
|---|---|---|---|
| **Time translation** | $\boxed{\Rightarrow}_{t_c} g(t)$ | $= M_{t_c,0,0,0,0}\, g(t)$ | $g(t-t_c)$ |
| **Frequency translation** | $\boxed{\uparrow\uparrow}_{t_c} g(t)$ | $= M_{0,f_c,0,0,0}\, g(t)$ | $e^{j2\pi f_c t} g(t)$ |
| **Time dilation /Frequency dilation** | $\boxed{\Leftrightarrow}_{a} g(t)$ | $= M_{0,0,a,0,0}\, g(t)$ | $e^{-a/2} g[e^{-a}(t-a)]$ |
| **Time shear** | $\boxed{\square}_{p} g(t)$ | $= M_{0,0,0,p,0}\, g(t)$ | $(-jp)^{-1/2} e^{j\pi \frac{1}{p} t^2} * g(t)$ |
| **Frequency shear** | $\boxed{\downarrow\uparrow}_{q} g(t)$ | $= M_{0,0,0,0,q}\, g(t)$ | $e^{j2\pi \frac{q}{2} t^2} g(t)$ |

Based on this idea, a new dimension consulted box dimension is proposed in this paper, namely chirplet-fractal dimension, by describing the results of chirplet transform with the concept of fractal dimension, that is, the fractal characteristics of signals is expressed by chirplet-fractal dimension.

Chirplet-fractal dimension is defined as follows:

*1)* Decompose time sequence signals with orthogonal chirplet transform, then high frequency coefficients are results of measuring signals with chirplet ruler. And note the coefficients with different scales as $\{D_{j,k} | k = 1, 2, \cdots M_j\}$, where $j$ is the level of decomposition, and $M_j$ is the sample length of high frequency sequence in chirplet transform.

*2)* Take the sum of the absolute values of modified coefficients as a measurement of decomposition result obtained by chirplet transform, note it as:

$$S_j = \sum_{k=1}^{M_j} |D_{j,k}|^2 \tag{6}$$

*3)* Plot the curve of $\ln j \sim \ln S_j$, then get Richardson curve in a similar way as box dimension, and divide the curve into three regions A, B and C in terms of different slopes, and regard region B as the scale-free region. Suppose the start point and the end point of this region are $j_1$ and $j_2$, respectively, then $\ln j \sim \ln S_j$ satisfies linear regress model:

$$\ln S_j = -D_W \ln j + b, j_1 \le j \le j_2 \tag{7}$$

The slope $D_W$ is defined as chirplet-fractal dimension of discrete sequence, and can be gotten through least squares method:

$$D_W = -\frac{(j_2 - j_1 + 1)\sum \ln j \ln S_j - \sum \ln j \sum \ln S_j}{(j_2 - j_1 + 1)\sum \ln^2 j - \left(\sum \ln j\right)^2} \tag{8}$$

where $j_1 \le j \le j_2$.

# 3 Test and Results

## 3.1 Test Model

The chirplet-fractal dimension of LC on HV insulators is gotten from the method proposed in this paper. And LC is gotten from the following artificial test of contaminated insulator flashover: the test is under the condition of clean fog, constant voltage and contaminating beforehand with solid pollution method. The fog room is 4m×4m×5m in size, the transformer is non-free transformer with specification of 125kVA, and its rated capacity, rated current and rated voltage is 125/250kVA, 312.5/0.5A and 0.4/250kV, respectively. The tested insulator is XP-70. Among 7 pieces of insulator string in 110kV transmission line, the first insulator endures the highest voltage under the condition of without grading ring. So only a single insulator is tested, and the test voltage is 26kV, NSDD is 2.0$mg/cm^2$, ESDD are 0.10$mg/cm^2$, 0.15$mg/cm^2$, and 0.20mg/$cm^2$, respectively.

The detecting devices include a homemade LC sensor and a conditioning circuit. The acquisition card is Yanhua 9111DG. An industrial computer samples LC, temperature and humidity at the same time.

### 3.2 Results

The chirplet-fractal dimension of denoised LC is calculated through the above method. Figures 4 to 9 are LC curves of the artificial test and the corresponding chirplet-fractal dimensions, and their ESDD are 0.10, 0.15 and 0.20mg/cm2, respectively.

The chirplet-fractal dimension of LC decreases along with the increasing of the LC amplitude. In initial stage, because pollution layer is being humidified in low humidity, the impact of LC is very small with sporadic spark and glow discharge, and the corresponding chirplet-fractal dimension changes hardly. Little arcs begin to

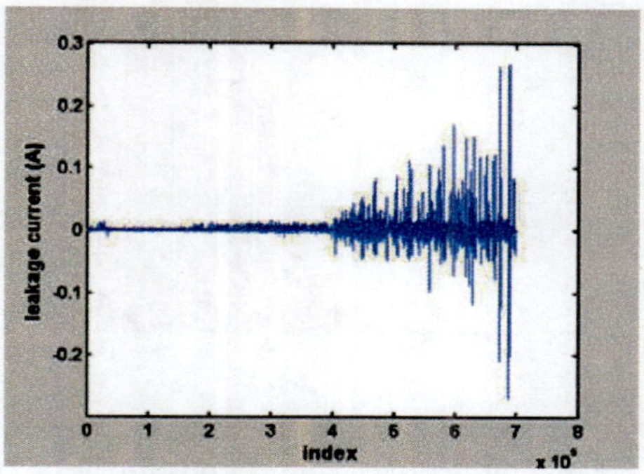


**Fig. 5.** Leakage current (ESDD=0.10)

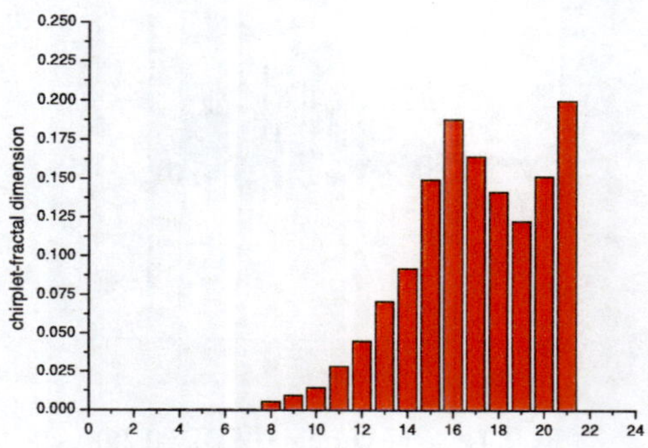


**Fig. 6.** Chirplet-fractal dimension of figure 5

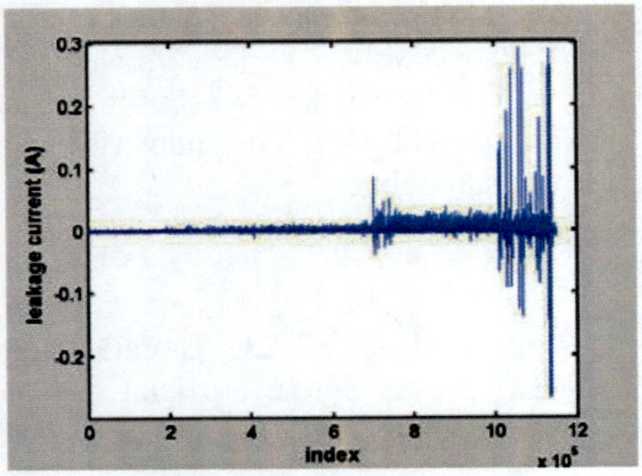


**Fig. 7.** Leakage current (ESDD=0.15)

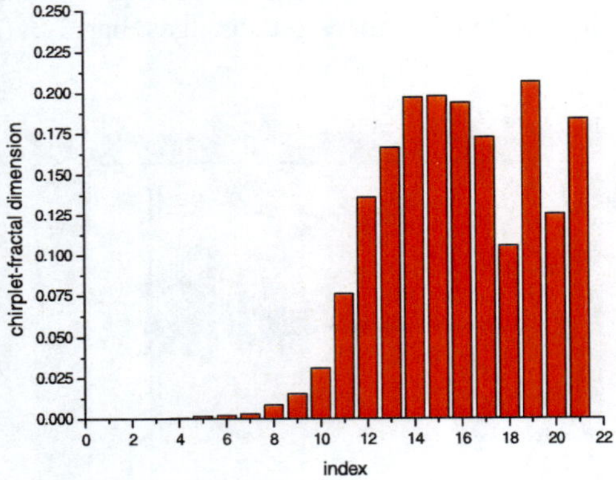


**Fig. 8.** Chirplet-fractal dimension of figure 7

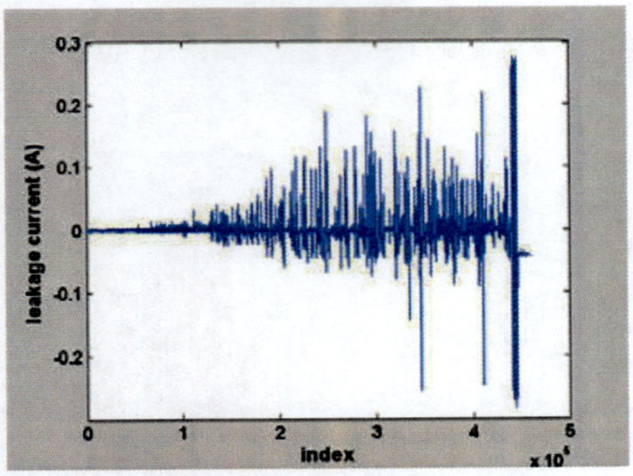


**Fig. 9.** Leakage current (ESDD=0.20)

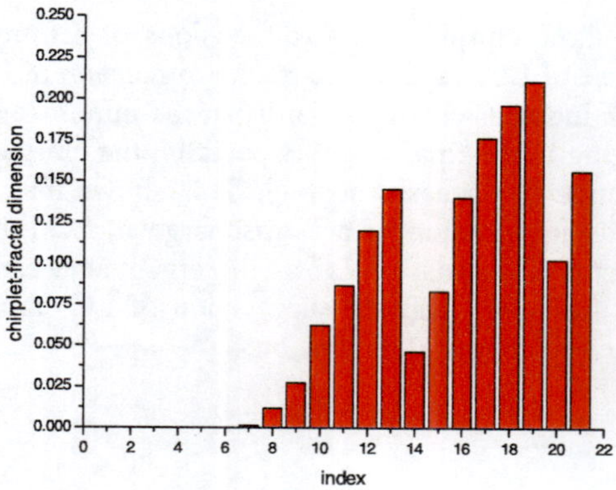


**Fig. 10.** Chirplet-fractal dimension of figure 9

discharge along with sufficiently humidified pollution layer, which leads to large impacts of LC. But owing to a few distinguish among intensities of little arcs and multiple little arcs existing at one time, the impact of arcs discharge counteract each other, which results in some fluctuation of LC[10], and the corresponding chirplet-fractal dimension increases evenly. And then a big arc occurs and develops to the primary arc[11], meanwhile, LC fluctuates obviously[12]. While the primary arc runs through the whole insulator string and the flashover was brought on, LC presents a very large impact, shown as the final impact current close to 0.3A in Figures 5, 7, and 9, and the chirplet-fractal dimension concusses acutely. At the moment of flashover, the chirplet-fractal dimension decreases heavily, shown as the time of 19 in Fig.6, 18, 20 in Fig.9, and 14, 20 in Fig.11.

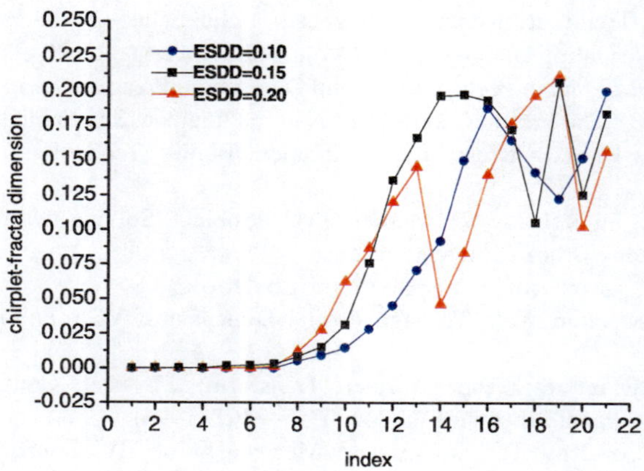


**Fig. 11.** Comparison of chirplet-fractal dimension of LC

The comparison of three chirplet-fractal dimensions of different LC is shown in Fig.10. With the increase of LC, the chirplet-fractal dimension of LC presents obvious increasing trend, which includes changing hardly in the initial stage, increasing in the middle stage, and surging in the final stage. Especially, the chirplet-fractal dimension decreases heavily when LC concusses acutely. This shows that, the chirplet-fractal dimension describes the information of arcs discharge in LC efficiently. Therefore, taking the chirplet-fractal dimension of LC as the eigenvalue can reflect the change law of LC during the flashover, and the supervision of LC can be realized exactly through the change of chirplet-fractal dimension.

## 3 Conclusion

A new idea called chirplet-fractal dimension, using chirplet as the ruler to define fractal dimension, is proposed in this paper, and the algorithm is given too.

The change rule of chirplet-fractal dimension of LC in artificial test of contaminated insulators is studied. The results show that the chirplet-fractal dimension can efficiently describe the change law of LC: it ascends along with the ascending of LC, and falls quickly when LC impacts heavily. Thus, chirplet-fractal dimension is a valid method to depict feature of signals, and it is also a good eigenvalue for flashover discriminations and risk predictions.

## References

1. Qiu, S., Shi, W., Zhang, W.: High Voltage Engineering. Xi'an Jiaotong University Press, Xi'an (1995)
2. Yang, B.: High Voltage Engineering, vol. I. Chongqing University Press, ChongQing (2002)
3. Yan, Z.: New Trend of Insulation Diagnosis - Sidelights on the 3rd Japan-Chinese Conference of Insulation Diagnosis. High Voltage Engineering 21, 90–92 (1995)
4. Mandelbrot, B.B.: Fractal: Form, Chance and Dimension. Freeman, San Francisco (1977)
5. Mandelbrot, B.B.: The Fractal Geometry of Nature. Freeman, San Francisco (1983)
6. Lin, H., Li, Y.: Fractal—Irregularity exploration. Beijing Technology University Press, Beijing (1992)
7. Simanca, S.R., Sutherland, S.: Mathematical Problem Solving with computers. The University at Stony Brook (Summer 2002), `http://www.math.sunysb.edu/~scott/Book331`
8. Kay, B.H.: A Random Walk Through Fractal Dimensions. VCR Publishers, New York (1989)
9. Mann, S., Haykin, S.: The Chirplet Transform: Physical Considerations. IEEE Transactions on Signal Processing 43(11), 2745–2761 (1995)
10. He, B.: Study of Key Technology and Mechanism of HV Contaminated Insulators Flashover for Online Monitoring and Risk Prediction, Ph.D. dissertation, School of Automation, Northwest Polytechnical University, Xi'an, China (July 2006)
11. Alston, L., Zoledzioqski, S.: Growth of Discharges on Polluted Insulation. IEEE Proceedings 10, 1260–1266 (1963)
12. De La, A., Gorur, R.S.: Flashover of Contaminated Nonceramic Outdoor Insulators in a Wet Atmosphere. IEEE Trans. On Dielectrics and Electrical Insulation 5, 814–823 (1998)

# An Supply Chain Network Evolving Model Based on Preferential Attachment of Path and Degree

Peihua Fu and Yanchu Liu

College of Computer and Information Engineering, Zhejiang Gongshang University
Hangzhou, China
fph666@hotmai.com, yanchu_liu@126.com

**Abstract.** Supply chain network is a complex giant system, and its complexity is determined by the structure of network. Empirical studies have been shown: the supply chain network had the scale-free and cluster characteristics, but today supply chain network models could not describe cluster characteristic of real network. This paper introduced a concept of path, and presentes a supply chain network model DPPA, which based on path and degree preferential attachment mechanism. This model depictes the scale-free feature of supply chain network, while also reflecting the cluster characteristic. The average cluster coefficient of this model with 5000 vertices in the network reached 0.5754, which can be controlled by the adjusting parameters of DPPA model.

**Keywords:** supply chain network, complex network, path and degree, preferential attachment, evolving model.

## 1 Introduction

Suppliers, manufacturers and distributors are interdependent in strategic, resources and capacity, and constitute the supply-product-distribution network, which is the supply chain network [1]. Supply chain network is staggered by a number of supply chains, whose internal exists wide range of complex material, information and financial flow, is a complex adaptive system [2]. Because of its ability to reveal the hidden laws of large-scale economic and social complex systems, complex network theory rapidly became one of these researchers' effective tool and an international frontier [3-4].

At present, using complex network theory to study the supply chain network has attracted many scholars' interestes. They have studied the information amplification effect, material supply network, scale-free distribution viability and robustness of supply chain network, but basic research on supply chain network evolution model is few [5-8]. In paper [9-11] they have proposed the local-world model and fitness model to simulate the supply chain network evolution, but there were a large gap between model and real supply chain network. Compared with the current complex network models, supply chain network has its own characteristics: cluster, hierarchy and dynamic. Therefore, the supply chain network evolution model can be a good way to study its mechanism and macro-behavior, and this has a good reference for such a complex system. Therefore, we need a model to simulate real supply chain networks vividly.

## 2 Supply Chain Network

Supply chain network is an open dynamic complex networks [12], including a large number of member firms which can be abstracted as a vertex, can be described as a graph. If the volume of transaction between two member firms is over a certain threshold, the edge between two vertices are connected. Today in the rapid development of global economic situation, supply chain networks have a large number of members and complex relationships between them, performancing the following characteristics:

Dynamics. These member firms in supply chain network are linked by logistics, information and capital flows, and their relationships are uncertain in the changing market environment. The network have to keep updated dynamically to meet the changing market demands, so there are some member firms to exit while other new member firms to join in, thus the supply chain network is dynamics open.

Complexity. Supply chain network has many types of member firms, which have their own corporate culture and value preference. They make decisions according to their preferences, but should be interacted with each other while impacted by complex market environment. And relationships between member firms are complex, because it is decided by external and internal environment.

Cluster. By the development of supply chain network, some core firms must appears, which is surrounded as star-shaped by other member firms. Of course, these core firms are changing because of the supply chain network has metabolic mechanism.

Supply chain network has scale-free and small-world characteristics. Tsutomu Nakano have studied a supply chain network in Tokyo with over 8000 member firms, and found that the network had a power-law distribution and preferential attachment characteristics [15]. As be shown in Figure 1, there was a power-law in the in-degree distribution. Li [16] have studied the competition network of automotive components enterprises in China, which had 9298 member firms, and the Average-Path-Length(APL) was 2.476 and the Average-Cluster-Coefficient(ACC) was 0.843.

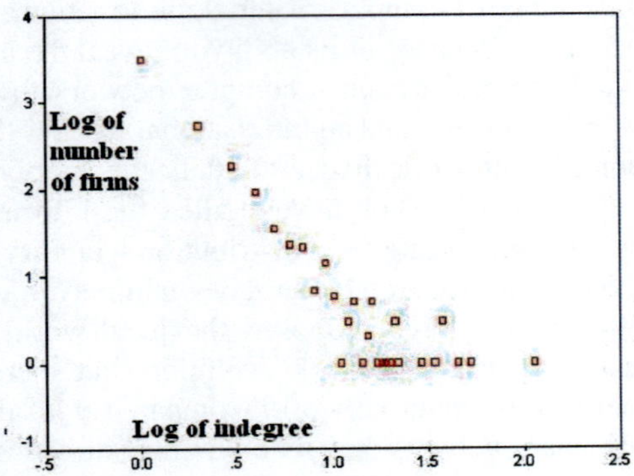


**Fig. 1.** A Power Law in-degree distribution of Tokyo

## 3 Supply Chain Network Model

A real supply chain network can be abstracted as a graph $G = (V, E)$, consisted by a vertex set V and an edge set E. The number of vertex is denoted as $N = |V|$, and the number of edge is denoted as $M = |E|$, while one edge of set E are corresponded with one vertex of set V. Matrix A represents the adjacency matrix of supply chain network, and the value of matrix element $a_{ij}$ represents the trade volum between two member firms. No trade happen between member firms and no edge connected between two vertexes when the matrix element $a_{ij} = 0$. Take into account the mutual of trade, where the supply chain network is undirected network. This paper measures the supply chain network in the topology structure based on the complex network theory.

The degree of vertex $i$ is denoted as $k_i$, that is the number of edges connect with vertex $i$, and the degree distribution of network is represented by the distribution function $P(k)$. The supply chain network has the power-law characteristic as other real networks [13], even in the supply chain networks [15-16]. The APL is the average distance of any two vertices in the network. So the smaller the APL is, the faster the flows in the supply chain network.

The clustering coefficient measures the degree of the network group through quantifying the concept of neighbourhood in the graph. The clustering coefficient of supply chain network is the Average-clustering-coefficient(ACC) of all the vertices, which can show the degree of concentration of the network. Empirical studies have been shown: the real supply chain networks have big ACC and small APL, that is the cluster and small-world characteristics.

In the traditional BA model, there are two characteristics of growth and preferential attachment, and the probability to connect new vertices based on the degree [13]. Some scholars have used the complex network model to simulate the supply chain network model. Paper [9][11] have proposed supply chain network model based on the Local-World model, and paper [10] introduced the linear combination of APL and risk measure as the probability of preferential attachment. These studies can delineate the scale-free feature of supply chain network, but with small ACC, which is common in real supply chain networks. So this paper introduce DPPA model, whose preferential attachment based on path and degree.

In real supply chain network, the parameter for member firms to choose to connect is not only the vertex degree. For example, an automotive components manufacturer Firm A, who locates in Hangzhou, plan to join the supply chain network of automotive components. Following the mechanism of BA model, Firm A firstly have to trade with the world largest automobile manufacturer (Toyota). However, this is not true. Because Firm A want to become a Toyota supplier must have adequate technology base and product advantages, which need to spend a lot of cost and time. There, Firm A first select the object to trade with should have paths to get it.

We introduce a new parameter $p_i$, that is the probability of the new vertex connect vertex $i$. If there is a vertex connects new vertex and vertex $i$, then $p_i = 1$. Otherwise, $p_i = 0$. Priority in the supply chain network connection, not only to consider the degree to the value of the node, but also consider whether it is connected to the path, you may have a very large trade costs without thought of path, more losses than gains. Therefore, we need to form an evolving model based on the concept of math, following is the evolving model of supply chain network (DPPA model):

- Initial: Begins at a network with $m_0$ vertex and full connected at $\mathrm{T}=0$ moment. Each time interval $t$ to add a new vertex and connect to the vertices already existed.
- Preferential attachment: The probability of new vertex connects exist vertex $i$ is $\Pi_i$, and $\Pi_i = \alpha \frac{k_i}{\sum_j k_j} + (1-\alpha)\frac{p_i}{\sum_j p_j} (0 \le \alpha \le 1)$ (1)
- At $\mathrm{T} = \mathrm{n}t$ moment, generates a supply chain network with $(m_0 + n)$ vertices.

In the DPPA model, the parameter $\alpha$ is a constant. When $\alpha = 1$, it is a BA model.

## 4 Numerical Simulation Analysis

Taking the above DPPA model to simulate a supply chain network with 5000 vertices, which begins at $m_0 = 5$ and $m = 3$. That it began at 5 member firms and each has trade with others, and have generated 10 networks because of the value of $\alpha$. As be shown in Figure 2, the degree distribution of network is power-law, that is supply chain network has scale-free characteristic. Figure 3 show the curve of the clustering coefficient varies with the adjusting parameter $\alpha$, and ACC can be determined by it. When $\alpha = 0$, that is the probability of preferential attachment based on the path $p_i$, the ACC of network reached 0.5754. When $\alpha = 1$, the network's preferential attachment is based on degree which is called BA model, have a small value of ACC. In Figure 4, obviously be seen that the preferential attachment based on path have big ACC, that is the combination of path and degree can simulate the cluster characteristic of supply chain networks.

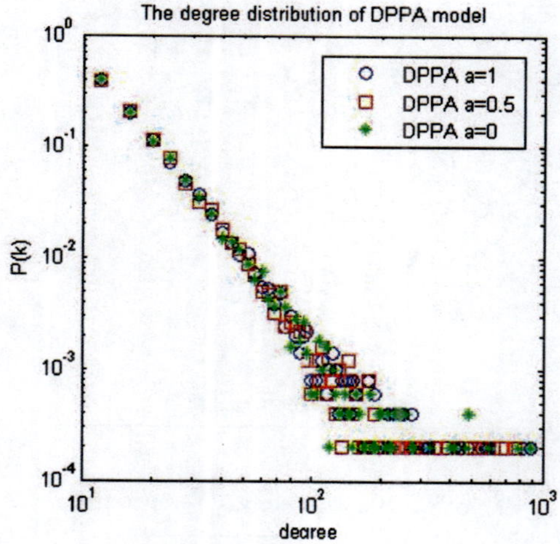


**Fig. 2.** Degree Distribution of DPPA

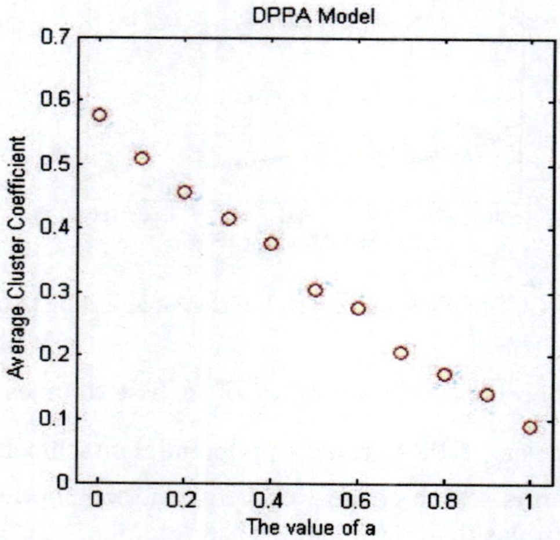


**Fig. 3.** ACC of DPPA

The ACC, that is average clustering coefficient of network, is an important measured parameter in supply chain network.

$$ACC = \frac{1}{N}\sum_{i=1}^{N} C_i = \frac{2}{N}\sum_{i=1}^{N}\frac{T_i}{k_i(k_i - 1)} \tag{2}$$

Here $\mathrm{T}_i$ is the neighborhood triples number of vertex $i$. Suppose the degree distribution $\mathrm{P}(k)$ and size $\mathrm{N}$ of DPPA is the same as the BA model, so $\mathrm{M}_i = \mathrm{P}(k) * \mathrm{N}$ and $\mathrm{D}$ is the biggest degree number.

$$ACC = \frac{2}{N}\sum_{i \in D} M_i \sum_{j \in M_i} T_{ij}\left(\frac{1}{k_i - 1} - \frac{1}{k_i}\right)$$
$$= \frac{2}{N}\sum_{i \in D} \frac{M_i}{k_i(k_i - 1)} \sum_{j \in M_i} T_{ij} \tag{3}$$
$$= \frac{2}{N}\sum_{i \in D} \frac{P(k)N}{k_i^2} \sum_{j \in M_i} T_{ij}$$

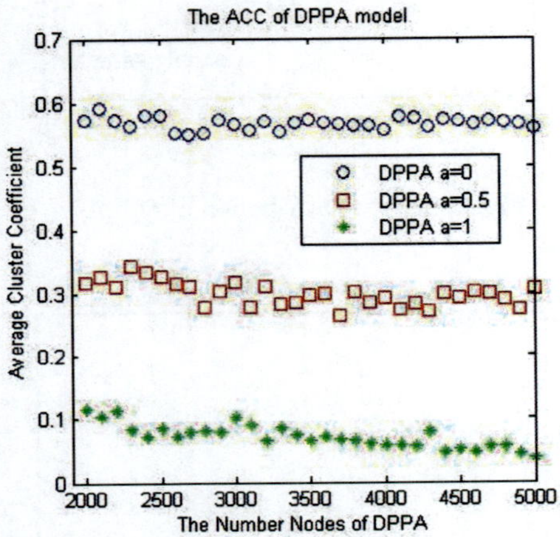


**Fig. 4.** The ACC of DPPA models with different adjusting parameter

Here $T_{ij}$ is the neighborhood triples number of the j-st vertexes in the group with their degree are $k_i$. Because DPPA model's preferential attachment is based on path and degree, and the path is search connecting vertex in neighborhood, so the DPPA model generates more triples than BA model. Therefore, the ACC of DPPA model is bigger than BA model, that is the DPPA can generate network with cluster characteristic.

## 5 Conclusion

The DPPA model could simulate the supply chain network and evolving mechanism . At the same time, it is able to simulate a network with cluster characteristic, in line with the reality of the basic characteristics of supply chain network, which further study of network control and coordination of new ideas and theoretical basis. Because it is difficult to get the data of real supply chain network, so further study about the supply chain network model should be given attention. Next we will research the importance of vertices and edge in the supply chain network, and the factors such as remove of vertex and edge can impact the evolving behavior of supply chain network.

**Acknowledgement.** This research was supported by the Innovation Fund of Zhejiang Gongshang University under Grant 1130XJ1509046, Supply chain network optimization study based on complex network theory.

## References

1. Fu, R.L., Michael, J.S.: Reengineering the Order Fulfillment Process in Supply Chain Networks. The International Journal of Flexible Manufacturing Systems 10, 197–229 (1998)
2. Amit, S., Soundar, K., Mark, G., et al.: Supply chain networks: a complex adaptive systems perspective. International Journal of Production Research 43(20), 4235–4265 (2005)
3. Barabasi, A.L., Bonabeau, E.: Scale - free networks. Scientific American 5(1), 60–69 (2003)
4. Newman, M.E.J.: The structure and function of complex networks. SIAM Review 45(2), 167–256 (2003)
5. Helbing, D.: Information and material flows in complex networks. Physica A 363(1), 6–16 (2006)
6. Kuhnert, C., Helbing, D.: Scaling laws in urban supply networks. Physica A 363(1), 89–95 (2006)
7. Thadakamalla, H.P., Raghavan, U.N., Kumara, S., et al.: Survivability of multiagent based supply networks: a topological perspective. Intelligent Systems and Their Applications 19(5), 24–31 (2004)
8. Snyder, L.V., Daskin, M.S.: Reliability models for facility location: The expected failure costcase. Transportation Science 39(3), 400–416 (2005)
9. Zhang, J.Y., Xu, J.Q.: Study on Complex Network Model for Adaptive Supply Chains. Chinese Journal of Management Science 17(2), 76–79 (2009)
10. Yu, H.S., Zhao, L.D., Lai, X.H.: Trade Quantity-Based Supply Chains Networks Evolving Model. Chinese Journal of Management 6(2), 187–191 (2009)
11. Qi, X., Yanjun, L., Tiejun, W.: A local2world network model based on inter-node correlation degree. Physica A 378, 561–572 (2007)
12. Harland, C.M., Lamming, R.C.: Developing the Concept of Supply Strategy. International Journal of Operations and Productions Management 19(6), 650–673 (1999)
13. Barabasi, A.L., Albert, R.: Emergence of scaling in random network. Science 286(5439), 509–512 (1999)
14. Holme, P., Kim, B.J., Yoon, C.N., et al.: Attack vulnerability of complex networks. Phys. Rev. E 65(5), 056109 (2002)
15. Nakano, T., White, D.R.: Network Structures in Industrial Pricing: The Effect of Emergence Roles in Tokyo Supplier-Chain Hierarchies. Structure and Dynamics 2(3), 130–154 (2007)
16. Li, J.M., Zhang, N.: Analysis to Competition Network Topological Structure of Automotive Components Enterprises in China. Complex Systems and Complexity Science 5(2), 72–78 (2008)

# Researches on Content Delivery Policies for Interactive Video Network

Dao-yi Wang[1], Wenan Zhou[2], and Yuanan Liu[1]

[1] School of Electronic Engineering, Beijing University of Posts and Telecommunications
Beijing, P.R. China
wangdy1972@263.net, yuliu@bupt.edu.cn

[2] School of Computer Science & Tech. Beijing University of Posts and Telecommunications
Beijing, P.R. China
zhouwa@bupt.edu.cn

**Abstract.** Interactive video network, such as Video-on-Demand (VoD) or IPTV system, is the primary part of the Next Generation Network (NGN), in which the application server resources are typically shared by multiple end users. Content delivery policy is one of the most important issues in such kind of network due to the huge storage and bandwidth required by video streaming. While the deployment of commercial interactive video network, we developed a series of content delivery policies, among them the mathematical definitions of object popularity, the measure methods for the inequality of object popularity, and a Dynamic-Resource-Allocation (DRA) based flash crowd mitigation algorithm are presented in this paper. The simulation results show that our researches are helpful to deploy QoS guaranteed video network in a cost-effective way.

**Keywords:** Interactive video network, DRA, flash crowd mitigation, content popularity, cable network.

## 1 Introduction

Interactive video network is one of the most important systems in the Next Generation Network (NGN), in which the application server resources are typically shared by a large number of end users. There are two main factors making content delivery policies become one of the most important issues in such kind of network. Firstly, compared to the traditional voice and data communication, video streaming sessions need tens to hundreds times of application server storage capacity and network transport bandwidth. Secondly, the interactive video networks are typically deployed to meet the expected peak demands based on normal time-of-day effect. Such an over-provisioning approach is not robust to flash crowd scenarios, where the streaming load is increased much higher than the expected peak load. For instance, on September 11, 2001, the load on the CNN website doubled every 7 minutes to reach a peak of almost 20 times the normal load [1].

We developed a systematic method for researches on content delivery policies and several key aspects are presented in this paper, including the definition of objects popularity, the measurement methods for the inequality of the object popularity, and a Dynamic-Resource-Allocation (DRA) based flash crowd mitigation algorithm. The simulation results show that our researches are helpful to deploy QoS guaranteed video network in a cost-effective way.

The remainder of this paper is organized as follows. In section 2, we provide the background of the related work. In section 3, the definition of object popularity and its inequality are brought forward. In section 4, the DRA based flash crowd mitigation algorithm is proposed, and the simulation results are given as well. In section 5, the conclusion of this paper is given.

## 2 Related Work

The inequality of the object popularity is the basic principle of the content delivery mechanism. In the related work [2-6], the object popularity distribution characteristics were observed. However, there are still lack of the standard object popularity definition and the quantitive measurement method for the popularity inequality. In this paper the mathematic definitions of object popularity and several popularity inequality measurement methods are proposed, which are helpful for the research and development of the interactive video systems.

Due to the unpredictability of flash crowd, the systems designed based on normal peak load are likely to fail under sustained overload conditions. In recent years, several approaches for handling unexpected workload surges have been proposed as follows:

- To use the techniques of service differentiation, such as admission control [7].
- To borrow resources from other applications [8, 9] based on the dynamic resources allocation.
- To borrow resources from other systems based on overlay network techniques, such as invoking P2P mode when C/S mode fails to fulfill the increasing requests [10].

Related work [7-10] focused on website applications, and the general idea can be utilized in our interactive video system, including admission control and overlay architecture. A DRA based flash crowd mitigation algorithm for the multicast and unicast hybrid cable network is developed in this paper, which is different from the related work in several aspects. Firstly, the website applications provider can only control the server-side resources, and the cable operator network can control both server-side and network-side resources. Secondly, our work here brings forward a DRA mechanism for commercialized interactive video services by taking into account the system robustness, cost-effectiveness and quality of service.

## 3 Object Popularity and Its Inequality

### 3.1 Definition

The set of the content objects stored in the video server cluster can be denoted by

$$C = \{c_i \mid i = 1,2,\ldots,N\}. \tag{1}$$

The elements in $C$ are ordered according to its popularity in descending order. The access probability of $c_i$ is denoted by $p_i$. Obviously, $p_i \geq p_j$ for all $1 \leq i < j \leq N$.

Therefore, the object popularity distribution of $C$ can be defined as

$$P = \{p_i \mid i = 1,2,...,N\} \tag{2}$$

The object concentrated-popularity distribution of $C$ can be defined as

$$s_i = P(X \leq i) = \sum_{j=1}^{i} p_j, \quad i = 1,2,...,N \tag{3}$$

$s_i$ can be used to measure the total access probability of the $i$ most popular objects.

As in (3), let $q = i/N$, then we get

$$f(q) = s_{qN} = \sum_{j=1}^{qN} p_j, q = \frac{1}{N}, \frac{2}{N}, ..., 1. \tag{4}$$

Obviously, $f(q)$ is the total access probability in percentage of the most $q$ part popular objects, which is called percentile-popularity distribution in this paper.

Let x-axis and y-axis denote $q$ and $f(q)$ respectively, we can get a curve, denoted by L(f(q), q), which is called Lorenz curve of object popularity distribution. As shown in figure 1, the curve L(f(q), q) is for the typical content objects set and L($f_E$(q), q) is for the perfect equality distribution.

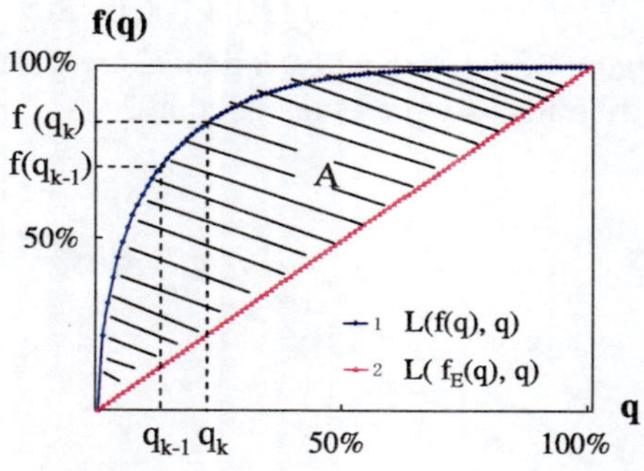


**Fig. 1.** Lorenz curve of object popularity distribution

The area enclosed by L(f(q), q) and L($f_E$(q), q), labeled by *A* in figure 1, can be used to measure the inequality of the object popularity. By the theories on statistics, we can get the Gini coefficient (*G*), Coefficient of variation (*CV*), and Theil index (*TI*). These variables are given as follows:

$$G = 2A = 2\left(\int_0^1 f(q)dq - 1/2\right) \tag{5}$$

$$CV = \frac{\sigma}{\mu} = \frac{\sqrt{\frac{1}{N}\sum_{i=1}^{N}(p_i - \frac{1}{N}\sum_{i=1}^{N}(p_i))^2}}{\frac{1}{N}\sum_{i=1}^{N}(p_i)} \tag{6}$$

$$TI(X) = -\sum_{i=1}^{N}\frac{1}{N}\log\frac{1}{N} - (-\sum_{i=1}^{N} p_i \log p_i) \tag{7}$$

By the method proposed by Brown M [11], $G$ can be got by the following equation.

$$G_b = 1 - \sum_{k=1}^{N}(q_k - q_{k-1})(2 - f(q_k) - f(q_{k-1})) \tag{8}$$

### 3.2 Simulation and Analysis

In this section, we use Zipf-like [12] law to verify our proposed methods for object popularity and its inequality. The popularity distribution of the content object set following Zipf-like law, denoted by $Zipf(\alpha, N)$, is given as follows:

$$P(i;\alpha,N) = \frac{1/i^{\alpha}}{\sum_{n=1}^{N} 1/n^{\alpha}} \tag{9}$$

The Lorenz curve of $Zipf(0.6,100)$, $Zipf(0.8,100)$, $Zipf(1,100)$ and Perfect Equality are shown in figure 2, the inequality variables are given in table 1, and the percentile-popularity distributions are presented in table 2.

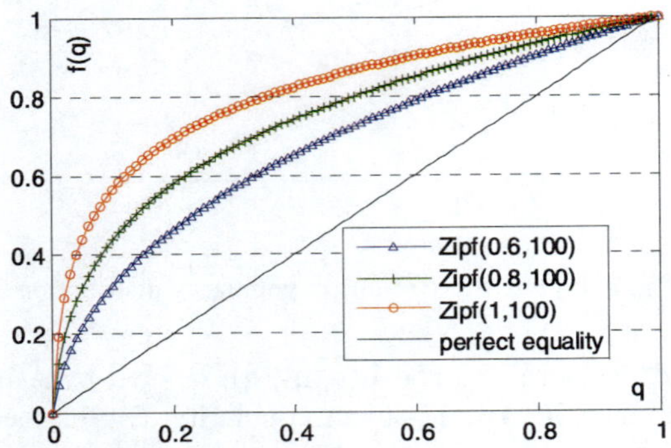


**Fig. 2.** Lorenze curve of Zipf-like sample

**Table 1.** Popularity Inequality

| Distribution Sample | Popularity Inequality | | |
|---|---|---|---|
| | *G* | *CV* | *TI* |
| Zipf(0.6,100) | 0.355 | 0.937 | 0.115 |
| Zipf(0.8,100) | 0.493 | 1.515 | 0.234 |
| Zipf(1,100) | 0.624 | 2.253 | 0.401 |
| Equality | 0 | 0 | 0 |

**Table 2.** Percentile popularity Distribution

| Distribution Sample | Percentile Popularity Distribution | | |
|---|---|---|---|
| | *10%* | *20%* | *30%* |
| Zipf(0.6,100) | 32% | 46% | 51% |
| Zipf(0.8,100) | 44% | 58% | 61% |
| Zipf(1,100) | 56% | 69% | 72% |
| Equality | 10% | 20% | 30% |

From figure 2, table 1 and table 2, we can see that our proposed method can measure the object popularity and its inequality correctly.

## 4 Dra Based Flash Crowd Mitigation Algorithm

### 4.1 Flash Crowd Model

The event that catches the attention of a large number of people will trigger flash crowd, which causes unexpected and overloading surge of traffic. According to related work [8, 10], only a small part of the objects are involved under flash crowd condition. In this section we use the following parameters to define a flash crowd, and the flash crowd model is given by figure 3.

The normal traffic rate ($\lambda_0$): $\lambda_0$=10 requests/min.

- The flash crowd traffic rate ($\lambda_f$): $\lambda_f=(\lambda-\lambda_0)$ is the traffic rate that exceeds $\lambda_0$.
- The peak traffic rate ($\lambda_p$): $\lambda_p=16\lambda_0=160$ requests/min.
- Flash crowd start time: $T_0=24\Delta T=480$ min, i.e. 8:00 AM.
- Flash crowd peak start time: $T_1=28\Delta T=560$ min, i.e. 9:20 AM.
- Flash crowd peak end time: $T_2=60\Delta T=1200$ min, i.e. 20:00 PM.
- Flash crowd end time: $T_3=64\Delta T=1280$ min, i.e. 21:20 PM.

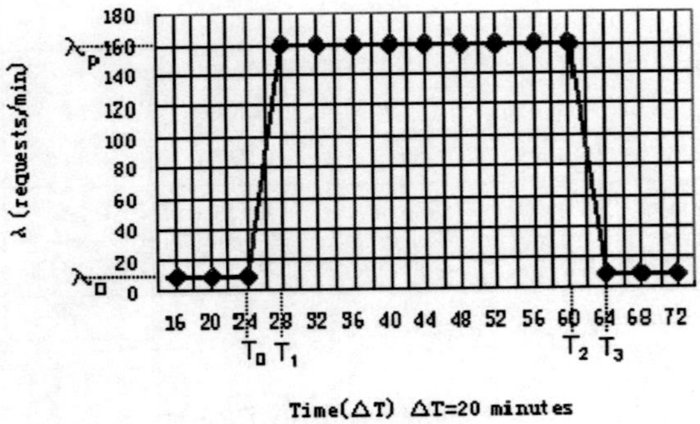


**Fig. 3.** Flash crowd model

As shown in figure 3, $\lambda_0$ is for the video objects from 1 to 1000, and $\lambda_f$ is for the 10 most popular objects, i.e. from 1 to 10.

### 4.2 DRA Based Flash Crowd Mitigation Algorithm

Let $T_r$ denote the request times of a popular video object during its play duration $T_L$, which is the time period of 20 minutes in this paper. Let $T_w$ denote the maximum number of waiting minutes allowed. Let $M_n$ denote the number of multicast streaming sessions used to serve all the requests of a video object under flash crowd condition.

*1) Flash crowd detection.* The central node keeps tracing the access characteristic of every video object and calculates $T_r$. The flash crowd condition of a popular object can be recognized if

$$T_r > T_L / T_w \tag{10}$$

Then, the transition from flash crowd status to normal traffic load condition can be recognized if $T_r$ decreases to be less than $T_L / T_w$.

2) *DRA algorithm.* Under flash crowd condition, the central node sets up a group of multicast streaming sessions for the flash crowd object, which can be denoted by $(M_1, M_{2,} ..., M_n)$. The play time offset between $M_i$ and $M_{i+1}$ is $T_w$. The central node redirects all the incoming requests for the flash crowd video object to this group of multicast streaming. Therefore, all the requests for the flash crowd video can share the limited number of streaming resources, and the number is given by the following equation.

$$M_n = T_L / T_w \tag{11}$$

When the flash crowd event ends, all the multicast streaming resources will be released.

*3) QoS guarantee.* Obviously, the maximum waiting time for the request to the flash crowd objects is $T_w$. For trick mode operation, such as Fast Forward and Fast Rewind, the central node can guide user to jump to another multicast streaming.

In this paper, the number of multicast streaming the central node needs to set up is

$$M_{total} = \sum M_n(i) \quad , \text{i=1, 2, ... , 10.} \tag{12}$$

As in (12), $M_n(i)$ denotes the number of multicast stream for video object labeled by $i$, and $M_{total}$ is only relative to the number of flash crowd objects but not the traffic load. Moreover, the parameter $T_w$ and the threshold of $T_r$ are configurable. This mechanism is helpful to get the perfect balance between the maximum multicast streaming resources and the maximum waiting time their customer can tolerate.

### 4.3 Simulation and Analysis

The simulation environment is given as follows:

- The interactive video network consists of one central node and ten edge nodes. The central node is responsible for session management and content delivery to the edge nodes and the edge nodes are responsible for streaming to the end users [13].
- Flash crowd model is given by figure 3.

The simulation result is shown in figure 4 and figure 5. At the time $T_0$, flash crowd event occurs and the traffic load increases rapidly, which is shown by the time period from 24(ΔT) to 28(ΔT) in figure 4. While increasing of the traffic load, the capacity of the edge node can not handle all the requests, and some of the requests are refused based on call admission control policy. After a small period of time, the system detects that $T_r$ exceeds its threshold and the requests for the flash crowd objects are switched to multicast streaming resources.

At the time $T_2$, flash crowd event ends, and the traffic load drops to normal level gradually, which is shown by the time period from 60(ΔT) to 64(ΔT) in figure 5. When $T_r$ decreases to be less than its threshold, the multicast streaming resources are released, and the following requests to these popular objects are served by setting up interactive video sessions. As shown in figure 5, the dynamical transition between interactive streaming session and multicast session can handle flash crowd traffic smoothly.

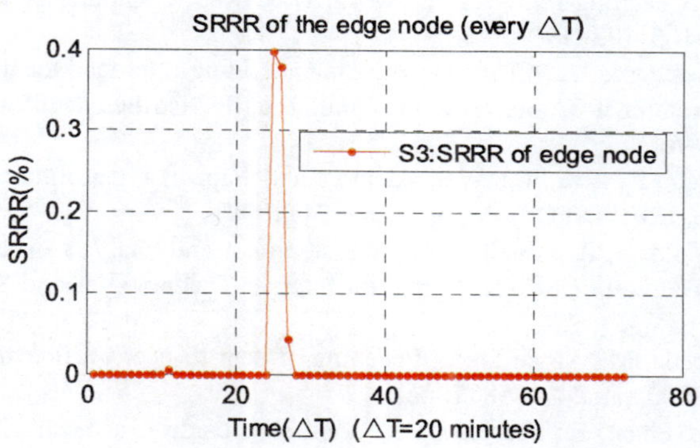


**Fig. 4.** SRRR of the edge node

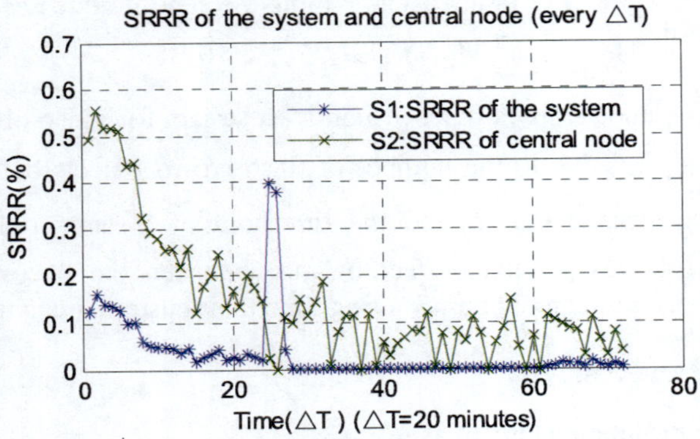


**Fig. 5.** SRRR of the system and central node

## 5 Conclusion

In this paper, two key aspects on content delivery policies for interactive video network are presented. The simulation on content object popularity and its inequality shows that the popularity definition and the inequality measurement methods are helpful for the general researches on content delivery policies. The simulation on DRA based flash crowd mitigation algorithm shows that the proposed algorithm can handle flash crowd event in a robust and cost-effective way. The configurable parameters can help the operator to get the perfect balance between resource cost and QoS level.

**Acknowledgment.** This work is part of the network research and planning of CITIC Network Co. Ltd. and Henan Cable Network Co. Ltd.

## References

1. LeFebvre, W.: CNN.com: facing a world crisis. In: Invited Talk, USENIX Annual Technical Conference (June 2002)
2. Glassman, S.: A caching relay for the world wide web. Computer Networks and ISDN Systems 27(2), 165–173 (1994)
3. Almeida, V., Bestavros, A., Crovella, M.: Characterizing reference locality in the WWW. In: IEEE International Conference in Parallel and Distributed Information Systems, pp. 92–103 (1996)
4. Breslau, L., Cao, P., Fan, L.: Web caching and Zipf-like distributions: evidence and implications. In: IEEE INFOCOM, pp. 126–134 (1999)
5. Chesire, M., Wolman, A., Voelker, G.: Measurement and analysis of a streaming media workload. In: The 2001 USENIX Symp. on Internet Technologies and Systems, pp. 5–12 (2001)
6. Saleh, O., Hefeeda, M.: Modeling and caching of peer-to-peer traffic. In: IEEE Computer Society ICNP 2006, pp. 249–258 (2006)
7. Cherkasova, L., Phaal, P.: Session-based admission control: a mechanism for peak load management of commercial web sites. IEEE Transactions on Computers 51(6), 669–685 (2002)

8. Chandra, A., Shenoy, P.: Effectiveness of dynamic resource allocation for handling internet flash crowds. Technical Report TR03-37, Department of Computer Science. University of Massachusetts Amherst (November 2003)
9. Lassettre, E., Coleman, D., Diao, Y.: Dynamic surge protection: An approach to handling unexpected workload surges with resource actions that have lead times. In: First Workshop on Algorithms and Architectures for Self-Managing Systems, pp. 82–92 (June 2003)
10. Pan, C., Atajanov, M., Hossain, M.B.: FCAN: flash crowds alleviation network using adaptive P2P overlay of cache proxies. IEICE Transactions on Communications E89-B(4), 1119–1126 (2006)
11. Brown, M.: Using Gini-style indices to evaluate the spatial patterns of health practitioners; theoretical considerations and an application based on the Alberta data. Social Science and Medicine 38(9), 1243–1256 (1994)
12. Joos, M.: Review of G.K. Zipf. The psychobiology of language. Language, 196–210 (1936)
13. Wang, D.Y., Liu, Y.A.: An improved time-window-based content caching algorithm. In: IEEE ICPCA 2008, pp. 375–379 (October 2008)

# Study on the Distribution and Computer Simulation of Metro Rail Potential

Yu-qiao Wang, Wei Li, Xue-feng Yang, Guo Ye, Qi-gao Fan, and Li-ping Zhang

College of Mechanical and Electrical Engineering
China University of Mining & Technology
Xuzhou 221008, China
cumtwangyuqiao@163.com

**Abstract.** Through simplifying and limiting the practical metro models, the bilateral power supplying system has been studied by using the resistance network model and a theoretical analytic model of the rail potential distribution in the bilateral power supplying system has been established. Computer simulation of rail potential distribution in different traction current is carried out. The results show that the rail potential is affected by substation pitch, the amount of current taken by locomotive, railway portrait resistor, rail-to-ground transition resistor and other factors, the distribution of rail potential is provided from a macro on the qualitative. This gives theoretical basis on solving high rail potential and how to set the distance of substation. Ultimately it provides a good solution to reduce the stray current corrosion.

**Keywords:** stray current, rail potential, metro.

## 1 Introduction

At present, urban rails(metro, light rail) basically use DC traction power supply of walking rail backflow, which can greatly reduce investment, but railway resistance causes voltage drop when current flows, railway has track potential to the ground. The railway can not be completely insulated to the ground, causing current to flow into the ground by the track and form stray current leakage, this leakage causes serious corrosion to buried metal steel, and high track potential may threaten personal safety. Factors influencing distribution of track potential are complex, we need to analyze the distribution combining with the actual situation.

## 2 Model Building of Track Potential Distribution

Supposing each resistor is uniformly distributed, we can consider the circuit to be linear in the circuit analysis[1]. By superposition principle, complex circuits containing many substations and load can be substituted by several simple circuits, we deduce formula of bilateral power supply system. To simplify the problem, we only study one power area, and without loss of generality, we deduce and simulate situation with drainage net.

Supposing:(1) portrait resistors of walking rail are uniformly distributed;(2) rail-to-drainage net transition resistor and soil resistor are uniformly distributed;(3) drainage net resistors are uniformly distributed;(4)at a certain time, current taken by train finally flows into two substations adjacent to the train;(5) Feeder circuit impedance r can be ignored.

*R* is railway portrait resistor, Ω/km; $R_P$ is drainage net portrait resistor, Ω·km; $R_g$ is rail-to-drainage net transition resistor,Ω/km ; $R_{g1}$ is soil resistance from the drainage net to the ground, Ω·km; *I* is current taken by locomotive, A; $I_1$ is returning current of substation A,A; $I_2$ is returning current of substation B,A;*L* is distance between substation A and substation B, km; $L_1$ is distance between substation A and the train, km.

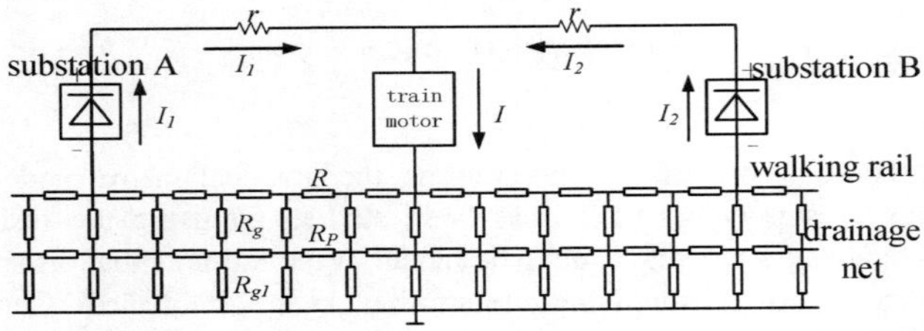


**Fig. 1.** Resistance distribution network of railway-buried mental-ground structure

According to hypothesis (1) (2) (3) (4),

$$I_1 = \frac{L - L_1}{L} I \tag{1}$$

$$I_2 = \frac{L_1}{L} I \tag{2}$$

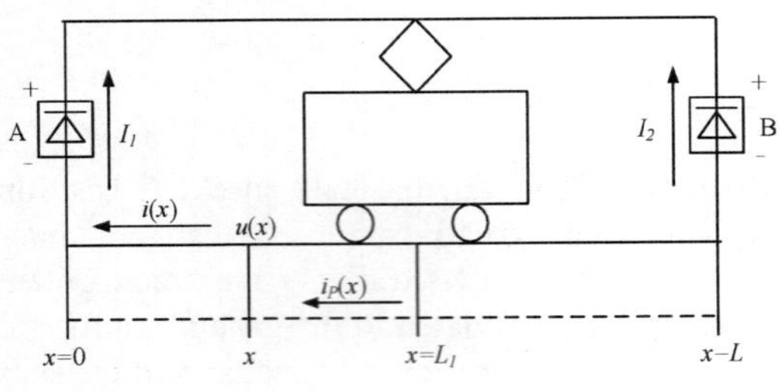


**Fig. 2.** Scheme diagram of the train power supply

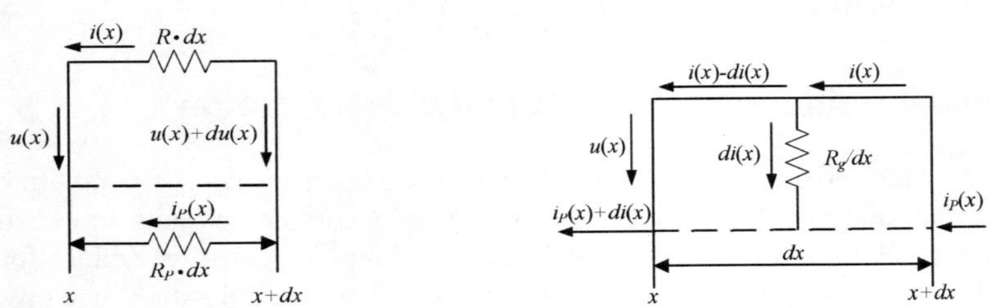


(a) Figure of voltage node

(b) Figure of current node

**Fig. 3.** Figure of voltage node and current node at the point of 3x

Fig.2 shows simplified power supply model diagram. $u(x)$ is voltage of walking rail at $x$ ,V; $i(x)$ is current of walking rail at $x$, A; $i_s(x)$ is stray current at x ,A; $i_P(x)$ is current flowing on the drainage net at $x$, A; $x$=0 at substation A.

For 0～$L_1$, Fig.3 (a) shows voltage situation at $x$, Fig.3 (b) shows current situation at $x$.

According to kirchhoff's first law $\left(\sum U=0\right)$, and Fig.3 (a) we get:

$$i(x)\cdot R\cdot dx+u(x)-i_p(x)\cdot R_p\cdot dx-[u(x)+du(x)]=0 \tag{3}$$

Simplifying equation (3) we get:

$$\frac{du(x)}{dx}=i(x)\cdot R-i_p(x)\cdot R_p \tag{4}$$

According to Fig.3 (b): $u(x)=di(x)\cdot R_g/dx$ that is:

$$di(x)=\frac{u(x)}{R_g}\cdot dx \tag{5}$$

According to hypothesis(4): $I_1=i(x)+i_P(x)$, that is :

$$i_p(x)=I_1-i(x) \tag{6}$$

Substituting equation (6) into equation (4) we get:

$$\frac{du(x)}{dx}=i(x)\cdot(R+R_P)-I_1\cdot R_P \tag{7}$$

From the deviration of x in equation (7), we get:

$$\frac{d^2u(x)}{dx^2}=\frac{di(x)}{dx}\cdot(R+R_p) \tag{8}$$

Substituting equation (6) into equation (8) we get:

$$\frac{d^2u(x)}{dx^2}-\frac{R+R_p}{R_g}\cdot u(x)=0 \tag{9}$$

Let be $\alpha=\sqrt{\dfrac{R+R_p}{R_g}}$, $\beta=\dfrac{R}{R+R_p}$, $\gamma=\dfrac{R_p}{R+R_p}$, from equation (9) we get:

$$\frac{d^2u(x)}{dx^2}-\alpha^2u(x)=0 \tag{10}$$

General solution of equation (10) is

$$u(x)=A\operatorname{ch}\alpha x+B\operatorname{sh}\alpha x \tag{11}$$

A, B are undetermined constants, from the deviration of x in equation (11), we get

$$\frac{du(x)}{dx}=\alpha\left(A\operatorname{sh}\alpha x+B\operatorname{ch}\alpha x\right) \tag{12}$$

Let be $Z=\sqrt{R_g\cdot(R+R_p)}$, Substituting equation (12) into equation (7) we get:

$$i(x)=\frac{A}{Z}\text{sh}\,\alpha x+\frac{B}{Z}\text{ch}\,\alpha x+I_1\gamma \tag{13}$$

According to hypothesis(4) we know, train returning current flow back to two adjacent substations respectively, and current flowing back to substation A is $I_1$ ; current flowing back to substation B is $I_2$ .At the train, that is at $x=L_1$ ,we get:

$$\lim_{x\to L_1^-} i(x)=I_1 \qquad \lim_{x\to L_1^+} i(x)=I_2$$

At $0\sim L_1$, under the boundary condition that $x=0$ , $i(x)=I_1$ ;When $x=L_1^-, i(x)=I_1$.We substitute the two conditions into equation (13):

$$A=-\frac{R}{\alpha}\cdot I_1\cdot\text{th}\frac{\alpha L_1}{2} \qquad B=\frac{R}{\alpha}\cdot I_1$$

Substituting A, B into equation (11) and equation (13) we get:

$$u(x)=-\frac{R}{\alpha}I_1\,\text{th}\frac{\alpha L_1}{2}\text{ch}\,\alpha x+\frac{R}{\alpha}I_1\,\text{sh}\,\alpha x \tag{14}$$

$$i(x)=-I_1\beta\,\text{th}\frac{\alpha L_1}{2}\cdot\text{sh}\,\alpha x+I_1\beta\cdot\text{ch}\,\alpha x+I_1\gamma \tag{15}$$

Similarly, for $L_1\sim L$, we get:

$$u^{'}(x)=A^{'}\,\text{ch}\,\alpha x+B^{'}\,\text{sh}\,\alpha x \tag{16}$$

$$i^{'}(x)=\frac{A^{'}}{Z}\text{sh}\,\alpha x+\frac{B^{'}}{Z}\text{ch}\,\alpha x+I_2\gamma \tag{17}$$

$A^{'}$ 、 $B^{'}$ are undetermined constants, when $x=L_1^+$ , $i(x)=I_2$ ;When $x=L, i(x)=I_2$.according to these two boundary conditions we get:

$$A^{'}=-\frac{R}{\alpha}\cdot I_2\cdot\frac{\text{ch}\,\alpha L_1-\text{ch}\,\alpha L}{\text{sh}\,\alpha(L_1-L)} \qquad B^{'}=\frac{R}{\alpha}\cdot I_2\cdot\frac{\text{sh}\,\alpha L_1-\text{sh}\,\alpha L}{\text{sh}\,\alpha(L_1-L)}$$

Substituting $A^{'}$ $B^{'}$ into equation (16), we get:

$$u^{'}(x)=-\frac{R}{\alpha}\cdot I_2\cdot\frac{\text{ch}\,\alpha(L-x)-\text{ch}\,\alpha(x-L_1)}{\text{sh}\,\alpha(L_1-L)} \tag{18}$$

Combining the equations above, and simplifying equation (14) and equation (18), we get metro rail potential distribution of bilateral power supplying system:

$$u(x)=\begin{cases}\dfrac{R}{\alpha}\cdot\dfrac{I_1}{ch\dfrac{\alpha L_1}{2}}\cdot sh\alpha\left(x-\dfrac{L_1}{2}\right), & 0\le x\le L_1\\[2ex] \dfrac{R}{\alpha}\cdot\dfrac{I_2\left[ch\alpha(x-L)-ch\alpha(x-L_1)\right]}{sh\alpha(L-L_1)}, & L_1<x\le L\end{cases} \tag{19}$$

# 3 Computer Simulation of Metro Rail Potential Distribution

According to equation (19), supposing railway portrait resistor R=0.026Ω/km, the drainage net portrait resistor $R_P$=0.01Ω/km, $R_g$=15Ω·km, we use MATLAB to simulate it, and to find the change law of potential distribution with other parameters in different traction current.

## 3.1 Influence of Substation Pitch on Rail Potential Distribution

Supposing the train is running at the distance of 0.8km from substation A, when the amount of current taken by locomotive is changing within the range of 1000A~4000A, we simulate rail potential changing with substation pitch. Fig. 4 shows rail potential changing situation when substation pitch L is 1.2km and 2.4km.

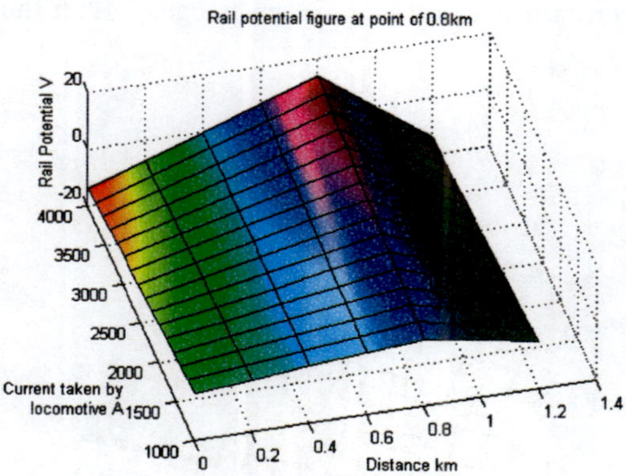


(a) Substation pitch $L$=1.2km

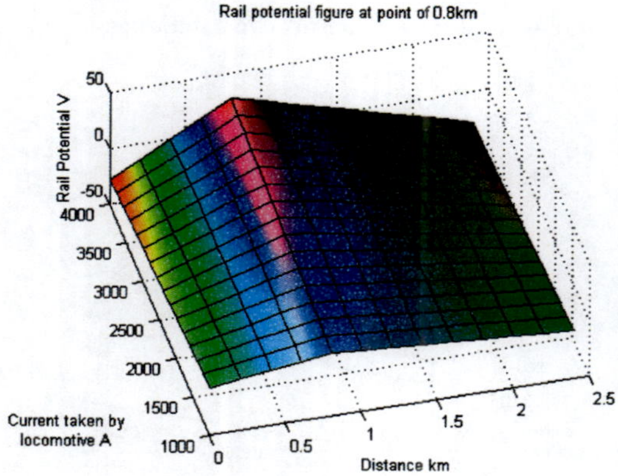


(b) Substation pitch $L$=2.4km

**Fig. 4.** Rail potential distribution changing with substation pitch in different traction current

From Fig. 4 we know, maximum rail potential in positive direction is at the train, and increases with substation pitch, maximum rail potential in negative direction is at substation return point, and the absolute values of positive maximum and negative maximum are the same. At the middle position between the train and the substation, rail potential is zero. Rail potential increases with the increase of current taken by locomotive, and the larger the current taken by locomotive, the higher the rail potential changing rate.

### 3.2 Influence of Railway Portrait Resistor on Rail Potential Distribution

Supposing substation pitch L=2.0 km, the train is running at the distance of 0.8km from substation A, when the amount of current taken by locomotive is changing in 1000A~4000A,we simulate rail potential changing with railway portrait resistor. Fig. 5 shows the rail potential situations of two points at the distance of 0.6km and 1.5km from substation A, when railway portrait resistor changes within the range of 0.01~0.04Ω/km,

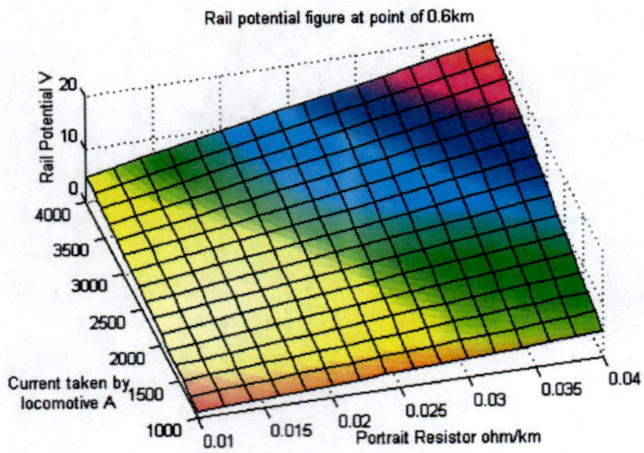


(a) At the distance of x=0.6km from substation A

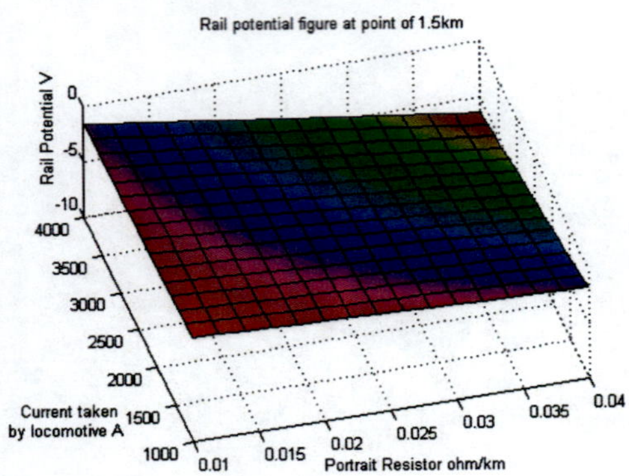


(b) At the distance of x=1.5km from substation A

**Fig. 5.** Rail potential distribution changing with portrait resistor at different currents

From Fig. 5 we know, at the same current taken by locomotive, the larger the railway portrait resistor at point less than locomotive, the higher the rail potential; The larger the railway portrait resistor at point more than locomotive, the smaller the rail potential, the rising or decreasing slope of rail potential becomes larger and larger with currents taken by locomotive increasing, so the influence of railway portrait resistor on rail potential becomes greater and greater with current taken by locomotive increasing.

### 3.3 Influence of Rail-to-Ground Transition Resistor on Rail Potential Distribution

Supposing substation pitch $L=2.0$ km, the train is running at the distance of 0.8km from substation A, when the amount of current taken by locomotive is changing in 1000A～4000A, we simulate rail potential changing with rail-to-ground transition resistor. Fig.6 shows the rail potential situations of two points at the distance of 0.6km and 1.5km from substation A, when rail-to-ground transition resistor changes within the range of 0.1～20Ω·km.

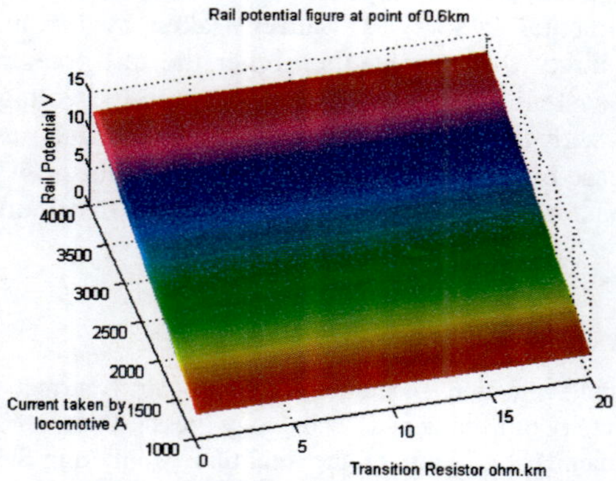


(a) At the distance of $x$=0.6km from substation A

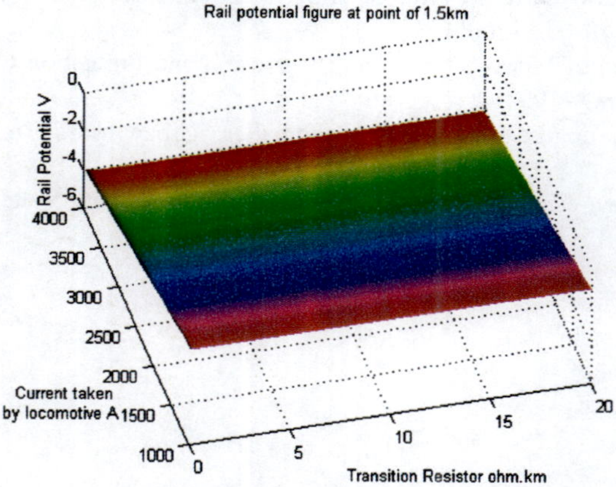


(b) At the distance of x=1.5km from substation A

**Fig. 6.** Rail potential distribution changing with rail-to-ground transition resistor at different currents

From Fig. 6 we know, when transition resistor changes within the range of 0.1～1Ω·km, the larger the rail-to-ground transition resistor and the rail potential are, the more greatly the transition resistor affects rail potential. In addition, transition resistor affects rail potential more obviously with the increase of current taken by locomotive. When rail-to-ground transition resistor is larger than 1Ω·km, rail potential becomes a relatively fixed value with the change of transition resistor.

## 4 Conclusion

Through simplifying and limiting the practical metro current circuit models, using the resistance network theory, this paper deduced the simplified analytic model of rail potential, and carried out its computer simulation, we derived the following conclusion:(1) substation pitch affects rail potential and rail current directly, and decrease of substation pitch can reduce the maximum of rail potential. Therefore, in designing urban rail systems, substation pitch should be reduced within permission of economic and environmental factors; (2) Current taken by locomotive affects rail potential greatly. The larger the current, the higher the rail potential, so in the DC tractive power supply system, we should use higher tractive voltage to reduce load current of locomotive with the same power;(3) railway portrait resistor affects rail potential greatly; decrease of railway portrait resistor can reduce rail potential;(4) rail-to-ground transition resistor has little effect on rail potential distribution.

## References

1. Li, W.: Monitoring and Protection Technique for Metro Stray current Corrosion, vol. 10, pp. 9–48. China University of Mining and Technology Press (2004)
2. Li, Y.-G., Li, J.-I., Yuan, H.-M.: Study on the Real-time Monitoring System of Laboratory Simulation of Metro Stray Current. China Railway Science 9, 26(5), 119–122 (2005)
3. Zhu, X.-X.: Metro Stray current Corrosion and Protection. Development and Application of Materials 9, 26(5), 119–122 (2005)
4. Technical Regulation of Metro Stray current Corrosion and Protection CJJ49-92(S). China Planning Press, Beijing (1993)
5. Wangyuqiao: Study on the rail potential anomalies. Urban Mass Transit 8(12), 35–37 (2009)
6. Moody, K.J.: Stray Current Characteristics of DC Transit Systems. Cathodic & Anodicprotection 7, 15–19 (1994)

# Study on Method of Data Fusion in Metro Stray Current Monitoring

Yu-qiao Wang, Wei Li, Xue-feng Yang, Guo Ye, Xi-yun Ge, and Xiu-ping Su

College of Mechanical and Electrical Engineering
China University of Mining & Technology
Xuzhou 221008, China
cumtwangyuqiao@163.com, suxiuping@xznu.edu.cn

**Abstract.** According to the characteristics and demands of metro stray current corrosion monitoring and protection, the monitoring parameters are reasonably selected and on-line monitoring methods of various parameters are analyzed. Data fusion model is established to make the data more reliability and accurate. The data fusion methods for measurements of stationary potential of reference electrode and rail potential are proposed and verified with the testing data. The result shows that the data fusion method can make the monitoring data more reasonable and effective, and it is important for the overall comprehensive analysis of metro stray current corrosion.

**Index Terms:** rail potential, data fusion, stationary potential, metro.

## 1 Introduction

A D.C. electrified system is widely adopted in metro, On a D.C. electrified system, low resistance between the traction return rails and the ground allows a significant part of the return current to leak into the ground. This is normally referred to as leakage current or stray current. Parts of the stray current will enter underground metal structures which form electrodes with the soil as the other electrolyte. The current point into the structure is known as a cathodic zone with the other current point back into the ground as an anodic zone. The current from anodic zone can enter another structure or return to a substation. Stray current corrosion at anodic zones will put the structure at risk, so measures should be taken to monitor stray current and protect the buried structure from stray current corrosion.

## 2 Monitoring Method for Stray Current Corrosion

### 2.1 The Selection of Monitoring Parameters

According to metro stray current Corrosion Protection Provisions [1], polarization potential of the buried structure steel is the key parameter of stray current monitoring. Studies have shown that the leakage of metro stray current is greatly influenced by rail potential. Rail potential can directly affect polarization potential of the structure steel, so the rail potential measurement is also very important. The monitoring parameters of stray current should include stationary potential, polarization potential of structure steel buried and rail potential.

### 2.2 Principle of Metro Stray Current Monitoring

*1) Method for measurements of polarization potential and stationary potential*

During the course of stray current monitoring in metro, there is IR drop due to the effect of stray current. Near reference method is used in the measurement of polarization potentials to reduce IR ohmic drop. Schematic diagram is shown as Fig. 1.

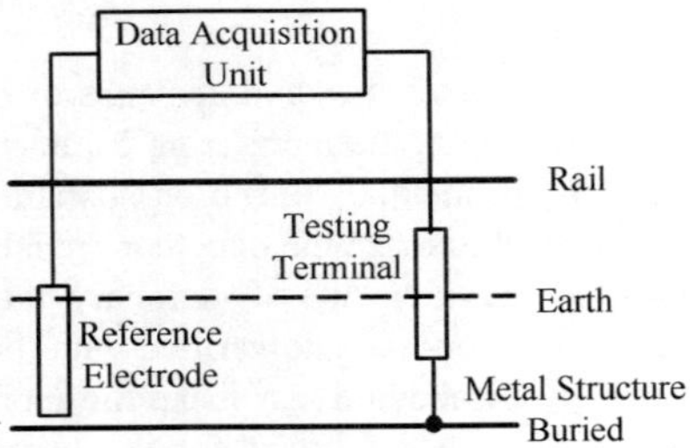


**Fig. 1.** Schematic diagram of polarization potential monitoring

Because of stationary potential of reference electrode, the sampled data are not values of polarization potentials; they are the potential differences between metal structure and reference electrode. Suppose stationary potential of reference electrode is $V_1$, and polarization potentials of metal structure is $V_2$. When the subway stops, metal structure is not influenced by the stray current, $V_2 = 0$, the potential difference between metal structure and reference electrode ($V'$) is the stationary potential of reference electrode, i.e. $V_1 = V'$.

When subway runs, if there is stray current leaking from metal structure, the polarization potential is not equal zero ($V_2 \neq 0$), and the potential difference between metal structure and reference electrode is the superposition of polarization potential and stationary potential, i.e. $V = V_2 - V_1$, so the polarization potential of metal structure is $V_2 = V + V'$.

So the polarization potential of metal structure can be calculated provided the differences between polarization potential and stationary potential under subway's stop and run conditions is measured.

*2) The measurement of rail potential*

During the stray current testing, rail potential is the difference between running rail and metal structure. The measuring principle is shown as Fig. 2.

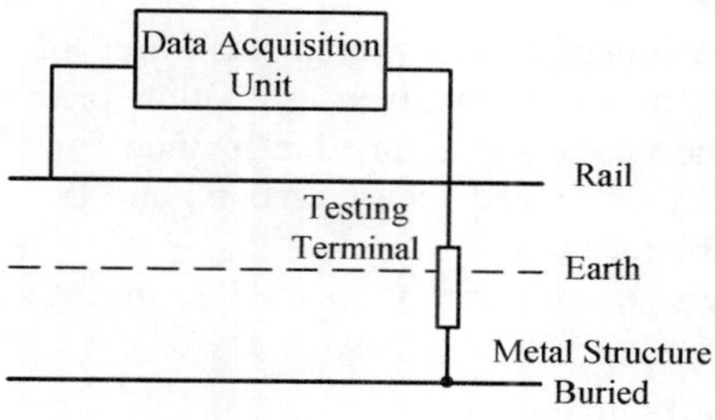


**Fig. 2.** Schematic diagram of rail potential monitoring

# 3 Data Fusion Model of Metro Stray Current Monitoring

Based on the analyses of metro stray current monitoring parameters and method, Data fusion model of metro stray current monitoring is built as Fig. 3.

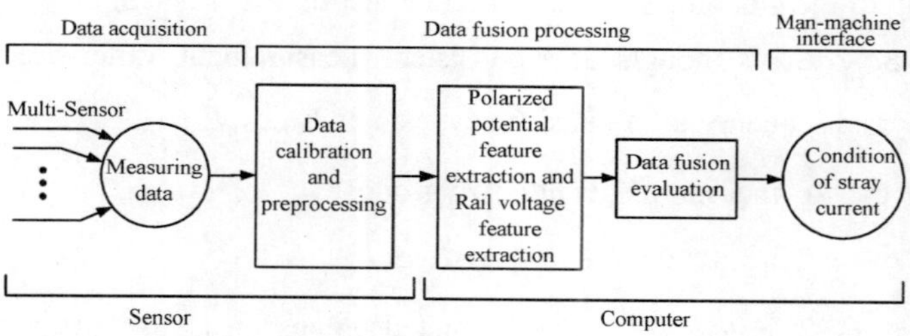


**Fig. 3.** Function model of stray current data fusion

Fig. 3 shows data fusion of stray current monitoring mainly includes data acquisitions, data fusion processing and man-machine interface. Data acquisitions consist of two parts: the acquisition of metal structure polarization potential and the acquisition of rail potential. The man-machine interface is used to display the results of data fusion processing. The following is the study on the method of data fusion processing of stray current monitoring.

# 4 Data Fusion Methods for Monitoring Parameters

## 4.1 Data Fusion Method for Stationary Potential Measurement

The measurement of polarization potential is based on reference electrode, the measurement accuracy of whose stationary potential directly influences the accuracy of polarization potential, so the data fusion method is introduced to improve the processing accuracy. During the stray current monitoring, $CuSO_4$ is commonly used as reference electrode. Experiments[2] show $CuSO_4$ reference electrode is characterized by steady performance in the course of measurement, then it can be considered as the value of stationary potential is constant during a short time measurement, and the multiple measurements are of normal distribution characteristic; arithmetic mean and statistic estimation method can be adopted to handle data. Study [3] shows statistic estimation method can obviously improve the accuracy and repeatability of limited measurements, and the obtained data is more reliable than the arithmetic mean.

Suppose the $N$ data of stationary potential during a measurement are $E_1, E_2, E_3, \cdots, E_n$.

The arithmetic mean of sampling data:

$$\overline{E} = \frac{1}{N}\sum_{i=1}^{N} E_i \tag{1}$$

$$\sigma = \sqrt{\frac{1}{N-1}\sum_{i=1}^{N}(E_i - \overline{E})^2} \tag{2}$$

According to the guidelines $3\sigma$, consistency test on the $N$ sampling data is done, if $\left|E_k - \overline{E}\right| \le 3\sigma$, $E_k$ is thought as a consistent measurement; otherwise the data is rejected. The data sequence after consistency test is: $E_1, E_2, E_3, \cdots, E_M$.

To divide the $M$ measurements into two groups:

$E_{11}, E_{12}, E_{13}, \cdots, E_{1n}$ and $E_{21}, E_{22}, E_{23}, \cdots, E_{2m}$

Where, $m, n \ge 2$, $m + n = M$, The estimated values of sample mean and sample variance of two sets of measurements respectively are:

$$\overline{E}_1 = \frac{1}{n}\sum_{i=1}^{n} E_{1i} . \overline{E}_2 = \frac{1}{m}\sum_{i=1}^{m} E_{2i} \tag{3}$$

$$\sigma_1^2 = \frac{\sigma_{01}^2}{n} . \sigma_{01}^2 = \frac{1}{n}\sum_{i=1}^{n}(E_{1i} - \overline{E}_1)^2 \tag{4}$$

$$\sigma_2^2 = \frac{\sigma_{02}^2}{m} . \sigma_{02}^2 = \frac{1}{m}\sum_{i=1}^{n}(E_{2i} - \overline{E}_2)^2 \tag{5}$$

According to statistic estimate theory, the estimated value of stationary potentials is

$$\hat{E} = \frac{\sigma_2^2 \overline{E}_1 + \sigma_1^2 \overline{E}_2}{\sigma_1^2 + \sigma_2^2} \tag{6}$$

The square deviation is

$$\sigma = \sqrt{\frac{\sigma_1^2 \sigma_2^2}{\sigma_1^2 + \sigma_2^2}} \tag{7}$$

The Equation (6) is the result of data fusion based on the arithmetic mean and statistic estimate.

The stationary potential data from Guangzhou metro No.1 is fused. The testing data of one testing point power supply sector between Fangcun and Changshoulu is shown in Table 1.

**Table 1.** Stationary potential measurements data

| | | | | | | | |
|---|---|---|---|---|---|---|---|
| time | 1 | 2 | 3 | 4 | 5 | 6 | 7 |
| value(V) | 0.123 | 0.121 | 0.124 | 0.129 | 0.119 | 0.121 | 0.124 |
| time | 8 | 9 | 10 | 11 | 12 | 13 | 14 |
| value(V) | 0.125 | 0.128 | 0.123 | 0.124 | 0.12 | 0.121 | 0.117 |
| time | 15 | 16 | 17 | 18 | 19 | 20 | 21 |
| value(V) | 0.122 | 0.151 | 0.12 | 0.123 | 0.122 | 0.122 | 0.117 |
| time | 22 | 23 | 24 | 25 | 26 | 27 | 28 |
| value(V) | 0.123 | 0.119 | 0.121 | 0.123 | 0.125 | 0.125 | 0.121 |
| time | 29 | 30 | 31 | 32 | 33 | 34 | 35 |
| value(V) | 0.124 | 0.122 | 0.124 | 0.164 | 0.125 | 0.121 | 0.122 |
| time | 36 | 37 | 38 | 39 | 40 | 41 | 42 |
| value(V) | 0.123 | 0.121 | 0.116 | 0.118 | 0.122 | 0.124 | 0.123 |
| time | 43 | 44 | 45 | 46 | 47 | 48 | 49 |
| value(V) | 0.121 | 0.122 | 0.122 | 0.123 | 0.124 | 0.122 | 0.119 |

The distribution of stationary potential is shown in Fig. 4. The action must be taken in the measurement because there are obviously errors in the testing data.

In the traditional stray current monitoring system, arithmetic mean is used as estimated value of stationary potential. In this measurement, the estimated value of stationary potential is: $\overline{E} = 0.1235V$ the mean square deviation of the estimated value is: $\sigma = 0.0075$

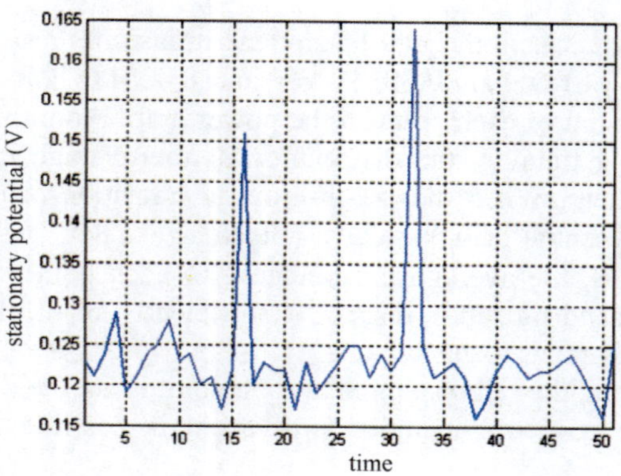


**Fig. 4.** Curve of stationary potential measurement data

To estimate the measurements by employing arithmetic mean and statistic estimate. Firstly, consistency tests on the measurements are done, and the 16th, the 32nd measurements are rejected; secondly, the measurements are divided into two groups. The first data set is the first 24 data, and the second is made up of the latter 25 data. The estimated fusion value based on the arithmetic mean and the statistic estimation is: $\hat{E} = (\sigma_2^2 \overline{E}_1 + \sigma_1^2 \overline{E}_2)/(\sigma_1^2 + \sigma_2^2) = 0.1221\text{V}$.

The mean square error $\hat{\sigma} = \sqrt{\dfrac{\sigma_1^2 \sigma_2^2}{\sigma_1^2 + \sigma_2^2}} = 0.0019 < \sigma$

So the accuracy of the measurement based on the data fusion of arithmetic mean and the statistic estimate is obviously improved.

## 4.2 Data Fusion Method of Rail Potential in Stray Current Monitoring [4, 5]

*i. Analysis of rail potential characteristic*

The rail potentail should be linear in one power area when the rail is in good working condition. In addition, the rail potentail will achieve maximum value at the train location, and negative maximum value at the reflux point of substation, the rail potential is zero in the middle location between the train and the substation. In theory, if the zero rail potential location deviates from the middle location between the train and the substation, it indicates that insulation in the zero rail potential location is damaged, but because the train is a moving load, the zero rail potential is always moving and changing, and because of signal interference, it is difficult to give a reasonable assessment of the instantaneous value of the rail potential , it is quite necessary to study the rail potential distribution through making the average rail potential as its characteristic value.

In order to simplify the questions, we consider the distribution of the average rail potential in one power area. Supposing substation pitch is 2.5 $km$, railway portrait resistor R = 0.025 $\Omega/km$, drainage net portrait resistors $R_p$ = 0.01 $\Omega/km$, current taken by locomotive is 1000A. We use MATLAB to simulate the distribution of the average rail potential when rail-to-grouand transition resistor $Rg$ changes during running of the train.

Fig. 5 shows that when the rail-to-grouand transition resistor is respectively 20 $\Omega \cdot km$, 3 $\Omega \cdot km$, 0.01 $\Omega \cdot km$, 0.001 $\Omega \cdot km$ and 0.0001 $\Omega \cdot km$, the distribution of the average rail potential of each point in the power area. From the figure we can see that when the train is running, the distribution of the average rail potential in one power area is not linear. When the rail-to-grouand transition resistor is greater than 3 $\Omega \cdot km$, the rail potential will basically not change a lot, but as rail-to-grouand transition resistor falls, the average rail potential of each point decreases gradually, when the rail-to-grouand transition resistor is sufficiently small, the average value of rail potential gradually tends to zero.

In the metro system, the rail insulation at each point is not the same due to different working conditions, in metro operation process, it is possible that in some place

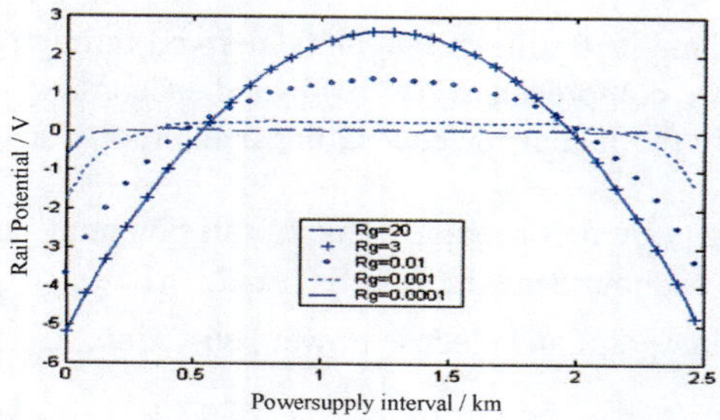


**Fig. 5.** Distribution curve of rail potential average value

insulation is damaged while other places are in good insulation. Fig. 6 shows distribution curve of the average rail potential at partial-damaged rail insulation, Fig. 6 (a) is distribution of the average rail potential in one power supply interval when rail insulation is good. Fig. 6 (b) is distribution of the average rail potential when insulation is damaged at $x = 0.3$km, $x = 1.2$km, and $x = 1.8$km.

We can see from the figure, when the rail insulation in some place is damaged, this will affect the rail potential distribution of the entire power area, and in the location of damaged insulation, the absolute value of the average rail potential will reduce. Therefore, in the monitoringop of the rail potential, if the average rail potential of the rail somewhere is closer to zero than the average potential around the rail, then the rail insulation is likely to be damaged, and measures should be taken for processing.

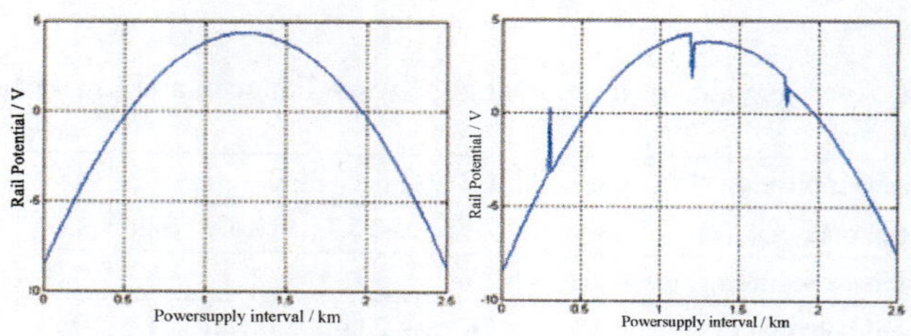


(a) No damage of rail insulation (b) Partial damage of rail insulation

**Fig. 6.** Distribution curve of average rail potential at partial-damaged rail insulation

*ii. Data fusion method of rail potential in stray current monitoring*

The working condition of running rail can be judged by the variation of rail potential average in a power supply sector, but the sensors of monitoring system are scattered, and the insulation of running rails anywhere may drop. Therefore the collected rail potential averages need to be processed. Interpolation can give an overall similar description to the discrete points with a simple expression. The corresponding relation between rail potential and its testing point can be derived through interpolation, and the rail potential can be comprehensively evaluated according to the interpolation curve trend. To simply the matter, piecewise linear interpolation are used in the actual design.

The data of testing point and average value of rail potential can be measured in one measurement at half an hour and the data is $(x_i, U_i)(i = 0,1,\cdots,m)$, where $x_i$ is the distance between testing point and electric power substation, $U_i$ is the rail potential of testing point $x_i$.

Processed by interval linear interpolation respectively in $[x_i, x_{i+1}]$

$$P_i(x) = y_i \frac{x - x_{i+1}}{x_i - x_{i+1}} + y_{i+1} \frac{x - x_i}{x_{i+1} - x_i} \quad x \in [x_i, x_{i+1}](i = 0,1,\cdots,n-1) \tag{8}$$

$P_i(x)$ is one-degree polynomial in $[x_i, x_{i+1}]$, it meets: $P_i(x) = y_i$, $(i = 0,1,\cdots,n)$, and continuous at testing point $x_i$.

According to Equation (8) the distribution curve of the average value of rail potential in one power supply sector can be derived. If the average value of rail potential somewhere changes abnormally, which means the running rail insulation may be destroyed, and corresponding measures should be taken to solve the problem.

Based on the above analysis, we spatialy fuse the data of rail potential from stray current monitoring system of the first Guangzhou Metro Line, when the train is running, we take the average of half-hour measured data. Table 2 shows sensor installation position and data of the average rail potential in one interval of power supply between Fangcun road and changshou road.

**Table 2.** Average value of rail potential between Fangcun and Changshoulu

| | | | | | | |
|---|---|---|---|---|---|---|
| sensor position (km) | 0 | 0.2 | 0.4 | 0.6 | 1.1 | 1.6 |
| rail potential (V) | -36.3 | -13.2 | -15.7 | 18.4 | 26.9 | 12 |
| sensor position (km) | 2.1 | 2.6 | 3.1 | 3.6 | 3.8 | 4.0 |
| rail potential (V) | 34.1 | 26 | 23.4 | 14.7 | 8.4 | -9.9 |

We use MATLAB to process measured data in Table 2 through linear interpolation. The interpolated curve is shown in Fig. 7. We can see from the figure, there exists rail potential mutations at 0.5km and 1.5km, thus, the rail insulation near the location may become bad, measures need to be taken to deal with.

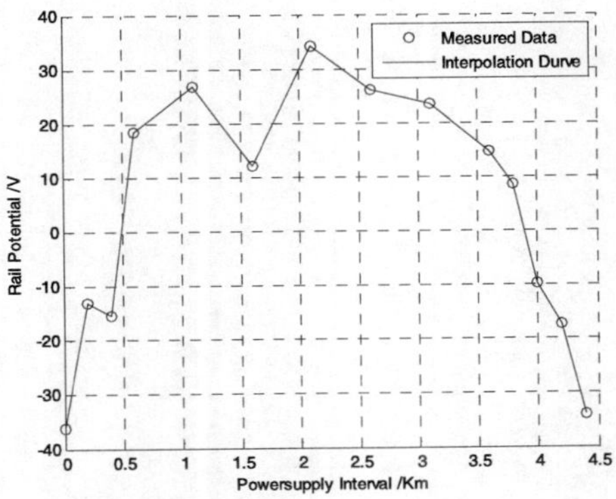


**Fig. 7.** Distribution curve of average rail potential between Fangcun and Changshoulu

## 5 Conclusions

The data fusion method is presented based on feature analysis of monitoring parameters of stray current corrosion, and verified through stray current monitoring data from Guangzhou metro No.1. The results show that the accurate and reasonable evaluation on stray current corrosion state can be derived with the data fusion method.

## References

1. VDV RECOMMENDATION 501/1 4/93. Reduction of the Corrosion Danger due to Stray Currents in Tunnels of DC Traction Systems with Return Current via Running Rails—Provisions and Bases for Calculation
2. Li, W., Wang, A.-B., Wang, Y.-Q.: Research of Experiment on Bluestone Electrode Applied in Metro Stray Current Corrosion Measuring. Journal of China University of Mining & Technology 34, 62–66 (2005)
3. Cai, F.-N., Liu, Q.-X.: Single sensor data fusion and analysis of effectiveness. Journal of Transducer Technology 24, 73–74 (2005)
4. Sun, S.-L., Deng, Z.-L.: Multi-sensor Optimal Information Fusion Criterion in Linear Minimum Variance Sense. Science Technology and Engineering 4, 334–336 (2004)
5. Zhou, J., Wang, Z.-S., Zhou, F.-Q.: The theory of multi-sensor system data fusion based on linear least square estimation. Journal of Astronautics 24, 364–367 (2003)

# How to Effectively Manage the Cost and the Time of Logistics Quick Response

Junbo Li

Management School, Guilin University of Technology
Guilin, China
junboli@glite.edu.cn

**Abstract.** Quick Response has become an important strategy to increase competitiveness in logistics. Variables affecting logistics quick response capability have been widely studied from the perspectives of assessment on logistics performance or supply chain performance. This paper analyzes the key variables to assess logistics quick response capability in China. It analyzes empirical data collected from 50 logistics experts and professionals to identify the importance sequence of the variables. The results indicate that, in the Chinese context, the three most important variables affecting logistics quick response capability are customer service, time management and logistics cost. It's critically important to manage the latter two variables in order to achieve improved customer service. A methodology of non-linear cost-profit balance is proposed for effective logistics quick response implementation.

**Keywords:** logistics, quick response, cost, time management.

## 1 Introduction

Quick Response(QR), proposed by Kurt Salmon Associates as a strategy to strengthen the competitive advantages of the American Apparel Manufacturers Association in the 20th century, is a method that allows manufacturers or retailers to communicate inventory needs for their shelves or assembly lines in near-real time. The generic QR schema requires a company to share what could be seen as proprietary data about its sales and manufacturing operations with its partners.

As a result, QR has been applied widely in the manufacturing industry over the past two decades. The objective is to utilize real-time customer and business partner information to increase speed of response in the supply chain. When implemented appropriately, QR can be an effective approach for the creation of more lean and efficient supply chains. QR has been unanimously applied in the domain of logistics, and successful logistics companies like FedEx, UPS and DHL take Quick Response as an effective strategy to improve relative competitive advantages.

The challenges to realize the QR potentials will be difficult and will take time, while there is obviously enormous potential for growth in China logistics market. A main issue is concerned with how to justify logistics quick response correctly. With the support from the social science fund [08BJY024] and natural science fund

[0991019] in Guangxi Autonomous Region in China, we are commissioned to do the investigation and research to assess and justify logistics quick response capability.

As a key component of the supply chain, improvements in logistics response times lead directly to improvements in the ability to meet evolving customer demands in terms of overall supply chain performance. This paper explores the application of QR to the logistics industry in China, investigates the key variables affecting logistics quick response ability, and studies the improvements of logistics quick response.

## 2 Variables of Logistics Quick Response

A number of studies have been carried out to identify the variables and mechanisms affecting logistics quick response from various perspectives. Flynn [1] emphasizes that customer loyalty is directly affected by logistics service quality. Harrison [2] perceives time as the main variable to be considered affecting logistics quick response with other variables such as customer service and logistics cost also being key. Paul Dooley [3] pointes to the limitations to logistics QR from the so-called Bullwhip effect in the supply chain. Kumia and Johnston [4] argue that collaboration and trust between partners are key variables affecting logistics cost, profit and risk. In the context of China, Na Li [5] concludes that the lack of well accepted standards by supply chain partners is a key variable affecting logistics quick response capabilities. Zhangyi Xu [6] believes the inefficient flow of information is the key variable to limiting logistics QR, and Xuping Wang [7] lists a number of salient variables including the ready availability of information, time, resources and facilities.

From the above discussion it is evident that there is a need to organize the multifarious variables already identified as affecting logistics QR. The paper reports results from ongoing empirical research towards this aim. The research correlates the opinions of 50 logistics industry experts using a methodological framework that taxonomies the variables as being broadly internal or external in their relevance to QR. Section III below provides a brief overview of the application of this methodological framework to key logistics QR variables in China.

## 3 Analysis of Logistics QR Variables

### 3.1 Variables Affecting Logistics Quick Response

As in many other contexts, logistics provides a number of services including warehousing, transportation, distribution, packaging, assembly, fulfillment and information processing. The various logistics functions come together to form the totality of logistics support. With this backdrop, the study, derived from our investigation and from the body of literature on logistics systems theory with reference to supply chain management (see for example [1], [2], [3], [4], [5], [6], [7] etc), conceptualizes logistics services as an integrated system [9] comprised of variables affecting its QR ability, broadly classified into two categories: external and internal (table 1).

**Table 1.** The Variables Affecting Logistics QR

| Internal Variables | | | | | | | External Variables | | |
|---|---|---|---|---|---|---|---|---|---|
| $u_1$ | $u_2$ | $u_3$ | $u_4$ | $u_5$ | $u_6$ | $u_7$ | $u_8$ | $u_9$ | $u_{10}$ |
| $u_1$=logistics hardware, $u_2$=time management, $u_3$=logistics cost, $u_4$=customer service, $u_5$=information system, $u_6$=coordination control, $u_7$=staff quality | | | | | | | $u_8$=structural variable, $u_9$=social variable, and $u_{10}$=natural variable. | | |

## 3.2 Analysis of the Variables

A total of 50 responses were received from selected experts who were each invited to fill questionnaires with objective of justifying the importance sequence upon 10 listed variables in table 1. Delphi fuzzy Arithmetic was used to determine the importance sequence coefficient $a_i$ (i =1, 2,..., m) of the variables through the collection of data from the experts. The relevant procedures of the methodology are as follows:

*1) The Importance Sequence Calculation*

*a) The Variable Importance Sequence Value:* $e_i \in \{1, 2,..., m\}$ is the importance sequence value of variable $u_i$, for less important variables, $e_i$=m, and for the most important variable, $e_i$ =1. The selected experts are invited to do the evaluation by questionaries with specific values $e_i$ (k), which is the importance sequence value provided by expert 'k' to evaluate variable $u_i$.

**Table 2.** The $e_i$ Evaluation by Expert 'K'

| **No.** | $u_1$ | $u_2$ | ...... | $u_m$ |
|---|---|---|---|---|
| $e_i(k)$ | $e_1$ (k) | $e_2$ (k) | ...... | $e_m$ (k) |

*b) To Edit the Preference Credit Table:* The analysis and statistics procedures on $e_i$ are as follows:

$$\text{when } \frac{e_j(k)}{e_i(k)} > 1, \quad A_{ij}(k) = 1$$

$$\text{when } \frac{e_j(k)}{e_i(k)} < 1, \quad A_{ij}(k) = 0 \tag{1}$$

The added $A_{ij}(k)$ from a number of 'n' experts is:

$$A_{ij} = \sum_{k=1}^{n} A_{ij}(k) \quad (\text{i=1, 2, ..., m; j=1, 2, ..., m}) \tag{2}$$

Table 3 summarizes the number (m×n) of statistic value of $A_{ij}$ preference credit. Here, $A_{12}$ refers to the comparison of variable $u_1$ to variable $u_2$ and $A_{21}$ refers to the comparison of variable $u_2$ to variable $u_1$, therefore $A_{12} + A_{21} = n$.

**Table 3.** m×n Statistic Value $A_{ij}$ Preference Credit

| No. | $u_1$ | $u_2$ | …… | $u_m$ |
|---|---|---|---|---|
| $u_1$ | $A_{11}$ | $A_{12}$ | | $A_{1m}$ |
| · | · | · | …… | · |
| $u_m$ | $A_{m1}$ | $A_{m2}$ | | $A_{mm}$ |

*c)* $\sum A_i$ *calculation:* $\sum A_i$ is the aggregate accumulation of $A_{ij}$ in a row of Table 3.

$$\sum A_i = \sum_{j=1}^{m} A_{ij}, \quad (j=1, 2, \ldots, m) \tag{3}$$

And $\sum A_{max} = \max\{\sum A_{1,} \sum A_{2,} \ldots \sum A_m\}$,

$$\sum A_{min} = \min\{\sum A_{1,} \sum A_{2,} \ldots \sum A_m\}$$

Here, the variable relevant to $\sum A_{max}$ is the most important one, and the variable relevant to $\sum A_{min}$ is the least important one.

*d) To Calculate the Grade Difference 'd':*

if $a_{max}=1$, $a_{min}=0.1$, then

$$d = \frac{\sum A_{max} - \sum A_{min}}{a_{max} - a_{min}} \tag{4}$$

*e) the Variable Importance Coefficient '* $a_i$ *':*

$$a_i = \frac{\sum A_i - \sum A_{min}}{d} + 0.1, \text{ or}$$

$$a_i = \frac{\sum A_{max} - \sum A_i}{d} \tag{5}$$

Thus, we can get the variable importance sequence 'A'.

$$A = (a_1, a_2, \ldots a_m)$$

*2) the Results From the Calculation*

Statistic work is done upon the collected data from the $e_i$ value provided by individual expert to calculate the $A_{ij}$ by formula (2). Table 4 summarizes the preference credit $A_{ij}$, and $\sum A_i$ (i=1, 2, …, 10) is the aggregate accumulation of the $A_{ij}$ in each row and is listed in column 12 correspondingly.

Continuous calculation of the grade difference 'd' by formula (4) and the variable importance coefficient '$a_i$' by formula (5) results respectively as presented in table 4:

**Table 4.** Preference Credit $A_{ij}$

| $u_i$ | $u_1$ | $u_2$ | $u_3$ | $u_4$ | $u_5$ | $u_6$ | $u_7$ | $u_8$ | $u_9$ | $u_{10}$ | $\sum A_i$ | $a_i$ |
|---|---|---|---|---|---|---|---|---|---|---|---|---|
| $u_1$ | * | 16 | 20 | 18 | 24 | 32 | 39 | 45 | 49 | 44 | 287 | 0.791 |
| $u_2$ | 34 | * | 32 | 26 | 37 | 39 | 42 | 47 | 49 | 47 | 353 | 0.979 |
| $u_3$ | 30 | 18 | * | 25 | 39 | 44 | 43 | 50 | 50 | 49 | 348 | 0.965 |
| $u_4$ | 32 | 24 | 25 | * | 38 | 45 | 46 | 50 | 50 | 50 | 360 | 1.000 |
| $u_5$ | 26 | 13 | 11 | 12 | * | 39 | 45 | 50 | 49 | 47 | 292 | 0.805 |
| $u_6$ | 18 | 11 | 6 | 5 | 11 | * | 41 | 47 | 49 | 45 | 233 | 0.635 |
| $u_7$ | 11 | 8 | 7 | 4 | 5 | 9 | * | 34 | 48 | 42 | 168 | 0.449 |
| $u_8$ | 5 | 3 | 0 | 0 | 0 | 3 | 16 | * | 34 | 36 | 97 | 0.246 |
| $u_9$ | 1 | 1 | 0 | 0 | 1 | 1 | 2 | 16 | * | 24 | 46 | 0.100 |
| $u_{10}$ | 6 | 3 | 1 | 0 | 3 | 5 | 8 | 14 | 26 | * | 66 | 0.157 |

$$d=\frac{\sum A_{\max}-\sum A_{\min}}{a_{\max}-a_{\min}}=\frac{360-46}{1.000-0.100}=348.89$$

$$a_i=\frac{\sum A_i-\sum A_{\min}}{d}+0.1=\frac{\sum A_i-46}{348.89}+0.1$$

The final calculation of the fuzzy sets 'A' by utilizing the calculated 'd' and '$a_i$' is:

$$A=(0.791,0.979,0.965,1.000,0.805,0.635,0.449,0.246,0.100,0.157)$$

The result (fuzzy sets 'A') indicates the importance sequence of variables affecting logistics quick response in China. This sequence, from the most important variable to the least important one, can be listed as follows: customer service, time management, logistics cost, information system, logistics hardware, coordination control, staff quality, structural variable, natural variable, and social variable. Thus, customer service, time management and logistics cost are at one end of the spectrum whilst at the other are external variables.

## 4 To Manage the Cost and the Time Management

Based on the above quantitative study, we can find out the three most important variables affecting logistics quick response are customer service, time management, logistics cost. This is reasonable and acceptable because the main goal of logistics QR is improved customer satisfaction through time-based logistics management with logistics cost well under controlled.

However, variable time management ($u_2$) and variable cost ($u_3$) of logistics quick response can not be usually achieved simultaneously, because logistics quick response generally saves time at the price of increased logistics cost. Therefore, it's critically important to effectively manage these two variables while implements logistic quick response strategy.

The cost and the time of logistics quick response have a nonlinear relation [10]. This means that logistics quick response is a double-edged sword which can produce good returns for an organization at a substantial cost. There exist such dilemma situations as over-response and under-response. Over-response appears when the profit is less than the excessive input cost to implement logistics quick response, while under-response refers to that when negative profit is yielded from less input cost towards logistics QR strategy. However, there exists a maximum point where the profit is the maximum while the input cost is the minimum for logistics quick response implementation. This can be illustrated in figure 1.

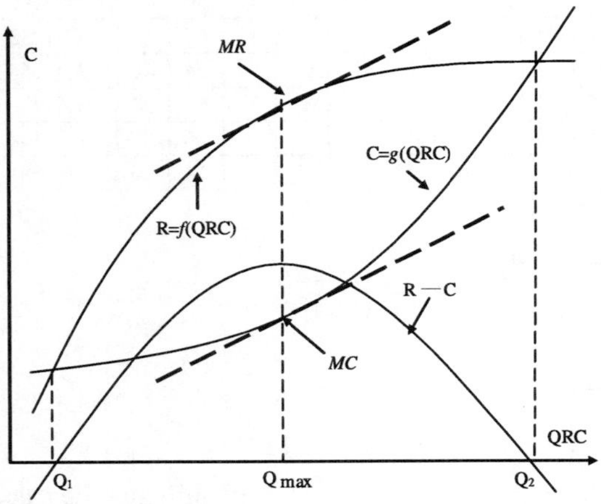


**Fig. 1.** Logistics Quick Response Non-Linear Cost-Profit Balance

In figure 1, QRC=*h* (t) refers to logistics quick response capability, C=*g* (QRC) and R=*f*(QRC) represents respectively the cost and the profit derived from logistics quick response implementation. The first derivative of function *g* and function *f* is the marginal cost *MC* and marginal profit *MR* of logistics quick response. Here:

$$MC = \frac{dC}{dg(QRC)} \geq 0 \qquad MR = \frac{dR}{df(QRC)} \leq 0$$

The maximum point $Q_{max}$ appears when the value of (A=R-C) produce a maximum value which occurs while the tangent slope is the same to function *g* and function *f* curves.

$$\frac{dA}{dA(QRC)} = 0, \Rightarrow \frac{dC}{dg(QRC)} = \frac{dR}{df(QRC)} \Rightarrow MC = MR$$

There also exist two balance points where the profit equals the input cost at $Q_1$ and $Q_2$. Organizations implementing logistics quick response lose money at situations when $Q<Q_1$ and $Q>Q_2$, while make profit only when $Q_1<Qmax<Q_2$. The best logistics quick response profitability occurs at $Q=Qmax$.

## 5 Conclusions

Based on the above quantitative study, we can conclude that the most important variables affecting logistics quick response capability, in the Chinese context, are customer service, time management and logistics cost. It's critically important to coordinate the latter two variables and manage them effectively while implementing logistic quick response strategy in order to achieve improved customer service. A methodology of non-linear cost-profit balance is a good tool for logistics quick response implementation.

## References


1. Flynn, B.B., Schroeder, R.G.: The impact of quality management practices on performance and competitive advantage. Decision Sciences 26, 659–692 (1995)
2. Harrison, A.: Logistics management and strategy, pp. 87–143. Pearson Education Limited (2002)
3. Dooley, P.: Automated unattended B2B replenishment. Ascet (4), 97–112 (2002)
4. Kumia, S., Johnston, R.B.: Adoption of efficient consumer response: The issue of mutuality. Supply Chain Management 6, 230–241 (2001)
5. Li, N., Hu, J.: Discussion of agile logistics development in China. Transportation and Delivery (4), 16–18 (2006)
6. Xu, Z.: Optimization of enterprise supply chain, pp. 145–189. Qinghua University Publication (2006)
7. Wang, X.: QR mechanism of emergency logistics. China Soft-Science, 127–131 (2005)
8. Yusuf, Y.Y., Gunasekaran, A., Adeleye, E.O.: Agile supply chain capabilities: determinants of competitive objectives. European Journal of Operational Research 59, 379–392 (2004)
9. Li, J., Gao, Y., Ambaye, D.: A Systematic View of the Logistics Quick Response Factors and the Importance Sequence in China. In: Proceedings of 2008 IEEE International Conference on Service Operations and Logistics, and Informatics, Beijing, China, October 12-15, pp. 2166–2170 (2008)
10. Alard, R., Schonsleben, P.: System-oriented supply chain risk management. Production Planning & Control 20, 343–361 (2009)
11. Devlin, J., Yee, P.: Trade logistics in developing countries: the case of the Middle East and North Africa. The World Economy 28, 435–456 (2005)
12. Bajgoric, N.: Web-based information access for agile management. International Journal of Agile Management Systems 2, 121–129 (2000)
13. Lin, C.-T., Chiu, H., Chu, P.-Y.: Agility index in the supply chain. International Journal of Production Economics 10, 285–299 (2006)
14. Mentzer, J.T., Konrad, B.P.: An efficiency/effectiveness approach to logistics performance measurement. Journal of Business Logistics 12, 33–52 (1992)
15. Ralescu, D.A., Adams, G.: Fuzzy integral. Journal of Mathematical Analysis and Applications 75, 562–570 (1980)
16. Kaufmann, A., Gupta, M.M.: Introduction to fuzzy arithmetic: theory and application. Van Nostrand Reinhold, New York (1991)

# The Application of SOM and GRNN in Creativity Measurement for Middle School Students

Jiayuan Yu

Department of Psychology, Nanjing Normal University
Nanjing, China
yujiayuanwx@163.com

**Abstract.** In this paper, creativity measurement data was analyzed by neural networks. Williams Creativity Test B(WCTB) and Adolescent Scientific Creativity Scale(ASCS) were used to measure the creative affective and scientific creativity for 550 middle school students. The data was clustered with SOM neural network, and three categories were obtained. There were significant differences for creativity factors except imagination among the categories. In 550 students, 70% of them were used as modeling group, and the other as testing group. Generalized regression neural network (GRNN) and multivariable linear regression (MLR) were used for modeling and testing. Risk-taking curiosity, imagination and complexity scores used as input and independent variable, scientific creative scores used as output and dependent variable. The result showed the predictive error of GRNN was lower than the error of MLR. The neural networks could analyze creativity measurement data very well.

**Keywords:** lneural networks, clustering, creativity measurement.

## 1 Introduction

### 1.1 Creative Affective and Scientific Creativity

Williams defined creativity in relation to four cognitive factors and four affective factors (risk-taking, curiosity, imagination, and complexity). Scientific creativity is the ability in learning scientific knowledge and solving scientific problem.

In this research, the creativity measurement data of middle school students were clustered. The creative affective factors were used to predict the scientific creativity.

### 1.2 Som Network Based Clustering

Self-Organizing Feature Map (SOM) is a competitive learning network [1]. It adjusts weight vectors and divides them into groups.

### 1.3 Generalized Regression Neural Network Based Predictive Validity

Multivariable linear regression (MLR) is often used for validity test [2]. MLR requires that predictive factor and criterion are interval variables, and there is a linear relationship between them. However, it is hard to satisfy these assumptions for psychological measurement data. Neural networks have been tried to apply in psychological measurement and solve this problem.

# 2 Method

## 2.1 Test Instrument

*1) Williams Creativity Test B (WCTB)*

WCTB is a 50-item creativity assessment instrument that provides scores for risk-taking, curiosity, imagination and complexity.

*2) Adolescent Scientific Creativity Scale (ASCS)*

ASCS could measure seven aspects of scientific creativity.

## 2.2 Subject

550 middle school students' data were analyzed in this study.

# 3 Using SOM to Cluster the Creativity Measurement Data

## 3.1 SOM Clustering

SOM was used to cluster11 variables of creativity measurement (including 4 factors of WCTB and 7 aspects of ASCS). The results showed that students were divided into A, B, and C categories. There were 36,194,320 students in each category respectively. Their average scores of every variable in creativity measurement were in Table 1.

**Table 1.** The average scores of every variable for each category students

| | *X1* | *X2* | *X3* | *X4* | *X5* | *X6* | *X7* | *X8* | *X9* | *X10* | *X11* |
|---|---|---|---|---|---|---|---|---|---|---|---|
| A | 26.44 | 34.44 | 29.64 | 30.31 | 27.58 | 23.71 | 14.76 | 18.08 | 11.32 | 12.43 | 14.80 |
| B | 25.64 | 33.73 | 28.53 | 29.16 | 13.40 | 15.21 | 8.83 | 12.12 | 5.96 | 9.27 | 13.63 |
| C | 25.06 | 32.66 | 28.46 | 28.34 | 6.81 | 5.83 | 4.41 | 5.02 | 3.41 | 6.03 | 10.86 |

In above table, X1 to X4 were risk-taking, curiosity, imagination, and complexity in WCTB. X5 to X11 were object use, problem giving, product improvement, science fancy, problem solving, science experiment and product design in ASCS.

## 3.2 The Difference Significant Test of Creativity Scores for Categories

ANOVA was used to test the difference of creativity scores among the categories. The results showed there were significant differences for these factors except imagination.

## 4 Comparing with Neural Networks and Multiple Regression

The score of scientific creativity was tried to predict by the scores of four factors of WCTB.

The data was divided into A and B groups. 70% of the dada was in group A which was used as modeling. Others of the data were in group B which was used as testing.

### 4.1 Predicting with Neural Networks

Generalized regression neural network (GRNN) was used in this research. Following are the steps for GRNN modeling:

1) Pretreatment of the raw data;
2) Determining the node numbers of input and output layer according to training sample;
3) Determining the SPREAD parameter, it is related with the fitness and curve smooth;
4) Testing the accuracy of the network.

The root mean square error RMSE1=1.031.

### 4.2 Predicting with Statistics Regression Method

The WCBT scores of group A was used as independent variable X1, X2, X3 and X4, the ASCS scores of them were used as dependent variable Y. The regression equation was set up as following:

$$Y=23.151+0.042X1+0.963X2+0.543X3 +0.739X4,$$

Then the WCBT scores of group B were put into this equation, and the Y values were got. They were compared with the ASCS total scores of group B students. The root mean square error RMSE2=1.763. It was larger than RMSE1. It meant the prediction ability of neural networks was higher than statistics regression method.

## 5 Conclusion

Above research gave following conclusion:

*A.* SOM network could cluster the creativity measurement data very well. Meaningful information could be got form the clustering results.

*B.* The students of group A have higher creativity scores. It is pity there are only 6.5% of total students in group A. The students of group C have lower creativity scores, and they are 58.2% of the total students. It is very urgent to train the students' creativity.

*C.* Neural networks could be used to predict the scientific creativity and has higher accuracy comparing with statistical method.

## References

1. Dong, C.: Matlab Neural Networks and Its Application. Defense Industry Publishing House (2005)
2. Cyders, M., Smith, G., Spillane, N.: Integration of impulsivity and positive mood to predict risky behavior. Psychological Assessment 19, 107–118 (2007)

# A General Throughput Model for Parallel Cluster Tools

Zheng Xiuhong[1,2,3], Yu Haibin[1], and Hu Jingtao[1,2]

[1] Key Laboratory of Industrial Informatics, Shenyang Institute of Automation, CAS, Shenyang 110016, China
[2] Graduate School of the Chinese Academy of Sciences, CAS, Beijing 100039, China
[3] School of Computer Science and Technology, Shenyang Institute of Chemical Technology, Shenyang, 110142, China
zhengxiuhong@sia.cn, yhb@sia.cn, hujingtao@sia.cn

**Abstract.** To effectively evaluate and analyze performance of cluster tools in semiconductor manufacturing, the paper presents a throughput model for parallel cluster tools .For developping this model, the timing diagram is applied to describe the usage of the chamber and robot in parallel processing of cluster tools. Finally, the theoretical model which indicates the relationship between the number of chamber, processing time, transport time, and throughtput is deduced. This model can analyze and computer the wafer period and the batch period fast and accurately.

**Keywords:** parallel, cluster tools, throughput, model, timing diagram.

## 1 Introduction

A cluster tool is an integrated, environmentally isolated semiconductor manufacturing system consisting of process chambers, transport robots, and loadlocks, where the wafers enter and leave the system. Figure 1 shows a 4-chamber cluster tool. The proliferation of cluster tools in semiconductor manufacturing motivates an understanding of how basic cluster tool parameters, such as processing time, number of chamber, rate of robot, and configuration, determine the performance of cluster tools. Throughput is one of the common performance measures. The deterministic models developed here attempt to predict the throughput of a single-arm *N*-chamber parallel cluster tool. The model provides general insight into the impact of cluster tools parameters on performance.

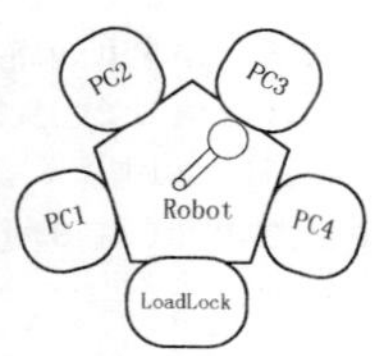


**Fig. 1.** 4-chamber cluster tool

## 2 Related Work

A lot of researches are focused on issues of cluster tools modelling and simulation. Basic performance analysis and throughput model development have been done in [1] and [2]. [3] extends the model in [1] to a cluster tool with redundant chambers. In the case of cluster tools with dual-blade robots, throughput studies are presented in [4].

Petri nets have been used to model cluster tools. Srinivasan first employs Petri nets to evaluate the performance of cluster tools [5]. Zuberek uses timed Petri nets to model the flow of wafers and sequences of actions performed by robots in [6] and [7]. [7] also investigates the steady-state performance of cluster tools with chamber revisiting. For a complicated configuration, Petri nets modelling and analysis are more difficult.

Simulation also plays an important role in performance analysis of cluster tool. [8] presents a flexible and sufficiently accurate cluster tool simulation model. [9] develops a simulation model to verify the above-mentioned model in [1] and [3]. But simulation does not reveal the relationships between the cluster tools parameters and performance.

There are two special configurations of cluster tools. They are referred as serial and parallel configurations. Consider a cluster tool with *N*-chamber in a serial configuration, all of the chambers are different from each other, and each wafer visits each chamber once. [1] develops a throughput model of a generic single-arm *N*-Chamber cluster tool in a serial configuration. On the other hand, in a parallel configuration, all of the modules are the same. The parallel cluster tool performs only one step on a wafer. This paper only considers the parallel configuration of the cluster tool with *N*-chamber.

## 3 Assumptions and Preliminaries

Before discussing the throughput model, we need to make some assumption and give a general description of the parallel cluster tools.

- At beginning, the robot is assumed to be idle at the loadlock.
- Every chamber's processing time is the same.
- The robot transport time between any two chambers or between a chamber and the loadlock is a constant.

When a batch of wafers arrives at an empty parallel cluster tool, it is placed in the loadlock which is then typically pumped down to vacuum. All the time required to do this is denoted as $T_{load}$.

The robot transports the first wafer and loads it into the first chamber, and then returns to the loadlock to transport another wafer and loads it into the second chamber. After there are wafers in all chambers, it returns to the first chamber and unload it. This is called the initial transient state and the time during the state is donated as $T_s$.

After that, the steady state begins. The robot unloads the processed wafer from a chamber, transports and loads it into the loadlock, and unloads anther wafer from the loadlock, transports and loads it into the empty chamber. Then it processes a wafer in next chamber in the same way. When the number of raw wafers in loadlock is less than or equal to *N*, the steady state ends. The time during the steady state is donated as $T_{steady}$. In the steady state, the time between two subsequent processed wafers arriving at the loadlock is defined the wafer fundamental period and denoted as $WFP$.

The final transient state comes after the steady state. The robot transports the wafers one by one to the loadlock after they are processed. The time during the final transient state is donated as $T_e$.

When all wafers return to the loadlock, the loadlock is raised to atmosphere pressure and the batch is removed. The time during this period is denoted as $T_{unload}$.

In general, the time to process a batch, denoted as $TB$, is the sum of $T_{load}$, $T_s$, $T_{steady}$, $T_e$ and $T_{unload}$.

$$TB = T_{load} + T_s + T_{steady} + T_e + T_{unload} \tag{1}$$

# 4 The Model

There are two types of schedules in the cluster tool operation, called transport-bound schedule and process-bound schedule. In the transport-bound schedule, the robot is always busy. When a process chamber finishes the wafer process, it has to wait the robot to unload the wafer. In the case of the process-bound schedule, the robot is idle sometimes and waits for a process chamber to complete the wafer process.

## 4.1 Transport–Bound Schedule

Figure 2 shows a timing diagram of a cluster tool during the initial transient state in the transport–bound mode. The timing diagram is a Gantt charting of the chambers and robot operations in the cluster tool. Here, filled boxes depict chamber busy periods, denoted as $P$. The arrows depict robot movements, and each arrow represents the time required to transport a wafer. The transfer of a wafer involves: (1) the robot without a wafer arriving at a chamber or the loadlock ($t_e$); (2) unloading a wafer from the chamber or the loadlock($t_l$); (3) transporting the wafer to another chamber or the loadlock ($t_w$); (4) loading the wafer into the chamber or the loadlock($t_l$). In figure, $T_1 = t_w + 2t_l$, $T_2 = t_w + t_e + 2t_l$.

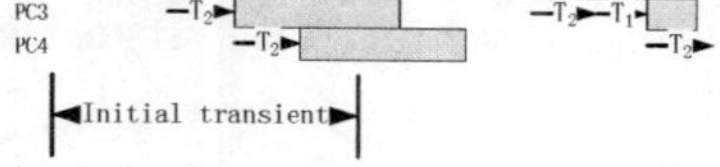


**Fig. 2.** Initial transient in transport-bound

As figure 2 shows, the robot operations in initial transient state are as follows:

- Unload the first raw wafer from the loadlock, transport it to chamber 1 and load it; chamber 1 starts its process;
- Move to the loadlock, unload the next raw wafer, transport it to chamber 2 and load it; chamber 2 starts its process;
- Load chamber 3 and 4 in the same way;
- Move to chamber 1, unload the processed wafer, transport it to the loadlock and load it;

With reference to Figure 2, the initial transient state time $T_s$ is

$$T_s = T_1 + 3T_2 + T_2 = T_1 + 4T_2 \tag{2}$$

For a parallel cluster tool with $N$-chamber, the following equation can be derived:

$$T_s = T_1 + (N-1)T_2 + T_2 = T_1 + NT_2 \tag{3}$$

After the initial transient state, the cluster tool enters the steady state. As shown in Figure 3, the robot operations in each cycle are as follows:

- Unload next raw wafer from the loadlock, transport it to chamber 1 and load it; chamber 1 starts its process;
- Move to chamber 2, unload the processed wafer, transport it to the loadlock and load it;
- unload next raw wafer from the loadlock, transport it to chamber 2 and load it; chamber 2 starts its process;
- process chamber 3 and 4 in the same way;
- Move to chamber 1, unload the processed wafer, transport it to the loadlock and load it;

From Figure 3, the wafer foundation period $WFP$ is $T_1 + T_2$, it is independent of the number of process chamber $N$. Furthermore, the wafer foundation period of a $N$-chamber cluster tool in steady state can be derived:

$$WFP = T_1 + T_2 \tag{4}$$

The time spent in steady state is

$$\begin{aligned} T_{steady} &= M * WFP \\ M &= \left\lfloor \frac{NW - N}{N} \right\rfloor * N \end{aligned} \tag{5}$$

$M$ is the number of wafers processed in a steady state. $NW$ is the number of wafers in a batch. Operator $\lfloor\ \rfloor$ represents rounding down.

As shown in figure 3, in the transport-bound schedule, the chamber process time needs to obey the condition

$$P < K = 3*(T_1 + T_2) \tag{6}$$

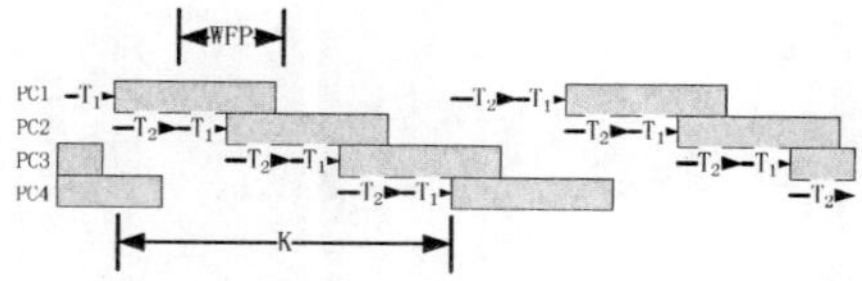


**Fig. 3.** Steady state in transport-bound

For a $N$-Chamber cluster tool, the condition of transport-bound is

$$P < (N-1)*(T_1 + T_2) \tag{7}$$

Changes of the process time within the limit K does not alter $WFP$. When the process time is more than K, the cluster tool enters the process-bound region.

The final transient timing diagram is shown as Figure 4. The robot operations are slightly different from that in the steady state. When there are no raw wafers in the loadlock, the robot always unloads wafers from chambers and loads them into the loadlock.

The final transient state time in this example is

$$T_e = 3T_2 + 3(T_1 + T_2) \tag{8}$$

As shown in figure 4 , the time of the final transient depends on both transport time and the remainder of $N$ divided by $NW$. For a general parallel cluster tool with $N$-chamber, the following equations can be derived:

$$\begin{aligned} T_e &= (N-1)*T_2 + m*(T_1 + T_2) \\ m &= \mathrm{mod}(NW, N) \end{aligned} \tag{9}$$

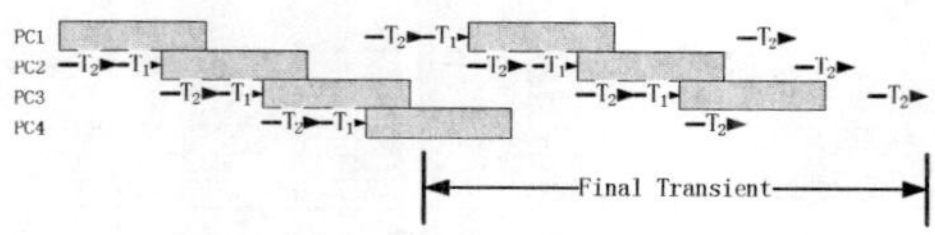


**Fig. 4.** Final transient in transport-bound

### 4.2 Process-Bound Schedule

Based on the condition (7), when the process time is more than $(N-1)*(T_1 + T_2)$, the cluster tool enters process-bound region. The timing diagram is shown as figure 5. The time of initial transient state is

$$T_s = P + T_1 + T_2 \tag{10}$$

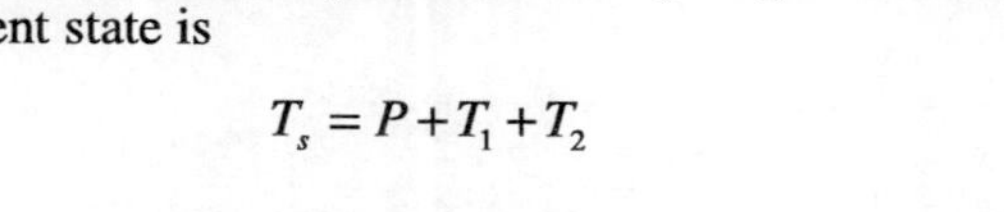


**Fig. 5.** Initial transient in process-bound

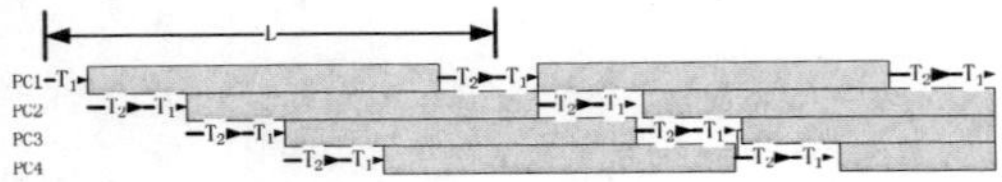

**Fig. 6.** Steady state in process-bound

The number of process chamber has no impact on this time. Thus the initial transient time of a *N*-chamber cluster tool is the same as (10).

In process-bound mode, the timing diagram of the cluster tool operating in the steady state is shown in figure 6. The steady state can be divided into several cycle L. As figure 6 shown, there are 4 wafers outs in each cycle L for a 4-chamber cluster tool. The wafer foundation period is

$$WFP = (P + T_1 + T_2)/4 \tag{11}$$

For a $N$-chamber cluster tool, there are N wafers outs in each cycle L. The wafer foundation period can be derived:

$$WFP = (P + T_1 + T_2)/N \tag{12}$$

Figure 7 depicts the timing diagram of a 4-chamber cluster tool in which chamber's process time are different from each other. Here, filled boxes represent chamber busy periods and empty boxes in dashed lines denote allowable process limits. Changes in the process time within these time limits do not alter *WFP* of the cluster tool. Thus the assumption that all chambers' processing time is the same can be removed. That is, when chamber's process times are different, the maximum process time defines the wafer foundation period.

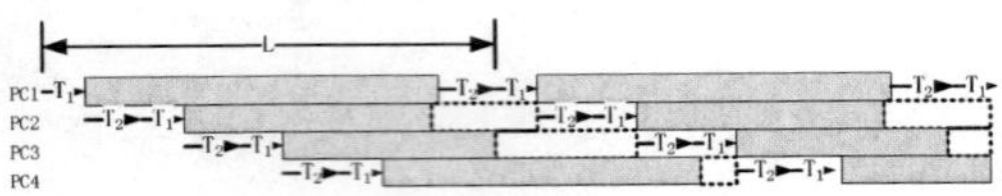

**Fig. 7.** Variable process time cluster

Figure 8 illustrates the final transient state operation of a 4-chamber cluster. The time of the final transient state is

$$T_e = P + 3 * (T_1 + T_2) \tag{13}$$

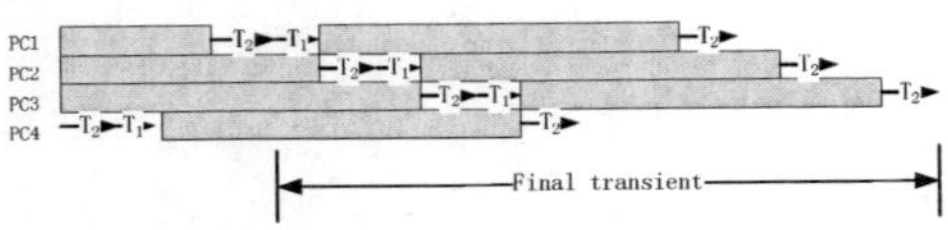


**Fig. 8.** Final transient in process-bound

For a parallel cluster tool with *N*-chamber, the following equation of the final transient time can be derived:

$$T_e = \begin{cases} (N-1)*(T_1+T_2) & m=0 \\ P+m*(T_1+T_2) & m\neq 0 \end{cases} \tag{14}$$
$$m = \mathrm{mod}(NW, N)$$

Combining equation (3), (4), (9), (10), (12), (14), the complete model can be derived:

$$TB = \left\lfloor \frac{NW-N}{N} \right\rfloor * N * WFP + T_{load} + T_{unload} + T_s + T_e \tag{15}$$

## 5 Performance Analysis Software

Based on the throughput model of the parallel cluster tool presented in section IV, this paper develops the performance analysis software for the *N*-chamber cluster tool. Figure 9 is the running interface of the software. The software can output accurately transient state time, wafer foundation period, and batch period based on input parameters.

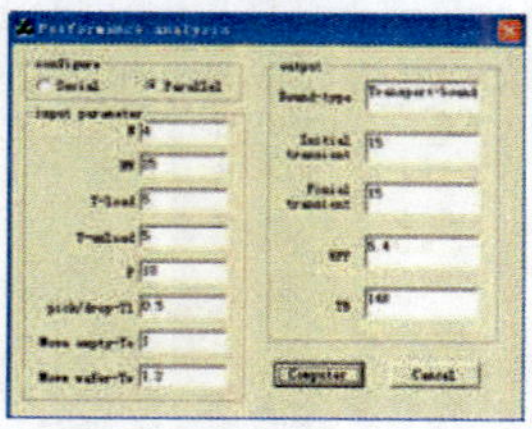

**Fig. 8.** Performance analysis software of cluster tools

## 6 Conclusion

Throughput analysis of cluster tools is critical for production improvement and cost reduction. This paper deduces a throughput model of a general single-arm $N$-Chamber parallel cluster tool. In transport-bound schedule, throughput depends on the number of chambers, transport time, and the remainder of $N$ divided by $NW$, but changes of process time have not affect throughput. When the processing time is more than $(N-1)*(T_1+T_2)$, the cluster tool comes into process-bound region. Throughput in this region also depends on process time. This model is helpful to compute and analyse performance of cluster tools.

## References

1. Perkinson, T.L., MacLarty, P.K., Gyurcsik, R.S., Cavin III, R.K.: Single-wafer cluster tool performance: An analysis of throughput. IEEE Trans. Semicond. Manufact. 7, 369–373 (1994)
2. Wood, S.C., Tripathi, S., Moghadam, F.: A generic model for cluster tool throughput time and capacity. In: 1994 IEEE/SEMI Advanced Semiconductor Manufacturing Conference, pp. 194–199 (1994)
3. Perkinson, T.L., Gyurcsik, R.S., MacLarty, P.K.: Single-wafer cluster tool performance: An analysis of the effects of redundant chambers and revisitations sequences on throughput. IEEE Trans. Semicond. Manufact. 9, 384–400 (1996)
4. Venkatesh, S., Davenport, R., Foxhoven, P., Nulman, J.: A steady state throughput analysis of cluster tools: Dual-blade versus single-blade robots. IEEE Trans. Semicond. Manufact. 10, 418–423 (1997)
5. Srinivasan, R.S.: Modeling and performance analysis of cluster tools using Petri nets. IEEE Trans. Semicond. Manufact. 11, 394–403 (1998)
6. Zuberek, W.M.: Timed Petri nets in modeling and analysis of cluster tools. IEEE Transactions on Robotics and Automation 17, 562–575 (2001)
7. Zuberek, W.M.: Cluster tools with chamber revisiting-modeling and analysis using timed Petri nets. IEEE Trans. Semicond. Manufact. 17, 333–344 (2004)
8. LeBaron, H.T., Hendrickson, R.A.: Using emulation to validate a cluster tool simulation model. In: Proceedings of the 2000 Winter Simulation Conference Proceedings, vol. 2, pp. 1417–1422 (2000)
9. Gupta, A.K., Lendermann, P., Sivakumar, A.I., Priyadi, J.: Simulation analysis of cluster tool operations in wafer fabrication. In: Proceedings of the 2008 Winter Simulation Conference Proceedings, Austin, TX, pp. 2141–2142 (2008)

# Two Level Strategies for Supply Chain Knowledge Innovation*

Wei-hua Wang and Nie Qing

International Business School
Shanghai Institute of Foreign Trade, SHIFT, Shanghai, China
Wilson_0011@163.com, n5678g@hotmail.com

**Abstract.** Analyzed the strategy of knowledge innovation in supply chain level after discussing the Characteristic of supply chain knowledge innovation, presented enterprise level strategies according to difference in market station and technical character of product.

**Keywords:** Supply chain, Knowledge innovation, Strategy.

## 1 Introduction

The vital importance of innovation in today's competitive climate has been widely canvassed. "In today's business environment, there is no executive task more vital and demanding than the sustained management of innovation and change…to compete in the new environment, companies must create new products, services and processes…they must adopt innovation as a way of corporate life"[1]. While the competition between enterprises becomes competition between supply chains in the 21st centaury [2], innovation with no doubt plays a critical role for the success of supply chain operation. Many researchers are devoted to the problem solving of supply chain knowledge management [3][4].

## 2 The Characteristic of Supply Chain Knowledge Innovation

Knowledge Innovation is developed by interaction of individuals and their knowledge no matter for supply chain or for enterprise, but the structure of supply chain makes the knowledge innovation of supply chain has its own characteristic:

- The complexity of the products in supply chain that they often composed of many kinds of components whose technique may either be mature or under development determines that the knowledge innovation of supply chain cannot be made in a same mode.

* Sponsored by 085 Project (08509009), Research Innovation Fund (10YS167) of Shanghai Committee of Education, ChenGuang Project (09CG62) of Shanghai Educational Development Foundation.

- The connection for different sub-system of products is dissimilar. Some of them are closely connected, others are loosely connected, and they should be treated distinctively in knowledge innovation.
- The difference of the components in the supply chain that some of them are advanced in technique and market and some of them are not makes it necessary to handle the knowledge innovation diversely.

Strategies designing for Supply Chain Knowledge Innovation should take these characteristics into consideration. We will discuss it from the supply chain level and the enterprise level:

## 3 The Knowledge Innovation Strategy for Supply Chain Level

The design of knowledge innovation strategy for supply chain level should take the systematic characteristic of knowledge in supply chain into account, strategies may include:

### 3.1 Strategy to Modularize and Manage Level by Level

The interaction of the knowledge makes the knowledge management very complex because changes in part of the system will bring forth a series of changes in the whole system. Measures should be taken to reduce the complexity and the interaction scope. By managing the whole system level by level, the complexity can be reduced. Meanwhile, the connection rules are regulated between different modules, the selection scope for designers is restricted and the repeated interaction is decreased.

Modularization and managing the knowledge innovation level by level makes the knowledge interaction in the supply chain well-ordered. The interaction of the knowledge is no longer in the whole system but in the levels and modules and between them.

The intensity difference of interaction between knowledge system and subsystems in the supply chain should also be taken into account. Charles Perrow puts forward a framework to analyze from two aspects: (1) task variability, that is frequency of unexpected and novel events that occur in conversion process (2) problem analyzability, that is task can be reduced to mechanical steps or specified with a computational procedure. The combination of these two dimensions yields Perrow's framework which classified four kinds of technologies that is craft, routine, nonroutine and engineering, as shown in figure 1.

The organizing and coordinating ways vary with the kind of technology, the more routine the technology is, the less interaction level we needs, vice versa.

### 3.2 Strategy to Intensify Interaction between Similar Knowledge Modules

Many of the knowledge modules in supply chain are closely linked or similar to each other such as knowledge needs for suppliers provide similar products. Managers should make well arrangements to intensify interaction between these modules because:

First, the more closely linked between different knowledge modules, the more value the interaction will be, this kind of interaction will give resonance to each other and yield many new ideas.

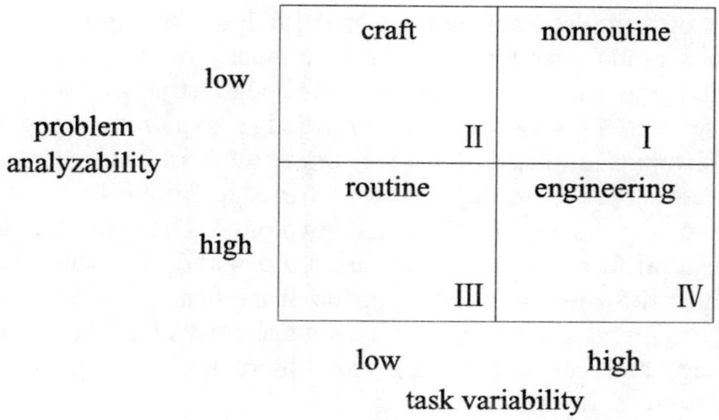


**Fig. 1.** Perrow's framework

Second, the interaction in supply chain needs coordinated knowledge level for participant parts. Parts with low knowledge innovation level weaken the competition level of the whole supply chain. By intensifying interaction between similar knowledge modules the low level parts can get help from high level parts. But the high level parts should be put a premium on, for example, more portfolios to encourage knowledge innovation efforts and avoid hitchhiking.

The interaction modes involved in the supply chain knowledge innovation are numerous, to make the strategy for supply chain level works well, the management system should be consistent with these modes.

# 4 The Knowledge Innovation Strategy for Enterprise Level

The knowledge innovation strategy should consider not only interaction for supply chain level, but also arrangement for enterprise level. Market position and evolution phase of technology can be taken into account for the strategy design at this level.

Many aspects of knowledge innovation have been analyzed, for example, the types of knowledge (tacit or explicit, general or specific, etc.), how to acquire knowledge (internally acquired vs. externally acquired orientation), and the stability of knowledge system (whether they need to update and develop new knowledge continuously). The relationship between a company's strategy for knowledge management and its competitive strategy has also been discussed.

Hansen, Nohria &Tierney [5] concluded two kinds of strategy called "codification strategy" and "personalization strategy". For the former strategy, knowledge is carefully codified and stored in databases, where it can be accessed and used easily by anyone in the company, for the latter strategy, knowledge is closely tied to the person who developed it and is shared mainly through direct person- to- person contacts, the chief purpose of computers is to help people communicate knowledge, not to store it.

Companies follow a standardized product (sell products that do not vary much or mature product) strategy should consider the codification strategy, and companies follow a customized or innovative product strategy should consider the personalization strategy.

At the aspect of attitude, Zack noticed two kinds of strategy called aggressive strategy and conservative strategy.[6] Firms oriented toward exploiting internal knowledge exhibit the most conservative knowledge strategy, while unbounded innovators (those who closely integrate knowledge exploration and exploitation without regard to organizational boundaries) represent the most aggressive strategy. Firms taking a conservative strategy view knowledge primarily as an objectified proprietary asset to be protected and financially exploited. They create tacit knowledge internally, explicate it to ease its transfer and reuse within the firm, and attempt to create barriers to its diffusion or transfer outside of the firm into the industry at large. For firms taking a aggressive strategy, their advantage comes from being able to absorb external knowledge and integrate it with their internal knowledge to develop new insights faster than the competitor.

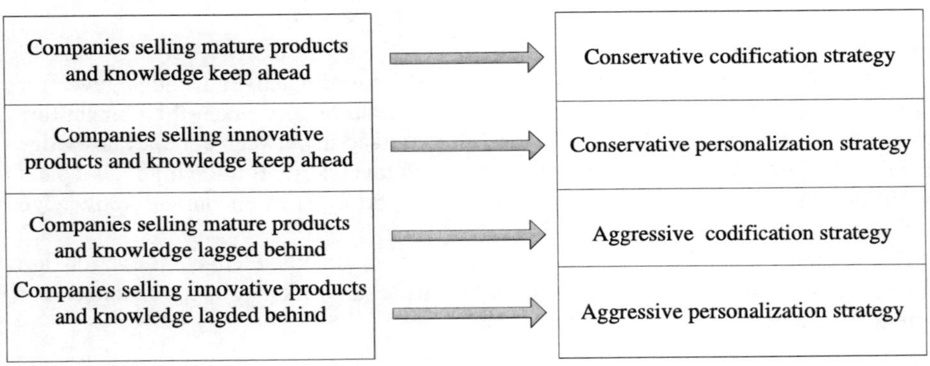


**Fig. 2.** The Knowledge Innovation Strategy for Enterprise Level

Market position and evolution phase of technology of companies in the supply chain vary from each other. Some are selling mature products, while others selling innovative products. As for the market position, knowledge of some of them keep ahead, others may lag behind. Taking different requirements of these companies into account, knowledge innovation strategy may design as follows:

Companies selling mature products and their knowledge keep ahead can take conservative codification strategy. For companies whose knowledge keeps ahead, an conservative strategy will be required to protect its core knowledge, and the codification strategy opens up the possibility of achieving scale in knowledge reuse and thus of growing the business.

Companies selling innovative products and their knowledge keep ahead can take conservative personalization strategy. The firm can exploit internal knowledge while people in companies seeking innovation by sharing information that would get lost in document form, use person to person approach to facilitate utilizing and developing tacit knowledge.

Companies selling mature products and their knowledge lagged behind can take aggressive codification strategy. These companies need to aggressively acquire knowledge to improve their competition level, and at the same time develop elaborate ways to codify, store, and reuse knowledge.

Companies selling innovative products and their knowledge lagged behind can take aggressive personalization strategy. Absorb new knowledge to create differentiate competition strength; cooperate with contract to communicate tacit knowledge with other company.

Aggressive or conservative, codification or personalization, Companies may pursue one strategy predominantly and use the second strategy to support the first. In knowledge intensive industries, it's a wise choice to pursue an aggressive knowledge strategy.

## 5 Conclusion

The design of knowledge innovation strategy for supply chain should take into consideration the complexity of the products in supply chain, the connection for different sub-system of products and the different characteristic of the member's technology. For strategy designing in supply chain level, modularization and managing level by level and intensifying interaction between similar knowledge modules can make the knowledge interaction more efficient. For strategy designing in enterprise level, whether to pursue aggressive or conservative, codification or personalization strategy varies according to a company's market position and evolution phase of technology.

## References

1. Storey, J.: The management of innovation problem. International Journal of Innovation Management 4(3), 3 (2000)
2. Christopher, M.: Logistics and Supply Chain Management – Strategies for Reducing Cost and Improving Service, vol. (1), pp. 82–85. Electronic Industries Press, Peking (2003)
3. Hall, R., Andriani, P.: Management Focus Analyzing intangible resources and managing knowledge in a supply chain context. European Management Journal 16(6), 685–697 (1998)
4. Holsapple, C.W., Singh, M.: The knowledge chain model: activities for competitiveness. Expert Systems with Applications 20, 77–98 (2001); 28 (1999)
5. Hansen, M.T., Nohria, T., Tierney, T.: What's your strategy for managing knowledge. Harvard Business Review, 106–116 (March-April 1999)
6. Zack, M.: Managing codified knowledge. Sloan Management Review 40(4), 45–58 (1999)

# Analysis and Design of RFID Tag Ownership Transfer Protocol

Wang Shaohui

College of Computer, Nanjing University of Post and Telecommunication NanJing, 210046, China
wangshaohui@njupt.edu.cn

**Abstract.** In some radio frequency identification (RFID) applications, the tag may change owner during the course of its lifetime. In the conference RFIDsec 2008, Song proposed an RFID tag ownership transfer protocol. In this paper, we point out that the protocol can not satisfy the security requirement of Old Owner Privacy, i.e., the new owner can track the previous communication between the old owner and the tag. Based on a recently proposed efficient secure mutual authentication protocol [5], we present an efficient tag ownership transfer protocol, and the analysis show the protocol can reach the security requirement of privacy, authentication, old owner privacy and new owner privacy.

**Keywords:** RFID, Ownership Transfer Protocol, Mutual Authentication, Privacy.

## 1 Introduction

Radio Frequency Identification (RFID) is an automated object identification technology. RFID systems at least consist of two main components: tags and readers. An RFID tag consists of an integrated circuit for storing and processing data, and an antenna for communicating via an RF signal with RFID readers. RFID readers are typically connected to a back-end server with a database containing information associated with the RFID tags that it manages.

While being promising in a wide range of applications such as supply chain management, inventory tracking and libraries, RFID also raises privacy and security concerns [1]. Authentication and privacy are two significant problems confronted by RFID systems. Because of the limited computing power and source of RFID tags, the protocols used in RFID system are required to be efficient. However, the security features needed in RFID systems are almost the same as needed in other systems: authenticity, integrity, confidentiality, untraceability and availability. A considerable volume of papers have been published to provide solutions to protect RFID communications [2,3,4,5].

Tag ownership transfer is one of the possible requirements for some RFID applications, in which an RFID tag may change its owner a number of times during its lifetime. When the ownership of a tag is transferred from one owner to another, the new(old, respectively) owner should not have the capability to trace the tag interactions happening before (after, respectively) the ownership transfer. Molnar et al.[6] proposed the first paper explicitly to deal with ownership transfer using tree structure. The protocol is based on a three-party model, which involves a Trusted Center (TC), so this scheme is rather restrictive. Saito, Imamoto and Sakurai[7] suggested two approaches for reassigning an RFID tag's key for ownership transfer. The first one is based on three-party model, and the second (SIS-2 scheme) uses a two-party model. SIS-2 scheme is based on the premise that the backward channel (i.e. the communication channel from the tag to the reader) is more secure against eavesdropping than the forward channel (i.e. the communication channel from the reader to the tag), which is questionable. Other two-party ownership transfer protocols either provide no old owner privacy[8] or are vulnerable to tag tracking attacks, or replay attacks[9,10,11].

In 2008, Song[12] proposed an RFID tag ownership transfer protocol using a two-party model. The protocol utilizes secure channel to protect the secret transferred between the new and old owner. The protocol is claimed to provide all the necessary security properties and is efficient in terms of storage and communication requirements. But Cai et.al.[13] discovered the protocol is vulnerable to some active attacks including server impersonation, tag impersonation and de-synchronization.

In this paper, we give a further security analysis on Song's tag ownership transfer protocol, and point out the protocol can not satisfy old owner privacy, i.e. the new owner can deduce the previous secret shared between the old owner and the tag, thus he can track the previous communication. In addition, we present an efficient secure ownership transfer protocol based on the mutual authentication protocol recently proposed by Zhao Yunlei et. al.[5].

The rest of the paper is organized as follows. In section 2, we review the security requirements for tag ownership transfer, and give the notations used in the protocols. The attack on Song's protocol is given in section 3. In section 4, we propose our new protocol, and give the security analysis. Section 5 concludes the paper.

## 2 Security Requirements and Notations

An adversary can use different strategies to undermine an RFID system. The classes of attacks are divided as passive attacks and active attacks. The aim of a well-suitable RFID authentication protocol is to thwart these attacks. The RFID protocol should mainly resist the following threats[12]:

- Tag information leakage. If an unauthorized reader can obtain a tag identifier, then it may be able to access the private information related to the tag held in the server database.
- Replay attacks. An adversary can intercept messages exchanged between a reader and a tag, and replay them.
- Desynchronization attack. This is a kind of active attack. The adversary tries to destroy the synchronization between the tag and Reader through interrupting the authentication messages.
- Tag tracking attack. The adversary traces the tags from their protocol flows.
- Man-in-the-middle attacks. An adversary could insert or modify messages sent between a reader and a tag without being detected.

- Tag impersonation. In this case, an adversary would like to impersonate a legitimate tag to fool a legitimate reader. It would communicate with a reader instead of the tag and be authenticated as the tag.

- Server impersonation. If an adversary compromises a reader, then it might be able to impersonate a legitimate server to the tag using knowledge of the tag internal state.

In a tag ownership transfer protocol, except for the above security requirements, the following requirements should be satisfied in order to protect the privacy of the tag:

- New owner privacy: after ownership transfer, only the new owner should be able to identify and control the tag. The old owner should no longer be able to identify or trace the tag.

- Old owner privacy: When ownership of a tag has been transferred to a new owner, the new owner of a tag should not be able to trace past interactions between the tag and its old owner.

**Notations:** Here we give the notations used in the following protocols. Let $l$ and $l'$ be two integers, and notations $\oplus$, >>, <<, || and $\leftarrow$ mean XOR, right circular shift, left circular shift, concatenation and substitution operator respectively. $f_k$ is a keyed hash function with $f_k : \{0,1\}^* \; * \{0,1\}^l \rightarrow \{0,1\}^l$, and $h$ is a hash function with $h : \{0,1\}^* \rightarrow \{0,1\}^l$. Let $E_k : \{0,1\}^{2\kappa} \rightarrow \{0,1\}^{2\kappa}$ be a lightweight block cipher such as PRESENT [14], and $D_k$ is the corresponding decryption algorithm. $E_k^0$ (respectively $E_k^1$) is the $\kappa$-bit prefix (resp., suffix) of the output of $E_k$. $pad_1$ and $pad_2$ are two predetermined padding strings with the bit length $l'$, and $ctr$ is the counter of length $2\kappa - l'$ with the initial value 1. $I$ is the index number with the initial value being $E_K^0(1 \,||\, pad_1)$ \$.

## 3 Attack on Song's Protocol

We first review Song's ownership transfer protocol, then we will show the protocol can not satisfy the security requirement of old owner privacy.

Song's protocol is divided into two parts: ownership transfer protocol (**P1**) and secret update protocol (**P2**). Through the first protocol, the old owner will transfer the information related to the tag to the new owner via a secure channel, and the new owner will establish new secret with the tag in the second protocol. The tag has the secret $t$, and the old owner has secret of $(t, s)$ with $t = h(s)$. Here we only depict the protocol **P1** as follows.

1. The new owner generates and sends to the tag a random string $r_1$ of $l$ bits.

2. The tag generates a random string $r_2$ of $l$ bits, computes $M1 = t \oplus r_2$ and $M2 = f_t(r_1 \oplus r_2)$, and then sends $M1$ and $M2$ to the new owner.

3. The new owner forwards $r_1$, $M1$ and $M2$ to the old one, with a request for ownership of Tag (RT).

4. After receiving the message, the old owner performs the following steps:

(a) If the received request RT is valid, it searches through its database for a pair $(t, s)$ for which the value of $t$ satisfies $M2 = f_t(r_1 \oplus M1 \oplus t)$.

(b) If such a pair $(t, s)$ is found, it sets $r_2 = M1 \oplus t$ and computes $M3 = s \oplus (r_2 >> l/2)$. Otherwise, the session stops.

(c) The old owner computes current secrets as $s' \leftarrow (s << l/4) \oplus (t >> l/4) \oplus r_1 \oplus r_2$ and $t' = h(s')$.

(d) The old owner sends $M3$ to the new owner, and transfers the updated secrets $(t', s')$ and the other necessary information for the tag ($Info$) via a secure channel.

5. The new owner stores $(t', s')$, $Info$ in its database, and forwards $M3$ to the tag.

6. The tag then computes $s = M3 \oplus (r_2 >> l/2)$, and checks that $h(s) = t$. If the verification succeeds, the tag has authenticated the new owner as an legitimate server, and updates $t$ as $t' \leftarrow h((s << l/4) \oplus (t >> l/4) \oplus r_1 \oplus r_2)$. Otherwise, the session stops.

**Attack on Song's Protocol.** Our attack bases on thefollowing two simple facts about the operators:

**fact 1:** for any integers $a, b, c$, Operation $\oplus$, >> has the following property: $(a \oplus b) >> c = (a >> c) \oplus (b >> c)$, which is also satisfied for the operator <<.

**fact 2:** As to the operation >> and <<, for any integer $a$ with the bit length $l$, the value of $(a >> l/2) << l/4$ is equal to $(a >> l/4)$.

In the above protocol, the new owner has received the messages $M1$, $M3$ and $s'$ from the tag and the old owner as follows:

$$\begin{cases} M1 = t \oplus r_2 \\ M3 = s \oplus (r_2 >> l/2) \\ s' = (s << l/4) \oplus (t >> l/4) \oplus r_1 \oplus r_2 \end{cases}$$

From the first equation, the new owner can get $t = M1 \oplus r_2$, and he can obtain $s = M3 \oplus (r_2 >> l/2)$ from the second equation. Utilizing these two expressions and the two facts, the new owner can first compute the random number $r_2$ which is chosen by the tag, then deduce the previous secret $(t, s)$ between the old owner and the tag easily:

$s' = (s << l/4) \oplus (t >> l/4) \oplus r_1 \oplus r_2 = ((M3 \oplus (r_2 >> l/2)) << l/4) \oplus ((M1 \oplus r_2) >> l/4) \oplus r_1 \oplus r_2 = (M3 << l/4) \oplus ((r_2 >> l/2) << l/4) \oplus (M1 >> l/4) \oplus (r_2 >> l/4) \oplus r_1 \oplus r_2 = (M3 << l/4) \oplus (M1 >> l/4) \oplus r_1 \oplus r_2$, thus the random number $r_2$ can be calculated as: $r_2 = s' \oplus (M3 << l/4) \oplus (M1 >> l/4) \oplus r_1$. From the first and second equations, the new owner can get the secret $(t, s)$, so the previous communication of the tag can be tracked, and the requirement of old owner privacy does not satisfy.

## 4 A New Tag Ownership Transfer Protocol

In this section, we first present a new efficient tag ownership transfer protocol based on the mutual authentication protocol proposed in [5]. Then we give a security analysis briefly.

### 4.1 New Ownership Transfer Protocol

The new protocol follows the same structure of Song's protocol, i.e. it is divided into two sub-protocols: Ownership transfer Protocol (**P1**) and Key Update Protocol (**P2**).

**Ownership transfer Protocol (P1).** Through this protocol, the old owner will transfer the information related to the tag to the new owner, and the new owner can convince that the tag and the new owner are all legitimate. The old owner has the secret information of $(I, k, ctr, ID)$ for all the tags, and the tag has the secret of $(k, ctr)$. In order to take ownership of the tag, the new owner will communicate with the old owner and the tag as follows:

1. The new owner will choose randomly a challenge string $c \in \{0,1\}^{\kappa}$, and sends it to the tag;

2. After receiving the challenge, the tag will compute $I = E_k^0(ctr \| pad_1)$, $(r_0, r_1) \leftarrow E_k(c \| I)$, $r_\tau \leftarrow r_0 \oplus (ctr \| pad_2)$. The tag sends $I \| r_\tau$ to the new reader, and update its counter $ctr$ as $ctr + 1$.

3. The new owner forwards $c \| I \| r_\tau$ to the old owner, together with a request for ownership transfer of the tag $R_T$.

4. On receipt of the message, the old owner will perform the following operations:

(a) If the request is valid, the old owner first searches its database for a tuple $(I, k, ctr', ID)$, computes $(r_0, r_1) \leftarrow E_k(c \| I)$, and checks whether $ctr' \| pad_2 = r_0 \oplus r_\tau$. If no such tuple exists, the tag will check whether $(I', k, ctr', ID)$ exists that satisfies $ctr \| pad_2 = E_k^0(c \| I) \oplus r_\tau$ and $E_k^0(ctr \| pad_1) = I$.

(b) If such tuple exists, the old owner will choose a random secret key $k'$, and compute $r_R = E_k(r_1 \| k')$. Otherwise the old owner will stop this session.

(c) The old owner sends $r_R$ to the new owner, and transfers the new secret key $k'$ and necessary information for the tag ($Info$) via a secure channel.

5. The new owner stores $k'$ and $Info$ in its database, and send $r_R$ to the tag.

6. After receiving the message $r_R$, the tag computes $r' \| k' \leftarrow D_k(r_R)$, if $r_1' = r_1$, The tag will update its secret as $k'$. Otherwise, the session stops.

**Secret Update Protocol (p2).** Through this protocol, the new owner will establish a new secret with the tag in order to prevent the old owner from tracing the tag. The protocol **P2** should be performed at a distance from any readers connected to the old owner. Protocol **P2** is proceeded as follows:

1. The new owner randomly chooses a challenge $c$, and sends it to the tag;

2. Then the tag will compute $I = E^0_{k'}(ctr \| pad_1)$, $(r_0, r_1) \leftarrow E_{k'}(c \| I)$, $r_\tau \leftarrow r_0 \oplus (ctr \| pad_2)$. The tag sends $I \| r_\tau$ to the new reader, and update $ctr$ as $ctr + 1$.

3. On receipt of the message, the new owner will check whether a tuple $(k', ID)$ exists that satisfies $ctr \| pad_2 = E^0_{k'}(c \| I) \oplus r_\tau$ and $E^0_{k'}(ctr \| pad_1) = I$. If such tuple exists, the new owner will choose a random secret key $k_{new}$, and send $r_R \leftarrow E_{k'}(r_1 \| k_{new})$ to the tag. In addition, the owner sets $ctr = 1$, stores $I = E^0_{k_{new}}(1 \| pad_1)$. Otherwise the owner will stop this session.

4. After receiving the message $r_R$, the tag computes $r_1' \| k_{new} \leftarrow D_{K'}(r_R)$, if $r_1' = r_1$, The tag will update its secret as $k_{new}$, and $ctr$ as 1. Otherwise, the session stops.

## 4.2 Security Analysis

The protocols **P1** and **P2** are efficient in terms of storage, communication and computation. In **P1**, the computation is at most three lightweight block cipher computations in the tag and the old owner; and in **P2**, the computation is also three lightweight block cipher computations in the tag and the new owner.

The protocol is based on a mutual authentication protocol[5] which is proved to be adaptive completeness, mutual authentication and zero-knowledge privacy. In **P1**, the old owner and the tag should authenticate mutually; and so should the new owner and the tag in the protocol **P2**. Along with the authentication in both sub-protocols, the owner sends $r_R = E_k(r_1 \| k')$ (or $r_R = E_{k'}(r_1 \| k_{new})$) in order to update the secret key, which does not leak any information of the secret.

The messages exchanged between the server and tag are computed using randomly chosen challenge $c$, and a lightweight block cipher. Because the adversary does not know the secret $k$, without which the adversary can not counterfeit legitimate messages, so the protocol can resist some passive and active attack, such as replay attack, tag(reader) impersonation, man-in-the-middle attack. And because each time the tag's response is anonymous and unlinkable, the schemes can protect against tag location tracking attack.

In addition, in the protocol **P1**, the new owner obtains a randomly chosen string $k'$ as a secret key, and what he received is only ciphertexts which do not leak any information of the previous key $k$, without which the new owner can not track the previous communication. So the protocol can satisfy the requirement of Old owner privacy. In the same way, In the protocol **P2**, the new owner changes the secret $k'$ to $k_{new}$. Because we premise that the old owner can not eavesdrop the communication in this sub-protocol, the old owner can not trace the tag any more without the knowledge of $k_{new}$. Thus the protocol satisfy New owner privacy.

## 5 Conclusion

In some RFID applications it is necessary to allow for transfer of tag ownership. Using the properties of the operators of XOR and circular shift operators, we presented an analysis on the Song's tag ownership transfer protocol. The analysis showed the new owner can deduced the previous secret of the old owner and the tag, thus the previous communications can be tracked.

Based on a recently proposed efficient mutual authentication protocol, we propose a novel efficient tag ownership transfer protocol, which is divided into two sub-protocols **P1** and **P2**. In **P1**, after the mutual authentication between the old owner and the tag with the participation of the new owner, the old owner give the authorization to the new one. And in **P2**, after the mutual authentication between the new owner and the tag, the secret key will be updated in order to protect the future communication from the old owner.

**Acknowledgment.** This work was supported by National Natural Science Funds (Grant No.60903181) and Nanjing University of Post and Telecommunication Funds (Grant No.NY208072).

## References

1. Juels, A.: RFID security and privacy: A research survey. IEEE Journal on Selected Areas in Communications 24, 381–394 (2006)
2. Peris-Lopez, P., Hernandez-Castro, J.C., Tapiador, J.M.E., Ribagorda, A.: Advances in Ultralightweight Cryptography for Low-Cost RFID Tags: Gossamer Protocol. In: Chung, K.-I., Sohn, K., Yung, M. (eds.) WISA 2008. LNCS, vol. 5379, pp. 56–68. Springer, Heidelberg (2009)

3. Juels, A., Weis, S.A.: Authenticating Pervasive Devices with Human Protocols. In: Shoup, V. (ed.) CRYPTO 2005. LNCS, vol. 3621, pp. 293–308. Springer, Heidelberg (2005)
4. Cichoń, J., Klonowski, M., Kutyłowski, M.: Privacy Protection for RFID with Hidden Subset Identifiers. In: Indulska, J., Patterson, D.J., Rodden, T., Ott, M. (eds.) PERVASIVE 2008. LNCS, vol. 5013, pp. 298–314. Springer, Heidelberg (2008)
5. Deng, R.H., Li, Y., Yao, A.C., Yung, M., Zhao, Y.: A new framework for RFID Privacy. ePrint 2010/059, http://eprint.iacr.org/curr/
6. Molnar, D., Soppera, A., Wagner, D.: A Scalable, Delegatable Pseudonym Protocol Enabling Ownership Transfer of RFID Tags. In: Preneel, B., Tavares, S. (eds.) SAC 2005. LNCS, vol. 3897, pp. 276–290. Springer, Heidelberg (2006)
7. Saito, J., Imamoto, K., Sakurai, K.: Reassignment Scheme of an RFID Tags Key for Owner Transfer. In: Enokido, T., Yan, L., Xiao, B., Kim, D.Y., Dai, Y.-S., Yang, L.T. (eds.) EUC-WS 2005. LNCS, vol. 3823, pp. 1303–1312. Springer, Heidelberg (2005)
8. Lim, C., Kwon, T.: Strong and Robust RFID Authentication Enabling Perfect Ownership Transfer. In: Ning, P., Qing, S., Li, N. (eds.) ICICS 2006. LNCS, vol. 4307, pp. 1–20. Springer, Heidelberg (2006)
9. Osaka, K., Takagi, T., Yamazaki, K., Takahashi, O.: An Efficient and Secure RFID Security Method with Ownership Transfer. In: Wang, Y., Cheung, Y.-m., Liu, H. (eds.) CIS 2006. LNCS (LNAI), vol. 4456, pp. 778–787. Springer, Heidelberg (2007)
10. Fouladgar, S., Afifi, H.: A simple privacy protecting scheme enabling delegation and ownership transfer for RFID tags. Journal of Communications 2(6), 6–13 (2007)
11. Fouladgar, S., Afifi, H.: An efficient delegation and transfer of ownership protocol for RFID tags. In: First International EURASIP Workshop on RFID Technology, Vienna, Austria (September 2007)
12. Song, B.: RFID tag ownershio transfer. In: The 4th Workshop on RFID Security (RFIDsec 2008), Budaperst, Hungary (July 2008)
13. Cai, S., Li, Y., Li, T., Deng, R.: Attacks and Improvements to an RFID Mutual Authentication Protocol and its Extensions. In: The 2nd ACM Conference on Wireless Network Security (WiSec 2009), Zurich, Switzerland, March 16-18 (2009)
14. Bogdanov, A.A., Knudsen, L.R., Leander, G., Paar, C., Poschmann, A., Robshaw, M.J.B., Seurin, Y., Vikkelsoe, C.: PRESENT: An Ultra-Lightweight Block Cipher. In: Paillier, P., Verbauwhede, I. (eds.) CHES 2007. LNCS, vol. 4727, pp. 450–466. Springer, Heidelberg (2007)

# Image Detection to Vehicle Queue Length of Crossroad Based on DSP

Lu Weina, Wang Haifang, Ma Yuquan, Zhang Lihong, and Wang Qingzhu*

Department of Mechanics & Electronics Hebei Normal University of Science & Technology
Qinhuangdao, Hebei Province, China, 066004
haibian016@yahoo.com.cn

**Abstract.** An image detection to vehicle queue length at crossroad based on DSP is presented in this paper. It describes the hardware structure and the main program of the image detection system, as well as the vehicle queue length extraction method. The system is applied to the city traffic signal control machine, and it shows the system is better than the means of using inductance coil on precision, fixing and etc.

## 1 Introduction

It is needed to obtain road traffic volume when the signal lamps of crossroad are controlled by urban traffic signal controllers. Presently, in most domestic and international traffic signal control systems, the main detecting technology of traffic volume is using annular inductance coil [1, 2]. Such traffic detection device must be buried in the road and usually damage road surface. In addtion, it is not very satisfying on precision, stableness, installation and maintenance. However, capturing image on field and measuring the vehicle queue length with image detection can be described as another better way. This method avoids the disadvantage of the inductance coil. As the amount of data is very large, most of the image processing is executed on microcomputer platform. But the microcomputer has some instability in the system, such as computer crash, and the cost price of this system is high. Using the suitable DSP chips to construct the embedded system can solve this problem perfectly. The DSP chips adopt Harvard Bus Architecture and Pipelining. The hardware multiplication-accumulator MAC is specially designed for various digital signal processing algorithms. In this paper, the system is constructed with the master chip TMS320VC5402. Thus, it is realized to detect the vehicle queue length at crossroad by image processing techniques and obtain the traffic information on field.

## 2 Hardware Structure and Working Principle of System

According to the actual needs, the detection system of vehicle queue length is mainly composed of TMS320VC5402, SAA7111A, DT72V235, CY7C1021and CY7C1041 chips.

* Corresponding author.

TMS320VC5402 is high-performance digital signal processor from TI Company. It adopts Harvard Bus Architecture and special instruction system. In the TMS320VC5402 chip, there are eight buses, CPU, on-chip memory and on-chip peripheral circuits. The instruction cycle is 10ns and the computation capacity is 100MIPS [3, 4]. Also, the advantages are low power consumption, highly parallel operation and fast calculating speed.

SAA7111A is an image decoding chip from Philips Company. It is used to decode the image signals from analog CCD. The chip provides four analog input channels, two analog signal processing channels. It has several kinds of image data output stream like 16-bit, 12-bit or 8-bit. All kinds of image data, such as YUV4:1:1, YUV4:2:2, RGB(5,6,5) and RGB(8,8,8), are output from SAA7111A in the form of odd-even field. In the chip, there are also 24 internal registers initialized through I2C bus [5]. Because of the high operating frequency of TMS320VC5402 (more than 100MHz) and the low output clock frequency of SAA7111A (only 13.5MHz), a synchronous FIFO chip is introduced between the two chips as data cache. The selected type of FIFO chip is IDT72V235. Its storage capacity and data width are 2K × 18bit. The maximal read-write speed is 67MHz [6].

CY7C1021 and CY7C1041 from CYPRESS Company are respectively used for space expansion of data and program. Such expansion is necessary, because the available storage space in TMS320VC5402 chip is only 16K DARAM. The data space expansion chip is 64K×16bit CY7C1021 and the program memory expansion chip is 256K×16bit CY7C1041. The two chips are both CMOS-based asynchronous static RAM. Their accessing speed is 12ns and the maximal operating frequency can reach to 80MHz. All of these capabilities have satisfied the requirements of system [7, 8].

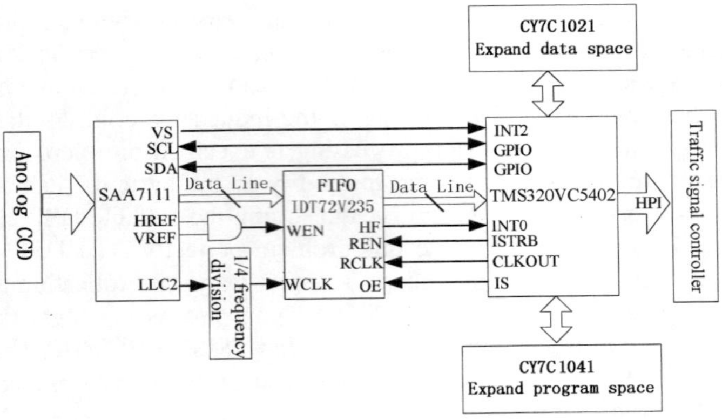


**Fig. 1.** The hardware structure of system

The hardware structure of the system has been shown in Fig.1. The analog image signals are captured by CCD and input to SAA7111A. After A/D conversion and horizontal & vertical sync separation, SAA7111A outputs YUV signals or RGB signals (Either signal to output can be decided by the initialized configuration of SAA7111A ), as well as some other sync signals. After the analog image is converted to digital image through SAA7111A, the image is sent to the FIFO chip IDT72V235. Later, while 1k data has been stored into the chip, FIFO begins to apply for DSP interruption to read the data. Following the interrupt, DSP can go back again to execute previous programs. In this way, image data can be read into DSP data space again and again through interrupt mode.

Here, another particular thing has to been explained. That is, the data space of TMS320VC5402 is only 64KB, while a full captured picture is $720 \times 576 = 405$KB, so we can see that TMS320VC5402 data space is far from able to save a complete image. However, our system requirement is only to extract the vehicle queue length. It is not very high to the image resolution ratio. Therefore, only the odd field is taken and the even field is discarded. Thus, the image size becomes just $405 \div 2 = 202.5$KB. Although the image resolution ratio is not very high, it is enough to detect the vehicle queue length on the image which can be stored in the data space.

The initialization of SAA7111A is to configure its 24 registers through the two general-purpose I/O of TMS320VC5402 simulating I2C-bus. I2C bus is a two-line bidirectional serial communication bus which is developed and patented by Philips. Its data transmission requires only two lines. One is the clock line SCL, and the other is the data line SDA. The initialization routine is prepared in accordance with the following programming formats: begin→ write slave address → respond → sub address → respond → 24 bytes → respond → end.

## 3 Software Design of System

### 3.1 Main Program Design of System

The main program flowchart of the system is shown in Fig.2. First, initialize the DSP, including rational configuration of status register ST0, status register ST1, processor mode control register PMST, software wait state register SWWSR, partitioned conversion control registers BSCR, clock mode register CLKMD and etc. It also includes the configuration of .cmd, vectors.asm and other documents, so that DSP works normally.

Second, determine the size of recorded images in conformity to the application requirements of this system. It is unnecessary to take a complete image. The size of image captured from analog CCD is $720 \times 576$. That is too large. Actually, just $180 \times 288$ is already enough for image processing. Therefore, LLC2 of SAA7111A is divided into four frequency parts and then connected to the WCLK of FIFO. Thus, the data is written into FIFO by taking one pixels at intervals of three pixels and taking a total of 180 pixels to a row. RTS0 of SAA7111 judges whether odd or even field. Through RTS0, HREF and VREF connecting with WEN of FIFO, only one field of pixels can be taken, that means 288 rows. The size of each adopted image is accordingly $180 \times 288 = 51840$ pixels. In each interruption, the half-fullness of FIFO is 1024(1K) (standard mode) and 1024 pixels are taken every time; hence the number of interrupts needs to be 51840/1024=50.625. Therefore, set up the variable INT_NUM=50, INT_VARIABLE=0.

In addition, to ensure that each image starts from the first pixel, the vertical synchronizing VS signal is connected to the external interrupt INT2 of DSP. While inside the interrupt service program, make XF = 0, delay, XF = 1. Since XF is joined with reset pin RST of FIFO, the interrupt program clears FIFO and brings the read pointer and write pointer back to the starting point. After the falling edge of VS, once VREF and HREF are effective, it begins to read the first pixel. The program goes into INT2 subroutine at first, reset FIFO and then close INT2. (This is because in the process of reading this image, as long as the first pixel point is taken, this interrupt is no longer needed unless taking the next image). When the INT0 interrupt is coming again, it means the half-fullness signal of FIFO. Take data, close interrupt and return. Comparing INT_VARIABLE with INT_NUM, if they are unequal, continue to wait; if they are equal, the acquisition of one image is finished.

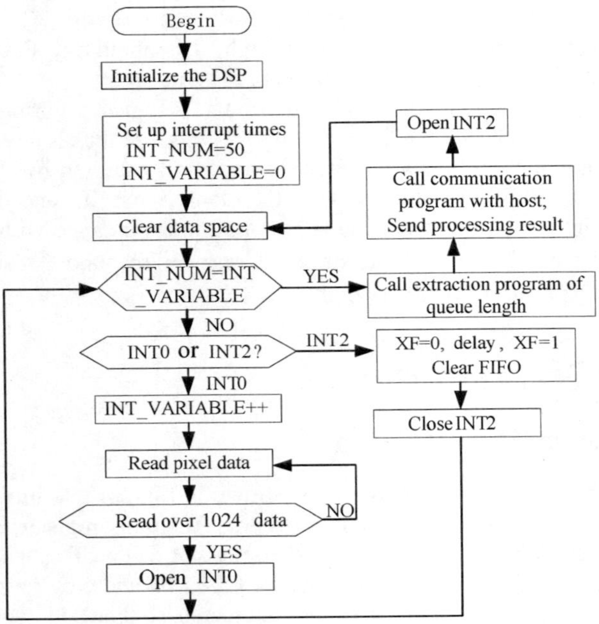


**Fig. 2.** The flowchart of main program

### 3.2 The Extraction Program of Vehicle Queue Length

The extraction of vehicle queue length is to recognize the vehicles from the scene image containing vehicles. The key is to directly find an appropriate threshold for image binarization. Besides, since each scene is different and susceptible to natural conditions, the threshold is not fixed but changed with the actual scene image. In order to search the best segmentation threshold, a variety of threshold selection algorithms have been developed at the present time [9], such as the P parameter method, double-peak method, maximum between-cluster variance method (OTSU method) and etc. Each method has its own characteristics and applications. The maximum between-cluster variance method (OTSU method) is considered as one of the optimal means to select the threshold automatically [10]. It uses the gray level histogram to dynamically determine the image segmentation threshold, according to the fact that the variance between target and background is maximal. The performance of image segmentation can be obviously improved by OTSU method.

As OTSU is based on the image merely consisting of brightness (background) and darkness (target), the between-cluster variance is not only considering variance but also utilizing target and background probability as weight coefficients. The probability is actually the proportion of the target or background in the whole images. It reflects the size of target or background. Therefore, while doing image segmentation of vehicles with OTSU, if the car with a certain color (such as white) occupies a small proportion in the whole image (that is, the image is not only composed of the gray cars and the road), there will be so much background to be wrongly segmented as the target, or the target to be segmented as the background. Thus, it is unable to recognize the vehicles effectively. Since the colors of cars in grey image are generally two colors, white and grey, we can use multiple segmentations method to process images. The flowchart of vehicle length extraction program is shown in Fig3.

First of all, the optimal thresholds for extracting light and dark regions of the image are calculated by OTSU method. The threshold extracting light region is used for image binarization processing, so that the white vehicles can be detected; then the dark region is similarly extracted from the image and the grey vehicles are detected. After reversing the first image colors and uniting it with the second image, both the light and dark vehicles can be detected and displayed synchronously.

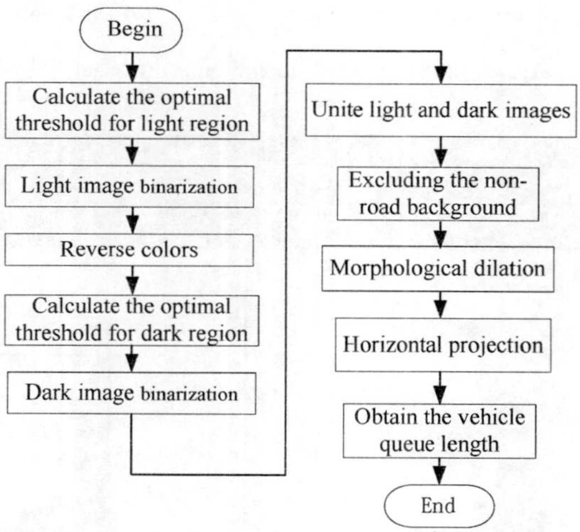


**Fig. 3.** The flowchart of vehicle length extraction program

However, the roadside buildings is often remained in the combined image because of its gray levels are little different with the vehicles'. So it is necessary to filter these roadside buildings from the image (Here the scope of filtration should be determined according to the erected location of CCD). Afterwards, the black edge region of the vehicles is expanded through morphological dilation. The processed picture is shown in Fig4. It can be seen that resting vehicles are obviously separated from the moving ones. And then, the average gray value of each row is calculated, so that the horizontal projection is formed. As the vehicles in the queue are sequential, the black continuous region after projection is the region of the vehicles queue (Fig.5). After the length of the black region is converted into the actual distance, the vehicle queue length is obtained finally.

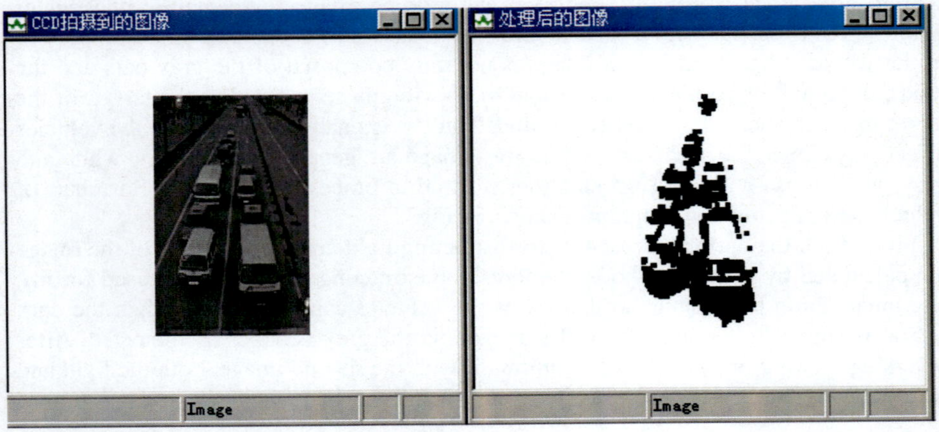


(a) The original image captured by CCD (b) The processed image

**Fig. 4.** The images before and after processing

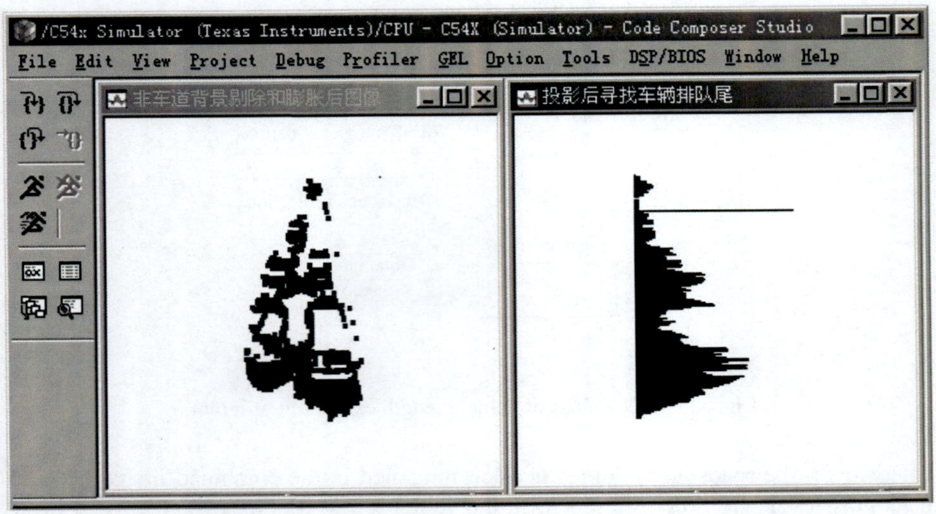


**Fig. 5.** Effect picture of determination on the vehicle queue tail

## 4 Conclusion

The paper has designed and realized an image detection system based on DSP for the vehicle queue length at urban crossroad. The system has been successfully applied to traffic signal real-time controller based on image. The erected analog CCD at crossroad captures the field images. And then the images are processed in DSP chips, so that the length of vehicle queue is extracted. In this way, the traffic data are provided for urban traffic signal controller to scheme real-time signal cycle and green split at intersection. The system can replace the devices such as inductor coils to catch the field traffic data, with its advantage of reducing road surface damage, improving collection accuracy, being more flexible and convenient. It has great significance for the planning and development of Intelligent Transportation Systems (ITS).

## References

1. Min, G., Bin, M.: The application of England Traffic Control Technology. Journal of Foreign Highway 17(6), 33–36 (1997) (in Chinese)
2. Chang, L.S.: Road Traffic Control. Hunan Science & Technology Press, Changsha (1995) (in Chinese)
3. TI Incorporated. TMS320C54x DSP Reference Set. CPU and Peripherals, vol. 1 (April 1999)
4. TI Incorporated. TMS320VC5402 fixed-point signal processor (January 2000)
5. Philips Semiconductors. SAA7111A Enhanced Video Input Processor (EVIP) (May 15, 1998)
6. IDT Corporate. IDT72V235 Data Sheet (February 2002)
7. Cypress Semiconductor Corporation. CY7C1041 Data Sheet (October 1999)
8. Cypress Semiconductor Corporation. CY7C1021 Data Sheet (August 31, 2001)
9. Qi, H.S., Lei, W.: A Survey of Thresholding Methods for Image Segmentation. Engineering and Electronics 24(6), 91–94 (2002) (in Chinese)
10. Otsu, N.: A Threshold Selection Method From Gray-Level Histograms. IEEE Transactions on SMC (9), 62–67 (1979)

# SI-NS Diagram Program Design and Development Method

Si Chuangen, Si Ximing, and Zhang Chunjuan

Hangzhou ThinkinPic Software Technology Co. Ltd, Hangzhou, China
Tel.: 86-571-88859763
cgsi@sohu.com

**Abstract.** His paper attempts to exemplify SI-NS diagram technology and its related program design and development method. With this method programmers can always focus on the SI-NS diagram, which is much easier to be understood than the abstruse program itself. This new development model could guarantee the quality of the developed programs and a high success rate of development, which naturally shortens the software development cycle and reduces the software costs.

**Keywords:** Software Engineering, SI-NS diagram, program design, development method.

## 1 Introduction

Exercise: Write a program to read in a string that is in the format of XXX…X<A>B. A and B are integers between 2 and 10, while X is a number between 0 and 9, and XXX…X is a Base A Number System integer. It is required to converse XXX…X into a Base B Number System integer and then output it.

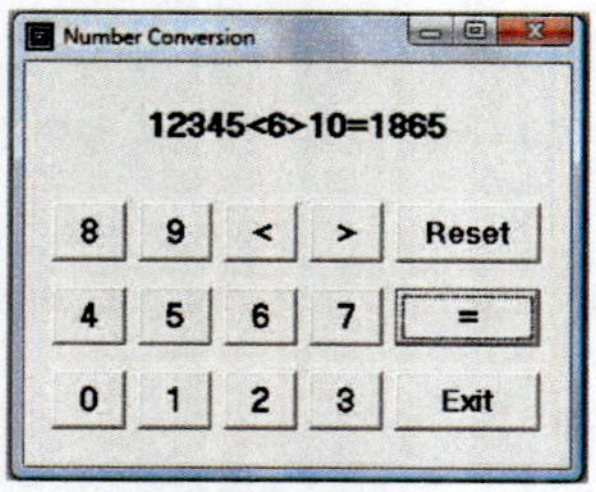


**Fig. 1.** Conversion among different Base Numbers Systems

For example, the input string is 178<10>6.

This string means to change a decimal integer 178 into a senary integer and then output the senary integer string.

# 2 Solution Analysis

This seems to be a simple question, but related to the syntax and semantic analysis of the string that is read in. Here we will apply the SI-NS diagram as the program design and development method, which is to augment the program through a top-down method, so as to break the problem above into the following steps or problems that are much easier to be solved.

*1) Use the syntax chart to define the syntax of the string to be input.*

*2) Use the SI-NS diagram to design a syntax analysis program, and test the program.*

*3) Augment the program to identify semantically invalid string errors, for instance "17<2>8", and test the program.*

*4) Augment the program to converse numeric strings among different Number Systems that are below the decimal system, and test the program.*

# 3 Problem Resolving Process

## 3.1 Use the Syntax Chart to Define the Syntax of the String to Be Input

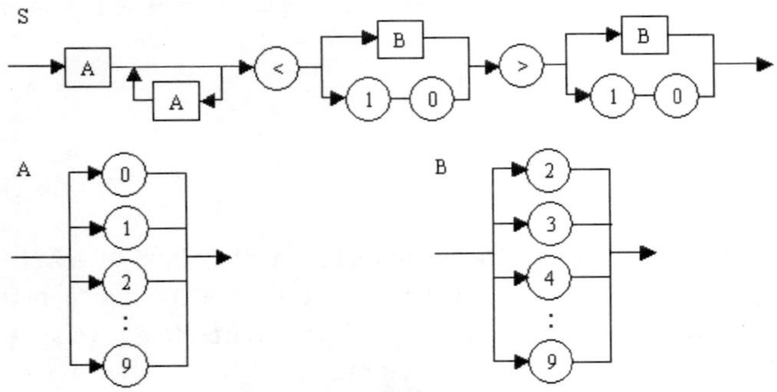


**Fig. 2.** Syntax Chart

## 3.2 Use the SI-NS Diagram to Design a Syntax Analysis Program, and Test the Program

1) Based on the syntax chart, design the syntax analysis program using the SI-NS diagram. (Please use pencils, easy to modify)

2) Execute the program to verify the programmed logics (white box testing) of the well-designed SI-NS diagrams. The syntax analysis program can identify the syntax errors that exist in the strings read into the program, such as 2J<8>、 8<18>3, etc, but cannot distinguish those errors that satisfy the syntax rules but fail to meet the semantics, for example 17<2>8.

3) Based on the well-designed SI-NS diagram, use the keyboard to input the corresponding code. (Notes: Will become skilled with a little training)

4) The program is converted to SI-NS diagrams immediately. (Use the Software Tool developed by our Company)

5) Compare the auto-generated SI-NS diagram with the manually designed SI-NS diagram to check whether their structures are consistent, and then confirm the correctness of the code entered.

The above process can be presented as Fig. 3.

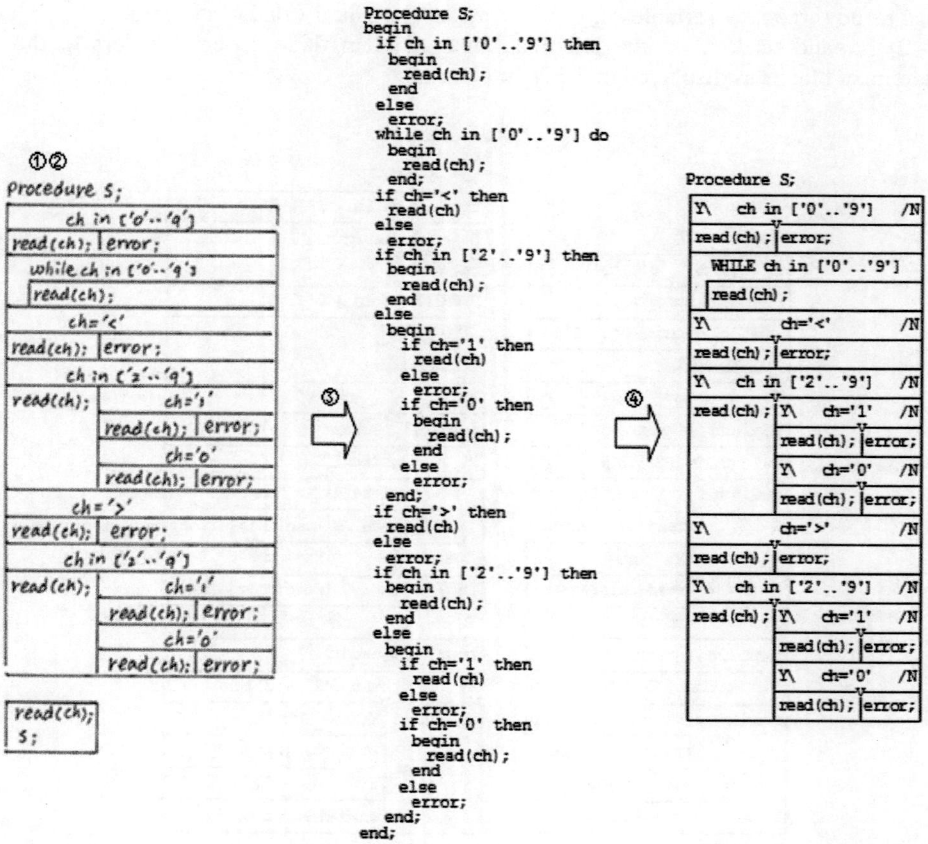


**Fig. 3.** Design Syntax Analysis Program

6)Add in related variables and procedure specifications to ensure that the syntax analysis program can be compiled and executed.

7)Test the program with a computer. If any issue occurs during testing, first analyze and modify the generated SI-NS diagram, and then modify the corresponding code. Then convert the modified code into SI-NS diagram. This way, to analyze the SI-NS diagram is actually the same as to analyze the code itself. To put it another way, to modify the code is to update the SI-NS diagram. In a word, the software engineers only have to focus on the SI-NS diagram, rather than exert all their efforts to understand and analyze the code itself.

### 3.3 Augment the Syntax Analysis Program to Identify Semantically Invalid String Errors, for Instance "17<2>8", and Test the Program

The algorithm is to read in every single character from the XXX…X<A>B string. If there is an X character greater than or equal to the A character, then a semantic error occurs.

The following augments are implemented:

1) Add three new variables, i, j and Array A. The initial value of i is zero.

2) Expand read(ch) in the left diagram marked in three different colors to the statement blocks as displayed in the right diagram.

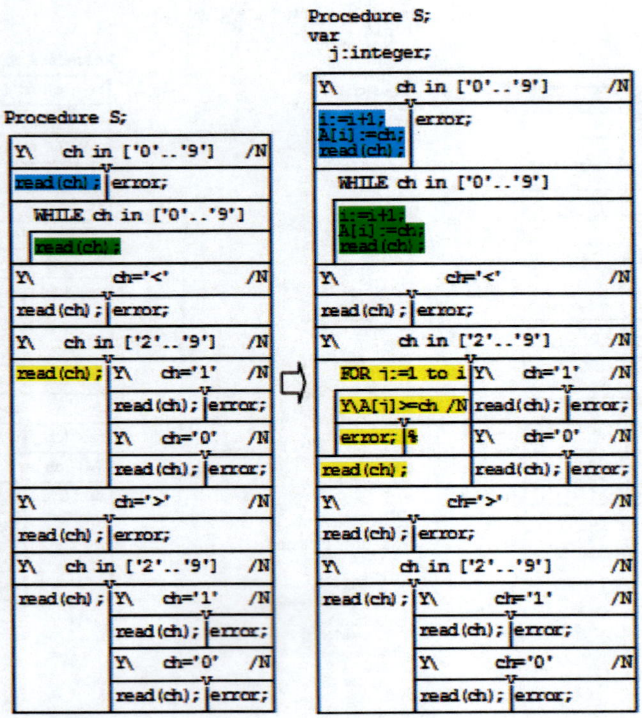


**Fig. 4.** New Syntax Analysis Program Expanded by Semantic Analysis

### 3.4 Augment the Program to Convert Numeric Strings among Different Base Number Systems That Are Below the Decimal System, and Test the Program

The algorithm is to

1) First convert "XXX…X" in string XXX…X<A>B from the Base A Number System to the decimal variable Iten, as displayed in green color in Fig.5.

```
var
  Form1: TForm1;
  st:string;
  ch:char;
  k,i:integer;
  A:array[1..20]of char;

procedure TForm1.B EaqualClick(Sender: TObject);
procedure error;

beep;
showMessage('Syntactic or Semantic Error in Input String! ');
halt;

(*-----------------------------------------*)
procedure read(var ch:char);

k:=k+1;
ch:=st[k];

(*-----------------------------------------*)
Procedure S;
var
  j:integer;
  IB1,Iten,IB2:integer;
  C:array[1..20]of integer;
```

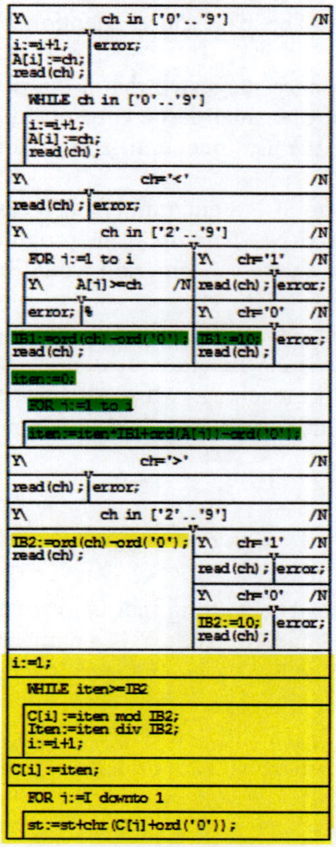


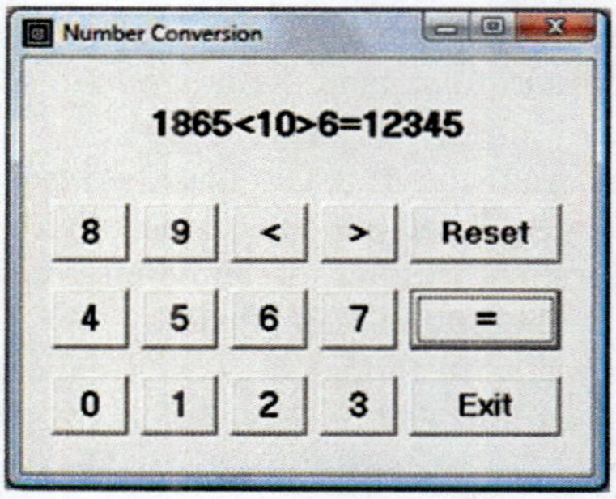


```
Label1.Caption:=Label1.Caption+'=';
st:=st+'=';
k:=0;
read(ch);
s;
Label1.Caption:=st;
```

**Fig. 6.** Final Test Results after the Conversion and Expansion among Different Base Number Systems

2) Second divide Iten by B to get both division and reminder, and then covert the results to a Base B system number as displayed in yellow color in Fig.5. (See next page).

Fig. 5 displays the SI-NS diagram program by pressing the equal key after the Conversion and Expansion among Different Base Number Systems, and the final test results.

# 4 Conclusions

1) To evaluate the problem solving capability of software engineers (students), a great indicator is the capability of them to make good use of the decomposition method to break down complex problems into several small issues and steps, which are much easier to be solved. This is worth to be paid close attention by us (in teaching).

2) During problem-solving process, it is strongly suggested that a top-down, step by step method be adopted. That is to say, after one small issue is resolved, first test it, and then start to expand the program to solve the next one, until all the small issues are solved.

3) Compared with pure text based program, the SI-NS diagram is much easier to be understood, expanded and maintained. With the SI-NS diagram as the program design and development method, it can be ensured that every sub-program would be robust, optimized, reliable and maintainable.

4) The exact methods, steps and experience can only be shared during coding complex programs that have a certain size. The SI-NS diagram technology can play a positive role as to solving the issue that exists in teaching “Most program examples are less than one page, therefore teaching can only impart knowledge but not cultivate capability”.

# Exercise

Use the above method to write a program, to read in a string that is in the format of XXX…X<A>B. A and B are integers between 2 and 16, while X is a number between 0 and 9, or A and F, and XXX…X is a Base A Number System integer. It is required to converse XXX…X into a Base B Number System integer and then output it.

For example, the input string is 17A<16>6.

This string means to converse a hexadecimal integer 17A into a senary integer and then output it.

The syntax of the string to be input is as Fig.6.

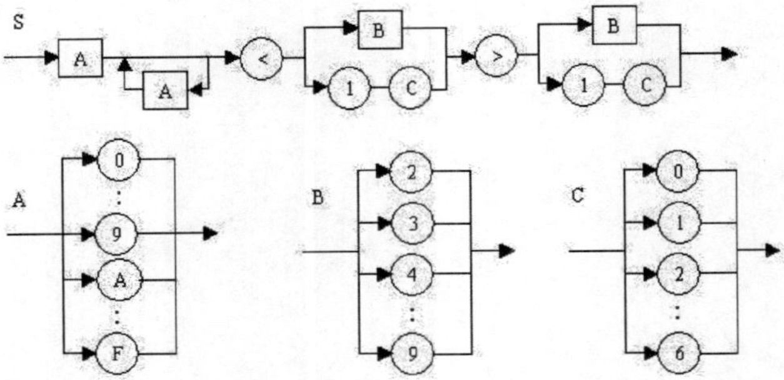


**Fig. 6.** Syntax Chart

# References

1. Chuangen, S.: Auto-Conversion between PASCAL Program and SI-NS Structured Diagrams. Chinese Journal of Computers 9(3), 237–240 (1986)
2. Chuangen, S.: Compile Design and Development Technolgoy. Tsinghua University Press, Beijing (2003)

# Linear Detection Research of Remote Sensing Image Based on Morphological Filter and Hough Transformation

Gaohua Chen, Zhicheng Zhao, and Tiemei Yang

School of Electronic Information Engineering, Taiyuan University of Science & Technology
Taiyuan, P.R. China
1978cgh@163.com

**Abstract.** Because remote sensing image is characterized by abundant grayscale change, great amount of information, complicated background noise, some methods can successfully detect binary image, however, detection of these methods in the remote sensing image is difficult to meet actual needs. The paper summed up the key problems of Hough transformation, and a modified algorithm for peak value cluster round phenomena in the peak value extraction of Hough transformation is proposed in this paper, the experiments have shown that peak value extraction algorithm proposed in this paper can greatly increase the accuracy of parameter estimation. Combining the modified Hough transformation and morphological filter in the linear detection of remote sensing image, the paper put forward a linear detection method based on morphological filter and Hough transformation, comparing to the simulation results of classical Hough transformation, the algorithm in this paper can achieve good results in the detection speed and the target location precision.

**Index Terms:** Remote sensing image, Morphological filter, Hough transformation, Linear detection.

## 1 Introduction

The straight line or line segment detection is an important course in the computer vision region, which is a representative task in the pattern recognition. Many objects have line outline, and detection and localization of these line outlines can provide conditions for further identification and analyses [1]. Remote sensing image has abundant noise because the influence of height, sensor precision, weather and temperature. Some methods can successfully detect binary image, however, detection of these methods in the remote sensing image is difficult to meet actual needs.

Because Hough transformation is characterized by part dearth insensitivity, stochastic noise robustness, parallel process flexibility and practical application etc., it is widely used by scholars in the image processing, pattern recognition and computer vision.

The mathematical morphology is a new science established on the strict mathematical theory, which is a good instrument in the analysis and description of geometry morphology [2]. The mathematical morphology method has some obvious advantages compared to the other image processing methods in the space or frequency domain. The morphology operator could filter noise effectively, reserve the original information of image and emphasize the geometrical property of image [3]. So, it is convenient for further analysis of image.

A modified algorithm for peak value cluster round phenomena in the peak value extraction of Hough transformation is proposed in this paper, the experiments have shown algorithm of the paper can greatly increase the accuracy of parameter estimation. Combining the modified Hough transformation and morphological filter in the linear detection of remote sensing image, the paper put forward a linear detection method based on morphological filter and Hough transformation, comparing to the simulation results of classical Hough transformation, the algorithm in this paper can achieve good results in the detection speed and the target location precision.

## 2 Linear Detection Principle of Hough Transformation

Duda and Hart put forward a method by using Hough transformation to extract line, which has realized mapping from pattern space to parameter space. The basic idea of Hough transformation is point-linear duality. In general, Hough transformation was carried out through the polar equation of a straight line.

$$\rho = x\cos\theta + y\sin\theta$$

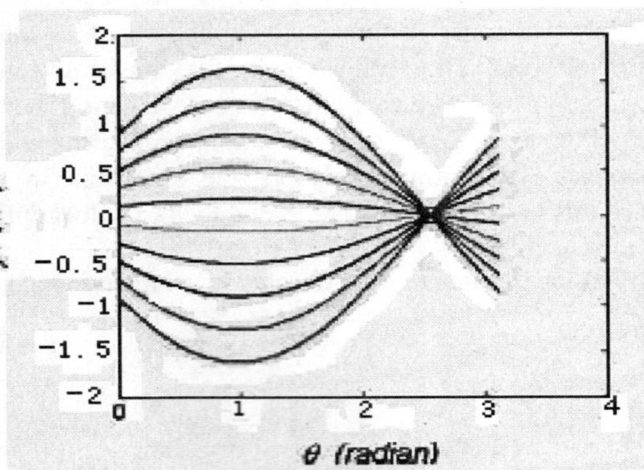


(a) Parameter space map

| | | | | | | | | |
|---|---|---|---|---|---|---|---|---|
| | | | | | | | | |
| | | | | | | | | |
| 1 | | | | | | | 1 | 1 |
| 2 | 3 | 1 | | | | 1 | 2 | 2 |
| 2 | 3 | 5 | 8 | 10 | 9 | 6 | 3 | 2 |
| 3 | 3 | 4 | 2 | | 1 | 2 | 2 | 2 |
| 2 | 1 | | | | | 1 | 2 | 2 |
| | | | | | | | | 1 |
| | | | | | | | | |

(b) Accumulator matrix

**Fig. 1.** Parameter space map and accumulate matrix without any disturbance

It is using sinusoid as the point of straight line in the pattern space. If there are $n$ pixels on a straight line in the image space, $n$ strip sinusoids would have a public intersection in the Hough space.

In the practical application, the parameter space into a discrete accumulator matrix through Hough transformation, then according to Eq.(1), every point $(x, y)$ in the image is mapped to a series of accumulator in the parameter space, and the corresponding accumulator value would add 1. If image space contains a straight line, parameter space would have a local maximum in the corresponding accumulator. Through detecting the local maximum, it is easy to ascertain a pair of parameters $(\rho, \theta)$ corresponding to the straight line, thereby straight line would be detected. It is shown as Fig.1.

## 3 Key Problems in Hough Transformation

Through a large number of simulation experiments, found the existing problems in Hough transformation has a strong impact on its performance in the actual application. Key problems as follows:

*1) Calculation and storage.*

As a batch method, Hough transformation takes both time and space into consideration, and inevitably has the shortcomings such as great calculation and large storage.

*2) Parameter quantized interval selection.*

It could be found that in the linear detection of Hough transformation, first, variable parameters need to be quantified. If parameter quantized interval is undersize, the calculation is bigger; if parameter quantized interval oversize, the parameter spatial accumulation effect is not good, and it is incapable of finding the parameter values of accurately describing the curve. Therefore, selecting parameter quantized interval is important.

*3) Peak value accumulation.*

By using Hough transformation, the emphasis is how to detect out voting accumulation peak value of public voting region in the parameter space even more accurately and more effectively. Therefore it is an important problem to detect out the part accumulation peak value and seek the appropriate accumulation mode more effectively.

*4) Peak value extraction.*

After the parameter space proof accumulation, Hough transformation usually ascertain alternative estimate parameter through adopting setting threshold. But because the characteristics of Hough transformation, the extracted alternative estimate parameters more than the practical parameters. Therefore, in order to keep the accurate detection, Hough transformation needs to dispose the alternative estimate parameters when ascertain them ultimately.

## 4 Modified Hough Transformation

For obtaining estimate parameters of the straight line, setting threshold is a common method used to get the alternative accumulation unit. After setting threshold, it is necessary to detect entire accumulation matrices, extract and record accumulation unit meeting the threshold condition. After the data points of the same straight line proceed voting to parameter space through Hough transformation, the accumulation value in each accumulation unit is relatively bigger in the major neighborhoods of accumulation peak value. These led to some alternative accumulation units extracted by setting the threshold are completely adjacent. Whereas the accumulation value of the adjacent alternative accumulation units is the outcome of the voting in the same straight line. In other words, the data points voted through these alternative accumulation units are belong to the same straight line. And only one parameter corresponds to veritable parameter, and other parameters can not really reflect straight line detected.

In Fig.1, it is shown that accumulation peak value of parameter space is 10, accumulation value of its adjacent accumulate units is 8 and 9, If the threshold is 7 and the extract algorithm is not adopted, it is possible to detect three straight lines. Whereas there is only one straight line in the image space.

Based on above-mentioned peak value cluster round phenomena, the paper put forward a modified peak value extract algorithm:

1) Through the method of setting threshold, accumulate unit of meeting threshold condition is extracted, and their position coordinates in the entire accumulate matrices is recorded, and is kept as two-dimensional array, it is called as the alternative accumulation unit position matrices in the paper. The first column element of matrices refers to the number of rows of accumulation unit of meeting threshold condition. The second column element refers to the number of columns of the accumulation unit of meeting threshold condition.

2) Analysis to the position matrices of the alternative accumulation unit in the previous step.

a. Column vector can be obtained through the first column element subtracts the first one element of this column in the alternative matrices. If this column vector exist elements of absolute value less than $C_1$, then the number of the rows of these elements should be recorded to a vector *row*, which is for the alternative accumulation matrices.

b. Use the second column of alternative matrices subtracts the first one element of this column, and detects absolute value of the row of corresponding *row*. If absolute value of corresponding position is less than $C_2$ in the column vector, relevant number of rows should be recorded as *rowfinal*. It is thought parameters determined by these accumulation units should correspond to the same straight line. The value of $C1$ and $C2$ can be determined according to actual needs. $C1=3$, $C2=5$ in this paper.

3) Analysis to the accumulation unit matrices in the previous step.

a. Extract maxima of every row in the accumulation unit matrix, record maxima to vector *rowm*, and record the column of maxima to vector *pos*.

b. Extract maxima of vector *rowm*, record maxima and number of row of maxima in the vector *rowm* to *[e, p]*, give $p$ value to $a$, give column value of maxima to $b$. and then treat determinate parameter of this accumulate unit as straight line parameter.

c. Set the corresponding position element in the matrix of the accumulate unit as negative infinity.

4) Dispose the alternative accumulation unit matrix use the same way, but set the second element as subtrahend until the end of detection.

Fig.2 shows parameter extraction adopting peak value extraction algorithm and without adopting any extraction algorithm.

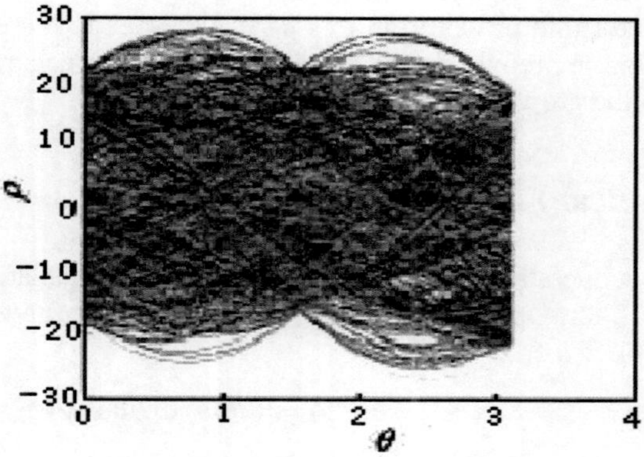


(a) Parameter space maps

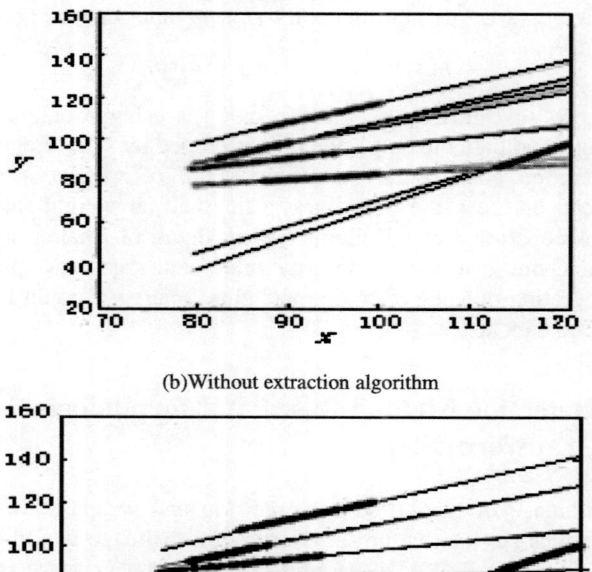


(b)Without extraction algorithm

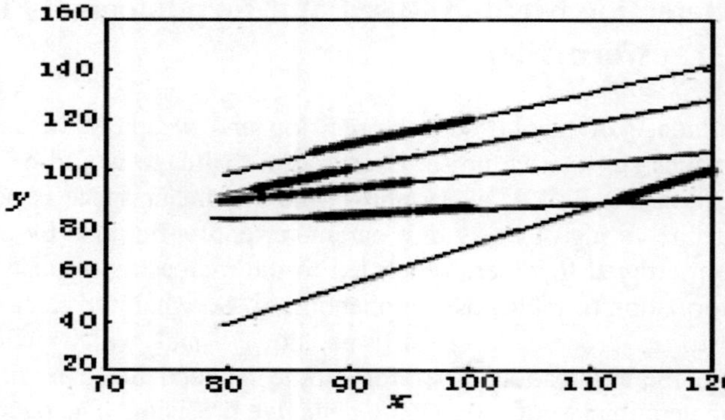


(c) The paper extraction algorithm

**Fig. 2.** Parameter extraction comparison

Fig. 2(a) is a parameter space maps, Fig. 2(b) is a parameter extraction diagram without adopting any peak extraction algorithm, Fig. 2(c) is a parameter extraction diagram adopting extraction algorithm proposed in this paper. It is shown that peak value extraction algorithm proposed in this paper can greatly increase the accuracy of parameter estimation, moreover, in the same conditions, this accuracy with the growth of the data used in the growth.

## 5 Morphological Filter

Combining opening operation and close operation in the mathematical morphology can form morphology filter. Opening operation of structure unit function *b (i, j)* to input image *f(x, y)* is defined as:

$$f(x, y) \circ b(i, j) = [(f \Theta b) \oplus b](x, y) .$$

Equation (2) shows that gray-scale opening operation is using structure unit function to erode input image, and erosion outcome is bulged by structure unit function.

Close operation of structure unit function *b (i, j)* to input image *f (x, y)* is defined as:

$$f(x, y) \bullet b(i, j) = [(f \oplus b) \Theta b](x, y) .$$

Equation (3) shows that gray-scale close operation is using structure unit function to bulge input image, and then the bulge outcome is eroded by structure unit function.

Opening operation could eliminate bright detail of smaller size comparing to structural element, but keep the whole gray value and large light space of the image unchanged. Close operation could eliminate dark detail of smaller size comparing to structural element, but keep the whole gray value and large dark space of the image unchanged. Integrating opening operation and close operation could filter all noises in the light space and dark space.

## 6 Linear Detection Method Based on Morphological Filter and Hough Transformation

Because the influence of height, sensor precision and weather, the remote sensing image always makes substantive noise. Every point of image would be transferred in the Hough detection, the poll of every unit in the parameter space is determined not only by straight line voting of every unit parameter entity, but also by probable value voting of random arrayed together, which led to the increasing of false accumulation points and the formation of multi-inspect phenomena. So when the calculation is larger in Hough transform, process speed will be slower, and which will lead to the inconvenience in the application. Therefore some isolated noise in linear detection should be eliminated, in case of influencing detective precision. The paper put forward a linear detection method based on Hough transformation and morphological filter. This method can filter some noise information in the course of detection, and can increase detection speed. The algorithm is as follows:

1. Design open-close morphology filter. First, make opening operation to input image *f(x, y)*, and the working equation is $f_1 = f \circ B = (f \Theta B) \oplus B$ , And then *f(x, y)* operate close operation to the outcome of opening operation, the working equation is: $f_2 = f_1 \bullet B = (f_1 \oplus B) \Theta B$ .

2. Edge detection of the filtered image made by *Sobel* operator.

3. Linear detection to the image made in the previous step by using the modified Hough transformation

4. Output image.

## 7 Simulation Results and Analysis

Fig 3(a) is a highway remote sensing image randomly selected, Fig 3(b) is the edge detection results by the *Sobel* operator, Fig 3(c) is the straight line detection results by the classical Hough transform, Fig. 3(d) is the straight line detection results by algorithm proposed in this paper.

It is shown that classical Hough transformation algorithm needs large storage, large calculation, and is slow and inefficient, and exists false detection phenomena. The algorithm proposed in this paper could effectively increase detection speed, furthermore, Fig. 3 shows that the algorithm is inexistent false detection phenomena, and target location is accurate.

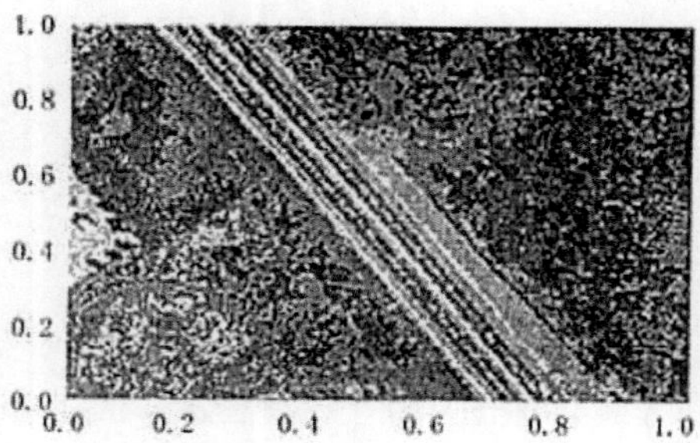


(a) Original image

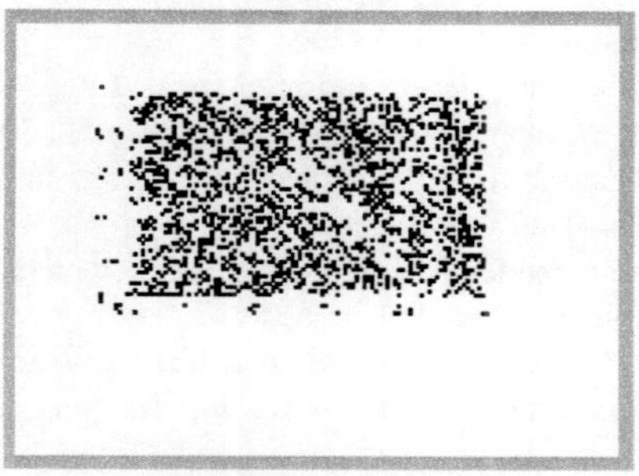

(b) Edge extraction figure

**Fig. 3.** Simulation results figure

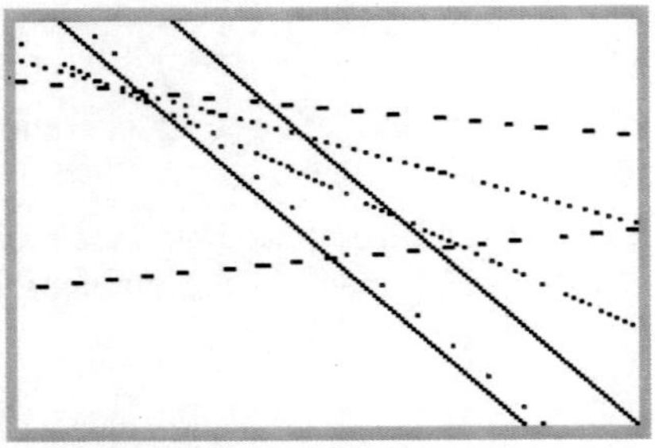

(c) Classical Hough transformation detection result

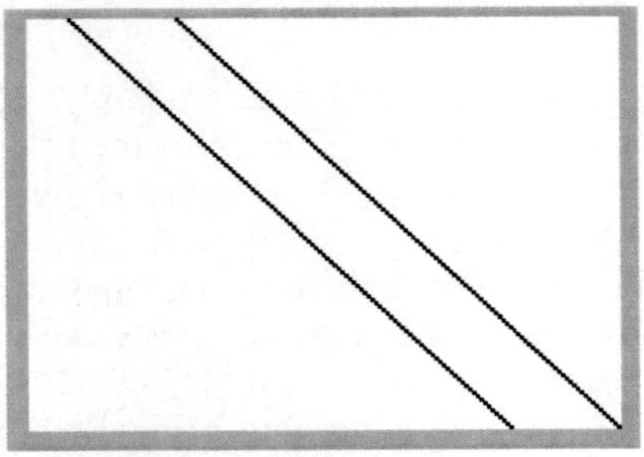

(d)The paper algorithm detection result

**Fig. 3.** (*continued*)

**Acknowledgment.** This work is partially supported by university technology development project of Shanxi province, China (20091024).

## References

1. Gao, X., Chen, G., Li, Y.: A Hough transform based method for the damage detection. Journal of Computer-Aided Design & Computer Graphics 21(2), 268–274 (2009)
2. Wang, S.: Application of mathematical morphology in image processing. Computer Engineering and Applications (32), 89–92 (2004)
3. Song, J., Delp, E.J.: The analysis of morphological filters with multiple structuring elements. Computer Vision Graphics and Image Processing 50, 308–328 (1990)
4. Dai, Q., Yu, Y.: The advances of Mathematical morphology in image processing. Control Theory & Applications 18(4), 478–482 (2001)
5. Xu, X., Dong, G.: Simple model based line detection with Hough transform. Chinese Journal of Scientific Instrument 25(4), 13–15 (2004)
6. Zheng, Y., Qiu, Z., Wu, J.: A palm-print principal line extraction based on morphological filter and Hough transformation. Journal of Beijing Jiaotong University 32(5), 107–120 (2008)

# The Research on How Social Capital Facilitates Knowledge Sharing between Individuals

Jin Hui and Chen Yijia

School of Economics and Management
Jiangsu University of Science and Technology
Zhenjiang, China
hui0409236@126.com, 513706708@qq.com

**Abstract.** As the knowledge activity is always embedded in networks of social context, so the net structure relationship quality and common social cognition have an important impact on knowledge sharing effect. Based on reviewing the theory of social capital, we researched on how social capital facilitates knowledge sharing between individuals. Conclusions from theoretical study are as following. (1) Each Social capital variable (namely structure variable, relationship variable and cognitive variable) has significant influence on knowledge sharing. (2) There are inter-relationships among the variables of social capital. Structure variable has positive effect on relationship variable and cognitive variable. Cognitive variable has positive effect on relationship variable. (3) The knowledge tacit moderates the impact of social capital on knowledge sharing effect.

**Index Terms:** knowledge sharing, social capital, structure variable, relationship variable, cognitive variable, knowledge tacit.

## 1 Introduction

The knowledge-based theory of the firm thought the enterprise core ability came from knowledge sharing among internal members promoted by the organization, but not knowledge creation [1]. Multitudinous empirical experience indicated that knowledge sharing plays an important role in researching team collaborative operation, proliferation of core manufacturing technology, new products development, learning speed, cooperation of trans-departmental and business unit and many other management backgrounds activities.

On the part of occurrence mechanism of knowledge sharing, scholars found 'selective transferring phenomenon' in knowledge sharing process [2], namely that the knowledge recipient have the tendentious choice to seek and transfer knowledge with some specious knowledge sources. This is because knowledge sharing is always embedded in certain social network relations. Therefore explaining the occurrence mechanism of knowledge sharing from the social capital theory angle becomes one of the knowledge sharing studying focus in recent years. In the research of the relationship between social capital and knowledge sharing, social capital plays an important role in promoting the exchange and combination of enterprise internal intelligence capital [3]; the forms of organizational social capital will affect the flowing of its internal knowledge.

Although social capital theory starts a new horizon, there are a few articles systematically discussing the impact mechanism of the various social capital variables on knowledge sharing and the interaction influence among the various social capital variables. Based on that, we focused on the influence mechanism of social capital promoting personal knowledge sharing, and inspected three questions specifically: (1) How do the various social capital variables promote knowledge sharing; (2) Whether the each social capital variables is independent, or there is a interaction influence; (3) Whether knowledge tacit degree has moderate impact on the influence of the various social capital variables on knowledge share effect.

## 2 Reviewing of Social Capital Theory and Construction of Theoretical Model

### 2.1 Reviewing of Social Capital Theory

The concept of social capital theory first appeared in Hanifan's book "The community center" in the 20 years of the 20 Centuries, and then it was introduced to economic, management and other fields. In the numerous researching of the definition about social capital, the most authoritative one is defined by the French society scholar Bourdieu (1980) who defined social capital as 'aggregation of actual or latent resourcewhich are related with enduring network holding relationship of the institution relations that is familiarized or acknowledged by everyone'[4]. In the researching of social capital dimensions, Nahapiet and Ghoshal (1998) proposed social capital dimensions model [3], which has historical significance. They pointed out that social capital should included three dimensions, namely structure variable (reflects the social interaction condition of both sides), relationship variable (reflects the quality of both sides' social relations) and cognitive variable (reflects the individual's cognition to the colony where he is in, including common values and knowledge, etc).

### 2.2 Construction of Theoretical Model

Although knowledge management experts approved that the research of social capital can effectively explain the reason of knowledge sharing. But the original research mostly focused on the independent mechanism of various social capital variables on knowledge sharing effect, lacking integrity and systematic characteristic. There were only a few papers studied the function relationship among the various social capital variables, and there was almost no research comprehensive considering the moderate impact of knowledge's characteristic on the way of various social capital variables to knowledge sharing.

We think social capital is a compound concept which is combined by structure variable, relationship variable and cognitive variable, and the three variables is not independent with each other; there is a certain degree interaction relationship. On the other hand, knowledge tacit degree will hinder the occurrence of knowledge sharing, and the higher the degree is, the higher the hindrance is, so knowledge tacit degree has moderate impact on knowledge sharing. Based on this, based on the Nahapiet and Ghoshal (1998) research frame, we integrated many scholars' researching results of relation intensity, net centrality, trust, common language, common values and so on, constructed this paper's theoretical model.

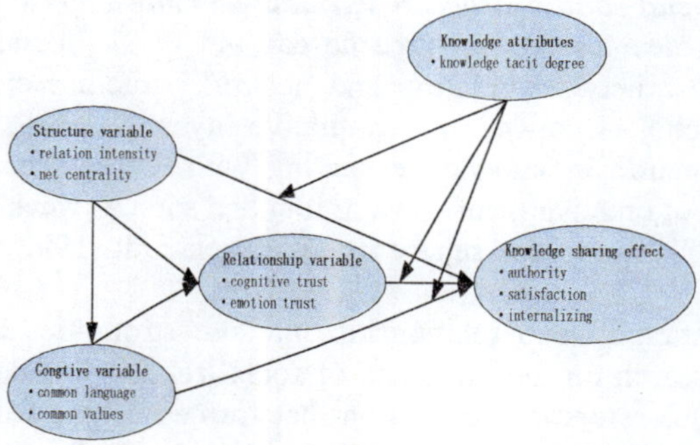


**Fig. 1.** Theoretical model of the influence of various social capital variables on knowledge sharing effect

## 3 Dependent Variable: Knowledge Sharing Effect

We use knowledge sharing effect as dependent variable to examine the positive and negative impact of social capital on knowledge sharing. But there are a lot of controversies on measuring knowledge sharing effect. These studies had gone through three stages. In the initial stage, scholars first focused on 'the quantity of knowledge sharing', it became one of the most convenient operation methods to make the use of the number of common knowledge to measure knowledge sharing effect. The following scholars found 'the quantity of knowledge sharing ≠ the quality of knowledge sharing', and they put forward that knowledge sharing effect should be divided into two separate dimensions: sharing quantity and sharing quality. But regardless of measuring the quantity of knowledge sharing or the quality of knowledge sharing, both of them couldn't express 'the reconstruction process of knowledge sharing' [5]. Knowledge sharing is not merely transmitting knowledge from the knowledge owner to the acceptors, it also includes knowledge absorption and digestion by acceptors, and namely knowledge sharing should include a reconstruction process. Therefore, there were a few scholars began to focus on knowledge recreation degree in the recently research stage on measuring knowledge sharing output [6].

Based on the above analysis, we choose three dimensions (authority of individual knowledge acquisition, satisfaction of individual knowledge acquisition, internalizing of individual knowledge acquisition) to measure knowledge sharing effect. Authority of individual knowledge acquisition reflects the quantity of the knowledge; satisfaction of individual knowledge acquisition reflects the reliability and usefulness of the knowledge; internalizing of individual knowledge acquisition reflects the extras effort when individual practice the knowledge to work (degree of knowledge re-creation).

# 4 Independent Variable: Structure Variable, Relationship Variable and Cognitive Variable

## 4.1 Influence of Social Capital Structure Variable

*1) Influence of social capital structure variable on knowledge sharing effect*

Personal position decides the resource he can get [7-8]. Social capital structure variable focus on how network relation and network structure characteristic affect personal behavior. Scholars conducted quantitative investigation to the social capital structure variable function in knowledge sharing with the society network analysis method and numbers of creative theories were founded such as weak-ties superiority of Granovetter (1973) [7], strong-ties superiority of Kraekhardt (1992) [9], net centrality of Tsai(1998) [10].

In 1973, Granovetter divided relationship into the strong-ties and the weak-ties which started the research on the influence of social relation intensity on knowledge sharing [7]. Strong-ties refers to the relation that formed by the frequent interaction; weak-ties refers to the loose relation between individuals. Researchers have three different opinions in the influence of social relation intensity to knowledge sharing. Some researchers pointed out the strong-ties superiority theory which emphasizes that the strong interpersonal relation can promote the knowledge transfer easily [2, 9]. But scholars as Granovetter who hold the weak-ties superiority thought weak-ties is easier to initiate novel idea and knowledge dissemination [7, 11]. Other scholars who hold the weak-ties and Strong-ties correlation theory no longer separately emphasize these two relations' function instead of their correlation function in knowledge sharing [1].

"Net centrality" theory believes that every actor in social net structure has his position which affects the resource he can get. Using net centrality in knowledge sharing mechanism research reveals some nodes in knowledge sharing network have more direct relations than others; there is an "authority" relationship between the core node and the non-core ones. Scholars such as Tsai (1998) [10], Perry-smith (2003) [12] verified if the individual is closer to the center position, he can control more related resource and share wider benefits, knowledge sharing effect will be more obvious. Based on the above analysis, we hypothesize:

Hypothesis 1: structure variable has significant positive influence on knowledge sharing effect.

Hypothesis 1a: relation intensity has significant positive influence on knowledge sharing effect.

Hypothesis 1b: net centricity has significant positive influence on knowledge sharing effect.

*2) Influence of social capital structure variable on social captial relationship variable*

Burt (1992) thought that knowledge exchange was different from economical exchange, but a society exchange and it can't have the equal repayment immediately, but will become a "human sentiment" [8]. According to society exchange theory, dealers should have good anticipation to each other, because repayment in social exchange won't come immediately. After society exchange succeeds many times, trust relations will be established naturally [13]. Thus society exchanges density (structure dimension) will influence the production and promotion of both exchange sides' trust (relation dimension) [10].

McAllister thought interpersonal trust could be divided into cognitive trust and emotion trust, and exchange frequency is an important prerequisite to produce these two trusts [14]. During knowledge sharing process, exchange intensity will play two important roles. First, frequent exchange of sharing main bodies is helpful in understanding each other, reducing conflicts, promoting emotion attachment, then bringing emotion trust. Second, because cognitive trust is based on the cognitive judgment of the one's ability or reliability, opening close exchange is helpful for both trust sides collecting enough information to distinguish whether the other side has the ability to share knowledge, then cognitive trust will be produced.

On the other hand, the network centricity is also good at promoting the trust of both exchange sides. First, People are willing to believe a person who is in the center of the net has profound knowledge and remarkable ability, and it is easy for him to gain cognitive trust of other individuals. Second, the one who is in the center of the net is the important link to relate others. He keeps close relationship with other individuals, and it is easy for him to gain others emotion trust [13]. Based on the above analysis, we hypothesize:

Hypothesis 2: structure variable has significant positive influence on relationship variable.

Hypothesis 2a: relation intensity has significant positive influence on relationship variable (emotion trust and cognitive trust).

Hypothesis 2b: net centrality has significant positive influence on relationship variable (emotion trust and cognitive trust).

*3) Influence of social capital structure variable on social captial cognitive variable*

Cannon (2001) found one of the most important reasons that a team can operate in a complex, dynamic, the fuzzy situation effectively is that team members have a common understanding which displays as team members' common value and cognitive way [15]. Social capital structure variable is an effective variable that can explain how the community can form and share serious of goals and values. Van Maanen (1979) observed that social interaction is helpful for individual studying organizational values and common knowledge [16].

Deep and frequent communication is the basement of establishing wider common knowledge between both knowledge sharing sides in knowledge sharing domain. Kogut (1996), Henderson (1994) thought relation intensity was the important way to form sharing linguistic environment [17-18]. On the other hand, net centrality can promote individuals forming common viewpoint. Individual who has higher net centrality can apperceive organizational values and common knowledge by frequent commutation with other members. Besides, the individual who is in the center of the net has the ability and chance to gather other individuals around himself. This characteristic will urge him to transfer organizational values and common knowledge to other individuals intentionally or innocently, and that will increase other personal approval and cognition to organization. Based on the above analysis, we hypothesize:

Hypothesis 3: structure variable has significant positive influence on cognitive variable.

Hypothesis 3a: relation intensity has significant positive influence on the production of network's common knowledge and values.

Hypothesis 3b: net centricity has significant positive influence on the production of network's common knowledge and values.

## 4.2 Influence of Social Capital Relationship Variable

Social capital relationship variable as total sum of emotion relation formed by long-term interactions which focus on describing intimate relations established among individuals. High quality social relations will eliminate distrust and fear in knowledge sharing process and impel knowledge transferring positively, while bad quality social relations will reduce contact and communication and even disturb knowledge sharing. In personal relations quality research, scholars focused on "trust" and pointed out that "trust" was the first choice variable in studying social capital relations.

In order to analysis the influence mechanism of trust to knowledge sharing, we took McAllister' bisection method to personal trust dividing trust into two types: emotion trust and the cognitive trust [14]. Emotion trust is based on the emotion connection of both trust sides. Cognitive trust is based on the judgment of the ability or credibility of the person who is trusted. Different trust relationship plays different role on knowledge sharing. In knowledge research stage, in order to avoid "offensive" or exposing "knowledge ignorance", knowledge recipient is always willing to send out knowledge sharing request to emotion trust members. And it can avoid the risk of be ridiculed as well as improve the possibility of obtaining help. Knowledge source will believe the recipient won't bring harm to him and is willing to contact and communicate with the recipient. However, uncertainty of knowledge transferring and many detail problems will cause misunderstanding of knowledge sharing sides in knowledge transferring stage. At this time, based on cognitive trust of knowledge sources, knowledge recipient has confidence on knowledge of source and is willing to spend more time listening, understanding viewpoint of knowledge source. Meanwhile, based on cognitive trust of knowledge recipient, knowledge source will treat the knowledge recipient's inconsistent behavior understandingly, try his best to help the recipient to understand and accept the knowledge. Based on the above analysis, we hypothesize:

Hypothesis 4: relationship variable has significant positive influence on knowledge sharing effect.

Hypothesis 4a: emotion trust has significant positive influence on knowledge sharing effect.

Hypothesis 4b: cognitive trust has significant positive influence on knowledge sharing effect.

## 4.3 Influence of Social Capital Cognitive Variable

Nahapiet and Ghoshal defined social capital cognitive variable as those collective resource that can provide common statement, interpretation and values among members [3]. Organizations don't have human brain, but they have cognitive systems and memories, organizations retain the memory of certain behaviors, mental maps, norms and values with time going.

*4) Influence of social capital cognitive variable on knowledge sharing effec*

Social capital cognitive variable as organizational "public goods" has special significance in knowledge sharing. Personal cognitive structure implicitly determines personal intelligent activity patterns such as knowledge selection, knowledge acquisition and knowledge absorption which are key links of knowledge sharing process. But in the previous social capital studying, cognitive variable has not been give sufficient attention. A few scholars try to use common knowledge and values to describe the influence mechanism that cognitive variable on knowledge sharing effect.

Reagans (2003) who thought the most important way for people to learn new knowledge was going with their existing knowledge. This means if there is common knowledge among the knowledge source and the recipient; the knowledge sharing would occur more easily [2]. Nonaka (1994) also approved that "if there isn't any common knowledge, it is very difficult to share new knowledge" [19]. In knowledge sharing process, if knowledge source has more common knowledge with the recipient, not only he can be clearer about what kind of knowledge the recipient really need, but also he can express and explain the knowledge to the recipient with a kind of language that the recipient can understand. If knowledge recipient has more common knowledge with the knowledge source, he can more easily understand the knowledge provided by knowledge source; improve the efficiency of knowledge absorption.

"Common values" emphasis the mutual recognition and the sense of cohesion among individuals. Individuals who have been in an organization with a long time will coincidently understand and percept organizational visions and strategies. Based on the sense of identity and belonging to the organization, they can form "common values" [10]. When the two sides of knowledge sharing have common values, they will have a common goal and behavior direction. It can stimulate personal driving force and participation of knowledge learning and sharing; otherwise, in the case of lacking common values, individuals are often self-centered, learning only for themselves, and tending to master knowledge for their own unique. Based on the above analysis, we hypothesize:

Hypothesis 5: cognitive variable has significant positive influence on knowledge sharing effect.

Hypothesis 5a: common knowledge has significant positive influence on knowledge sharing effect.

Hypothesis 5b: common values have significant positive influence on knowledge sharing effect.

*5) Influence of social captial connitive variable on social captial relationship variable*

"Common values" are considered as kinds of societal force that are based on organizational identification. In an organization, individuals share the same values and life experiences, and naturally consider the organizational members as the "insiders". For the same recognition of common values, individuals tend to associate with the "insiders" who have the same social attributes with themselves. This emotional trust accompanied with common values has a positive effect in knowledge sharing process. Knowledge sharing among individuals is not necessarily for a clear return, but in order to respond to the collective norms, to achieve collective goals, or start from the reciprocal admission of staffs' common preferences. On the other hand, "common knowledge" can help the knowledge sharing both sides to express or understand the information with each other using a familiar language. The similarity of cognitive structures will greatly reduce the friction, misunderstanding and conflict among interpersonal communication, enhance mutual understanding and recognition, and thereby bring rise to trust based on cognition. Based on the above analysis, we hypothesize:

Hypothesis 6: cognitive variable has significant positive influence on relationship variable.

Hypothesis 6a: common knowledge has significant positive influence on emotion trust.

Hypothesis 6b: common values have significant positive influence on cognitive trust.

## 5 Moderetor Variable: The Degree of Knowledge Tacit

Knowledge attributes are often regarded as the important factors that influence knowledge sharing. Although based on different attributes standards, knowledge can be classified in different ways, classification of tacit knowledge and explicit knowledge is the most popular knowledge classification structure in knowledge management field. Knowledge with high tacit degree is highly personalized, difficult to formalized, and plants deeply at individual experience, judgment and subconscious knowledge. Knowledge with low tacit degree can be codified and measured, and can be presented by words, formulas and computer programs. Because of the important influence of tacit knowledge on knowledge sharing, it is necessary to consider the moderate impact of tacit knowledge degree on the influence mechanism those social capital variables on knowledge sharing effect.

In the moderate impact of knowledge tacit degree to structure variable, knowledge with high tacit degree requires social interaction with high intensity to be shared; otherwise knowledge with high explicit degree can be shared easily in loose social relations. Research on the reliability of knowledge tacit degree and relation intensity verified this point. For example, Reagans (2003) confirmed the tacit of knowledge can adjust the relationship between the relation intensity and the ease of knowledge transferring and it indicated when knowledge tacit degree is higher, strong-ties is more easily to promote knowledge sharing than weak-ties [3].

In the research of the moderate impact of knowledge tacit degree on relationship variable field, some scholars noted that trust has different impact on the personal tacit or explicit knowledge sharing. For example, Szulanski (2004) pointed out that the influence of trust on knowledge sharing effect would be affected by knowledge ambiguity. This result indicated that trust would affect the accuracy of the knowledge obtained by the team members. Knowledge ambiguity moderates the impact of trust on knowledge sharing effect [20].

In the moderate impact of knowledge tacit degree on cognitive variable field, knowledge tacit degree will also influence the impact of social cognition on knowledge sharing. For one thing, the higher the shared knowledge tacit degree is, the more time and energy both knowledge sharing sides have to pay. At this time individuals with highly common values will positively take part in knowledge sharing behavior considering the organizational mission and goals. For another thing, when the tacit degree of the knowledge that needs to be shared is higher, it is easier to cause causal ambiguity in searching and transferring process. Only the knowledge sharing sides have more common language, can they succeed in realizing the entire process of knowledge sharing. Based on the above analysis, we hypothesize:

Hypothesis 7: knowledge attributes has moderate impact in the relationship between social capital variables and knowledge sharing effect.

Hypothesis 7a: compared with knowledge with low tacit degree, relation intensity and net centrality have big impact on knowledge sharing effect when the tacit degree of shared knowledge is higher.

Hypothesis 7b: compared with knowledge with low tacit degree, personal emotion trust and cognitive trust have big impact on knowledge sharing effect when the tacit degree of shared knowledge is higher.

Hypothesis 7c compared with knowledge with low tacit degree, common values and knowledge have big impact on knowledge sharing effect when the tacit degree of shared knowledge is higher.

## 6 Conclusion

Social capital plays an active role in promoting knowledge sharing between individuals. But the internal mechanism of how social capital influence knowledge sharing in knowledge sharing process has been covered in "veil" continuously. This study targets at the internal function mechanism of social capital to personal knowledge sharing, and comes to the conclusions as following: (1) each Social capital variable has independent promotion function on knowledge sharing; (2) structure variable affects knowledge sharing effect by the positive intermediary impact of relationship variable and cognitive variable; (3) cognitive variable affects knowledge sharing effect by the positive intermediary function of relationship variable; (4) knowledge tacit degree has moderate impact on the way of various social capital variables to knowledge sharing. Those conclusions have a deeper reflection to the mechanism of individual knowledge sharing, and provide some theoretical references to organizations in knowledge sharing practice.

## References

1. Grant, R.M.: Toward a Knowledge-Based Theory of the Firm. Strategic Management Journal 17(Special Issue), 109–122 (1996b)
2. Reagans, R., McEvily, B.: Network structure and knowledge transfer: The effects of cohesion and range. Administrative Science Quarterly 48(2), 240–267 (2003)
3. Nahapiet, J., Ghoshal, S.: Social capital, intellectual capital, and the organizational advantage. Academy of Management Review 23(2), 246–266 (1998)
4. Bourdieu, P.: The Social Capital. The Research on Sciences Socials 31, 2–3 (1980)

5. Davenport, T., Prusak, L.: Working Knowledge: How Organization Manage What They Know, pp. 23–24. Harvard Business School Press (1998)
6. Cummings, J.L., Teng, B.S.: Transferring R&D knowledge: The key factors affecting knowledge transfer success. Journal of Engineering &Technology Management 20(1-2), 39–68 (2003)
7. Granovetter, M.: The strength of weak ties. American Journal of Sociology 78(6), 1360–1380 (1973)
8. Burt, R.S.: Structural holes: The social structure of competition. Harvard University Press, Cambridge (1992)
9. Krakhardt, D.R.: The strength of strong ties: The importance of philos in the organizations. In: Nohria, N., Eccles, R. (eds.) Networks and Organizations: Structures Forms and Action, pp. 216–239. Harvard Business press, Boston (1992)
10. Tsai, W., Ghoshal, S.: Social capital and value creation: The role of intrafirm networks. Academy of Management Journal 41(4), 464–476 (1998)
11. Uzzi, B., Lancaster, R.: Relational embeddedness and learning: The case of bank loan managers and their clients. Management Science 49(4), 383–399 (2003)
12. Perry-Smith, J.E., Shalley, C.E.: The social side of creativity: A static and dynamic social network perspective. Academy of Management Review 28(l), 89–106 (2003)
13. Luo, J., Zhen, M., Xie, Z.: Influence of social capital in practical social group on knowledge sharing. Jiangxi Society Science (3), 131–141 (2007)
14. McAllister, D.: Affect and cognition-based trust as foundations for interpersonal cooperation in organizations. Academy of Management Journal 38(l), 24–59 (1995)
15. Carmon, M.D., Edmondson, A.C.: Confronting failure: Antecedents and consequences of shared beliefs about failure in organizational workgroups. Journal of Organizational Behavior 22, 161–177 (2001)
16. Van Maanen, J., Schein, E.H.: Toward of Theory of Organizational Socialization. Research in Organizational Behavior 1, 209–264 (1979)
17. Kogut, B., Zander, U.: What firm do: Coordination, identity, and learning. Organization Science 7(5), 502–515 (1996)
18. Henderson, R.M., Coekburn, I.: Measuring competence? Exploring firm effects in Pharmaceutical research. Strategic Management Journal 15(Winter special issue), 63–84 (1994)
19. Nonaka, I.: A dynamic theory of organizational knowledge creation. Organization Science 5(1), 14–35 (1994)
20. Szulanski, G., Cappetta, R., Jensen, R.J.: When and how trust worthiness matters: Knowledge transfer and the moderating effect of causal ambiguity. Organization Science 15(5), 600–613 (2004)

# Research on the Organization Performance Appraisal in the Information Era

Maofeng Li and Lunqu Yuan

Beijing Jiaotong University, School of Economics and Management
Beijing, China
Jeff.zyq@gmail.com

**Abstract.** Performance appraisal can make the organization strategy come true and keep the organization sustainable development. The information era characterized in more and more knowledge workers and corporations informationization has lead the new challenges for the current performance appraisal. Knowledge-based employees are playing increasingly important role in the modern organization. The working place is not only in office and the working time is unfixed any more. Some of the organizations depend on the informationization too much. All these changes are changing the way of the current performance appraisal. This paper suggests the performance appraisal should combine with the ability appraisal, individual appraisal combine with the team appraisal, eHR (electronic human resource) combine with the traditional appraisal, which can exactly assess the employ's contribution in the organization in the information era.

**Keywords:** Performance appraisal, Information era, eHR.

## 1 Introduction

Performance appraisal is an important tool to implement the organization strategy, and provide evidence for the compensation, awards and punishment. The main current performance appraisal methods implemented by most organizations are balanced scored card, 360 degree appraisal, and key performance index, etc. These methods stress the reifiable, measurable, achievement, reasonable and timeline (SMART) principle, which are pretty effective appraisal methods during the industry era. The human being came into the information ear from the end of 90s in the 20 century, especially when the information and electric industry became the pillar industry. The employees with more knowledge and the organization with more information are the characteristics under the background of the information era. Those characteristic significantly improve the efficiency of the corporation's operations and management. Meanwhile, they bring many new challenges for the organization's performance appraisal. More and more knowledge workers come out, the working place is multi-plots and working time is unfixed. All these changes have been challenging the current performance appraisal methodology.

## 2 New Challenges for the Current Performance Appraisal under the Background of Information Era

People have been being replaced by computer in the information era. The production, operation, and management have been informationized. More and more organizations oriented to the information and knowledge production. The informanization has tremendously improved the efficiency of production and management. More and more corporations has been utilizing eHR as the main management way, meanwhile, which has lead many new challenges to the current performance appraisal.

### 2.1 More and More Knowledge Workers

The famous American management scientist Peter Drucker has proposed the concept of the knowledge workers, who can explore modern technology to improve the efficiency on one hand, and have the relatively high learning and innovation ability as well [1] [2]. Drucker defines six factors for knowledge worker productivity (1999):

- Knowledge worker productivity demands that we ask the question: "What is the task?"
- It demands that we impose the responsibility for their productivity on the individual knowledge workers themselves. Knowledge workers have to manage themselves.
- Continuing innovation has to be part of the work, the task and the responsibility of knowledge workers.
- Knowledge work requires continuous learning on the part of the knowledge worker, but equally continuous teaching on the part of the knowledge worker.
- Productivity of the knowledge worker is not — at least not primarily — a matter of the quantity of output. Quality is at least as important.
- Finally, knowledge worker productivity requires that the knowledge worker is both seen and treated as an "asset" rather than a "cost." It requires that knowledge workers want to work for the organization in preference to all other opportunities [3].

The significant characteristic of the knowledge workers is innovation. Moreover, the main working way for the knowledge workers is thinking, and their knowledge is update according to the environment, which is quite flexible. The knowledge workers are knowledgeable, creative and flexible. The characteristics are following as in summer [4].

- More and more work is team work, multi-team work and even virtual teamwork
- The process of work is unobservable
- The performance is unmeasurable
- The research explained that the performance appraisal is inefficient in many corporations, which accounts for the proportion from 30% to 50%. The defects follow as:
- Performance appraisal's inaccuracy
- Performance appraisal's subjectivity
- Performance appraisal without feedback

### 2.2 Muti-working Plots and the Unfixed Working Time

During the information era, especially with the maturity of network meeting technology, more and more workers can finish their work online, even though staying home and they don't need to work from 9:00 am to 5:00 pm according to the schedule. This makes the performance appraisal more and more difficult, because you can't supervise how your employees work in their home or any else place.

### 2.3 Over-Informationization in Some Corporations

More and more corporations are using the management information system (MIS), or utilizing the enterprises relation plan (ERP) to integrate the human resource, materiel and capital in the corporation. ERP is one part of the MIS, which can reduce the production cost to improve the efficiency of production, operation and management. eHR is one part of the ERP as well, which is very important for the humane resource management. The eHR is a new human resource management based on the advanced software and hardware. eHR can reduce the management cost and improve corporations profit and employee's service through the centralized repository of information, automatic processing of information, employee self-service, external cooperation, as well as shared service. It's very hard to assess one employee by eHR completely, especially for the knowledge workers, since the computer can't totally replace the human being to appraise the employee.

## 3 The New Way of the Performance Appraisal under the Background of Information Era

In a word, information era is workers with more knowledge and organization with more informationization in essential. The performance should be adjusted according to these two characteristics.

### 3.1 Appraising the Performance Combined with Ability

How to manage knowledge workers is changing the way of the current organization performance appraisal. The traditional performance appraisal aims to the performance of the employees, for instance, the research institution appraise employees by their research achievement and papers published on public. If there is no good mechanism to evaluate the research achievement, it will lead the researchers pursuing the instant success and being impulsive, which is pretty harm for the research. In this case, the researcher can't get much innovation. Apart from appraising the performance of the knowledge workers, we also should appraise the employee's attitude including the self-motivation and responsibility, as well as the ability including the analysis and judgment, incorporation, flexibility and communication [5]. It's the only way to exactly appraise one knowledge employee by appraising both the performance and ability under the background of the information era. It can motivate them to work hard and keep the core employees staying in the organization if the organization has a good compensation institution in this case.

### 3.2 Appraising the Performance Based Both on the Individual and Team

The current performance appraisal stress the individual performance, however, the information era has changed the performance appraisal method. With the information expanding, the work becomes more and more complicated. The jobs depending only on individual become less and less, which has been being replaced by the team or multi-teams [6]. Therefore, it's very hard to appraise the individual's contribution by individual with the traditional performance appraisal method. Even more, many jobs are multi-teams, which lead more difficult to appraise the individual contribution among the teams.

### 3.3 Appraising Combined eHR and Traditional Human Resource Management

If we only explore the traditional human resource management to manage corporation, it won't fit the requirement of the information era and will fail in the market competition. If the organization only depends on the eHR method, they can't exactly appraise the employee's performance under the background of the information era. Both the traditional and eHR are the effective method to appraise the performance, but we can't only explore one and we should have the main appraisal method. Under the background of the information era, we should explore the eHR as the main performance appraisal method and the traditional way as the assistant method. We have to make full use of the computer and the initiative of human being as well.

## 4 Conclusions

Performance appraisal can make the organization strategy come true and keep the organization sustainable development. The information era characterized in more and more knowledge employees and corporation with more informationization have lead the new challenges for the current performance appraisal. Knowledge-based employees are playing more and more important role in the modern organization. The working place becomes not unique and the working time is unfixed any more. The organizations depend on the informationization too much. All these changes are changing the way of the current performance appraisal. This paper suggests the performance appraisal should combine with the ability appraisal, individual appraisal combine with the team appraisal, and eHR combined with the traditional appraisal, which can assess the employ's contribution in the organization under the information era background in order to keep the core employees for the corporations.

## References

1. Drucker, P.F.: Mangement: Tasks, Responsiblities, Practices, p. 839. Harper & Row, New York (1973)
2. Saaty, T.L.: The Analytic Hierarchy Process. Mc Graw Hill, Newyork (1980)
3. Drucker, P.F.: Mangement Challenges of the 21st Century. Harper Business, New York (1999)

4. Maofeng, L., Weizhong, Z.: The Analysis and Research of Performance Assessment of Knowledge Workers in Engineering Design Corporates. China Exploration and Survey (6), 49–51 (2008) (in Chinese)
5. Hong, W.: Research on the Examination and Critique Model of Knowledge Staff Achievements. Journal of Huaiyin Institute of Technology 16(6), 45–47 (2007) (in Chinese)
6. Jing, G.: Explore and Analyze the Strategy of Cadre Image Communication in Information Era. Gansu Theory Research (1), 108–112 (2010)

# Identity-Based Parallel Key-Insulated Encryption with Multiple Long-Term Keys

Yanli Ren, Shuozhong Wang, Xinpeng Zhang, and Zhenxing Qian

School of Communication and Information Engineering, Shanghai University, Shanghai 200072, China
{renyanli1982123,shuowang,xzhang,zxqian}@shu.edu.cn

**Abstract.** Parallel key-insulated encryption (PKIE) allows two independent long-term keys to be alternately used in short-term key update operations. At least half of short-term keys would be exposed and at least half of ciphertexts could be decrypted if one of the long-term keys is exposed. In this paper, we propose a new PKIE scheme with n long-term keys in the identity-based setting. If one of the long-term keys is exposed, only 1/n short-term keys would be exposed and 1/n ciphertexts could be decrypted, so the new PKIE scheme can greatly decrease loss due to key exposure. The scheme is adaptive-ID secure without random oracles, and it has a tight reduction. Moreover, its public key has a constant size, while sizes of ciphertexts and short-term keys depend on the number of long-term keys.

**Keywords:** identity-based, parallel key-insulated encryption, long-term keys, without random oracles.

## 1 Introduction

Key exposure is an important issue in practice. To deal with the problem, Dodis et al. introduced the key-insulated public key encryption (KIPE), which stores long-term keys in a secure but computationally limited device [1]. Short-term secret keys are kept in a powerful but insecure device, and needs to be updated frequently to reduce the risk of key exposure. This increases the risk of long-term key exposure, which is damaging as it typically requires the system to restart with a new public key. Hanaoka et al. [2] introduced the concept of parallel key-insulated encryption (PKIE), in which distinct independent long-term keys are alternately used in key updating operations. As we know, using two long-term keys may simultaneously increase security of both long-term keys and users by allowing frequent updates. In [2], the authors provided a PKIE construction in the random oracle model. At PKC'07, Libert et al. described an efficient PKIE scheme without random oracles [3].

Classical IBE schemes [4] rely on the assumption that secret keys are kept perfectly secure. In practice, it is easier for an adversary to obtain the secret key from a naive user than to break the computational assumption on which the system is based. Thus, it is a worthwhile task to deal with the key-exposure problem in IBE scenarios. Weng et al. proposed an identity-based PKIE (IBPKIE) scheme without random oracles [5].

However, the scheme has long public parameters and the reduction is loose. We have proposed an IBPKIE scheme that is adaptive-ID secure without random oracles [6]. This scheme has short public parameters and a tight reduction.

All PKIE schemes only allow two long-term keys to be alternately used in key update operations until now. It means that at least half of short-term secret keys would be exposed and at least half of ciphertexts could be decrypted if one of the long-term keys is exposed. Therefore, a PKIE scheme can greatly decrease loss due to key-exposure if it has $n$ long-term keys, where $n>2$. If one of the long-term keys is exposed, only $1/n$ short-term secret keys would be exposed and $1/n$ ciphertexts could be decrypted. Thus, Libert et al. [3] proposed an open problem that is to construct a PKIE scheme by increasing the number of long-term keys. In this paper, we present an IBPKIE scheme with multiple long-term keys based on an IBE scheme [7]. The scheme is adaptive-ID secure without random oracles and has constant size public parameters, and the sizes of short-term keys and ciphertext are dependent on the number of long-term keys.

## 2 Definitions

### 2.1 Bilinear Map

Let $\kappa$ be a security parameter and $p$ a large prime number of bit $\kappa$. $G$ and $G_T$ are two groups of order $p$, and $g$ is a generator of $G$. $e: G \times G \to G_T$ is a bilinear map with the following properties [8,9]:

1) Bilinear: For all $(u,v) \in G \times G$, and $a,b \in Z_p$,
$e(u^a, v^b) = e(u,v)^{ab}$.

2) Non-degenerate: $e(g,g) \neq 1$ for a generator $g \in G$.

### 2.2 Complexity Assumption

Assume $G$ and $G_T$ are groups of order $p$, where $p$ is a large prime number, then the augmented bilinear Diffie-Hellman exponent (ABDHE) problem in $(G, G_T)$ is defined as follows. Given $(g', (g')^{\alpha^{q+2}}, g, g^{\alpha}, g^{\alpha^2}, \cdots g^{\alpha^q}, Z) \in G^{q+3} \times G_T$ to decide whether $Z = e(g', g)^{\alpha^{q+1}}$ [7]. For an adversary $A$, the advantage in solving the ABDHE problem is

$$|\Pr[A(g', (g')^{\alpha^{q+2}}, g, g^{\alpha}, \cdots, g^{\alpha^q}, e(g', g)^{\alpha^{q+1}}) = 0] \\ - \Pr[A(g', (g')^{\alpha^{q+2}}, g, g^{\alpha}, \cdots, g^{\alpha^q}, Z) = 0]|,$$

where the probability is over random coins of $g, g', \alpha, Z$, and the random coins of $A$.

We say that the $(T, \varepsilon)$ − ABDHE assumption holds in $(G, G_T)$ if no $T$-time algorithm has advantage at least $\varepsilon$ in solving the ABDHE problem in $(G, G_T)$.

# 3 The Proposed IBPKIE Scheme

## 3.1 Setup

Let $G, G_T, e, g$ be defined as in Section 2.1. $h:\{0,1\}^* \rightarrow Z_p^*$ is a collision-resistant hash function, $f:\{0,1\}^* \times \{0,1\}^{1+\log_2 p} \times \{0,1\}^* \rightarrow Z_p^*$ is a pseudo random function. $PKG$ randomly chooses $\alpha \in Z_p^*$, $h_0, h_1, h_2, h_3$, $h_4 \in G$, and defines $F(ID,t) = (h_1 h_2^{-ID})^{h(t)} (h_3 h_4^{-ID})^{h(ID,t)}$, $g_1 = g^{\alpha}$. $\alpha$ is the master key of $PKG$, and the public parameters are $(g, g_1, h_0, h_1, h_2, h_3, h_4, f, h, F(ID,t))$.

## 3.2 Extract

For an identity $ID$, $PKG$ executes as follows:

1. Randomly choose $HK_{ID,0}, \cdots, HK_{ID,n-1}$, and define $r_{1-n} = f(ID, HK_{ID,n-1}, 1-n), \cdots, r_0 = f(ID, HK_{ID,0}, 0)$.
2. Randomly choose $r_{ID} \in Z_p^*$, and compute
   $d_{-n,0} = r_{ID}$, $d_{1-n,0} = (g_1 g^{-ID})^{r_{1-n}}, \cdots, d_{0,0} = (g_1 g^{-ID})^{r_0}$,
   $d_{1,0} = (h_0 g^{r_{ID}})^{1/(\alpha - ID)} F(ID, 1-n)^{r_{1-n}} \cdots F(ID,0)^{r_0}$.
3. Return $HK_{ID,0}, \cdots, HK_{ID,n-1}$, $TSK_{ID,0} = \{d_{-n,0}, \cdots, d_{1,0}\}$.

## 3.3 UpdH

Given $HK_{ID,j}$ and $t$, $j \in \{0,1 \cdots, n-1\}$, compute:

$$UK_{ID,t} = ((g_1 g^{-ID})^{r_t - r_{t-n}},\ F(ID,t)^{r_t} / F(ID, t-n)^{r_{t-n}}),$$

where $r_t = f(ID, HK_{ID,j}, t), r_{t-n} = f(ID, HK_{ID,j}, t-n)$.

## 3.4 UpdT

Given $UK_{ID,t}, TSK_{ID,t-1}$ and $t$, the user does:

1. Let $UK_{ID,t} = (h_t, h_t')$,
   $TSK_{ID,t-1} = \{d_{-n,t-1}, \cdots, d_{1,t-1}\}$.
2. $d_{1,t} = d_{1,t-1} \cdot h_t, d_{-n,t} = d_{-n,t-1}$, $d_{1-n,t} = d_{2-n,t}$,
   $\cdots$, $d_{-1,t} = d_{0,t-1}$, $d_{0,t} = d_{1-n,t-1} \cdot h_t'$,
   $TSK_{ID,t} = (d_{-n,t}, d_{1-n,t}, \cdots, d_{0,t}, d_{1,t})$,

where $d_{-n,t} = r_{ID}$, $d_{1-n,t} = (g_1 g^{-ID})^{r_{t-n+1}}, \cdots$,

$$d_{0,t} = (g_1 g^{-ID})^{r_t},\ d_{1,t} = (h_0 g^{r_{ID}})^{1/(\alpha - ID)}$$
$$\cdot F(ID, t-n+1)^{r_{t-n+1}} \cdots\cdot F(ID,t)^{r_t}.$$

### 3.5 Encrypt

Given $t$, randomly choose $s \in Z_p^*$, and compute

$$c_{-n} = e(g,g)^s,\ c_{1-n} = F(ID, t-n+1)^s,\ \cdots, c_0 = F(ID,t)^s,$$
$$c_1 = (g_1 g^{-ID})^s, c_2 = m \cdot e(g, h_0)^s,\ c = (t, c_{-n}, \cdots, c_1, c_2)..$$

### 3.6 Decrypt

The recipient decrypts:

$$\frac{e(c_1, d_{1,t})}{c_{-n}{}^{d_{-n,t}} e(c_{1-n}, d_{1-n,t}) \cdots e(c_0, d_{0,t})} = e(g, h_0)^s,\quad c_2 / e(g, h_0)^s = m.$$

## 4 Analysis of the New IBPKIE Scheme

**Theorem 1.** Assume that the $(T', \varepsilon', q)-$ABDHE assumption holds in $G, G_T$, and $h$ is a collision resistant hash function, then the new IBPKIE scheme is $(T, \varepsilon, q-1)-$IND-ID-KI-CPA secure for $T' = T + O(t_{\exp} \cdot qn)$, $\varepsilon' = \varepsilon - 1/p$, where $t_{\exp}$ is the average time required to exponentiate in $G$ .

**Proof.** Assume $A$ is an IND-ID-KI-CPA adversary, and we can construct an algorithm $B$ to solve the $q-$ ABDHE problem as follows. At first, $B$ is given a vector $(g', (g')^{\alpha^{q+2}}, g, g^{\alpha}, \cdots, g^{\alpha^q}, Z) \in G^{q+3} \times G_T$ to decide whether $Z = e(g', g)^{\alpha^{q+1}}$. $B$ flips a $Coin \in \{1,2\}$. If $Coin = 1$, $B$ plays $Game$ 1 with $A$ and else $Game$ 2.

**Game 1:** $B$ acts as a challenger expecting that $A$ will never corrupt the long-term keys with the challenged identity. $B$ interacts with $A$ as follows:

**Setup.** $B$ randomly chooses $I(x) = \sum_{i=0}^{q} a_i x^i$, and computes $g_1 = g^{\alpha}$, $h_0 = g^{I(\alpha)}$, where $a_i \in Z_p^*$. Let $h_1 = g_1^{u_1}, h_2 = g^{u_1}$, $h_3 = g_1^{u_2}, h_4 = g^{u_2}$, where $u_1, u_2 \in Z_p^*$ are two random numbers. Then $B$ sends the public keys $(g, g_1, h_0, h_1, h_2, h_3, h_4)$ to $A$.

**Phase 1.** $A$ adaptively issues queries.
Long-term key query. $A$ sends $< ID, j >$ to $B$, where $j \in \{0,1,\cdots,n-1\}$. $B$ maintains a list $HK^{list}$, which is initially empty. $B$ checks whether $HK^{list}$ contains $(ID, j, HK_{ID,j})$. If yes, return $HK_{ID,j}$ to $A$. Else, randomly choose $HK_{ID,j} \in \{0,1\}^{1+\log_2 p}$, add $(ID, j, HK_{ID,j})$ on $HK^{list}$, and send $HK_{ID,j}$ to $A$.

Extract query: $A$ sends $ID$ to $B$. $B$ does:

1. Issue long-term key queries on $< ID, j >$ to obtain $HK_{ID,j}$, where $j \in \{0,1,\cdots,n-1\}$.
2. Compute $d_{-n,0} = r_{ID} = -I(ID)$,

$$d_{1-n,0} = (g_1 g^{-ID})^{r_{1-n}}, \cdots, d_{0,0} = (g_1 g^{-ID})^{r_0},\ d_{1,0} = g^{I_{ID}(\alpha)} \cdot F(ID,1-n)^{r_{1-n}} \cdots F(ID,0)^{r_0},$$

where $I_{ID}(x) = \dfrac{I(x) - I(ID)}{x - ID}$, $r_{1-n} = f(ID, HK_{ID,n-1}, 1-n)$, $\cdots$, $r_0 = f(ID, HK_{ID,0}, 0)$,

$$TSK_{ID,0} = (d_{-n,0}, \cdots, d_{1,0}).$$

3. Return $(TSK_{ID,0}, HK_{ID,0}, \cdots, HK_{ID,n-1})$ to $A$.

It is a valid private key since

$$g^{I_{ID}(\alpha)} = g^{\frac{I(\alpha)-I(ID)}{\alpha - ID}} = (h_0 g^{r_{ID}})^{1/(\alpha - ID)},$$

$$d_{1,0} = (h_0 g^{r_{ID}})^{1/(\alpha - ID)} \cdot F(ID,1-n)^{r_{1-n}} \cdots F(ID,0)^{r_0}.$$

Short-term key query: $A$ sends $< ID, t >$ to $B$.

1. Issue long-term key queries on $< ID, j >$ to obtain $HK_{ID,j}$, where $j \in \{0,1,\cdots,n-1\}$.
2. Compute $d_{-n,t} = r_{ID} = -I(ID)$,

$$d_{1-n,t} = (g_1 g^{-ID})^{r_{t-n+1}}, \cdots, d_{0,t} = (g_1 g^{-ID})^{r_t},$$

$$d_{1,t} = g^{I_{ID}(\alpha)} \cdot F(ID, t-n+1)^{r_{t-n+1}} \cdots F(ID,t)^{r_t},$$

where $r_{t-n+1} = f(ID, HK_{ID,n-1}, t-n+1), \cdots,\ r_t = f(ID, HK_{ID,j}, t), j = t \bmod n$.

3. Return $TSK_{ID,t} = (d_{-n,t}, \cdots d_{0,t}, d_{1,t})$ to $A$.

**Challenge:** $A$ sends $< ID^*, t^*, m_0, m_1 >$ to $B$.

$B$ randomly chooses $m_\omega, \omega \in \{0,1\}$, and computes

$$c_1^* = (g')^{\alpha^{q+2} - (ID^*)^{q+2}},\ c_{-n}^* = Z \cdot e(g', g)^{\sum_{i=1}^{q+1} \alpha^{q+1-i} (ID^*)^i},$$

$$c_{1-n}^* = (g')^{(\alpha^{q+2} - (ID^*)^{q+2})(u_1 h(t^* - n+1) + u_2 h(ID^*, t^* - n+1))},$$

$$\cdots, c_0^* = (g')^{(\alpha^{q+2} - (ID^*)^{q+2})(u_1 h(t^*) + u_2 h(ID^*, t^*))},$$

$$c_2^* = m_\omega \frac{e(c_1^*, d_{1,t^*})}{(c_{-n}^*)^{d_{-n,t^*}} e(c_{1-n}^*, d_{1-n,t^*}) \cdots e(c_0^*, d_{0,t^*})},$$ where

$(d_{-n,t^*}, \cdots, d_{0,t^*}, d_{1,t^*})$ is a secret key of $ID^*$ at $t^*$.

$B$ sends $c^* = (t^*, c_{-n}^*, c_{1-n}^*, \cdots, c_1^*, c_2^*)$ to $A$.

Let $s^* = \log_g g' \dfrac{\alpha^{q+2} - (ID^*)^{q+2}}{\alpha - ID^*}$, where

$$\frac{\alpha^{q+2}-(ID^*)^{q+2}}{\alpha-ID^*}=\sum_{i=0}^{q+1}\alpha^{q+1-i}(ID^*)^i.$$

If $Z=e(g',g)^{\alpha^{q+1}}$, $c_1^*=(g_1g^{-ID})^{s^*}$,

$$c_{-n}^*=e(g',g)^{\sum_{i=0}^{q+1}\alpha^{q+1-i}(ID^*)^i}=e(g,g)^{s^*},$$

$$c_{1-n}^*=(g_1g^{-ID^*})^{s^*(u_1h(t^*-n+1)+u_2h(ID^*,t^*-n+1))}$$

$$=(h_1h_2^{-ID^*})^{s^*h(t^*-n+1)}(h_3h_4^{-ID^*})^{s^*h(ID^*,t^*-n+1)}$$

$$=F(ID^*,t^*-n+1)^{s^*},$$

$$\cdots,c_0^*=F(ID^*,t^*)^{s^*},\ c_2^*=m_\omega\cdot e(g,h_0)^{s^*}.$$

**Phase 2:** $A$ adaptively issues queries as Phase 1, and $B$ answers in the same way as Phase 1.

**Guess:** $A$ submits $\omega\in\{0,1\}$ and wins the game if $\omega'=\omega$ and the following conditions are satisfied:

(1) $<ID^*>$does not appear in extract queries;
(2) $<ID^*,t^*>$ does not appear in short-term key queries;
(3) $<ID^*,0>$, …, and $<ID^*,n-1>$ can not appear in long-term key queries.

**Game 2:** $A$ will be able to corrupt exactly one of the long-term keys with the challenged identity.

**Setup, Phase 1, Challenge, Phase 2:** As **Game 1**.

**Guess:** $A$ submits $\omega\in\{0,1\}$ and wins the game if $\omega'=\omega$ and the following conditions are satisfied:

(1) $<ID^*>$does not appear in extract queries;
(2) $<ID^*,t^*>$does not appear in short-term key queries;
(3) $A$ can not issue both short-term key query $<ID^*,t^*-1>$ and long-term key query $<ID^*,t^* \bmod n>$;
(4) $A$ can not issue both short-term key query $<ID^*,t^*+1>$and long-term key query $<ID^*,(t^*+1) \bmod n>$;
(5) Only one of $<ID^*,0>$, …, and $<ID^*,n-1>$ can appear in long-term key queries.

**Probability analysis:** Refer to [7].

**Time complexity.** Each query requires $O(n)$ exponentiations in $G$, and $A$ makes at most $q-1$ queries, so $t'=t+O(t_{\exp}\cdot qn)$. □

## 5 Conclusion

We have presented an IBPKIE scheme which allows multiple long-term keys to be alternately used in key updates. The scheme is adaptive-ID secure without random oracles, and it has a tight reduction. The sizes of private key and ciphertext depend on the number of long-term keys, we expect to solve this problem in the future research.

**Acknowledgement.** The work described in this paper was supported by the Natural Science Foundation of China (60872116, 60832010, and 60773079), and Shanghai Postdoctoral Science Foundation, China (10R21413200).

## References

1. Dodis, Y., Katz, J., Xu, S., Yung, M.: Key-Insulated Public Key Cryptosystems. In: Knudsen, L.R. (ed.) EUROCRYPT 2002. LNCS, vol. 2332, pp. 65–82. Springer, Heidelberg (2002)
2. Hanaoka, G., Hanaoka, Y., Imai, H.: Parallel key-Insulated Public Key Encryption. In: Yung, M., Dodis, Y., Kiayias, A., Malkin, T. (eds.) PKC 2006. LNCS, vol. 3958, pp. 105–122. Springer, Heidelberg (2006)
3. Libert, B., Quisquater, J.-J., Yung, M.: Parallel Key-Insulated Public Key Encryption without Random Oracles. In: Okamoto, T., Wang, X. (eds.) PKC 2007. LNCS, vol. 4450, pp. 298–314. Springer, Heidelberg (2007)
4. Shamir, A.: Identity-Based Cryptosystems and Signature Schemes. In: Blakely, G.R., Chaum, D. (eds.) CRYPTO 1984. LNCS, vol. 196, pp. 47–53. Springer, Heidelberg (1985)
5. Weng, J., Liu, S., Chen, K., Ma, C.: Identity-Based Parallel Key-Insulated Encryption without Random Oracles: Security Notions and Construction. In: Barua, R., Lange, T. (eds.) INDOCRYPT 2006. LNCS, vol. 4329, pp. 409–423. Springer, Heidelberg (2006)
6. Ren, Y., Gu, D.: CCA2-secure (hierarchical) identity based parallel key-insulated encryption without random oracles. Journal of Systems and Software 83(1), 153–162 (2010)
7. Gentry, C.: Practical Identity-Based Encryption without Random Oracles. In: Vaudenay, S. (ed.) EUROCRYPT 2006. LNCS, vol. 4004, pp. 445–464. Springer, Heidelberg (2006)
8. Boneh, D., Franklin, M.: Identity-based encryption from the weil pairing. SIAM Journal of Computing 32(3), 586–615 (2003)
9. Waters, B.: Efficient Identity-Based Encryption without Random Oracles. In: Cramer, R. (ed.) EUROCRYPT 2005. LNCS, vol. 3494, pp. 114–127. Springer, Heidelberg (2005)

# A Novel Approach to Calibrating the Line Scan Camera of Vehicle-Borne 3D Data Acquisition System

Youmei Han[1,2] and Liuzhao Wang[2]

[1] College of Geodesy and Geomatics
Shandong University of Science & Technology, Qingdao China
hanyoumei@126.com
[2] 3D Information Acquisition & Application Key Lab, Capital Normal University Beijing China
jzwlz@163.com

**Abstract.** It proposed a novel approach to calibrate the line scan camera (LSC) based on the characters of the line scan camera and the vehicle-borne 3D data acquisition system. Using the angle information from the laser scanner data, combing the principle of the line scan camera, it built a calibration model for LSC and designed some experiments for that. Then it computed high precision calibration parameters, which provides basis for the data peocessiong of the whole vehicle-boren 3D data aquisition system and gives references to the similar sensors calibration.

**Keywords:** City Virtual Reality and Visualization, Line Scan Camera, dynamic calibration, Laser Scanner, Vehicle-borne.

## 1 Introduction

Building surface texture data acquisition and real texture mapping have been a difficult problem which City Virtual Reality and Visualization should solve, and which is also one of the 3D city modeling key problems [1,2].Vehicle-borne 3D data acquisition system is one good system for City Virtual Reality and Visualization, which has accelerated the pace of urban three-dimensional modeling. Meanwhile in this 3D data acquisition system, using line scan camera to capture the true texture makes the city modeling more real.

Vehicle-borne 3D data acquisition system consist of line scan camera, laser scanner, GPS, Inertial Measurement Unit（IMU）et al. Line Scan Camera（LSC）has many advantages, such as high frequency, wide angle, which can save the data in time and don't lead to images missing which the planar camera couldn't reach. While it's non-metric and lens distortions are still the key problems to avoid it being used as a major sensor to capture the texture data. Traditional planar camera calibration methods, such as spatial resection, direct linear transformation, Vanishing Point method based on multi images, analytical plumb line, as well as self-test method [3-6],which always rely on the different images that captured on different direction then calculate the internal elements and their lens distortions based on co-linearity equation. Now regarding the line scan camera as a special case of the planar camera, its' special features need explore new ways to calibrate it. Until now, most of the calibration methods for the line scan camera are to fix the camera position and move the targets to get the calibration data [7-10], which have low precision because of the short distance between the camera and the targets.

Line Scan Camera （LSC） and Laser Scanner are the two important components of the vehicle-borne 3D data acquisition system. The former is used to obtain the building color texture information on both sides of the street and the latter can obtain the corresponding high-precision three-dimensional coordinates of the building. They have a common feature: the laser scanner data can be seen as lines consist of discrete points, and line scan camera data are RGB color lines with one pixel width. So it presented a novel approach which is different from traditional planar camera calibration methods and current line scan camera calibration methods. It can provide reference for similar linear array CCD sensors calibration.

## 2 Calibration Parameters of the lsc

The line scan camera CV-107CL made by Japan JAI company is used by Vehicle-borne 3D data acquisition system, which has been designed on the system. This camera uses the prism spectroscopic technique, which has good color reproduction. The white balance function button is on the back of the camera which is very convenient to use. The camera frequency can reach 19048 lines per seconds. It also has high resolution with 14μm pixel size which can get very real colorful images. But it should be calibrated to eliminate mechanical error till it can fusion the image data with the point clouds to get good information for the City Virtual Reality and Visualization.

The errors of the LSC mainly come from the lens distortion and the CCD itself. This LSC uses a Nikon AF lens (the focus is 14mm). The error caused by the lens comes from lens design, production and assembly which lead to image point's deviation from its correct location. It is also named as the optical distortion, which includes radial distortion and eccentric radial distortion. Eccentric distortion is always less than one-third of one pixel, so here it only considers the radial distortion [3, 11].

The CCD errors of LSC always come from CCD placement, unevenness of CCD planar and the distortion of each CCD unit. The CCD placement can result in the distortion of the principal point; the unevenness of CCD planar, which can only be calibrated by special device directly, has nothing to do with photography. In fact it is too small to influence the whole LSC calibration [3].

## 3 The Calibration of the lsc

### 3.1 Summary

Line Scan camera should move around the target objects then the normal images of the objects can be grabbed. However, when it is moving the position and the relative attitude is changing over the time. If it uses the traditional plane-array camera calibration methods to calibrate LSC, the relative attitude could not be captured because of the high frequency of the camera itself, which is the key content to get the orientation elements by traditional methods. Until now many domestic and foreign experts have provided some calibration methods: R. Horaud designed a special graph for the LSC in 1993, which was made of several lines [7], some scholars comes from Capital Normal University have also done some similar experiments [8]. In 2008, Beijing Information Science and Technology University also invented a LSC calibration target which also was made by a number of equally spaced vertical and horizontal lines [9]. Tianjin University has proposed a two-steps method to calibrate line CCD, which is also based on similar principles [10]. Their common feature is that based on this kind of specific targets, analyzed the images of the special lines on the targets to calibrate LSC' elements of interior orientation by the lines' relative positions relationship. If the target was far from the camera the image of it would be very small on the whole image. In order to get the target image on the edge of the frame, very big target should be designed which is impossible and not convenient to do this. So the accurate of these kinds of methods is relatively low.

### 3.2 The Idea of the Novel Calibration Method Based on the Laser Scanner

Regarding the middle of the line image (named as O) as the original point, with the positive direction pointing to the top of the camera, one image plane coordinate system was built (fig.1). In this paper, the main calibration content is camera inner orientation elements, which mainly includes the projection of the principal point on the image coordinate system ($x_0$), the camera focal length $f$ and the radial distortion.

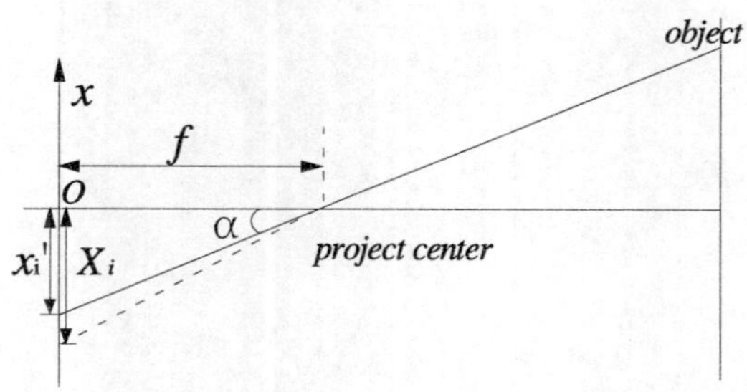


**Fig. 1.** The Imaging Principal of the Line Scan Camera

From Fig. 1 it can conclude that:

$$x_i = \tan \alpha_i * f \tag{1}$$

Where: $x_i$ is the theory value of the object on the image coordinate system; $f$ is the focal length; $\alpha_i$ is an angle between the line from some pixel (here name it as Z) to the principle point and the axis (it is also called incident angle).

Above all, the real image coordinate of the objects only has to do with incident angle and the focal length. As long as these two parameters could be determined accurately, it also could get the real image coordinate then get interior orientation elements of the LSC accurately.

In the vehicle-borne 3D data acquisition system, the laser scanner captures the 3D coordinate data in high frequency and it has a relatively high accuracy angle measure unit. The nominal accuracy of the laser scanner is ±20″ [12] .The normal focal length of the CV-L107CL lens is 14mm (equals to 1000 pixel size, each pixel size is 14μm).

Formula (1) , $x_i = \tan\alpha_i * f$ ,when $\alpha$ is a small value, it can be written as $x_i = \alpha_i * f$

$$M_x = f * m_a = \pm 0.097 \text{ (pixel)}$$

Where $m_a$ is 20″, and *f* is 1000. That is to say the influence to the image coordinate comes from the angle measure is less than 0.1 pixel size. So the angle distortion can be ignored during the calibration of the LSC. The LSC was fixed on the lasers scanner by precision machining (fig.2), and the installation error can be controlled within ±1cm. When the distance between the objects and the camera is farer than 25m, according to formula (1), it can compute the eccentric error of these two sensors, which is about 0.4 pixel sizes. So this error could be ignore, which is a very important primary condition to propose this novel calibration approach.

**Fig. 2.** The rigid construction of the LSC and LS

Fixed the LSC on the top of the laser scanner strictly, and made the working faces of these two sensors closed to parallel which can be regarded as a whole rigid body and the intersecting angle was adjusted within ± 5" by precision feeler gauge (fig 3). This rigidity parallel structure makes they have the same attitudes during moving. Moved this rigid body along a precision guide rail, then it can get the images and the geometrical information (cloud points) of the objects. It used a building wall which has much special characteristic information as the object. Extract characteristic points and the corresponding incident angle of them from the cloud points by programming. Obtain the corresponding points' pixel from the line scan images and make it group with the cloud points extracted above. It needed to observe some groups of data which distribute on the wall evenly. Then it used these two kinds of information as the original data to compute the LSC parameters by precision adjustment computations. That is the main idea of this novel approach.

### 3.3 The Calibrate Model for This Approach

According to the geometrical optics, the LSC $y = 0$, the radial distortion of its lens $\Delta r$ can be written as a homogeneous equation as following:

$$\Delta_r = x(k_0 r^2 + k_1 r^4 + k_2 r^6 + ....) \tag{2}$$

Where $\Delta r$' unit is μm, $k_i (i = 0,1,2,....)$ is the coefficient to express the lens of the camera, $r$ is the radius vector of some pixel .The $\Delta r$ is very little, so it can be written as:

$$r = \sqrt{(X' - x_0)^2} \tag{3}$$

Where $x_0$ and $X'$ are the image coordinate of the original point and this point on the image.

For most of the lens, three $k$ coefficients are enough to describe the deviation curve. However two is enough for some good lens [4].

Using (1), (3) and (2) ,it constructed the calibrate model for the LSC:

$$\tan \alpha_i * f = (X_i' - x_0) + k_0 (X_i' - x_0)^3 + k_1 (X_i' - x_0)^5 + k_2 (X_i' - x_0)^7$$

Because of $f$ is related with $x_0, k_0, k_1, k_2$, it need to compute these parameters in two steps in order to get high precision results.

First, compute $x_0, k_0, k_1, k_2$

Carrying on the linearization to this model it can get (4) as following,

$$v_{x_i} = (x_i) - x_i + (1 + 3k_0 (X_i' - x_0)^2 + 5k_1 (X_i' - x_0)^4 + 7k_2 (X_i' - x_0)^6) \Delta x_0 + (X_i' - x_0)^3 \Delta k_0 + (X_i' - x_0)^5 \Delta k_1 + (X_i' - x_0)^7 \Delta k_2 \tag{4}$$

With a lot of observation value, formula (4) can be written as a matrix: $V = AX - L$

here :

$$A = [(X_i' - x_0)^3, (X_i' - x_0)^5, (X_i' - x_0)^7, 1 + 3k_0 (X_i' - x_0)^2 + 5k_1 (X_i' - x_0)^4 + 7(X_i' - x_0)^6]^T$$

$$X = [\Delta k_0, \Delta k_1, \Delta k_2, \Delta x_0]^T$$

$$L = x_i - (x_i)$$

After iterative computation it gets $x_0, k_0, k_1, k_2$.

Second, compute *f*.

Carrying on the linearization to this model it can get (5) as following,

$$v_{x_i} = (x_i) - x_i + tga_i \cdot \Delta f \tag{5}$$

With a lot of observation value, (5) can be written as a matrix: $V = AX - L$

Where, $A = tga_i$, $X = \Delta f$, $L = x_i - (x_i)$

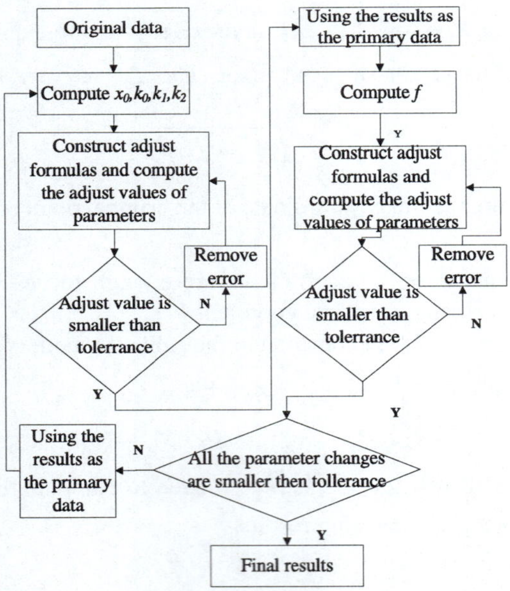


**Fig. 3.** The flow chart of computing parameters

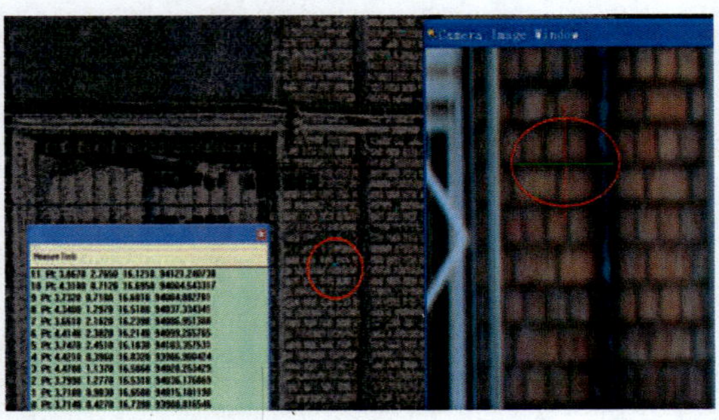

**Fig. 4.** The corresponding of the cloud points and the Image

Using the results computed by the first step as the primary data, with iterative computation $f$ can be obtained. Then use this $f$ result as the primary data to compute the parameters of the first step for a second time. Till to all the parameters can get a convergent and stable result, then the final results will have a high precision. The details flow char of this idea is in the Fig. 3.

## 4 Experiments and the Results

Fixed the rigid body onto a lift platform (fig.2), then move the platform in some speed to capture the image and cloud points of the wall of a building. Fig. 4 is a section of the experiment wall in our programming, with many brick-work joints (characteristic points) in it.

A programming was written for computing the calibrate parameters with an error removing algorithm in it. For each pair of data,

When: $$|\Delta x_i| > 3\sqrt{\frac{\sum \Delta x_i * \Delta x_i}{n-1}}$$

this pair would be regarded as wrong data and be removed from the original data. Then the rest data is used to join the iterative computation.

Three groups of original data were observed for this experiment, using them to compute the calibrate parameters respectively. Then used the results to calibrate the camera and got the corresponding distortion curve, which were draw in the Fig. 5 following. Comparing these three curves, they have the similar size and shape, so this calibrate model is right for calibrating the LSC. And the results have high precision.

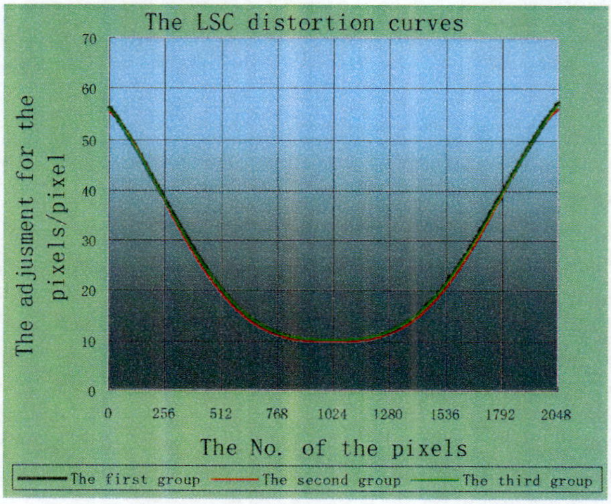


**Fig. 5.** The curves of the distortion

The results of the three group data are in table 1.It also computed the root mean square errors respectively.

$$m_{xi1} = \pm\sqrt{vv/(n-1)} = \pm 0.41$$

$$m_{xi2} = \pm\sqrt{vv/(n-1)} = \pm 0.51$$

$$m_{xi3} = \pm\sqrt{vv/(n-1)} = \pm 0.37$$

The pixel size of CV-L107CL is 14μm. Each group original data has about 150 pairs of data and the maximize root mean square error is 0.51×14=7.14μm, which has enough precision to be fusion with other data and provide data to City Virtual Reality and Visualization.

Analyzing the three results, it got the root mean square error of the principle point and the focus length as following.

$$m_{x0} = \pm\sqrt{vv/(n-1)} = \pm 0.16$$

$$m_{\mathrm{f}} = \pm\sqrt{vv/(n-1)} = \pm 0.54$$

All the above root mean square error'(RMSE) unit is one pixel size (14μm), that is to say the root mean square error of principle point is 0.16×14=2.24μm,the focal length one is 0.54×14=7.56μm.

**Table 1.** The Results and the Accurate of the Three Group Data

| groups | $x_0$ | $k_0$ | $k_1$ | $k_2$ | *f*(pixel) | RMSE(pixel) |
|---|---|---|---|---|---|---|
| 1 | 10.23 | 9.35x10-8 | 5.82x10-14 | 9.27x10-21 | 1024.38 | ±0.41 |
| 2 | 9.86 | 8.50x10-8 | 3.43x10-14 | 9.80x10-21 | 1023.25 | ±0.51 |
| 3 | 9.96 | 9.75x10-8 | 5.57x10-14 | 13.41x10-21 | 1024.42 | ±0.37 |

## 5 Conclusions

Analyzing the results, this novel approach has high calibration precision. The lens distortion root mean square error is reach to 7.14 μm, the principal point one is 2.24μm and the focal length one is7.56μm. After calibrate by this approach the image data has totally enough precision to fusion with the cloud points. The high precision and stability of this approach makes it can provide reference to calibrate the similar line sensors. This method had also been test by a TVI LSC and got a well result. In order to improve the precision in the future, the distance between the LSC can the objects could be farer. Decreasing the influence of the laser scanner angle error and the machine error would also be better for precision.

**Acknowledgment.** The authors would like to thank Professor Xianlin Liu and the leaders of 3D Information Acquisition and Application Key Lab for providing the experiment data.

## References


1. Bo, W., Aiwu, Z.: Design and implementation of a vehicle-borne system of 3D data acquisition and processing. Chinese Journal Of Stereology And Image Analysis 13, 30–33 (2008)
2. Yunfang, C., Zetian, Y.: Research on mobile data collection system based on multi-sensor. Transducer and Micro-system Technologies 25, 23–25 (2006)
3. Wenhan, X., Zuxun, Z.: Camera calibration based on vanishing points of multi image. Acta Geodaetica et Cartographica Sinica 33, 335–340 (2004)
4. Wenhao, F.: Close range photogrammetry, pp. 185–193. Wuhan Universe Press, Wuhan (2002)
5. Zongjian, L., Hongxia, C.: Research on the digital camera distortion calibration. Geomatics and Information Science of Wuhan University 30, 122–125 (2005)
6. Dong, W., Wenhao, F., Xiushan, L.: Camera Calibration Of Nikon D1X. Science of Surveying and Mapping 32(2), 33–34 (2007)
7. Horaud, R., Mohr, R., Lorecki, B.: On single-scan line camera calibration. IEEE Transactions On Robobtics And Automation 9, 71–74 (1993)
8. Wenqian, Q., Yi, Y., Xiaole, H.: The application of single scan line camera in vehicle borne 3D information data acquisition system. Journal of Capital Normal University 29, 9–11 (2008)
9. `http://zhuanli.baidu.com/pages/sipo/20081016/75/3e3c374255cffcb788653f34d810af80_0.html`
10. Hongtao, Z., Fajie, D.: Study on calibration of linear CCD based on two steps. Acta Metrologica Sinica 28, 311–313 (2007)
11. Liuzhao, W.: Small digital and aerial photogrammetry system. KunMing China, University of Science and Technology, 5–6 (June 2006)
12. Liuzhao, W., Laichao, Q.: The calibration of RA-360 laser scanner. In: The 6th International Symposium on Digital Earth, pp. 102–105 (September 2009)

# The Bartered-Type Return Policy Based on Strategic Customer Behavior*

Hong Jiang, Ershi Qi, Yanfang Huo, and Daojian Yang

School of Management
Tianjin University
Tianjin City, China
jianghong_helen@yahoo.com.cn

**Abstract.** Considering strategic customer behavior (SCR), this paper studies the bartered-type return policy, and builds an extended newsvendor model with the bartered-type return policy based SCR. By analyzing our model we obtain the expression of the seller's optimal price and inventory. The optimal price is related to the distribution function of the value of the product, and the optimal inventory is related to the price of the replaced product when the cost and the salvage price is given. At last we simulate the actual data using the software matlab, and clearly illustrates the application of our model in the real condition.

**Keywords:** Bartered-type, Return Policy, Strategic Customer Behavior, Newsvendor Model.

## 1 Introduction

Nowadays there are more and more strategic customers in the markets. Maximizing self-benefits is their code of conduct. Their presence seriously challenges the enterprises how to make the correct decision [1]. Some scholars have developed the research on strategic customer behavior. These researches first appeared in the field of the marketing, which are used to choose the best marketing strategy [2-3]. Then the studies on strategic customer behavior have achieved a great success in the marketing field. Recently with the development of the supply chain, some scholars began to pay attention to strategic customer behavior in the supply chain distribution channels [4-5]. After that strategic customer behavior are introduced formally into the field of the supply chain operation for solving the problem of enterprises' optimal production decisions. The current study in this area is still in its beginning stage, and the researches are mainly focused on the price decision [6-9]. In other literatures of the operational areas of the supply chain, strategic customer behavior has rarely been mentioned.

This paper studies enterprises' optimal return policy with no quality defects considering the existence of the strategic customers in the retail market. The early relevant researches are mainly focused on how to avoid returning, but they all disregarded the strategic customer behavior [10-13]. Su studied the strategic customer's

* Supported by National Natural Science Foundation of China (70971095).

return behavior, and he had an assumption that the cash is returned to the customers directly [14]. However in our daily life many stores offer a wide selection of the different products, and consumers who wouldn't like to keep the product may be permitted to change another cheaper one for free. In fact, when consumers return unwanted products, some stores offer store credit only. So with regards to this practical phenomenon our paper studies the bartered-type return policy considering the strategic customer behavior, which means the customers who want to return one must buy another instead. This paper aims to help enterprises make the optimal bartered-type return policy and maximize their own profits.

## 2 Problem Description

The basis of this paper is the classic newsvendor model. We assume that there are a seller and numerous strategic customers in the market. In the sales period they estimate their own surplus before buying the product and if their expected surplus is more than zero, they will buy the product at the full price, otherwise they will not purchase anyone or wait for the salvage period. Then, because of the deviation of the product value estimation, a reassessment will be held after the strategic customers buy the product. In this process if the product value is more than its price, they will keep it, otherwise they will consider whether to change it for the cheaper one. In the salvage period the seller salvages all leftover units at the salvage price. These leftover units include the units that were unsold as well as that were sold but returned. Here the customers can't be permitted to change the product any more. We take the symbol as follows:

$X_i$ : random market demand of product *i*;

$F_i(\cdot)$ : the distribution function of the $X_i$ , its density function is $f_i(\cdot)$ ;and $\overline{F}_i(\cdot)=1-F_i(\cdot)$ ;

$q_i$ : seller's inventory of product $i$ ;

$V_i$ : random value of product $i$ in the customer's mind;

$G_i(\cdot)$ : the distribution function of the $V_i$ , and its density function is $g_i(\cdot)$ , and $\overline{G}_i(\cdot)=1-G_i(\cdot)$ ;

$p_i$ : full price of product $i$ ;

$c_i$ : cost of product $i$ ;

$s_i$ : salvage price of product $i$ ;

$H_{v_1,v_2-v_1}(\cdot)$ : the probability of $V_1 \le v_1$ and $V_2 - V_1 \le v_2 - v_1$ , and its density function is $h_{v_1,v_2-v_1}(\cdot)$ .

## 3 Model

For the simplicity of our model we assume that the customers can only be permitted to barter the product just once. Namely, the customer changes product 1 for product 2 (namely i=1,2), and the price of product 1 is higher than the product 2. The process of the customer decision is like this: first, he will decide whether to buy or not. When his expected surplus is more than zero ( $E\Pi_C \geq 0$ ), he will buy the product at the full price. If he purchases the product 1, he will re-estimate it and choose whether to return it. If the value of product 1 is lower than its price ( $V_1 \leq p_1$ ), he will have the wishes of returning. And when he has the wishes of returning if the surplus brought by bartering is greater than keeping it, he will choose to barter the product; otherwise he will still keep it. His surplus of keeping the product 1 is $V_1 - p_1$.and the surplus of changing it for the product 2 is $V_2 - p_1$ First, the customer bartered behavior is decided by the rational factor, namely:

$$\max(V_1 - p_1,\ V_2 - p_1) \tag{1}$$

When the value $V_1 - p_1 \leq V_2 - p_1$ (namely $V_1 \leq V_2$ ), the customer will choose to barter the product rationally. Besides, because of $p_1 \geq p_2$ the customer's decision is also influenced by the psychology of the price-loss. So his decision is also dependent on :

$$\max\{V_1 - p_1,\ V_2 - p_1 - (p_1 - p_2)\} \tag{2}$$

If the value $V_1 - p_1$ is not more than $V_2 - p_1 - (p_1 - p_2)$ (namely $V_2 - V_1 \geq p_1 - p_2$ ), the customer will choose to barter the product irrationally. According to the rational and emotional factors, we can draw a conclusion as follow:

*Proposition 1: If the customer decides to return and change the product, the value* $V_i$ *and the price* $p_i$ *should be characterized by:*

$$\begin{cases} V_1 \leq p_1 \\ V_2 - V_1 \geq p_1 - p_2 \end{cases} \tag{3}$$

The first inequality represents "the wishes of returning", and the second one represents "the wishes of bartering". When the customer has the wishes of returning ( $V_1 \leq p_1$ ) and the value difference of the two products is more than the difference of their prices ( $V_2 - V_1 \geq p_1 - p_2$ ), the customer will decide to change the product 1 for the product 2. We assume that the joint distribution function of $v_1$ and $v_2 - v_1$ is

$$\begin{aligned} H_{v_1, v_2 - v_1}(\cdot) &= P\{V_1 \leq v_1, V_2 - V_1 \leq v_2 - v_1\} \\ &= P\{V_1 \leq v_1, V_2 \leq v_2\} \end{aligned} \tag{4}$$

Because of the value $V_1$ and $V_2$ are independent, the joint density function is $h_{v_1, v_2 - v_1} = g_1(v_1) \cdot g_2(v_2)$ . Therefore the customer's surplus function is

$$\begin{aligned} E\Pi_C &= \underbrace{\int_{p_1}^{+\infty} (v_1 - p_1) g_1(v_1) dv_1}_{(a)} + \underbrace{\int_0^{p_1} \int_0^{2p_1 - p_2} (v_1 - p_1) g_1(v_1) g_2(v_2) dv_1 dv_2}_{(b)} \\ &\quad + \underbrace{\int_0^{p_1} \int_{2p_1 - p_2}^{+\infty} (v_2 - p_1) g_1(v_1) g_2(v_2) dv_1 dv_2}_{(c)} \\ &= EV_1 - p_1 + \int_0^{p_1} \int_{2p_1 - p_2}^{+\infty} (v_2 - v_1) g_1(v_1) g_2(v_2) dv_1 dv_2 \end{aligned} \tag{5}$$

In the equation (5), the parts (a), (b) and (c) represent respectively "keeping", "willing to return but not to barter" and "bartering". When the value and the price satisfy $V_2 - p_1 = 0$ and $V_2 - p_1 - (p_1 - p_2) = 0$ (namely $V_2 = p_2 = p_1$), it means the seller offers full refunds. Here the customer's expected surplus may be expressed by $E\Pi_C = \int_{p_1}^{+\infty}(v_1 - p_1)g_1(v_1)dv_1$, which is consistent with the literature in [14]. By analyzing the formula (5) we can find that the surplus $E\Pi_C$ is decreasing by the price $p_1$ only when the surplus brought by bartering ($\int_0^{p_1}\int_{2p_1-p_2}^{+\infty}(v_2 - v_1)g_1(v_1)g_2(v_2)dv_1dv_2$) is decreasing by $p_1$. Also we draw the following conclusion:

*Proposition 2: i) The customer will choose to buy the product 1 in the "full price" period, only when the price* $p_1$ *is characterized by:*

$$p_1 - \int_0^{p_1}\int_{2p_1-p_2}^{+\infty}(v_2 - v_1)g_1(v_1)g_2(v_2)dv_1dv_2 \le EV_1 \tag{6}$$

*ii) In order to induce the customers to buy at full price, when the bartered surplus is decreasing by* $p_1$*, the seller's optimal price* $p_1$ *is given by:*

$$p_1 - \int_0^{p_1}\int_{2p_1-p_2}^{+\infty}(v_2 - v_1)g_1(v_1)g_2(v_2)dv_1dv_2 = EV_1 \tag{7}$$

The above proposition tells the seller that he should fix the price $p_1$ satisfying the difference between the price $p_1$ and the surplus by bartering is less than the expected value $EV_1$ in order to make the customer buy the product 1 at the full price.

The seller's expected profit function is

$$\begin{aligned} E\Pi_R &= \underbrace{p_1\overline{G}_1(p_1)E\min(X_1,q_1)}_{\text{not return}} + \underbrace{(p_1 - p_2 + s_1)G_1(p_1)\overline{G}_2(2p_1 - p_2)E\min(X_1,q_1)}_{\text{barter}} \\ &\quad + \underbrace{p_1G_1(p_1)G_2(2p_1 - p_2)E\min(X_1,q_1)}_{\text{not barter}} + s_1(q_1 - E\min(X_1,q_1)) - c_1q_1 \\ &= [p_1 - s_1 - (p_2 - s_1)G_1(p_1)\overline{G}_2(2p_1 - p_2)]E\min(X_1,q_1) - (c_1 - s_1)q_1 \end{aligned} \tag{8}$$

By $\partial E\Pi_R/\partial p_1 = 0$ and $\partial E\Pi_R/\partial p_2 = 0$, we can obtain the optimal prices $p_1$ and $p_2$ as follow:

*Proposition 3: In order to maximize the seller's profit, the price* $p_1$ *and* $p_2$ *should both satisfy the expression as follow:*

$$G_1(p_1) = \frac{p_1}{A_2} + \frac{A_1}{A_2}(p_2 - s_1) \tag{9}$$

$$\overline{G}_2(2p_1 - p_2) = \frac{A_2}{p_2 - s_1} \tag{10}$$

From the above proposition, we can find the optimal price $p_i$ is related to the value distribution function $G_i(\cdot)$. If we know the certain distribution of the product value (for example: $G_i(\cdot)$ obeys uniform distribution) and the salvage $s_1$, we can obtain the optimal price $p_i$. And the constants $A_1$ and $A_2$ can be calculated by some specified $p_i$ and $G_i(\cdot)$ in practice.

For the same reason by $\partial E\Pi_R / \partial q_1 = 0$, we can find out the optimal inventory.

*Proposition 4: The seller's optimal inventory satisfies the following expression :*

$$\overline{F}(q_1^*) = \frac{c_1 - s_1}{(A_1 - 1)s_1 - A_1 p_2} \tag{11}$$

The equation (11) tells us that the optimal inventory is related to the cost $c_1$, the salvage $s_1$ and the product 2's price $p_2$. When the $c_1$ and $s_1$ is given, the product 1's optimal inventory $q_1^*$ is only related to the $p_2$. When the $A_1$ is positive, the $q_1^*$ is decreasing by the $p_2$, which means in this situation the decreasing number of customers who buy the product 1 at full price $p_1$ plays a leading role, so the seller should reduce the $q_1$; when the $A_1$ is negative, the $q_1^*$ is increasing by the $p_2$, which means in this situation it plays a leading role that customers buy the product 1 and keep it, so the seller should increase the $q_1$.

## 4 Example and Analysis

In order to illustrate the actual application of our model, we use the software matlab to simulate the practical problems. We assume the value distribution function obeys the uniform distribution in the region $[0, \max V_i]$, and the cost $c_1$ is equal to 4, and the salvage price $s_1$ is equal to 2.

Suppose that the seller can get a approximate group data from the actual world, for example $p_1 = 8$, $p_2 = 6$, $G_1 \sim U(0,10)$ and $G_2 \sim U(0,12)$. Then by calculating the equation (9) and (10), we can obtain the equations as follow:

$$\begin{cases} \dfrac{p_1}{v_1} = \dfrac{3p_1}{2} - \dfrac{14}{5}(p_2 - 2) \\ \dfrac{2p_1 - p_2}{v_2} = 1 - \dfrac{2}{3(p_2 - 2)} \end{cases} \tag{12}$$

Using the equation (12) we can obtain a group of the optimal prices which is shown inTab.1, Fig.1 and Fig.2. In this example the minimum of the price $p_i$ is respectively 4.4 and 4, because below the two prices the seller will lose some profit.

**Table 1.** The optimal price of the product

| $\max V_1$ | $p_1$ | $\max V_2$ | $p_2$ | $\overline{F}(q_1^*)$ |
|---|---|---|---|---|
| 15 | 11.5 | 17 | 7.9 | 0.330 |
| 14 | 10.8 | 16 | 7.5 | 0.347 |
| 13 | 10.1 | 15 | 7.1 | 0.366 |
| 12 | 9.4 | 14 | 6.8 | 0.386 |
| 11 | 8.7 | 13 | 6.4 | 0.409 |
| 10 | 8 | 12 | 6 | 0.436 |
| 9 | 7.3 | 11 | 5.6 | 0.466 |
| 8 | 6.6 | 10 | 5.2 | 0.501 |
| 7 | 5.9 | 9 | 4.8 | 0.543 |
| 6 | 5.1 | 8 | 4.4 | 0.595 |
| 5 | 4.4 | 7 | 4 | 0.662 |

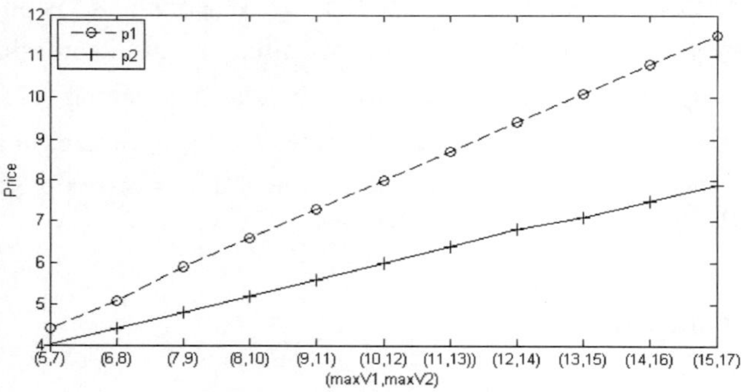


**Fig. 1.** The optimal price of the products

From Fig.1 we can see clearly the optimal price $p_i$ is approximately linear. And with the increasing of the value of the products, the optimal price is increasing, too. Although the difference between the $\max V_1$ and $\max V_2$ is the constant "2", the value of $p_1$ minus $p_2$ (namely $p_1 - p_2$) is increasing gradually.

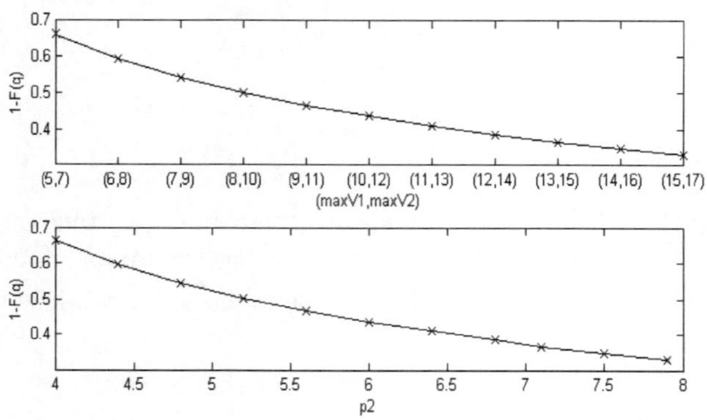


**Fig. 2.** The distribution value of the optimal inventory

Fig.2 shows us the distribution function of the optimal inventory of the ( $\max V_1$ , $\max V_2$ ) and the $p_2$. In our example because the constant $A_1$ is negative, the distribution function $\overline{F}(q_1^*)$ is decreasing by the $p_2$. However the optimal inventory $q_1^*$ is increasing by the $p_2$ (the function $\overline{F}(q)$ is decreasing by $q$ ).

# 5 Conclusion

This paper mainly studies the bartered-type return policy based on the classic newsvendor model considering strategic customer behavior. We build a model of bartered-type return policy. Through analyzing the customer's expected surplus and the seller's profit we will find out some conclusions as follow:

1) When the seller offers the bartered-type return policy, he should fix the price of the product which satisfies that the difference between the price and the surplus brought by bartering is less than the strategic customers' expected value of product 1. Otherwise the strategic customers will not buy the prodcut at the full price.

2) In order to maximize the seller's profit, the optimal price is related to the value distribution function of the product. Using some actual optimal point we can obtain the seller's optimal price in other situation.

3) The seller's optimal inventory is related to the price of the product 2. In the different situations the optimal inventory can be affected by the price of the product 2 oppositely.

However, in reality the supply chain is much more complicated than the newsvendor model, which usually has a network structure. So the research on the bartered-type return policy in the decentralized supply chain is a main field of study. Furthermore, some sellers offer a time-limit refund policy in the early period and a bartered-type return policy in the next period. Then the research of how to fix a optimal mixed return policy combined above two return policies is a very meaningful direction.

# References


1. Su, X.: Intertemporal Pricing with Strategic Customer Behavior. Management Science 53, 726–741 (2007)
2. Coase, R.H.: Durability and Monopoly. Journal of Law and Economics 15, 143–149 (1972)
3. Bulow, J.I.: Durable-Goods Monopolists. Journal of Political Economy 90, 314–332 (1982)
4. Desai, P., Koenigsberg, O., Purohit, D.: Strategic Decentralization and Channel Coordination. Quantitative Marketing and Economics 2, 5–22 (2004)
5. Arya, A., Mittendorf, B.: Benefits of Channel Discord in the Sale of Durable Goods. Marketing Science 25, 91–96 (2006)
6. Su, X., Zhang, F.: Strategic Customer Behavior, Commitment, and Supply Chain Performance. Management Science 54, 1759–1773 (2008)
7. Aviv, Y., Pazgal, A.: Optimal Pricing of Seasonal Products in the Presence of Forward-Looking Consumers. Manufacturing and Service Operations Management 10, 339–359 (2008)

8. Cachon, G.P., Swinney, R.: Purchasing, Pricing, and Quick Response in the Presence of Strategic Consumers. Management Science 55, 497–511 (2009)
9. Lai, G., Debo, L.G., Sycara, K.: Impact of Posterior Price Matching on Profit with Strategic Consumers. Manufacturing and Service Operations Management (unpublished)
10. Hess, J.D., Chu, W., Gerstner, E.: Controlling Product Returns in Direct Marketing. Marketing Letters 7, 307–317 (1996)
11. Chu, W., Gerstner, E., Hess, J.D.: Managing Dissatisfaction: How to Decrease Consumer Opportunism by Partial Rrefunds. Journal of Service Reseach 1, 140–155 (1998)
12. Davis, S., Gerstner, E., Hagerty, M.: Money Back Guarantees in Retailing: Matching Products to Consumer Tastes. Journal of Retailing 71, 7–22 (1995)
13. Yabalik, B., Petruzzi, N.C., Chhajed, D.: An Integrated Product Returns Model with Logistics and Marketing Coordination. European Journal of Operational Research 161, 162–181 (2005)
14. Su, X.: Consumer Returns Policies and Supply Chain Performance. Manufacturing and Service Operations Management 11, 595–612 (2009)

# Runtime-Based Boot Components Re-measurement Scheme for Trusted Platform

Peng Weiping[1,2,3,4], Zhou Yajian[1,3,4], Wang Cong[1,3,4], and Yang Yixian[1,3,4]

[1] Key Laboratory of Network and Information Attack & Defence Technology of MOE, Beijing University of Posts and Telecommunications, Beijing, 100876, China
[2] School of Computer Science and Technology, Henan Polytechnic University, Jiaozuo Henan, 454001, China
[3] Information Security Center, State Key Laboratory of Networking and Switching Technology, Beijing University of Posts and Telecommunications, Beijing, 100876, China
[4] National Engineering Laboratory for Disaster Backup and Recovery, Beijing University of Posts and Telecommunications, Beijing, 100876, China
pwp@bupt.cn

**Abstract.** Integrity measurement is just one of the most important features in a trusted computing platform to provide evidence of platform configuration and behavior. Each set of software instructions (binary code) is measured and recorded before it is executed. Especially, all of boot components which be recorded in pre-boot stage can't be re-measured until the platform is rebooted. However, the reality is that many components belong to pre-boot stage, such as BIOS, Boot Loader and MBR, are easily modified or updated in runtime environment. If these updated results of components are treated as configuration requirements for a remote server-side to verify the access request, the remote network access connection will fail. It must restart the host to re-measure there components, and the result of the new measure is not guaranteed right absolutely. We solve this problem by embedding Integrity Measurement Agency Component (IMAC) which has the capability of monitoring transformation, verifying credibility, updating Platform Configuration Registers and recording integrity measurement logs in Linux kernel. The result of performance analysis demonstrates that the method is feasible and credible.

**Keywords:** Trusted Computing, Boot Components, PCR, Integrity measurement Agency Component.

## 1 Introduction

Trusted computing technology acts as a relatively new approach to secure systems is brought forward by the Trusted Computing Group (TCG)[1] in 2003. Trust in terms of the Trusted Computing Group means that an entity can be trusted if it always behaves in the expected manner for the intended purpose. Its main idea is to equip computer systems with a device that can be trusted by both remote communication peers and the user of the platform herself. This device, called the Trusted Platform Module (TPM)[2,3,4], is acted as core root of Trust for measurement, or CRTM, can measure additional system attributes and then later verifiably report them as a basis for determining the overall trustworthiness of a platform. Integrity measurement is just one of the most important features in a trusted computing platform to provide evidence of platform configuration and behavior. The TCG standardizes the measurement and reporting of attributes covering trust establishment into the boot process of a system. A trusted platform, starting from a root of trust in hardware, performs a series of measurements that record summaries of software that has executed (or is executing) on a platform. By these means, each set of software instructions (binary code) is measured and recorded before it is executed. Especially, the recording which be recorded during pre-boot stage can't be undone until the platform is rebooted. However, the reality is that many components belong to Pre-boot stage, such as BIOS, option ROMs, GURB and MBR, are easily modified or updated during runtime environment. If these results of components or configurations which be update or revise by user are treated as evidences for a remote server-side to verify whether the access requester meet the configuration requirements, the remote network access connection will fail. It must restart the host to re-measure the various components, and we can not guarantee the right of the new measure absolutely.

We solve this problem by embedding Integrity Measurement Agency Component (IMAC) in Linux kernel which has the ability of monitoring transformation, verifying credibility, updating Platform Configuration Registers (PCRs) and recording Integrity Measurement Logs (IMLs).

This paper is structured as follows. We first introduce the basic technology of integrity measurement in Section 2, and then present the design principle, system architecture and each components of IMAC in Section 3. We discuss the feasibility and credibility of our scheme in Section 4 and finally we conclude in Section 5.

## 2 Configuration Measurement

The TCG has defined a set of specifications that are intended to provide an open set of security related building blocks for enhancing the trust associated with a computing platform. One of the services provided by TPM is a computer measures the configuration of hardware and software and keeps chain of trust which is passed to BIOS, Boot Loader (i.e. GRUB-IMA), kernel (i.e. Linux-IMA[5]), and applications, and reports precise values of a measurement to a third party as reliable information.

The configuration measurements are stored in a PCRs inside of the TPM. The TPM offers 24 PCRs that allow extending 160 bit numbers into them. The first 16 PCRs are reset to 0 while the remaining 8 PCRs are reset to 1 whenever the system is reset (e.g. reboot). The first 8 PCRs (PCR0-PCR7) are used for attesting the booting steps, the remaining 8 PCRs (PCR8 – PCR15) are allocated for use by the booted system [6]. TPM uses a specific process called PCR extension to renew the register in record mode. This process is indicated in the following equation.

$$NewPCR_iValue{=}HASH\ (OldPCR_iValue \parallel Additional\ Value) \tag{1}$$

Here, the suffix "$i$" is an index identifying each PCR register. During a boot sequence the platform continuously executes extension operations for each register using a hash value of the software component executed sequentially as an additional value. This information is valuable to an individual who needs to know the reliability of the platform as a service provider.

## 3 Design of IMAC

IMAC plays a primary role in the client-side architecture during operation system runtime environment by providing policy to collect updated information of pre-boot stage, pre-computing integrity values, formatting integrity values and implementing remote attestation request. The difference of IMAC and Platform Trust Services (PTS)[7] is that PTS collects the integrity status of Trusted Network Connection (TNC)[8,9] components after OS loaded while IMAC collects the integrity status of pre-boot components.

### 3.1 Principle of IMAC

Each component will be measured in accordance with the order of the measurement list after the computer be powered. The details of each component will be saved in the IML and measurement values should be extended into PCR. The purpose of inserting the IMAC measure into the process of OS_Loader measures OS is to embed IMAC credibility into the trust chain. Once the credibility of IMAC was destroyed, the trust chain would be destroyed and the platform will become trustless. Some core components (e.g. POST, BIOS, GRBU, MBR, etc.) can be changed or updated normally by application software during OS running. If exist arbitrary operation to firmware or software, IMAC should collect relevant information and make corresponding operation according to the strategy.

If and only if there operations be limited to changes, IMAC collects metric parameters, determines the PCR index, and calculates the new measure by calling TPM commands. Then, IMAC sends a platform attestation request to the remote reference server or TTP. After receives the response, they communicate with each other to examine the credibility of this measure. If the result is match with the reference value, the platform can still be trusted after this update or modification. Otherwise, it is failure. The user must download a new patch or re-configuration the platform. Normal network access services unavailable until the credibility of certification adopted.

### 3.2 Architecture

The system architecture is depicted in Figure 1. The IMAC acts as a system component which can collect measurement information from the BIOS, Options ROMs, OS Loader, OS components and some additional software or applications that are loaded onto the platform (i.e., software and applications that are either part of the Trusted Computing Base (TCB) or can potentially subset the TCB). Metric information may be stored in an IML where they may be retrieved for later use. The Measurement Log also contains the Transitive-trust Chain which be created by the RTM and other measurement agents that may have executed prior to IMAC execution. Ideally, part of the Trusted OS should compute the IMAC measurement and append it to the Transitive-trust chain prior to passing execution to the IMAC. The V_IMAC is used to maintain the digest of conference value which is a non-volatile storage area out of the TPM.

IMAC is an intermediate component between Trusted Software Stack (TSS) and TPM, and intercommunicate with TSS by using TCG Device Driver Library Interface (TDDLI) and TPM by using TPM Device Driver Interface (TDDI).

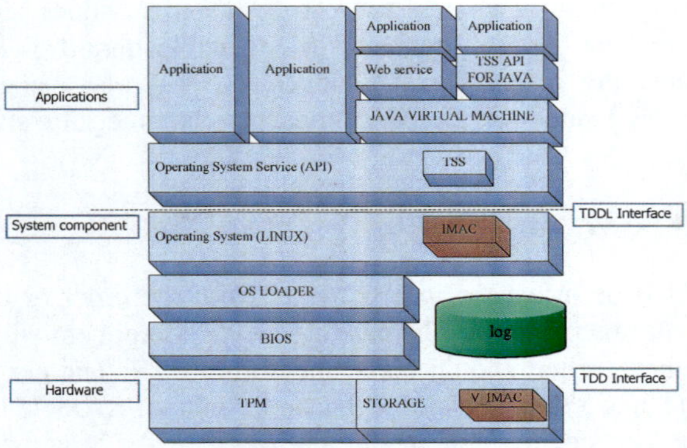


**Fig. 1.** System Architecture

## 3.3 IMAC Structure

The IMAC is consisted of Information Collector (IC), Strategy Files (SF), Event Analysis Module (EAM), Platform Attestation Request Module and TPM_Commands Trigger. The system architecture is depicted in Figure 2.

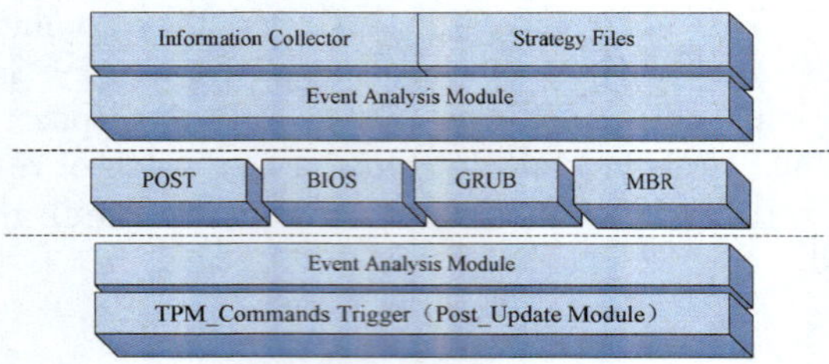


**Fig. 2.** IMAC Structure

*1) Strategy Files*

Static policy information provides detailed parameters to identify different components. There parameters include component character, trigger function, Interrupt Request (IRQ), Interrupt function number and operation information. The policy information is specified and managed by a Trusted Third Party (TTP) which is assumed to provide absolute security policy information. Figure 3 shows the format of Policy Parameters.

| | |
|---|---|
| Component Label | Reserved |
| Interrupt Number | Function Number |
| Operations Style | |

**Fig. 3.** Policy Parameters

♦ Component Label: It can be used to identify different component.
♦ Reserved: It is reserved for future use.
♦ Interrupt Number: System call interrupt is the only interface for applications to use the operating system resources. It is set to int128 (0x80).
♦ Function Number: It is used to indicate what function is.
♦ Operations Style: It includes the operation of reading, modification and inserting.

*2) Information Collector*

Information Collector is primarily used by the IMAC to monitor the change of pre-boot components, determine whether this change affect the chain of trust and gather the alternate state in the form of policy structure.

The collected information of IC includes the identity information of user and platform, the integrity information of platform configuration and system availability information. The required standard for information collection is depending on the integrity measurement computation. The measurement information should include both static information and dynamic information according to the implementation of trusted platform based on TPM [10]. The static information is easily to be obtained, for the platform identity information has been stored in the TPM and the platform configuration integrity information has been stored in the IML. Simultaneously the result will be extended hash and stored in the PCR. As to the dynamic information measurement, it mainly refers to the OS component class. The dynamic information also includes the identity information, integrity information and system availability information, which is stored in the form of log of PTS in the protected storage region according to a specific schema. Meanwhile the result should be extended hash and the final hash value will be stored in the corresponding PCR in TPM too.

*3) Event Analysis Module*

EAM can determine whether to continue the next process according to the collected and monitored information.

EAM must meet the following requirements:

♦ If the event information is insufficient to modify or update the core components, such as read-only, it is only to take conventional treatment, such as discarded this event.
♦ If the event leads to the interruption to update the core components, IMA should determine PCR index and the IML file by querying the policy file.
♦ Provide interface for Platform Attestation request module.

*4) Platform Attestation Request Module*

Similar to the majority applications, the Remote Access Requests will be sent to the Remote Authentication Server (RAS)firstly. The RAS determines whether to provide relative services through the request information. If response, the server will send a demand of verification platform configurations and reliability. At the same time, in order to prevent the middle attacks and replay attacks, along with sending random nonce values. According to the information feedback from server, the local host needs to send back the appropriate PCR, log and the key certificate.

The server is maintained by a relatively credible entity, such as product manufacturer or service provider, and can provide standard reference values of different versions and models.

*5) Post_Update Module*

According to the returned information from the upper platform authentication request module, PUM should call the corresponding TPM Commands to extend PCR and update Integrity Measurement Log. PCRS can be altered only by *TPM_Startup, TPM_Extend and TPM_SHA1ComputeExtend,* and read using functions of *TPM_Quote* and *TPM_PCRRead.*

## 3.4 V_IMAC Operation

*1) Generation*

The *V_IMA*C is a non-volatile storage area outside of the TPM. It is used to maintain the digest of conference value. Typically, this reference value is generated by the platform administrator in kernel mode. We use a specific process to seal the conference value. This process is indicated in the following equation.

$$V\_IMAC = TPM\_Seal_{keyhandle}\ (SHA1\ (TPM\_MV\ (IMAC))) \quad (2)$$

*TPM_MV* is a function which can measure the entity and return the measurement value. *keyhandle* is a loaded key that be used to digital signatures, and *TPM_Seal* is used to encrypt private objects that can only be decrypted using *TPM_unSeal. V_IMAC* is the ultimate reference value.

The generation process of conference value as follows:

*a)* *Step1:* The operator must act as the role of platform administrator or TPM_Owner

*b)* *Step2:* Before the first time of IMAC loaded, PTS generates the measurement value of IMAC by using some interfaces or functions of TSS.

*c)* *Step3:* Get the digest of measurement value of IMAC by using SHA1.

*d)* *Step4:* The operator sends the request of Key generation and AuthData to TPM.

*e)* *Step5:* TPM generates the pairs of key. The public key is used to seal the digest in step3 and the private key is used to unseal and verify the sealed data which resided in the TPM.

*f)* *Step6:* store the value.

*2) Verification*

This unseal process is indicated in the following equation.

$$V\_IMAC' = TPM\_Unseal_{parenthandle}\ (V\_IMAC) \quad (3)$$

Where *V_IMAC'* represents the digest of standard conference value, and *TPM_Unseal* decrypts the structure internally and checks the integrity of the result. *Parenthandle* is a loaded key that can unseal the data.

The verify process of integrity as follows:

*a) Step1:* IMAC is loaded as a component service of OS after OS loaded.

*b) Step2:* Before starting the IMAC, PTS will measure the component integrity and extend PCR. IMAC captures the intermediate value and calculate the SHA1.

*c) Step3:* unseal the V_IMAC.

*d) Step4:* compare between sha1_value and V_IMAC'.

*e) Step5*: return the result according to comparison.

# 4 Performance Analysis

In this section, we discuss the feasibility and credibility of the new re-measurement scheme.

## 4.1 Feasibility

1)Linux is an open-source operating system and ereryone can reuse the system core components. So, technical problems are not exist in the process of IMAC components be embedded in it.

2)Some components belong to pre-boot stage are easily modified or updated during runtime environment. These operations are implemented by calling interrupt functions which be provided by the Linux kernel interface, and can be monitored and captured.

3)TSS provides fundamental resources to support the TPM. It can Supply one entry point for applications to the TPM functionality, provide synchronized access to the TPM and manage TPM resources. There are two methods: Local Procedure Call (LPC) or Remote Procedure Call (RPC). The LPC is a direct call from one application, module, or component to another within the same process or address space allocated by the OS. In an LPC the calling routine can resolve directly (compile time) or indirectly (run time) the local address of the routine to call or pass control to. So, the IMAC can call these functions smoothly through the interface.

4)The TCG TPM 1.2 Main specification has defined TPM Main Part 3 Commands specification which provides method to call TPM Commands. IMAC can call these commands such as TPM_Extend, TPM_Quote, TPM_PCRRead and TPM_SHA1ComputeExtend.

## 4.2 Credibility

TPM can provide the credibility and validation. Both data integrity registers defined in TPM1.1 specification and the non-volatile Storage in TPM1.2 specifications, the aim is to provide a storage area to store reference values. IMAC is involved in the trust chain After be embedded into Linux kernel,. The comparison result of measurement value of IMAC and the reference value demonstrates the credibility of IMAC. The credibility of IMAC is related with the initial platforms. So, The IMAC would be trusted while its platform is credible in the initialization phase.

## 5 Conclusions and Future Work

This paper focuses on how to implement integrity measurement and credibility assessment about changed platform in the conditions of un-reboot. We solve this problem by embedding IMAC in Linux kernel which has the capability of monitoring transformation, verifying credibility, updating PCRs and recording integrity measurement logs. The result of performance analysis demonstrates that our proposed method is feasible and credible.

Based on the work proposed in this paper, we will investigate how to improve all functions of IMAC and test the proposed method in real environment, as well as in-depth studying the mechanism of Trusted Network Connections.

**Acknowledgments.** This work is supported by National 863 (No. 2009AA01Z430), National Natural Science Foundation of China (No. 60972077, 60973146) and National S&T Major Program (2009ZX03004-003-03).

## References

1. TCG: TCG Specification Architecture Overview. TCG Specification Version 1.2, The Trusted Computing Group (TCG), Portland, Oregon, USA (April 2003)
2. TCG: TPM Main, Part 1: Design Principles. TCG Specification Version 1.2 Revision 94, The Trusted Computing Group (TCG), Portland, Oregon, USA (March 2006)
3. TCG: TPM Main, Part 2: TPM Data Structures. TCG Specification Version 1.2 Revision 94, The Trusted Computing Group (TCG), Portland, Oregon, USA (March 2006)
4. TCG: TPM Main, Part 3: Commands. TCG Specification Version 1.2 Revision 94, The Trusted Computing Group (TCG), Portland, Oregon, USA (March 2006)
5. Sailer, R., Zhang, X., Jaeger, T., van Doorn, L.: Design and Implementation of a TCG-Based Integrity Measurement Architecture. In: Thirteenth Usenix Security Symposium, pp. 223–238 (August 2004)
6. Garfinkel, T., Pfaff, B., Chow, J., Rosenblum, M., Boneh, D.: A Virtual Machine-Based Platform for Trusted Computing. In: 9th ACM Symposium on Operating System Principles (2003)
7. TCG: TCG Infrastructure Working Group Platform Trust Services Interface Specification (IF-PTS), Specification Version 1.0, The Trusted Computing Group (TCG), USA (November 2006)
8. TCG: TCG Trusted Network Connect TNC IF-IMC, Specification Version 1.0, The Trusted Computing Group (TCG), USA (May 2005)
9. TCG: TCG Trusted Network Connect TNC IF-IMV, Specification Version 1.0, The Trusted Computing Group (TCG), USA (May 2005)
10. Chen, L., Zhang, H., Zhang, L., Cai, L.: A New Information Measurement Scheme Based on TPM for Trusted Nerwork Access. In: 2007 International Conference on Computational Intelligence and Security Workshops, pp. 506–509 (2007)

# Multi-invariance MUSIC Algorithm for DOA Estimation in Acoustic Vector-Sensor Array

Weiyang Chen, Zhongwei Sun, and Xiaofei Zhang

Department of Electronic Engineering
Nanjing University of Aeronautics & Astronautics
Nanjing, 210016, China
{weiweigenes,szwsg}@126.com, fei_zxf@163.com

**Abstract.** This paper discusses the problem of direction of angle (DOA) for acoustic vector-sensor array, and derives multi-invariance MUSIC (MI-MUSIC) algorithm therein. The proposed MI-MUSIC, which only requires one-dimension searching, can avoid the high computational cost within two-dimension MUSIC algorithm (2D-MUSIC). Our proposed algorithm can resolve the DOAs estimation of acoustic vector-sensor array. We illustrate that MI-MUSIC algorithm has better performance than ESPRIT algorithm, and it also can be viewed as a generalization of MUSIC. Simulation results verify the usefulness of our algorithm.

**Keywords:** Acoustic vector-sensor array, DOA estimation, multi-invariance MUSIC.

## 1 Introduction

Direction of arrival (DOA) estimation [1-6] is a key problem in the underwater acoustic signal processing field. Recently, this problem has particularly attracted considerable attention, and it has been used for wireless location and emergency service [7]. Obviously, a precise DOA and time delay estimation results in a more accurate location estimate. The elements of a sensor array are often been arranged in a very regularity, structured geometry. Linear, circular, and rectangular arrays are common examples. These special geometries can be exploited in developing computationally efficient algorithms for DOA estimation. Eigen-decomposition based methods, including multiple signal classification (MUSIC) and estimation of signal parameters via rotational invariance techniques (ESPRIT) [8-9], have high resolution DOA estimation performance, but they fail to work in coherent signal condition. ESPRIT is a closed-form eigen structure-based parameter estimation technique which requires data model with invariance structures. An alternative eigen-decomposition based method to estimate DOAs is MUSIC algorithm [10-11], which uses the noise-subspace eigenvectors of the data correlation matrix to form a null spectrum and yields the corresponding signal parameter estimates. Notably, it has also concerned that MUSIC matches some kind of irregularly-spaced array with high popularity [12].

In this paper, we propose a novel algorithm for exploiting arrays with multiple translated subarrays, while our work links the acoustic vector-sensor array signal detection problem to it. It can be viewed as a generalization of MUSIC [13], and thus, we refer to this algorithm as MI-MUSIC, which distinctively reduces the complexity for DOA estimation. It estimates the DOAs using only a one-dimensional (1-D) search, and it is able to enforce the constraint that the subarray responses for a given source are related by a scalar multiplier, which is lying on the unit circle. In addition, we compare the root mean squared error (RMSE) and the cost of computational complexity against those of ESPRIT. The proposed MI-MUSIC algorithm can have better DOA estimation performance than ESPRIT algorithm. Our proposed algorithm can overcome the shortcomings of EPSPIT method.

***Notation*:** $(.)^*$, $(.)^T$ ,$(.)^H$ and $\|\ \|_F$ are denoted as the complex conjugation, the matrix transpose, the matrix conjugate transpose and the Forbenius norm, respectively; $\circ$ is the Khatri–Rao product; $\otimes$ is the Kronecker product, it is also denoted that $I_P$ is a $P \times P$ identity matrix.

## 2 Problem Formulation

### 2.1 Data Model

A total of $K$ narrowband plane waves impinge on an array containing $M$ acoustic vector sensors with half-wavelength spacing between adjacent sensors. We define $\boldsymbol{\theta}_k = [\phi_k, \varphi_k]^T$, which is the DOA of the $k$th source. In this space scenario, the output of an acoustic vector sensor which is located at $\boldsymbol{r}$, is the four-element vector

$$\begin{bmatrix} y_p(t) \\ y_v(t) \end{bmatrix} = \sum_{k=1}^{K} \begin{bmatrix} 1 \\ u_k \end{bmatrix} b_k(t) \tag{1}$$

where $y_p(t)$ is the pressure part of the measurement. $y_v(t)$ is the velocity part of the measurement. $b_k(t)$ is the transmit signal of the $k$th source. $u_k$ is the unit vector at the sensor pointing towards the source

$$u_k = \begin{bmatrix} \cos\phi_k \cos\varphi_k \\ \sin\phi_k \cos\varphi_k \\ \sin\varphi_k \end{bmatrix} \tag{2}$$

### 2.2 The Received Signal Model for Acoustic Vector-Sensor Array

In this paper, the output of a uniform linear array containing $M$ acoustic vector sensors located at $r_1, r_2, \ldots, r_M$ is shown

$$\mathbf{y}(t) = \boldsymbol{\Psi}(\boldsymbol{\theta})\mathbf{b}(t) + \mathbf{n}(t) \tag{3}$$

where $\mathbf{b}(t)$ contains $K$ source signals, n(t) is the received noise. The $k$th signal $b_k(t)$ is arriving from direction $(\phi_k, \varphi_k)$, where $\varphi_k$ and $\phi_k$ stand for the elevation angle and the azimuth angle, respectively. $\boldsymbol{\Psi}(\boldsymbol{\theta})$ is shown $\boldsymbol{\Psi}(\boldsymbol{\theta}) = \begin{bmatrix} \boldsymbol{\psi}(\boldsymbol{\theta}_1) & \boldsymbol{\psi}(\boldsymbol{\theta}_2) & \cdots & \boldsymbol{\psi}(\boldsymbol{\theta}_K) \end{bmatrix}$, and

$$\boldsymbol{\psi}(\boldsymbol{\theta}_k) = \mathbf{a}(\varphi_k) \otimes \mathbf{h}(\boldsymbol{\theta}_k) \tag{4}$$

where $\mathbf{a}(\varphi_k)$ is the $M \times 1$ steering vector of an acoustic pressure sensor array with the same geometry as the acoustic vector sensor array for the $k$th signal. Define $\mathbf{h}(\boldsymbol{\theta}_k) = \begin{bmatrix} 1 & \mathbf{u}_k^T \end{bmatrix}^T$ ,which is the bearing vector of the $k$th source. Considering that channel state information is constant during $L$ transmitting symbols. The received signal of acoustic vector-sensor array is

$$\begin{aligned} \mathbf{X} &= [\mathbf{A} \circ \mathbf{H}]\mathbf{B}^T + \mathbf{N} \\ &= \begin{bmatrix} \mathbf{H}D_1(\mathbf{A}) \\ \mathbf{H}D_2(\mathbf{A}) \\ \vdots \\ \mathbf{H}D_M(\mathbf{A}) \end{bmatrix} \mathbf{B}^T + \mathbf{N} \end{aligned} \tag{5}$$

where $\mathbf{A} = [\mathbf{a}(\varphi_1), \mathbf{a}(\varphi_2), ..., \mathbf{a}(\varphi_K)] \in \mathbb{C}^{M \times K}$ is the direction matrix, $\mathbf{H} = [\mathbf{h}(\boldsymbol{\theta}_1), \mathbf{h}(\boldsymbol{\theta}_2), ..., \mathbf{h}(\boldsymbol{\theta}_K)] \in \mathbb{C}^{4 \times K}$ is the bearing matrix. $\mathbf{B} \in \mathbb{C}^{L \times K}$ is the source matrix for $L$ samples. $D_m(.)$ is to extract the $m$th row of its matrix and constructs a diagonal matrix out of it. $\mathbf{N}$ is the noise matrix.

For the signal model in (5), the covariance matrix $\mathbf{R}_x$ can be given by

$$\mathbf{R}_x = \mathbf{X}\mathbf{X}^H = \mathbf{E}_s \mathbf{D}_s \mathbf{E}_s^H + \mathbf{E}_m \mathbf{D}_m \mathbf{E}_m^H \tag{6}$$

where $\mathbf{D}_s$ stands for a $K \times K$ diagonal matrix whose diagonal elements contain the largest $K$ eigenvalues and $\mathbf{D}_m$ stands for a diagonal matrix whose diagonal entries contain the smallest $4M - K$ eigenvalues. $\mathbf{E}_s$ is the matrix composed of the eigenvectors corresponding to the largest $K$ eigenvalues of $\mathbf{R}_x$, while $\mathbf{E}_m$ represents the matrix including the rest eigenvectors. Note that $\mathbf{E}_s$ and $\mathbf{E}_m$ can be regarded as the signal subspace and the noise subspace, respectively.

# 3 Multi-invariance Music (mi-music) Algorithm for doa Estimation Acoustic Vector-Sensor Array

According to Eq.(6), we construct the 2D-MUSIC spatial spectrum function in this form

$$f_{2dmusic}(\theta)=\frac{1}{[\mathbf{a}(\varphi)\otimes h(\theta)]^H\mathbf{E}_m\mathbf{E}_m^H[\mathbf{a}(\varphi)\otimes h(\theta)]} \tag{7}$$

Here we have the $K$ largest peaks of $P_{MUSIC}(\theta)$ taken as the estimates of the DOAs for the sources. Since 2D-MUSIC requires an exhaustive two-dimension search, this approach is normally inefficient due to high computational cost. In the following subsection, we present the multi-invariance MUSIC algorithm, which qualify for the DOA estimation just through one-dimension searching.

## 3.1 Multi-invariance Music (mi-music) Algorithm for Angle Estimation in Acoustic Vector-Sensor Array

MI-MUSIC algorithm can be viewed as a generalization of MUSIC and has been proposed for DOA estimation with the exploitation of array invariance [14]. In this subsection, the idea of MI-MUSIC has been adopted to estimate DOA in acoustic vector sensor array. Assuming that no noise is presented, the signal subspace $\mathbf{E}_s$ in Eq.(6) can be denoted as

$$\mathbf{E}_s=\begin{bmatrix} HD_1(\mathbf{A}) \\ HD_2(\mathbf{A}) \\ \vdots \\ HD_M(\mathbf{A}) \end{bmatrix}\mathbf{T}=\begin{bmatrix} H \\ H\mathbf{\Phi} \\ \vdots \\ H\mathbf{\Phi}^{M-1} \end{bmatrix}\mathbf{T}=\mathbf{\Lambda}\mathbf{T} \tag{8}$$

$$=[\mathbf{a}(\varphi_1)\otimes h(\theta_1),\mathbf{a}(\varphi_2)\otimes h(\theta_2),...,\mathbf{a}(\varphi_K)\otimes h(\theta_K)]\mathbf{T}$$

where $\mathbf{T}$ is a $K \times K$ full-rank matrix, and $\mathbf{\Phi}= diag\{[\exp(-j\pi\sin\varphi_1)\ \exp(-j\pi\sin\varphi_2)\ \cdots,\exp(-j\pi\sin\varphi_K)]\}$ is the rotation matrix. The signal in Eq.(8) is with multi-invariance characteristic, and then we can use MI-MUSIC algorithm [14] for angle estimation in acoustic vector sensor array. According to (8), $\mathbf{\Lambda}=\mathbf{E}_s\mathbf{T}^{-1}$ can be easily obtained, then the signal subspace fitting is given in this form $\hat{\mathbf{T}},\hat{\mathbf{\Lambda}}=\arg\min_{\mathbf{T},\mathbf{\Lambda}}\left\|\mathbf{\Lambda}-\hat{\mathbf{E}}_s\mathbf{T}^{-1}\right\|_F^2$ , where $\hat{\mathbf{E}}_s$ is the estimate of $\mathbf{E}_s$ ,which is also denoted as

$$\hat{\mathbf{T}},\hat{\mathbf{\Lambda}}=\arg\min tr(\mathbf{\Lambda}^H\Pi_{\hat{\mathbf{E}}_s}^{\perp}\mathbf{\Lambda}) \tag{9}$$

where $\Pi_{\hat{\mathbf{E}}_s}^{\perp}=\mathbf{I}_{MN}-\hat{\mathbf{E}}_s(\hat{\mathbf{E}}_s^H\hat{\mathbf{E}}_s)^{-1}\hat{\mathbf{E}}_s^H$, $tr(.)$ denotes the sum of the elements of the principal diagonal of the matrix. For

$$\Lambda = [\mathbf{a}(\varphi_1) \otimes h(\theta_1), \mathbf{a}(\varphi_2) \otimes h(\theta_2), ..., \mathbf{a}(\varphi_K) \otimes h(\theta_K)] \tag{10}$$

By Eq. (10), the minimization of Eq. (9) becomes

$$\mathbf{a}(\varphi), h(\theta) = \arg\min \sum_{k=1}^{K} [\mathbf{a}(\varphi_k) \otimes h(\theta_k)]^H \Pi_{\hat{\mathbf{E}}_s}^{\perp} [\mathbf{a}(\varphi_k) \otimes h(\theta_k)] \tag{11}$$

Also, the minimization for Eq.(11) can be attained by searching for the deepest $K$ minimum values in the following criterion, which is shown as

$$\begin{aligned} \mathbf{V}(\theta) &= [\mathbf{a}(\varphi) \otimes h(\theta)]^H \Pi_{\hat{\mathbf{E}}_s}^{\perp} [\mathbf{a}(\varphi) \otimes h(\theta)] \\ &= h(\theta)^H [\mathbf{a}(\varphi) \otimes \mathbf{I}]^H \Pi_{\hat{\mathbf{E}}_s}^{\perp} [\mathbf{a}(\varphi) \otimes \mathbf{I}] h(\theta) \\ &= h(\theta)^H \mathbf{Q}(\varphi) h(\theta) \end{aligned} \tag{12}$$

where $\mathbf{Q}(\varphi) = [\mathbf{a}(\varphi) \otimes \mathbf{I}_4]^H \Pi_{\hat{\mathbf{E}}}^{\perp} [\mathbf{a}(\varphi) \otimes \mathbf{I}_4]$, Eq.(12) is the problem of quadratic optimization. $\mathbf{I}_4$ is the 4 × 4 identity matrix. We also consider the constraint of $\mathbf{e}_1^T h(\theta) = 1$, where $\mathbf{e}_1 = [1, 0, ..., 0]^T \in \mathbb{C}^{4\times1}$, which has been added to eliminate the trivial solution $h(\theta) = \mathbf{0}$. The optimization problem can be reconstructed with the linear constraint minimum variance solution, for which we have

$$h(\theta) = \frac{Q(\varphi)^{-1} e_1}{e_1^T Q(\varphi)^{-1} e_1} \tag{13}$$

$$\hat{\varphi} = \arg\min_{\varphi} \frac{1}{\mathbf{e}_1^T Q(\varphi)^{-1} \mathbf{e}_1} \tag{14}$$

The DOA $\varphi$, $k = 1, 2, \ldots, K$, are estimated via a 1-D search. The solution for $\varphi$ is

$$\hat{\varphi} = \arg\max_{\varphi} \mathbf{e}_1^T Q(\varphi)^{-1} \mathbf{e}_1 \tag{15}$$

Searching $\varphi \in [-90^\circ, 90^\circ]$, we find the $K$ largest peaks of the (1, 1) element of $Q(\varphi)^{-1}$. The $K$ largest peaks should correspond to DOA $\varphi$, $k = 1, 2, \ldots, K$.

### 3.2 Major Steps of Mi-Music

Till now, we have achieved the proposal for multi-invariance MUSIC-based algorithm for DOA estimation in acoustic vector sensor array. We show the major steps of MI-MUSIC algorithm as follows:

**Step 1.** Perform the eigen-decomposition operation for covariance matrix $\mathbf{R}_x$ in order to get the signal subspace $\hat{\mathbf{E}}_s$, and calculate $\Pi_{\hat{\mathbf{E}}_s}^{\perp}$;

**Step 2.** By searching $\varphi$, we find the $K$ largest peaks of the (1, 1) element of $\mathbf{Q}(\varphi)^{-1}$ with respect to Eq.(15) to get the estimation of DOAs.

### 3.3 Complexity Analysis

In contrast, the MI-MUSIC algorithm that we presented could have heavier computational load than ESPRIT. ESPRIT requires $O((4M)^2L + (4M)^3 + K^3)$. For MI-MUSIC algorithm, the computational complexity of formation of the covariance matrix is $O((4M)^2L)$ and calculation of its eigen-decomposition requires $O((4M)^3)$, and it employs a 1-D criterion that must be searched for $K$ local maximums, thus the computational load is determined by the complexity of evaluating the criterion for each $\varphi$. MI-MUSIC requires the formation of the matrix $\mathbf{Q}(\varphi)$ for each $\varphi$, which is an $O((4M)^2)$ operation, followed by calculation of the (1, 1) element of $\mathbf{Q}^{-1}(\varphi)$, which requires $O((4M)^2)$.

## 4 Performance Evaluation

In order to assess the angle estimation performance of our algorithm, we present Monte Carlo simulations and set its trial numbers as 1000. Define the root mean squared error (RMSE) as $RMSE = \frac{1}{K}\sum_{k=1}^{K}\sqrt{\frac{1}{1000}\sum_{n=1}^{1000}(\hat{\varphi}_k^n - \varphi_k)^2}$, where $\hat{\varphi}_k^n$ is the estimated DOA of the $n$th Monte Carlo trial, $\varphi_k$ is the perfect DOA. Noted that: $L$ is the number of snapshots; $K$ and $M$ are the number of receive sources and array elements, respectively.

In the following simulations, we normally adopt the acoustic vector sensor array with $M$ acoustic vector sensors and half-wavelength spacing between adjacent sensors. We assume that there are three non-coherent sources located at angles $10^\circ$, $20^\circ, 30^\circ$, respectively.

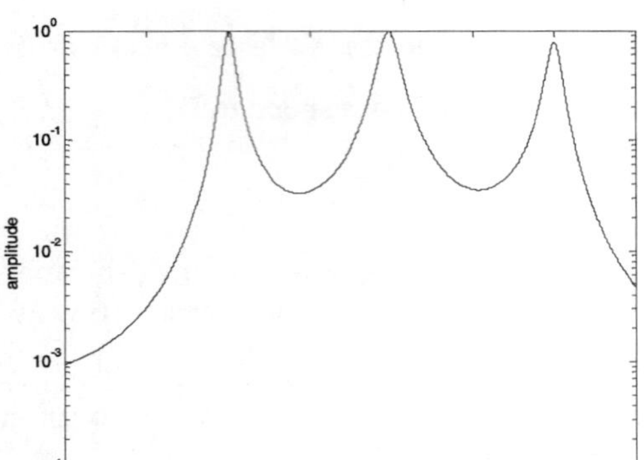


**Fig. 1.** DOA estimation of MI-MUSIC with SNR=15dB

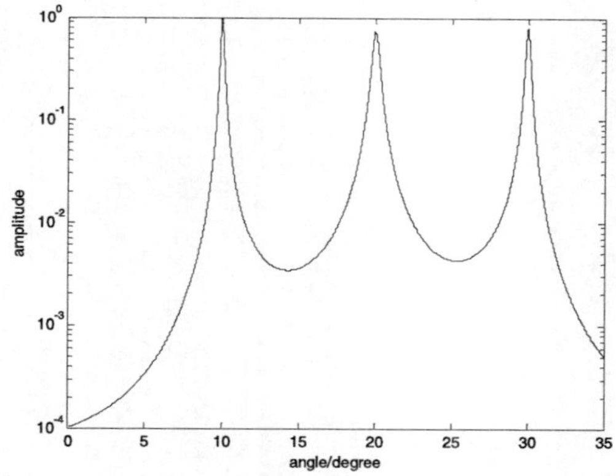


**Fig. 2.** DOA estimation of MI-MUSIC with SNR=25dB

Fig. 3 shows the DOA estimation performance comparison for three sources ($M$=8, $K$=3, $L$=200), where we compare MI-MUSIC algorithm with ESPRIT method. It is indicated in Fig. 4 that between the two algorithms, MI-MUSIC has better performance.

Fig. 4 depicts the algorithmic performance comparison for three sources where MI-MUSIC has been adopted, and the simulation is shown with different array elements $M$ ($K$=3, $L$=200). It is indicated that the performance of angle estimation for the acoustic vector sensor array becomes better in collaboration with $K$ increasing. It is clearly shown that the estimation performance of MI-MUSIC for the acoustic vector sensor array is gradually improving with the number of antennas increasing. Multiple antennas improve DOA estimation performance because of diversity gain.

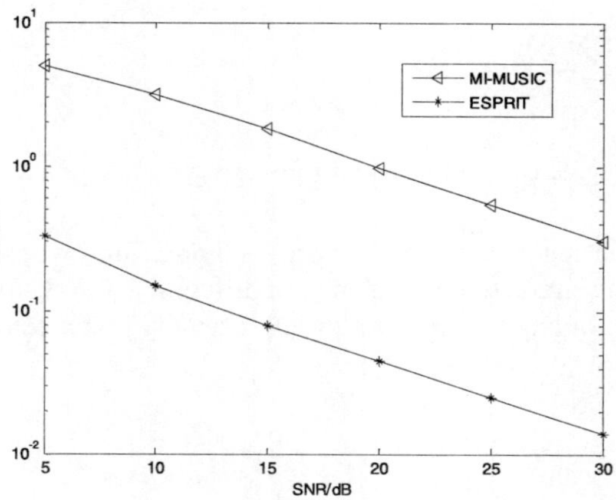


**Fig. 3.** DOA estimation performance comparison

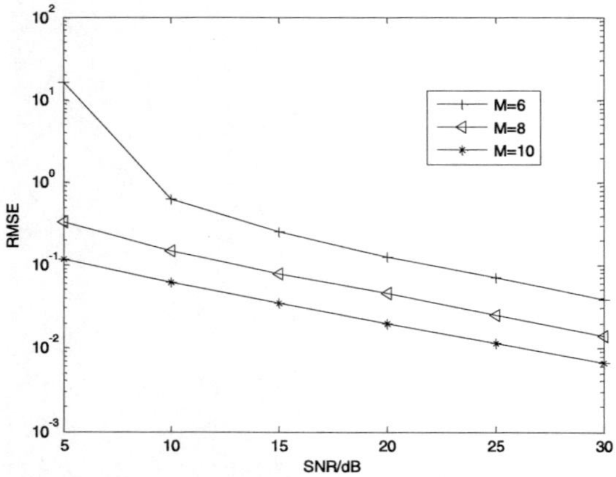


**Fig. 4.** DOA estimation performance with different *M*

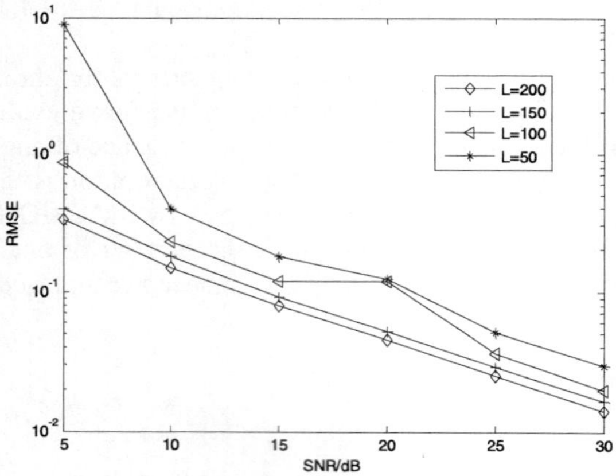


**Fig. 5.** DOA estimation performance with different *L*

Fig. 5 depicts the algorithmic performance comparisons where MI-MUSIC has been adopted, and the simulation is shown with different $L$ ($M$=8, $K$=3). It is indicated that the performance of angle estimation for the acoustic vector sensor array becomes better in collaboration with $L$ increasing.

# 5 Conclusion

We have presented MI-MUSIC algorithm for DOA estimation in the acoustic vector sensor array in this paper. The new MI-MUSIC approach is able to exploit the full invariance structure of the array, and it estimates the DOAs using a simple 1-D search, then, as such, enjoys a significant computational advantage over ESPRIT. We demonstrate that MI-MUSIC can have much better performance for DOA estimation in contrast to ESPRIT, and it may enjoy better robustness than ESPRIT in difficult threshold scenarios. MI-MUSIC algorithm typically outperforms other suboptimal estimators for the same problem.

## References

1. Lizzi, L., Viani, F., Benedetti, M., Rocca, P., Massa, A.: The M-DSO-esprit method for maximum likelihood DOA estimation. Progress In Electromagnetics Research, PIER 80, 477–497 (2008)
2. Gu, Y.-J., Shi, Z.-G., Chen, K.S., Li, Y.: Robust adaptive beamforming for steering vector uncertainties based on equivalent DOAs method. Progress In Electromagnetics Research, PIER 79, 277–290 (2008)
3. Lie, J.P., Ng, B.P., See, C.M.: Multiple UWB emitters DOA estimation employing time hopping spread spectrum. Progress In Electromagnetics Research, PIER 78, 83–101 (2008)
4. Mukhopadhyay, M., Sarkar, B.K., Chakrabarty, A.: Augmentation of anti-jam GPS system using smart antenna with a simple DOA estimation algorithm. Progress In Electromagnetics Research, PIER 67, 231–249 (2007)
5. Harabi, F., Changuel, H., Gharsallah, A.: Direction of arrival estimation method using a 2-L shape arrays antenna. Progress In Electromagnetics Research, PIER 69, 145–160 (2007)
6. Changuel, H., Harabi, F., Gharsallah, A.: 2-L-shape twodimensional arrival angle estimation witha classical subspace algorithm. Progress In Electromagnetics Research, PIER 66, 301–315 (2006)
7. Rappaport, T.S., Reed, J.H., Woerner, B.D.: Position location using wireless communications on highways of the future. IEEE Commun. Mag. 34(10), 33–41 (1996)
8. Van Der Veen, A.-J., Vanderveen, M.C., Paulraj, A.J.: Joint angle and delay estimation using shift-invariance properties. IEEE Signal Processing Letters 4(5), 142–145 (1997)
9. Picheral, J., Spagnolini, U.: Angle and delay estimation of space-time channels for TD-CDMA systems. IEEE Transactions on Wireless Communications 3(3), 758–769 (2004)
10. Vanderveen, M.C., Papadias, C.B., Paulraj, A.: Joint angle and delay estimation (JADE) for multipath signals arriving at an antenna array. IEEE Communications Letters 1(1), 12–14 (1997)
11. Wang, Y.Y., Chen, J.T., Fang, W.H.: TST-MUSIC for joint DOA-delay estimation. IEEE Trans. on Signal Processing 49(4), 721–729 (2001)
12. Wong, K.T., Zoltowski, M.D.: Self-initiating MUSIC-based direction finding and polarization estimation in spatio-polarizational beamspace. IEEE Trans. Antennas Propagat. 48(8), 1235–1245 (2000)
13. Schmidt, R.: A signal subspace approach to multiple emitter location and spectral estimation. Ph.D. dissertation, Stanford Univ., Stanford, CA (1981)
14. Swindlehurst, L., Stoica, P., Jansson, M.: Exploiting arrays with multiple invariances using MUSIC and MODE. IEEE Trans. Signal Process. 49(11), 2511–2521 (2001)

# The Design of the Internal Combustion Engine Based on the Intelligent PID Control Algorithm

Wang Dongyun and Wang Kai

Department of Electronic&Information Zhongyuan Institute of Technology,
Zhengzhou Henan Province, China
wdy1964@yahoo.com.cn, wkai_85@163.com

**Abstract.** At present, the technology of the traditional relay controller and honeywell digital controller are adopted in many combustion machine control systems. But the two technologies show some shortcomings in practice. A new combustion machine control system based on PLC was presented in which the intelligent PID control algorithm was adopted. In the new proposed system, the temperature is controlled by the analog outputs. The practical application results show that the proposed system with properties of high accuracy, good stability and high reliability. The design of software and hardware of the new system is introduced in this paper.

**Index Terms:** PLC, combustion machine, temperature control, PID.

## 1 Introduction

Now, the technology of the traditional relay control and the technology of honeywell controller are adopted in many combustion machine control systems. The former has many disadvantages because of using the hardware which is fixed wiring to realize the logic control of the control system. For example, the volume increases, more power, inefficient, easy to malfunction and cannot guarantee the normal industrial production. The latter requires a combustion machine must be equipped with a honeywell controller. If there are many combustion machines need to be controlled, the construction cost will be high. The PLC shows its unique advantages in logic control, sequence control and system protection with the continuous improvement of the technology of PLC. In addition, the industrial automatic system is more practical, more economical, more flexible and more reliable due to its high reliability, flexibility, simple programming, convenient maintenance, strong anti-jamming capability and many other features.

The new system is developed by our department and a combustion company named ZhongCheng. In the system, the honeywell controller is replaced by PLC. We composed a PID temperature closed loop system by using the A/D and D/A functions of the PLC. The temperature closed loop system can accurately control the temperature of the combustion machine.

## 2 The Basic Structure and Working Principle of the System

### 2.1 Basic Structure

The basic structure of the combustion machine control system is shown in Fig.1, which consists of PLC Master Control System, HMI (Human Machine Interface), combustion furnace, temperature sensor, flame sensor, A / D and D / A conversion modules.

### 2.2 Working Principle

First, the temperature of the combustion furnace will be transformed into PLC identifiable digital signals when the flame is detected by the flame sensor. Then, the system will do PID operation after the signals be input into PLC. And the operation results will be transformed into analog information which can be used to control the opening degree of the electric-air valve of the combustion machine. The regulation of the furnace temperature is implemented.

## 3 Hardware Configuration

In this system, we adopted PLC (EC-O8MO8R -O4NO4B) which is produced by KEWEI automation Co., LTD. The PLC has 8 opto-coupler inputs, 8 relay outputs, 4 arbitrary indexing thermocouple inputs and 4 analog outputs(4～20mA). And the system has the CAN network and RS485 network function. The PLC has many advantages. Such as high reliability, flexible configuration, high cost performance and so on.

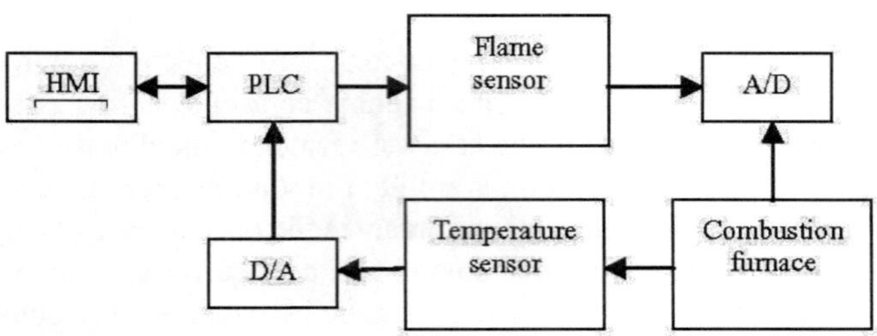


**Fig. 1.** The structure of the control system

A PLC (EC-O8MO8R-O4NO4B) can control 2 combustion machines at the same time. In addition, we can utilize the CAN network function or the 485 network function for saving more cost by composing combustion machine control groups. The PLC control system links with external devices and the condition of connection is shown in TABLE 1 and Fig.2.

# 4 The Software Design of the System

## 4.1 Working Procedure of the Combustion Machine Control System

First, the flame will be detected if exist or not when the system be started. If there is no flame, the system continues to work normally. Otherwise, the alarm will be started and all the outputs will be closed.

When no flame is detected, the igniter and the ignition valve work simultaneously. If there still no flame is detected after the igniter working for 10 seconds, the alarm will be started and all the outputs will be closed.

The igniter will be closed, if the flame is detected in 10 seconds or less. The electric-air valve(4～20mA) and the main air valve will be opened. At the same time, the temperature signals are detected by the temperature sensor. Then they will be transmitted to PLC for the PID operation after the A/D conversion. The operation results can be transformed into analog information(4～20mA) which can be used to control the opening degree of the electric-air valve of the combustion machine. The furnace temperature is kept

**Table 1.** The Address Assignment of I/O of The Combustion Furnace Control System

| Input | Address | Output | Address |
|---|---|---|---|
| Startup | X000 | Alarm | Y000 |
| Detect Flame | X001 | Igniter | Y001 |
| Reset | X002 | Ignition Valve (I.Value) | Y002 |
| Detect Temperature | AI0+ AI0- | Main Air Valve (M.Value) | Y003 |
| | | Electric-air Valve | IO+ VI0- |

If the flame disappears when the main air valve is working, the main air valve will be closed at once and all the switches will be reset. After that, the system will start again. The steps of the system are shown in fig.3.

## 4.2 The Design of the Temperature Control System

According to big lag and time-delay of the nonlinear system, an intelligent PID algorithm is proposed which is based on the conventional PID control algorithm. The new algorithm is effective, stability and strong anti-jamming ability. It can be devided into many different sections. In the sections, the system can be analyzed by using different algorithms. The intelligent PID algorithm is more applicable in practice than others because it doesn't depend on the mathematical model of the system and isn't sensitive to the change of the system parameters.

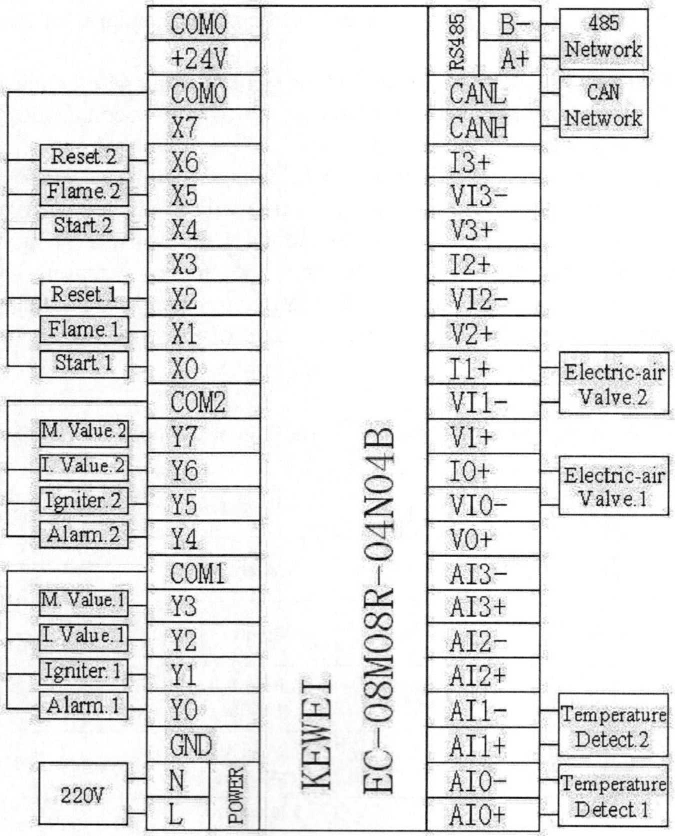


**Fig. 2.** The PLC control system links with external devices

According to the characteristics of the system and the database of PID, we can choose this data: $e(k)$, $u(k)$, $e_{\max}$, $e_{\mathrm{m}}$, $u_{\max}(k)$. $e(k)$ is the error of the actual value and the target value; $u(k)$ is controlled variable; $e_{\max}$ is the maximum allowable deviation; $e_{\mathrm{m}}$ is allowable deviation; $u_{\max}(k)$ is the maximum control variable. The data can be used to build knowledge base. The knowledge is independent respectively. In any case, you can find the machine conditions. And the corresponding control rules will be triggered. The control rules are described as follows:

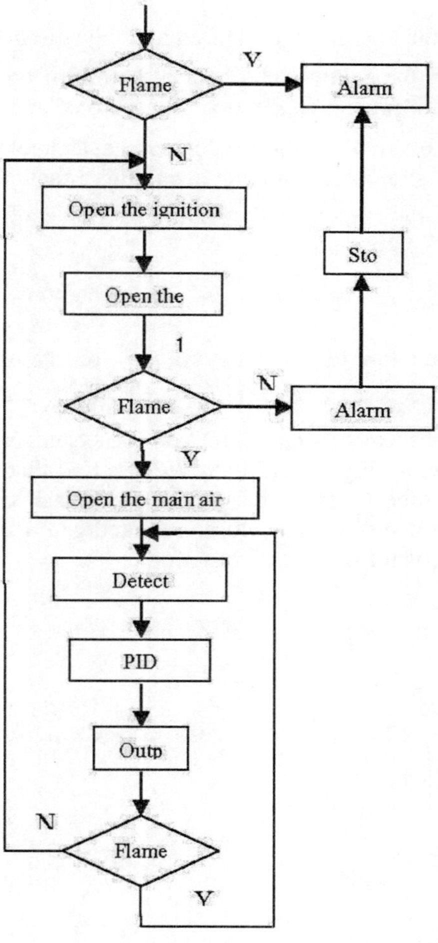


**Fig. 3.** Flow chart of the combustion machine control system

1) If $e(k) > -e_{\max}$, then $u(k) = u_{\max}(k)$. The result shows that the error is too big, we must decrease the deviation as soon as possible.

2) If $-e_{\max} \leq e(k) < -e_{\mathrm{m}}$, then $u(k) = k_p \bullet u_{\max}(k)$. $k_p$ is the proportionality factor. In order to prevent the overshoot too big, the control variable must be reduced properly when the temperature is close to the target value.

3) If $|e(k)| \leq e_{\mathrm{m}}$, the traditional PID operation is working. And $u(k)$ is the traditional PID operation result. The overshoot of the system will be decreased and the system will be stability faster, if the traditional PID operation is executed in a small scale.

4) If $e_{\mathrm{m}} < e(k) < e_{\max}$, then $u(k) = 0$. At this moment, the overshoot reaches ceiling value. We must close the control variable to make the system be cooled.

5) If $e(k) \geq e_{\max}$ , the alarm will be started and the system will be stopped. When the temperature reaches the ceiling value, the system must be stopped and the alarm devices will be started for protecting safety of the personal and the equipments.

The temperature control system which adopted the technology of intelligent PID required to conduct on-site commissioning according to actual situations to gain the properly parameters of $e_{\max}$ and $e_{\mathrm{m}}$ for getting the best control results.

## 5 Simulation

We next present simulation to illustrate efficacy of the intelligent PID control algorithm. When the $K_p = 1$ , $T_i = 1$ , $T_d = 1$ , $e_{\max} = 0.5$ , $e_{\mathrm{m}} = 0.2$ , the simulation results of the conventional PID control system and the intelligent PID control system as shown in Fig.4 and Fig.5. We can find that the overshoot, the rise time and the settling time of the intelligent PID control system is less than the conventional PID control system. Therefore, the intelligent PID control system is far superior to the conventional PID control system.

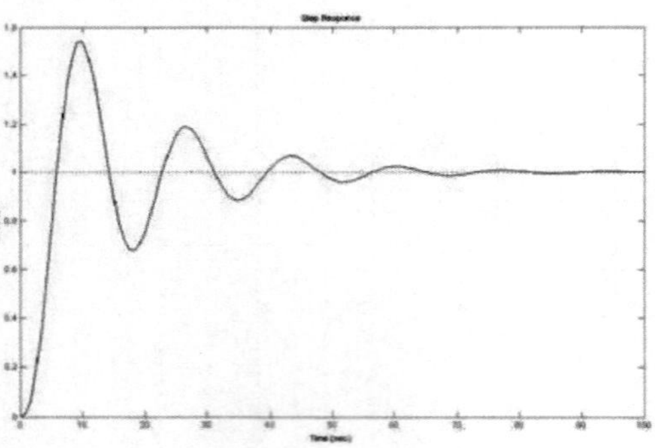

**Fig. 4.** The simulation of the conventional PID control system

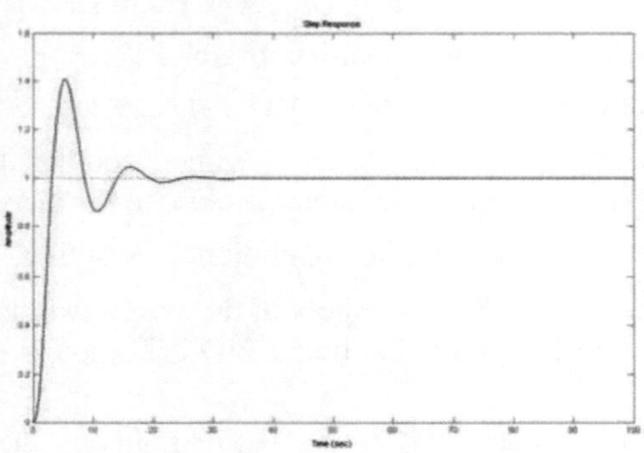

**Fig. 5.** The simulation of the intelligent PID control system

## 6 Conclusion

The PLC is used to replace the traditional relay controller and honeywell digital controller in this system. This design does not only simplify the system, improve the reliability and stability of the equipment, but also improve the efficiency of combustion greatly. The new system is far superior to the conventional PID control system.

The internal combustion control system based on PLC is developed by our department and ZhongCheng combustion company. In practice, it is putted into operation. The actual operation result shows that the system runs stably, safe, reliable and energy saving.

**Acknowledgement.** The authors would like to thank graduate student Li Luowei and Zeng Ping at Zhongyuan Institute of Technology for their experimental contribution to this work.

## References

1. Wang, L., Wu, W.: Application of Intelligent PID Control Algorithm for Variable Frequency and Constant Pressure Water Supply Control System. Chinese Journal Of Electron Devices 32(3), 693–695 (2009)
2. Li, W.: The application of intelligent PI D algorithm in the control of lubricating tank. Machinery Design & Manufacture (11), 82–83 (2008)
3. He, D., Wang, W.: Temperature Control System for Chemical Reaction Kettle Based on S7-200. Electric Drive 37(2), 43–46 (2007)
4. Lin, C., Du, W., Zhou, X.: The Application of PLC System in an Experimental Electric-heated Boiler. Industry Control and Applications 25(5), 22–25 (2009)
5. Chen, Z., Qin, J., Qiu, X.: The Design of Boiler Combustion Control System Based on PLC. Programmable Controller & Factory Automation (3), 61–62 (2009)
6. Hu, L., Li, H., Fang, Z., Zhuang, S.: PID Parameters Auto-tuning Based on PLC. Control Engineering of China 15(S2), 14–15, 65 (2008)

# Research on Knowledge Management from the Perspective of Human Resource Management

Liu Fengwei

School of Economics and Management
Zhengzhou University of Light Industry
Zhengzhou, P.R. China
liufengwei@zzuli.edu.cn

**Abstract.** Based on the theory of knowledge management and human resource management, this paper explored the relationship between knowledge management and human resource management and analyzed the problems in knowledge management in China's enterprises. In addition, this paper put forward some suggestions to improve the level of knowledge management according to the China's practice. The paper argued that the integration of human resource management and knowledge management can achieve effective management of knowledge innovation. On the one hand, the knowledge management can improve the knowledge innovation through more effective system and method of human resource management; on the other hand, combining with human resource management can dispel misconceptions of knowledge management, and knowledge management can achieve further development by taking people as the core of management.

**Index Terms:** knowledge management, human resource management, knowledge innovation, integration.

## 1 Introduction

Transferring the management focus to the human capital management is a qualitative leap in the development process of management, which marks the management, has entered an advanced stage. The initial performance of this stage is that human resource management is becoming increasingly important and gradually becomes the core of organizational management. In recent years, the worldwide emergence of knowledge management indicates the human capital management in organization has entered a new historical period. Human resource management in organization already has a long history, but knowledge management is understood and accepted by organizations with the emergence of knowledge economy. Although the common issues of human resource management and knowledge management is to realize the maximum utilization of human capital, but there is a clear difference in the historical stages of development, the objects, goals, methods and contents of management between them. It is why there are so many organizations confusing the relationship between them in the management practice, which not only wastes resources, but also makes the developments of them be very much restricted. This paper discusses the relationship between them, aiming at finding a coordinated development approach and providing a reference for organizations in effectively increasing the utilization efficiency of human capital in the knowledge-based economy.

# 2 The Relationship between Human Resource Management and Knowledge Management

## 2.1 The Arrival of Knowledge-Based Economy era Brings Forward a New Challenge to Human Resource Management

The application of human resource management in organizations has a long history, and many managers have more emphasis on the human factor, through improving working conditions, shorter working hours, providing workers with a variety of welfare and other methods to improve the motivation of workers. In the "management movement" of United States, Taylor, the "father of scientific management", proposed that enterprises should select and appoint qualified staff for different positions and conduct training to improve efficiency, and he established the professional personnel management department in the enterprise for the first time. By the late 1950s, with the deepening of psychological research on motivation, needs, group dynamics, etc. and coupled with the rapid progress of science and technology, the needs and expectations of staff is also undergoing profound changes. These objective factors prompted managers to re-explore the positive ways to motivate staff, so human resource management became an important part of organizational management. Compared with the traditional personnel management, modern human resource management has significant progress, changing the view of taking the man as a cost and a "tool", beginning to develop and use the man as a "resource", and under the guidance of this idea, there has established a modern human resource management model which takes talent assessment, performance evaluation and salary incentive system as the core. However, for the central elements of human capital - the knowledge, the passive situation of management does not be changed.

In the era of agricultural and industrial economy, what the people mainly provide in the organization are manual labor and professional technical activities. For these elements, organizations can carry out effective control through controlling the number of labor, the worker's professional and technical background, etc.. For example, organizations can change the quantity of labor force to meet the physical requirements of the organization, hire workers with different professional backgrounds or conduct specialized training for members to suit the needs of different expertise. They also can determine the organization's competitive advantage and development direction through the number of labor force, professional structure, and the time of engaging in one profession. In such a case, the human resource management is in the active position in the management of human, manual work and technical activities provided by the man, the contents of which also focused on talent assessment, performance evaluation and salary incentive.

The most important factor to decide the success of organization is people's knowledge in the knowledge-based economy, especially the capability of knowledge innovation. The capability of knowledge innovation is unique to staff in the organization, but it is not just the processing and structuring of information grasped by a person, as well as a higher level things inside, such as the imagination and judgment. Therefore, the modern human resource management methods and systems based on the physical control can not effectively control the knowledge innovation capability of members. This makes modern human resource management concentrate on the people themselves, through introducing, motivating, maintaining and controlling the people who have the knowledge innovation capability to achieve knowledge management. As we all know, much of people's knowledge is implicit, while the innovation and creativity of knowledge are high, so that it is very difficult to screen the talents with the knowledge innovation capability. This will not only cause difficulties to the introduction of intellectual in human resource management, but also be not conducive for organization to assess their own knowledge innovation capability. In addition, the movement of modern organization members is frequent, which also makes the passive management of organization to knowledge resources greatly increase organizational risk. In short, knowledge-based economy has caused a huge impact on the modern human resource management of organization, so organizations urgently need a new management approach to effectively control the knowledge resources of organization.

### 2.2 Bringing Knowledge Management into the Framework of Human Resource Management is an Inevitable Choice for Human Resource Management Adapting to Knowledge-Based Economy

In the wave of knowledge-based economy, knowledge innovation takes the place of other factors and becomes an important component of the core competitiveness of organization, and in this context, the knowledge management starts to enter people's horizons, and is understood and used for the organization. The purpose of knowledge management lies in that the organization guides, supports, coordinates and controls the knowledge innovation of their knowledge workers through the provision of new environment, concept, system and method, and ultimately improves the core competitiveness of organization. The combination of human resource management and knowledge management will have an impact on human resource management in the following areas: First of all, knowledge management can convert the tacit knowledge of organization members into the sharing knowledge of enterprise before them retire or leave, so this can greatly reduce the risk of human resource management of organization. In addition, knowledge management enhances the rapid emergence and accumulation of personal experience, which actually is the distillation of code-based, tacit knowledge and explicit knowledge, and can provide more convenient conditions for the human resource development of organization. Second, knowledge management through the integration of knowledge within the organization, not only can provide better services for the various departments within organization, but also can directly service for consumers and customers, so human resource management walks forward from the background, and becomes the real axis of the entire organization management. Again, the knowledge management can put forward new theoretical knowledge in terms of development, use, performance improvement, performance appraisal and incentive mechanisms of human resources.

And human resource management mainly concerns the relationship between the organization and people, while knowledge management pays more attention to the relationship between organization and knowledge innovation, so the combination of both makes the knowledge management pay more attention to the development of knowledge innovation of people, and can provide a better organizational environment for tapping people's potential. In addition, the knowledge management broadens the boundaries of human resource management, and transfers management focus from the latter to the former in the knowledge and non-knowledge business. Finally, knowledge management also puts forward quantitative problem of knowledge for human resource management, and makes the past single performance evaluation rise to a comprehensive knowledge assessment, which improves the knowledge innovation of organization, thereby enhances the core competitiveness of organization. In short, bringing the knowledge management into framework of human resource management could achieve the innovative and proactive intervention through knowledge management and change the passive situation of human resource management in knowledge management, and thus make the human resource management become the core of modern organizational management.

### 2.3 The Integration of Human Resource Management and Knowledge Management Is the Only Way for the Further Development of Knowledge Management

Many organizations confuse the knowledge management with information management in understanding, they believe that the knowledge management is to firstly break out of knowledge, converts the tacit knowledge in people's minds into explicit knowledge that can be seen, and then conducts management using specific IT technology, so that the knowledge can be shared within the organization to enhance its knowledge innovation. This understanding has encountered great difficulties in practice, mainly because that: First, many knowledge is difficult to explicitly state, and a number of knowledge will lose a lot of features in the refining process, simultaneously these features may be more valuable than the common things. Second, the computer can not achieve the knowledge innovation. And only human beings can play a central role in the process of knowledge innovation, regardless of how much of the information processing ability of computers, they are nothing but a tool of people. And knowledge will always be innovative and creative, but now the computer still could not produce anything creative, in this regard, we still need to use the human brain for creative thinking.

The wrong understanding of knowledge management not only greatly has reduced the role of knowledge management, but also has affected the continued development of knowledge management. This is why the introduction of concept of knowledge management into our country for years but the response is only a handful, and understanding is also very vague, many people believe that knowledge management is the OA System (Office Automation system), and even provider's products of some well-known knowledge management software or platforms also achieve the office automation as the primary function. Is office automation system the knowledge management for which we have high expectations? We know that raising the organization's knowledge innovation is the ultimate goal of knowledge management, in this sense, the knowledge management does not equal to the information management, and it must takes human as the core of management. Among an organization, department of human resource management is one section which directly manages members of the organization, and the same as knowledge management, both of their goals are to maximize the use of human resources. Both advocate using means of good culture to create a relaxed atmosphere, as well as focus on learning team-building to achieve their respective goals. And human resource management is also closely related to various departments and each member of the organization, and has some more in-depth understanding to organization's members, which is very helpful to the implementation of the concept of knowledge management and the development of knowledge management work. Therefore, the integration of human resource management and knowledge management is the only way for the further development of knowledge management.

## 3 Integrated Approaches of Human Resource Management and Knowledge Management

### 3.1 To Increase the Management of Knowledge Map in Human Resource Management

Traditional management conducts knowledge management taking knowledge as an object, breaks out of the knowledge using the idea of decomposition-combination, like the textbook, and converts it into explicit knowledge, and conducts management using a computer program, then mix it up when re-uses the knowledge. And now to combine with human resource management is to use knowledge map. We know that knowledge exists in the human brain, when you need some kind of knowledge, you can find these people through some means (telephone, interviews, project team, BBS, virtual studio, etc.), and achieve these reusable knowledge, which is the use of knowledge map. Specifically, what we should do is to integrate information in human resource management, for example, what are the personnel in the organization, do they have any kind of professional background, what kind of training they have received, what projects they have specifically engaged in, what kind of experience they have, and other information, then when the customer has particular need, you can find timely the relevant personnel. Through a period of accumulation, you can find people with relevant experience for every new project, and then let them participate in, focus on all resources of organization, and provide the best solution. This is knowledge management, or human resource management system oriented capacity or work experience.

### 3.2 To Create Human Resource Management System of Knowledge-Sharing

To create an atmosphere to encourage and promote exchanges between members through human resource management system, that is knowledge-sharing. For example, to set up "two entrances" in the human resource management system of organization, one is project management system storing specific project information, including description of project, team members and the performance of each person during the course. This gives other members an opportunity to learn how others complete a project step by step. In this sense, this system is one "garden" of generating and sharing knowledge, another entrance is the file system of members, except for storing the personal data of members, the system will also update the completion of work of members, for example, what project the member has completed, and so on. Through the comprehensive use of information in these two systems, leaders of the organization can arrange the project for member quite reasonable, and human resource managers can also receive the latest information of the member performance, and conduct rewards and punishments accordingly on that basis.

### 3.3 To Build Knowledge-Precipitation Platform in Human Resource Management

Human resource managers should have an overall resource management thinking and concept of business process, they also should understand which aspect the difficult issues may arise, could the knowledge be precipitated when brain drain, can its successor has possibility of using previous experience, knowledge and method. To resolve this problem, we must have a knowledge management approach, method and work platform, and let employees precipitate the knowledge of work plan, summary and successful cases then filter such information using comment mechanism, so that the valuable personal knowledge can become enterprise knowledge, so as to maintain a sustainable competitiveness.

## 4 Conclusions

For the integration of human resource management and knowledge management, human resource management can achieve the active and effective management of knowledge innovation, which is an important factor of human resource. And on the one hand, the knowledge management can improve the knowledge innovation of organization through more mature and effective system and method of human resource management; on the other hand, the combination with human resource management can also dispel misconceptions of knowledge management, and knowledge management is different from information management, knowledge management can achieve further development only taking people as the core of management.

## References

1. Li, S.-Y.: Dialysis of human resource development mechanism from knowledge management. China Science and Technology Forum (2) (2007)
2. Li, C.-P., Shi, K.: The human resource management strategies of promoting knowledge management. Journal of China Agricultural University (3) (2003)
3. Zeng, J.-Q., Zheng, P.-E.: On the human resource management in knowledge economy age. Journal of Management Sciences (3) (2001)
4. Guan, X.-L.: On the development and management of knowledge economy and human resources. Journal of Shanxi Radio & Amp; TV (5) (2005)

5. Arthur, J.: Effect of human resource systems on manufacturing performance and turnover. Academy of Management Journal 37(3), 670–687 (1994)
6. Baron, J.N., Kreps, E.D.: Strategic Human Resources. John Wiley and Sons, New York (1999)
7. Appelbaum, E., Bailey, T., Berg, P., Kalleberg, A.L.: Manufactuing Advantage: Why High Performance Work Systems PayOff. Cornell University Press, New York (2000)
8. Delery, J.E., Doty, H.D.: Modes of theorizing in strategic human resources management: Tests of universalistic, contingency, and configurational performance predictions. Academy of Management Journal 39, 802–835 (1996)
9. Guthrie, J.P.: High-involvement work practices, turnover, and productivity: Evidence from New Zealand. Academy of Management Journal 44, 180–190 (2001)
10. Ichniowski, C., Shaw, K.: The effects of human resource management system on economic performance: An international comparison of U.S. and Japanese plants. Management Science 45(5), 704–721 (1999)

# Design and Application of Fieldbus OPC DA Server

Jianchun Xing, Qiliang Yang, Ping Wang, and Ronghao Wang

Engineering Institute of Engineering Corps, PLA University of Science and Technology
Nanjing, China
{xjc,yql,wp,wrh}@893.com.cn

**Abstract.** OPC is one of the most important communication specification in the industrial control community. Taking one kind of LonWork fieldbus products developed by our self as one case study, we implement an OPC DA Server for the fieldbus based on the given APIs of the device driver. The implementation principles of four important parts: real-time database, GUI, OPC data access, I/O read and write, are probed in detail. This opc server, together with several popular SCADA software such as iFix and KingView, has successfully applied in many intelligent buildings' projects.

**Keywords:** OPC Specification, OPC DA Server, Fieldbus.

## 1 Introduction

OPC, OLE for Process Control, means OLE technologies applied in industrial process control and is one of the most important specifications in the community of industrial control [1], [2]. OPC defines a series of standard OLE interfaces with the COM/DCOM technologies. Using these interfaces, clients can communicate to various software or hardware products under the universal data access standards, and so they make the integration of automation-control systems very easy. OPC has many specifications; Data Access Specification is the most important one among them [3]. OPC Data Access Server (usually shortened as OPC Server) is the software implementation of OPC Data Access Specification.

Fieldbus is the communication and network technology in the industrial automation community. It plants CPU into traditional meters, which makes the meters communicate with other computing devices using open protocols. FF, Profibus, LonWorks, etc. are the main international standards of Fieldbus. Many big companies such as SIEMENS are all develop OPC Servers (e.g. S7-200 PC Access PPI OPC Server) for their own Fieldbus products to enhance the communication abilities of them. However, most of the OPC Servers[4][5] only provide data-exchange functions and lack such advanced managing capabilities as simulation, communication fault detection, error recording.

In the paper, taking the LonWork fieldbus developed by our self as one case study, we implement one OPC DA Server for the fieldbus. Comparing with other opc servers, our OPC Server has the good features: friendly user interface, device-communication-state diagnosis and error automatic log, device simulation, defining scan periods of devices with f rates the application of the OPC server. We summarize the paper in Section 6. reedom, supporting string data type. Using the server, any industrial control software that supports OPC standards can access the LonWorks fieldbus, which greatly broadens the application area of the fieldbus.

The rest of the paper is organized as follows. Section 2 gives an overview of our LonWork fieldbus and describes the motivation of the implementation of the OPC server. The general architecture of the OPC server is discussed in Section 3. Section 4 presents the implementation of key parts of the server such as constructing the real-time database, implementing the read-write thread and so on. Section 5 illust.

## 2 Backgrounds

LonWorks fieldbus, initiated and supported by Echelon Inc., is a very popular and open fieldbus control network and widely applied in the world. It allows other users to developed their own various LonWorks products according with the core LonTalk protocols. We have developed a kind of LonWorks control network using the general technologies of LonWorks. Our LonWorks network includes a serial of measure-control modules such as LM1101, LM1204, LM1205 to support analogy input/output and digital input/output. Previously, we have to access this network using APIs, which needs tired programming and make the process of data access complicated. User applications call the APIs which are provided with the OS (Operating System) driver of the LonWorks network to access the modules in the network. The APIs are all encapsulated in the .dll file-Lmdrv.dll.

Here are the important APIs related with the I/O access:

- *LmPort:* assigning the port address of the LonWorks communication cards,
- *LmState:* reading the states from the digital I/O channels in digital modules,
- *LmNum*: reading the values from the analogy I/O channels in analogy modules,
- *LmCmd*: sending control commands from the up-level computers to the network.

This paper intends to implement an OPC Server using OPC DA Specification on the base of the given APIs of this kind of LonWorks so as to change the traditional data exchange style. Fig.1 illustrates this intention.

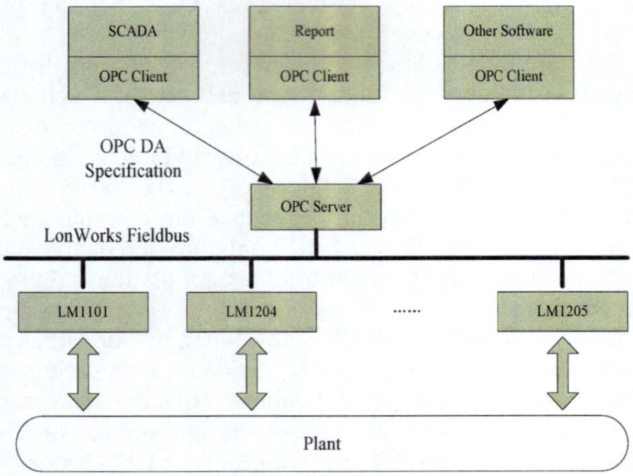


**Fig. 1.** Easy access to the LonWorks fieldbus using the OPC server

We can learn from Fig.1 that through the opc server, any software supporting OPC DA Specification can easily access our LonWorks fieldbus without any programming.

## 3 Architecture of the OPC DA Server

The LonWorks OPC server is completely compatible to the communication protocol of the LonWorks network, and is developed on the basis of the API-based driver of the network.

Fig.2 describes the design architecture of the server. The OPC server is made up of the following software modules.

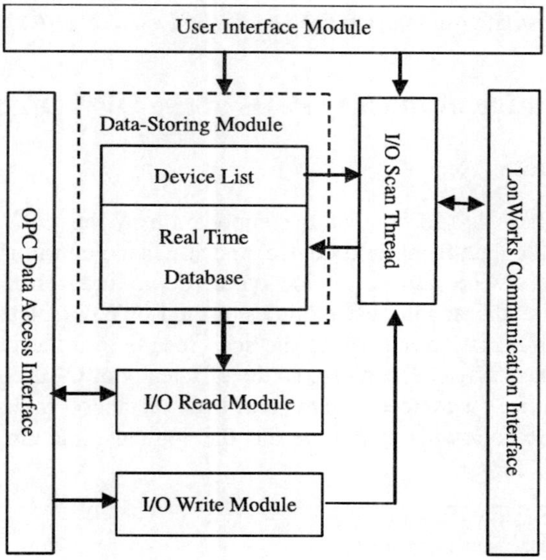


**Fig. 2.** The software Architecture of the Lonworks OPC Server

*User interface module.* This module is a graphic window to interface with users. The window has a standard document/view structure with explore-style user interface.

*OPC DA interface module.* This module implements all objects and interfaces defined by the OPC Data Access 2.03 specification. These objects involve OPC Server object, OPC Group object and OPC Item object. The OPC interfaces of the LonWorks OPC server support COM interfaces and automation interfaces.

*Data-storing module.* This module consists of the device list and the real-time database. The device list is used to store configuration information such as device names, device addresses, device types, device-scanning periods and so on.

*I/O scan thread module.* According to the device list configured before, this module scans the LonWorks network periodically under the pre-defined intervals of devices and save the scan results to the real-time database. During the scan of the thread, if the thread find write commands coming into the write-command queue, it will process the write commands firstly.

*I/O read module.* This module processes the data-read packets from OPC clients and read values from the real-time database in terms of the packets. With being changed from common data types to the variant type, the values read from the real-time database are set to the corresponding OPC items. After being stamped with the time and quality mark, the OPC items will be returned to OPC clients.

*I/O write module.* This module parses the write command packets from OPC clients, and convert the packets to the command strings according with the LonWorks protocol.

*LonWorks communication interface module.* This module uses a series of API functions that encapsulated in lmdrv.dll to access the LonWorks I/O device directly.

# 4 Implementation of the Key Parts of the OPC DA Server

## 4.1 Real-Time Database

The real-time database is the core of the application of the OPC server. It timely stores and updates the data from field device, and stores the configuration information of devices such as device names, device types, device addresses and so on. The background scan thread scans the I/O modules of the LonWorks network according to the configuration information and saves the scan results into the real-time database. The real-time database stays in memory, so the efficiency of OPC clients reading data from the database is guaranteed. The real-time database is implemented as a structure array. The following codes define the logical structure of the real-time database in the OPC server.

```
//Real-Time Database
typedef struct DeviceRTValue
{
    CString deviceType;          //device type
    long address;                //device address
    float AIRTValue[80];         //real-time values of analogy
    long DIRTValue[240];         // real-time values of digital
    WORD ComIsErr;               //communication state of device
    long LastScanTime;           //the last scan time of device
    long ScanPeriod;             //the scan period of device
    BOOL IsSim;                  //simulate or not
} DEVICERTVALUE;
```

The real-time database is as the global database for the OPC server application, and makes all data be shared among the threads in the application.

## 4.2 Graphic User Interface

General OPC Server can be divided into out-process server and in-process server. The former server exists in the style of an executable file (EXE) and the later is in the style of a dynamic-link library (DLL). Out-process servers can be accessed by clients over processes or networks, but in-process servers can not. The LonWorks OPC server in this paper belongs to out-process servers and has a separate process space, so the server will be of a stability performance when it exchanges data with OPC clients.

The main user interface of the OPC server has a document-view architecture, so it is easily operated by users.

This server uses three level objects, devices, groups and tags to manage data. Devices mean the LonWorks I/O control modules, tags represent I/O channels of devices and groups are the logical partitions for tags. Devices, groups and tags are displayed with explore styles in the main interface. A tree-view ActiveX control is used to display devices and groups, and an ActiveX list-view control is used to display tags and their properties. For devices and tags, property dialogs are designed for users editing and modifying their properties. Fig.3 is the main graphic interface of the OPC server.

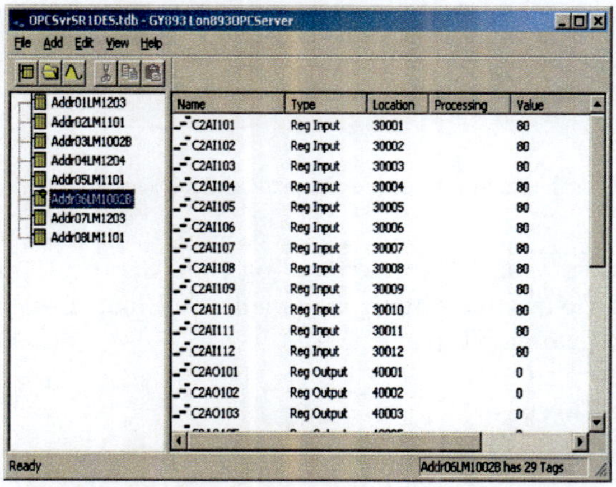


**Fig. 3.** The main user interface of the LonWorks OPC Server

The main graphic interface provides command toolbars and menus to support user operations such as adding, deleting, editing, copying and so on. Using the graphic interface, users can save the configuration into files, and the files can be loaded automatically when the OPC server starts.

### 4.3 OPC Data Access Interface Based on the Object-Oriented Technology

OPC objects include server objects, group objects and item objects. Server objects maintain relative information of themselves and act as containers for OPC group objects to dynamically create or release group objects. Similarly, group objects not only maintain relative information of themselves but also provide a suit of containing mechanisms to logically manage item objects. Item objects represent data links of OPC servers. From the viewpoint of common interfaces, item objects are not COM objects[6], which can be directly operated by OPC clients, so there are no interfaces defined for the objects. All operations to item objects are through the group objects that contain the items. Server objects and group objects are true COM objects and form an aggregation relation. Namely, after creating group objects, server objects transfer the pointers of the group objects to OPC clients to let the clients operate the group objects directly. Fig.3 illustrates the relations between objects in OPC Server and Client.

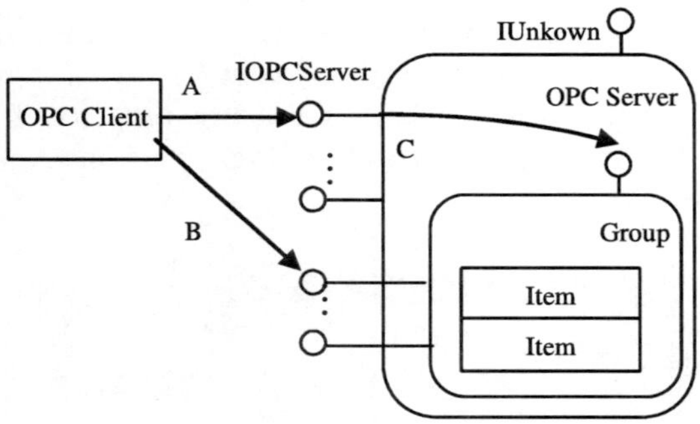


**Fig. 4.** The relations between objects in OPC server and client

In Fig.4, A means calling the interfaces of OPC Server; B means calling the interfaces of OPC Goup after getting the pointer of Group from Server; C means creating Group and returning its pointer.

### 4.4 Read-Write Threads

The rational process for read-write of I/O device is the key to promote the efficiency of the OPC server.

There is a specific thread designed to implement the read and write of I/O devices in the server. The thread is created at the time when a user-configuration document is loaded. This thread circularly scans I/O channels of devices with the information from the user-configuration document and transfers the scanned results to the real-time database. In the duration of the scan, the thread checks the writing flag: WriteCmdIsComing. If the value of the flag is TRUE, which illustrates that there are writing commands in the message queue, the thread will read the command strings from the queue by the first-in-first-out rule and sent the command strings to the LonWorks network one by one.

Writing commands are processed firstly. Only the process of write commands is complete, the thread will enter the read loop. This mechanism insures that writing commands can be responded to timely. Fig.4 is the flow chart of the I/O scan thread.

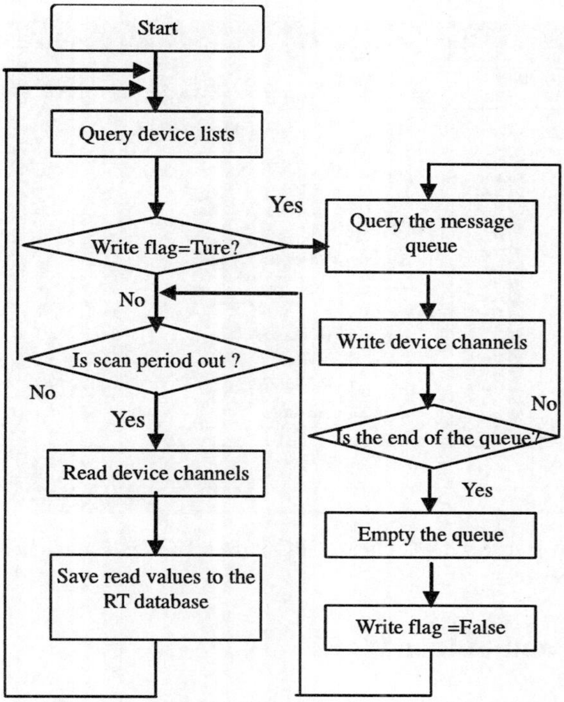


**Fig. 5.** The flow chart of the I/O scan thread

In Fig.5, the main part of the reading-writing thread is a dead loop. The dead loop scans devices ceaselessly. When writing commands come, the thread will stop reading to process writing commands; after the writing commands are completely processed, the thread continues to read.

# 5 Applications of the OPC Da Server

The API-based OPC server has been successfully developed and applied in many intelligent buildings' projects. In these projects, several popular SCADA software such as Proficy HMI SCADA iFIX and KingView all can easily connect the opc server to exchange data with our LonWorks fieldbus using their own OPC clients. Fig.6 shows that iFix OPC client, OPC Power Tool. is successfully communicating with our Lon893OPCServer application.

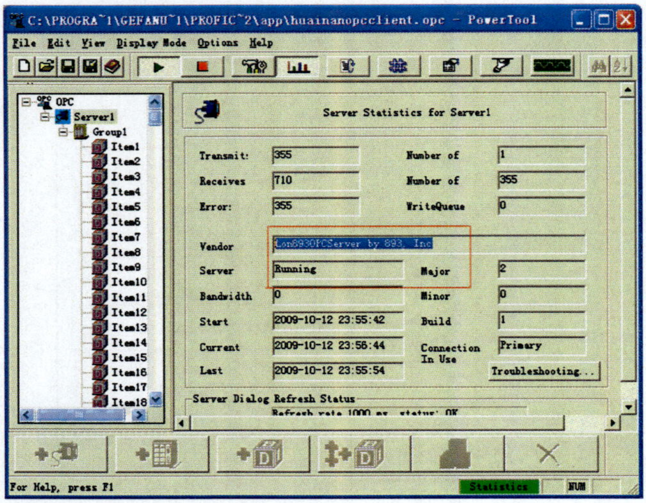


**Fig. 6.** The scenario that iFix OPC client, OPC Power Tool, is communicating with the opc server Lon893OPCServer

# 6 Summary and Futuer Work

With the aim to make our own LonWorks fieldbus products more open and more easily be accessed, we develop the OPC server Lon893OPCServer based on OPC Data Access Specification. With comparisons to other OPC servers, this OPC server has much more managing strengths such as device simulation, device-communication-state diagnosis and so on. This paper presents the design principles and technologies of the OPC server. The implementation of several key parts of the server as real-time database, GUI, OPC data access, I/O read and write, are probed in detail. The API-based approaches for developing the Lon893OPCServer can also be used for realizing other fieldbus OPC DA Server.

In the future work, we will continue to perfect the simulation function of the OPC server, which make it be able to not only simulation the values of devices, but also be able to simulate the control logics of devices.

# References

1. OPC Foundation. Data Access Custom Interface Standard Version 2.04 [EB/OL] (September 5, 2000), http://www.OPCFoundation.org
2. OPC Foundation. Data Access Custom Interface Standard Version 2.02 [EB/OL] (February 5, 2000), http://www.OPCFoundation.org
3. Xing, J.C., Wang, L., Wang, P.: Summarizations of the Co-operation Standard of Industrial Control Software. Industrial Control Computer (January 2000)
4. Siemens Energy & Automation, Inc., S7 200 PC Access User's Manual (Online Help) (2004)
5. ABB, Inc., EIB OPC Server 2.0 User's Manual (Online Help) (2004)
6. Pan, A.M.: COM Principle and Applications. China Tsinghua University Press, Beijing (1999)

# Security Enhancement on an Authentication Method for Proxy Mobile IPv6

Qi Jiang[1], Jianfeng Ma[1], Guangsong Li[2], and Ayong Ye[3]

[1] Key Laboratory of Computer Networks and Information Security (Ministry of Education), Xidian University, Xi'an 710071, P.R. China
jiangqixdu@gmail.com, jfma@mail.xidian.edu.cn
[2] Department of Information Research
Zhengzhou Information Science and Technology Institute
Zhengzhou 450002, P.R. China
guangsongli@yahoo.com.cn
[3] Key Lab of Network Security and Cryptology, Fujian Normal University, Fuzhou, 350007, P.R. China
yay@fjnu.edu.cn

**Abstract.** Proxy Mobile IPv6 (PMIPv6) is an emerging network-based localized mobility management scheme. Recently, Zhou, Zhang and Qin proposed an authentication method for PMIPv6. However, this paper shows that their scheme fails to achieve mutual authentication between the Mobile Node (MN) and network. It performs only unilateral authentication of the MN to network, and the MN has no information about the authenticity of the network, thus it is susceptible to network spoofing attack and malicious MAG attack. Then, an improved scheme is proposed to remedy these security flaws.

**Keywords:** authentication, handover, proxy mobile IPv6.

## 1 Introduction

As a representative solution of network-based mobility management, Proxy Mobile IPv6 (PMIPv6) [1] is being actively standardized by the Network-based Localized Mobility Management (NETLMN) working group of IETF [2]. PMIPv6, which provides mobility services without the involvement of the MN and reduces mobility signaling over wireless links, is being adopted as part of several wide-area wireless network (e.g., 3GPP, 3GPP2, WiMAX) and local area network environments.

In PMIPv6, the core mobility service provisioning entities include the Mobile Access Gateway (MAG) and the Local Mobility Anchor (LMA). The MAG, typically running on the access router, performs mobility management on behalf of the MN. The LMA maintains reachability state of the MN in a PMIPv6 domain. Accordingly, an MN with in PMIPv6 does not require protocol stack modification in order to support PMIPv6. The MAG and LMA manage all mobility signaling and establish a bidirectional tunnel for all traffic sent to and from the MN. In view of the MN, the entire PMIPv6 domain appears as its home network.

MN authentication is a prerequisite of registration for network access, that is, the MN should be authenticated mutually with the network in the first place. When an MN attaches to a network, the MN wants to be confident that it is not a malicious network. Meanwhile, the network wants to ensure that the MN is who he or she claims to be and that the MN is authorized to access the service. However, the current specification [1] provides no authentication scheme.

Zhou et al. [3] proposed an authentication method for PMIPv6 and adopted pre-authentication to reduce handover delay. The scheme is a symmetric encryption based challenge-response mechanism. However, a detailed security analysis is absent in their paper.

In this paper, we show that their scheme fails to achieve mutual authentication between the MN and network, i.e., the scheme only achieves unilateral authentication of the MN to network, and the MN has no information about the authenticity of the network, thus it is susceptible to network spoofing attack and malicious MAG attack. To overcome the security flaws of Zhou et al.'s scheme, we propose an effective remedy which provides the missing security provisions that are necessary for PMIPv6 access authentication.

The remainder of this paper is organized as follows. Section II gives some related work on PMIPv6 security. Section III briefly reviews Zhou et al.'s scheme and elaborates on the weakness of their scheme. Section IV presents our improved authentication scheme. Section V discusses the security analysis of our scheme, and at the end, Section VI concludes the paper.

## 2 Related Work

MIPv6 security has gained considerable research effort in the literature [4], whereas PMIPv6 security has only recently emerged.

Security threats to PMIPv6 on the two interfaces of the architecture, i.e., the interface between an LMA and an MAG, and the interface between an MN and its current MAG, are discussed in [5].

To solve the inefficiency of ordinary authentication mechanisms such as Extensible Authentication Protocol (EAP) and Kerberos when applied to PMIPv6, Lee et al. [6] propose a ticket-based authentication mechanism to optimize handover authentication during handover process. They also provide the proof of the security of the scheme using BAN logic.

To minimize handover delay, in [7], a variant of the Diffie-Hellman key agreement is used to eliminate time required for re-authentication by AAA server when handover occurs. A pre-shared key based authentication scheme for exchanging mobility signaling is proposed in [8]. Reference [9] describes a certificate-based public key authentication mechanism for PMIPv6, which is based on EAP-TLS. The authentication mechanism involves initial authentication and handover authentication, and is analyzed by BAN logic. Lee and Chung [10] introduce three authentication methods that are based on security level, i.e., plaintext based, hash function based and shared secret key based. The authentication methods are compared in terms of security level and signaling cost.

## 3 Review of Zhou, Zhang, and Qin's Scheme

The authentication scheme proposed by Zhou et al. [3] provides both initial authentication and handover authentication. The former occurs when the MN accesses to the PMIPv6 domain at the first time, while the latter occurs when the MN changes its point of attachment in the same authentication domain. The notations involved are listed in Table I.

### 3.1 Assumption

It is assumed that each MN is identified by a globally unique network access identifier (NAI) and that the authentication, authorization and accounting (AAA) server in each domain possesses profile of each MN and shares a long-term key with each one. The AAA server is also assumed to share a pre-established key with each network entity, i.e., each MAG and LMA, and to maintain the trust relationship through DIAMETER [11] protocol.

### 3.2 Initial Authentication

The message sequence for initial authentication is depicted in Fig. 1. We describe its steps in detail as follows.

**Table 1.** Notations

| Notation | Quantity |
|---|---|
| NAI | The identity of an MN |
| MAGID | The identity of an MAG |
| LMAID | The identity of an LMA |
| LC | Local challenge |
| CR | Credential |
| RPI | Replay protection indicator |
| HNP | Home network prefix |
| KAAA | Pre-shared key between an MN and the AAA |
| KAM | Pre-shared key between an MAG and the AAA |
| KAL | Pre-shared key between the AAA and an LMA |
| KML | A dynamic key between an MAG and an LMA |
| SPML | Security parameters for constructing KML |
| $E_K(M)$ | Encryption of a message M using a key K |
| HMAC-SHA1(K, M) | A keyed hash of a message M using a key K |
| \|\| | Bit string concatenation |

**Step 1.** (MN→MAG): AS
MN sends an attendant solicit (AS) message to initiate a new session.
**Step 2.** (MAG→MN): AA= {LC}

MAG responds with an attendant advertisement (AA) message including LC, which is a random number generated by the MAG.

**Step 3.** (MN→MAG): AReq= {NAI, MAGID, RPI, LC, CR}

MN computes CR=$E_{KAAA}$(LC), then, sends authentication request (AReq) message to MAG.

**Step 4.** (MAG→AAA): AMR= {NAI, MAGID, RPL, LC, CR}

MAG sends AA-MAG-request (AMR) message to AAA. The content of AMR is the same as that of AReq.

**Step 5.** (AAA→LMA): ALR= {NAI, MAGID, $E_{KAL}$(SPML)}

AAA computes CR'=$E_{KAAA}$(LC), and compares CR' with CR received in AMR. If they are equal, MN is successfully authenticated. Then, the AAA server generates a nonce SPML, and creates KML=HMAC-SHA1(SPML, NAI||MAGID||LMAID) to secure bi-directional tunnel between the MAG and LMA. Finally, the AAA server sends AA-LMA-request (ALR) message to LMA. Note that the above KML computation differs slightly from the specification in [2], where it is stated that KML=HMAC-SHA1(KAAA, SPML||NAI||MAGID||LMAID). This change has been made because the specification in [2] would appear to be an error, since the LMA and MAG cannot construct KML as specified in [2].

**Step 6.** (LMA→AAA): ALA= {NAI, LMAID}

LMA decrypts $E_{KAL}$(SPML) and constructs KML in the same way as AAA. Then it replies to AAA with an AA-LMA-answer (ALA) as confirmation.

**Step 7.** (AAA→MAG): AMA= {NAI, LMAID, HNP, $E_{KAM}$(SPML)}

AAA sends to MAG AA-MAG-answer (AMA) message to notify the MAG of the authentication result.

**Step 8.** (MAG→MN): ARep= {NAI, HNP}

MAG decrypts $E_{KAM}$(SPML) and constructs KML in the same way as AAA. Moreover, MAG sends to MN the authentication reply (ARep) message.

**Step 9.**

When receiving ARep from MAG, MN is informed of the authentication result.

### 3.3 Handover Authentication

As illustrated in Fig. 2, handover authentication only differs from initial authentication in that the handover authentication procedure is performed in the new MAG while the MN is still attached to its old MAG. Therefore, the detailed description of handover authentication is omitted due to limited space.

### 3.4 Security Weakness

We now demonstrate that Zhou, Zhang and Qin's scheme fails to achieve mutual authentication between the MN and network. It only achieves unilateral authentication of MN to network. Therefore, there exist two attacks in the scheme. We describe these two attacks as follows.

1) Network spoofing attack

Assume that *A* is an attacker, *A* can always impersonate as a legitimate network to MN. Specifically, when an MN sends AS to initiate a new session, *A* can send AA as a response. After receiving AReq, *A* can reply with forged ARep. Therefore, MN has no awareness of the attack.

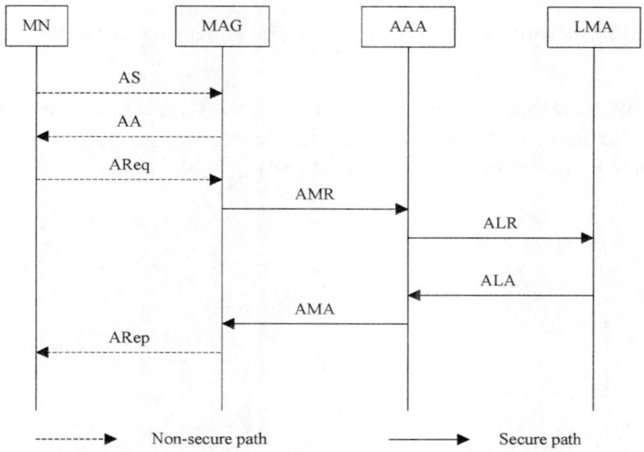


**Fig. 1.** Initial authentication of Zhou, Zhang and Qin's scheme.

2) Malicious MAG attack

Even when layer-2 security mechanisms are enabled, a malicious MAG *M* can impersonate as the intended MAG which the MN wants to attach. In specific, assuming an MN wants to attach an MAG with the identity of MAGID. After the MN has sent AReq, *M* intercepts the AReq, and modifies it to AReq'= {NAI, MAGID', RPI, LC, CR}, where MAGID' is identity of *M*. Then *M* follows the authentication process. Finally, *M* sends ARep to the MN in the name of the MAG. *M* have successfully impersonated as the intended MAG to the MN without being noticed by the MN.

# 4 Our Improved Scheme

## 4.1 Security Requirements

Before we present the improved scheme, we re-examine the security requirements of PMIPv6. The main goal of the authentication scheme is to mutually authenticate the MN and network to each other, and to achieve integrity, replay protection. An additional requirement, specific to PMIPv6 under consideration, is identification of the network path on which the session is taking place. Specifically, the scheme requires supporting the following:

**S1 – MN Authentication.** Allow the AAA to authenticate and authorize the MN with rights to establish a security association with and receive mobility services from, the MAG and LMA in the PMIPv6 domain.

**S2 – AAA server authentication.** Allow the MN to assure that it is authenticating to a trusted AAA server.

**S3 – Integrity.** Prevent attackers from altering protocol messages.

**S4 – Replay protection.** Prevent attacks from replaying messages from previous protocol runs.

**S5 – Path authentication by the AAA server and MN.** Allow the AAA server and MN to verify the identity of the network entities along the path from MN to AAA server, particularly, that of the intended MAG of the MN.

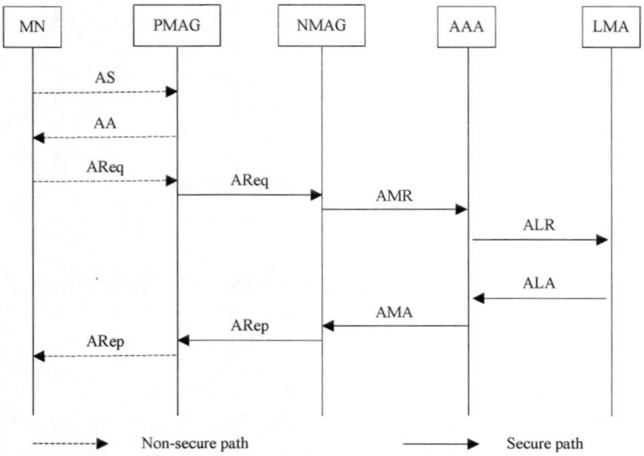


**Fig. 2.** Handover authentication of Zhou, Zhang and Qin's scheme

### 4.2 Our Improved Scheme

We propose an improved scheme to accomplish mutual authentication between MN and MAG. The differences between our improved scheme and Zhou et al.'s scheme are described below.

**Step 2'.** (MAG→MN): AA= $\{LC_{MAG}\}$

MAG responds with an attendant advertisement (AA) message including $LC_{MAG}$, which is a random number generated by the MAG.

**Step 3'.** (MN→MAG): AReq= $\{NAI, MAGID, RPI, LC_{MAG}, LC_{MN}, CR_{MN}\}$

MN generates a random number $LC_{MN}$ and computes $CR_{MN}= E_{KAAA}(NAI||MAGID||RPI||LC_{MAG}||LC_{MN})$, then, sends authentication request (AReq) message to MAG.

**Step 4'.** (MAG→AAA): AMR= $\{NAI, MAGID, RPL, LC_{MAG}, LC_{MN}, CR_{MN}\}$

MAG sends AA-MAG-request (AMR) message to AAA. The content of AMR is the same as that of AReq.

**Step 5'.** (AAA→LMA): ALR= $\{NAI, MAGID, E_{KAL}(SPML)\}$

AAA computes $CR'_{MN} = E_{KAAA}(NAI||MAGID||RPI||LC_{MAG}||LC_{MN})$ and compares $CR'_{MN}$ with $CR_{MN}$ received in AMR. If they are equal, MN is successfully authenticated. Then, AAA checks if the MAG sending AMR is with the identity of

MAGID. If so, AAA generates a nonce SPML, and creates KML=HMAC-SHA1(SPML, NAI||MAGID||LMAID) to secure bi-directional tunnel between MAG and LMA. Finally, AAA sends AA-LMA-request (ALR) message to LMA.

**Step 7'.** (AAA→MAG): AMA= {NAI, LMAID, HNP, $CR_{AAA}$, $E_{KAM}$(SPML)}

AAA computes $CR_{AAA}$= $E_{KAAA}$(NAI||MAGID||RPI||$LC_{MN}$||$LC_{MAG}$), and sends to MAG AA-MAG-answer (AMA) message to notify the MAG of the authentication result.

**Step 8'.** (MAG→MN): ARep= {NAI, HNP, CRAAA}

MAG decrypts EKAM(SPML) and constructs KML in the same way as AAA. Moreover, MAG sends to MN the authentication reply (ARep) message.

**Step 9'.** When receiving ARep from MAG, MN computes $CR'_{AAA}$ = $E_{KAAA}$(NAI||MAGID||HNP||$LC_{MN}$||$LC_{MAG}$), and compares $CR'_{AAA}$ with $CR_{AAA}$ received in ARep. If they are equal, the MAG is successfully authenticated.

## 5 Security Analysis

In this section, we provide an in-depth security analysis and discussion of the proposed scheme.

**S1** – MN authentication is achieved in Step 5' by validating the authenticator $CR_{MN}$. The nonce $LC_{MAG}$ acts as a challenge to the MN to assure the AAA server that it possesses the pre-shared key KAAA.

**S2** – AAA server authentication is achieved in Step 9' by validating the authenticator $CR_{AAA}$. The nonce $LC_{MN}$ acts as a challenge to the AAA server to assure the MN that it possesses the pre-shared key KAAA.

**S3** –Integrity follows from the fact that the messages are protected by $CR_{MN}$ and $CR_{AAA}$. If an attacker alters the messages, the verification of $CR_{MN}$ and $CR_{AAA}$ will fail because of the property of the encryption function.

**S4** – Replay protection is due to the randomness and freshness of the nonces $LC_{MN}$ and $LC_{MAG}$. $LC_{MN}$ ($LC_{MAG}$) assures the MN (AAA server) that $CR_{AAA}$ ($CR_{MN}$) is fresh for every session.

**S5** – Path authentication by the AAA server and MN comes from the fact that MAGID is included in the computation of $CR_{AAA}$ and $CR_{MN}$ which precludes any MAG other than the intended MAG from posing as the intended MAG. If a malicious MAG *M* modifies the MAGID included in the AReq, however, it could not obtain a valid value $CR_{MN}$, AAA will observe it and abort the current authentication process.

## 6 Conclusion

We have pointed out that two security attacks, i.e., network spoofing attack and malicious MAG attack, exist in the scheme of Zhou, Zhang and Qin. Then, an improved scheme is proposed to overcome the identified problems. Our ongoing work includes a formal proof of the security of the improved scheme and performance evaluation.

**Acknowledgment.** This work is supported by the National High Technology Research and Development Program of China (863 Program) (2007AA01Z429), the National Natural Science Foundation of China (60633020, 60872041, 60803150), and funds of Key Lab of Fujian Province University Network Security and Cryptology (No.09A001).

## References

1. Gundavelli, S., Leung, K., Devarapalli, V., Wichorus, Chowdhury, K., Patil, B.: Proxy mobile IPv6, IETF Internet Draft, deaft-ietf-netlmm-proxymip6-18.txt (May 2008)
2. Kong, K., Lee, W., Han, Y., Shin, M., You, H.: Mobility Management for All-IP Mobile Networks: Mobile IPv6 vs. Proxy Mobile IPv6. IEEE Wireless Communications 15(2), 36–45 (2008)
3. Zhou, H., Zhang, H., Qin, Y.: An authentication method for proxy mobile IPv6 and performance analysis. Security and Communication Networks 2(5), 445–454 (2009)
4. Elgoarany, K., Eltoweissy, M.: Security in Mobile Ipv6: A Survey. Information Security Tech. Report 12(1), 32–43 (2007)
5. Vogt, C., Kempf, J.: Security threats to network-based localized mobility management (NetLMM). IETF RFC4832 (April 2007)
6. Lee, J., Lee, J., Chung, T.: Ticket-based Authentication Mechanism for Proxy Mobile IPv6 Environment. In: Proceedings of the Third International Conference on Systems and Networks Communications (ICSNC 2008), pp. 304–309. IEEE Computer Society (2008)
7. Kim, H., Oh, B.: Secure and low latency handoff scheme for proxy mobile IPv6. In: Proceedings of the International Conference on Mobile Technology, Applications & Systems (Mobility Conference 2008). ACM Press (September 2008)
8. Mun, Y., Kim, M., Kim, G.: Mutual Authentication Scheme in Proxy Mobile IP. In: Proceedings of International Conference on Computational Sciences and Its Applications, ICCSA 2008, pp. 65–72. IEEE Computer Society (2008)
9. Park, S., Lee, J., Chung, T.: Authentication Analysis Based on Certificate for Proxy Mobile IPv6 Environment. In: Gervasi, O., Taniar, D., Murgante, B., Laganà, A., Mun, Y., Gavrilova, M.L. (eds.) ICCSA 2009. LNCS, vol. 5592, pp. 885–896. Springer, Heidelberg (2009)
10. Lee, J., Chung, T.: A Traffic Analysis of Authentication Methods for Proxy Mobile IPv6. In: Proceedings of International Conference on Information Security and Assurance (ISA 2008), pp. 512–517. IEEE Computer Society (2008)
11. Calhoun, P., Loughney, J., Guttman, E., Zorn, G., Arkko, J.: Diameter base protocol. IETF RFC 3588 (September 2003)

# Software Piracy Prevention: Punish Buyer or Seller?

Hong Ding and Xiaofeng Zhang

School of Computer Science and Technology
Nantong University
Nantong, China
ntuzxf@163.com

**Abstract.** Software piracy is illegal and the government should adopt effective measures against it. After many years of fighting against it, software piracy does not die out, but rather becomes more serious. In this paper, the phenomenon of software piracy is investigated from the perspective of game theory. An evolution model is proposed and several simulations of the software market trend are followed this model. They simulated the market without punishment, with punishment on buyers, on sellers and on both of them. Through these analyses, a conclusion can be drawn that punishing the buyers is a necessary measure against software piracy.

**Keywords:** software piracy, punishment mechanism, game theory.

## 1 Introduction

Software piracy is an illegal copying of computer software. It is a global problem and causes enormous losses[1-3]. Banners have appeared on the media urging citizens not to buy them, but the effect is so faint that software piracy becomes more. The pirated software sellers are fined too, but software piracy still exists. So it is important to establishing and improving the framework of laws and rules on copyright software piracy.

In order to prevent the losses, many researchers investigate the phenomenon of software piracy[1, 3, 4]. They show that software piracy is influenced by many factors such as gender, age and experience etc. These theories are foundation of laws and rules. Some researchers [2] investigate the use of equity theory (fairness) as a deterrent to software piracy. But all these methods can not utterly destroy software piracy. We need a feasible plan to fight against software piracy.

In this paper, we use evolutionary behavior[5] to imitate the action tendency of the software market. We use penalties to change the market tendency. From the results of the simulations, it is obvious that inflicting penalty on buyers can achieve a better result than on sellers.

## 2 The Model

We use hunting deer game model to mimic the software market, this model was introduced by Rousseau in his book *Discourse on Inequality*(1755)[6]. Each buyer or seller can choose authorized or pirated software in every trading. Their payoffs can be represented by matrix in table 1.

**Table 1.** Game Matrix Between Buyer and Seller

| | | buyer | |
|---|---|---|---|
| | | *A* | *P* |
| seller | *A* | $(A_s, A_b)$ | (0,0) |
| | *P* | (0,0) | $(P_s, P_b)$ |

There are two cases in the matrix. One is that the requirements of the buyer and the seller do not match each other. In this case their payoffs are 0. The other case is that the seller provides what the buyer wants. If it is an authorized edition, they get the payoffs $A_s$ and $A_b$ separately. If it is a pirated edition, they get the payoffs $P_s$ and $P_b$ separately. Generally, the functions of a pirated and an authorized edition are almost the same, but the price of a pirated one is much lower, so $P_b >> A_b$. That is the buyer will get more if he buy pirated software. To a seller, there is no exact relations between $A_s$ and $P_s$, but the sales can rise because of the low price of pirated software. If it can rise $n$ times sales, $nP_s$ must higher than $A_s$. Then the seller tends to provide pirated software to get more payoffs. In fact, even if $A_s > P_s$ and $n=1$, there are sellers to provide pirated software. Suppose all buyers and sellers choose authorized software at the beginning, average payoff of sellers is $A_s \times n_1 / n_2$ ($n_1$ is the number of buyers and $n_2$ is the number of sellers). Now there is one seller come to provide pirated software, then all buyers will buy software form this seller for $A_b << P_b$. This seller's payoff will be $P_s \times n_1$, while the other sellers will get 0. When there is no significant differences between $A_b$ and $P_b$, this seller's payoff will be $n_2$ times(for a real market $n_2 > 100$ is normal) and others sellers' payoffs will fall sharply. So this sharp contrast of payoffs is a strong temptation to sellers and will force the other sellers to sell pirated software. Of course it is a vicious cycle.

According to Nash Equilibrium strategy theory[7], the optimization strategies are the requirements of the buyer and the seller are the same editions. So there are two Nash Equilibrium strategies in this game model. One is the buyer and the seller choose pirated software, another is they choose authorized software at the same time.

In fact, the buyer and the seller will choose the pirated editions for higher payoffs. That is because $P_b >> A_b$ and pirated software can make a high sales. Then the whole market becomes a chaos.

In order to strictly standardizing market order, punishments should be adopted. If the punishments reduce the pirated software's payoffs to a minus value, the buyers and sellers will not choose pirated software any more. According to the method of repeated deleting strict bad strategy, there is only one Nash Equilibrium strategy which is all buyers and sellers choose authorized editions. Though it is impossible to punish every trade of pirated software, the existence of punishments is a caution to the trade of pirated software. In the next paper, we will propose a model and discuss the difference between punishing buyers and sellers. We will see that the difference of punishment effect and target may cause different results.

Suppose in a market there are $n_1$ buyers and $n_2$ sellers. To a buyer, his probability of choosing an authorized edition is $p_{bi}(t)$ and his probability of choosing an pirated edition is 1-$p_{bi}(t)$. To a seller, his probability of providing an authorized edition is $p_{si}(t)$ and his probability of providing an pirated edition is 1- $p_{si}(t)$. If pirated editions are selected, the sales will be $n$ times, that is the seller will sell $n$ times editions and the buyer will buy $n$ times editions. Suppose that the buyer and the seller can draw lessons form current trade and may change his choose if his payoff is lower than average payoff.

We consider an action of market is all buyers finish a trade. So in an action of a market, every buyer must buy a software edition(If the buyer's requirement is not agree with a seller's, he will change to another seller. If the buyer's requirement is not agree with all sellers', he will change his requirement) and a seller's sales may be 0 or more than one. If they choose authorized editions, a buyer's payoff is

$$R_{Ab}=A_b \tag{1}$$

and a seller's payoff is

$$R_{As} = A_s \times m_1 , \tag{2}$$

where $m_1 = \dfrac{\Sigma p_{bi}(t)}{\Sigma p_{si}(t)}$. If they choose pirated editions, a buyer's payoff is

$$R_{Pb}=nP_b \tag{3}$$

and a seller's payoff is

$$R_{Ps} = nP_s \times m_2 , \tag{4}$$

where $m_2 = \dfrac{\Sigma(1-p_{bi}(t))}{\Sigma(1-p_{si}(t))}$.

If there is no punishment in the market, the payoffs of pirated software are must more. So a buyer's probability of choosing an authorized edition next time is

$$p_{bi}(t+1) = p_{bi}(t) - \Delta p_{bi} , \tag{5}$$

where $p_{bi}(t+1)) \in (0,1)$ and $\Delta p_{bi} > 0$ . $\Delta p_{bi}$ is the buyer's largest adjustment strength per time step. $-\Delta p_{bi}$ means the buyer tends to buy a pirated edition. And a seller's probability of providing an authorized edition next time is

$$p_{si}(t+1) = p_{si}(t) + F(\max\{R_{As}, nR_{Ps}\})\Delta p_{si} , \tag{6}$$

where $p_{si}(t+1) \in (0,1)$ , $\Delta p_{si} > 0$ and $F(\ ) \in \{-1,0,1\}$ . $\Delta p_{si}$ is the seller's largest adjustment strength per time step. $F(\ )$ is a function with three values -1, 0 and 1, $F(R_{As}) = 1$ , $F(nR_{Ps}) = -1$ , if $R_s = nR_{Ps}$ , $F(\ ) = 0$ . In most occasions $R_s < nR_{Ps}$ , that is $F(\max\{R_{As}, nR_{Ps}\}) =$ $F(nR_{Ps}) = -1$ , so $-\Delta p_{si}$ means the seller tends to sell a pirated edition.

In order to put an end to pirated software, punishment should be added. If the seller is punished by a penalty of *C*, then the seller who provides pirated software can get a payoff

$$R_{Ps} = nP_s \times m_2 - C\,. \tag{7}$$

The seller's $p_{si}(t+1)$ can still be described by (6). Generally this punishment mechanism will decline the amount of sellers who sell pirated software. But the actual result is not perfect because of the competition relations between sellers. If the buyers are punished by a least penalty of confiscation the pirated editions, the payoff of the buyer will be

$$R_{Pb} = -nP\,, \tag{8}$$

where *P* is the price of pirated software. In addition, we can add extra penalties if we can not punish every trade of pirated software. Now $R_{Pb} < R_{Ab}$, so a buyer's probability of choosing an authorized edition next time is

$$p_{bi}(t+1) = p_{bi}(t) + \Delta p_{bi}\,, \tag{9}$$

and buyers will turn to authorized software because of the punishment.

## 3 Result and Discussion

We use a large market and suppose there are $n_1$=$10^5$ buyers and $n_2$=$10^3$ sellers, $A_s$=10, $A_b$=10, $P_s$=5, $P_b$=50, $P$=5, $n$=1, $\Delta p_{bi} = 0.01$ and $\Delta p_{si} = 0.01$. We use the above model and imitate the trends of the market in next cases.

### 3.1 Condition without Punishment

Now we discuss an orderly market with little pirated software. We do not inflict any penalties on buyers or sellers. At the beginning, all buyers choose authorized editions with firm convictions( $p_{bi} = 1$ ), most sellers(90%) provide authorized editions with firm convictions( $p_{si} = 1$ ) and a part of sellers(10%) sell pirated software(their $p_{si} = 0$ ).

If there are no punishments, the market trend is shown in Fig. 1.

In Fig. 1, the vertical axis is the average probability of choosing an authorized edition, the horizontal axis illustrates the time, the "buyer" stands for the buyers' average probability( $p_b = \sum p_{bi}(t)/n_1$ ) of choosing an authorized software and the "seller" stands for the sellers' ( $p_s = \sum p_{si}(t)/n_2$ ). The buyers' trend follows (5) and the sellers' trend follows (6). The buyers will trend to pirated software because of its high payoff. Even if there is one seller provides pirated software, all buyers will only choose pirated software. The sellers' average $p_s$ rises at first and drops down fast then, till all sellers provide pirated software. At the beginning, the market is orderly. Few buyers choose pirated software and the market of pirated software is little. In this condition, the payoff of selling pirated software is less than the payoff of selling authorized software, so a part of sellers turn to authorized software again. With the amount of choosing pirated software buyers increasing, pirated software market grows. Then the amount of providing pirated software sellers increases till all sellers only provide pirated software.

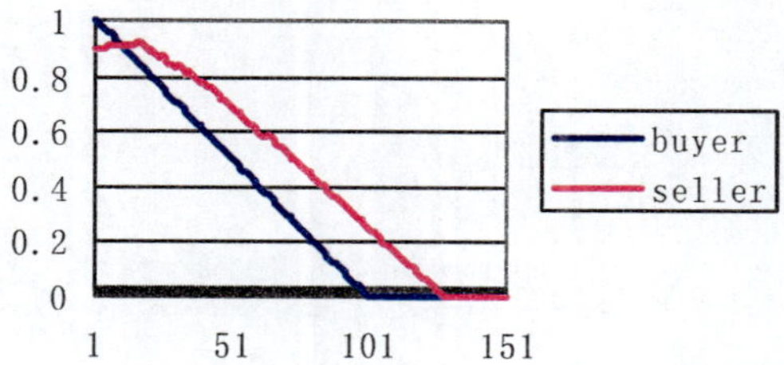


**Fig. 1.** Market trend without punishment

So in a market without punishment, once a seller provides pirated software, the whole market will chooses pirated software only.

### 3.2 Condition with Punishment on Sellers

At the beginning, we suppose the condition is the same as part A. If there is no punishment, all buyers and sellers will choose pirated software. If don't punish pirated software buyers and there are sellers providing pirated software, the buyers' average $p_b$ will follow (5) and be the same as Fig. 1. Fig.2 illustrates the sellers' average $p_s$ when the penalty is 0, 1000, 2000 or 10000.

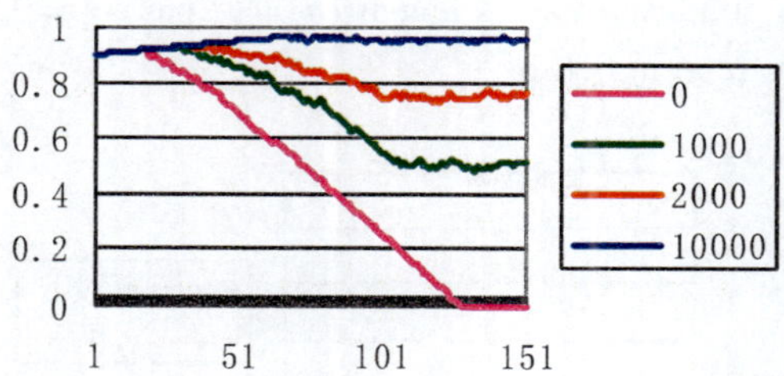


**Fig. 2.** Market trend with punishments on sellers

From (6) we know that when the market reaches a dynamic equilibrium[5]

$$p_{si}(t+1) = p_{si}(t) . \tag{10}$$

Then we can have next derivative process,

$$F(\max\{R_{As}, R_{Ps}\}) = 0 , \tag{11}$$

$$R_{As} = R_{Ps} , \tag{12}$$

$$A_s \times m_1 = nP_s \times m_2 - C , \tag{13}$$

where $m_1 = \frac{\Sigma p_{bi}(t)}{\Sigma p_{si}(t)}$, $m_2 = \frac{\Sigma(1-p_{bi}(t))}{\Sigma(1-p_{si}(t))}$, $n$=1。 At last $p_{bi}(t)=0$ (see Fig. 1), so

$$P_s \times m_2 = C, \tag{14}$$

$$m_2 = C/P_s, \tag{15}$$

$$\frac{n_1}{n_2(1-p_s)} = C/P_s, \tag{16}$$

$$p_s = 1 - \frac{n_1}{n_2} \cdot \frac{P_s}{C}. \tag{17}$$

If $P_s$=5 and $C$=1000, $p_s = 0.5$; If $P_s$=5 and $C$=2000, $p_s = 0.75$; If $P_s$=5 and $C$=10000, $p_s = 0.95$ (see Fig.2). That is even if the penalty is very high(10000 is greater than the average payoff of selling pirated software which is $5\times10^5/10^3$=500), it can not make all sellers provide authorized editions and there are still a few sellers(when $C$=10000, it is about 5%) providing pirated software.

So we can conclude that it is not a practicable plan with punishments only on sellers.

### 3.3 Condition with Punishment on Buyers

If the buyer's pirated software will be expropriated, the payoff $R_{Pb}$ will be less than $R_{Ab}$ and the buyer's average $p_b$ will follow (9) till $p_b = 1$.

First we start with condition like A and B( $p_b = 1$ and $p_s = 0.9$ ). Follow (6) and (9), we get the market trend in Fig. 3.

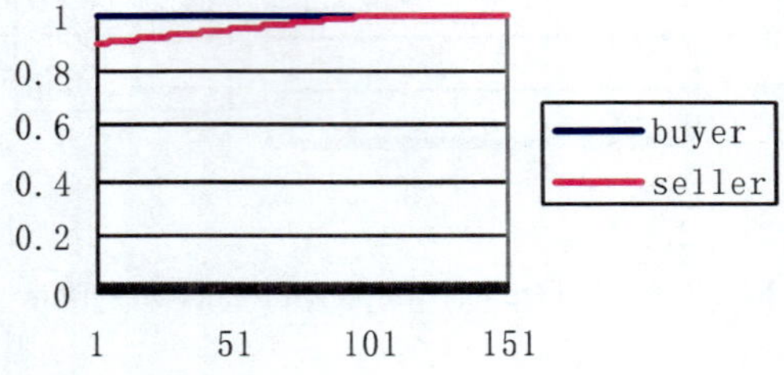


**Fig. 3.** Order market trend with punishments on buyers

Because of the punishment, the buyer's average $p_b$ does not decline. The seller's sales of pirated software will be 0 and he has to turn to authorized editions. So the seller's average $p_s$ also increases monotonically. We can see that it becomes an easy problem to prevent software piracy when we inflict punishment on buyers.

Then we start with a bad condition( $p_b = 0$ and $p_s = 0.9$ ). Follow (6) and (9), we get the market trend in Fig. 4.

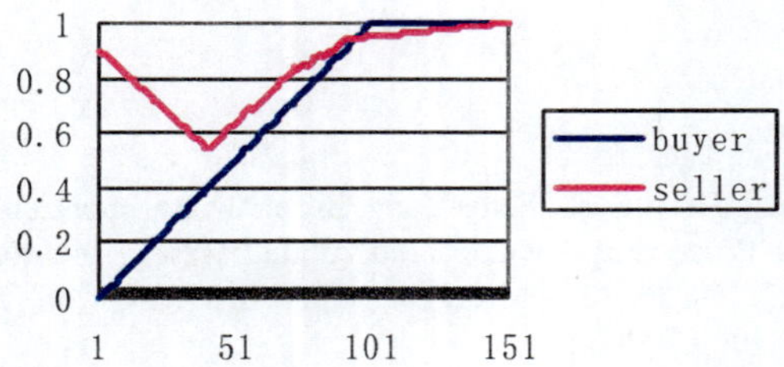


**Fig. 4.** Chaotic market trend with punishments on buyers

At first it is a big pirated software market and most buyers choose pirated software. So a part of sellers turn to pirated software for more payoffs. With the amount of the pirated software buyers declining, the sellers have to provide authorized software again. This time punishment on buyers also works well.

So it is a practical plan to inflict punishment on buyers.

### 3.4 Condition with Punishment on Buyers and Sellers

We use the condition like Fig. 4. If we punish the two sides, the market trend will be in Fig. 5(the penalty of seller is 1000).

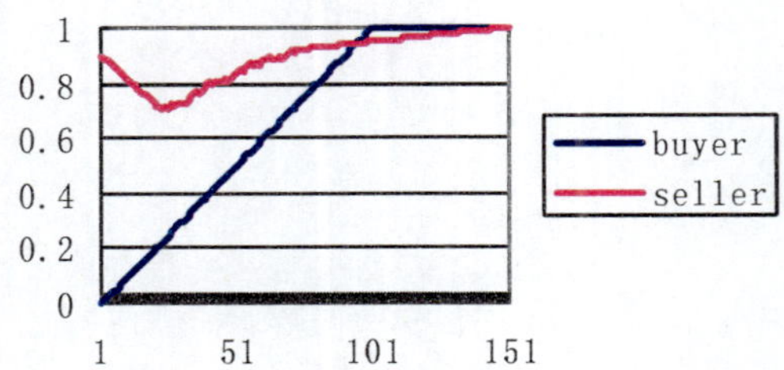


**Fig. 5.** Market trend with punishments on two sides

In Fig. 5 we can see the decline of the sellers' average $p_s$ is less than in Fig. 4. So punishment on two sides may help the market become orderly rapidly.

## 4 Conclusion

From above simulations, we can conclude that the punishment on sellers is less powerful than we expected. Even with high penalty, a few sellers will provide pirated software. And the punishment on buyers works well. It is an easy way to make all market orderly. So if we want to prevent the software piracy completely, we should not forget to put the punishment on pirated software buyers.

**Acknowledgment.** We acknowledge the support from Nantong Application Research Plan(No. K2009045) and Nantong University Natural Science Foundation (No. 09Z026).

## References

1. Tung-Ching, L., Meng Hsiang, H., Feng-Yang, K., et al.: An intention model-based study of software piracy. In: Proceedings of the 32nd Annual Hawaii International Conference on Systems Sciences, HICSS-32. Abstracts and CD-ROM of Full Papers (1999), doi:10.1109/HICSS.1999.772932
2. Douglas, D.E., Cronan, T.P., Behel, J.D.: Equity perceptions as a deterrent to software piracy behavior. Information & Management 44(5), 503–512 (2007)
3. Yang, D., Sonmez, M., Bosworth, D., et al.: Global Software Piracy: Searching for Further Explanations. Journal of Business Ethics 87(2), 269–283 (2009)
4. Mishra, A., Akman, I., Yazici, A.: Software piracy among IT professionals in organizations. International Journal of Information Management 26(5), 401–413 (2006)
5. Xu, C., Hui, P.M., Yu, Y.Y., et al.: Self-organized cooperative behavior and critical penalty in an evolving population. Physica a-Statistical Mechanics and Its Applications 388(20), 4445–4452 (2009)
6. Rousseau, J.J., Cranston, M.W.: Discourse on Inequality. Penguin Group (1984)
7. Myerson, R.B.: Refinements of the Nash equilibrium concept. International Journal of Game Theory 7(2), 73–80 (1978)

# Two Propositions about Geographical Distributions of BitTorrent File Resources*

Ming Chen, Lidong Yu, and Huali Bai

Institute of Command Automation
PLA University of Science and Technology
Nanjing, China
{yu.lidong,mingchennj,bhl0916}@163.com

**Abstract.** Recently, much attention has been paid to applying underlay information in optimizing BitTorrent (BT) systems. However, most of these approaches have taken such an assumption that BT file resources are distributed uniformly on the earth, which directly results in performance degradation on BT. In this paper, we study the geographical distribution of BT file resources. By measuring and analyzing BT systems, BT file resources are found to be non-uniformly distributed both in country level and AS level. Consequently, two propositions about the characteristics of geographical distribution of BT files resources are derived. These propositions overthrow the foundation based on which many P2P locality-based algorithms used to optimize cross ISPs traffic, i.e., BT file resources were thought to follow uniform distributions geographically. Finally, a general and adaptive traffic optimizing algorithm called GeoDTO is proposed and analyzed.

**Index Terms:** BitTorrent, file resource, geographical distrib-ution, locality, network measurement.

## 1 Introduction

BitTorrent (BT) [1] is a popular file resources sharing system. Due to its ignorance of the underlying topology of Internet and link costs of Internet Service Provider (ISP), one may choose some remote peers as its neighbors in BT, which potentially results in huge cross ISPs traffic in the Internet and increments of the operating costs of ISPs [2],[3].

At present, underlying network information [2]-[8] were widely used to localize P2P traffic to decrease consumptions of the network resources. However, among these optimizing algorithms of BT, a precondition was supposed that BT file resources are distributed uniformly on the earth. If this assumption does not correct, i.e., BT file resources are distributed in some specific areas following certain rules. Once there are a crowd of peers nearby, it is abundant excellent file resources that can be selected by P2P users. So, in this scenario, the locality-based strategies can be successful. Contrarily, it is hard to optimize traffic in the scenarios of few peers

* This work was partly supported by National High Technology Research and Development Program of China (No.2007AA01Z418) and the Natural Science Foundation of Jiangsu Province (No.BK2009058).

nearby. Therefore, it is necessary to study what geographical characteristics of BT file resources worldwide are.

## 2 Geographical Characteristics of BT File Resoures

It is convenient for discussion to define the following terms.

**Definition 1 File Resources** are defined as the shared files interested by peers in BT systems, which consist of many attributes such as filename, author, size, language, storage location, and format. In fact, only seeds have the whole parts of the file, while most P2P users in the system have parts of it.

**Definition 2 Geographical Distributions of File Resources** are defined as the distributions of physical storage locations of BT file resources worldwidly, which can be represented by the information such as country/city, the longitude and latitude, or the ISP/AS. In our paper, we approximatively assume that the geographical distributions of the file resource are equal to the distribution of the hosts storing the file resources.

**Definition 3 Native Countries of File Resources** are defined as the countries which take the language of BT file resources as their official language. People are usually interested in those files, the languages of which are comprehensible.

To explore the geographical distributions of BT file resources, we first collect lots of data by network measurement, then analyze the statistical data thoroughly, and finally derive the rules of the geographical distributions of BT file resources.

The detailed processes are described as follows.

1) By extending Vuze client [10], a plugin called gVuze is developed. gVuze is configured to retrieve the list of peers' IP addresses from tracker servers in every 20s.

2) To observe the distribution rules of the file resources in detail, from August to December 2009, a total of four months, about 600 popular video .torrent files are investigated. In the environment of China Education and Research Networ (CERNET), each .torrent file is added into gVuze clients and all peers' IP address lists returned from tracker server and other peers are recorded during the whole observing period. During the whole experiments, all IP addresses are stored in the local hash table. Considering the copyright issues, we do not download any file data during the whole experiment.

3) For the 600 sets of peers' IP address collected, the following processing methods are applied: (1) The country information of each peer's IP address is obtained by using GeoIP API functions from Maxmind [11] to access GeoLite Country database. (2) The AS number corresponding to each IP address is obtained by analyzing Prefix-to-AS mapping table from iPlane [12].

### 2.1 Geographical Distributions of Files Resources in Country Level

In the experiments, among all the 600 files, there are 280 Chinese files, 90 Japanese files and 230 English files. By our analysis, we find the following interesting phenomena: there are 90% ~ 95% of IP addresses belonging to China, 92% ~ 98% of IP addresses belonging to Japan, and 85% ~ 90% of IP addresses belonging to Europe and North America in the swarms of Chinese, Japanese and English files respectively. Apparently, the language types of file resources have strong correlations to the geographical distributions of BT file resources.

According to whether the languages of file resources are coincident with their official languages of countries in which peers reside, we classify file resources into two categories: native and non-native countries' file resources. As shown in Fig. 1 (a), more than 90% file resources locate in their native countries and 8.7% of the total resources distribute in their non-native countries. Furthermore, according to the size of peers belonging to various countries, we rank the top 3 countries in the swarms of Chinese files in Fig.1 (b), where there are about 90.8% of IP addresses reside in China, 3.2% belonging to Europe, 2.2% belonging to America, and 3.8% belonging to other countries. Statistically, about 9% of the total IP addresses located in non-native countries, which might be relevant to Chinese immigrants.

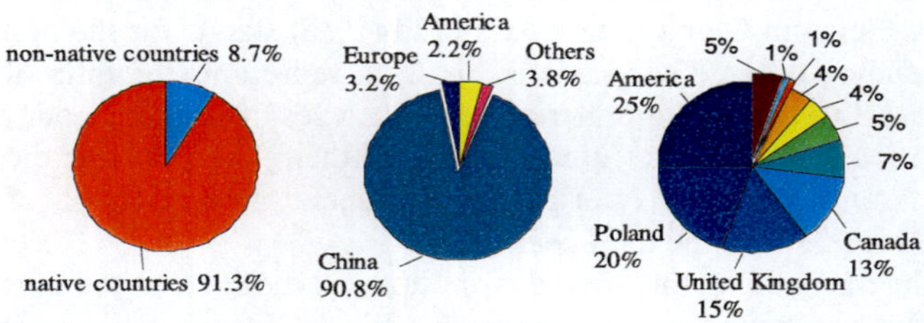


**(a) Total files** **(b) Chinese files** **(c) English files**

**Fig. 1.** Distributions of file resources in the interesting area

Fig. 1(c) shows the file resources are distributed in top 10 countries in the swarms of the English files, which show as follows: America occupies 25%, Poland 20%, United Kingdom 15%, Canada 13%, Australia 7%, Norway 5%, India and Holland 4% respectively, China and Malaysia 1% respectively. Statistically, there are 93% of the IP addresses located in their native countries such as North America, Europe and Asia etc, where English is their official language. While only 7% of the IP addresses locate in their non-native countries. The results must be relevant to the fact that English is one of the most popular languages worldwide.

According to our experiments, BT file resources are highly located in their native-countries, and the language of file resources is one of the most important factors affecting their geographical distributions. Thereat, Proposition 1 is derived.

**Proposition 1.** *File resources in BT systems are non-uniform distributed geographically worldwidely, i.e., majority of file resources are distributed in their native countries, while less in their non-native countries.*

## 2.2 Geographical Distributions of File Resources in as Level

To better understand the geographical distributions of BT file resources in finer granularity, the geographical distributions of BT file resources in AS level in their native countries is investigated further.

Since the Internet is divided into thousands of distinct regions of administrative domain, each of which possesses one or several ASes. Each AS in Internet is represented by a unique 16-bit AS number and administered by one ISP. Intuitively, the distributions of BT file resources in AS level should be related to some factors such as the degree of ISPs' support for P2P technology, the preference of P2P users, and issues of politics and copyrights. Supposed that P2P users have the same preference of file resources in their native-countries, the more users belong to an ISP/AS, the higher probability of file resources are distributed in this ISP/AS.

Taking Chinese files and English files for example, we first prefer the IP addresses belonging to China and America respectively. And then, we group the IP addresses into clusters according to their AS locations. The proportion between the number of P2P users in each AS and the number of the total system users is illustrated in Fig. 2(a), 2(b), 2(c) and 2(d) respectively.

Fig.2 (a) and Fig.2 (b) show the geographical distributions of Chinese file resources in AS level in China. The x-axis in Fig.2 (a) stands for the rank of ASes in China in descending order. The y-axis in Fig.2 (a) represents the ratio of the number of peers in the AS to the total number of peers in ASes of China. In our experiments, there are 60 ASes belonging to China. Among of them, AS4134 has the most peers, which accounts for 43% of the total peers. The second most peers of AS is AS4837, which accounts for 23% of the total peers.

Fig.2 (b) illustrates the cumulative distributiosn of the ratio of the number of peers in the AS to the total number of the peers in China. Clearly, the ratio about 80% of ASes in China is approximately 0, and there are more than 80% of the total peers locating in less than 10% of the ASes of China. According to the number of peers in the AS from our experiments, Table I presents the rank of the top 10 ASes in term of the number of peers belonging to China in the swarms of Chinese files.

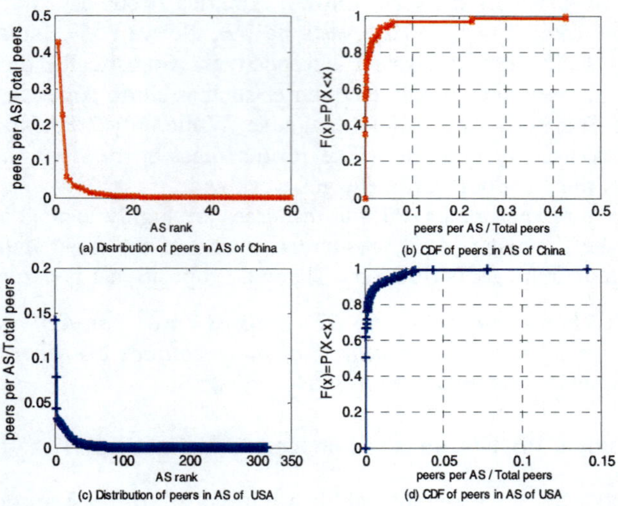


**Fig. 2.** Distributions of BT interesting file resources in ASes of China and US

The "AS rank globally" in Table I is obtained from CAIDA project [13], the goal of which is to provide the ISP ranking based on business relationships between ASes. Their approach is rooted in economic AS relationships, ranking each AS as a function of the number of IP prefixes advertised by this AS. The AS ranking results in Table 1 is derived by the four metrics in the following order: the number of /24s in their customer cluster, the number of prefixes in their customer cluster, the number of ASes in their customer cluster; and the AS degrees. Interesting enough, the AS rank of CAIDA is surprisingly in accord with the rank of the number of Chinese P2P users in the ASes of China in the rough except little exceptions.

Similarly, Fig.2 (c) and Fig.2 (d) illustrate the geographical distributions of English file resources in AS level in United States. There are 315 ASes belonging to US in our experiments. Among of them, the largest AS having the most peers, which accounts for 14% of the total peers. The second largest AS accounts for 8% of the total peers. Fig.2 (d) gives the cumulative distribution of the ratio of the number of peers in the AS to total number of the peers in US. We also surprisingly find the similar results that the ratio about 80% of ASes in US is approximately 0, and there are more than

**Table 1.** Top 10 ASes in China

| Rank | AS Number | Peers | AS Name/ISP | AS rank globally[13] |
|---|---|---|---|---|
| 1 | 4134 | 25715 | CHINANET-BACKBONE | 122(CN) |
| 2 | 4837 | 13768 | CHINA169-BACKBONE CNCGROUP China169 Backbone | 165(CN) |
| 3 | 4538 | 3843 | ERX-CERNET-BKB China Education and Research Network Center | 199(CN) |
| 4 | 4808 | 3073 | CHINA169-BJ CNCGROUP IP network China169 Beijing Province Network | 240(CN) |
| 5 | 9394 | 1863 | CRNET CHINA RAILWAY Internet(CRNET) | 224(CN) |
| 6 | 4812 | 1857 | CHINANET-SH-AP China Telecom (Group) | 232(CN) |
| 7 | 3462 | 1752 | HINET Data Communication Business Group | 227(TW) |
| 8 | 17816 | 1225 | CHINA,169-GZ CNCGROUP IP network China,169 Guangzhou MAN | 409(CN) |
| 9 | 4760 | 975 | HKTIMS-AP PCCW Limited | 580(HK) |
| 10 | 17785 | 962 | CHINATELECOM-HA-AS-AP asn for Henan Provincial Net of CT | 712(CN) |

**Table 2.** The Top 10 ASes in US

| Rank | AS Number | Peers | AS Name/ISP | AS rank globally[13] |
|---|---|---|---|---|
| 1 | 209 | 10905 | Qwest Communications Corporation | 2(US) |
| 2 | 7132 | 6099 | AT&T Internet Services | 31(US) |
| 3 | 22773 | 3126 | Cox Communications Inc. | 50(US) |
| 4 | 19262 | 2508 | Verizon Internet Services Inc. | 201(US) |
| 5 | 20115 | 2347 | Charter Communications | 166(US) |
| 6 | 6389 | 2228 | BellSouth.net Inc. | 176(US) |
| 7 | 11426 | 2207 | Road Runner HoldCo LLC | 250(US) |
| 8 | 12271 | 1306 | Road Runner HoldCo LLC | 394(US) |
| 9 | 33287 | 1301 | Comcast Cable Communications, Inc. | 386(US) |
| 10 | 33651 | 1164 | Comcast Cable Communications, Inc. | 443(US) |

70% of the total peers locating in less than 10% of the ASes of US. Table 2 presents the rank of the top 10 ASes in term of the number of peers belonging to US in the swarms of English files.

As shown in Table 2, the AS rank of CAIDA is approximately in accord with the rank of the number of the peers in the ASes of US. There is an exception that the number of peers in the ASes ranged from the rank 3 to the rank 10 is roughly same.

In summary, BT file resources have strong locality features in AS level, i.e., the higher the rank of ASes, the more file resources. Consequently, the rank of the ASes is another important factor affecting the geographical distributions of BT file resources worldwide. So, we derive Proposition 2.

**Proposition 2.** *In their native-countries, file resources in BT systems are non-uniform distributed geographically as well. The number of file resources in their ASes is in direct proportion to the global ASes ranking approximately and the number of P2P users in their AS.*

However, we also note that some ASes having higher rank are not presented in Table 1 and Table 2. It is possible that some of these ASes taking charge the inter-connections of Internet backbone have few P2P users. Or some P2P applications are blocked by the ASes. Still lots of P2P users are connected to Internet via limited wireless bandwidth provided by the ASes. The actual reasons will be explored in our future work.

Proposition 1 and Proposition 2 imply that there might exist some faults when executing recent traffic locality-based strategies [2~8], if few peers nearby.

## 3 When Executing Traffic Locality-Based Strategies

Consequently, we will discuss the effects of characteristics non-uniform geographical distributions of BT file resources on traffic optimization in the Internet.

**Scenario 1:** When a user locates in the non-native country of file resources, there may be few peers nearby. So country-locality based or ISP/AS-locality based strategies [3] may not only degrade the performance of BT system, but also make the BT systems don't work. In such scenario, the performance of locality-based optimizing strategies may degrade the performance of original BT.

**Scenario 2:** When a user locates in the non-native country, but it belongs to the AS having a lower rank globally. Due to few peers nearby, ISP/AS-locality based strategies may not improve the performance of BT system, and still worse, make BT system don't work either.

**Scenario 3:** When a user locates in the native country of file resources, and it belongs to the AS having a prior rank. Due to enough peers locate in the AS, ISP/AS-locality based strategies can not only improve the performance of BT system but also decrease traffic across ISPs dramatically. In this scenario, the number of peers within the AS is so large enough that there is no need to explore more peers in extended region, which may results in unnecessary traffic.

According to the proposition 1 and proposition 2, the scenario 3 is a more general case that majority of users, located in certain large ASes in their native countries, are interested in the file resources of native language. However, most of recent locality-based mechanisms [2]~[8] have paid less attention to the scenario 1 and scenario 2, which directly results in poor performance of BT. So it is necessary to develop an adaptive mechanism adapted in most scenarios to optimize traffic in locality-based fashion.

Fig.3 gives the flow chart of our traffic optimizing algorithm called GeoDTO (Geographical Distribution-based Traffic Optimizing).

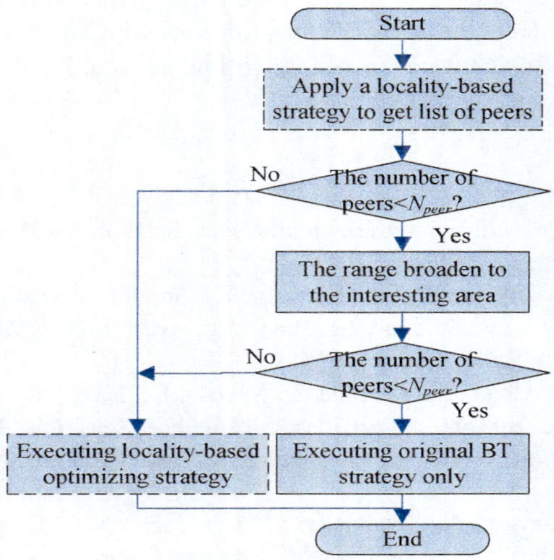


**Fig. 3.** Flow chart of GeoDTO

GeoDTO is an adaptive and simple algorithm utilizing the geographical distribution characteristics of BT file resources. Firstly, GeoDTO applies certain traffic locality-based strategy to get a list of peers, and then count the number of peers within the region. If the number of the peers in this set is less than $N_{peer,}$ the threshold of the peer number, the region is widened to its native country; otherwise, going on to execute the traffic locality-based strategy. And then, only the original BT strategy is adopted if the number of the peers in this set is still less than $N_{peer}$. For supplement, in the first step of GeoDTO, the locality-based strategy can be any one of existing locality-based optimizing algorithms[2]-[6].

Obviously, GeoDTO is a more general algorithm extending the range of applying various locality-based optimizing traffic mechanisms in most scenarios. It will loose the foundation of locality-based optimizing traffic if the number of peers is too small, so that it is necessary to broaden the range to their native countries. The best choice is only the original BT strategy if the number of peers still too small. Therefore, using GeoDTO and given the threshold $N_{peer}$, the general optimizing traffic algorithm base on the geographical distribution characteristics can be used without putting one's foot in troubles. Recent analysis and experiments show us that the threshold $N_{peer}$ may be chosen as 24, which will be discussed in detail later.

## 4 Conclusions

By measuring BT systems and analyzing the data, the non-uniform geographical distribution characteristics of BT file resources both in country and AS level are discovered, and two Propositions are derived. These propositions overthrow the foundation based on which many P2P locality-based algorithms used to optimize traffic. A general and adaptive traffic optimizing algorithm called GeoDTO is proposed. Further work includes the Propositions will be verified in more situations worldwide and the performance of GeoDTO will be analyzed.

## References

1. Cohen, B.: Incentives build robustness in bittorrent. In: Proc. 1st Workshop on Economics of Peer-to-Peer Systems, P2PEcon (2003)111
2. Karagiannis, T., Rodriguez, P., Papagiannaki, K.: Should internet service providers fear peer-assisted content distribution? In: Proc. 5th ACM SIGCOMM Conf. on Internet Measurement (IMC), Berkeley, CA (2005)
3. Bindal, R., Cao, P., Chan, W., Medved, J., Suwala, G., Bates, T., Zhang, A.: Improving Traffic Locality in BitTorrent via Biased Neighbor Selection. In: Proc. 26th IEEE International Conf. on Distributed Computing Systems (ICDCS ), Lisboa, Portugal (July 2006)
4. Aggarwal, V., Feldmann, A., Scheideler, C.: Can ISPs and P2P Systems Cooperate for Improved Performance? ACM SIGCOMM Computer Communications Review 37(3), 29–40 (2007)
5. Xie, H., Yang, Y.R., Krishnamurthy, A., et al.: P4P: Provider Portal for Applications. ACM SIGCOMM Computer Communications Review, 351–362 (August 2008)
6. Choffnes, D.R., Bustamante, F.E.: Taming the Torrent: A Practical Approach to Reducing Cross-ISP Traffic in Peer-to-Peer Systems. ACM SIGCOMM Computer Communications Review, 363–374 (August 2008)
7. Agarwal, S., Lorch, J.R.: Matchmaking for Online Games and Other Latency-Sensitive P2P Systems. ACM SIGCOMM Computer Communications Review, 315–326 (August 2009)
8. Wang, H., Liu, J., Xu, K.: On the Locality of BitTorrent-based Video File Swarming. In: Proc. International Workshop on Peer-to-Peer Systems (IPTPS ), Boston, MA (April 2009)
9. Mieghem, P.V.: Performance Analysis of Communications Networks and System, pp. 267–270. Cambridge University Press, New York (2006)
10. Azureus BitTorrent Client, `http://www.azureus.com/`
11. Maxmind, MaxMind GeoIP Country® Database, `http://www.maxmind.com/app/country`
12. iPlane: An Information Plane for Distributed Service, `http://iplane.cs.washington.edu/data/origin_as_mapping.txt`
13. The Cooperative Association for Internet Data Analysis. Introduction to Relationship-based AS Ranking, `http://www.caida.org/research/topology/rank_as/`

# The Strategic Decisions and Evaluation System of Chinese Logistics Agency Enterprises in Green Logistics Development

Xu Deling and Geng Fengchun

Business Management and Strategy Development Institute
Tianjin Normal University
Tianjin, China
tgxdl@163.com, gengfengchun@126.com

**Abstract.** Green logistics is an inevitable trend of international economic development. But Chinese logistics agency enterprises in green logistics development are restricted by external environment and internal condition. With SWOT analysis, the paper analyses the opportunities, threats, strengths and weaknesses of Chinese logistics agency enterprises in green logistics development. On the basis, the paper puts forward the strategic decisions of Chinese logistics agency enterprises in green logistics development, and puts forward the corresponding countermeasures in both hardware and software and evaluation system supporting the strategic decisions.

**Keywords:** green logisticst, logistics agency, strategye, countermeasure.

## 1 Introduction

With the development of international economy, green logistics has drawn attention of international logistics agency enterprises. It is important for China's logistics agency enterprises to decide strategies to develop green logistics. In order to China's logistics agency enterprises to achieve the strategies of green logistics, we must evaluate their behaviors to develop green logistics so that they improve international competitiveness.

## 2 Overview of the Research on Green Logistics

Early in 1990s, foreign scholars began to put forward the concept of green logistics (Green Logistics or Environmental Logistics) and began to research the theory of green logistics. Szymanklewicz,Jao(1993) proposed the enterprises must make the logistics environment green through survey of logistics enterprises[1].Remko I.vanHoek,(1999) pointed out that environmental management should be integrated into the business strategy from the view of the reverse logistics and green supply chain[2].Paul R.Murphy;Riehard F.Poist(2000) pointed out that enterprises should pay attention to environmental issues and develop a formal policy to protect environment from the view of green logistics strategy[3].Paul R.Murphy; Riehard F.Poist(2003) pointed out that environmental problems can affect many functions of the logistics by comparing logistics situation of the U.S enterprises and non-US enterprises[5]. In the 21st century, Wang changqiong (2002),Chinese scholar, proposed business strategy should be combined with environmental protection[6].He Baocheng(2004) thought green logistics is the objective requirements of enterprises participating in international competition[7].Guo Qiang and Zeng Jiaoyan(2004) discussed the development of green logistics from the view of strategy and practice[8].Liu Na(2009) researched the development situation of China's green logistics and put forward the development strategy of green logistics suited to China by SWOT analysis[9].Zhao Lijun(2008) proposed and constructed a comprehensive benefit evaluation model of green logistics by analytic hierarchy process(AHP) and fuzzy and comprehensive evaluation[10].Shen Jianjun(2009) pointed out that establishment of principal-agent model can improve government effectiveness and the level of effort of enterprises implementing green logistics[12]. Based on these studies, we put forward the strategic decisions suited to China's logistics agency enterprises developing green logistics. In addition, we promote the corresponding countermeasures in both hardware and software and evaluation system supporting the strategic decisions.

## 3 SWOT Analyses of China's Logistics Agency Enterprises Developing Green Logistics

In resent years, China's logistics agency enterprises rapidly develop and have certain advantages and strength. In addition, China has strongly advocated the development of green logistics which provides external conditions for logistics agency enterprises.

### 3.1 Strength

- After development of many years, China's logistics agency enterprises have a solid foundation and advantages in hardware such as transportation, storage, network, packaging, distribution processing and resource recycling, which offers solid material foundation for the development of green logistics. For example, the containers transportation as modern transportation means has been widely used; large modern warehouse facility has been built; three-dimensional storage and automated machinery have also been widely adopted. These resources and technologies ensure green logistics in traditional green logistics agency enterprises.

- China's logistics agency enterprises have a solid foundation for green logistics in software including organization, information, human and culture. China's logistics agency enterprises have partly reorganized enterprises' structure, which saves the logistics resources and reduces pollution of environment. Green logistics technologies have been applied such as standardization of information, collection, process and integration of logistics information and automation equipment.

## 3.2 Weakness

- Small-scale enterprises and decentralized operation exist. In developed countries, logistics agency has been well developed and application of logistics agency in enterprises has been more than 75%.But it is still less than 30%. Because logistics agency needs many fixed capital, many logistics agency enterprises in China have strong risk to expand the scale and integrate competitors. It is difficult to create economies of scale.
- The ideas and knowledge are short and technique is backward. China's green logistics agency enterprises shortly know the ideas and knowledge of green logistics. Importantly, China's logistics agency enterprises mainly engage in traditional transportation and warehousing business. Many enterprises just begin to apply green technique. Compared with the international advanced enterprises, they are backward.
- Low level of information and shortage of resource integration are in many enterprises. Information resource integration needs technique, including network, communication software and hardware. However, in China, application of GPS and GIS in large-scale enterprises reaches 23%, in large-scale logistics enterprises only reaches 12.5%, and in small enterprises have almost not.

## 3.3 Opportunity

- China's good in economic environment. In 2009, when the global financial crisis emerges and national economy recess, many enterprises are out of the market. However, China's economy stayed up the backdrop of economy and maintained sustained growth, which provides good economy environment for logistics agency enterprises.
- Governments pay great attention to sustainable development and develop policies. Government adheres to sustainable development and "low-carbon economy". In 2010, "two sessions" puts the "green" on the new schedule. To limit the polluting of enterprises and encourage enterprises to save energy, government introduced a large number of laws and regulations on environmental protection.
- Adequate infrastructure exists. China improves many transportation modes such as railways, highways, airports, ports and pipelines. In addition, China has built a large number of warehouses.
- Potential green logistics market is large. With the concept of sustainable development expanding, the awareness to protect environment is growing, and people strongly concern enterprises' environmental reputation and green products and services.
- The advanced experiences of foreign logistics agency enterprises provide support. Logistics agency enterprises in U.S, Europe and Japan have advanced techniques and concepts to develop the green logistics. China should learn their valuable experience.

### 3.4 Threats

- The concept of green logistics has not spread. In the golden age of economic growth, Chinese mostly concern the economic development and life improvement, so popularization of green logistics has to be improved.
- The foreign logistics agency enterprises threat our ones. With the opening of China's logistics industry, a large number of foreign logistics enterprises come to China. Their advanced logistics techniques and management threat China's logistics agency enterprises.
- Green logistics policies and regulations are incomplete. Since 90s in the 20th century, China has been formulated policies and regulations of environmental protection. But the policies and regulations in the logistics industry are not perfect and are badly caught out.

### 3.5 The Strategic Decisions of China's Logistics Agency Enterprises Developing Green Logistics

Based on the above analysis, we construct the SWOT analysis matrix and make strategic decisions.

**Table 1.** Swot Analyses

| **SWOT analysis of China's logistics agency enterprises developing green logistics** | **Strength**<br>1.The hardware has a solid foundation such as transportation, storage, network, package, distribution process and resources recycling<br>2.The software has a solid foundation such as organization, information, human and culture | **Weakness**<br>1. Enterprises' small-scale and decentralized operation<br>2. The ideas and knowledge are short and technique is backward<br>3. Low level of information and shortage of resource integration in many enterprises |
|---|---|---|
| **Opportunity**<br>1. China's good economic environment<br>2. Government's emphasis on sustainable development and support of policies<br>3. Adequate infrastructure<br>4. Potential green logistics market is large 5. The advanced experience of foreign logistics agency enterprises | **SO Strategy**<br>1.Make full use of China's logistics infrastructure and integrate external resources to develop green logistics<br>2.The enterprises use own hardware and software to develop green logistics<br>3.Innovatively use the foreign advanced experience to expand green logistics market | **WO Strategy**<br>1.Expand the business scale, increase the green investment and improve inter-firm cooperation<br>2. Promote the concept and knowledge of green logistics<br>3.Develop techniques and management of green logistics enterprises<br>4. Study foreign advanced experiment |
| **Threat**<br>1. The concept of green logistics has not spread<br>2. Threat of foreign logistics agency enterprises<br>3. Green logistics policies and regulations are incomplete | **ST Strategy**<br>1.Subdivide logistics market and open the new areas of green logistics<br>2.Expand enterprises own advantages and meet the new challenges | **WT Strategy**<br>Develop own international competitiveness, promote the culture of green logistics and make logistics system perfect |

## 4 Green Logistics Valuation Index System of Logistics Agency Enterprises Developing Green Logistics

To support and evaluate strategic decisions of China's logistics agency enterprises developing green logistics, we design and construct the following evaluation index system by learning foreign and domestic scholars' researches.

**Table 2.** Evaluation Index System

| Target layer | Criteria layer | Indicator layer |
|---|---|---|
| Strategic evaluation level of green logistics (A) | 1 Green logistics comprehensive strength $B_1$ | C11Capital<br>C12Capital profit margins<br>C13 Green investments (hardware and software) accounted for the proportion of capital<br>C14 Utilization level of alternative sources<br>C15 Logistics network, information, e-degree<br>C16 The extent of promotion of green culture and system |
| | 2 Green transport distribution $B_2$ | C21 Complex container utilization rate<br>C22 The proportion of the common distribution<br>C23 Vehicle exhaust emission rate to the standard<br>C24 Actual loading rates of vehicles<br>C25 Unit energy consumption of transport |
| | 3 Green package and storage $B_3$ | C31 The use proportion of new packaging materials<br>C32 The average cargo storage cycle<br>C33 Package recycle rate<br>C34 The level of packages modeling and reducing |
| | 4 Green distribution process $B_4$ | C41 The proportion of centralized processing<br>C42 Comprehensive utilization rate of raw materials |
| | 5 Reverse logistics and waste logistics $B_5$ | C51 Utilization rate of renewable resources<br>C52 Waste concentrated process and recycling rate<br>C53 Regeneration rate of returned product |

## 5 Strategic Evaluation Level of Green Logistics

We determine the weight of evaluation index and evaluate the strategic level of China' logistics agency enterprises by applying analytic hierarchy process and fuzzy comprehensive evaluation method.

*A. Determine the weight of evaluation index*

The template is designed so that author affiliations are not repeated each time for multiple authors of the same affiliation. Please keep your affiliations as succinct as possible (for example, do not differentiate among departments of the same organization). This template was designed for two affiliations.

*1) Construct comparison matrix*

The following comparison matrix is judged by the experts.

*a) Layer criteria to judge the target layer matrix (A-B):*

| A-B | B1 | B2 | B3 | B4 | B5 |
|---|---|---|---|---|---|
| B1 | 1 | 2 | 3 | 3 | 6 |
| B2 | 1/2 | 1 | 2 | 3 | 4 |
| B3 | 1/3 | 1/2 | 1 | 1 | 3 |
| B4 | 1/3 | 1/3 | 1 | 1 | 2 |
| B5 | 1/6 | 1/4 | 1/3 | 1/2 | 1 |

*b) Layer indicators to determine the criteria matrix (B-C):*

| B1-C | C11 | C12 | C13 | C14 | C15 | C16 |
|---|---|---|---|---|---|---|
| C11 | 1 | 3 | 1 | 8 | 2 | 5 |
| C12 | 1/3 | 1 | 1/3 | 5 | 6/7 | 5/4 |
| C13 | 1 | 3 | 1 | 6 | 3 | 4 |
| C14 | 1/8 | 1/5 | 1/6 | 1 | 2/5 | 4/3 |
| C15 | 1/2 | 7/6 | 1/3 | 5/2 | 1 | 3 |
| C16 | 1/5 | 4/5 | 1/4 | 3/4 | 1/3 | 1 |

| B2-C | C21 | C22 | C23 | C24 | C25 |
|---|---|---|---|---|---|
| C21 | 1 | 4/3 | 9/7 | 5/6 | 5/3 |
| C22 | 3/4 | 1 | 7/6 | 1 | 5/7 |
| C23 | 7/9 | 6/7 | 1 | 5/6 | 1 |
| C24 | 6/5 | 1 | 6/5 | 1 | 5/3 |
| C25 | 3/5 | 7/5 | 1 | 3/5 | 1 |

| B3-C | C31 | C32 | C33 | C34 |
|---|---|---|---|---|
| C31 | 1 | 2 | 1 | 3/5 |
| C32 | 1/2 | 1 | 5/7 | 1/2 |
| C33 | 1 | 7/5 | 1 | 6/7 |
| C34 | 5/3 | 2 | 7/6 | 1 |

| B4-C | C41 | C42 |
|---|---|---|
| C41 | 1 | 6/7 |
| C42 | 7/6 | 1 |

| B5-C | C51 | C52 | C53 |
|---|---|---|---|
| C51 | 1 | 1/3 | 4/5 |
| C52 | 3 | 1 | 5/3 |
| C53 | 5/4 | 3/5 | 1 |

*2) The level of single-sort.*

AHP software used can be calculated in relation to the target second-tier and third tier layer relative to the weight of the second layer:

*c) The second level in relation to the target layer weights*

A-B : $\lambda_{11}=0.413$ $\lambda_{12}=0.266$ $\lambda_{13}=0.141$ $\lambda_{14}=0.120$ $\lambda_{15}=0.060$ $\lambda_{max}=5.059$ CI=0.015 RI=1.120 CR=0.013<0.1 Therefore, through the consistency test.

*d) Compared to the third level of the weight of second-tier:*

$B_1$ -C : $\lambda_{111}=0.321$ $\lambda_{112}=0.118$ $\lambda_{113}=0.316$ $\lambda_{114}=0.042$ $\lambda_{115}=0.137$ $\lambda_{116}=0.066$ $\lambda_{max}=6.152$ CI=0.030 RI=1.240 CR=0.024<0.1 Therefore, through the consistency test.

$B_2$-C : $\lambda_{221}$=0.235 $\lambda_{222}$=0.180 $\lambda_{223}$=0.176 $\lambda_{224}$=0.236 $\lambda_{225}$=0.173 $\lambda_{max}$=5.073 CI=0.018 RI=1.120 CR=0.016<0.1 Therefore, through the consistency test.

$B_3$-C : $\lambda_{331}$=0.252 $\lambda_{332}$=0.157 $\lambda_{333}$=0.252 $\lambda_{334}$=0.339 $\lambda_{max}$=4.035 CI=0.012 RI=0.900 CR=0.013 <0.1 Therefore, through the consistency test.

$B_4$-C : $\lambda_{441}$=0.462 $\lambda_{442}$=0.538 $\lambda_{max}$=2.000 CI=0.000 RI=0.000 CR=0.000 <0.1 Therefore, through the consistency test.

$B_5$-C : $\lambda_{551}$=0.197 $\lambda_{552}$=0.524 $\lambda_{553}$=0.279 $\lambda_{max}$=3.015 CI=0.007 RI=0.580 CR=0.012<0.1 Therefore, through the consistency test.

*e) The level of total order:*

$\lambda_1$=0.133 $\lambda_2$=0.049 $\lambda_3$=0.130 $\lambda_4$=0.017 $\lambda_5$=0.057 $\lambda_6$=0.027 $\lambda_7$=0.063 $\lambda_8$=0.048 $\lambda_9$=0.047 $\lambda_{10}$=0.063 $\lambda_{11}$=0.046 $\lambda_{12}$=0.036 $\lambda_{13}$=0.022 $\lambda_{14}$=0.036 $\lambda_{15}$=0.048 $\lambda_{16}$=0.055 $\lambda_{17}$=0.065 $\lambda_{18}$=0.012 $\lambda_{19}$=0.031 $\lambda_{20}$=0.017

The total level of sort, there are: CI＝0.019 RI=0.972 CR=0.020<0.1 Therefore, through the consistency test.

According to AHP and the fuzzy comprehensive evaluation determining of weights of individual indicators, we can evaluate green logistics strategy of logistics agency enterprises. (Fuzzy comprehensive evaluation is omitted.)

## References


1. Szymankiewicz, J.: The logistics dilemma. Logistics Information Management 6(3), 36–44 (1993)
2. van Hoek, R.I.: From reversed logisties to green supply chains. Supply Chain Management 4(3), 129 (1999)
3. Murphy, P.R., Poist, R.F.: Green logistics strategies: An Analysis of usage pattems. Transportation Journal 40(2), 15–16 (2000)
4. Langnau, L.: A new shade of green for reverse logistics. Material Handling Management 56(3), 2 (2001)
5. Murphy, P.R., Poist, R.F.: Green perspective and practices: A "comparative Logistics", study. Supply Chain Management 8(2), 122–131 (2003)
6. Wang, C.: Study on green logistics management based on sustained development. Science & Technology Progress and Policy 2, 12–13 (2002)
7. He, B.: Green logistics: In the 21st century, the new trend of logistics development. Port Economy 5, 57–58 (2004)
8. Guo, Q., Zeng, J.: The macro-strategies and special practice of green logistics. Logistics Management (9), 69–71 (2004)
9. Na, L.: The SWOT analysis on the stratagem of Chinese green logistics development. Commercial Research (2), 175–178 (2009)
10. Zhao, L.-j.: Research on the evaluation of environmental logistics based on fuzzy comprehensive evaluation. Journal of Wuhan University of Technology (6), 174–177 (2008)

11. Hui, L.: Construction and perfecting of modern logistics agency system. China Logistics & Purchasing (6), 58 (2004)
12. Shen, J.-j.: An incentive mechanism study on green logistics based on principal-agent theory. Logistics Engineering and Management (3), 18–20 (2009)
13. Xu, X.: Construction of evaluation index system on Circular Economy and green logistics. Zhejiang Statistics (3), 22–24 (2003)
14. Sbihi, A., Eglese, R.W.: Combinatorial optimization and green logistics. Annals of Operations Research 175(1), 159–175 (2010)
15. Xu, L., Shang, J.-c., Wang, Y.-m.: Theories and methods of strategic environmental assessment of modern logistics development. Chinese Geographical Science 15(2), 145–150 (2005)

# Research on Methods of Optimal Road Transportation Routing for Dangerous Goods

Huo Hong[1] and Xiong Zhenyu[2]

[1] Harbin University of Commerce, Vice Director of Logistics Institute
Harbin, China
huohong1963@126.com
[2] Harbin University of Commerce, Postgraduate in Logistics Institute
Harbin, China
xzy-shz@163.com

**Abstract.** The transportation of dangerous goods is increasing greatly with the development of economy; the risks of it, to our human being's safety and health, are expanding, the consequence is very serious once the accident arises. So the problems that how to choose an optimal path for it to ensure transport safety and arrive at their destinations quickly need to solve urgently. The essay takes the special nature of dangerous goods into account, builds and optimizes a model, ultimately selects a transportation route of lower risk and cost.

**Keywords:** Transportation of dangerous goods, routing optimization, Risk Assessment.

## 1 Introduction

Dangerous Goods usually refers to something that is explosive, flammable, toxic, corrosive and radioactive chemicals as well as they are as a raw material made from a variety of products, they play an important role in the national economy.

However, the accidents of dangerous goods are different from the general cargo transport accidents. For example, in the process of road transportation, due to management reasons, human error, the deficiencies of vehicles, packaging, equipment and facilities, traffic and environment reasons, The holder that contains the combustive, explosive, toxic and dangerous goods may be broken or damaged because of the traffic accident, Leaking out the flammable, explosive, toxic chemicals, threatening to the residents, pedestrians, vehicles and polluting the air, water and soil, and even result in a broader range of casualties and property losses. We must monitor and manage the transportation of dangerous goods strictly in order to protect the economic construction and the safety of lives. The paper builds a transport model, on the one hand pursuing for lowest risks in transit and on the other hand pursuing for the shortest transport routes.

## 2 Description and Analysis of the Problem

The risk factors of dangerous goods, in the road transport system, are tremendous; therefore, the randomness and uncertainty of route are relatively large. Generally speaking, we can identify and analyze from the number of people affected, the total accident rate, risk costs, transport costs and so on.

We can not only get the number of people affected from the radius of personnel risk, cargo transport distances and population density; but also get the total accident rate from accident rate in statistics of this section; as well as get the transportation cost from the product of cargo freight per kilometers and total traffic. In this article, the total cost of risk can be get from the number of people affected, the annual accident rate, mortality and loss of life.

## 3 A Comprehensive Model for Transportation of Dangerous Goods

### 3.1 Modeling

Choosing route for the transportation of dangerous goods is, actually, looking for the route, which is not only Minimum risk but also minimum transportation costs, meanwhile, satisfying the Basic requirements of road.

According to the above analysis, we can build a bi-objective constrained optimization model for selection lines, based on risk costs and transportation costs [7].

$$\text{Min } Z = \lambda_1 fm \sum_{i=1}^{n} l_i + \lambda_2 \eta VAL \sum_{i=1}^{n} n_i v_i$$

$$n_i = \rho_i l_i d\ ; \quad \lambda_1 + \lambda_2 = 1\ ; \quad \lambda_i \in [0,1]$$

$$S.t \begin{cases} \sum_{i=1}^{n} l_i \le L \text{ ------ } (1) \\ \sum_{i=1}^{n} u_i \le U \text{ ------ } (2) \\ \sum_{i=1}^{n} c_i \le C \text{ ------ } (3) \end{cases}$$

### 3.2 Explanation of Model Symbols

Z--the sum of risk costs and transportation expenses

$\lambda_i$--the weight of risk costs and transportation expenses

$f$--Freight rate, ¥/ (tonkm)

$m$--the total freight volume of dangerous goods

$l_i$--The length of i section

$\eta$--Mortality rate of the Accident Area

VAL--Loss of life value (/million)

$n_i$--the total number of the affected people of i section

$v_i$--Annual accident rate of i section (based on accident statistics over the years)

$\rho_i$--Population density of i section

r--Risks associated with radius

$L$--the total length of transportation of cargo

U--the overall accident rate being acceptable

C--the cost of acceptable risk

## 4 Example Analysis

10ton dangerous goods have been transported from A to D. The risk radius of influenced people is 5km as supposed.

freight rate is 1￥per (ton km), Loss of life value (/million) is 500,000￥per a person [3], Mortality rate of the Accident Area is 0.002, Also, based on accident statistics over the years and the forecast of expert, the weight of risk costs and transportation expenses is regarded as $\lambda_1 = 0.2, \lambda_2 = 0.8$. Requirements: the total length L is not more than 25km, the overall accident rate U is lower than $8\times10^{-6}$/km [5], And the Risk costs C is not more than 800 RMB. Try to use model to figure out the Transport routes which possesses minimum risk and minimum transportation costs from A to D.

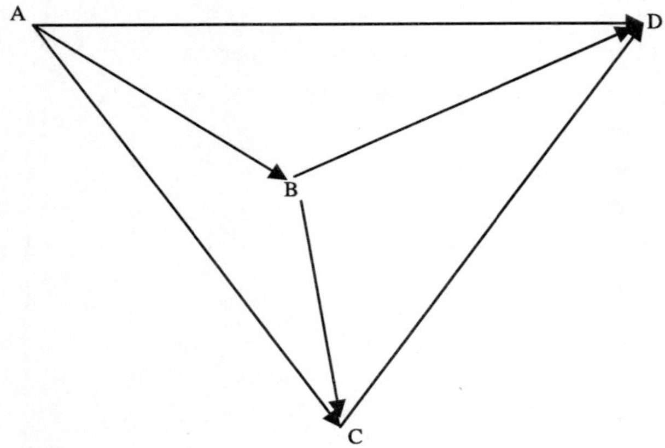


Chart 1

| | Length (km) | Annual accident rate ($\times 10^{-6}$/km) | Population density (person/km$^2$) |
|---|---|---|---|
| $l_1$:A-B | 5 | 7 | 2000 |
| $l_2$:A-C | 15 | 1 | 1200 |
| $l_3$:A-D | 30 | 3 | 400 |
| $l_4$:B-C | 5 | 8 | 1000 |
| $l_5$:B-D | 15 | 1 | 500 |
| $l_6$:C-D | 10 | 1 | 700 |

**Fig. 1.** According to known conditions, we can construct model

$$\text{Min } Z = 0.2 \times 1 \times 10 \sum_{i=1}^{n} l_i + 0.8 \times 0.002 \times 500000 \sum_{i=1}^{n} n_i v_i$$

$$n_i = \rho_i l_i d\ ;\quad \lambda_1 + \lambda_2 = 1\ ;\quad \lambda_i \in [0,1]$$

$$S.t \begin{cases} \sum_{i=1}^{n} l_i \le 25 \text{ - - - - - - - - } (1) \\ \sum_{i=1}^{n} u_i \le 8 \times 10^{-6} \text{ - - - } (2) \\ \sum_{i=1}^{n} c_i \le 800 \text{ - - - - - - } (3) \end{cases}$$

From table one; there have been 4 routes from A to D in all

(1) A——D
(2) A——B——D
(3) A——B——C——D
(4) A——C——D

Calculating the total transportation costs. the number of people who are affected and the risk costs.

(1) A——D

The total length of transportation: $\sum_{i=1}^{n} l_i = l_3 = 30km$

The overall accident rate: $\sum_{i=1}^{n} u_i = \sum_{i=1}^{n} n_i l_i$

$= 3\times10^{-6}\times30 = 9\times10^{-5}$ $\sum_{i=1}^{n} n_i = \sum_{i=1}^{n} \rho_i l_i d$

The total number of the affected people:

$= 400\times5\times2\times30 = 120000$ person

Transport fee: $F = fm\sum_{i=1}^{n} l_i = 11030 = 300$¥

The total risk costs: $\sum_{i=1}^{n} u_i = \eta VAL\sum_{i=1}^{n} n_i v_i$

$= 0.002\times500000\times120000\times3\times10^{-6} = 360$¥

so: $Z=0.2\times1\times10\sum_{i=1}^{n} l_i + 0.8\times0.002\times500000\sum_{i=1}^{n} n_i v_i$

$=0.2\times300+0.8\times360=348$¥

(2) A——B——D

The total length of transportation: $\sum_{i=1}^{n} l_i = 20\text{km}$

The overall accident rate: $\sum_{i=1}^{n} u_i = \sum_{i=1}^{n} n_i l_i$

$= 7\times10^{-6}\times5 + 15\times1\times10^{-6} = 5\times10^{-5}$

The total number of the affected people:

$\sum_{i=1}^{n} n_i = \sum_{i=1}^{n} \rho_i l_i d$

$= 2000\times5\times5\times2 + 500\times15\times5\times2$

$= 1750000$person

Transport fee: $F = fm\sum_{i=1}^{n} l_i$

=10×1×（5+15）=200￥

The total risk costs: $\sum_{i=1}^{n} u_i = \eta VAL \sum_{i=1}^{n} n_i v_i$

$0.002 \times 500000 \times (100000 \times 7 \times 10^{-6} + 75000 \times 1 \times 10^{-6})$

$= 775$￥

so:

$Z=0.2 \times 1 \times 10\sum_{i=1}^{n} l_i + 0.8 \times 0.002 \times 500000\sum_{i=1}^{n} \mathrm{n}_i v_i$

=0.2×200+0.8×775=660 元

（3）A——B——C——D

The total length of transportation: $\sum_{i=1}^{n} l_i = 20\text{km}$

The overall accident rate: $\sum_{i=1}^{n} u_i = \sum_{i=1}^{n} n_i l_i$

$= 7 \times 10^{-6} \times 5 + 5 \times 10^{-6} \times 8 + 10 \times 1 \times 10^{-6} = 8.5 \times 10^{-5}$ $\sum_{i=1}^{n} \mathrm{n}_i = \sum_{i=1}^{n} \rho_i l_i d$

The total number of the affected people:

$= 2000 \times 5 \times 5 \times 2 + 1000 \times 5 \times 5 \times 2 + 700 \times$

$10 \times 5 \times 2 = 220000$person

Transport fee: $F = fm\sum_{i=1}^{n} l_i$

$= 10 \times 1$ $(5+5+10) = 200$￥ The total risk costs: $\sum_{i=1}^{n} u_i = \eta VAL \sum_{i=1}^{n} n_i v_i$

$0.002 \times 500000 \times$

$(100000 \times 7 \times 10^{-6} + 50000 \times 8 \times 10^{-6} + 7000010^{-6}) = 1170$￥

so:

$Z=0.2 \times 1 \times 10\sum_{i=1}^{n} l_i + 0.8 \times 0.002 \times 500000\sum_{i=1}^{n} \mathrm{n}_i v_i$

$= 0.2 \times 200 + 0.8 \times 1170 = 9400$￥

(4) A——C——D

The total length of transportation: $\sum_{i=1}^{n} l_i = 25\text{km}$

The overall accident rate: $\sum_{i=1}^{n} u_i = \sum_{i=1}^{n} n_i l_i$

$= 1\times10^{-6}\times15 + 1\times10^{-6}\times10 = 2.5\times10^{-5}$

The total number of the affected people: $\sum_{i=1}^{n} \text{n}_i = \sum_{i=1}^{n} \rho_i l_i d$

=1200×15×5×2+700×10×5×2=250000 person

Transport fee: $F = fm\sum_{i=1}^{n} l_i$ =1×10× (15+10)

=250￥

The total risk costs: $\sum_{i=1}^{n} u_i = \eta VAL \sum_{i=1}^{n} n_i v_i$

$0.002\times500000\times$

$(180000\times1\times10^{-6} + 70000\times10^{-6}) = 250￥$

So:

Z =0.2×250+0.8×250=250￥.

| | A—D | A—B—D | A—B—C—D | A—C—D |
|---|---|---|---|---|
| **The total length L(km)** | 30 | 20 | 20 | 25 |
| **The total number of the affected people (person)** | $1.2\times10^{-5}$ | $1.75\times10^{-5}$ | $2.2\times10^{-5}$ | $2.5\times10^{-5}$ |
| **The overall accident rate ($\times10^{-6}$)** | 9 | 5 | 8.5 | 2.5 |
| **Transport fee (￥)** | 300 | 200 | 200 | 250 |
| **Risk costs (￥)** | 360 | 775 | 1170 | 250 |
| **Z** | 348 | 660 | 9400 | 250 |

**Fig. 2.**

## 5 Conclusion

The transport accident of dangerous goods is a typical “low probability but high consequence” event; the result of quantitative risk assessment model is very sensitive to the path optimization. if the starting and end point are given, the optimal path of different risk assessment model vary greatly, usually it is the path of minimize the risk of casualties and transport of environmental risk, without considering the other higher transportation costs, but also reflecting the special nature of dangerous goods themselves, unlike the general line selection and planning issues. So we should strengthen the safety management, working out the factors that are likely to cause the accident and eliminating the negative effects.

## References

1. Zhou, C.-l., Tian, H.: Research and Realization of Route Optimization of Dangerous Goods Transporting. Industrial Safety and Environmental Protection 34, 23–25 (2008)
2. Ren, C.-x., Wu, Z.-z.: Probe into Methods of Optimal Road Transportation Routing for Hazardous Materials. China Safety Science Journal 16, 12–13 (2006)
3. Zhang, M., Yang, C., Yang, J., Ma, Y.-F.: Research on Combined Location-routing Problems in Integrated Hazardous Materials Logistics Systems. Journal of Management Sciences in China 11, 45–46 (2008)
4. Song, J.-p., Ma, T.-s., Shao, H.-p., Wei, L.-f.: Study on the Dynamic Route Choice Model for Hazardous Material Road Transportation in Case of Well-informed. Journal of Satety and Environment 2, 58–59 (2009)
5. Kai, Y.-x., Wang, H.-y.: Research on optimization of transportation mode and route for hazardous materials transportation network. Journal of Safety Sscience and Technalogy 15, 124–127 (2009)
6. Song, J., Ding, Y., Meng, L.: Hazardous Material Transportation Network Design Based on Bi-level Programming. Journal of Shanghai Maritme University 2, 74–77 (2006)
7. Ren, C.-x., Wu, Z.-z.: Study on Route-choice Analysis of Hazardous Materials Transportation. China Safety Science Journal 8, 158–159 (2006)
8. Ma, C.-x., Guang, X.-p., Wu, F., Zhao, Y.-c.: Highway Transportation Route Decision-Making of Hazardous Material in Developed Transportation Network. Journal of Transportation Systems Engineering and Information Technology 9, 110–111 (2009)
9. Ma, C.-x., Wen, J.-j., Li, C.-h.: Highway Transportation Route Multiple Attribute Decision-making of Hazardous Material Under the Certain Environment. Journal of Lanzhou Jiaotong University 19, 127–130 (2009)
10. Shi, L.-c., Wei, L.-j., Wu, Z.-z.: Study on Multi-objective Route Optimization for Hazardous Material Road Transportation. Journal of Safety Science and Technology, 23–25 (February 2006)

# Design of Active Disturbance Rejection Controller for Main Drive System of Rolling Mill with Uncertainty Parameters

Zhang Ruicheng and Chen Zhikun

College of Computer and Automatic Control
Hebei Polytechnic University
Hebei Tangshan 063009, China
rchzhang@yahoo.com.cn, zkchen@heut.edu.cn

**Abstract.** The output and quality of metal plate and strip are heavily affected by performance of the main drive system of rolling mill. Aim at the characteristics of main drive system of rolling mill, such as uncertain external load disturbance and unmodelled dynamics, the model of main drive system of rolling mill is established, where uncertain external load disturbance and unmodelled dynamics are regarded as one integrated disturbance. Then, using extended states observer (ESO), a new observer is proposed to observe and compensate this integrated disturbance, and a robust controller of speed system of rolling mill is designed based on active disturbance rejection control (ADRC) technology. This controller has very strong robustness not only to external load disturbance, but also to unpredictable plant parameter variations such as inertia of load、resistance of armature. Its effectiveness is verified by the simulated result.

**Keywords:** Rolling mill, Speed control system, Uncertainty parameters, Active disturbance rejection control, Robustness.

## 1 Introduction

The increasing demand for higher quality steel strips and higher strip thickness precision in the steel industry, has lead to more demanding performance requirements from rolling mill drives. Howere, the mechnical vibrations inherent in rolling mill drives remain a big obstacle for achieving high-performance speed control loop with higher bandwidths and higher accuracy. The major problems of the speed control loop are: the mechanical vibration due to the elastic shafts and couplings of the mechanical system, the uncontrolled load disturbances and the plant uncertainties (parameters variation and unmodeled dynamics).

High starting torque, extended speed range, and ease of control in DC machines have ensured their continued used in particular industrial applications, such as traction and steel plants. There are two important requirements for high performance controlled the main driver of the rolling mill: one is the steady –state and dynamic tracking ability to set point changes and the other is the ability to recover from load disturbance. Conventional PI controllers for such drives are designed on the basis of local linearization about an operating point. These controllers are very effective if the load changes are small and the operating conditions do not force the system too far away from the linearizing equilibrium point. In order to compensate the load disturbance, a observe of the load diturbance is designed[1] and a robust speed control of a DC motor based on the equivalent load observer is design[2]. Howere, it has been known that these observes of the load diturbance depend on the system model. In fact, the systems have no precise mathematical models. To overcome these difficulties, using extended states observer（ESO）, a new observer is proposed to observe and compensate this integrated disturbance, and a robust controller of speed system of rolling mill is designed based on active disturbance rejection control(ADRC) technology. This controller has very strong robustness not only to external load disturbance, but also to unpredictable plant parameter variations such as inertia of load、resistance of armature. Through simulation, it was demonstrated that this is a promising new technology. It is a more powerful method than PID and is easy to use.

The active disturbance rejection control (ADRC) approach was first proposed in [3]. It is a nonlinear control law. To improve performances of the controller, ADRC adopts two important innovations. (1) Through simple nonlinear algorithm, the system states and their differential coefficients are controlled. (2) The extended states observer (ESO) is proposed for the on-line estimation of the total dynamics, lumping the internal nonlinear dynamics and the external disturbance. That is to say, ADRC controls not only the system states, but also their differential coefficients and allows the dynamic compensation of disturbance. So the controller has very strong stability and robustness. Therefore, ADRC has found wide industrial applications [4-5].

## 2 The Model of the Main Drive System of the Rolling Mill

### 2.1 Linear State-Space Model

Under the assumption of a homogenous magnetic field, the direct current (DC) motor is modeled as a linear transducer from motor current to electrical torque. The classical model of the DC motor is composed of a coupled electrical and a mechanical subsystem. The system block diagram is shown in Figure 1. The inertia time constant of the three-phase bridge rectifier unit is very small, so it is

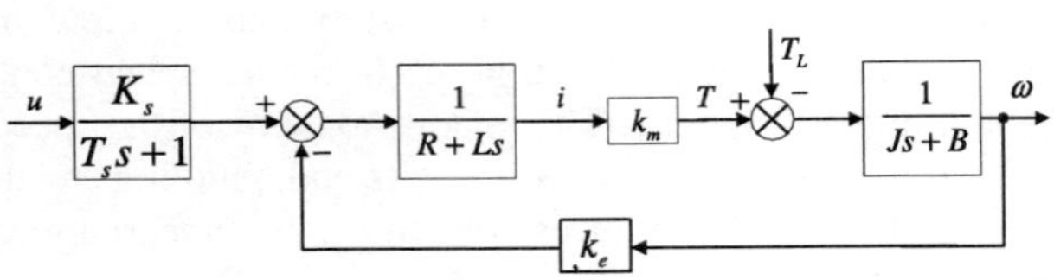


**Fig. 1.** Dynamic configuration of speed system of rolling mill

approximative proportion componemt. The motor dynamics are described by:

$$\begin{cases} \dfrac{di}{dt} = \dfrac{K_s}{L}u - \dfrac{R}{L}i - \dfrac{1}{L}k_e\omega \\ \dfrac{d\omega}{dt} = \dfrac{1}{J}T - \dfrac{1}{J}T_L - \dfrac{B}{J}\omega \end{cases} \tag{1}$$

where $i$ denote the armature current, $\omega$ denote the motor angular velocity, $u$ denote the input voltage, $L$ and $R$m represent the inductance and the total connection resistance of the motor, $K$s denote the magnification coefficient of the thyristor, $J$ denotes the mechanical inertia including the motor armature and shaft, $k$m, $k$e and $B$ denote the motor torque, the back-electromotive force (EMF) and the damping constants, the motor torque is $T=k_m i$ , The system load is denoted by $T_L$.

The block diagram figure 1 is transformed as figure 2. The signal $u_1(s) = \dfrac{T_L(s)}{P_1(s)}$, where $P_1(s) = \dfrac{K_s k_m}{R + Ls}$, and the controlled system transfer function is given as $P_2(s) = \dfrac{K_s k_m}{(R+Ls)(Js+B)+k_m^2}$. Then the model can be tansformed as (2).

$$\begin{cases} \ddot{\omega} = -\dfrac{JR+BL}{JL}\dot{\omega} - \dfrac{RB+k_m^2}{JL}\omega + \dfrac{K_s k_m}{JL}(u - u_1) \\ u_1 = \dfrac{L}{K_s k_m}\dot{T}_L + \dfrac{R}{K_s k_m}T_L \end{cases} \tag{2}$$

Where $u_1$denotes the input of the disturbance.

Let $x_1 = \omega$, $x_2 = \dot{\omega}$, $a_1 = \dfrac{RB + k_m^2}{JL}$, $a_2 = \dfrac{JR + BL}{JL}$, $b = \dfrac{K_s k_m}{JL}$

Then equation (2) can be transformed as (3).

$$\begin{cases} \dot{x}_1 = x_2 \\ \dot{x}_2 = -a_1 x_1 - a_2 x_2 + b(u - u_1) \end{cases} \tag{3}$$

## 2.2 The Internal Disturbance and the Variation Parameter

The main drive system of the rolling mill has a lot of disturbance. These disturbance have significantly effect to the dynamic behavior of the modeled system. So the load disturbance and the variation parameter must be included.

(1) The load disturbance

Because the gear gap, coupling gap, roll eccentricity cause the periodic oscillation during operation, the rolling load is given as (4).

$$T_L = T_{L0} + T_{L1}\sin\omega_r t \tag{4}$$

Where $T_L$ denotes the load torque, $T_{L0}$ denotes the rated load torque, $T_{L1}$ denotes the amplitude of the disturbance load torque, $\omega_r$ denotes the frequency of the disturbance load torque.

(2) variation parameter

The variation parameters of the main drive system of the rolling mill include the mechanical parameters and electromagnetic parameters. So the state-space model of the main drive system of the rolling mill is given as (5).

$$\dot{x} = (A_m + \Delta A)x + (B_m + \Delta B)u + f(\dot{T}_L, T_L) \tag{5}$$

Where $A_m = \begin{bmatrix} 0 & 1 \\ -a_1 & -a_2 \end{bmatrix}, B_m = [0, b]^T$

$$f(\dot{T}_L, T_L) = \left[ 0, -\frac{1}{J}\dot{T}_L - \frac{R}{JL}T_L \right]^T$$

$\Delta A$ and $\Delta B$ denote the variation parameter. Let

$$F_w = \Delta A x + \Delta B u + f(\dot{T}_L, T_L) \tag{6}$$

So the equation (5) can be transformed as (7).

$$\dot{x} = A_m x + B_m u + F_w \tag{7}$$

The parameters are substituted into the equation (7), and let $c_1 = \frac{R}{JL}$, $c_2 = \frac{1}{J}$, the state-space model is given as (8).

$$\begin{cases} \dot{x}_1 = x_2 \\ \dot{x}_2 = -a_1 x_1 - a_2 x_2 + bu + F_{w2} \end{cases} \tag{8}$$

Where
$$F_{w2} = \Delta a_1 x_1 + \Delta a_2 x_2 - c_1 T_L - c_2 \dot{T}_L \tag{9}$$

## 3 Design of the Active Disturbance Rejection Controller for the Rolling Mill Main Drive System

The ADRC is based on the idea that in order to formulate a robust control strategy. Although the linear model makes it feasible for us to use powerful classical control techniques such as frequency response based analysis and design methods, it also limits our options to linear algorithms and makes us overly dependent on the mathematical model of the plant. Instead of following the traditional design path of modeling and linearization and then designing a linear controller, the ADRC approach seeks to actively compensate for the unknown dynamics and disturbances in the time domain. This is achieved by using an extended state observer (ESO) to estimate the total dynamics, lumping the internal nonlinear dynamics and the external disturbances. Once the external disturbance is estimated, the control signal is then used to actively compensate for its effect, then the system becomes a relatively simple control problem. More details of this novel control concept and associated algorithms can be found in [3]-[5]. A brief introduction is given below.

### 3.1 The Transform of System Model

In order to control the main drive system of the rolling mill using active disturbance rejection control strategy. Let

$$a(t) = -a_1 x_1 - a_2 x_2 + F_{w2} \tag{10}$$

$$b = K_s k_m / JL \tag{11}$$

the equation (8) can be transformed as (12).

$$\begin{cases} \dot{x}_1 = x_2 \\ \dot{x}_2 = a(t) + bu \end{cases} \tag{12}$$

From (12), if the integrated disturbance $a(t)$ can be estimated. The control signal is used to actively compensate for its effect, then the main drive system of rolling mill becomes a linear system. The active disturbance rejection control can solve this problem. The architecture of ADRC is shown in Figure 2. It consists of three components: nonlinear tracking-differentiator, the ESO and the Nonlinear PD.

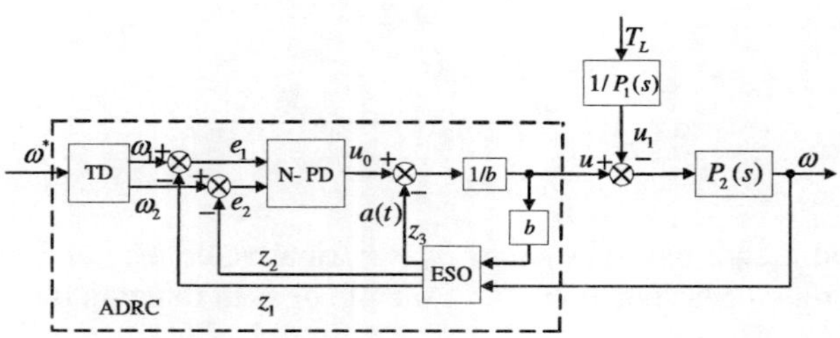


**Fig. 2.** Schematic diagram active disturbance rejection controller of speed system of rolling mill

### 3.2 Nonlinear Tracking-Differentiator

In many practices, the performances of the controlled system are limited by how to pick out the differential signal of the non-continuous measured signal with stochastic noise. In practice, the differential signal is usually obtained by the backward difference of given signal, which is very noisy and limits the overall performance. Han and Wang [6] developed a nonlinear tracking-differentiator (TD) to solve this problem effectively. It is described as

$$\begin{cases} \dot{\omega}_1 = \omega_2 \\ \dot{\omega}_2 = -Rsat(\omega_1 - \omega^* + \frac{|\omega_2|\omega_2}{2R}, \delta_0) \end{cases} \tag{13}$$

Where $R$ —tracking parameter, $\delta_0$—control parameter, $\omega_1$ tracking $\omega^*$.

The function of *sat* as following

$$sat(A, \delta_0) = \begin{cases} sign(A) & |A| > \delta_0 \\ \frac{A}{\delta_0} & |A| \le \delta_0, \delta_0 > 0 \end{cases} \tag{14}$$

### 3.3 The Extended State Observer (ESO)

The extended state observer (ESO) proposed by Prof. Han [7] is a unique nonlinear observer designed to estimate $a(\cdot)$, First, $a(\cdot)$ is augmented as a state for the system:

$$\begin{cases} \dot{x}_1 = x_2 \\ \dot{x}_2 = x_3 + b(t)u \quad x_3 = a(t) \\ \dot{x}_3 = h(t) \\ y = x_1 \end{cases} \tag{15}$$

Where $a(\cdot)$ is unknown. Then, the following nonlinear observer is used to estimate $x$ and $a(\cdot)$.

$$e = z_1 - y \tag{16}$$

$$\begin{cases} \dot{z}_1 = z_2 - \beta_{01} g_1(e) \\ \dot{z}_2 = z_3 - \beta_{02} g_2(e) + b_0 u(t) \\ \dot{z}_3 = -\beta_{03} g_3(e) \end{cases} \tag{17}$$

where $\beta_{01}$,$\beta_{02}$ and $\beta_{03}$ are observer gains, $b_0$ is normal value of $b$ and $g_1(\cdot)$ , $g_2(\cdot)$ and $g_3(\cdot)$ are linear or nonlinear functions. An example of such function is

$$g_i(e, \alpha_i) = fal(e, \alpha_i, \delta) = \begin{cases} |e|^{\alpha_i} sign(e) & |e| > \delta \\ e/\delta^{1-\alpha_i} & |e| \le \delta \end{cases} \tag{18}$$

### 3.4 The Nonlinear PD Control Law

Once the design of TD and ESO is accomplished, the general error and its change between the reference and the estimated states can be defined as $e_1$ and $e_2$ . The nonlinear proportional derivative (N-PD) law is used to synthesis the preliminary control action, which can be described as

$$\begin{cases} e_1 = \omega_1 - z_1 \\ e_2 = \omega_2 - z_2 \\ u_0 = \beta_1 fal(e_1, \alpha_1, \delta_1) + \beta_2 fal(e_2, \alpha_2, \delta_1) \end{cases} \tag{19}$$

Which reduces the plant to a double integrator, which in turn is controlled by the nonlinear PD controller. Where $\beta_1$ and $\beta_2$ are the gains of PD controller.

Plus the control action to cancel out the external disturbance, then the total control action of ADRC can be determined as follows

$$u(t) = \frac{1}{b}[u_0 - a(t)] \tag{20}$$

## 4 Simulation Study

In order to confirm the effectiveness of the control of main drive system of rolling mill based on ADRC, several simulations are carried out. The parameters of the main drive system of a 2030mm tandem mill [8] are as follows: $R$=0.0314Ω, $L$=0.0003H, $J_1$=1542kg·m$^2$, $J_2$=1552kg·m$^2$, $B$=0.0064, $K_s$=152, $k_m$=29N·m/A, $k_e$=29N·m/A.

The simulations are done in the condition as follows:

(1) $J=J_0$, the load disturbance $T_L$=14500+2910$sin(\pi t)$ is exerted at $t$=5.0 second.

(2) $J=J_0$, $\omega^* = 27.3 + 2.73\sin(2\pi t)$.

(3) $J=10J_0$, $T_L$=14500+2910$sin(\pi t)$.

Condition 1 simulate the bite steel at low speed. The speed responses is shown in Figure 3 when $\omega^*$=27.3$rad/s$. The key to the success of the ADRC is the ESO, which enables the controller to actively compensate for the changes in the dynamics (inertia) or the disturbances. To see the effectiveness of the ESO, The input and estimate value of load disturbance are shown in Figure 4. The solid line is the real value of load disturbance $T_L$, and the broken line is the estimate of load disturbance using ESO. It is clearly shown that they are very closely.

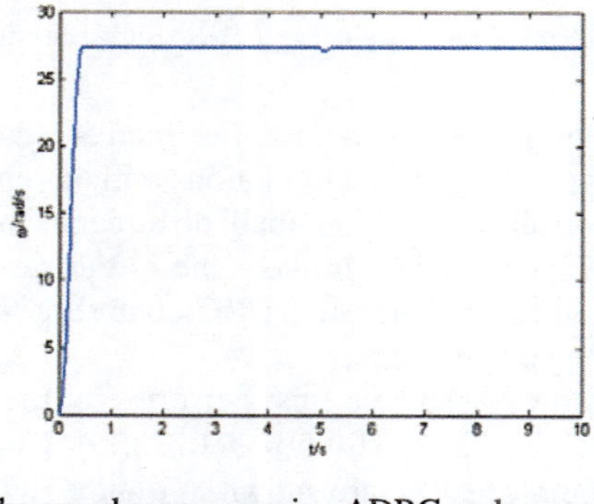


**Fig. 3.** The speed response using ADRC under condition (1)

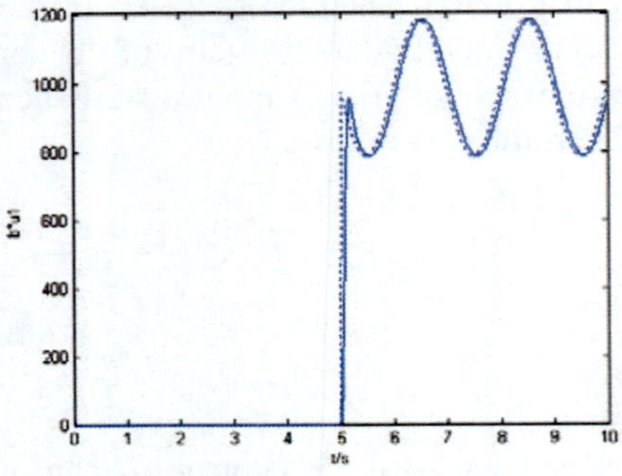

**Fig. 4.** The input and estimate value of load disturbance

Condition 2 simulate the tracking ability for time-varying reference input signal. the load disturbance $T_L$=14500+2910$sin(\pi t)$. The speed responses is shown in Figure 5. It demonstrates that the controller performs well and has not overshoot.

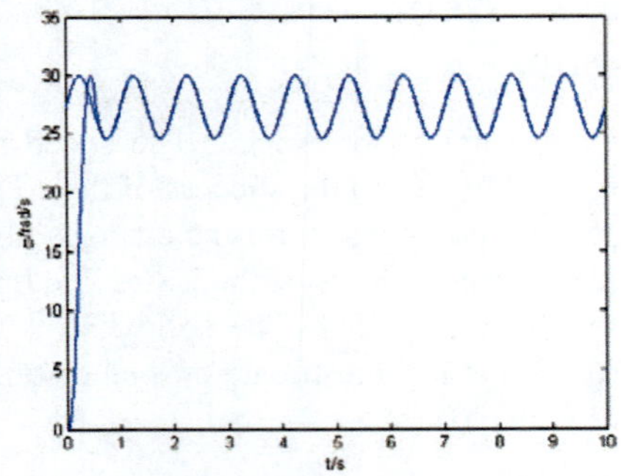

**Fig. 5.** The speed response using ADRC under condition (2)

Condition 3 simulate the robust the ADRC. we increase the inertia constant $J_L$ of the load by 100% and repeat the control simulations without changing any parameters of the ADRC and a controller based on load disturbance observer[1]. The speed responses are shown in Figure 6. The broken line is the result of load disturbance observer, and the solid line is the result of ADRC controller. It demonstrates that the ADRC controller has stronger robustness.

In addition, to test how these systems perform in the presence of external disturbances, a load disturbance $T_L$=14500+2910sin($\pi t$) Nm is exerted at $t$= 5.0 second. The responses of motor speed are shown in Figure 7. The curve 2 is the result of PI controller, and the curve 1 is the result of ADRC controller. Once again, the ADRC performs much better than PI. PI performed poorly in this case because, in order to avoid overshoot, the integral gain used is very small.

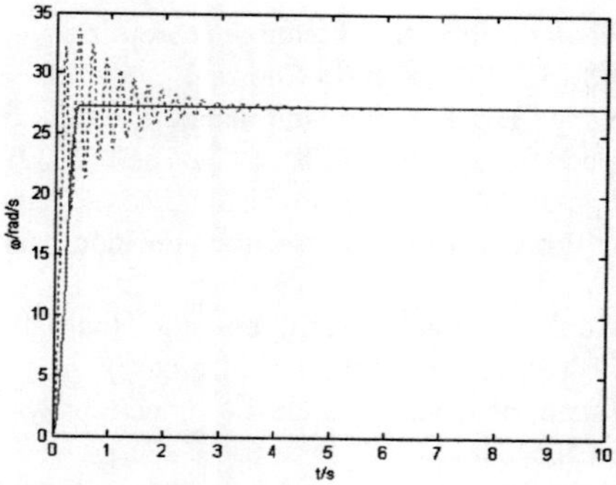


**Fig. 6.** The speed response using ADRC under condition(3)

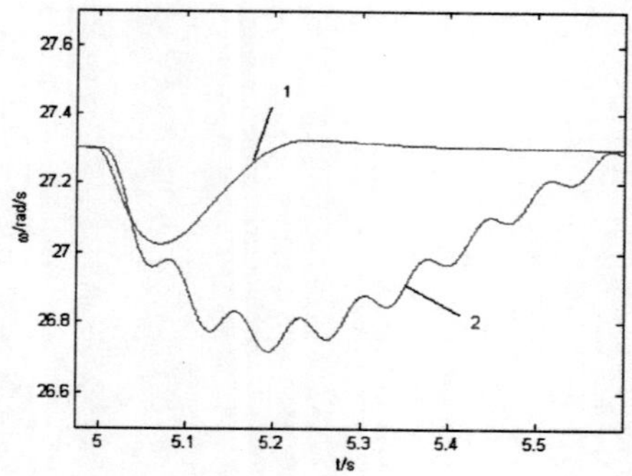


**Fig. 7.** Dynamic speed drop under load disturbance

# 5 Conclusion

In this paper, a new control concept is proposed for the drive system of rolling mill in order to achieve the robustness and the disturbance rejection. The unknown states and the load torque disturbance is estimated using an extended state observer and feedforward compensation for load torque in the system. The robustness of the ADRC to plant uncertainty and external disturbance is inherent in its structure. So the proposed controller guaranteed the stability in case of mechanical parameters are changed (the variation of 100% of $J_L$), and its validity and superiority is verified in comparison with the conventional PI controller. The simulation results showed the controller effectively improved the performances of command following, suppressed electromechanical vibration, reduced the dynamic speed fall brought by load disturbance.

# References

1. Li, C., Duan, W.: The rolling mill drive AC speed electromechanical vibration control. Metallurgical industry press, Beijing (2003) (in Chinese)
2. Fang, Y., Wang, L., Wang, Y.: The robust speed control of a DC motor based on the equivalent load observer. Electric Machines and Control 6(4), 297–300 (2002) (in Chinese)

3. Han, J.: From PID technique to active disturbance rejection control technique. Control Engineering of China 9(3), 13–18 (2002) (in Chinese)
4. Li, S.L., Yang, X., Yang, D.: Active disturbance rejection control for high pointing accuracy and rotation speed. Automatica 45(8), 1854–1860 (2009)
5. Wu, D., Chen, K.: Design and Analysis of Precision Active Disturbance Rejection Control for Noncircular Turning Process. IEEE Transactions on Industrial Electronics 56(7), 2746–2753 (2009)
6. Han, J., Wang, W.: Nonlinear tracking-differentiator. Journal of Systems Science and Mathematical Sciences 14(2), 177–183 (1994) (in Chinese)
7. Han, J.: The "extended states observer" of a class of uncertain system. Control and Decision 10(1), 85–88 (1995) (in Chinese)
8. Zou, J., Xu, L.: Tandem mill vibration control. Metallurgical industry press, Beijing (1998) (in Chinese)

# Design of Active Disturbance Rejection Controller for the Rolling Mill Automatic Position Control Stopping System

Zhang Ruicheng[1] and Li Guide[2]

[1] College of Computer and Automatic Control
Hebei Polytechnic University
Hebei Tangshan 063009, China
[2] Engineering Agency Tangshan Municipal Construction Corporation
Hebei Tangshan 063000, China
rchzhang@yahoo.com.cn, ligdxwq@163.com

**Abstract.** A nonlinear model of the rolling mill automatic position control system is derived based on analyzing the friction between roll and metal plate、the flexibility and the roll eccentricity force. The nonlinear characteristic of the automatic position control (APC) system is analysed. The result shows the APC system has chaotic characteristics. Aim at the characteristics of the APC system,such as nonlinear and variable parameter,a robust controller of the APC system is designed based on active disturbance rejection control(ADRC) technology. The simulation results show that the ADRC controller has very strong robustness to the system chaos and the large load disturbance, and rapid and smooth control process and high steady precise performances can be implemented.

**Keywords:** Automatic Position Control(APC), Chaos, Active Disturbance Rejection Controller(ADRC), Robustness.

## 1 Introduction

Automatic Position Control(APC) is a very important technology in the modern rolling process, knowledge of the precise width and the position of the strip edge is crucial for precise mill control. In particular the wedge-shape of slit strip causes the strip position in the rolling mill to deviate more and more from one pass to another unless it is corrected. In extreme cases, this may require that the rolling process be stopped and the coil be rewound. Automatic Position Control helps to optimize strip flatness during the cold rolling process.But for the same rolling mill, sometimes, the strip can not accurate stop, and the system will produce negative damping phenomenon when the external conditions are basic same. The motion of APC system may become very complex and appear chaos[1]. Because of the unknown load disturbance and variable parameter, the traditional PID control and adaptive control methods are often difficult to meet the control requirements. It can be seen from above that, it is of great significance to find a robust controller of the APC system, and the controller can adapt to a variety of operating conditions.

The active disturbance rejection control (ADRC) approach was first proposed in [2]. It is a nonlinear control law. To improve performances of the controller, ADRC adopts two important innovations. (1) Through simple nonlinear algorithm, the system states and their differential coefficients are controlled. (2) The extended states observer (ESO) is proposed for the on-line estimation of the total dynamics, lumping the internal nonlinear dynamics and the external disturbance. That is to say, ADRC controls not only the system states, but also their differential coefficients and allows the dynamic compensation of disturbance. So the controller has very strong stability and robustness. Therefore, ADRC has found wide industrial applications [3-5]. However, it is rare in the domain of chaos control. In this paper, a nonlinear model of the rolling mill automatic position control system is derived based on analyzing the friction between roll and metal plate、 the flexibility and the roll eccentricity force. The nonlinear characteristic of the automatic position control (APC) system is analysed. The result shows the APC system has chaotic characteristics. Then a robust controller of the APC system is designed based on active disturbance rejection control(ADRC) technology. The simulation results show that the ADRC controller has very strong robustness to the system chaos and the large load disturbance, and rapid and smooth control process and high steady precise performances can be implemented.

## 2 APC System Model and It's Chaos Phenomenon

The model of automatic position control is shown in Fig.1. Let the displacement of the strip relative to the centerline is $x$. In the rolling process, the strip sways aroud the centerline because of the self-tuning of the APC system. The roll is affected by the damping force, elastic force and eccentric force because of the friction and deformation resistance between the roll and strip. The damping force is proportional to speed. Howere, the roll are weared, so the damping force is nonlinear. It is givens as $k_1x^2$. Where, $k_1$ is the damping coefficient. It shows that the damping force is small in center and larger in edge. The elastic potential energy of the roll system is symmetrical. It is givens as $U= k_2x^2/2+ k_3x^4/4$, where $k_2$ and $k_3$ are the elastic coefficient. If $k_3$ compare with $k_2$, it is a very small number. So the elastic force is gives as $f=-dU/dx=- k_2x- k_3x^3$. The roll eccentric force is givens as $F=\sum F_i \cos(\omega_i t+\theta_i)$, here the first-harmonic $F\cos(\omega t)$ is considered. Where $F$ is the amplitude of roll eccentric force and $\omega$ is the angular frequency of roll eccentric force. For the APC system, motor torque is very large, so the motor with load can not brake. Therefore the stored energy of motor converts in force. It is givens as $k_4\dot{x}$, where $k_4$ is the energy storage coefficient. According to Newton's law, the motion equations of APC system is given as (1).

$$-k_1x^2\dot{x}-k_2x-k_3x^3+F\cos(\omega t)+k_4\dot{x}=m\ddot{x} \tag{1}$$

Let $k_1 = k_4$, so the equation (1) is given as (2).

$$\ddot{x}+\frac{k_1}{m}(x^2-1)\dot{x}+\frac{k_2}{m}x+\frac{k_3}{m}x^3-\frac{F}{m}\cos(\omega t)=0 \tag{2}$$

Where $k_1$ is the damping coefficient, $k_2$ and $k_3$ are the elastic coefficient, $F$ is the amplitude of roll eccentric force, $\omega$ is the angular frequency of roll eccentric force and $m$ is the quality of the strip.

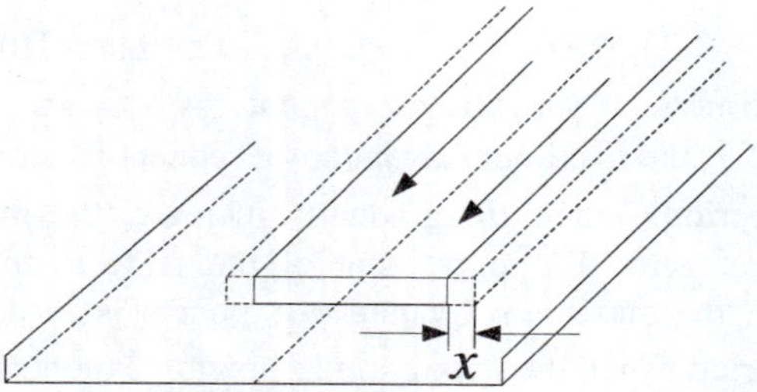


**Fig. 1.** The schematic diagram of the strip swaying on the path

Let $\tau = \omega_1 t$ , $\omega_1 = \sqrt{k_1 / m}$ , $\varpi = \omega / \omega_1$ , $\alpha = \omega_1^2$ , $\beta = k_2 \omega_1^2 / k_1$ , $\gamma = k_3 \omega_1^2 / k_1$, $\delta = F \omega_1^2 / k_1$, so the equation (2) is transformed as (3).

$$\ddot{x} + \alpha(x^2 - 1)\dot{x} + (\beta + \gamma \cdot x^2) \cdot x - \delta \cos(\varpi \cdot \tau) = 0 \tag{3}$$

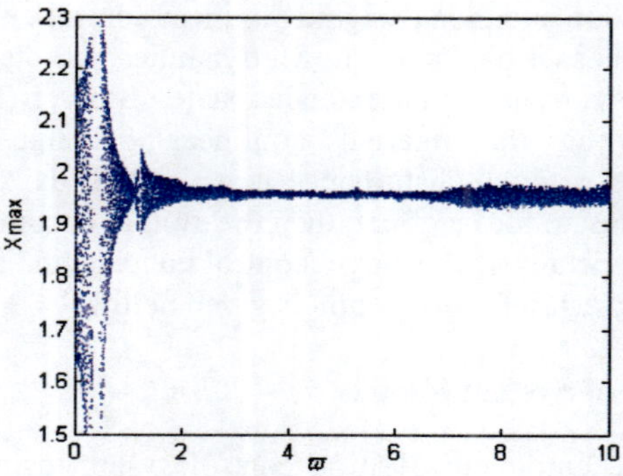


(a) Bifurcation diagrams

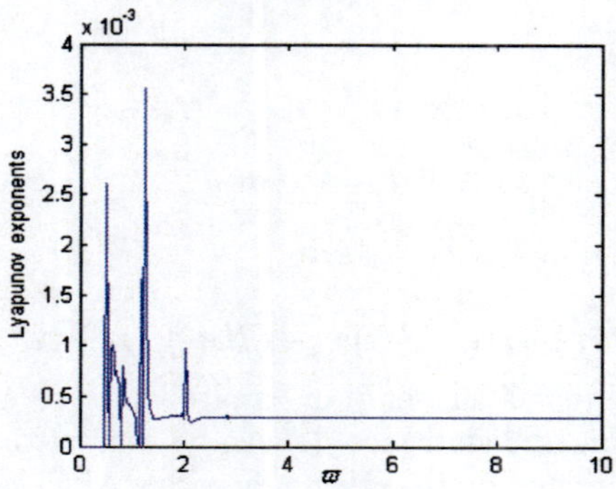


(b) Maximum Lyapunov exponent

**Fig. 2.** When $\varpi \in (0.005, 10)$, bifurcation diagrams and maximum Lyapunov exponent

Where $\alpha$ =0.05, $\beta$ =0.08, $\gamma$ =0.03, $\delta$ =0.05, $\varpi \in [0.005,10]$ , the bifurcation diagrams and maximum Lyapunov exponent is shows in Fig.2. When $\varpi \in [0.005, 0.48095]$, the maximum Lyapunov exponent is zero. The movement of the APC system is period. When $\varpi$ gradually increase, the maximum Lyapunov exponent is more than zero. The movement of the APC system is chaos. When $\varpi \in [1.2074, 1.2325]$, the maximum Lyapunov exponent is zero. The movement of the APC system is period. Then the maximum Lyapunov exponent is more than zero. The movement of the APC system is chaos.

# 3 Chaos Control of the APC System Based on ADRC

The ADRC is based on the idea that in order to formulate a robust control strategy. Although the linear model makes it feasible for us to use powerful classical control techniques such as frequency response based analysis and design methods, it also limits our options to linear algorithms and makes us overly dependent on the mathematical model of the plant. Instead of following the traditional design path of modeling and linearization and then designing a linear controller, the ADRC approach seeks to actively compensate for the unknown dynamics and disturbances in the time domain. This is achieved by using an extended state observer (ESO) to estimate the total dynamics, lumping the internal nonlinear dynamics and the external disturbances. Once the external disturbance is estimated, the control signal is then used to actively compensate for its effect, then the system becomes a relatively simple control problem. More details of this novel control concept and associated algorithms can be found in [2]-[6]. A brief introduction is given below.

## 3.1 The Transform of System Model

In order to control the automatic position control system using active disturbance rejection control strategy. In the (3), let $x_1 = x, x_2 = \dot{x}$, the system model can be transformed as (4).

$$\begin{cases} \dot{x}_1 = x_2 \\ \dot{x}_2 = f(t) + b(t)u \\ y = x_1 \end{cases} \tag{4}$$

Where $f(t) = \delta \cos(\varpi \tau) - \alpha(x_1^2 - 1)x_2 - (\beta + \gamma \cdot x_1^2) \cdot x_1$ , $u$ is input of the controller. $b$ is the regulation parameter of the controller.

From (4), if the integrated disturbance $f(t)$ can be estimated. The control signal is used to actively compensate for its effect, then the APC system of rolling mill becomes a linear system. The active disturbance rejection control can solve this problem. The architecture of ADRC is shown in Figure 3. It consists of three components: nonlinear tracking-differentiator, the ESO and the Nonlinear PD.

### 3.2 Nonlinear Tracking-Differentiator

In many practices, the performances of the controlled system are limited by how to pick out the differential signal of the non-continuous measured signal with stochastic noise. In practice, the differential signal is usually obtained by the backward difference of given signal, which is very noisy and limits the overall performance. Han and Wang [6] developed a nonlinear tracking-differentiator (TD) to solve this problem effectively. It is described as

$$\begin{cases} \dot{v}_1 = v_2 \\ \dot{v}_2 = -Rsat(v_1 - v_0 + \dfrac{|v_2|v_2}{2R}, \delta_0) \end{cases} \tag{5}$$

Where $R$ —tracking parameter
$\delta_0$—control parameter
$v_1$ tracking $v_0$
The function of *sat* as following

$$sat(A, \delta_0) = \begin{cases} sign(A) & |A| > \delta_0 \\ \dfrac{A}{\delta_0} & |A| \le \delta_0, \delta_0 > 0 \end{cases} \tag{6}$$

### 3.3 The Extended State Observer (ESO)

The extended state observer (ESO) proposed by Prof. Han [7] is a unique nonlinear observer designed to estimate $f(\cdot)$, First, $f(\cdot)$ is augmented as a state for the system:

$$\begin{cases} \dot{x}_1 = x_2 \\ \dot{x}_2 = x_3 + b(t)u \quad x_3 = f(t) \\ \dot{x}_3 = h(t) \\ y = x_1 \end{cases} \tag{7}$$

Where $h$(t) is the derivative of $f(\cdot)$ and is unknown. Then, the following nonlinear observer is used to estimate $x$ and $f(\cdot)$.

$$e = z_1 - y \tag{8}$$

$$\begin{cases} \dot{z}_1 = z_2 - \beta_{01} g_1(e) \\ \dot{z}_2 = z_3 - \beta_{02} g_2(e) + b_0 u(t) \\ \dot{z}_3 = -\beta_{03} g_3(e) \end{cases} \tag{9}$$

where $\beta_{01}$,$\beta_{02}$ and$\beta_{03}$ are observer gains, $b_0$ is normal value of $b$ and $g_1(\cdot)$ , $g_2(\cdot)$ and $g_3(\cdot)$ are linear or nonlinear functions. An example of such function is

$$g_i(e, \alpha_i) = fal(e, \alpha_i, \delta) = \begin{cases} |e|^{\alpha_i} \operatorname{sign}(e) & |e| > \delta \\ e/\delta^{1-\alpha_i} & |e| \le \delta \end{cases} \tag{10}$$

Remarks:

1) The use of nonlinear function $g_i(\cdot)$ in (9) can make the observer more efficient. The nonlinear function (10) was selected based on experiments. Intuitively, it is a nonlinear gain function where smaller error corresponds to higher gain. This technique is used widely in industrial applications.
2) If $g_i(\cdot)$ $i$=1,2 …, are chosen as linear functions, then (9) is equivalent to the well known Luenburger observer found in linear system theory.
3) Similar to linear observers, the observer equation (9) reflects our best knowledge of the system. The information on the range of $b$ in (4) is needed for the selection of $b_0$ in the ESO. If we know more about the plant, such as the friction, then it should be incorporated into the observer to make it more efficient.
4) Note that there is a linear segment in (10) in the neighborhood of the origin. It was discovered experimentally that the nonlinear function in (10) makes the observer converge faster than its linear counterpart and the linear section in (10) makes the output of the observer smoother.
5) The proper selections of the gains and functions in (9) are critical to the success of the observer. One approach is to design the linear observer first and then gradually increase the nonlinearity to improve the performance. It was discovered that once the ESO is properly set up, the performance of the observer is quite insensitive to the plant variations and disturbances.

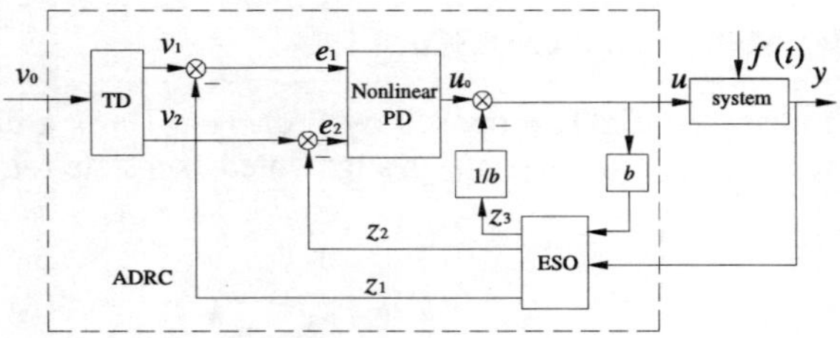


**Fig. 3.** Chaos control of the APC system based on ADRC

### 3.4 The Nonlinear PD Control Law

Once the design of TD and ESO is accomplished, the general error and its change between the reference and the estimated states can be defined as $e_1$ and $e_2$ . The nonlinear proportional derivative (N-PD) law is used to synthesis the preliminary control action, which can be described as

$$\begin{cases} e_1 = v_1 - z_1 \\ e_2 = v_2 - z_2 \\ u_0 = \beta_1 fal(e_1, \alpha_1, \delta_1) + \beta_2 fal(e_2, \alpha_2, \delta_1) \end{cases} \tag{11}$$

Which reduces the plant to a double integrator, which in turn is controlled by the nonlinear PD controller. Where$\beta_1$ and $\beta_2$ are the gains of PD controller.

Plus the control action to cancel out the external disturbance, then the total control action of ADRC can be determined as follows

$$u(t) = \frac{1}{b_0}[u_0 - f(t)] \tag{12}$$

## 4 Simulation Study

In order to confirm the effectiveness of chaos control of the APC system based on ADRC, several simulations are carried out. The parameters of the ADRC controller are as follows $R$=10, $\delta_1$=0.001, $\beta_{01}$=100, $\beta_{02}$=65, $\beta_{03}$=80, $\beta_1$=100, $\beta_2$=10.

**Step response tests**

When $\alpha$=0.05, $\beta$=0.08, $\gamma$=0.03, $\delta$=0.05, $\varpi$=0.4, The movement of the APC system is period. The final value of step response is 0.02, the ADRC is imposed at $t$=80s, the response curve is shown in Figure 4. It demonstrates that the controller performs well and the overshoot is small.

When $\alpha$=0.05, $\beta$=0.08, $\gamma$=0.03, $\delta$=0.05, $\varpi$=4, The movement of the APC system is chaos. The final value of step response is 0.02, the ADRC is imposed at $t$=80s, the response curve is shown in Figure 5. It demonstrates that the controller performs well and the system is not chaos.

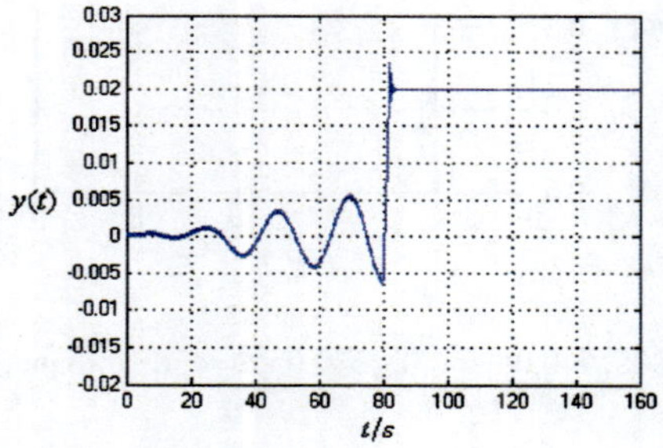


**Fig. 4.** When $\alpha$=0.05, $\beta$=0.08, $\gamma$=0.03, $\delta$=0.05, $\varpi$=0.4, step response curve

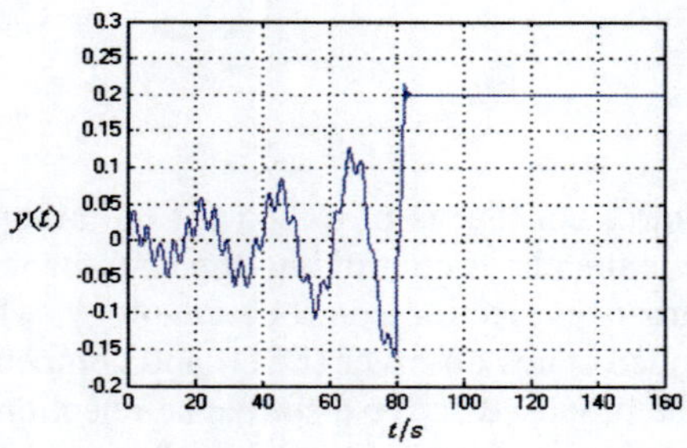


**Fig. 5.** When $\alpha$=0.05, $\beta$=0.08, $\gamma$=0.03, $\delta$=0.05, $\varpi$=4, step response curve

**Disturbance Rejection Tests**

When $\alpha$=0.05,$\beta$=0.08,$\gamma$=0.03,$\delta$=0.05,$\varpi$=0.4, The movement of the APC system is period. To test how these systems perform in the presence of external disturbances, a unit step load disturbance is exerted at $t$=6.0 second. The responses of APC system is shown in Figure 6. It is clearly shown that the result of ADRC is rapidly stable.

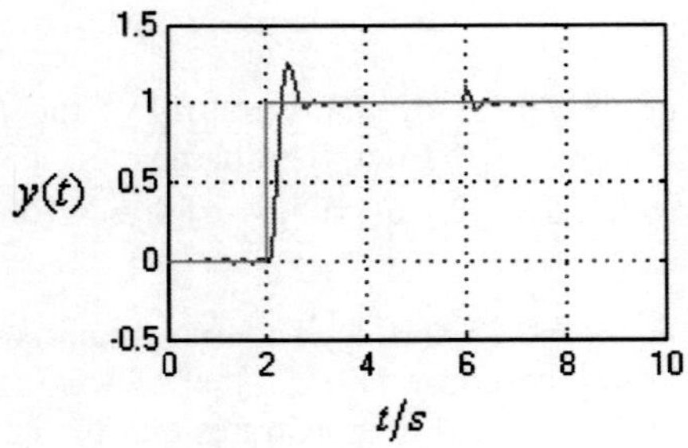


**Fig. 6.** When $\alpha$=0.05,$\beta$=0.08,$\gamma$=0.03,$\delta$=0.05,$\varpi$=0.4, disturbance response curve

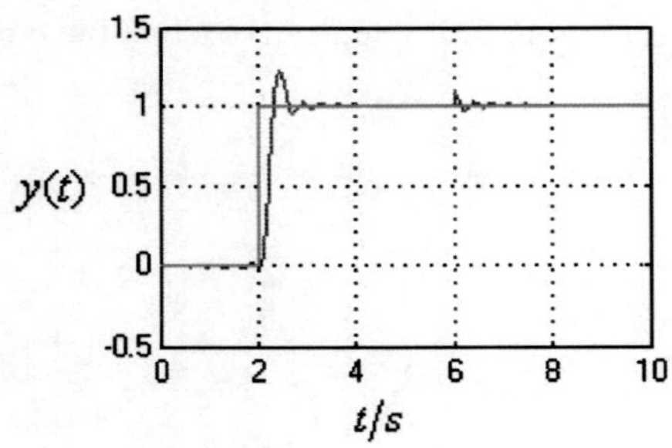


**Fig. 7.** When $\alpha$=0.05,$\beta$=0.08,$\gamma$=0.03,$\delta$=0.05,$\varpi$=4, disturbance response curve

When $\alpha$=0.05,$\beta$=0.08,$\gamma$=0.03,$\delta$=0.05,$\varpi$=4, The movement of the APC system is chaos. a unit step load disturbance is exerted at $t$=6.0 second. The responses of APC system is shown in Figure 7. It is clearly shown that the result of ADRC is rapidly stable.

# 5 Conclusion

In this paper, a new control concept is proposed for the automatic posotion control system in order to achieve the chaos control and the disturbance rejection. It is based on a unique active disturbance rejection control technology, where the disturbance is estimated using an extended states observer (ESO) and compensated for the dynamic change in the system. The proposed active disturbance rejection controller (ADRC) is designed without an explicit mathematical model of the plant. Hence the controller is inherently robust against plant variations. The improvement in steady state error is also quite evident.

## References

1. Zhang, L.: Chaotic Phenomena Research on Process Control. University of Science and Technology, Beijing (2003) (in Chinese)
2. Han, J.: From PID technique to active disturbance rejection control technique. Control Engineering of China 9(3), 13–18 (2002) (in Chinese)
3. Xia, Y., Shi, P., Liu, G.P., et al.: Active disturbance rejection control for uncertain multivariable systems with time-delay. The Control Theory and Applications 1(1), 75–81 (2007)
4. Li, S.L., Yang, X., Yang, D.: Active disturbance rejection control for high pointing accuracy and rotation speed. Automatica 45(8), 1854–1860 (2009)
5. Wu, D., Chen, K.: Design and Analysis of Precision Active Disturbance Rejection Control for Noncircular Turning Process. IEEE Transactions on Industrial Electronics 56(7), 2746–2753 (2009)
6. Han, J., Wang, W.: Nonlinear tracking-differentiator. Journal of Systems Science and Mathematical Sciences 14(2), 177–183 (1994) (in Chinese)
7. Han, J.: The "extended states observer" of a class of uncertain system. Control and Decision 10(1), 85–88 (1995) (in Chinese)
8. Yu, X., Zhu, L.: Dynamic parameter setting and application of active disturbance rejection controlle. Journal of North China Electric Power University 32(6), 9–13 (2005) (in Chinese)

# Wire Communication System Based on DSP in Feeder Automation

Yanfang Wang[1], Yongjun Liu[1], Xiaoping Wang[1], and Shenlin[2]

[1] Department of Computer & Electric Engineering, Shijiazhuang Tiedao University
Shijiazhuang, China
[2] Beijing Tiantan Biological products CO.LTD,
Beijing, China
{wxpyf,wxp1501,tulipsl_007}@163.com, hit101@126.com

**Abstract.** The requirements of feeder automation are that the communication system had the characteristic of lower cost, stronger anti-interference ability, higher reliability and much more real time ability. So wire (twister-pair) relay communication manner is chosen in the feeder automation system in the design, and the communication distance can reach to 10km. In the system, OFDM(orthogonal frequency division multiplexing)technology is applied to transfer data. The OFDM signal is produced and demodulated by way of IFFT (Inverse Fast Fourier Transformer) and FFT(Fast Fourier Transformer). A general DSP (Digital signal process) chip TMS320C54 act as the key of digital signal process to perform almost all the function including arithmetic calculation, modulating and demodulating the data. The OFDM signal Modem, designed basing on the DSP chip, had met the need of the communication system.

**Keywords:** feeder automation, modems, orthogonal frequency division multiplexing, digital signal processing, communication system.

## 1 Introduction

Feeder automation is the automation of distribution line of the electrified wire net, and it is the important part of the distribution automation system. The function of the feeder automation is monitoring the performance mode and the load of the feeder wire. When some trouble happened, it can exactly make sure where the failure is, and then rapidly insulate the trouble section to let the non-failure part resume normal power supply.

The reliability of the communication system of feed automation will be weakened due to it working in high EMI (electromagnetism interference) circumstance. Therefore, the communication system of feed automation must have very high anti-interference ability.

In fact, the investment in the communication system of a feed automation system is hugeness. So the designer should try his best both reducing the cost and meeting the need of the communication system. Considering its feasibility and all of the above characteristics of the feed automation system, we choose wire (twisted-pair) relay communication mode which has effective performance cost.

## 2 The Principle of the OFDM

The modulation and demodulation procedure is equivalent to discrete Fourier transform in the OFDM system. If using digital signal processor and FFT(Fast Fourier Transform) arithmetic, we need no pectinate structure filter group, so it is easier to achieve the modulation and demodulation process in OFDM[1].

## 3 The Configuration of the Communication Network

The communication network schematic diagram is shown as Fig.1. The communication between FTU(Feeder Terminal Unit) and substation is carried out by each Receiver/Transmitter. And the communication between Receiver/ Transmitter and FTU is achieved by RS-232 interface. The Receiver/Transmitter consists three parts: modem, line driver and a bypass set. For digital signal modulating, the modem transforms low-pass digital signal, including rich direct component, into band-pass sine wave in order to transmit on the line. On the other hand, the modem can demodulate sine wave from the transmission line into digital signal for CPU's recognizing and processing. The line driver amplifies the transmitting signal to achieve long distance transmission. The function of the bypass set is avoiding the communication system break long time when a node is failure or a equipment is being maintained.

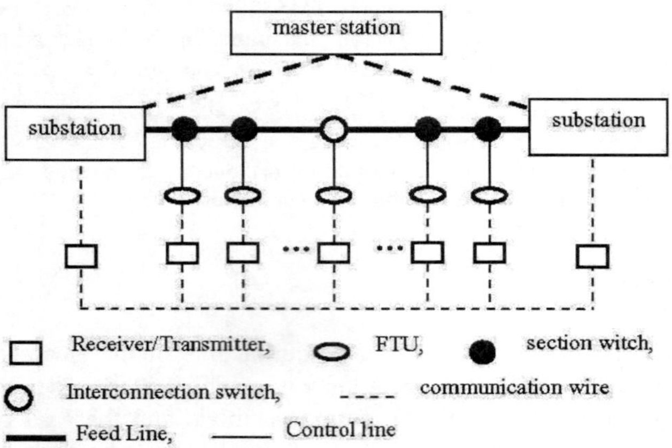


**Fig. 1.** Schematic diagram of feeder automation system Communication network

In Fig.1, a substation is defined as main substation and the other substation as the extra substation. Their primary communication operations are the self-reporting information of the line Receiver/Transmitter, getting in contact with each (sub-)stations, and touring or checking tasks sent by the main substation. The principle of the information transmitting is that since the transmission of each kind of information uses different address, separate out properly and can be commendably done in the communication procedure. They will do their best to avoid task conflict between each Receiver/Transmitter self-reporting, to prohibit task conflict between self-reporting and each station/substations. In general, the system firstly checks whether the channel busy before each self-reporting operation, and the data is sent until affirming the channel in idle state. Even if colliding between self-reporting information, substation could find the trouble according to the fuzzy fragment information of the self-reporting, and then the system goes into trouble query state.

# 4 The Design of the Modulator

The structural diagram of modulator is shown as Fig.2. The modulator and intellective FTU(Feeder Terminal Unit) is fixed on the same pole. Each FTU collect the switch running state on the pole, such as load, voltage, power information and the switch's position etc. The FTU then send the state information into the modulator in binary system via a RS-232 interface. Firstly, the serial input binary-data should be done error correction coding in BCH (Bose, Chaudhuri & Hocquenghem Type of code). Secondly, it should be sent into a 16-QAM (Quadrature Amplitude Modulation) encoder to form imaginary number sequence. After serial-parallel conversion, the imaginary sequence data together with pilot symbol act as a part of IDFT in 64 dot D(m). And treat D(m) with IDFT (inverse discrete Fourier transform) to get d(n). CP(Cyclic Prefix) is inserted into Guard space at the behind of each OFDM symbol. After expanding and winding, the d(n) is put into a D/A convertor to get OFDM signal via a low-pass filter.

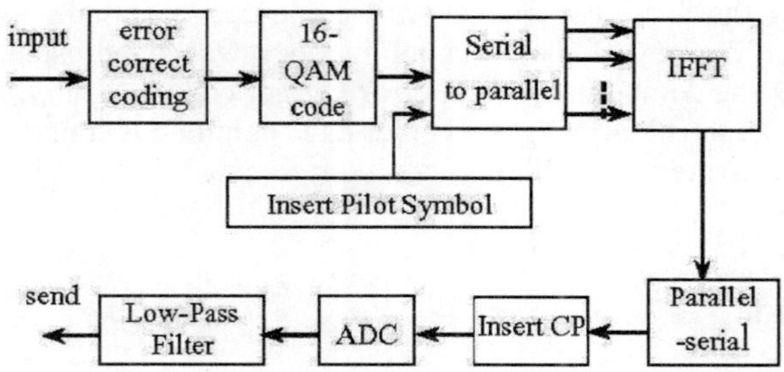


**Fig. 2.** The structural diagram of modulator

All the jobs before D/A are accomplished by TMS320C5402 device with software programming. And the sampling clock of the D/A convertor also supplied by the device TMS320C5402 through software. The D/A convertor and the analogue low pass filter are actualized by TLC32044.

# 5 The OFDM Demodulator

The structural diagram of demodulator is show as Fig.3.

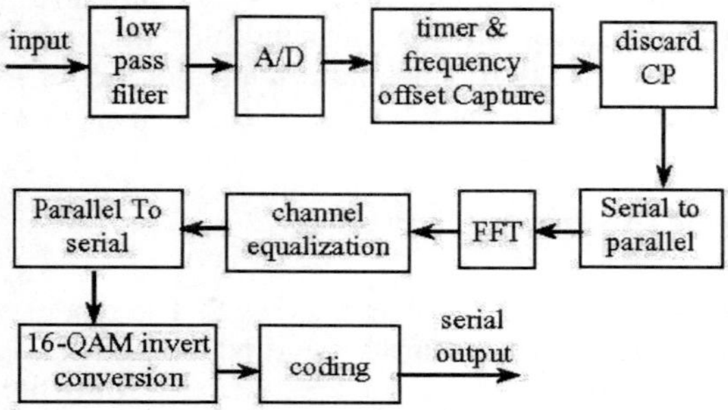


**Fig. 3.** The structural diagram of demodulator

When receiving form communication line, the data is sent into TLC32044 chip to filter off the disturbance and the noise portion by switch capacitance low-pass filter inside the chip. The data then is sent into A/D converter to get digital data. The converting precision is 14 bit, and its sampling speed could be adjusted as same as modulator's sampling speed, 16KHz, with software. After that, the digital data is sent into TMS320C5402 for demodulating. All the function behind A/D converter in Fig.3 is realized by software.

Before demodulating, the frame of capturing signal should first carry through real signal FFT. Using the 4 pilot frequency will compute Local sampling clock delay, according to which the next frame of sampling clock will be adjusted. parallel-serial conversion, 16- QAM demodulation and error correct coding then will be put up. At last, using amplitude threshold judges whether have continuous frames. And when the 4 pilot frequency average amplitude less than a fixed value, the demodulation procedure finishes.

## 6 Conclusion

In the paper, MODEM is designed to use wire relay communication method in the feed automation system, and use OFDM transmission and DSP technique. It has the characteristics of lower cost, higher transmission rate, higher anti-interference, and it can meet the needs of communication in feed automation system.

## References

1. Orthogonal frequency division multiplexing, U.S. patent 3 488 445 (issued January 6, 1970)
2. Li, Z., Mammela, A.: An all digital frequency synchronization scheme for OFDM. In: Proc. IEEE Int. Symp. PIMRC 1997, Helsinki, Inland, pp. 327–331 (September 1997)
3. Landstrom, D., Wilsom, S.K., Van de Beek, J.-J., Borjesson, P.O.: Symbol time offset estimation in coherent OFDM systems. IEEE Transaction on Communication 50(4), 545–549 (2002)

# Influence of Supply Voltage of Node on RSSI-Based Localization Performance and Calibration Technique

Zhang Rongbiao, Guo Jianguang, Chu Fuhuan, and Song Yongxian

School of Electrical Engineering
Jiangsu University
Zhenjiang, China
zrb@ujs.edu.cn, jianguang_72@163.com

**Abstract.** Location accuracy is one of the key indicators in measuring the performance of wireless sensor localization systems. Apparently, the localization performance is largely determined by ranging precision in the distance-based localization system. In the RSSI-based distance estimation techniques, the accuracy of their estimation is vulnerable to the change of RSSI measured values. As is known, the Received Signal Strength Indicator (RSSI) is environment dependent in certain extent. In addition, the received signal strength mainly depends on the emitting power of the sender. In this paper, the RSSI measured values are investigated when the sending power gets changed due to low supply voltage of the sender, and then a calibrated method is proposed. Experiments show that the proposed method can significantly mitigate distance estimation errors caused by low supply voltage, thus it can be widely used in battery powered wireless sensor localization systems.

**Keywords:** localization, supply voltage, received signal strength, calibration technique.

## 1 Introduction

As is known, one of the wireless sensor network (WSN) applications is to estimate locations of sensors with an acceptable accuracy, which are used in the fields such as precision agriculture, bush fire surveillance, environmental monitoring and so on [2]. Practically, it is important in these applications that every sensor need to know its approximate location. Usually, it is difficult to program the position of each device in advance for a large-scale wireless sensor network. Hence, a great number of localization techniques have been presented one after another, which can be classified technically into three categories: Triangulation, Scene Analysis and Proximity [1]. In range-based localization techniques like triangulation, the improvement of one dimensional range estimated values in reliability and quality is achieved, so does the position estimation. In general, the process of ranging involves transmission and reception of signals between two communication nodes of a system. Compared with other ranging methods such as (angle of arrival) AOA [4], [5], (time of arrival) TOA [3], utral wide band (UWB) [13] and (time difference of arrival) TDOA [8], the (received signal strength indicator) RSSI [7], [9], [10] does not require a special or a complicated hardware. For these reasons, the RSSI has attracted a lot of attention in position estimation. However, the accuracy of range estimation that is based on the received signal strength indicator (RSSI) is highly sensitive to environment factors, such as multipath, fading and non-line of sight (NLOS) of the environment. These factors may result in large ranging errors. Furthermore, the emitting power of a node could directly affect the received signal strength, for most of nodes in the WSN (wireless sensor network) are battery powered. This implies that the supply voltage of the node may influence its emitting power in a certain extent. Some researches [6], [11], [14] have attempted to deal with the location errors caused by multipath effects, channel fading and so on. In the research [12], the author has mitigated the problem of low voltage supply of the node by modeling the variance of the battery status as a normal distributed random variable. Unfortunately this method is not fit for those position systems with highly accurate demand.

In this paper, to improve location accuracy, we will focus on making the most use of the actual RSSI measured values instead of novel positioning techniques. We adopt the most widely used propagation model proposed in [6], and design an accurate voltage calibration procedure which could reduce RSSI measurement errors caused by low supply voltage. In addition, we evaluate the localization performance of our proposed method in an indoor environment with an IEEE 802.15.4 wireless sensor network. The results show that the localization system with low voltage calibration performs better than that without calibration in terms of the accuracy of ranging and location estimation.

This paper is organized as follows. Section II shows the measuring principles for locations and the positioning algorithms used in this paper. In section III, we investigate the impact of supply voltage of the node on the RSSI measured values. A simple scheme is proposed to calibrate the RSSI measurement errors caused by bad battery status in section IV. Section V presents the experiment specification and details the experimental results of the research. Finally, in section VI, conclusions of the research are summarized.

# 2 Principle of Position Estimation

## 2.1 Principle of RSSI-Based Range Determination

Existing RSSI-based distance estimation methods use radio signal attenuation properties to model the distance between two communication nodes as a function of the received signal strength indicator (RSSI). In this paper, we will improve the widely accepted propagation model proposed in [6], which can be stated as (1).

$$P_R(d)(dBm) = A - 10n\lg\left(\frac{d}{d_0}\right) + X_\sigma + RSS_{VC} \tag{1}$$

where $P_R(d)$ is the received power in dBm at distance d from a sender, A is the received power at the reference distance $d_0$. n is the path loss exponent, typically at 2 to 4, which measures the rate at which the received signal strength decreases with distance increasing. The value of n depends on the specific propagation environment. RSSvc is the compensated value of RSSI measurement errors caused by low voltage supply. In this research, n can be calculated through two beacons nodes during experiments, and is assumed to be a constant in our experiment. In addition, $d_0$ and A, which are the known constants, can be obtained from a priori calibration of the wireless device. $X_\sigma$ can be used to approximate the RSSI measurement errors caused by multipath and channel fading.

### 2.2 The Principle of Position Estimation

In most of distance-based localization algorithms, to estimate the position of the blind node, three beacon nodes at least are needed to measure the blind node's signal strength in the network. When the distance between the blind and the beacon node is estimated through the measured RSSI, then the possible position of the blind node can be represented by a circle at the radius of estimated distance $d_i$. As depicted in Fig. 1, by means of trilateration, the intersection point of three circles gives the blind node location (white node).

As shown in Fig. 1, it is assumed that the location of the i th beacon node could be expressed as $(x_i, y_i)$, $i \leq n$ $(n \geq 3)$, the blind node with unknown location can be expressed as (x, y). The distance between the i th beacon and blind node $d_i$ could be estimated from the corresponding RSSI measured values, then the location of the blind node can be calculated by the following overdetermined nonlinear equations [16]:

$$\begin{cases} (x-x_1)^2+(y-y_1)^2=d_1^2 \\ (x-x_2)^2+(y-y_2)^2=d_2^2 \\ \vdots \\ (x-x_n)^2+(y-y_n)^2=d_n^2 \end{cases} \tag{2}$$

Linearizing (2) gives proper linear equations in the form of (3)

$$AX = B \tag{3}$$

where

$$A = \begin{bmatrix} 2(x_1 - x_n) & 2(y_1 - y_n) \\ \vdots & \vdots \\ 2(x_{n-1} - x_n) & 2(y_{n-1} - y_n) \end{bmatrix} \quad B = \begin{bmatrix} x_1^2 - x_n^2 + y_1^2 - y_n^2 + d_n^2 - d_1^2 \\ \vdots \\ x_{n-1}^2 - x_n^2 + y_{n-1}^2 - y_n^2 + d_n^2 - d_{n-1}^2 \end{bmatrix},$$

$$X = \begin{bmatrix} x \\ y \end{bmatrix}$$

The Equation (3) can be solved by using the standard least-squares method

$$\hat{X} = (A^T A)^{-1} A^T B \tag{4}$$

Obviously, in multilateration scheme, the estimated distance $d_n$ with low accuracy takes over the major source of localization errors. Therefore, in order to improve the range-based localization accuracy, we mainly focus on improving the accuracy of distance estimation.

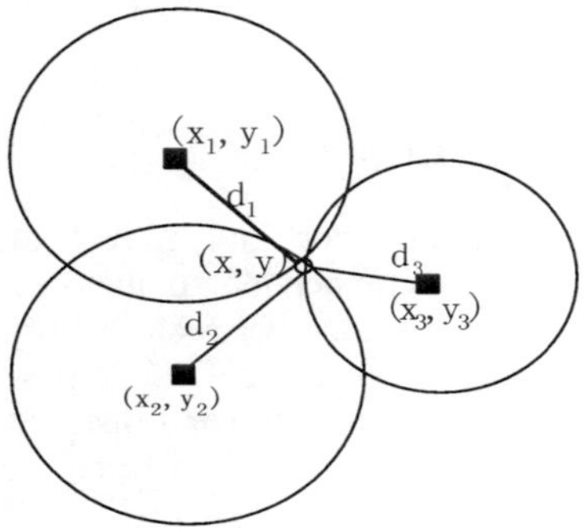


**Fig. 1.** The principle of trilateration

## 3 Impact of Voltage Supply of Node on RSSI Measurement

In this part, we will verify the impact of battery status of nodes on RSSI measured values. The details of the experiment testbed and relevant parameters of our devices will be shown in section V. In the first experiment, the same sender and receiver are placed 3.6 meters apart with the height of 0.2 m to the floor. The antennas of two nodes on the floor are pointing to the ceiling. The sender transmits a packet to the receive tags 40 times at the power level of -16 dBm, and the receiving node records RSSI values each time. When the process mentioned above is finished, we change the battery voltage of the sender, then repeate the experiment. The results shown in Fig. 2 indicate that the received signal strength at a given distance varies significantly when the supply voltage of the sender is less than 2.1V.

## 4 Solution

In order to correct the measured errors due to low supply voltage of nodes, the experiment No. 2 has been conducted in the testbed. Before the experiment is carried out, several preparation steps are necessary. Firstly, the sender is equipped with 3.3V batteries. Secondly, positions, heights and antennas of the two nodes are kept the same as the experiment No. 1, so does the sending power. During this experiment, the sender transmits a packet in which the supply voltage and ID of the sender are included, and then the listening node will record both RSSI measured values and the sender's battery status. The process continues until the battery is out of power.

As illustrated in Fig. 3, the trend of RSSI measured values indicates that the received signal strength still keep relatively stable until the supply voltage of the sender is below the threshold of 2.3V. The solid line is fitting curve of the experimental RSSI measured values by least square method.

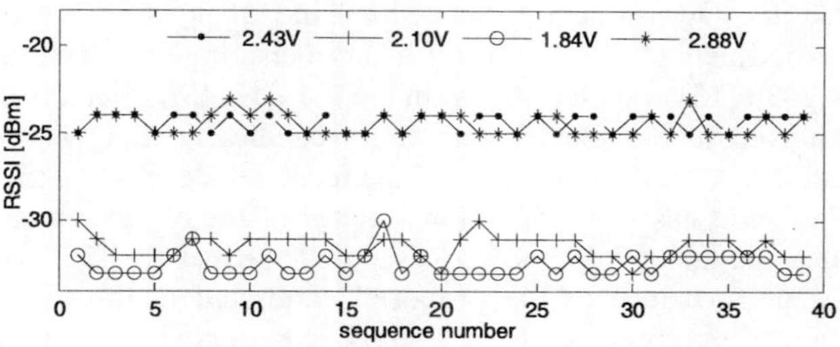


**Fig. 2.** Comparison of measured RSSI with the same distance but different battery status of the sender

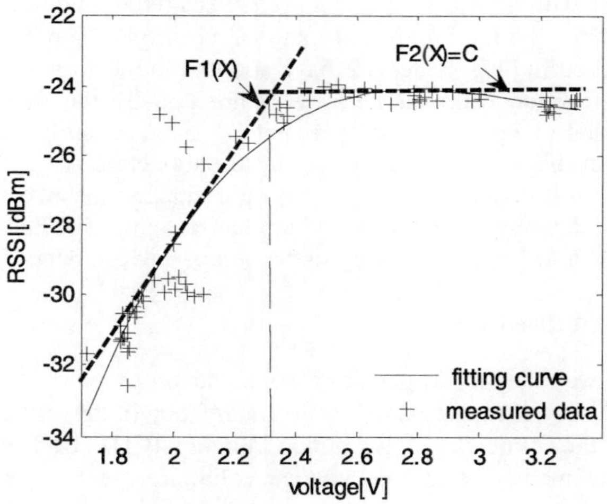


**Fig. 3.** RSSI measured values versus supply voltage of the sender

To reduce the complexity of calibration, we have deployed the linear function F1(x) to approximate the curve when the supply voltage is less than the threshold, and a const C is used to approximate the curve when the voltage is above the threshold. Therefore, the RSSvc can be computed according to different voltage V by means of (5)

$$RSS_{VC}=\begin{cases} K(2.3\text{-}V) & 1.8 \le V \le 2.3 \\ 0 & 2.3 < V \le 3.3 \end{cases} \tag{5}$$

where K denotes the slope of the straight line F1(x).

# 5 Experiment and Results

## 5.1 Experimental Setup

All experiments have been carried out in an indoor environment to test the influence of battery status on RSSI measured values and validate the proposed approach mentioned in previous sections. Our student office with (line of sight) LOS environment is selected as the experiment testbed. The physical dimension of the testbed is 15m×10m. The chipcon CC2430 [15], which operates in the 2.4 GHz ISM band, is the centerpiece of the node deployed in our experiment. It is specifically designed for low power wireless applications. A built-in indicator of the received signal strength gives an 8-bit digital value. The node has a length of one quarter of the electrical wavelength (λ/4) monopole antenna and the emitting power is set to -16dBm.

To verify the performance of the proposed calibrated method in mitigating the RSSI-based localization errors which are caused by battery status, the experiment No. 3 is conducted in the testbed. In this experiment, we deploy 3 reference nodes whose supply voltages are lower than 2.3V in the localization network. In addition, all nodes with antennas pointing to the ceiling in the testbed are placed 0.2 m above the ground. Twelve blind nodes with normal supply voltage are scattered randomly in the testbed. In contrast, reference nodes are put to known positions. The process of RSSI collection is proposed in [17], which we have adopted in the experiment No. 3, then the distance and position estimation can be computed by the method mentioned above. As mentioned in Section II, the parameters A and n can be estimated before localization, thus in this experiment the estimated parameters A= –20dBm ($d_0$=1m), n=2.5. However, in this paper we mainly focus on mitigating the errors caused by low supply voltage of the reference nodes rather than ranging and localization errors owing to multipath fading and shadowing in the indoor environment.

## 5.2 Experimental Results

In this subsection we analyze the performance of our proposed calibrated method in improving the accuracy of ranging and location estimation in the testbed. Fig. 4 shows the comparison of the Cumulative Distribution Function (CDF) of distance estimation errors between the method with low voltage calibration and the method without

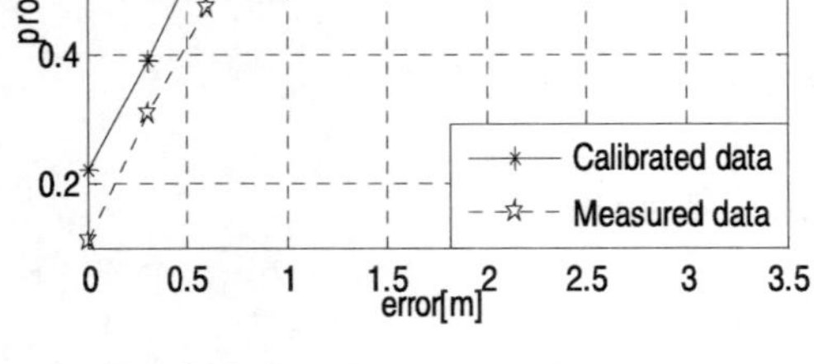


**Fig. 4.** Comparison of CDF of the distance estimation error

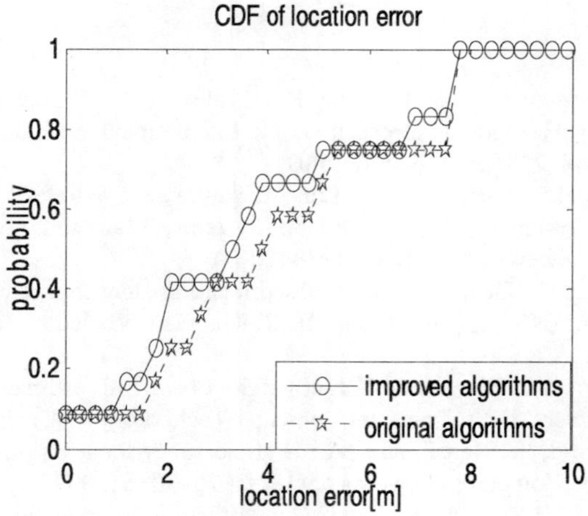


**Fig. 5.** CDF of the location error

calibration. Clearly, as shown in Fig.4, without increasing calculation complexity, our proposed ranging method outperforms the traditional one in term of the accuracy of distance estimation.

As depicted in Fig. 4, the CDF of the distance estimation errors obtained from low voltage calibration firstly reaches the peak at the error of 2.2 m, while the other one marked pentagram does not reach to 1 until the distance estimation errors increases to 3 m. Apparently, it is our proposed scheme that takes effect in mitigating the distance determination errors, even if the accuracy of ranging is not as high as UWB for the multipath effects in indoor environments.

As is known, in most of rang-based localization approaches, the accuracy of one dimensional range measurement is the key point in improving the performance of the localization system. Fig. 5 shows the localization results computed by trilateration. It can be seen that the proposed scheme performs well in mitigating the location errors caused by low voltage supply of the sender.

# Conclusion

In this research, we consider the impact of battery status of the sender on the accuracy of distance and location estimation in the RSSI-based indoor wireless localization system. Besides, a novel experiment-based calibrated scheme is proposed to decrease distance estimation errors due to the abnormal battery status. The concrete calibrated data are calculated according to the current supply voltage of the sender. Several experiments are conducted to verify the efficiency of the proposed scheme in reducing the location errors. In conclusion, without increasing the complexity of ranging and position estimation, we obtain a significant effect in mitigating the location error due to low battery voltage. Therefore, the calibrated method can be widely used in battery powered wireless sensor localization systems.

## References


1. Garcia, E.M., Bermudez, A., Casado, R., Quiles, F.J.: Wireless Sensor Network Localization using Hexagonal Intersection. In: International Federation for Information Processing IFIP, vol. 248, pp. 155–166 (2007)
2. Liu, H., Darabi, H.S., Banerjee, P., Liu, J.: Survey of Wireless Indoor Positioning Techniques and Systems. IEEE Trans. on Systems, Man and Cybernetics, Part C: Applications and Reviews 37(6), 1067–1080 (2007)
3. Wang, X., Wang, Z., O'Dea, B.: A TOA-based location algorithm reducing the errors due to non-line-of-sight (NLOS) propagation. IEEE Trans. on Vehicular Technology 52, 112–116 (2003)
4. Klukas, R., Fattouche, M.: Line-of-sight angle of arrival estimation in the outdoor multipath environment. IEEE Trans. on Vehicular Technology 47(1), 342–351 (1998)
5. Boushaba, M., Hafid, A., Benslimane, A.: High accuracy localization method using AoA in sensor networks. Computer Networks 53(18), 3076–3088 (2009)
6. Ladha, C., Sharif, B.S., Tsimenidis, C.C.: Mitigating propagation errors for indoor positioning in wireless sensor networks. In: Proc. IEEE Mobile Adhoc and Sensor Systems (MASS), Pisa, Italy, pp. 1–6 (2007)
7. Bahl, P., Padmanabhan, V.N.: VNRADAR: An In-building RF-Based User Location and Tracking System. In: Proc. 9th IEEE Computer and Communications Societies, vol. 2, pp. 775–784 (2000)
8. Chen, C.K., Gardner, W.A.: Signal-Selective Time-Difference-of-Arrival Estimation for Passive Location of Man-Made Signal Sources in Highly Corruptive Environments, Part II: Algorithms and Performance. IEEE Trans. on Signal Processing 40(5), 1168–1184 (1992)
9. Alippi, C., Vanini, G.: A RSSI-based and calibrated centralized localization technique for Wireless Sensor Networks. In: Proc. 4th IEEE Int. Conf., Pervasive Computing and Communications Workshops (PERCOMW 2006), pp. 301–305 (2006)
10. Chung, W.Y., Lee, B.G., Yang, C.S.: 3D virtual viewer on mobile device for wireless sensor network-based RSSI indoor tracking system. Sensors and Actuators B: Chemical 140(1), 35–42 (2009)
11. Bergamo, P., Mazzini, G.: Localization in sensor networks with fading and mobility. In: 13th IEEE International Symposium on Personal, Indoor and Mobile Radio Communications, vol. 2, pp. 750–754 (2002)
12. Zhou, G., He, T., Krishnamurthy, S., Stankovic, J.A.: Models and Solutions for Radio Irregularity in Wireless Sensor Networks. ACM Transactions on Sensor Networks (TOSN) 2(2), 221–262 (2006)
13. Gezici, S., Tian, Z., Giannakis, G., Kobayashi, H., Molisch, A., Poor, H., Sahinoglu, Z.: Localization via ultra wide band radios: a look at positioning aspects for future sensor networks. IEEE Signal Processing Magazine 22(4), 70–84 (2005)
14. Barsocchi, P., Lenzi, S., Chessa, S., Giunta, G.: A Novel Approach to Indoor RSSI Localization by Automatic Calibration of the Wireless Propagation Model. In: Proc. 69th IEEE Vehicular Technology Conference (VTC), pp. 1–5 (Spring 2009)
15. A True System-on-Chip solution for 2.4 GHz IEEE 802.15.4/ZigBee, `http://focus.ti.com/docs/prod/folders/print/cc2430.html`
16. Savvides, A., Han, C.C., Strivastava, M.B.: Dynamic fine-grained localization in Ad-Hoc networks of sensors. In: Proc. 5th IEEE Int. Conf. on Mobile Computing and Networking, Rome, Italy, pp. 166–179 (2001)
17. Patil, M.M., Shaha, U., Desai, U.B., Merchan, S.N.: Localization in Wireless Sensor Networks using Three Masters. In: Proc. IEEE Int. Conf. on Personal Wireless Communications (ICPWC 2005), pp. 384–388 (2005)

# Weak Signal Detection Method under the Strong Noise Background

Zhang Rongbiao, Chu Fuhuan, Ran Li, and Guo Jianguang

School of Electrical Engineering
Jiangsu University
Zhenjiang, China
zrb@ujs.edu.cn, chuchu1986115@163.com

**Abstract.** In response to the instability phenomenon of the weak signal detection process based on the single chaotic oscillator, an improved method which detects the weak signal by adjusting the system to reach a stable chaotic state is proposed. This method combines the chaotic attractor and double differential oscillator detection, and decides the critical value from chaos to period by the double differential oscillators at the same time. The simulation shows that this method can avoid misjudgment effectively and has a good real-time performance, which is beneficial to weak signal detection under the strong noise background, and it provides a referential method for practical engineering application.

**Keywords:** weak signal detection, double oscillators, chaos, stability.

## 1 Introduction

There are many relatively sophisticated methods for weak signal detection in time domain, such as correlation detection [1], wavelet transform [2] and so on. But in traditional detection methods, the minimum working signal-to-noise ratio (SNR) threshold is only -10dB, which limits the widespread application of these methods [3]. Since Birx used chaotic oscillators for detecting weak signal in 1992, the method has attracted much attention of domestic and foreign scholars in the defense industry, biomedical engineering, materials testing, confidential communications, fault diagnosis and other areas [4-8]. Weak signal detection method based on chaotic oscillators makes use of the feature that the oscillator is immune to noise but sensitive to the weak signal having the same frequency with the system period driving force to achieve the detection of the weak signal drowned in noise [9]. But during simulation, we find that the system has to go through a period of time before transiting to the final steady state. If the noise is added during the transition process, sometimes it will have a certain influence on the system and even change the final state of the system, thus the detection performance of the system is reduced due to the failure of distinguishing that which one really takes effect for the system phase transformation, noise or weak signals [10], [11]. Therefore, it is necessary to find a method that can eliminate the impact of noise on the transition process, prevent misjudgment, and improve the stability of the system detection process. In this paper, an improved method, on the basis of the single chaotic oscillator detection, is proposed. In this method, weak signals are detected by adjusting the oscillators to reach a stable chaotic state. Using two oscillators with a slight difference in driving force simultaneously, the depth of the system in periodic and chaotic state can be determined by the differential output amplitude, the movement state of the system can also be identified quickly and easily without blindly increasing the simulation time, and the critical chaotic value of the system can effectively be determined. The simulation results show that the method can effectively detect weak signals and has a good real-time performance. Furthermore, it is more stable than the detection method based on a single chaotic oscillator.

## 2 Single-Oscillator Chaos Detection Method

### 2.1 Detection Model

The chaotic oscillator is immune to noise and the signal having larger frequency difference from the system periodic force but sensitive to the weak signal having the same frequency with system periodic force. Adding the weak periodic signal as a disturbance of the chaotic system, the system will immediately transit from the chaotic state to the large-scale periodic state. Through the system identification, we can detect the weak signal by determining whether there are specific frequency signals. The Homes-Duffing equation, as listed in (1), is seen as a relatively mature weak signal chaotic detection model.

$$\ddot{x} + k\dot{x} - x + x^3 = r\cos(t) \tag{1}$$

where k is the ratio of damping, typically k = 0.5, $-x+x^3$ is the non-linear recovery force, and rcos(t) is the periodic driving force.

$$s(t) = a\cos(t) + n(t) \tag{2}$$

It is assumed that the weak signal to be measured is acos(t), and white noise is n(t), so mixed-signal to be measured is (2). When s(t) is added to the right of (1), (3) can be got. The right side of (3) denotes the system periodic driving force.

$$\ddot{x} + k\dot{x} - x + x^3 = r\cos(t) + (a\cos(t) + n(t)) \tag{3}$$

While ω=1, (4) is the corresponding state equation.

$$\begin{cases} \dot{x} = y \\ \ddot{x} + k\dot{x} - x + x^3 = r\cos(t) + (a\cos(t) + n(t)) \end{cases} \tag{4}$$

Based on above state equations, the system detection model is set up as shown in Fig. 1.

In the case of the fixed k and the given initial situation, the system will quickly transit from the homoclinic orbit and bifurcation state to chaos with the amplitude of periodic driving force r gradually increasing from zero, which will go into the large-scale periodic state after a long time of the chaotic state. Adjust the system to the critical state which is the transition from the chaotic state to the large-scale periodic state, add the signal to be measured, and observe the state of the system to determine whether there are useful weak signals. If the system keeps chaotic status, but the edge of the system trajectory has a strong disturbance, then it illustrates that the signal to be measured contains only noise. If the system immediately transits into the periodic state, it means that it contains a useful weak signal. Then reduce the amplitude of the driving force gradually until the system re-enters the chaotic state. So it is easy to calculate the amplitude of weak signals by comparing two time amplitudes of the system driving force, and then the weak signal can be detected from the strong noise background. Therefore, this method determines the existence of a useful weak signal mainly by observing whether the system generates phase transformation.

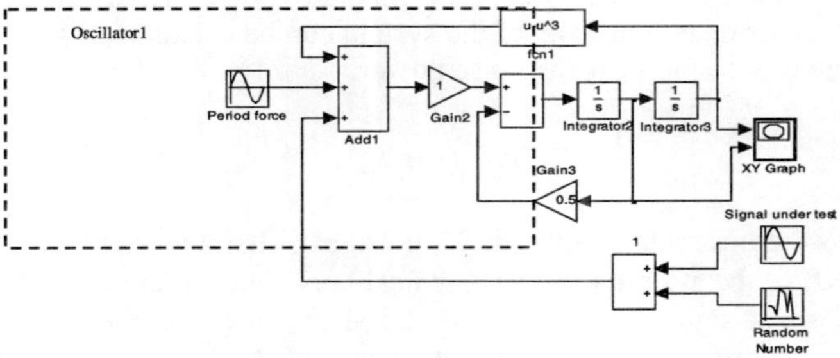


**Fig. 1.** Single-oscillator chaos detection model

In order to improve the general adaptability of the system detection model, we can do $t=\omega\tau$ scale transform for the Duffing equation (4) while $\omega\neq 1$. To facilitate the description, this article only describes the weak signal detection when $\omega=1$.

## 2.2 Impact of Noise on the System Detection Performance

In most cases, single-oscillator chaotic detecting model is able to detect weak signals from the strong noise background. However, the system can easily be noise-driven because the critical state of the single-oscillator chaotic detection system is unstable. During the simulation, we find that the adding of certain intensity noise has an impact on the phase state of the system. We enable the system to reach a chaotic critical state by adjusting period driving force $r=0.8260084$. It is assumed that a fixed step length is $h=0.01$, simulation time is $1000\times\pi$, and the initial value is $(x(0),x'(0)=(0,0))$. When the simulation starts, the white noise with variance of 0.001 and an average of 0 is added. The Equation (4) can be solved by using fourth-order Runge-Kutta method. Clearly, when the system is running to the 1800s, the envelope of x time-domain waveform is nearly a straight line, and the phase diagram presents a periodic state, thus it shows that the adding of the noise makes the system generate phase transformation. The x time-domain waveform of the system and the corresponding phase diagram are shown in Fig. 2.

Thus, the detection model, in a certain noise level, makes the system generate phase transformation. This situation makes us mistakenly believe that there is a useful weak signal, thereby the performance of the weak signal detection is greatly reduced.

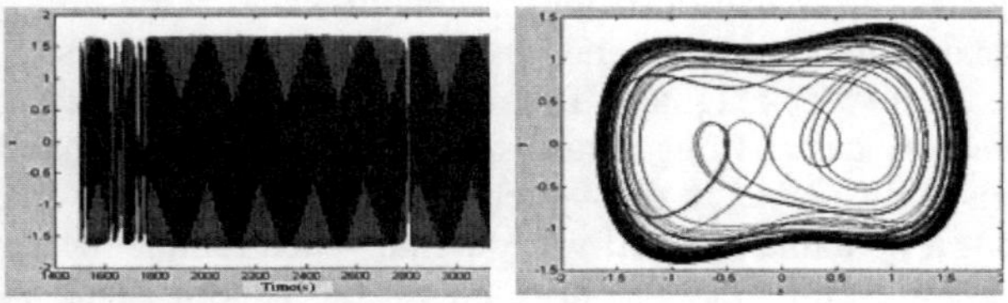

**Fig. 2.** x time-domain waveform of the system added noise with variance of 0.001 and the corresponding phase diagram

Separately adding a small signal in several different noise intensities, the minimum signal amplitude that drives the system to generate phase transformation can be observed, and then the detection SNR of the system can be calculated by (5), where $P_s$ denotes the signal power and $P_n$ is the noise power.

$$SNR = 10\lg \frac{P_s}{P_n} \tag{5}$$

Table 1 shows the minimum amplitude of the signals that the system can detect and the detection SNR of the system in several typical noise intensities. The symbol '--------' denotes that weak signals can not be detected in those noise intensities for the phase transformation of the system caused by noise. And we can see that single chaotic oscillator detection model can not effectively detect weak signals under certain noise level.

# 3 Improvement of the Detection Model

## 3.1 Improved System Detection Model

When the system is very close to the critical state, it is easily driven by noise, and the noise with pretty low amplitude would lead to the phase transformation of the system, therefore the stability of the system detection process will be greatly reduced. In order to improve the stability of the system detection process, we adjust the system to reach a stable state when the noise is added. There will be a transition process before the system enters the final steady state, but the adding of the noise would have an impact on the transition process and even change the final motion state of the system. In order to eliminate the noise-induced system phase transformation, the weak signal can be detected by adjusting the chaotic state of the system, and the signal to be measured is added when the system reaches a stable motion state. The improved detection model can be expressed as (6).

$$\begin{cases} x'' + kx' - x + x^3 = r\sin(t) + s(t)U(t - t_0) \\ x'' + kx' - x + x^3 = R\sin(t) + s(t)U(t - t_0) \end{cases} \tag{6}$$

$$s(t) = a\sin(t) + n(t) \tag{7}$$

where the signal to be measured is (7). The period driving forces of two chaotic oscillators, which can be regulated to make the system reach a critical chaos state, respectively, are R and r, R=λ×r, setting λ = 1.00001. U(t-$t_0$) is a step signal in which the value of $t_0$ is generally larger than that of the time of the system transition process. It is used to regulate the adding time of the signal to be measured, which ensures that the system has been in the stable chaotic state when noise is added. The created block diagram of the system detection model is shown in Fig. 3.

**Table 1.** The Detection Performance of the Method with a Single-Oscillator

| Noise input variance | Signal Amplitude Threshold | SNR |
|---|---|---|
| 0.1 | 0.02 | -27dB |
| 0.01 | 0.002 | -37dB |
| 0.001 | -------- | -------- |
| 0.0001 | 0.00005 | -69dB |
| 0.00001 | -------- | -------- |

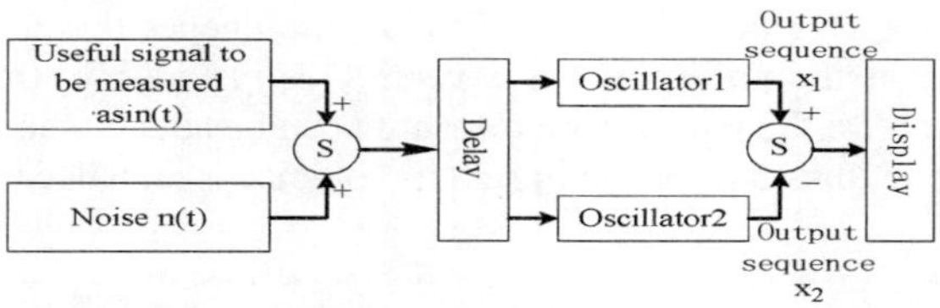


**Fig. 3.** Improved system detection block diagram

Fig. 3 shows that the detection model consists of double chaotic oscillators, delay module, addition module, subtraction module and display module. The specific detection model of oscillator 1 is the same as what it is in the dashed box in Fig. 1. The driving force amplitude of oscillator2 is λ-fold as that of the oscillator1, and other parameters are the same with oscillator1's. Firstly, the amplitude of the system driving force is adjusted to the chaotic critical state, and then the mixed-signal to be measured is added to the double oscillators system after a delay of $t_0$. Two-oscillator system is solved by using fourth-order Runge-Kutta method and the output sequences of the chaotic system, $x_1$ and $x_2$ are obtained respectively. The output sequences are processed through the subtraction module, and then the differential output sequence is got, which will be displayed in the scope module finally. As system period driving forces of double oscillators have only a tiny difference, such double oscillators have the following properties, in the chaos and large-scale periodic state, the outputs of two graphics have a very clear distinction, and the system state can be easily determined in a short time. Taking difference operation between two oscillators can fully suppress the period state and highlight the chaotic state, eliminate common mode interference, and effectively improve anti-noise performance of the system. Through observation of the output differential sequence, we find, in the chaotic state, that the output amplitudes of two oscillators are greatly different and irregular, and the waveform fluctuates seriously after difference operation. In the period state, the output amplitudes of two oscillators are slightly different and relatively stable, and the waveform remains smooth with small amplitude after difference operation. Automatic identification of weak signals using computers can be achieved by setting a threshold value. And setting a time threshold, as long as the amplitude of the differential waveform is greater than the setting threshold value continuously within the setting time, it can be considered that the system is in a chaotic state, and vice versa, in a periodic state. The differential graphics of the system in chaotic state and period state and the corresponding phase diagram are shown in Fig. 4.

By comparing the camera picture and the differential graphic, we find that there is a corresponding relationship between them, so the phase transformation of the system can be determined by the differential output. In comparison with the way that directly uses "XY phase diagram" module, this method saves a lot of machines memory, increases computing speed, and effectively improves the real-time performance of the detection system. The movement state of the system can be observed by judging the shape of the differential graph, thus more accurate critical value is obtained without blindly increasing the simulation time. Meanwhile, the chaotic depth of the system can be determined through the differential graph, the greater the amplitude is, the deeper the chaotic depth of the system. Also the level of immunity to noise is different due to the difference in depth of chaos, the deeper the chaos depth is, the smaller the impact of the noise, and vice versa [12]. Therefore, through observing the differential graphics, the signal to be measured can be added when the system is in the deep chaotic state, thus the system immunity to noise is enhanced.

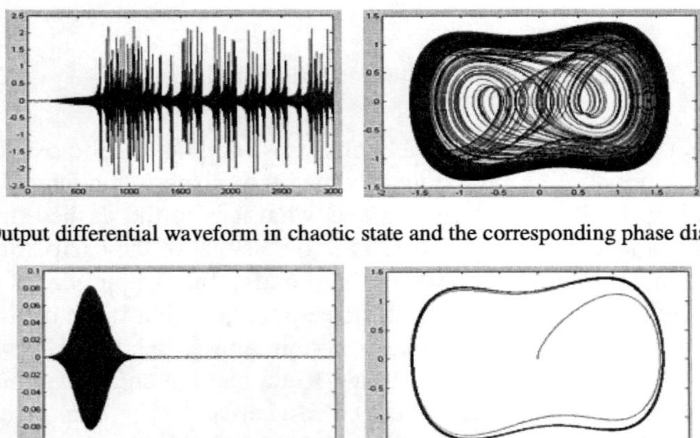

(a) Output differential waveform in chaotic state and the corresponding phase diagram

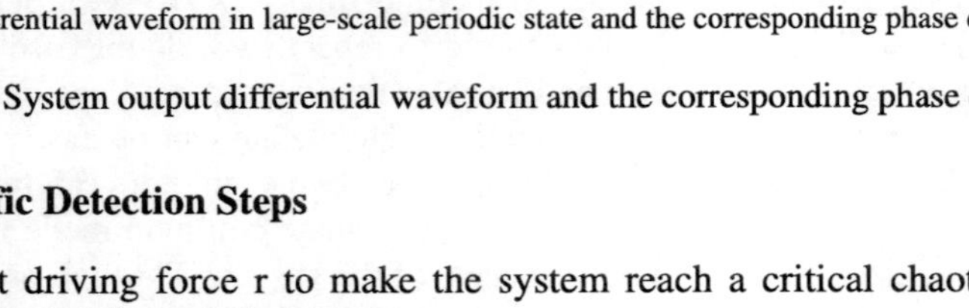

(b) Output differential waveform in large-scale periodic state and the corresponding phase diagram

**Fig. 4.** System output differential waveform and the corresponding phase diagram

### 3.2 Specific Detection Steps

1. Adjust driving force r to make the system reach a critical chaotic state from chaos to a large-scale period state.

2. Select a appropriate delay of $t_0$. In the simulation, we find that the system probably ends the transition process around the 700s and reaches the final steady state, which can also be seen in Figure 4. Before the 700s, the amplitude of the waveform after difference operation is relatively large and unstable, indicating that the system is in the transition process. After the 700s, a stable waveform with small

amplitude is obtained after differential operation, indicating that the system has already been in a steady period state. So we choose $t_0 = 300\times\pi$, which means that the signal to be measured is added when the system enters the 150th cycle.

3. Add the signal to be measured and observe the differential output of the system to detect the weak signal. If the system transits from the chaotic state to the large-scale periodic state, the signal to be measured contains the weak signal having the same frequency with period driving force. On the other hand, if the system state does not change, the signal to be measured is a pure noise.

4. If the system generates phase transformation, the amplitude of weak signals can be calculated by adjusting r, and the calculation method is the same with the detection based on a single oscillator.

# 4 Analysis of Weak Signal Detection Performance

## 4.1 Impact of Noise on the System

Take the same simulation conditions as the single-oscillator detection method, and add white noise with the variance of 0.001 and an average of zero when the system goes into the 150th cycle. We find the system does not generate phase transformation, which shows that the system has a certain degree of anti-noise. Obviously, this method has better anti-noise performance than the detection based on single oscillator. Output of differential Graph and the corresponding phase diagram are shown in Fig. 5.

As can be seen from Fig. 5, the system is still in the chaotic state, at this time, the amplitude change of the differential graphic is more intense and the envelope of the phase diagram is rough, which are the results of disturbance caused by the noise in the system. However, the immunity to noise is enhanced, for the noise is added when the system is in a steady motion state, so the added noise does not change the final motion state of the system.

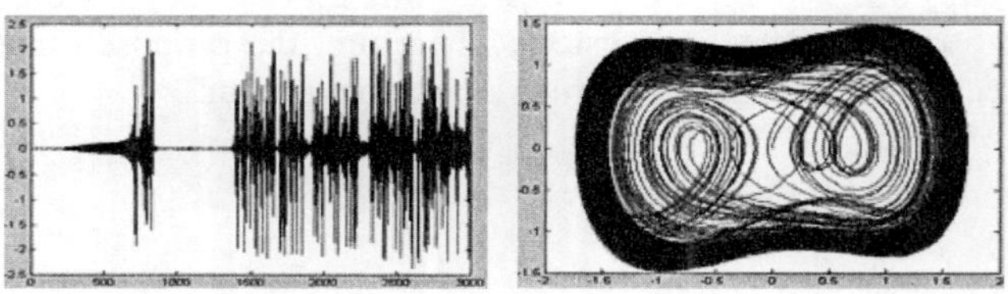

**Fig. 5.** The differential graph after adding noise and the corresponding phase diagram

## 4.2 Analysis of Detection Performance

Take the same condition in the simulation as the single-oscillator detection method, and we can get the critical chaotic value of r=0.8260084 after repeated simulation. Take $t_0$=300×π when the signal to be measured is added, and then observe the state of the system. Table 2 lists the signal amplitude threshold which the system can detect in several typical noise backgrounds and the corresponding SNR. Comparing Table 1

**Table 2.** Detection Performance of Improved Methods

| Noise input variance | Signal Amplitude threshold | SNR |
|---|---|---|
| 0.1 | 0.009 | -34dB |
| 0.01 | 0.002 | -37dB |
| 0.001 | 0.0003 | -43dB |
| 0.0001 | 0.00005 | -69dB |
| 0.00001 | 0.00004 | -41dB |

with Table 2, it is clear that, in the improved method, the detection SNR is decreased, the immunity to noise of the system is enhanced, and detection stability is greatly improved under certain noise level.

## 5 Conclusion

The conclusion of noise immunity, which is derived from theory of stochastic differential equations, is a study result of noise statistical properties, and it could not determine its impact on the detection system. In this paper, through analyzing the weak signal detection method based on a single chaotic oscillator, a method that detects weak signals by adjusting the chaotic state of the system is proposed. The signal to be measured is added after the transition process, thus this method avoids the influence of the noise on the phase state of the system, effectively enhances the immunity to noise, and improves the stability of the system testing process. Meanwhile, the system detects the weak signal by using double differential oscillators, the motion state of the system can be easily observed without blindly increasing the simulation time, and the real-time performance of the detection system is enhanced. Compared with methods such as traditional correlation detection, wavelet transform and other methods, this approach can detect the weak signal under much lower SNR. Furthermore, it enhances the immunity to noise more significantly than the single-oscillator detection system. Therefore, the proposed method is more suitable for detecting weak signals in strong noise background.

## References


1. Xiao, H.M., Wu, J., Chen, C.G., Su, X.Z.: Correlation Detection of Weak Laser Pulse Signals. Optics & Optoelecrtonic Technology 2(1), 61–64 (2004)
2. Zhang, R.B., Hu, H.Y., Feng, Y.B.: Study on weak signal detection method based on wavelet entropy. Chinese Journal of Scientific Instrument 28(11), 2078–2084 (2007)
3. Li, Y., Yang, B.J.: On detection of chaotic oscillator Citation, pp. 49–55. Publishing House of Electronics Industry, Beijing
4. Zheng, S.Y., Guo, H.X., Li, Y.A., Wang, B.H., Zhang, P.Y.: A new method for detecting line spectrum of ship-radiated noise using Duffing oscillator. Chinese Science Bulletin 54(14), 1906–1912 (2007)
5. Dai, Z.C., Wang, B.W.: Secure communication based on chaotic synchronization. System Engineering and Electronics 29(5), 699–702 (2007)

6. Li, C.S., Qu, L.S.: Fault Diagnosis of Rolling Bearings Based on Chaos and Symbol Sequence Statistics. Journal of Xi'an Jiaotong University 39(3), 261–265 (2005)
7. Wang, Y.C., Zhao, Q.C., Wang, A.B.: Extraction of periodic signals in chaotic secure communication using Duffing oscillators. Chinese Physics B 17(7), 2372–2376 (2008)
8. Wang, W., Li, Q., Zhao, G.J.: Novel approach based on chaotic oscillator for machinery fault diagnosis. Journal of the International Mesurement Conference 41(9), 904–911 (2008)
9. Wang, G.Y., Chen, D.J., Lin, J.Y., Chen, X.: The Statistical Characteristics of Weak Signal Detection Based on Duffing Oscillator. Acta Electronica Sinica 26(10), 38–44 (1998)
10. Wang, Y.S., Ma, X.L., Wei, Y., Fan, H.D.: A New Method of Weak Signal Detection Using Chaos Phase Change. In: 8th International Conference on Electronic Measurement and Instruments (ICEMI), pp. 812–816 (2007)
11. Xie, T., Wei, X.Y.: Study on reliability of chaotic oscillator in weak signal detection. Chinese Journal of Scientific Instrument 29(06), 1265–1269 (2008)
12. Zhang, H.X.: Extraction of Partial Discharge Signal Feature with Chaos Oscillator and Adaptive Lifting Wavelet. Sichuan Chong Qing, Xihua University (2008)

# Design of Rapid Control Prototyping Platform Based on RTX

Zhi Xu and Shuo Tang

College of Astronautics
Northwestern Polytecnical University
Xi'an, China
xuzhi@nwpu.edu.cn, suotang@ nwpu.edu.cn

**Abstract.** The rapid control prototyping(RCP) system that based on Simulink model and RTX real time system was designed. The RCP system can rapidly convert the Simulink model into the RTX real time model and realize: the automatic compilation and downloading of the Simulink model, the on-line modification of simulation parameters, real time storage of simulation data, automatic distribution simulation, configuration file administration. The simulation results, given in Fig. 8 and Table 1, show preliminarily that the RCP system designed by us is not only reliable but also gives good precision.

**Keywords:** component, rapid control prototyping, RTX real time system, hardware-in-the-loop simulation.

## 1 Introduction

Rapid control prototyping(RCP) that is rapid functional prototype test. RCP simulation and hardware-in-loop simulation (HILS) is the difference because the former is to realize the controller functions, while the latter is a device simulation or environment to realize a certain device or the entire external environment functions. There are now several typical RCP system such as DSPACE, xPC and RT-LAB. These RCP systems use host and target mode, the simulation model of the system has been developed in simulink environment by the host with WINDOWS operating system, while target for real-time simulation.

In this paper, the rapid prototyping systems is designed based on RTX(Windows Embedded Real-time System) and simulink environment for model development. The system can quickly convert simulink models into real-time model based on RTX environment, automatically compile and download the simulink model, on-line changes simulation parameters, on-line storage simulation data, automatic distributed simulation and simulation project file management functions. The system can achieve the closed-loop real-time simulation for complex system and can quickly achieve a single component prototype testing.

## 2 Rapid Prototyping System Support Environment

The support environment of Rapid prototyping system including software and hardware two parts. Software components including real-time operating system and modeling tools, hardware components including computer and other I/O interface boards. The system uses the design mode of RTX for real-time systems, simulink for model development tools, the x86-based Industrial control computer for hardware platform and software and hardware separation, so it was able to make the system with flexible and configurable, make user's models with modular, and maximize cost savings in order to improve design efficiency.

### 2.1 RTX Real-Time Architecture

RTX is designed to a real-time extension of subsystems for Windows. RTX provides the precise control for IRQ, I/O and memory in order to ensure real-time task execution with a 100% reliability. As as a full Windows extended system, RTX does not modify any of the Windows system architecture. RTX subsystem (RTSS) has a precise high-speed task scheduler, RTX scheduler can ensure that the thread switching time maintained at 500 nanoseconds to 2 microseconds. RTX clock resolution to 100 nanoseconds, and will not produce any drift. Its timer cycle can be done a minimum 100 microseconds. As an extension of Windows subsystem, in addition to using Windows resources, RTX can exist independently, even in the Windows blue screen when the system presented, RTX can still robust operation.

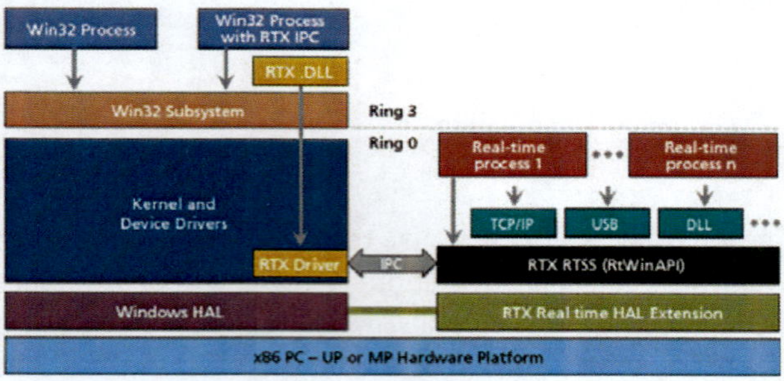


**Fig. 1.** RTX Architecture

### 2.2 Simulation Modeling Tools—Simulink/RTW

Simulink is a software package that used to modeling, simulation and analysis for dynamic systems, Simulink provides a library of modules to help users quickly create a dynamic system model. RTW can automatically generate code based on the simulink environment, directly can generate optimized, portable and personalized code from the simulink model, and automatically generate a variety of configurations depending on the target environment programs. The system model designed by the user can be directly converted to the original source by RTW that user need not write any program code.

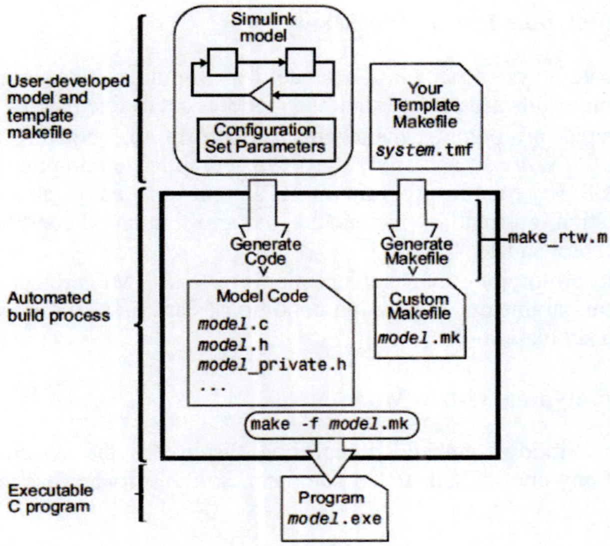


**Fig. 2.** Simulink/RTW Workshop executes

# 3 Rapid Prototyping Framework

The rapid prototyping system makes RTX as a real-time platform and simulink as the modeling tool. The non real-time tasks are separated from real-time task by the system, that is model parameters, model IO and simulation configuration separated from real-time tasks, in order to complete segmentation, compiler, loading, simulation and management of information changes for model. System architecture shown in Fig 1:

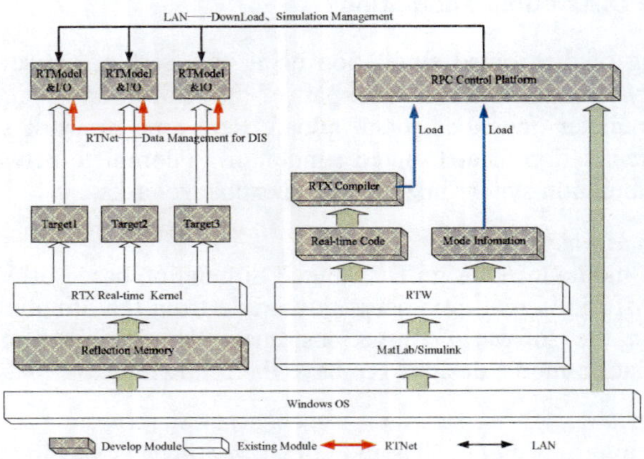


**Fig. 3.** Rapid Prototyping Framework

### 3.1 Rapid Prototyping System Workflow

- First, create a system simulation model by Simulink and confirm the digital simulation results are correct, then the model is set to fixed step.
- Apply rapid prototyping management platform to convert model to by Simulink/RTW tools, and apply the VC++ compiler to compile real-time code into RTSS Executable program in RTX real-time environment, the model configuration information (parameters, I/O and simulation conditions) generate the XML project file.
- The rapid prototyping management platform load XML project file to set the simulation parameters, simulation conditions, data files, and so set up, then it begins to simulation.

### 3.2 Rapid Prototyping System Mode

Since RTX is a Windows embedded real-time system, so the system can operate independently of any one RTX installed computer, so the software has three modes of operation.

- Stand-alone operation mode. RPC systems operate independently of any one target machine to compile, simulate and realize the control for the prototype device.
- Host operation mode. Running on the Host machine that does not simulation, and the completion of the split model, compile, download, on-line parameter modification and simulation management.
- Client operating mode. Running on the target machine to complete real-time simulation and controlled of model by the Host.

## 4 Design of the Basic Functions

### 4.1 Real-Time Distributed Simulation

Rapid prototyping of distributed simulation using two sets of network mechanisms. Ethernet is mainly used for non-real-time part of the realization of simulation models to download, parameter download, data upload. Real-time network system for real-time part of the realization of distributed simulation to complete between each of the target machine simulation synchronization, data exchange tasks.

*1) Simulation Mode.*

The system is modular in design Distributed Simulation based on VMIC5565 real-time network, VMIC5565 real-time network module from the Simulink model library (not integrated in the software) allows customers the flexibility to configure the Distributed Simulation mode, designed to have the following two kinds of distribution patterns

- Network interrupt mode. This mode that the timer is specified by the user in the simulink model in a target machine. The target send simulation command to others for each simulation step and other target machine, after receiving the instructions to complete the step simulation. The mode can achieve clock synchronization in each of the target machine, however, the model calls for simulation step size of each target to run the model must be consistent.

- Own timer mode. The mode that target machine has its own timer, each target machine to run model simulation step size can be inconsistent. Simulation starts, each of the target machine is running according to their own clock, the model can multi-sampling rate simulation.

*2) Distribution model configuration*

We must make the appropriate changes to the model to realize distributed simulation. To f14.mdl model as an example, as shown in Figure 4~7.

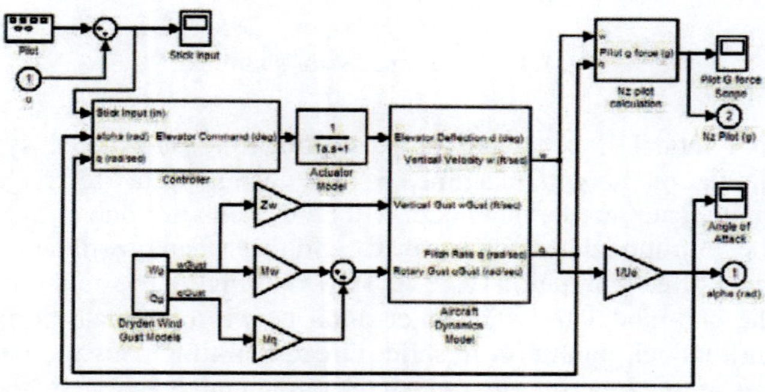


**Fig. 4.** f14.mdl model for simulink

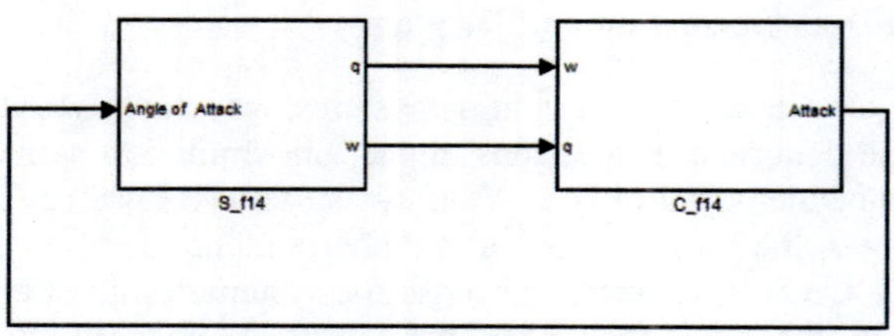


**Fig. 5.** f14 model divided into two target

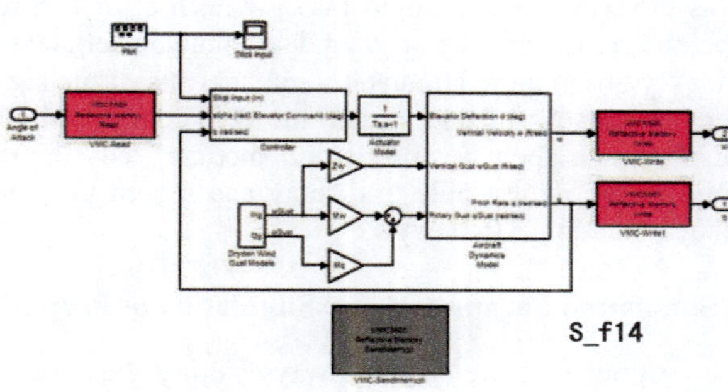


**Fig. 6.** S_f14 submodel for simulink

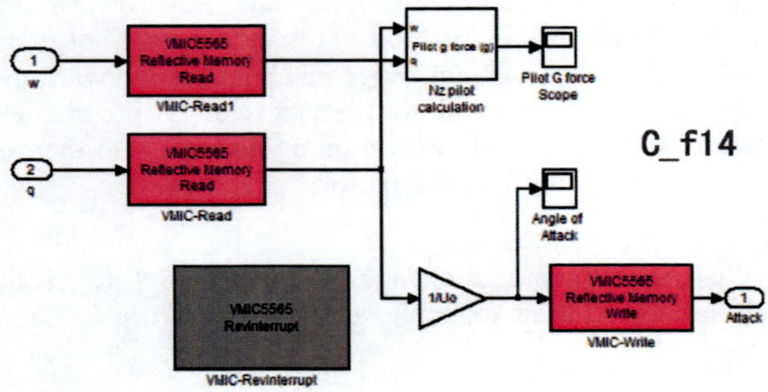


**Fig. 7.** C_f14 submodel for simulink

First, the root model is separated into the following sub-model system by the number of simulation target machine, if the users want to realize network synchronization interrupt mode, the model will be named after one of these sub-prefix is set to S_, other sub-models prefix set for C_, So that when downloading model, the system can be identified simulation type. As shown in Figure 3.

Secondly, the sub-model in which interaction between the data ports, by adding real-time network model, input port in series a read real-time network data model, the output port in series one to write real-time network data model, namely, the completion of the distribution of the model configuration. Application system software to complete the model segmentation, compile and download. As shown in Figure 4,5.

## 4.2 Simulation Data Recording and Display

As the simulation runs in real-time operating systems, real-time simulation of the hard disk read and write data records a serious impact on simulation performance. As the RTX real-time embedded system is a Windows system, its priority must be higher than any of Windows, the Windows operating environment on the hard drive does not have an impact RTX real-time performance, so the system design of a shared memory buffer to complete the data exchange between RTX and Windows. RTX each simulation step, the data stored in shared memory cache area, while the Windows program is detected continuously buffer zone and found that the data be written to immediately remove the hard disk is stored into a data file, or add to the real-time display oscilloscope, the same time, the original data in the buffer cleared.

To be able to quickly observe the simulation and analysis of the signal, the system can be mapped simulink blocks of data observations. Namely, Simulink, such as if there is Scope, Display data such as observation module, the simulation platform, load-time simulation model, is also able to display and record the same data, which greatly improved the simulation efficiency.

## 4.3 Modify the Simulation Parameters and Simulation of Project Management

Designers in real-time simulation, we should always modify the parameters to test the simulation results, so simulation is very important changes in the parameters, the system design eliminates the need to modify the Simulink parameters can be achieved through the platform of the parameter changes to the design of a set of shared memory mechanism to run a program on the Windows platform, and the RTX environment, run the program under the parameters of the completion of mapping the simulation before the start of all change the parameters downloaded to the RTX environment. On Windows platform, all parameters can be saved as a project XML file, the user can load it directly to the next simulation to improve the simulation efficiency.

### 4.4 Common I/O Interface

I/O devices are used to complete the data exchange with the prototype including AD / DA, RS232/422/485, DIO, 1553B, A429 and other commonly I/O protocol. Communication interface is packaged into a Simulink model library, users can directly build Simulink simulation system.

## 5 Simulation Analysis

To f14.mdl model as an example, set the simulation step size of 0.01 seconds, using fixed-step fourth-order Runge-Kutta integration algorithm and complete Simulink numerical simulation, rapid prototyping platform stand-alone real-time simulation, as well as the distribution of real-time simulation.

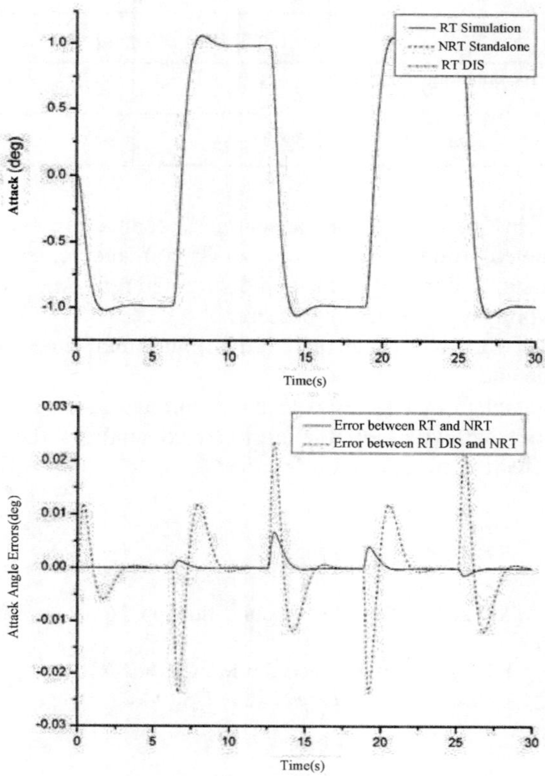


**Fig. 8.** Simulation data curves

Simulation results of the three kinds of angle of attack data comparison, as shown in Figure 8 and Table 1.

As can be seen from Figure 8 stand-alone real-time simulation and non real-time simulation are not fully consistent, mainly due to RTW generated real-time code, but the error is small and will not affect the simulation results. The error is slightly larger between Distributed simulation and real-time simulation, it is mainly caused because the model are divided, but the same does not affect the simulation results.

From Table 1 we can see the distribution of simulation can greatly increase the simulation speed, but requires the designer to balance the distribution model in order to achieve load balancing, maximize the use of computer resources. RTX with high timing accuracy, maintain microsecond-level clock jitter (maximum 6us), can fully satisfy the 100us simulation cycle real-time requirements.

**Table 1.** Simulation performance comparison table

| Simulation performance | Stand-alone RT | DIS RT |
|---|---|---|
| Step | 10ms | 10ms |
| minimum computing time for each step | 0.106ms | 0.083ms |
| maximum computing time for each step | 0.176ms | 0.13ms |
| average computing time for each step | 0.114ms | 0.089ms |
| number of waiting | 3000 | 3000 |

**Acknowledgment.** This paper presents a seamless connection between Simulink model, RTX real-time environment and Windows System and achieve the two modes prototype control for stand-alone real-time and distributed real-time simulation, which is other rapid prototyping systems do not possess.

The VMIC5565 network is used to realize distributed real-time simulation, clock synchronization mode and own timer mode.

Data caching mechanism can be used to store simulation data fully and display real-time oscilloscope that will help the designer to analysis the data results of simulation without affecting the simulation performance.

## References

1. Yang, D., Li, L.-t.: MATLAB/RTW and it's Application, pp. 163–246. Tsinghua University Press, Beijing (2002)
2. Opal-RT Company, " RT-Lab Quick Refence Guide", Canada (2002)
3. Ardence Company, " RTX 8.1 SDK Release Notes", US (2007)

# VxWroks-Based Mass Storage Driver Stack Framework and Application

Gong Yun[1] and Sun Li-hua[2]

School of Information Engineering, Nanchang University
Nanchang, China
gongyun757@163.com, slh52@163.com

**Abstract.** Universal Serial Bus (USB) driver is a module that has not been developed in VxWorks5.5. In this study, implementation of USB mass storage driver has been analyzed, including USB Driver (USBD) and Host Controller Driver (HCD) in USB1.1 protocol and Operation System Abstraction Layer (OSAL). Further, an application of USB block driver including configuration of components has been implemented.

**Keywords:** USB mass storage driver, USBD, HCD, OSAL.

## 1 Introduction

Besides the characteristic of direct hardware detail shielding and efficient resource management, embedded operation system has properties of high real-time, high-power and capability of being tailored as its special advantages[1]. VxWorks developed by Winder River System is an excellent embedded operation system. In this study, USB mass storage driver has been analyzed depending on the enhanced development ability, high-power kernel and user-friendly client development environment of VxWorks[2]. The analysis of Operation System Abstraction Layer (OSAL) and configuration of VxWorks components in this study is the most valuable theory.

## 2 USB Mass Storage Framework

Figure 1 shows VxWorks-based USB mass storage driver framework, including application module, I/O system module, file system module, USB host stack and hardware.

USB host stack is composed of USB protocol stack and USB Host Controller Driver (HCD). Protocol stack includes
software USB mass storage driver, USB HCD and OSAL.

Hardware includes USB Host Controller (HC) and USB device. They are separately analyzed in USB HCD and USB Device Driver.

### 2.1 USB Driver (USBD)

USBD, which is independent of hardware and above HCD, is the most important part of USB host.

- USBD is responsible for managing each USB device connected with host and providing path of communication between higher layer and USB device. For instance, root task tUsrRoot initializes HCD (OHCI in this study). Firstly, system calls the function *usrUsbHcdUhciAttach( )*. Then the function *usbdHcdAttach( )* is called to attach HCD to USBD. Finally, the task is passed to HCD by calling *usbHcdAttach( )*. USBD plays the role of the communication path between system and mass storage device.

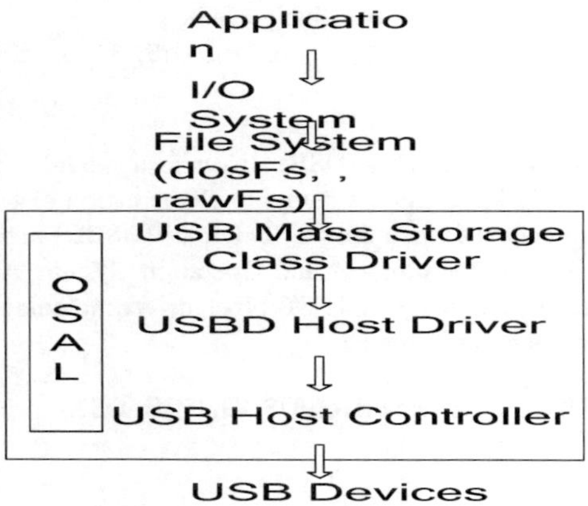


**Fig. 1.** USB mass storage driver framework

- USBD also processes USBD power bandwidth management automatically. Moreover, USBD manages USB hub and dynamic plug-play of device. *FncCurrentFrameGet( )* and *fncHubPortCountGet( )* are separately called to attain frame number and hub number on the bus, and in this way USB bandwidth and hub become manageable[3].

## 2.2 USB HCD

USB HC is the hardware of each USB device in the host system. Most manufactures produce USB HCs that conform to one of three major device specifications. These specifications include the Enhanced HC Interface (EHCI), the Open HC Interface (OHCI), and the Universal HC Interface (UHCI). USB host stack comes with EHCI, OHCI, and UHCI drivers. Choice of the type of controller depends on developers.

HCD is responsible for managing HC, distributing bandwidth and managing link list and root hub. HCD is also required to emulate the behavior of the root hub. It means HCD must intercept transfer requests intended for the root hub and synthesize standard USB responses to these requests. For instance, when a HC is first initialized, the root hub must respond at the default USB address 0, and its downstream ports must be disabled. The HCD recognizes which of these requests are intended for the root hub and responds to them appropriately. When transferred, data constitutes different link list according to the type of transmission. Data is submitted to the HC for processing. HC creates one or more transmission processing according to volume, and then, the data transmission with the device is completed. After the completion of transmission processing, the result is sent to USBD that will notify the client for processing [4].

Command sent by Application module is passed to the function in the library *usbHcdLib.c* by USBD. For instance, while initializing, system attains device descriptor. This process is implemented by sending command for calling *(*pNexus->hcdExecFunc) ((pVOID) &hrb)*. Then the information is sent to HC hardware by *fnclrpSubmit( )* in *usbHcdUhciLib* library, and finally to the device hardware. Actually, data is transmitted in *fnclrpSubmit( )*[5].

### 2.3 USB Device Driver

USB client module is on the top of USB protocol stack. USB class drivers are a typical instance of client module. They are responsible for managing different kinds of devices attached to USB and provide path to communicate with each device. Other instances of USB client are those applicable routines that communicate with devices using USBD. USB device drivers focus on the implementation of some special initialization of USB device and converting client applicable routines of higher layer to corresponding calls of USBD driver routines. This layer avoids details of implementation of USB by providing application programming interface functions to applicable layer. The initialization and attachment of mass storage are described in detail in mass storage device driver.

### 2.4 OSAL Modul

OSAL module is a module via which driver and VxWorks operating system interact with each other. It means that OSAL is used by VxWorks to implement efficient operation. OSAL includes task control and scheduling, task communication and interrupt handling of VxWorks operating system driver.

*1) Task control and scheduling:* Each task uses the structure task control block to convey its information such as ID, current status, system resource and 256 priority levels numbered 0 through 255. A task has four different states: ready, suspending, delaying and pending. There are two ways for Wind kernel task scheduling. One is default algorithm priority-based preemptive scheduling, the other is round-robin scheduling algorithm. Both algorithms rely on the task's priority, but the former one is for different-priority tasks, and the latter is for equal-priority tasks.

*2) Task communication:* VxWorks supplies a rich set of inter-task communication mechanisms including shared memory, semaphore, mutexes and condition variables, message queues and pipes, sockets and remote procedure calls and signals. Each of them has its own characteristic and usage, and semaphore and massage queue are the most frequently used ones. Because of its high-real time, semaphore is used in this study. For instance, during initializing USBD and *usbdInitialize( )* is called, *oss_sem_create (depth, 0, &pQueue->inUse)* produces a semaphore which is then stored in *semPoolQueue*, and then *ossSemGive( )*, *ossSemTake( )*, *ossMutexTake( )* and *ossMutexRelease( )* are used to attain and release binary or mutex semaphore. The function attains a semaphore can be processed and the one that does not need to wait for the semaphore. The high-optimized semaphore provides the fastest mechanisms for task communication to make task isochronous and mutually exclusive.

*3) Interrupt handling:* Many VxWorks facilities are available to Interrupt Service Routines (ISRs), but an ISR does not run in a regular task context and has no task control block. Therefore there are some important limitations. For instance, all ISRs share a single stack. ISRs cannot call any creation or deletion routines, and ISRs also must not perform I/O through VxWorks drivers. In mass storage driver, *intConnect* is used to set interrupt. *UsbPciIntConnect( )* function in *fncAttach* library is called to connect the routine to corresponding interrupt vector[6].

# 3 Mass Storage Device Driver

## 3.1 Block Drvier Fucntions

Complete block driver routine includes device driver routine install function, reading and writing operating functions, control function, creating block device and reading and writing functions[7].

*4) Install function:* The block install function is *usbBulkDrv( )*, which is in charge of initializing block device driver and configuring interrupt vector and controller number.

*5) Create block device: UsbBulkBlkDevCreate( )* creates block device. When initializing, system calls the function *usrUsbBulkDevInit( )* to initialize USB mass storage driver. The process includes two steps.

- *UsbBulkDevInit( )* is used to register block device, and *usbBulkPhysDevCreate( )* creates structure *USB_BLK_DEV( )* including setting and abstracting descriptor.
- *BulkMountDrive( )* is called to attach driver to dosFs. The function *usbBulkBlkDevCreate( )* included in it is used to initialize and return a pointer to *BLK_DEV( )* structure. Any operation on device depends on *BLK_DEV( )* structure. For instance, *pBulkDev->blkDev.bd_blkRd* is called to read from the device.

*6) Operate function: UsbBulkDevBlkRd( )* and *usbBulkDevBlkWrt( )* is responsible for reading from and writing to device. When reading from the device, *usbBulkFormScsiCmd* is firstly called to form SCSI command, and *usbBulkCmdExecute* is then called to read the command whose parameters all are *BLK_DEV*.

*7) Control function:* The controlling function of block device is *UsbBulkDevIoctl( )*, which is accessed by file system.

## 3.2 Configuration of USB Block Device

To complete USB driver, the board support package named lite5200bfinal has been created. After the completion of USB block device driver routine, a VxWorks Image Project that is compatible with lite5200bfinal should be created. The Image Project is called USB. And then, configuration of USB block device components is completed. The components, which should be built, are USB Hosts, USB Target and USB devices. The development board used here is MPC5200, whose HC is OHCI. So the component of OHCI and USB Host Stack located in hardware->buses->USB Hosts and the component of OHCI Init and USB Host Stack located in hardware->buses->USB Hosts->USB Host Init are selected in the USB Image Project. Because it belongs block device, Mass Storage-Bulk is included.

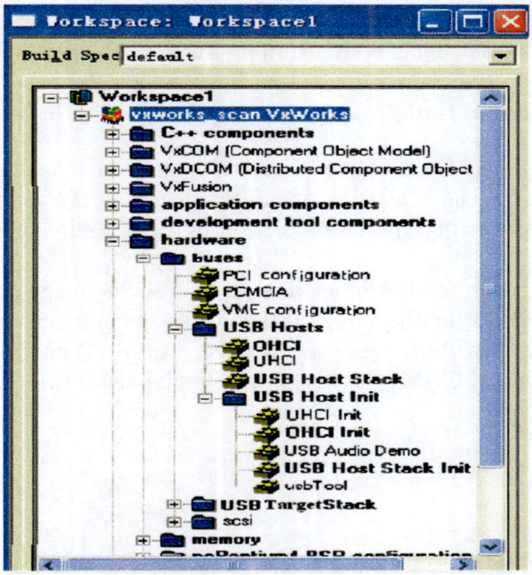


**Fig. 2.** Configuration of USB block driver components

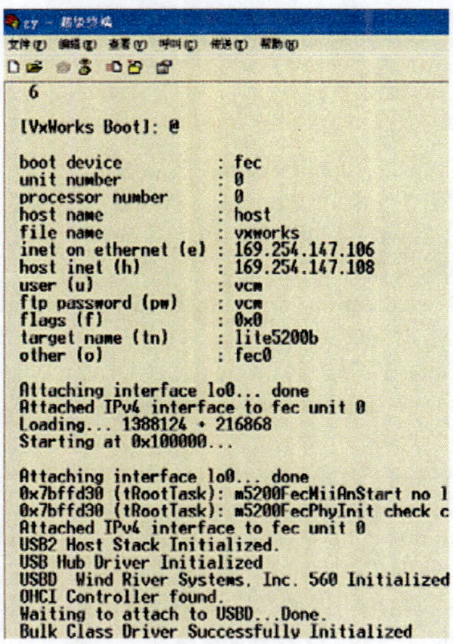


**Fig. 3.** USB device initialization simulation

Mass storage class driver is a type of block device driver. So a file system component should be included. And the file system can be dosFs, highly reliable file system and rawFs. To use a particular file system, the disk must be formatted under the chosen files system, the appropriate components required for that file system and with the event framework should be included. DosFs file system is chosen here, so components of dosFs file system are used with the USB host stack.

Configuration of components is shown in figure 2.

### 3.3 Implementation of Building and Testing

Tornado is an Integrated Development Environment, which is used for intersected software development. The real time operating system VxWorks5.5 is included in Tornado2.2. Building, testing and configuration of USB components are all done by Tornado2.2.

After the configuration of USB block device, build USB image project by Tornado. Download the VxWorks to the target after the image project has been built and a VxWorks image has been generated successfully. After booting VxWorks successfully, the terminal COM1 or COM2 will display the result. The result is shown in figure 3[8].

## 4 Conclusion

Innovations here are the mass storage driver routine and the implementation of mass storage device driver based on VxWorks. Mass storage device completes the function such as data transmission and preservation. Moreover, the configuration of USB driver components is described. Thus the period of other USB device drivers based on other operation system can be shortened.

## References

1. Brayton, R.K., Rudel, 1.R., Sangiovanni-Vincentelli, A.: AM ultiple-Level Logic Optimization System. IEEE Trans. on Computer-Aided Design 6, 1062–1081 (1987)
2. Wei, M.: Computer USB system and its host/slave design, pp. 1–8. Beijing University of Aeronautics and Astronautics, Beijing (2003)
3. Liao, Z.-d.: Design and Realization of Multi-USBScanner Device Driver in VxWorks System. Xidian University. Xi'an (2007)
4. Lin, B.-r., Zhang, F., Chen, Y.: Analysis on USB Driver Based on VxWorks Operation System. Journal of Chongqing Jianzhu University 27, 98–100 (2005)
5. Huang, F.-z., Chen, H.-m., Duan, X.-j., Wu, C.-f.: Implementation of USB Driver on VxWorks Real-Time Operation System. J. Measurement & Contorl Technology 27, 55–57 (2008)
6. Chen, M., Luo, y.-h.: An Embedded System Design for USB Device Driver. J. Tai Yuan Science & Technology, 78–79 (2009)
7. Wang, J.-g., Gong XiaoYang, X.-l., et al.: VxWorks-based Imbeded Real Time system design. Tinghua University, Beijing (2004)
8. Liu, Y.-W., Yu, G.-Q.: Research of the VxWorks-based USB Device Driver. J. Computer Knowledge and Technology 5, 7008–7017 (2009)

# An Improved Anomaly Detection Method Based on Fuzzy Association Rules

Zifen Yang

School of Informatics Linyi Normal University
Linyi, China
yzf_smile@163.com

**Abstract.** The discovery of association rule is a crucial aspect in database mining. In order to further improve the ability of fuzzy association mining in identifying the unknown intrusion data in anomaly detection, an improved algorithm called Hapriori using hash list on mining fuzzy association rules is proposed in this paper. Equivalence classes are used to search frequent items, so that the usual practice of reputedly database scanning can be avoided. At last, experimental results show that our proposed method to some extent increased recognition of the intrusion data.

**Keywords:** fuzzy association, hash list, anomaly detection.

## 1 Introduction

Intrusion detection is an important security technology of protecting information systems as well as firewall and data encryption techniques [1]. It has great capabilities in identifying the malicious behaviors in networks and responding to them. As a means of initiative defend, IDS has several problems to be solved. On the basis of retrospection the development of intrusion detection technology, the problems incurrent intrusion detection systems are analysis. Data mining techniques can get important data from a large number of information. So it is used widely in IDS [5]. However, data mining is usually able to deal with discrete data. So this often leads to "sharp boundary" problem .Thus researchers apply the fuzzy theory in data mining to solve this problem [6].

This paper will propose an improved fuzzy association algorithm to improve the capacity of their ability to identify the invasion of the data.

## 2 Related Work

### 2.1 Association Rules

The rule of association is a common way to discover the inter-relationship hidden in the data. . Let $D=\{t_1,t_2,....,t_k,...t_n\}$ , $t_k = (i_1, i_2... i_p)$ be one thing. The element $i_j$ of $t_k$ ( $j = 1, 2 ... p$ ) is transaction. Set $I = (i_1, i_2 ... i_j ... i_m)$ is composed of a collection of all the items of $D$ . Any subset such as $X$ of $I$ become an item set of $D$ . For each set of support is $\sup(X) = |X(T)|/|D|$ . In this formula, the item number of $X(T)$ is expressed as $|X(T)|$, and the item number of $D$ is expressed as $|D|$ . $X(T) = \{T \in |D| X \subseteq T\}$ , if it meets $\sup(X) \geq \min\sup$ , it is claimed frequent item sets. The confidence of assertion rules is $conf(X \Rightarrow Y)$ = $\sup(X \cup Y)/\sup(X)$ . An association rule of the form $X \Rightarrow Y$ type implication, in which $X \subseteq I$ , $Y \subseteq I$ , and $X \cap Y = \varnothing$ .

## 2.2 Fuzzy Association

Set the domain $U$ , the mapping $\mu_A : U \rightarrow [0,1], \forall X \in U$ $\mu_A(x)$ is the degree of membership for $A$ . For the two fuzzy sets $A$ and $B$ , $(A \cap B)(U) = \mu_A(X) \wedge \mu_B(X)$ $= \min\{\mu_A(X), \mu_B(X)\}$ $\forall \lambda \in [0,1]$ , we do the definition that $A_\lambda = \{\mu_A(x) \geq \lambda\}$ is the fuzzy set A of λ-set; if $A_\lambda = \{\mu_A(x) > \lambda\}$ then $A_\lambda$ is a λ-strong cut set[3].

$D = \{t_1, t_2, ...., t_k, ...t_n\}$ , $I =$ $(i_1, i_2 ... i_j ... i_m)$ is attribute of database $D$ 's attribute set. The fuzzy association rule will be found is in the form contains the following formula: $< X, A > \Rightarrow < Y, B >$ . Where, $X \cap Y = \varnothing$ $A = \{f_{x_1}, f_{x_2}, f_{x_p}\}$ is fuzzy sets of $X$ $B = \{f_{y_1}, f_{y_2}, f_{y_q}\}$ is fuzzy sets of $Y$ . And $< X, Y >$ in the properties of the corresponding set of fuzzy set $Y$ s. Sequence even set $< X, A >$ that the properties of $X$ , take the corresponding value of $A$ .

The basic Apriori algorithm finds frequent item sets for Boolean association rules, receiving as input a database $D$ of transactions and the minimum support for the rules [2]. It uses the Apriori property: if an item set I is not frequent, the item set $I \cup A$ is also not frequent; i.e. "all none empty subsets of a frequent item set must also be frequent".

The Apriori algorithm builds a set $C_k$ and $L_k$ to create frequent item sets of size *k+1*:

```
L1 = {frequent item sets}
k=2;
While (Lk! = ∅ {
        Ck =GenerateCandidates (Lk);
        for each transaction t in the database
                Increment the count of all candidates in Ck that belong to D
  Lk! =candidates in Ck with enough support
k ++;
}
Return (L=L1 ∪ L2 ∪ ....);
```

*GenerateCandidates*( ) returns a subset of the join operation of $C_k$ and $L_k$ ,pruning item sets that do not satisfy the Apriori property. Computing the support and confidence of all nonempty subsets of each frequent item sets generates the set of association rules.

## 3 The Improvement of Fuzzy Association Mining Algorithm Hapriori

However, bottleneck of Apriori algorithm for fuzzy association can be summarized as follows:

*1)* Do numerous visits to the entire database.
*2)* Pattern matching is used to identify frequent item sets and connect steps algorithm. So its efficiency is relatively low.
*3)* If we want to calculate the degree of support, we must repeat this calculation.

In this paper we propose a new algorithm called Hapriori to solve that problem. The main method of new algorithm is to use the new data structure hash list to store records that all the keywords with their synonyms in the same linear form [3]. It is convenient to check, because this can build a certain corresponding relationship between H (key) and key. The new algorithm will first visit to the database. Then we get *1*-frenquent item sets by calculating the properties of a sequence to be a frequent item sets. Every transaction in formations of *1*-frenquent item sets is stored in hash list. In this node contains two parts: the transaction No *Tid* and the degree of membership under this transaction $u$ . The transaction information is not stored, if this membership degree μ is zero or outside the scope of the cut-off set. If the membership degree $u$ set to zero, or outside the scope of the cut-off, the transaction information is not stored. The keyword of Hash list which we make for each item sets is the transaction number. We check whether $j$ is exist from hash list by scanning the keywords of hash list $i$ . If there is degree of membership to take min operation, or Hash list $i$ continue to scan until no keywords. $(k\text{-}1)$ frequent item sets is connected to generate k-candidate item sets, and we take min operations for the same degree of membership services. At the same time a Hash list will be build. The first node is set to connect the two items after $k$ - item sets, while the linked list node is the same in this transaction under the transaction number, and the minimum degree of membership. The sum of the minimum degree of membership is calculated through the temporary variable. If the result is not less than *minsup*, then the new list will be retained, otherwise it will be deleted. ($k$ -1) frequent item set list will be completely deleted when $k$ -frequent item sets is generated.

The above operation needs to access the entire database when we calculate *1*-frenquent item sets. And then we only need the information of $(k\text{-}1)$ frequent item sets in the database when we calculate the support of candidate item sets. With the increase $k$ se of $k$ , the number of ($k$ -1) frequent item sets is decreasing, therefore the amount of data which will be accessed has declined. In addition, double counting is avoided if we calculate support of $k$ - item sets to use the results of $k$ *-1*-frequent item set.

The optimization Steps will be done as follows. Equivalence classes with similar characteristics with the frequent sets, namely, a subset of equivalence classes must also equivalence class [4]. If an item set equivalence class then its higher level of item-set is not equivalence class. We let $R$ is a relationship which is defined based on set $A$ . And $R$ is reflexive, symmetric and transmission .For the formation of $R$ -equivalence classes namely, $[a]=\{b[k] \mid a[1:k-1] \quad =b[1:k-1] \mid\}$ . When $k=1$ , remove the orderly connection of the same attribute. When $k \geq 2$, Prefix $a$ string with $[a]$ collection, each element (items from the same string in a meta-equivalence class) will be $C_k = C_k \bigcup c$ connected to the formation of equivalence classes of candidate $C_k$ . i.e.: $[ab]=[a]\bigcap[b]$ .And then connects each element of the prefix string $ab$ and $[ab]$ to gain $C_k$ . This method avoids the connection of the pattern-matching as well as a property inspection item sets.

Based on the above thought, Hapriori algorithm are descript as follows:

---

Input: transaction database $D$ membership function; minimum support minsup.
$L_1 = \varnothing$ ; Establish hash list for each item sets.
for each item of $D$ {/ * generate a hash list and a frequent item set * /
int $\omega = 0$ ;
For each transaction in database $D$
  {Convert the data into fuzzy membership degree $\mu$ via membership function
    $F_k$
  if $\mu \geq \lambda$ insert ((Corresponding to the hash list item, transaction number,
    membership $\mu$ $conf(X \Rightarrow Y) = \sup(X \bigcup Y) / \sup(X)$ ) ;
    $\omega + = \mu$ ;
  }
 *if* $(\omega / n >= \min\sup) L_1 = L_1 \bigcup \{item\}$ *;*
 *else {Remove the hash list of the non-frequent item }*
*for* $(k=2; L_{k-1} \neq \varnothing;\ k++)$
  *{* $C_k$ *= Apriori_gen_equ (* $L_{k-1}$ *);*
  *for each* $c \in C_k + 1$ *{if(computer(* $c$ *,support)>=minsup)*
      $L_{k+1} = L_{k-1} \bigcup \{c\}$
      *else delete hash list c;*
  *}*
  *remove hash list of ( k -1) frequent item sets ;}*
*int compute(c, support) {/* calculate the degree of support item sets*/*
*Create (hash list c);*
*take pre k -1 item s , and the last k -1 item set j ;*
*for every key word in hash list s*
*{*
  *if (find key words in hash list j )*
  *Insert (hash list c, key word, min (the corresponding size of key word* $\mu$ *));*
 *Support+=min (the corresponding size of key word* $\mu$ *) ;*
 *}*

```
Return support=support/n;
  }
Apriori_gen_equ (L_{k-1} : (k-1)frequent item sets )
 {If k=2 connect item of L_{k-1} that the same attribution is deleted.
      else { L_{k-1} is divided to generate equivalence class L_{k-1} / R according to
 same prefix.}
 If k=3 save collection L_2 / R .
  for every equivalence class L_{k-1} / R
  {
          if k_i < = 1 continue ;
        else for(int i =0 ; i < k -1 ; i ++){
           [ C_k Prefix s ] = [equivalence class L_{k-1} / R ] ∩ [this prefix item i ];
             Connection String of s;
             C_k = C_k ∪ c ;
             }
}
```

# 4 Experiments and Analysis

## 4.1 Experiment

IDS are generally divided into two categories: anomaly detection and misuse detection. We use fuzzy association rules mining in anomaly detection according to the following method. First, set up normal mode system under the association rules $S_1$. Then we mine the association rule $S_2$ of the system in a transient mode. Then we calculate the similarity $(S_1, S_2)$ .We use the similarity to express the current status' degree of deviation from normal. Then we can sure weather the system is in abnormality.

We give two association rules as follows. $R_1 : X \rightarrow Y, c, s, R_2 : X' \rightarrow Y', c', s'$ . Similarity $(R_1, R_2) = \max[0,1, -\max[c - c'] / c, | s - s' | / s]$ is the degree between the two rules. The similarity between $S_1$ , $S_2$ is similarity ( $S_1, S_2$ ) $= \mu^2 / | S_1 || S_2 |$ in this formula $\mu = \sum similarity(R_1, R_2)$, $| S_1 || S_2 |$ is the amount of rules.

We analysis flow-stat LAN of our lab using Hapriori. We selected four attributes which are associated with network analysis. The proportion of TCP and UDP in all of data packets: $P_{tcp}$ and $P_{udp}$。 The average number of packets per second of network: Avg.packet/s. The average data bits per second: Avg.Mbit/sec. We sample one time every hour, and a total of ten times was obtained in Table 1.

From table 1, network data flows get membership function $F_k$ according to $F$ - Mean clustering method, and then we convert it into fuzzy data.

**Table 1.** Data of Experiments

| | $P_{tcp}$ | | $P_{ucp}$ | | Avgpacket/s | | AvgMbt/s | |
|---|---|---|---|---|---|---|---|---|
| *Tid* | *A.L* | *A.H* | *B.L* | *B.H* | *C.L* | *C.H* | *D.L* | *D.H* |
| 1 | 0 | 0.90 | 1 | 0 | 0 | 0.65 | 0 | 1 |
| 2 | 0 | 0.89 | 1 | 0 | 0 | 0.73. | 0 | 1 |
| 3 | 0 | 0.57 | 1 | 0 | 0 | 1 | 0 | 1 |
| 4 | 0 | 0.93 | 0 | 0.5 | 0 | 0.72 | 0 | 0.86 |
| 5 | 0 | 0.86 | 0.5 | 0 | 0.89 | 0 | 1 | 0 |
| 6 | 0 | 0 | 0 | 0.5 | 0 | 0.62 | 0 | 1 |
| 7 | 0 | 1 | 0.5 | 0 | 0 | 0.91 | 0 | 1 |
| 8 | 1 | 0 | 1 | 0 | 0 | 1 | 0 | 1 |
| 9 | 0.64 | 0 | 1 | 0 | 0.63 | 0 | 1 | 0 |
| 10 | 1 | 0 | 1 | 0 | 0 | 1 | 0 | 0.89 |
| sup | 0.26 | 0.51 | 0.7 | 0.1 | 0.15 | 0.66 | 0.2 | 0.78 |

Set $\lambda = 0.45, \min\sup = 0.25$ , we get $L_1 = (\{A.L\},\{A.H\},\{B,L\}\{C.H\},\{D.H\})$ hash list is connected in accordance with the order of through scanning frequent item to get $L_2$.We get the results as Table 2 as follows.

**Table 2.** The Results of Fuzzy Association Rules

| $L_2$ | Sup |
|---|---|
| $A.H,B.L$ | 0.32 |
| $A.H,C.H$ | 0.34 |
| $A.H,D.H$ | 0.46 |
| $B.L,C.H$ | 0.48 |
| $B.L,D.H$ | 0.52 |
| $C.H,D.H$ | 0.64 |

| | |
|---|---|
| $C.H,D.H$ | 0.64 |

→

| $L_3$ | Sup |
|---|---|
| $A.H,B.L,C._s H$ | 0.26 |
| $A.H,B.L,D.H$ | 031 |
| $A.H,C.H,D.H$ | 0.36 |
| $B.L,C.H,D.H$ | 0.42 |

## 4.2 Discussion

We can get equivalence classes from Table 2 $L_2 / R$ as follows:

$[A.H.B.L] \wedge [C.H] \rightarrow \{B.L.C.H.D.H\}$ $S$=26 % $c$ =74%

$[A.H] \wedge [DH] \rightarrow \{B.L, D.H\}$ $S$=25 % $c$ =73%

$[A.H] \wedge [B.L] = \{C.H, D.H\}$ $S$=42 % $c$ =53.2%

$[A.H] \wedge [C.H] = \{D.H\}$ $S$=28.6 % $c$ =52.3%

$[B.L, C.H] = [B.L] \wedge [C.H]$ $S$=28.6 % $c$ =40.1%

$C_3(\{A.H, B.L, C.H\}), \{A.H, B.L, D.H\}, \{A.H, C.H, D, H\}$, We set minconf=0.65, so $\{B, L, C, H, D, H\}$
we can get strong fuzzy rules through data mining as follows;

$AH \wedge B.L \rightarrow CH$ $s$ =21 % $c$ =76%
$A.H \wedge B.L \rightarrow D.H$ $s$ =27% $c$ =86%
$A.H \wedge D.H \rightarrow B.L$ $s$ =27% $c$ =73%

We get $A.H \wedge D.H \rightarrow B.L$ .That is if network flow in a high number of TCP packets and UDP packets is lower, then the data bits per second on average of the network is higher.

## 5 Conclusion

The cut-off operation of fuzzy set theory is applied in the paper, as well as Hash list Structure is used to achieve fast positioning search, thus reducing the degree of support for double-counting. And redundant pattern matching is avoided via the equivalence class optimization algorithms. On the contrary, building a new hash list itself is a definite increasing in space and time used. The next step is to identify the invasion event through mining the fuzzy time series association.

## References

1. Hu, D., Min, J.: Netwoek security, pp. 168–192. QingHua university PRESS, Beijing (2003)
2. Kuok, C., Fu, A., Wong, M.: Mining Fuzzy Association Rules in Database. SIGMOD Record 27, 41–46
3. Park, J.S., Chen, M.S., Yu, P.S.: Aneffective hash-based algorithm for mining assoation rules. In: Proc. 1995 ACM Int. Conf. Management of Data, San Jose, pp. 175–186 (1995)
4. Zhang, J., Gong, J.: An Association Ming Matrix Algorithm Based on Equivalence class. Computer Technology and Development 18, 36–40 (2008)
5. Wang, H., Shen, X.: An Anomaly Detection Method Based on Fuzzy Judgement. Journal of Computer Research and Development 40, 4–7 (2003)
6. Wang, W.D.: Genetic Algorithm Optimization of Membership Function for Ming Fuzzy Association Rules. In: International Joint Conference on Information Systems, Fuzzy Theory and Technology Conference, Atlantic City (2000)

# A Utility Quality of Service for Grid Resource Brokers

Jun Xie and Ming Chen

College of Information Engineering
Capital Normal University
Beijing 100048, China
xiejunsoftware@tom.com, chenming@cup.edu.cn

**Abstract.** Grid computing promise the ability to share geographically and organizationally distributed resource to increase effective computational power and resource utilization. However, for the grid computing to be successful, it is very important to provide middleware services that assist grid users to easily interact with grid environments. The paper presents a utility quality of service for grid resource brokers. Resource Brokers on the grid consult a number of distributed information service to select the best data source based on user requirements. The consultation task increases the design complexity of Resource Brokers. We have designed and implemented a utility quality of service for grid resource brokers to hide the underlying complexity of the grid resources from users and to meet not only grid user's requirements but also resource owner's policies. It focuses on the discovering and scheduling dynamic resources scattered across multiple organizations. Furthermore, it can be integrated with various scheduling services.

**Keywords:** Grid resource, QoS, Grid Resource Brokers, Distributed resource, dynamic resources.

## 1 Introduction

Grid computing infrastructures offer a wide range of distributed resources to applications [1]. To support application execution in the context of the grid, a Grid Resource Broker is desirable. Grid Resource Brokering is defined as the process of making scheduling decisions involving resources over multiple administrative domains [2]. The resource broker and job submission components are vital for any Grid computing infrastructure, as their functionality and performance to a large extent determine the user's experience of the Grid. In all, these components have to identify, characterize, evaluate, select, and allocate the resource best suited for a particular application. The brokering problem is complicated by the heterogeneous and distributed nature of the Grid as well as the differing characteristics of different applications. To further complicate matters, the broker typically lacks total control and even complete knowledge of the state of the resource.

Typically, resource brokers are closely integrated with, or at least heavily dependent on, some particular Grid middleware, with popular solutions ranging from brokering components being part of the job submission client to centralized Grid-schedulers not that different from traditional batch system schedulers [3,4,5]. Hence, it is normally non-trivial to migrate a broker from one middleware to another, or to adjust it to simultaneously work with resources running different middleware.

This contribution presents an architecture and an implementation of a utility quality of service for grid resource brokers. The service is general in the sense that it can be used with different Grid middlewares, with middleware-specific issues concentrated to minor components. These components are used for format conversions in interactions with clients and information systems as well as for middleware-specific interaction with resources.

The paper is structured as follows. In section 2, the Quality of Service for Grid resource brokers is presented. A prediction model that supports the implementation of quality of service is described in Section 3. In Section 4, the Quality of Service for Grid resource broker implementation is discussed. The paper concludes with a summary and discussion of future work.

## 2 T System Design

This section describes the design of a broker, which is enhanced by adding the functionality to enable adaptability. The Resource Broker performs a number of basic functions. The first step is the discovery and selection of resources that best fit the needs of the Grid application. The broker will then submit jobs in the application to the chosen machines. The broker thus handles submission of jobs but not how the job is actually executed on the resource, as that is part of the Resource Management system that resides on the resource involved. These actions are referred to as scheduling in the Resource Broker. Once jobs are being executed, the broker for that application monitors the resources and the progression of the jobs.

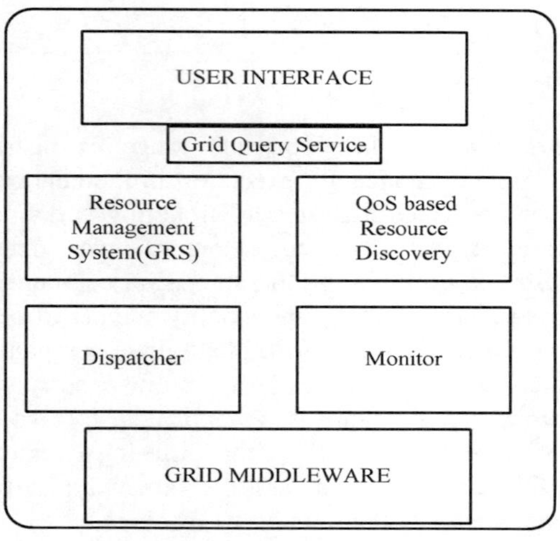


**Fig. 1.** Proposed Broker Design

Fig.1 depicts the proposed design of an adaptable broker. The broker is comprised of basic components implementing Grid Query Service, Resource Management System (GRS), Resource Discovery & Selection, Dispatching and Monitoring. In addition to this the broker includes an adapter manager.

The adapter manager controls migration, which is supported by job monitoring and enabled by rescheduling and check-pointing. The broker gathers dynamic information about the resources during the runtime like accessibility, system workload, performance, etc. Dynamic information is reported to the monitor.

The monitor provides predictive information to the adapter manager, which uses this information to make a decision as to whether job migration is required. The main task for the adapter manager is to ensure the job requirements are fulfilled.

## 2.1 Resource Management and Scheduling in Grid Computing

Computational Grids enable the coordinated and aggregated use of geographically distributed resources for solving large-scale problems in science, engineering, and commerce. In these environments, resource management and scheduling is always a complex undertaking due to the geographic distribution of resources that are often owned by different organizations having different usage policies and cost model, and varying loads and availability patters. And this has been widely studied by researchers and scholars across the world.

In the GlobusWORLD2004 [3] international workshop, the "community scheduler" is again proposed and emphasized, which servers as a broker in the grid environments between client applications and resource management, as shown in Fig.2.

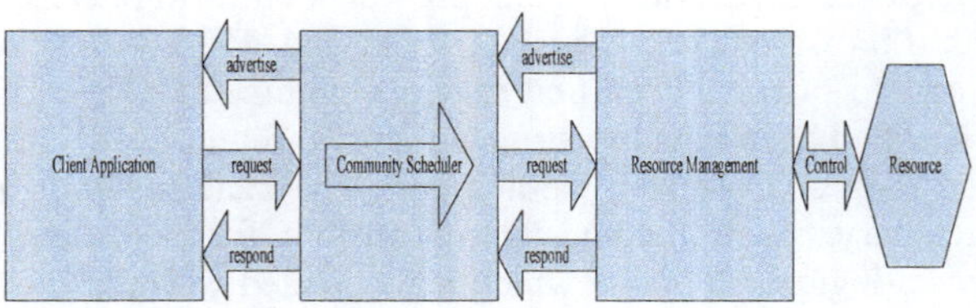


**Fig. 2.** Community Scheduler

## 2.2 QoS Based Resource Management System in Grid Resource Brokers

In dealing with the complexities of resource management and scheduling presented in Grid Resource Brokers, we investigate a QoS based resource management system in Brokers. In this architecture, the function of dynamic distributed resource management, real time task analysis and disposal are designed and implemented to satisfy customers' requirements, as shown in Fig 3.

Fig 3 shows the main components of our resource management system in Grid Resource Brokers, which include: QoS management, WS-Resource management, Grid Query Service and Grid Resource Scheduler.

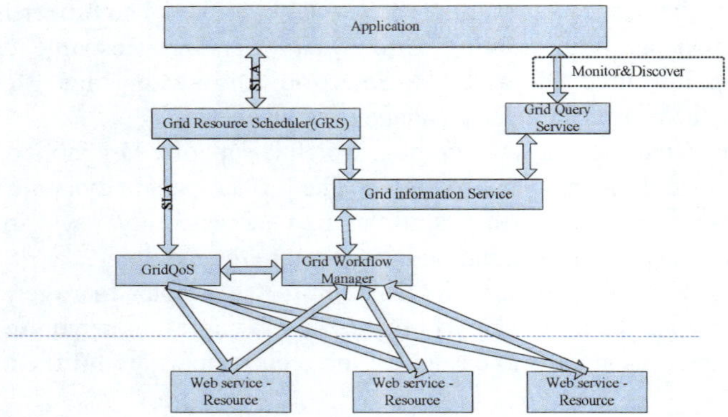


**Fig. 3.** QoS based Resource Management System in Grid Resource Brokers

## 2.3 Grid Query Service

Reducing the complexity of the Resource Broker design, increasing the overall performance of Grid based application, and abstracing over grid middleware versions are the main motivations for developing the Grid Query Service (GQS) [4]. It abstracts over Grid middleware, because it is based on OGSI grid service [5], and the overall performance is improved y encapsulating, in one framework, the grid information services that any Resource Broker needs. Therefore, the Resource Broker needs to consult only one service, the GQS, with user requirements to obtain information about data sources and/or computational resource. This work has developed both encapsulated services and a generic API that provide a uniform interface to the unified data services within its framework. These services and API are developed, based on OGSI-DAI services, to query the unified services within the GQS. A user portal to unified services has also been developed; detail can be found in [8]. In using the API, each query operation is invoked as a method call. The main methods are: mcsQuery, rliQuery, and lrcQuery. Each method is responsible for accessing a pre-defined database. It returns a logical file name which is associated with the provided metadata attributes. Further generic methods of the API are: mcs, rls, rli, lrc, rliHandle, and lrchandle. These methods use the main methods in a mix and match fashion to accomplish their tasks. Moreover, when new services are integrated within the GQS, Resource Brokers need only to know the newly developed API to interface to the newly added services.

## 2.4 GRS

Resource and tasks are tied up with quality of service. Each kind of resource has it own QoS properties or attributes to distinguish from each other, every task has its own specified QoS constraints to be satisfied. In order to deal with this quality of service problem during the product lifecycle management, we develop a Qos based Grid Resouce Scheduling system in Grid Resource Brokers to perform the application level resource scheduling functions; the process of GRS shown in Fig 4.

QoS Resource Discovery: the goal of this step is to identify a list of authorized resources that meet the task /subtask's requirements.

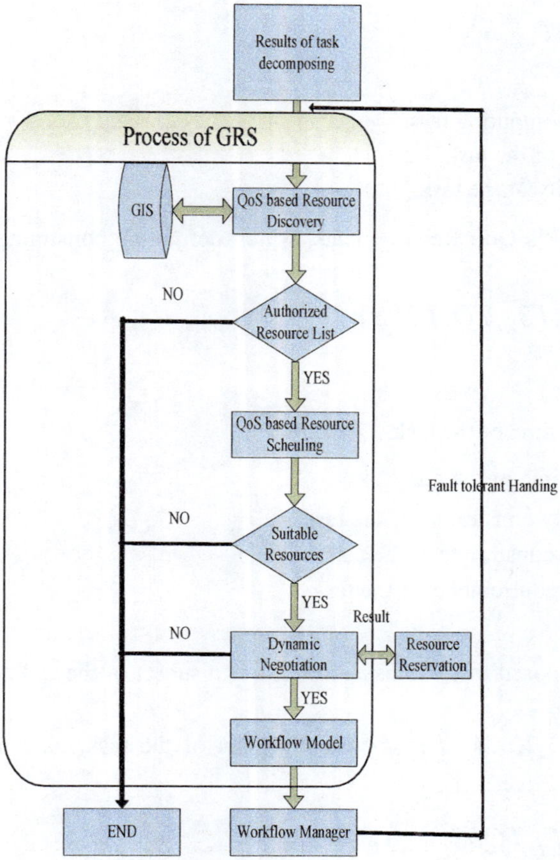


**Fig. 4.** Process of GRS

QoS based Resource Scheduling: Once the list of possible resource is known, GRS will select those resources that are expected to meet the QoS constraints imposed by customer.

Dynamic Negotiation: by interacting and negotiating with the selected resource, GRS determines the detailed target and contents for each resource, then, the negotiation results are delivered to Qos in Grid Resource Brokers (QGRB). QGRB will make sure the violation detection and penalty management solution, sign SLA with the resource, and then reserve it.

## 3 QoS for Grid Resource Brokers

### 3.1 Basic Definitions

Definition 1:
A task is usually composed of several working procedures in a predefined sequence. We define task as follows:

$$TS = \{TN, TL, TQoS\}$$

Where:

TS----The Grid computing task
TN----The name of the task
TL----The quantity of the task

TQoS--- The task's QoS Requirements, which define the constraint characteristics of the task:

$$TQoS = \{TS_S, TS_e, TQ, TC, S\}$$

Where:

$TS_S$ -- The start time of the task.
$TS_e$ -- The end time of the task.
$TQ$ -- The quality constraints of the task
$TC$ -- The price constraints of the task
S – The service requirements of the task

Definition 2:
Task can be decomposed into subtasks. Subtask is a subset of the task, which can be a component or a part.

$TS = \{TS_i\}$ i=1,2,….n, $TS_i$ means subtask i of the task. And the subtask can also be defined as:

$$TS_i = \{TN_i, TL_i, TQoS_i\}$$

Definition 3:
Grid Resource are physical elements to complete the fabric of a task. It is defined as follows:

$$R = \{RU, RK, RI, RD, RQoS\}$$

Where:
R --Grid Resource
RU --Resource UID
RK-- The kind or classification of the Resource
RI -- The input format of the Resource
RO -- The output format of the Resource
RD -- The dynamic information of the resource, which includes the current status of the resource.
RQoS -- QoS attributes of the resource.

We define it as:

$$RQoS = \{RT, RQ, RC, RS, RL, RY\}$$

Where:

RT -- The subset of the attributes relates to time
RQ -- The subset of the attributes relates to qualit
RC -- The subset of the attributes relating to cost
RS -- The subset of the attributes relates to service
RE -- The subset of the attributes relates to creditability
RL -- The subset of the attributes relates to reliability
RY -- The subset of the attributes relates to strategy

$$\beta_{ij} = \begin{cases} 1, \text{Re}\,source \ \ R_i \ \ is \ \ selected \ \ to \ \ task \ \ TS_j \\ \qquad 0 \quad Otherwise \end{cases}$$

Where, R={R}, (i=1,2….,m) is the resource list got from resource discovery component, and TS={TS}, (j=1,2,…n) is the subtask set decomposed from a task.

### 3.2 The Utilization Process of QoS for Grid Resource Brokers

First, receiving the resource lists (RL, $RL = \{R_1, R_2, ..., R_n\}, n \succ 1$ ) from QoS-based discovery component.

Second, constructing hierarchical structure with resource's QoS attributes as the evaluation criteria.

Third, choosing expert to be the decision making group.

Fourth, Constructing dimensionless fuzzy judgment matrix. Some of the resource's attributes are beneficial attributes, which means the larger the better, whereas, the others are the smaller the better, which called profitable attributes, and different attribute has different dimension and unit, so, it is difficult to build matrix.

Fifth, Adjusting the factors in the constructed fuzzy judgement matrix, so that the matrix is the fuzzy consistent judgment matrix. The adjustment principle is: comparing the corresponding factors belonging to two random specified rows, the balance is constant.

Sixth, Repeating the fourth dan fifth step until all the experts in the decision-making group give their judgment matrix.

Seventh, Aggregating all the experts' matrix into one fuzzy consistent evaluation matrix, according to the following definition and theorem: if $S = (s_{ij})_{m \times n}$ and $T = (t_{ij})_{m \times n}$ are two fuzzy consistent matrices, and $p_{ij} = \varpi_1 s_{ij} + \varpi_2 t_{ij}$ , $\varpi_1 + \varpi_2 = 1$, then $P = (p_{ij})_{m \times n}$ is a fuzzy consistent matrix.

Eighth, Ranking the alternative resources with additive global weights, and select the resource with the largest value as the suitable one for the specified subtask.

# 4 Conclusions and Future Work

In order to resolve resource scheduling problem to handle dynamic changes in availability of grid resources and user requirements, and to provide scalable, controllable, measurable, and easily enforceable policies for management of the resources. We propose a utility quality of service for grid resource brokers, which comprises of Grid Query Service, Qos-based resource discovery, Qos-based resource scheduling, and dynamic negotiation.

We have shown that a utility quality of service for grid resource brokers is a viable contender for use in future Grid implementations.

Future work will focus on developing an interactive job, where the user can change the attributes of the job during run time.

# References

1. Buyya, R.: Economic-based Distributed Resource Management and Scheduling for Grid Computing. PHD thesis, Monash University, Austrilia.2002
2. Phan, T., Ranganathan, K., Sion, R.: Evolving Toward the Perfect Schedule: Co-Scheduling Job Assignments and Data Replication in Wide-Area Systems using a Genetic Algorithm. In: Feitelson, D.G., Frachtenberg, E., Rudolph, L., Schwiegelshohn, U. (eds.) JSSPP 2005. LNCS, vol. 3834, pp. 173–193. Springer, Heidelberg (2005)
3. Foster, I., Kesselman, C., Tuecke, S.: Grid service for distributed system integration. Computer 35(6), 37–46 (2002)
4. Mahajan, R., Bellovin, S.M., Floyd, S., Ioannidis, J., Paxson, V., Shenker, S.: Controlling High Bandwidth Aggregates in the Network. Computer Communications Review 32(3), 62–73 (2002)
5. Frey, J., Tannenbaum, T., Livny, M., Foster, I., Tuecke, S.: Condor-G: A Computation Management Agent for Multi-Institutional Grids. Cluster Computing 5(3), 237–246 (2002)
6. Nielsen, N.R.: The allocation of computer resource: Is pricing the answer? Communications of the ACM 13(8), 467–474 (1970)
7. Waldspurger, C.A., Hogg, T., Huberman, B.A., Kephart, J.O., Stornetta, W.S.: Spawn: A Distributed Computational Economy. IEEE Transactions on Software Engineering 18(2), 103–117 (1992)
8. Foster, I., Tuecke, S.: Unger, J.: OGSA Data Services. Global Grid Forum (September 2003)
9. Luther, A., Buyya, R., Ranjan, R., Venugopal, S.: Peer-to-Peer Grid Computing and a .NET-based Alchemi Framwork. In: High performance Computing: Paradigm and Infrastructure. Wiley Press, USA (2005)
10. Rananathan, K., Foster, I.: Simulation studies of computation and data scheduling algorithms for data grids. Journal of Grid Computing 1(1), 53–62 (2003)
11. Rajasekar, R., Livny, M., Solomon, M.: Matchmaking: An extensible framework for distributed resource management. Cluster Computing 2(2), 129–138 (1999)
12. Foster, I., Kesselman, C., Tuecke, S.: The Anatomy of the Grid. International Journal on Supercomputer Applications (2001)

# Research and Realization of Streaming Media Digital Rights Management

Jun Xie and Chuanzhong Li

College of Information Engineering Capital Normal University
Beijing 100048, China
xiejunsoftware@tom.com, lichuanzhong@163.com

**Abstract.** Digital rights management (DRM) is becoming a key enabling technology to protect intellectual property of digital contents, control their dissemination and usage. Combining the actual application situation of the current stream license technology and the new generation of DRM technology, and then poses a concrete scheme for the digital right management system. What's more, it realizes the streaming media digital rights management system finally according to the scheme.

**Keywords:** DRM, Streaming Media, Digital Rights Management.

## 1 Introduction

With the rapid development of Internet and digital technology, streaming media is becoming a emerging network multimedia. Streaming Media refers to continuous time-base media which use streaming transmission technology in the internet / intranet, such as audio, video, or multimedia files. The emergence of streaming media promotes the rapid development of technologies, such as online and distance education, e-commerce, and online video on demand. As streaming media applications become increasingly wide, its problem of copyright is emerging increasingly. Streaming media content is wantonly stolen, transmitted, while content owners and creators of the legitimate rights and interests have not been duly protected, especially the secondary transmission (such as disk replication, network share, etc.) which take nothing into account, consume the resources of potential users at geometric progression rate. Thus, a new technical ------ digital rights management technology (Digital Rights Management, DRM) [1] was proposed. The purpose is to protect the intellectual property rights of digital content and to ensure the legitimate use of digital content and dissemination throughout the life cycle by technical means.

Digital rights management technology has been concerned by industry and academia, which is seen as the key technology of the spread and transactions of digital content. Calculation of many well-known internationally Machine companies and research institutions have launched their own products and systems, such as Microsoft, IBM, Real Networks, as well as SONY and other companies, these companies mainly engaged in the whole combination of DRM technology, communication systems, as well as the development of its components; MPAA (Motion Picture Association of American) and SDMI (Secure Digital Music initiative) and other organizations are focused on the development of some DRM systems which adapt to special requirements , such as DVD copy protection and digital music transmission protection. We can see that DRM technology has become one of the hot Internet technology to develop [2].

In this paper, we investigate and analyze the current multimedia digital rights management system in details, then put forward a common digital rights management framework. Based on the architecture, design and implement a prototype system SQDRM, with openness, scalability, and superior interoperability and so on. Vocational training in the Ministry of Labor remote network deployment and experimental results show that the system can be streaming media data in digital rights to effective protection.

## 2 Digital Copyright Protection Technology

Content providers to provide rich digital content at the same time, need to stop the illegal use of content, protection of intellectual property rights and interests of owners of the need in the digital world, the introduction of copyright protection technology. Design and implementation of DRM systems is a complex task that requires from the architectural point of view to know it. DRM architecture includes features, information, modules, etc., the most important is the function of the structure and information structure [3]. Functional structure defines the DRM system, high-level modules. DRM function of the system architecture as shown:

## 3 Streaming Media Copyright Protection of Critical Technology and Architecture

Streaming media copyright protection system is based on two key technologies: streaming media technology and digital copyright protection technology. In the traditional technology, the user in order to watch a film, need to reserve storage space in the local to the whole movie to download before playback, and streaming media technology will allow users to online play. Streaming Media refers to the network using streaming technology, transmission media. Transfer process can be described as: streaming media server to a continuous image and sound information is encapsulated into a separate data packets, according to certain protocols continuous, real-time to send to the client; the client to create and maintain a buffer zone, while the received data packets into the buffer, while the buffer available to the decoding of data inside the module. A buffer much smaller than the size of media files does not require a lot of storage space.

Streaming Media copyright protection of reducing storage pressure; users can download the edge while watching without having to wait for the entire file to download to a local. Streaming media technology can be widely applied into video ordering, online broadcast, distance learning, real-time video conferencing and other Internet applications, the network has brought a revolutionary change in the exchange of information, and they also bring tremendous business opportunities [4].

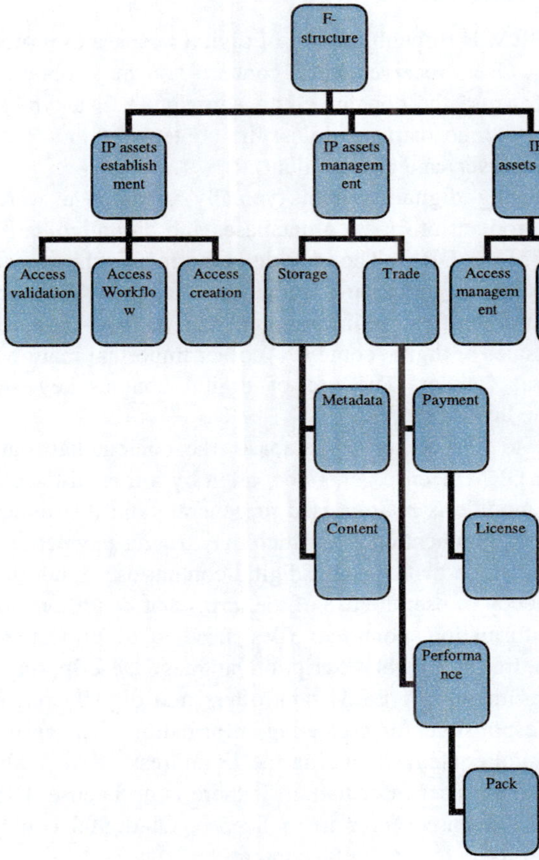


**Fig. 1.** Structure model of DRM

Digital copyright protection technology for digital content is the intellectual property protection, a series of hardware and software technologies for the protection of digital content throughout the life cycle of the legitimate use of digital content value chain, balancing the roles of the various interests and needs of the entire digital market development and dissemination of information [5]. In particular, DRM system for digital products to provide the following protection:

- Unauthorized users can not be a way to deceive and decryption using digital products online or offline content.
- Authorized user digital products can not be saved without the protection or distribution.

- Users do not permit limitations of digital products outside the scope of operation.
- Digital content providers can track their own digital products, digital products are found to be cracked and illegal use.

The core idea of DRM is through the use of digital licenses to protect the copyright of digital content [6]. Users receive digital content, you must obtain the appropriate number of licenses can use the content. Shown in Figure 2, a typical DRM system architecture consists of three main modules: the content server (content server), the license server (license server) and client (client).

Content server storing digital content typically includes a warehouse, storage product information product information databases and digital content packaging tool for the safe handling of the DRM. The module is the main achievement of the digital content encryption, such as processing and inserts a digital watermark processing results and the content identification information such as meta-data can be distributed along with packaged sales of digital content. Another important feature is the use of the right to create digital content. The use of digital content key and the right to information sent to the license server.

Includes the right to license server database, the content key database, the user identity database and DRM license generator, often by a trusted third-party module is responsible for. The module is mainly used to generate and distribute digital license, you can also achieve user authentication, which may trigger payment transactions such things. Digital license is one that contains digital content use rights (including the use of permissions, frequency of use, the use of the terms and conditions of use), licensors and their owners' information, computer files, used to describe the digital content licensing information, from the right description language Description.

Client-side consists mainly of DRM controllers and digital content using a tool. DRM controller is responsible for collecting information such as user identity and control the use of digital content. Including the IP address, MAC address, whether to join the blacklist and other user information. If there is no license, DRM controller is also responsible to the license server for a license. Of digital content using a tool mainly used to support users to use digital content.

## 4 Content Server Design

Content server storing digital content, including the contents of warehouses, storage product information product information databases and digital content packaging tool for the safe handling of the DRM. The module is the main achievement of the digital content encryption, such as processing and inserts a digital watermark processing results and the content identification information such as meta-data can be distributed along with packaged sales of digital content. Another important message is to create digital content use rights. The use of digital content key and the right to information sent to the license server.

## 5 License Server Design

Throughout the Streaming Media DRM system is the most central part of the license server system. Streaming media encoding encryption system and back office systems are for the permit system to provide the necessary authentication information, and ultimately control the flow of media rights management is done by the license server. Encoding encryption subsystem will first need of the management of streaming media encrypted according to a certain encoding rules, while encrypted information such as: Content ID, Key ID, public key, seed, etc. into the database; background management platform will be streaming users to encrypt applications for purchase of information and user information stored into the database. When users play protected streaming media files, they will apply for certification system through the network to broadcast licenses, the certification system through a database of all stored information, and access to user information, the dynamic management of user rights, and management information into database, so as to realize a wide range of streaming media copyright management.

License server by using the user rights are effectively manage dynamic, real-time response to a user's application and an application for leave feedback safe and reliable manner, and ultimately the entire DRM system, streaming media and its users to monitor the use of all forms of track management, thereby the entire Streaming Media DRM system has a new generation of DRM systems key features and functions.

## 6 Client's Interaction with the License Server

The certification process of streaming media is an information exchange process between client and license server in network.

When the user plays the protected streaming media file, the client through the streaming media in the URL information, connect to license server, while sending request information to obtain the certificate Challenge, server receives the message, will give the client feedback from a certified interface. Users only need authentication page, enter their authentication information that the user names, passwords, etc., and confirmed the authors. At the same time the client will send the client a variety of hidden information, including the user currently logged on time, IP address, MAC address and other information. Server-side processing through access to information and, based on user information stored in the database to determine the legitimacy of the user and the use of authority, after the adoption of the appropriate license will be sent to the client. Clients only need to store this license, and then can broadcast according to the permissions, set by license.

Principles of streaming media to achieve certification

In the process of streaming media to obtain the certificate, the server receives the client sends this information, how to complete the processing of information to the user's identity authentication and generation of broadcast license is the key technology throughout the certification process. In the license server receives the client sends all the information, the server will carry out the work of parsing the information from the information obtained to get streaming media file content ID, and the client info, and verify whether the private key of streaming media match with the public key of server. Combination of database analysis to determine whether user information for legitimate users, unauthorized users as well as the blacklist is not to permit a user-generated operations, failure to return the information to the client. After the identification information, the server based on the content ID, user information, and database information to generate authority objects; the last server based on the key, permission object, the client information, etc. to create the issue of a user-specified playback of streaming media files license, thus completing the whole process of streaming media obtaining the certificate.

Permission control

The key to permission control is permission analysis and validation. Permission analysis and validation is to analyze digital license, to verify whether the user's operation is permitted, prerequisite and restrictions is met with requirement. Right parsing and validation with the use of the rights description language is closely related to, from REL parser (interpreter) treatment. REL parser embedded in the DRM controller.

Figure 2 shows the processes of permission analysis and validation based on XrML. In the figure, user required to play a protected video files. Receive the user's playback request, playback control procedures XrML-enabled Application to collect user information, file information, and user operation information, to REL parser License Interpreter request. Parser to obtain the data corresponding to the license file and verify its effectiveness, confirm that the file has not been tampered with, and then parse the license file to check whether to include broadcast rights, and verify that the permissions are indeed owned by the current user. If the test is passed, then return the appropriate constraints and preconditions to the playback control procedures. Condition Validator is required to verify whether prerequisite and restrictions is met with the relevant conditions (such as whether the number of players to the extent permitted, etc.) by play control program. If the condition is satisfied, then the user can play the audio file.

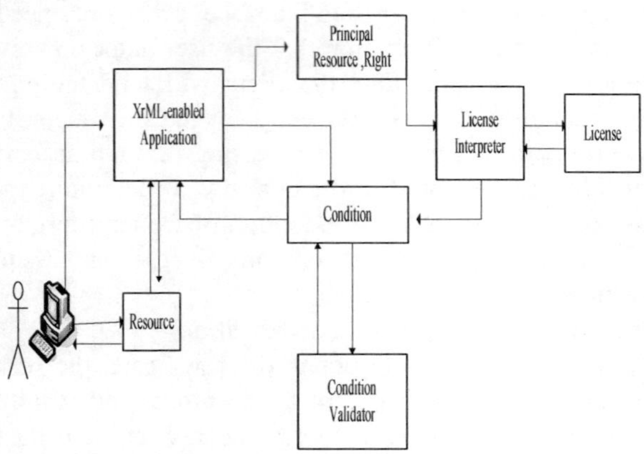


**Fig. 2.** Process of permission analysis and validation

# 7 Streaming Media Copyright Protection System

The packaged content server is responsible for the media data file encryption protection and copyright information input. Working process are: content owners to be specified to protect the streaming media file; content owners to protect streaming media files to be designated copyright information; system recognize the file type, if it is the type of system supports, continue the following operations; the use of RC4 stream encryption algorithms for video and audio data to the frame as the unit for encryption and decryption, key length of 128 bits. RC4 is essentially a pseudo-random number generator, and generating algorithms and data stream XOR. It is characterized by fast algorithms for real-time the impact of small; the copyright information documents to the license server signature. As shown in Figure 3, the content server-side device interface package.

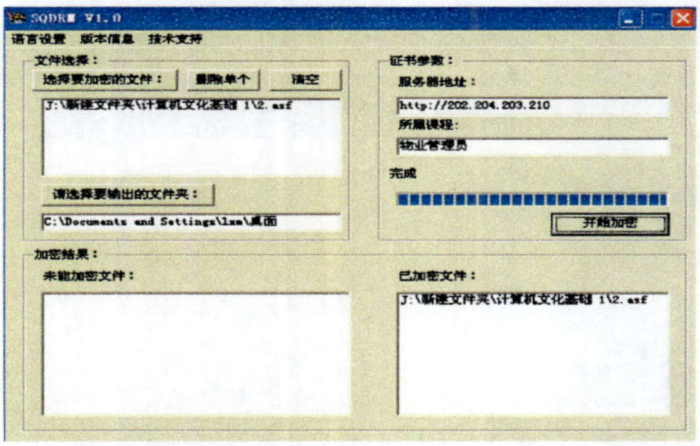


**Fig. 3.** Interface of content server packetizer

The licensing system of systems

Licensing system is built on open standards system, allowing clients to use the different services it provides, but also for the entire DRM system provides highly scalable, and certification services provided by this system is very flexible and can adapt to different The authentication of the certification requirements.

1) File playback authentication. This approach means that when a user plays the file automatically to the designated license server to obtain licenses, and requires the user identity and password information for authentication. Certification, then the authentication server automatic transmission license and stored in the user machine.

2) A single file permits the pre-transmission. The approach is that users do not have to play the file, but directly through the user identity and password authentication, as well as the product identifier to provide the necessary means to obtain a license. Certification, then the authentication server automatic transmission license and stored on the user's machine. The model generally used for users to have access to products, but without a permit.

3) Volume Documentation License pre-transmission. The approach is that users do not play the file, but directly through the user identity and password authentication to obtain a permit batch file method, certification, then the license server automatically

transfers the user to select all the media file playback license is stored in the user's machine. The model generally used for CD-ROM, courseware or network curriculum contains a lot of media documentation.

4) Document Play permit default. This approach means that when a user plays the file automatically to the specified authentication server to obtain the certificate, without need to enter authentication information, authentication server license will be automatically transferred and stored on the user's machine. The model generally used for sample preview, play a limited number. Certification of this system are four main methods of the above, making streaming media DRM systems can be built on a web-based services more flexible rights management patterns. And the system provides a friendly user-friendly interface, so to some extent also to improve the system's operability. Figure 4 shows the license management system interface

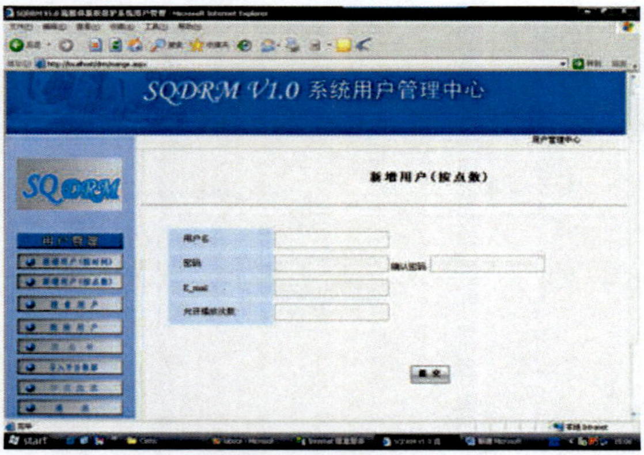


**Fig. 4.** Interface of license management system

# 8 Conclusions

In this paper, based on the study of streaming media digital rights, we proposed a universal streaming media system for digital copyright protection, and implements a prototype system of streaming media digital copyright protection, and conducted SQDRM system experiments and have a certain level of practical application. The results show that the system has capabilities of transparent supporting local playback and online play of streaming media, which effectively solves the key problem in the field of digital copyright and provides a standard interface for users to expand. The actual system deployment and initial experiments show that the system protects the streaming media data very well. Obviously, it is unrealistic to achieve absolute security, digital copyright protection solution will depend on a more secure software and hardware environment. We will have a further expansion of the system and improve the business model of the system in the future work.

## References


1. Trippe, W., Mooney, S.: Degital Rights Management: Business and Technology. M&T books, New York (2002)
2. Garnett, N.: Digital rights management, copyright, and napster. ACM SIGecom Exchanges 2(2), 1–5 (2001)
3. Zwollo, K.: Digital Document Delivery and Digital Rights Management. Information Services & Use (21), 9–11 (2001)
4. Neuman, B.C., Ts'o, T.: Kerberos: An Authentication Service for Computer Networks. IEEE Communications Magazine 32(9), 33–88 (1994)
5. Tueck, S., Welch, V., Thompson, E.D.: Internet X.509 Public Key Infrastructure Proxy Certificate Profile draft-ietf-pkix-proxy-10(work in progress). In: IETF (2003)
6. Welch, V., Foster, I., Kesselman, C., Mulmo, O., et al.: X.509 Proxy Certificates for Dynamic Delegation 3rd Annual PKI R&D Workshop (2004)

# The Attitude towards Translation Exposure and Its Hedging Techniques

Ying Zhang[1] and Yunlin Zheng[2]

[1] Department of Finance Hohai University
Nanjing, China
[2] Business School The University of South Dakota
Vermilion, U.S.A.
zoe511@hhu.edu.cn, Yunlin.Zheng@usd.edu

**Abstract.** It is highly debatable whether the Multinational Enterprises(MNEs) should hedge the translation exposure. The paper examined the stock market reaction to the translation adjustment gains and losses. Through reviewing the survey of current practices of hedging translation exposure, it indicated the management choice of whether to hedge translation exposure and which the most effective way to hedge is. The conclusion is that the existence of loan covenant, chief executive's commitment of profit and enjoy of a premium might force MNEs to hedge the translation exposure. Other factors discourage MNEs to engage in such hedging activities. And matching is still the most used internal hedging technique, while swap and forward contracts are prevailing but cursory external techniques.

**Keywords:** translation exposure, hedging techniques, multinational enterprises, financial reporting.

## 1 Introduction

Translation exposure arises from financial reporting requirements which is the potential of gains or losses resulting from the change in exchange rates. It is highly debatable whether the Multinational Enterprises (MNEs) should hedge the translation exposure.

The study is organized as follows: Section II discusses why firms hedge or not hedge their translation exposure. Section III examines the stock market reaction to the translation adjustment gains and losses. Section IV reviews the survey of current practices of hedging translation exposure and indicates the management choice of whether to hedge translation exposure and which the most effective way to hedge is. Section V is the conclusion.

## 2 The Reasons of Hedging or Not Hedging the Translation Exposure

Before talking about the feasibility of hedging the translation exposure, we need to examine the goals of hedging the translation exposures. According to Shapiro (2002), firms that engage in translation exposure hedging activities usually want to achieve the aim to maximize the shareholder value or, firm value. Therefore, firms need to balance off the benefits of hedging with its costs to evaluate whether it is worthwhile to hedge[1].

Hagelin (2003) summarized that, generally, the finance literature recommends not to hedge translation exposure for two reasons[2]. Firstly, translation adjustment gains or losses are listed separately from net income. Secondly, translation adjustment gains or losses could not properly estimate the firm value. Thus, hedging translation exposure is inefficient to reduce the firm's overall exposure.

For firms that have decided to hedge translation exposure, in Shapiro's (2002) view, they can only hedge against unexpected changes of exchange rates if financial markets are efficient. And he summarizes various kinds of costs of the basic hedging techniques, including the cash flow and human resource that MNEs spend on the hedging activities.

Companies with tight loan covenants usually consider translation exposure hedging. Through examining Swedish firms' use of financial hedges to hedge foreign exchange exposure by using survey data, Hagelin and Pramborg (2006) found that about 20% of the sampled firms hedge their translation exposure and that the existence of loan covenants and translation exposure hedging are positively related[3]. This is consistent with Butler's (2003) argument that firms hedge translation exposure in order to secure their access to funds under the loan covenants [4].

Furthermore, Haseltine (1981) found that, because the financial executives should responds to public demand, they have to minimize the size of reported losses even if a premium must be paid at times [5]. This premium might be the higher interest rate in home currency borrowing compared with local currency borrowing. The financial executives prefer to pay the premium rather than to show a foreign exchange loss in statements. Ensor and Muller (1981)'s finding supports this view [6]. Accounting and published results can have more influence than cash flow on the financial executives, despite the relevance of translation. Thus, Rodringuez (1981) stated that, for managers concerned with the influence of accounting result, they would first manage the translation exposure [7].

In addition, Moffett, Stonehill and Eiteman (2005) noted that tax consequence of foreign exchange rate exposure favors hedging translation exposure first. Generally, only realized gains or losses affect tax liability [8].

## 3 Stock Market Reaction to the Translation Adjustment Gains and Losses

If a company already has tight loan covenants, it should pay higher interest rate for additional capital to compensate the translation exposure. Also if chief executives cannot keep the earnings promise to the public due to translation exposure, the stock prices usually drop. And translation adjustment losses deteriorate the key financial ratios such as ROE, ROA and debt-to-equity. The extant accounting literature that focuses on the stock market reaction to the adjustment gains or losses compared the effect of SFAS No. 52 with that of SFAS No. 8.

Collins and Salatka (1993) found that, under SFAS No. 8, translation gains or losses included in the income statement adds significant "noise or garbling" to the reported earnings figures, while the translation method under SFAS No. 52 improve the informativeness of earnings signals[9]. Bartov's (1997) view[10] is consistent with them. In contrast, Soo and Soo (1994) found no significant change in valuation weights between these two standards, though the stock market incorporates translation gains or losses information into the stock prices [11]. Beside the above literature focusing on the unexpected return/earnings, Louis's (2002) study proved that the translation adjustment negatively related to the firm's value, i.e., a positive translation adjustment is associated with a loss of value. Thus, we may conclude here that, although the stock market may incorporate translation adjustment information when evaluating a firm, since there is negative relation between translation adjustment and a firm's value, that is, translation adjustment gains only add negative value to the firm, while translation adjustment losses add positive value and is good to firm, MNEs should not hedge the translation adjustment gains or losses.

## 4 Survey of Current Practices of Hedging Translation Exposure

Though theoretical and empirical studies suggest that MNEs should not to manage the translation gains or losses, what do investors choose in practice? In early 1990s, the treasury team at Touch Ross conducted a survey by interviews with senior managers from listed companies of UK. The survey's answers are classified into three categories. First, as for investment, 68% of the respondents didn't consider that they would simply invest in currency element, 72% would allocate UK quoted MNEs into their sterling portfolio, but all of the respondents considered whether by the foreign exchange fluctuations would affect their shareholdings. Second, as for translation exposure hedge, 63% of respondents preferred that companies not to hedge the translation exposure, 14% said they prefer hedge the foreign currency exposure by themselves, thereby companies' hedging could only duplicate their hedges. Third, as for hedging techniques, 77% of the respondents chose local currency borrowing as prudent hedging method, 32% investors selected swaps and options as their hedging instruments but still felt that these financial derivatives hedging is "cursory".

More recently, Nazarboland (2003) conducted a survey of financial reports to investigate the behavior of top UK corporations towards translation exposure [13]. And his conclusion is only a few of the surveyed corporations view the translation exposure as the primary currency risk, the other majority views it immaterial.

## 5 Conclusion

Except for the three special circumstances (the existence of loan covenant, chief executive's commitment of profit and enjoy of a premium) that might force MNEs to hedge the translation exposure, in most normal circumstances, other factors such as reduction of firm's value, hedging costs and the failure potential of hedging discourage MNEs to engage in such hedging activities. Currently, matching is still the most used internal hedging technique, while swap and forward contracts are prevailing but cursory external techniques.

## References


1. Shapiro, A.C.: Multinational Financial Management, 7th edn., pp. 421–583. John Wiley & Sons, Inc., Hoboken (2002)
2. Hagelin, N.: Why firms hedge with currency derivatives: an examination of transaction and translation exposure. Applied Financial Economics 13, 55–69 (2003)
3. Hagelin, N., Pramborg, B.: Empirical evidence concerning incentives to hedge transaction and translation exposures. Journal of Multinational Financial Management 16, 142–159 (2006)
4. Butler, K.C.: Multinational finance, 3rd edn., pp. 517–589. South-Western College Pub., Mason (2003)
5. Haseltine, J.B.: A Longer-term approach. In: Ensor, R., Muller, P. (eds.) The Essentials of Treasury Management, pp. 397–481. Euromoney Publications, London (1981)
6. Ensor, R., Muller, P.: The Essentials of Treasury Management, pp. 108–253. Euromoney Publications, London (1981)
7. Rodringuez, R.M.: Foreign exchange management in US Multinationals, pp. 36–72. University of Illinois at Chicago Circle (1981)
8. Moffett, M.H., Stonehill, A.I., Eiteman, D.K.: Fundamentals of Multinational Finance, 2nd edn., pp. 362–410. Pearson Education, Inc., Boston (2005)
9. Collins, D., Salatka, W.: Noisy accounting earnings signals and earnings response coefficients: The case of foreign currency accounting. Contemporary Accounting Research 10, 119–159 (1993)
10. Bartov, E.: Foreign currency exposure of multinational firms: Accounting measures and market valuation. Contemporary Accounting Research 14, 623–652 (1997)
11. Soo, B., Soo, L.: Accounting for the multinational firm: Is the translation process valued by the stock market? The Accounting Review 69, 617–637 (1994)
12. Louis, H.: The value relevance of the foreign translation adjustment. The Accounting Review 78, 1027–1047 (2002)
13. Nazarboland, G.: The attitude of top UK multinationals towards translation exposure. Journal of Management Research 3, 119–126 (2003)

# The Econometric Analysis of the Contribution Degree of Technological Progress in Bohai Rim Region*

Zhang Jingqiang[1], Zhou Hong[2], and Huang Zilang[1]

[1] School of Economics & Management
North China University of Technology
Beijing, China
School of Humanity and Law
[2] Beijing Institute of Civil Engineering and Architecture
Beijing, China
zjq811@163.com, zhh620@sohu.com, 335086019@qq.com

**Abstract.** In this paper, utilizing the econometric regression method, using the provincial panel data of Bohai Rim region, and selecting the factors of the technological progress in Bohai Rim region, the author created a space econometric regression model and analyzed the contribution degree of technological progress in Bohai Rim region. The results indicate that the effect of the domestic technology diffusion path spillover to the technological progress of Bohai Rim region is greater than the role of the foreign technology diffusion path spillover and the spillover of domestic R&D capital stock plays a leading role in the technological progress of Bohai Rim region.

**Keywords:** technological progress, econometric regression model, panel data, empirical study.

## 1 Introduction

Since the 30 years of reform and opening up, China's economy has shown rapid development. Various regional GDP maintained an average annual growth rate of 9%, which created a miracle in the history of China's economic development. However, the rapid growth of China's economy is mainly due to the input of capital, labor, energy and the extensive features of economic growth mode is still relatively pronounced. Under the new international economic situation, China's economic development must transform the economic growth mode from extensive development to intensive development. In the final analysis, we should depend on technological progress to promote regional economic growth.

Technological progress roots in innovation activities. The higher innovation efficiency is, the faster the rate of technological progress and the pace of economic growth. On the contrary, if there are no innovation actives in a country's economy, and

* This paper is supported by "Beijing's Undergraduate student Scientific Research and Entrepreneurial Action Plan Project (1018)"and "2010 Undergraduate student S&T Activities Project of North China University of Technology".

in the long run, this country's economy will fall into stagnation. In recent years, the research on the technological progress has become the key points of academic discussion. These scholars, for example, Fare. R(1994)[1], Chow and Lin(2002) [2], Wang and Yao(2003) [3], Yan Pengfei and Wang Bing(2004)[4], Zhao Wei(2005)[5], Yue Shujing and Liu Chaoming(2006)[6], studied the regional technological progress through various methods. The results show that the technological progress has played a certain role in promoting economic development, but the extent of the role and contribution of the technological progress to regional economic growth has been an academic discussion problem.

In this paper, from the effecting factors of the regional technological progress, we use spatial econometric analysis method to study the technological progress of China's Bohai Rim region, and establish the model of regional technological progress contribution degree, which has a strong reference and facilitating role on raising the technological progress level of Bohai Rim region and promoting regional economic growth.

## 2 Method and Model

### 2.1 Basic Model

In order to quantitatively study the technological progress contribution degree of Bohai Rim region, we use the trade spillovers model (CH model) to analyze regional technological progress [7]. The CH model is based on the "innovation-driven" growth theory model of Grossman & Helpman, and its aim was to evaluate foreign technological advantages on how to influence GDP growth, more precisely, to evaluate the technological spillover effects from the import of goods and services.

The expression form of CH model is as following:

$$\ln F_{it} = \alpha_{0i} + \alpha_{1i} \ln S_{it}^{d} + \alpha_{2i} \ln S_{it}^{f(CH)} + \varepsilon_{it} \tag{1}$$

Among which, $i = 1, 2, 3, \cdots\cdots, n$, represents the country; $F_{it}$ is the factor productivity of the $i$ th country in the $t$ period; $S_{it}^{d}$ is the domestic R&D capital stock of the $i$ th country in the $t$ period; $S_{it}^{f(CH)}$ is the foreign R&D capital stock of the $i$ th country through trade path spillover with the calculation of the CH method, and $S_{it}^{f(CH)} = \sum_{j \neq i} \frac{IM_{ijt}}{IM_{it}} * S_{jt}^{d}$, $IM_{ijt}$ is the import of the $i$th country from the $j$th country in the $t$ period, and $IM_{it} = \sum_{j \neq i} IM_{ijt}$, $S_{jt}^{d}$ is the domestic R&D capital stock the $j$ th country in the $t$ period; $\alpha_{0i}$ is the constant term; $\alpha_{1i}$ is the flexibility of the domestic R&D capital stock on total factor productivity of the $i$th country; $\alpha_{2i}$ is the flexibility of the foreign R&D spillovers stock on total factor productivity; $\varepsilon_{it}$ is the random interference term.

## 2.2 Model Expansion

In this paper, we mainly study the technological progress of Bohai Rim region, so we will redefine the variables of CH model and expand the CH model. First, we improve the domestic R&D capital stock through re-introducing the invention patent application variable under a single path of R&D capital stock; secondly, we improve the foreign R&D capital stock through re-introducing foreign direct investment (FDI) under a single path of trade import. In this paper, we mainly study the impact on the R&D stock at home and abroad to the technological progress of Bohai Rim region and this impact mainly manifested between the different individuals of the cross-section. Then, we assume that the time series parameters are homogeneous, and the parameters satisfy the time consistency, that is, the parameter values do not change over time.

Then, the expansion model is as following:

$$\ln TFP_{it} = \alpha_i + \beta_{1i} \ln S_t^{rd} + \beta_{2i} \ln S_{it}^{pat} + \beta_{3i} \ln S_{it}^{imp} + \beta_{4i} \ln S_{it}^{fdi} + \varepsilon_{it} \tag{2}$$

Among which, $TFP_{it}$ is the total factor productivity of the $i$th province in Bohai Rim region in the $t$ year, it measures generalized technology progress; $S_t^{rd}$ is China's R&D capital stock in the $t$ year, and we assume that it has been conducted among various regions; $S_{it}^{pat}$ is the invention patent applications of the $i$th province in Bohai Rim region in the $t$ year; $S_{it}^{imp}$ is the trade import of the $i$th province in Bohai Rim region in the $t$ year; $S_{it}^{fdi}$ is the foreign direct investment of the $i$th province in Bohai Rim region in the $t$ year; $\alpha_i$ is the constant term; $\beta_{1i}$, $\beta_{2i}$, $\beta_{3i}$ $\beta_{4i}$ are the coefficients; $\varepsilon_{it}$ is the random interference term.

## 2.3 Model Checking and Optimization

*1) The checkingof the spillover effects on FDI and trade imports*

There is likely to have lagged effect on the impact of FDI and trade import to the technological progress of Bohai Rim region, so we should separately check the impact of FDI and trade import to the total factor productivity of Bohai Rim region to determine its lag phases.

Then, we establish the regression equation and it is as following:

The current period:

$$\ln TFP_{it} = \alpha_0 + \beta_1 \ln S_t^{rd} + \beta_2 \ln S_{it}^{pat} + \beta_3 \ln S_{it}^{imp} + \varepsilon_{it} \tag{3}$$

$$\ln TFP_{it} = \alpha_0 + \beta_1 \ln S_t^{rd} + \beta_2 \ln S_{it}^{pat} + \beta_4 \ln S_{it}^{fdi} + \varepsilon_{it} \tag{4}$$

The lagged one period:

$$\ln TFP_{it} = \alpha_0 + \beta_1 \ln S_t^{rd} + \beta_2 \ln S_{it}^{pat} + \beta_3 \ln S_{it}^{imp}(-1) + \varepsilon_{it} \tag{5}$$

$$\ln TFP_{it} = \alpha_0 + \beta_1 \ln S_t^{rd} + \beta_2 \ln S_{it}^{pat} + \beta_4 \ln S_{it}^{fdi}(-1) + \varepsilon_{it} \tag{6}$$

The lagged two period:

$$\ln TFP_{it} = \alpha_0 + \beta_1 \ln S_t^{rd} + \beta_2 \ln S_{it}^{pat} + \beta_3 \ln S_{it}^{imp}(-2) + \varepsilon_{it} \quad (7)$$

First, we make F-test for various regression equations [8]. The F-test results are as shown in Table 1. From Table 1, the $F_2$ values of various equations are less than the critical value, which indicates that in all equations various provincial and municipal intercepts and coefficients are the same.

**Table 1.** F-test results of regression equation Under single path

| regression equation | (3) | (4) | (5) | (6) | (7) |
|---|---|---|---|---|---|
| $F_2$ | 0.2727* | 0.7114* | 0.3408* | 1.6701* | 1.8138* |

According to the F-test results, we can achieve the regression results of panel data under single path, as shown Table 2.

**Table 2.** Regression results of panel data under single path

| regression equation | (3) | (4) | (5) | (6) | (7) |
|---|---|---|---|---|---|
| $\alpha$ | -0.1683 (-0.08) | -0.5059 (-0.249) | -0.9796 (-0.617) | -0.4321 (-0.226) | -0.3534 (-0.151) |
| $\beta_1$ | 0.1491 (1.280) | 0.1656 (1.444) | 0.1646 (1.5306) | 0.1472 (1.280) | 0.1608 (1.306) |
| $\beta_2$ | 0.0305 (1.315) | 0.0256 (1.1076) | 0.0384 (1.6338) | 0.0352 (1.4550) | 0.0262 (0.1255) |
| $\beta_3$ | -0.0087 (-1.19) | | 0.0164 (2.261***) | | 0.0010 (0.126) |
| $\beta_4$ | | 0.0047 (0.591) | | -0.0043 (-0.539) | |
| $AR(1)$ | 0.9625 (35.9*) | 0.9604 (33.54*) | 0.9501 (29.49*) | 0.9562 (32.61*) | 0.9628 (30.26*) |
| $R$ | 0.9846 | 0.9845 | 0.9835 | 0.9828 | 0.9804 |
| $\overline{R}$ | 0.9841 | 0.9839 | 0.9829 | 0.9821 | 0.9796 |
| $F$ | 1826 | 1809 | 1596 | 1527 | 1249 |
| $SSE$ | 0.2946 | 0.2973 | 0.2713 | 0.2834 | 0.2797 |
| $DW$ | 1.83 | 1.84 | 1.83 | 1.89 | 1.88 |

Note: (1) the figures between brackets are t-test values;

(2) * Correlation is significant at the 0.01 level, ** Correlation is significant at the 0.05 level, *** Correlation is significant at the 0.1 level;

(3) All equations use mixed regression models.

*2) Econometric Regression Model*

From the regression results of TABLE II, the spillover effect of the trade import lagged one period and the FDI current period to the total factor productivity of Bohai Rim region is the most. So, we can establish the econometric regression model of the technological progress contribution degree of Bohai Rim region, as shown in (8).

$$\ln TFP_{it} = \alpha_0 + \beta_1 \ln S_t^{rd} + \beta_2 \ln S_{it}^{pat} + \beta_3 \ln S_{it}^{imp}(-1) + \beta_4 \ln S_{it}^{fdi} + \varepsilon_{it} \quad (8)$$

## 3 Empirical Calculation and Results Analysis

According to the technological progress contribution degree model of Bohai Rim region, we can achieve the regression model of panel data, as shown in (9).

$$\ln TFP_{it} = -0.984 + 0.167 \ln S_t^{rd} + 0.036 \ln S_{it}^{pat}$$

$$t = \; (-0.591) \quad (1.517) \qquad (1.503) \; + 0.016 \ln S_{it}^{imp}(-1) + 0.005 \ln S_{it}^{fdi} + \varepsilon_{it} \quad (9)$$

$$t = \quad (2.261^{***}) \qquad (0.548) \; [AR(1) = 0.952]$$

$$t = \quad (28.629^{*})$$

$$R = 0.983567 \; \overline{R} = 0.982791 \; SSE = 0.270583 \; DW = 1.83$$

Note: * Correlation is significant at the 0.01 level, *** Correlation is significant at the 0.1 level; $AR(1)$ is the estimated values of the $\rho_1$ in $\varepsilon_{it} = \rho_1 \varepsilon_{i,t-1} + \upsilon_{it}$, and $\hat{\rho}_1 = AR(1)$.

From the calculation results of the technological progress contribution degree of Bohai Rim region, the contribution degree of the R&D capital stock overflow to the technological progress is the most, but is not statistically significant. The domestic invention patent applications overflow to the improvement of the technological progress level of Bohai Rim region plays an important role.

The foreign two technology diffusion paths overflow to the technological progress of Bohai Rim region plays a certain role, and the overflow role of the trade import is still relatively statistically significant, which indicate that the impact of the trade import to the technological progress of Bohai Rim region is significant. Through comparing with the domestic and foreign technology diffusion path overflow, we find that the overflow role of the domestic technology diffusion path is significantly larger than the overflow role of the foreign technology diffusion path.

## 4 Conclusion

Based on the above empirical results, we can get the following conclusions:

First, the effects of the various technology diffusion overflow paths to the technological progress of Bohai Rim region are all positive, that is, these technology diffusion path overflows can promote the technological progress of Bohai Rim region.

Secondly, the effect of the domestic R&D capital stock spillover to the technological progress of Bohai Rim region is the greatest, followed by the overflow role of the domestic invention patent applications.

Thirdly, the domestic technology diffusion path for overflow of Bohai Sea region is greater than the impact of technological advances abroad, the role of technology diffusion path of the overflow.

**Acknowledgment.** I express heartfelt thanks for Mrs. Zhou Hong and Mrs. Huang Zilang. They collect and sort data for this paper, proofread the whole paper, and put forward many suggestions for the completion of this paper.

## References

1. Fare, R., Grosskopf, S., Norris, M.: Productivity growth, technical progress and efficiency change in industrialized countries. American Economic Review 84, 66–83 (1994)
2. Chow, G., Lin, A.: Accounting for economic growth in Taiwan and Mainland China: a comparative analysis. Journal of Comparative Economics 30, 507–530 (2002)
3. Wang, Y., Yao, Y.: Sources of China's economic growth 1952-1999: incorporating human capital accumulation. China Economic Review 14, 32–52 (2003)
4. Yan, P., Wang, B.: Technical efficiency, technological progress and productivity growth: an empirical analysis of DEA method. Economic Research, 55–64 (December (2004)
5. Zhao, W., Ma, R., He, Y.: Decomposition of changes in total factor productivity—based on the empirical analysis of the Malmquist productivity index. Statistical Research, 37–42 (July 2005)
6. Yue, S., Liu, Z.: Human capital and regional total factor productivity analysis. Economic Research, 90–97 (April 2006)
7. Coe, D.T., Elhanan, H.: International R&D spillovers. European Economic Review 39, 859–887 (1995)
8. Li, Z., Ye, A.: Advanced econometrics, pp. 147–151. Tsinghua University Press, Beijing (2007)

# The Application of Information Fusion and Extraction in Maize Seed Breeding

Ji Dan[1], Qiu Jianlin[1], Dai Xiaoyu[2], Gu Xiang[1], and Chen Li[1]

[1] School of Computer Science and Technology, Nantong University
Jiangsu, China
[2] School of Software, Harbin Institute of Technology
Heilongjiang, China
Qiu.jl@ntu.edu.cn, d_xiaoyu@126.com, Gu.x@ntu.edu.cn

**Abstract.** It has been the irresistible trend of agriculture informationization which uses information technologies to treat enormous data and finds out potential useful rules to direct the development and reformation of agriculture. Aiming at the specific application of maize seed breeding, this paper effectively integrates several data mining technologies and presents a new method called CA to analyze the whole maize information. The algorithm achieves transverse dimension reduction by combining PCA and other methods, and also achieves longitudinal dimension reduction by improving CURE and k-means. The decision-tree method of CA algorithm introduces three different classifiers in order to enhance the accuracy of trees. By comparing the results of improved algorithm with traditional methods, we can find that the new algorithm is better in performance and degree of parallelism.

**Keywords:** Dimension Reduction, Clustering, Decision Tree, Maize Seed Breeding.

## 1 Introduction

In maize seed breeding field, we have stored a large number of maize samples. Each corn is made up of many attributes like period, plant height, ear height etc. The whole corn dataset is a large-scale and high dimension one. So it is important for us to utilize data mining technologies to deal with it and find out potential useful rules to direct maize seed breeding.

Clustering is a common data analysis tool and is a kind of unsupervised learning. One of the prominent characteristics is clustering can deal with tremendous and complexity datasets, and also can be the pretreatment part of other methods. So we can consider the clustering result as the pretreatment of decision tree method. At present, researchers have present many different kind of clustering algorithms [1]-[3], and we can choose the most suitable one to deal with data according to the particular condition and applicable scope of certain algorithms. However, there also exist some kind of problems like sensitive degree of noise etc. So it is worth for us to consider how to select and improve original algorithms for obtaining the optimal results.

As one of the mostly used methods of data mining, decision tree method can deduce the classified rules in the form of a tree from a set of disorder and non-rule instances via modeling and can reflect some useful features and relationships syllabify which provide much convenience for users to deal with data and research information. There are quite a lot of traditional algorithms but also exist some problems more or less, such as the optimal feature selection, the pruning of trees and so on. At present, researchers have improved these methods in many ways [4]-[7]. However, to a certain extent, we can't obtain the real useful information by one and the computation costs are also high. So if we use decision tree method after preprocessing the original data, the result would be better.

The improved CA algorithm consists of three main parts: transverse dimension reduction, longitudinal dimension reduction and decision tree method. While the former one combines PCA (Principal Component Analysis) method and correlation coefficient matrix to find the most important attribute groups for analyzing datasets, and then obtains new feature sets according to the corresponding weight values which can reduce large feature dimensions. The latter one introduces grid partition, and then reforms k-means and CURE in order to pre-cluster the whole datasets to several clusters as the input of the decision tree. After estimating the cluster we needed, we can implement decision tree algorithm to it instead of dealing with the whole dataset, which can reduce the algorithm's calculation amount tremendously. Finally, improve the decision tree by introducing three different classifiers; we can build a simple and available tree.

The remainder of this paper is organized as follows. Section 2 introduces the whole CA algorithm and concretely describes important parts of CA which include longitudinal dimension reduction, and decision-tree part. Section 3 applies the CA algorithm to maize seed breeding, and give out the calculate steps and data results specifically. We also analyze the results of seed breeding, and compare them with other different methods. The experiment shows that the outcome of CA is better. Section 4 concludes this paper and point out CA's advantages and the problem that need to be settled.

## 2 Ca Algorithm

### 2.1 Transverse Dimension Reduction

With the development of computer technology and multimedia technology, more and more high-dimensional data have appeared in our lives. In low dimension, we can use Manhattan or Euclidean distance as the measurement of similarity between data samples. While, in high dimension, this kind of similarity do not exits any more. So it is important for us to reduce the high dimension and eliminate the redundancy in data. The approaches for dimension reduction generally can classify into several parts like the methods based on feature selection, linear or non-linear mapping and the methods combined with other domain knowledge. We often use Principle Component Analysis (PCA), kohonen matching, Sammon projection, the method based on rough set etc. But there still exits some problems, such as feature ordering method just consider the relations between features and the target, but do not think over the redundancy between features; linear mapping method may influence the initial distributing or sharp of data and lead to analysis deviation.

Aim at these problems, this paper combines PCA and correlation coefficient matrix to search associated features which would influence the target tremendously and eliminates the irrelevant redundancies in order to convert the high-dimensional data into low-dimensional data which can highly reduce compute consume and runtime. The whole procedures are shown below:

*1) Standardize the whole data set and computer the correlation coefficients between features.* Through analyzing the bigger coefficients in matrix, we can find out the features of which are highly related. Then group them together and mark M1. While the computation of correlation coefficient between feature X and Y is shown as formula (1), and $\bar{x}$, $\bar{y}$ are mean values:

$$r = \frac{\sum_{i=1}^{n}(x_i - \bar{x})(y_i - \bar{y})}{\sqrt{\sum_{i=1}^{n}(x_i - \bar{x})^2 * \sum_{i=1}^{n}(y_i - \bar{y})^2}} \tag{1}$$

*2) Execute PCA to the whole data set.* First calculate the eigenvalues and cumulative contribution rate to ensure the amount of principle components, then acquire the eigenvectors of every component and form a coefficient factor matrix. According to the magnitude of factors, we can find out highly associated features. Group them together and mark it as M2.

*3) View the features which do not appear in M1 and M2 as redundant features and disregard them.*

*4) Unite associated attribute groups M1 and M2.* And obtain the new feature set according to the weight values acquired by the coefficient factor matrix. While the weight value formula and new feature formula are shown below:

$$w_{ij} = \frac{p_{ij}}{p_i}; \tag{2}$$

$$W_i = (w_{ij})_{d\times 1}; \tag{3}$$

$$F_i = V * W_i; \tag{4}$$

While formula (2) is to compute the weight value of associated attributes in each principle component, and $p_{ij}$ is the coefficient factor of associated feature, $p_i$ is the summation of coefficient factors, $w_i$ is a matrix formed by d weight values. Formula (4) is to obtain a new feature set.

## 2.2 Longitudinal Dimension Reduction

Clustering is a common data analysis tool and belongs to a kind of unsupervised learning method. One of the prominent characteristics is clustering can deal with big datasets and can be the pretreatment part of other data mining methods. Traditional clustering can broadly be classified into the methods based on partition, hierarchy, density, grid and model. The common algorithms are k-means, BIRCH, CURE, DBSCAN, STING etc. But there exist many differences among the application conditions and clustering efficiency etc. For example k-means just can deal with the numerical values and is sensitive to the input parameters, but it is valid for small scaled datasets; DBCCAN shall be influenced by clustering centroids and also be sensitive to the exceptional samples, the efficiency is higher than k-means; k-means and BIRCH just can dig out convex and spherical sharps; CURE can cluster arbitrary datasets and performance is higher than BIRCH and STING. So this paper will improve the original algorithms based on CURE in order to cluster samples effectively.

The original CURE algorithm has two outstanding characteristics. One is to select a constant number of scattered points in space instead of single centroid or all points to represent clusters which can avoid traditional methods and apply to any geometry sharps. The other is to introduce shrink factor for reducing the impact of noise which can make the algorithm more haleness in handling outliers and recognize non-spheroid clusters.

Based on such advantages, we will import k-means and grid-partition theory to improve the original CURE algorithm. The procedures are shown below:

*5) Outlier analysis.* After standardization and transverse dimensional reduction, sample data are almost in the range of [-1.5,1.5]. Here we permit an error rate of 25% and extend the range to [-2.0,2.0]. If a data exceeds this area, we should view it as an outlier and eliminate it.

*6) Introduce grid-partition method.* Select appropriate partition factor and each dimension shall divide into the same parts according to this factor. We can obtain the partition grids of random dimension.

*7) Implement improved k-means[8] algorithm to the above samples in each grid which can cluster the whole dataset into many small clusters.* This method is improved in the confirmation of clustering centers and distance measurement. The advantage is it can enormously reduce iterations and compute consummations which can exert the application of k-means in small-scale dataset sufficiently.

*8) Adopt the clustering theory of CURE.* Make sure the number of representative points and the degree of shrink factor. Then cluster all the small clusters into big clusters of fixed number we needed.

### 2.3 Decision Tree Method

The decision tree is one of the widely used logical algorithms. It can reflect the classification results by modeling in the form of a tree which can intuitively tell us the relationships among samples and provide useful references for analyzing data. Traditional decision tree methods include ID3, C4.5, CART, CHAID, PUBLIC, SLIQ and SPRINT etc. and most of these methods are based on C4.5. These algorithms also exist some problems like feature deflection, the optimal partition threshold, non-increasing learning, non-global optimum results, the pruning of tree etc. At present, researchers have presented many improved algorithms [9]-[11]. It is unnecessary to go into details.

C4.5 algorithm applies information gain to select features which only considers the relationship between features and targets, but do not take into account the association among features. But in transverse dimension reduction part of CA, we have already thought over their relations which well make up C4.5's shortcomings. In order to make sure the correctness of feature selection, we introduce 3different classifiers. The measurement of classifiers is C4.5, gini-index, $\chi 2$ respectively. When more than one feature in classifiers are the same, we can choose them as tree node. Aim at the optimal partition threshold problem, we select the one of which has the smallest information entropy [12].

# 3 The Application of CA in Maize Seed Breeding

## 3.1 Sample Dataset

The whole dataset we should consider is made up of 51 corn samples, and each one has more than twenty features. Here we choose 9 features from the dataset, and they are period, plant height, ear height, ear length, ear diameter, thousand grain weight, seed rows per ear, seed number per row and output respectively. Now give out its scaled-down version, see Table 1, and f1-f9 in table corresponding to the features mentioned above.

**Table 1.** Sample Set of Maize Seed Breeding

| No. | f1 | f2 | f3 | f4 | f5 | f6 | f7 | f8 | f9 |
|---|---|---|---|---|---|---|---|---|---|
| Y1 | 100 | 194.8 | 78.9 | 15.63 | 4.19 | 200.8 | 15.8 | 38.1 | 6.73 |
| Y2 | 101 | 229.5 | 93.9 | 18.28 | 4.53 | 269.8 | 15.2 | 40.5 | 7.83 |
| Y3 | 99 | 270.1 | 114.7 | 16.34 | 4.65 | 28713 | 14.4 | 35.4 | 6.70 |
| ... | ... | ... | ... | ... | ... | ... | ... | ... | ... |
| Y51 | 102 | 245.4 | 103.3 | 18.00 | 4.23 | 250.3 | 13.4 | 41.0 | 7.49 |

## 3.2 Transverse Dimension Reduction

Transverse dimension reduction of CA combines PCA and correlation coefficients which can achieve reduction of features well. The specific steps are shown below:

*9) Standardize all the samples.* Compute the correlation coefficients of 9 features to form a correlation coefficient matrix R which is shown as formula (5).

$$R=\begin{cases} 1.0000 \\ 0.0502 & 1.0000 \\ 0.2038 & 0.7924 & 1.0000 \\ -0.1602 & 0.0794 & -0.0197 & 1.0000 \\ 0.1722 & 0.2256 & 0.2695 & -0.3415 & 1.0000 \\ 0.2250 & 0.2629 & 0.1296 & 0.1356 & 0.2498 & 1.0000 \\ 0.0213 & -0.0683 & -0.0741 & -0.3034 & 0.4356 & -0.3998 & 1.0000 \\ -0.0562 & 0.1921 & 0.2802 & 0.5785 & -0.2782 & -0.0238 & -0.4638 & 1.0000 \\ 0.1442 & 0.3648 & 0.2634 & 0.1046 & 0.3884 & 0.5488 & 0.0395 & 0.1583 & 1.0000 \end{cases} \tag{5}$$

After comparing the coefficients in matrix, we can obtain an associated feature group M1: {(2,3), (4,8), (6,9)}.

*10) Execute PCA to standardized dataset.* According to the rule: cumulative contribution rate should exceed 85%, we can confirm the number of principle component. Through computing the eigenvalues and equation set, we can obtain eigenvectors of every principle component and form a coefficient factor matrix which is formula (6). Finally we can obtain an associated feature group M2: {(2,3,9), (4,5,7,8), (3,6), (1), (1)}.

$$p=\begin{cases} 0.1769 & 0.1897 & 0.2317 & -0.6983 & 0.5867 \\ 0.4953 & 0.0428 & -0.3890 & -0.0136 & -0.2644 \\ 0.4728 & 0.0623 & -0.4780 & -0.2316 & -0.0842 \\ 0.1331 & -0.4887 & 0.0442 & 0.3369 & 0.3673 \\ 0.2312 & 0.4956 & 0.0346 & 0.2485 & 0.0899 \\ 0.3917 & 0.0125 & 0.6120 & 0.0176 & -0.2841 \\ -0.1626 & 0.4713 & -0.2690 & 0.3470 & 0.4210 \\ 0.2356 & -0.4869 & -0.1889 & 0.0468 & 0.3683 \\ 0.4396 & 0.1195 & 0.2841 & 0.4004 & 0.2053 \end{cases} \tag{6}$$

*11) Unite M1 and M2.* Achieve the final associated feature group (2,3,6,9), (4,5,7,8), (1). That is (plant height, ear height, thousand grain weight, output), (ear length, ear diameter, seed rows per ear, seed number per row), (period).

*12) Select corresponding principle component for the above feature groups and calculate its weight values for each associated feature.* After the product of multiplication, get a new feature set and get ready for the following data mining algorithm.

### 3.3 Longitudinal Dimension Reduction

Longitudinal dimension reduction of CA improves CURE algorithm and can achieve scale reduction excellently. The specific steps are below:

*13) Outlier analysis.* Search all the samples which exceed the range of [-2.0, 2.0] and view them as exceptions. Here we eliminate 5 outlier samples: Y1, Y30, Y33, Y35, Y45. From the original sample set, we know that Y1, Y30, Y45 are belonging to low yield of which all the feature values are very low, while Y33, Y35 are the high yield seed of which every feature values are very high.

*14) Introduce grid partition.* There are three new features which corresponding to three dimensions. Set the partition factor to 2 which means divide each dimension into two equal parts. Now we can obtain 8($2^3$=8) identical grids.

*15) Implement improved k-means to the samples in each grid.* By adjusting k, we can get multiple different small clusters.

*16) According to the clustering idea of CURE, we can cluster all the preceding small clusters into a few clusters.* Through selecting appropriate representative points and shrink factors, we finally obtain two big clusters.

Here we compare this result with the original method. First, we define the whole dataset as "INI", the standardized dataset as "STAD" and the dataset after CA's pretreatment as "CAS". Then we cluster the "INI", "STAD" by traditional algorithms and use CA's pre-clustering method to "CAS". Now compare these three cluster results by analyzing sensibility to outliers, the distances between clusters and mean squared errors as Table 2.

**Table 2.** Comparison of 3 Cluster Method

| Data set | Best k | Distance Between Clusters | Mean Squared Error |
|---|---|---|---|
| INI | 3 | 46.1595 | 1126.85 |
| STAD | 3 | 2.57985 | 102.425 |
| CAS | 2 | 1.5898 | 66.7633 |

From this table, we can know the preceding two methods are sensitive to the outliers when k is increased, while CA algorithm do not need this process in this part, and can cluster the samples directly. The distances between clusters and the mean squared errors also show that the clusters obtained by CA is much more compact.

## 3.4 Decision Tree Part

We implement improve C4.5 algorithm to the cluster we needed to consider, and can obtain a clear decision tree model for us to analyze. From this tree, we can know the classification of each corn sample and also can find out the optimal corn seed for breeding. The final decision tree is below, see Figure 1.

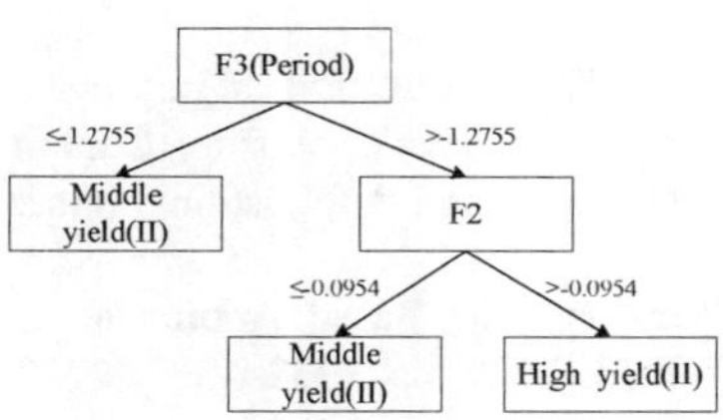


**Fig. 1.** Final Decision Tree

## 4 Conclusion

The application of improved CA algorithm in maize seed breeding can obtain much important useful information for us. We can figure out the important degree of every corn features by transverse dimension reduction, find out some exceptional maize seeds for us to investigate by outlier analysis, estimate input corns' assort by decision tree method etc. After comparing CA with original method, we can find that the efficiency of CA algorithm is much better than original method.

This improved algorithm is very useful in maize seed breeding, but there still exit many problems such as the incremental treatment of CA which need to handle the input data every time; the sensitivity of the parameters like cluster number k, grid partition g; the optimization of objective; the combination of other useful methods like genetic, fuzzy concept and so on. We should adjust the parameters in different applications or conditions to reach the best correct results.

**Acknowledgment.** This work was supported in part by the Science Foundation of JiangSu Education Department Grants 06KJB520090, 08KJB520009 and the Science Foundation of Nantong of JiangSu Grants K2006008, AL2007033, K2008008, K2009045, K2009057.

Correspondence to: Qiu Jian-lin, School of Computer Science & Technology, Nantong University, Nantong, China.

## References

1. Luo, J., Chen, D.: An Enhanced ART2 Neural Network for Clustering Analysis. In: Workshop on Knowledge Discovery and Data Mining, pp. 81–85 (2008), doi:10.1109/WKDD
2. Aslanidis, T., Souliou, D.: CUZ: An improved Clustering Algorithm. In: IEEE 8th International Conference on Computer and Information Technology Workshops, pp. 43–48 (2008), doi:10.1109/CIT.2008.Workshops.118
3. Shahriar Hossain, M., Angryk, R.A.: GDClust: A Graph-Based Document Clustering Technique. In: Shahriar Hossain, M. (ed.) Seventh IEEE International Conference on Data Mining – Workshops, pp. 417–422 (2007), doi:10.1109/ICDMW.2007.104
4. Shen, Y., Shao, H., Yang, L.: Privacy Preserving C4.5 Algorithm over Vertically Distributed Datasets. In: 2009 International Conference on Networks Security, Wireless Communications and Trusted Computing, pp. 446–448 (2009), doi:10.1109/NSWCTC.2009.253
5. Shao, Y., Chen, Q., Yin, W.: The Application of Improved Decision Tree Algorithm in Data Mining of Employment Rate: Evidence from China. In: 2009 First International Workshop on Database Technology and Applications, pp. 202–205 (2009), doi:10.1109/DBTA.2009.72
6. Marsala, C.: A Fuzzy Decision Tree Based Approach to Characterize Medical Data. In: Fuzzy Systems, FUZZ-IEEE 2009, pp. 1332–1337 (2009), doi:10.1109/FUZZY.2009.5277106
7. Ding, B., Zheng, Y., Zang, S.: A New Decision Tree Algorithm Based on Rough Set Theory. In: 2009 Asia-Pacific Conference on Information Processing, pp. 326–329 (2009), doi:10.1109/APCIP.2009.216

8. Bu, Y.-y.: Research of Clustering Algorithm based on k-means. Journal of Southwest University for Nationalities (Natural Science Edition) 35(1), 198–200 (2009)
9. Huang, A.-h., Chen, X.-t.: An Improved ID3 Algorithm of Decision Trees. Computer Engineering and Science 31(6), 109–111 (2009)
10. Wei, J.-M., Wang, S.-Q., Yu, G., et al.: A Novel Method for Pruning Decision Trees. In: 2009 International Conference on Machine Learning and Cybernetics, July 12-15, vol. 1, pp. 339–343 (2009)
11. Marsala, C.: A Fuzzy Decision Tree Based Approach to Characterize Medical Data. In: IEEE International Conference on FUZZ-IEEE 2009, August 20-24, pp. 1332–1337 (2009)
12. Zhao, J.-x., Ni, C.-p., Zhan, Y.-r.: Combined Optimization Decision Tree Algorithm Suitable for Large Scale Data-base. Systems Engineering and Electronics 31(3), 583–587 (2009)

# Wavelet Based Affine Projection Adaptive Filter

Wei-Wei Wu and Yan-Song Wang

Vehicle Engineering College
Shanghai University of Engineering Science
Shanghai, China, 201620
wwwu@sues.edu.cn, jzwbt@163.com

**Abstract.** A wavelet transform based affine projection algorithm is proposed, with a variable step-size scheme. The numerical simulations and engineering application verify the performance of the algorithm. On the same conditions, the proposed algorithm has a faster convergence and the same steady state MSE compare to normal affine projection algorithm. That means to achieve the same rate of convergence less order is needed by this algorithm, which counteract the computing expenses induced by wavelet transform. By means of variable step-size scheme, the proposed algorithm has a faster convergence rate and a lower steady state MSE or misadjustment than the corresponding normal variable step-size affine projection algorithms.

**Keywords:** affine projection algorithm, adaptive filter, wavelet transform, variable step-size.

## 1 Introduction

Adaptive filters are widely used to handle unknown environments or nonstationary signals. Basically, the linear adaptive filters can be generally divided into LMS (least mean square) and RLS (recursive least squares). Affine projection algorithm (APA) is introduced as a link between normalized LMS (NLMS) and RLS [1]. APA provides a much faster convergence than NLMS, especially when the reference input is colored, and its computing complexity is very low compare to RLS. In many applications, subband adaptive filters [6] have become a practical solution. In this paper, it is found that the convergence of APA can be enhanced by dividing the signal into subbands through wavelet transform, and the steady state MSE (mean square error) or misadjustment remains the same. The variable step-size scheme, which is used to balance the contradictory properties such as convergence rate and misadjustment, is also provided. The numerical simulations and engineering application verify the effective and performance of the proposed algorithms.

## 2 Wavelet Based Affine Projection Algorithm

The Mallat formula [4] of wavelet transform is

$$x_k^1 = \sum_n h_0(n-2k)x_n^0, \quad d_k^1 = \sum_n h_1(n-2k)x_n^0 \tag{1}$$

where $x_n^0$ is original signal, $x_n^1$ is approximate signal and $d_n^1$ detail signal one level below.

### 2.1 Transform Domain NLMS Algorithm

The wavelet transform NLMS algorithm [2] is actually a subband filter. Its principle is to decompose the reference signal $x(n)$ through wavelet transform, so to break down the signal into different frequency bands, then filter the subband signals with NLMS algorithm, and finally output the filtered signal. The convergence rate is improved because the frequency range of each subband signal is narrower than original. The simplified wavelet domain NLMS algorithm is shown in Fig. 1. This algorithm does not decompose the desired signal $d(n)$, but using vector $Z(n)=[z_1(n), z_2(n), \ldots z_N(n)]^T$, which is the wavelet transform of $X(n)$, as reference signal. The steps are:

$$\begin{aligned} e(n) &= d(n) - Z^T(n)W(n) \\ W(n+1) &= W(n) + \frac{\mu}{\|Z(n)\|^2} e(n)Z(n) \end{aligned} \tag{2}$$

where $0<\mu<2$ is the step size , and $\mu$ is set to 1 generally, $W(n)=[w_1(n), w_2(n), \ldots w_N(n)]^T$ is the weight vector. Obviously, compared with the standard subband filter algorithms, adaptive updating process of $W(n)$ described above includes estimation of the wavelet reconstruction, so its convergence rate is slowed down. But the wavelet decomposition and reconstruction is usually realized by FIR filter, and NLMS has a fast convergence rate in estimating FIR. So the algorithm shown by Fig. 1 is applied in many different cases because of its simple realization.

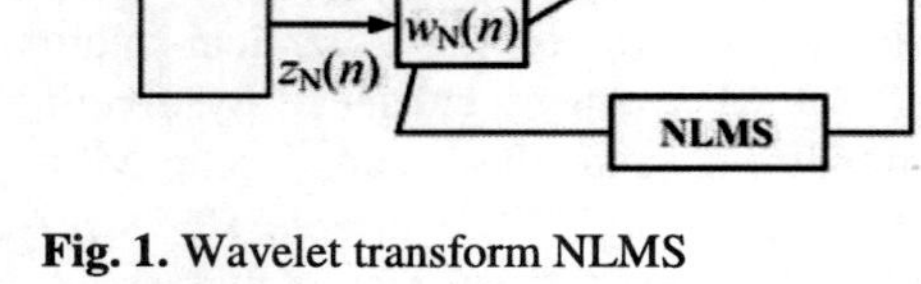


**Fig. 1.** Wavelet transform NLMS

### 2.2 Affine Projection Algorithm (APA)

The updating process of affine projection algorithm is

$$\begin{aligned} E(n) &= d(n) - U(n)W(n) \\ W(n+1) &= W(n) + \mu U^T(n)[U(n)U^T(n)]^{-1}E(n) \end{aligned} \tag{3}$$

where $U(n)=[X(n), X(n\text{-}1), \ldots X(n\text{-}M+1)]^T$ is $M\times N$ matrix, $X(n)=[x(n), x(n\text{-}1), \ldots x(n\text{-}N+1)]^T$ is input vector.

APA takes more computing expense than NLMS, but it has a faster convergence rate when used to handle colored data [3, 9].

To ensure high convergence rate and low misadjustment, variable step-size algorithms such as VSAPA [5] (variable step-size APA ) and VSS-NLMS [7] are proposed.

### 2.3 Wavelet Based Affine Projection Algorithm (WAPA)

WAPA decomposes the input signal into subband signals through wavelet transform, these subband signals are integrated into a vector $Z(n)$, and then using APA to update the weight vector $W(n)$ from $Z(n)$, $Z(n\text{-}1)$, … $Z(n\text{-}M\text{+}1)$. Here $M$ is the order of WAPA.

Specifically, WAPA can be represented as a constrained optimization problem: minimize the norm of weight vector variation $\delta W = W(n+1) - W(n)$ between two adjacent iterations, under the constraint of $d(n-k) = Z^T(n-k)W(n+1)$ for all $k$=0, 1, … $M$-1. Based on lagrange multiplier method, the constraint optimization problem above is turned to an unconstraint problem $J(n) = \|\delta W\|^2 + \sum_{k=0}^{M-1} \alpha_k [d(n-k) - W(n+1)Z(n-k)]$. Given $M \times N$ matrix $A(n)=[Z(n), Z(n\text{-}1), \ldots Z(n\text{-}M\text{+}1)]^T$, $M \times 1$ vector $D(n)=[d(n), d(n\text{-}1), \ldots d(n\text{-}M\text{+}1)]^T$, and $\lambda=[\alpha_1, \alpha_2, \ldots \alpha_{n\text{-}M\text{+}1}]^T$, the objective function can be expressed as $J(n) = \|\delta W\|^2 + (D(n) - A(n)W(n+1))^T \lambda$, from equation $\frac{\partial J(n)}{\partial W(n+1)} = 2(W(n+1) - W(n)) - A^T(n)\lambda = 0$, the updated weight vector can be got

$$W(n+1) = W(n) + \frac{1}{2} A^T(n)\lambda \cdot \quad (4)$$

From equation $D(n) = A(n)W(n+1) = A(n)W(n) + \frac{1}{2} A(n)A^T(n)\lambda$, the multiplier vector $\lambda$ can then be solved by $\lambda = 2[A(n)A^T(n)]^{-1}[D(n) - A(n)W(n)] = 2[A(n)A^T(n)]^{-1} E(n)$, so the updated weight vector in (4) can be computed as $W(n+1) = W(n) + A^T(n)[A(n)A^T(n)]^{-1} E(n)$.

From above, WAPA can be described as:

1) initialize $W(0)=\mathbf{0}$
2) for $n$=0, 1, … compute the wavelet transform of $X(n)=[x(n), x(n\text{-}1), \ldots, x(n\text{-}l\text{+}1)]^T$ and assembly the subband signal elements to get the vector $Z(n)$, then

$A(n)=[Z(n), Z(n\text{-}1), \ldots Z(n\text{-}M\text{+}1)]^T$
$D(n)=[d(n), d(n\text{-}1), \ldots d(n\text{-}M\text{+}1)]^T$
$E(n)= D(n)\text{-}A(n)W(n)$
$W(n\text{+}1)= W(n)+\mu A^T(n)[A(n) A^T(n)]^{-1}E(n)$.

Compared with normal APA, the computing expense of WAPA increases moderately, but the value of $M$ is usually not large, and the wavelet transform improves the convergence rate, which is tested in the following of the paper, so the performance improvement of WAPA deserves its computation increase. To reduce the misadjustment and improve the convergence rate further, VSWAPA (variable step-size WAPA) is proposed. The step-size of VSWAPA is a variable and defined as

$$\mu(n) = \mu_{\max} \frac{E\|P_n\|^2}{E\|P_n\|^2 + C} \tag{5}$$

where $P_n = A^T(n)[A(n)A^T(n)]^{-1}[E(n) - v(n)]$, and $C = M/\text{SNR}$ (signal-noise ratio). $P_n$ and $C$ can be statistically estimated during the iteration process. To reduce computing expense and storage, they are generally estimated by instant values computed in the iterations or their moving average, e.g. $\hat{P}_n = \alpha\hat{P}_{n-1} + (1-\alpha)A^T(n)[A(n)A^T(n)]^{-1}E(n)$, where $0<\alpha<1$ is smoothing factor.

## 3 Simulation Results

The performance of the WAPA and its variable step-size scheme is tested in simulations and engineering applications. In identification of IIR, which is the general application of adaptive filter, WAPA is compared with normal APA and NLMS. The usage of WAPA in engine vibration tests is also demonstrated.

### 3.1 IIR Identification

The IIR filter for test has the following poles: $0.46\pm0.6i$, $-0.16\pm0.72i$ and $-0.64\pm0.16i$, and the zeros are $-0.97\pm0.01i$, $\pm i$, $0.9\pm0.1i$. The output of testing IIR filter is mixed with measurement noise, which has mean square value as 0.01. The simulation of algorithm is performed in Matlab, and MSE is the average of 400 simulations.

For the convenience of comparison, following conditions are defined: weight vector length $l$, order $M$ and iteration step-size for APA and WAPA are identical; the length of input vector of NLMS filter is $l+M-1$, which means that the input data in each step is absolutely equivalent for all the tested algorithms. The wavelet used in the tests is Daubechies 4 [8] and the decomposition level is 3.

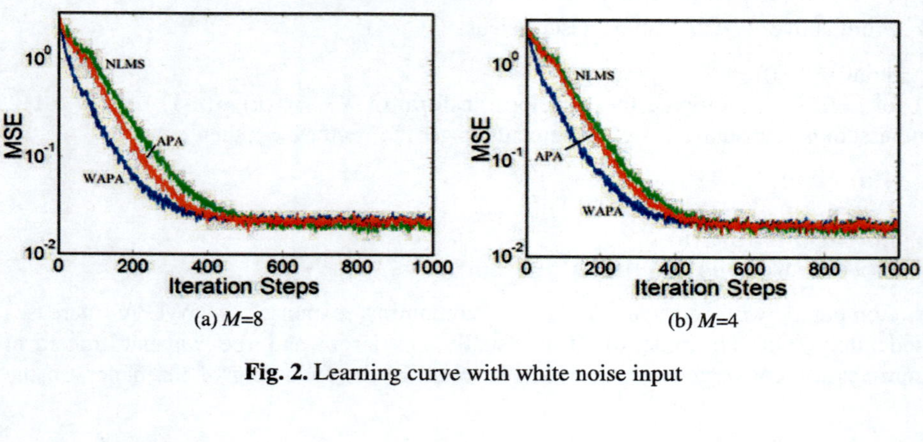


(a) $M$=8 (b) $M$=4

**Fig. 2.** Learning curve with white noise input

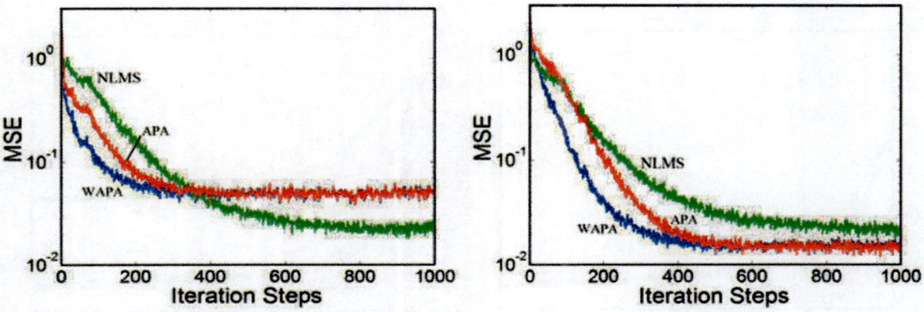


(a) Step-size=0.5 (b) Step-size=0.1

**Fig. 3.** Learning curves with colored input

Fig. 2 and Fig. 3 show the learning curves of algorithms when the input signals are white noise and colored signal. Obviously, as the *M* increases, the convergence rate of APA and WAPA is enhanced. Whatever the value of *M* is, WAPA has a faster convergence rate than APA and NLMS. In Fig. 2(b), when $M$ =4, APA does not have too much advantage over NLMS, but WAPA still maintains a relatively high convergence rate (it is even faster than APA with $M$=8 in Fig. 2(a)). This means to gain the same convergence rate, less order is needed by WAPA than APA, so the computing expense is reduced, which greatly counteract the computing increase induced by wavelet transform in WAPA.

In Fig. 3 the performance of the algorithms with colored input signals are compared in different step-size (the step-size of NLMS retained 1). Obviously, the convergence rate of WAPA is the fastest when the input signals are colored (the MSE of NLMS does not converge to the minimization even as iteration step reaches 1000). The convergence rate of APA and WAPA does not change greatly compare to Fig 2, and WAPA is still faster than APA. The misadjustment and convergence rate of APA is affected by step-size, which is also shown in Fig. 3. Small step-size will benefit misadjustment, but will slow down the convergence. Using the same step-size, WAPA and APA have the same minimal MSE or misadjustment.

Variable step-size scheme can balance the convergence rate and misadjustment. To test the performance of VSWAPA, IIR filter is kept the same as above, and mean square value of measurement noise is changed to $10^{-4}$. The test results with colored input are shown in Fig. 4. Obviously, the performance of VSWAPA is much better than VSAPA both at convergence rate and misadjustment, and variable step-size algorithms generally perform better than fixed step-size. When the step-size of APA and WAPA is small enough, their steady state MSE may be slightly lower than that of VSWAPA, this is because estimated values must be used in (5), which may be interfered by noise. But APA and WAPA converge much slower than VSWAPA in such cases.

### 3.2 Engine Vibration Signal Filtering

In the vibration tests of a four-cylinder engine, acceleration signals are measured at the cylinder head and near the supporter respectively. At the revolution speed of

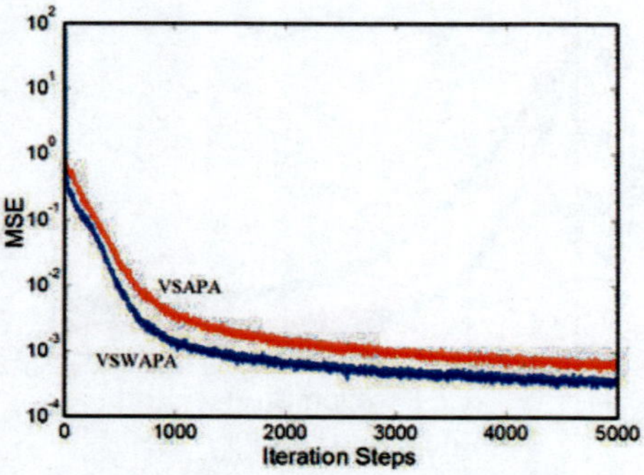


(a) Comparison of variable step-size algorithms

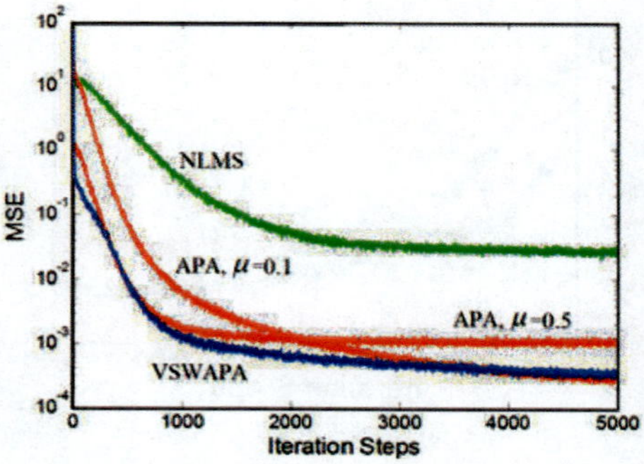


(b) Comparison of variable and fixed step-size algorithms

**Fig. 4.** Performance of variable step-size algorithms

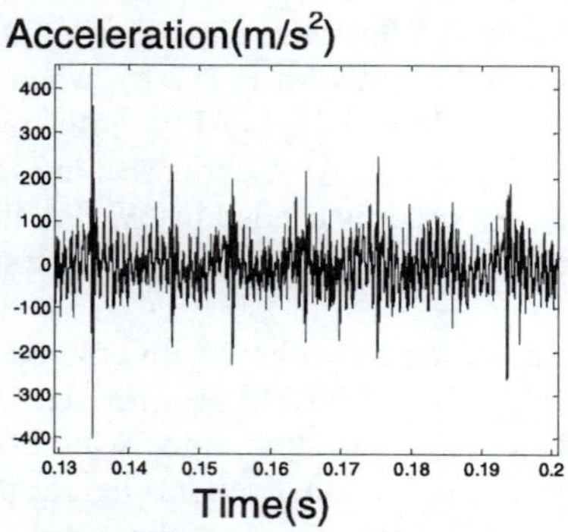


**Fig. 5.** Acceleration at cylinder head

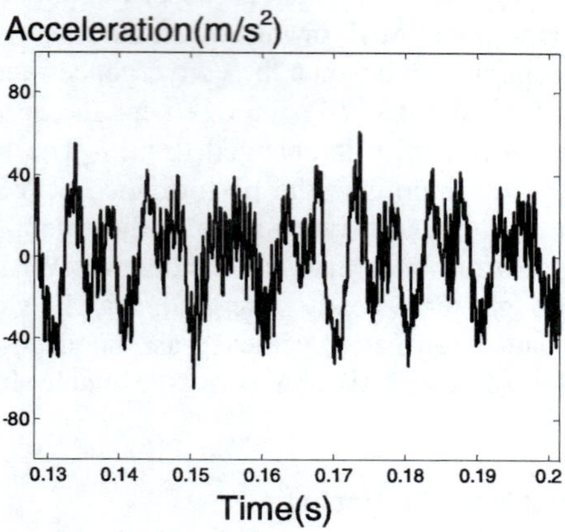


**Fig. 6.** Acceleration at supporter

3000 rpm, the measured acceleration signals are shown in Fig. 5 and 6. Acceleration signal at the cylinder head in Fig. 5 reflects the high-frequency impact during the engine works (the impact of combustion at time interval of about 0.01 second, the impact of valves closing, etc.).

In Fig. 5 there is still a clear low-frequency component, which is the effect of the global low-frequency vibration generated by excited elastic supporter. In order to study the working properties and surface vibration of the engine, the low-frequency vibration generated by elastic supporter should be removed. High-pass filter simply does not work well in this case. So the signal from the supporter, as shown in Fig. 6, is taken as reference input, and the low-frequency vibration components in Fig. 5 is removed by proposed WAPA. In Fig. 7, the result of filtering is shown. It can be seen clearly from the result time history and spectrum that the impact of combustion and valve switching is reserved, and the component of global low-frequency vibration has already been eliminated. The filtered signal satisfied the demands of further study on engine vibration.

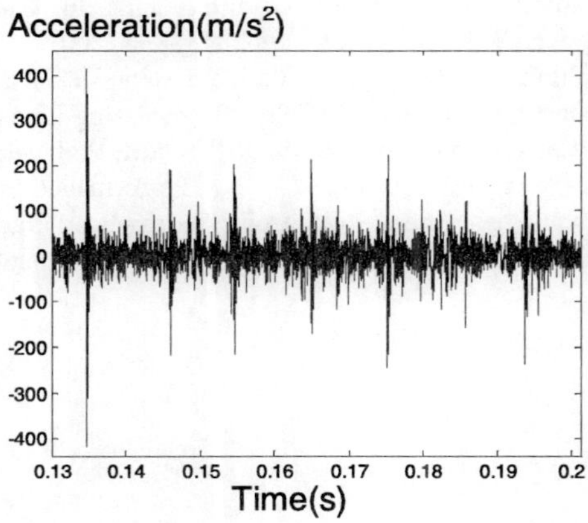


**Fig. 7.** Filtered signal

# 4 Conclusions

The proposed WAPA and its variable step-size scheme have excellent performance, which is confirmed by the numerical simulations and engineering tests. The main results are as following: 1) WAPA has higher rate of convergence than APA with the same order *M*, or equivalently speaking, compared to normal APA, WAPA with less order *M* can achieve equal convergence rate, which greatly counteract the computing expenses induced by wavelet transform; 2) Using the same fixed step-size, WAPA and APA have the same minimum steady state MSE or misadjustment; 3) By means of variable step-size scheme, the proposed VSWAPA has faster convergence rate and lower steady state MSE or misadjustment than the corresponding normal VSAPA and other fixed step-size algorithms.

**Acknowledgment.** Research is supported by The Program for Professor of Special Appointment (Eastern Scholar) at Shanghai Institutions of Higher Learning, and the Foundation of the 085 Knowledge Innovation Engineering (grant no. JZ0901).

## References

1. Haykin, S.: Adaptive filter theory, 4th edn. Prentice Hall, New Jersey (2002)
2. Erdol, N., Basbug, F.: Wavelet transform based adaptive filters: analysis and new results. IEEE Trans. Signal Processing 44(9), 2163–2171 (1996)
3. Shin, H., Sayed, A.: Mean-square performance of a family of affine projection algorithms. IEEE Trans. Signal Processing 52(1), 90–102 (2004)
4. Mallat, S.: A wavelet tour of signal processing. China Machine Press, Beijing (2002)
5. Shin, H., Sayed, A.: Variable step-size NLMS and affine projection algorithms. IEEE Signal Processing Letters 11(2), 132–135 (2004)
6. Abutalebi, H.R., Sheikhzadeh, H., Brennan, R.L., Freeman, G.H.: Affine projection algorithm for oversampled subband adaptive filters, icme. In: International Conference on Multimedia and Expo. (ICME 2003), vol. 1, pp. 833–836 (2003)
7. Paleologu, C., Ciochina, S., Benesty, J.: Variable step-size NLMS algorithm for under-modeling acoustic echo cancellation. IEEE Signal Processing Letters 15(10), 5–8 (2008)
8. Daubechies, I.: Ten Lectures on Wavelets, vol. 61. SIAM, Philadelphia (1992)
9. Zhu, Y.-G., Zhu, Y.-Y., Yao, F.-Q., Liu, Y.-X.: A Performance Study of Affine Projection Adaptive Equalizers in the Presence of Strong Narrowband Interference. In: 2009 WRI International Conference on Communications and Mobile Computing, vol. 2, pp. 386–390 (2009)

# Research of the New Principal Component Analysis Algorithm Based on the Dynamic Model

You Zhou[1], Qinglin Wang[1], Guangda Yang[2], and Dehui Qiu[3]

[1] School of Automation
Beijing Institute of Technology
Beijing, China
[2] National Laboratory of Electro-Optics System Technology
Sanhe, China
[3] School of Automation/School of Information Engineering
Beijing Institute of Technology/Capital Normal University
Beijing, China
zy21221@sina.com, wqlbit@gmail.com, copa9999@yahoo.com.cn,
qiudehui@126.com

**Abstract.** An extension algorithm, exponentially weighted dynamic principal component analysis (EWDPCA) algorithm is proposed to monitor the dynamic process, in particular for disturbances and set points changes in process. The technique is validated through simulation in the process of actual chemical polymerization. The results show that the EWDPCA algorithm can eliminate the linear relationship among variables and reflect the real-time information in dynamic process.

**Keywords:** principal component analysis (PCA), dynamic process, dynamic PCA (DPCA), exponential weighting.

## 1 Introduction

Modeling, monitoring and fault diagnose in dynamic process is one of the hot problems in recent international control fields, which is important in both theory and application. The multivariate statistical methods for the dynamic process are necessary to monitor the process performance and diagnose the faults.

The EWPCA method monitors the tendency overtime and variation of set point in time, which based on renewing the correlation matrix by the newest process data. It eliminates the diagnosis errors due to the random variations of process, provides the efficient results and decreases the misinformation. The sample variables always relate to each other in dynamic process. DPCA extracts the constraining linear relations between variables and diagnoses the fault by construct the Hankel matrix.

A new method called as EWDPCA is proposed to monitor the performance in dynamic process, and applied to deal with the sample data of the chemical polymerization process.

## 2 Multivariate Statistical Analysis Methods

### 2.1 Exponentially Weighted Principal Component Analysis

The concept of EWPCA was first introduced by Wold [1], and the idea of recursive algorithms for exponentially weighted projection to latent structures (EWPLS) was proposed by Dayal, MacGregor [2] and Qin [3] et. al.. The recursive EWPCA algorithm based on the achievements of predecessors was studied by Li et. al.[4]. In the algorithm, the model is updated when each new vector of process measurements becomes available. The updating of the model is carried out as follows [5]

$$(\mathbf{X}^{\mathrm{T}}\mathbf{X})_t = \lambda(\mathbf{X}^{\mathrm{T}}\mathbf{X})_{t-1} + (1-\lambda)(\mathbf{x}^{\mathrm{T}}\mathbf{x})_t \tag{1}$$

where $\mathbf{x}_t$ is the new vector of process measurements, $(\mathbf{X}^{\mathrm{T}}\mathbf{X})_{t-1}$ is the variance–covariance matrix formed from the exponentially weighted observations stored in the process memory, $\lambda$ is the weighting factor ( $0 \le \lambda \le 1$ ) and $(\mathbf{X}^{\mathrm{T}}\mathbf{X})_t$ is the updated matrix used to construct the process model.

The weighting factor $\lambda$ is calculated as follows [6]

$$\lambda_t = 1 - \frac{(1-T_t^2)SPE_t / n^2}{\sqrt{n_t - 1}} \tag{2}$$

where $n_t$ is the asymptotic memory length at time $t$, $n_t = \lambda_{t-1}n_{t-1} + 1, n_0 = 0$.

### 2.2 Dynamic Principal Component Analysis

The algorithm of DPCA was first proposed by Ku [7] in 1995. The matrix $\mathbf{X} = [\mathbf{x}(1), \mathbf{x}(2), \cdots, \mathbf{x}(N)]^{\mathrm{T}} \in \mathbf{R}^{N\times n}$ is composed by sample $\mathbf{x}$, where $N$ is the number of samples. The dynamic PCA algorithm extends the matrix $\mathbf{X}$ firstly to consider the dynamic relationships between $\mathbf{x}(k)$ in different sample times.

$$\begin{aligned}\mathbf{X}^* &= [\mathbf{x}^*(1), \mathbf{x}^*(2), \cdots, \mathbf{x}^*(N-d)]^{\mathrm{T}} \\ &= [\mathbf{X}_0, \mathbf{X}_{-1}, \cdots, \mathbf{X}_{-d}]\end{aligned} \tag{3}$$

where $\mathbf{x}^* = [\mathbf{x}_o^{\mathrm{T}}, \mathbf{x}_{-1}^{\mathrm{T}}, \cdots, \mathbf{x}_{-d}^{\mathrm{T}}]^{\mathrm{T}}$ , $\mathbf{X}_{-i}$ corresponds to the variable which shift sample $x$ with $i$ moment.

$$\begin{aligned}\mathbf{X}_{-i} &= [\mathbf{x}_{-i}(1), \mathbf{x}_{-i}(2), \cdots, \mathbf{x}_{-i}(N-d)]^{\mathrm{T}} \\ &= [\mathbf{x}(i+1), \mathbf{x}(i+2), \cdots, \mathbf{x}(N-d+i)]^{\mathrm{T}}\end{aligned} \tag{4}$$

Then $\mathbf{X}^* \in \mathbf{R}^{(N-d)\times n(d+1)}$ is expressed by the following

$$\mathbf{X}^* = \begin{bmatrix} \mathbf{x}^{\mathrm{T}}(1) & \mathbf{x}^{\mathrm{T}}(2) & \cdots & \mathbf{x}^{\mathrm{T}}(d+1) \\ \mathbf{x}^{\mathrm{T}}(2) & \mathbf{x}^{\mathrm{T}}(3) & \cdots & \mathbf{x}^{\mathrm{T}}(d+2) \\ \vdots & \vdots & \ddots & \vdots \\ \mathbf{x}^{\mathrm{T}}(N-d) & \mathbf{x}^{\mathrm{T}}(N-d+1) & \cdots & \mathbf{x}^{\mathrm{T}}(N) \end{bmatrix} \tag{5}$$

After the transforming from $X$ to $X^*$,the number of samples reduced from $N$ to $N-d$.

The key of the dynamic PCA algorithm is determine the parameters $d$. Ku [7] proposed to take $d = 1,2$ for general conditions.

## 3 Exponentially Weighted Dynamic Principal Component Analysis

A new algorithm for dynamic process monitoring and diagnostics, exponentially weighted dynamic principal component analysis (EWDPCA), is proposed to satisfy the real-time and overcome the dynamic properties. The idea of this approach constructs the Hankel matrix according with the DPCA, extracts the linear relationship among the variables in dynamic process, and then weights with the data matrix. The process model is updated when each new vector of process measurements becomes available.

The proposed algorithm of EWDPCA is as follows

*1) Initialization*

A nominal data $\mathbf{X}$ is first defined. The data set is then normalized to zero mean and unit standard deviation. The initial weighting factor $\lambda_0$ is given by

$$\lambda_0 = 1 - \frac{1}{m} \tag{6}$$

*2) Application of EWPCA to a new observation*

*a) For a new vector of currect process measurements* $\mathbf{x}_t$*, the updated variance-covariance matrix is calculated as*

$$\mathbf{X}_t = \lambda_{t-1}[\mathbf{X}^{\mathrm{T}} \mid \mathbf{x}_t]^{\mathrm{T}} \tag{7}$$

*b) Calculate the eigenvalues and eigenvectors for the updated matrix.*

*c) Calulate the number of principal components to include in the model.*

*d) Calculate the number of lags* $l$*, and construct the data matrix according with DPCA.*

*e) For the current vector of process measurements at time* $t$ *, calculate the principal component scores* $T_t^2$ *and* $SPE_t$ *.*

*f) Update the control limits for the* $T_t^2$ *and* $SPE_t$ *monitoring charts.*

*g) If both* $T_t^2$ *and* $SPE_t$ *values less than the action limits, go to 3); else go to 4).*

*3) Update the EWDPCA parameters*

*a) Update the weighting factor* $\lambda_t$ *and the asymptotic memory length* $n_t$ *.*

*b) Return to 2) and repeat the iteration for the next online vector for process measurements at sample point* $t+1$*.*

*4) Action incompatible with the operation*

If either the value of $T_t^2$ or $SPE_t$ are >action limits, store the vector of press measurements in a store matrix.

## 4 Simulation Analyses

The EWDPCA method is used to deal with the data sampled by actual chemical polymerization and prove the application in process monitoring.

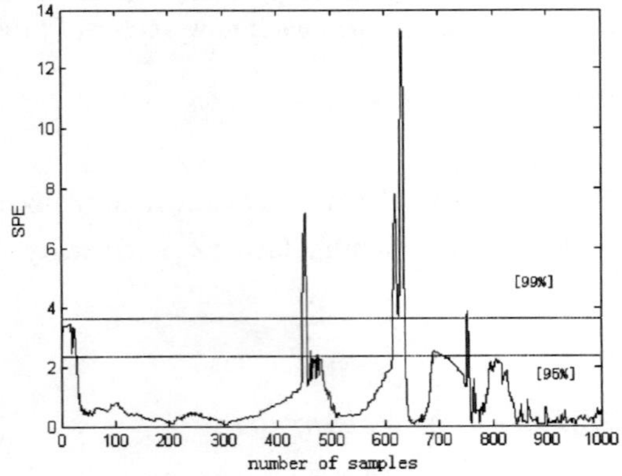


**Fig. 1.** SPE Monitoring plan in DPCA

Fig. 1 and Fig. 2 show that the EWPCA algorithm can regulate the failures occurred in the process and respond promptly by compared to the DPCA algorithm. The DPCA method is able to eliminate the linear correlation between variables quickly and reduce the impact of dynamic characteristics of the process by building a delay Hankel matrix. In sample process, the EWPCA method improves the system model continuously which reflected by the real-time changes.

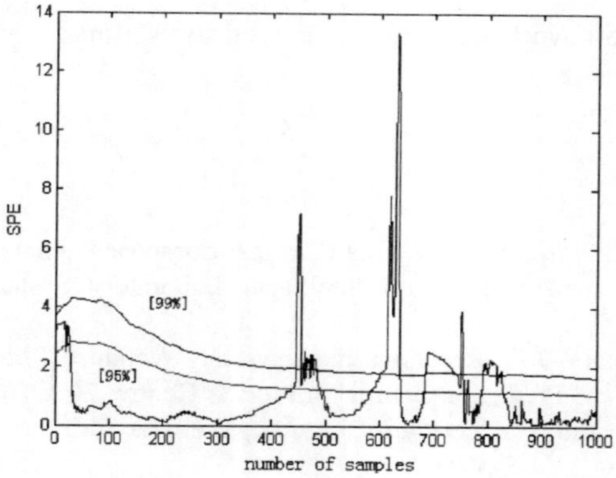


**Fig. 2.** SPE Monitoring plan in EWPCA

Fig. 3 shows that the EWDPCA approach not only can eliminate the linear relationship between variables in system, but also can enhance real-time performance and carry out the fault detection in dynamic systems more effectively. Comparing the result of Fig. 2 with that of Fig.3, there are high coincidence degrees according to the result in both technologies. It shows that both of them can be used in the error diagnose of dynamic system and the EWDPCA can determine the fault more quickly and accurately.

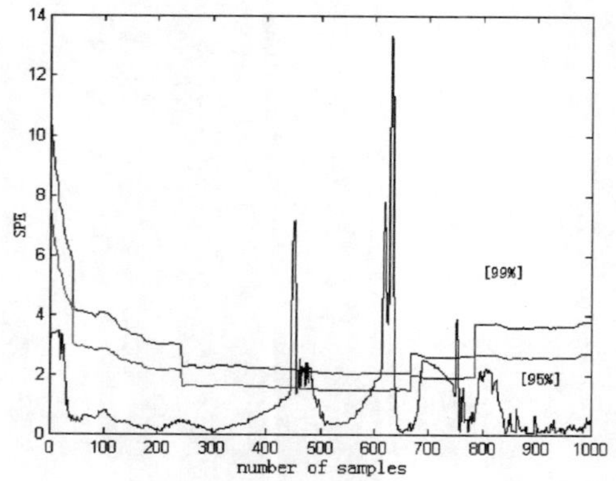


**Fig. 3.** SPE Monitoring plan in EWDPCA

# 5 Conclusion

A new exponentially weighting dynamic principal component analysis (EWDPCA) method based on DPCA and EWPCA method is presented. The advantage of this algorithm can not only eliminate the linear relationship between multi-variables, but also reflect the real-time information of process. The simulation results show that the EWDPCA approach can detect the abnormal situation in performance monitoring effectively.

**Acknowledgment.** This work was fully sponsored by National Natural Science Fund No. 60905051.

## References

1. Wold, S.: Exponentially Weighted Moving Principal Component Analysis and Projection to Latent Structures. Chemometrics and Intelligent Laboratory Systems 23(11), 149–161 (1994)
2. Dayal, B.S., MacGregor, J.F.: Recursive Exponentially Weighted PLS and Its Application to Adaptive Control and Prediction. Journal of Process Control 7(3), 169–179 (1997)
3. Qin, S.J.: Recursive PLS Algorithms for Adaptive Data Monitoring. Computer and Chemical Engineering 22(4-5), 503–514 (1998)
4. Li, W., Yue, H.H., Valle-Cervantes, S., et al.: Recursive PCA for Adaptive Process Monitoring. Journal of Process Control 10(5), 471–486 (2000)
5. Lane, S., Martin, E.B., Morris, A.J., Gower, P.: Application of exponentially weighted principal component analysis for the monitoring of a polymer film manufacturing process. Transactions of the Institute of Measurement and Control 25(1), 17–35 (2003)
6. Fortescue, T.R., Kershenbaum, L.S., Ydstie, B.E.: Implementation of Self-tuning Regulators with Variable Forgetting Factors. Automatica 17, 831–835 (1981)
7. Ku, W., Storer, R.H., Georgakis, C.: Disturbance rejection and isolation by dynamic principal component analysis. Chemometrics and Intelligent Laboratory Systems 30, 179–196 (19950020039

# Control Method Research of Three-DOF Helicopter Based on Active Disturbances Rejection Controller

You Zhou[1], Qinglin Wang[1], and Dehui Qiu[2]

[1] School of Automation
Beijing Institute of Technology
Beijing, China
[2] School of Automation/ School of Information Engineering
Beijing Institute of Technology/ Capital Normal University
Beijing, China
zy21221@sina.com, wqlbit@gmail.com, qiudehui@126.com

**Abstract.** The flight control system of helicopter is difficult to control for their performance of high-order, multi-inputs multi-outputs and strong non-linear. This paper proposed a stability control algorithm of three-DOF laboratory helicopter based on active disturbances rejection controller (ADRC). Controlling the motion of axis guaranteed the tracking control in degrees of helicopter respectively. The stability analysis is carried through Lyapunov method. The results show that the algorithm compensate the disturbance and improve the system performance effectively.

**Keywords:** active disturbances rejection controller, three-DOF helicopter, tracking.

## 1 Introduction

The flight control system of helicopter is a typical complex system which has the performance of high-order, multi-inputs multi-outputs and strong non-linear. The advantages of the helicopter are taking off and landing vertically, hovering in the air and multi-angle flight. But there are several problems such as the great vibration, difficulties in manipulation and the poor stability caused by the special flight condition. The serious aerodynamic coupling and that in axes between control channels impact the stability of helicopter in the low-speed flight and hovering. The high-accuracy manipulation will increase the burden on driver undoubtedly for overcome the inherent instability of the helicopter [1]. Therefore, the good control algorithm is needed to expand the flight features, improve the operational reliability and reduce the burden on drivers.

The algorithms is proposed to control the three-DOF helicopter experimental device which developed by Googol Technology Ltd. The motion equations in three degrees of freedom (pitch, lateral and rotation) can simulate the actual characteristics of helicopter partly which used to design the control methods and test the stability. The common algorithms in the helicopter model are PID control and linear quadratic regulator (LQR) control. The frequent changes of PID parameters deteriorate the control performance [2] when the object in an uncertainty environment. It is difficult to choose the states and control weighting matrix of performance in LQR control [3]. The other control methods have been used to helicopter control system [4] - [6]. In this paper, a new algorithm based on the active disturbances rejection controller is proposed to track the input signals separately and simulate the effects of the control process in the three-DOF helicopter system.

## 2 System Description

The experimental system of three-DOF helicopter is composed by base, balancing pole, balancing block, propellers and other components. The pitching of balancing pole (the base as the fulcrum) is caused by the lift from the two propellers and the rotation of balancing pole (the base as the axis) is induced by the speed difference of the two propellers [7]. The structural model of the helicopter is showed in Figure 1.

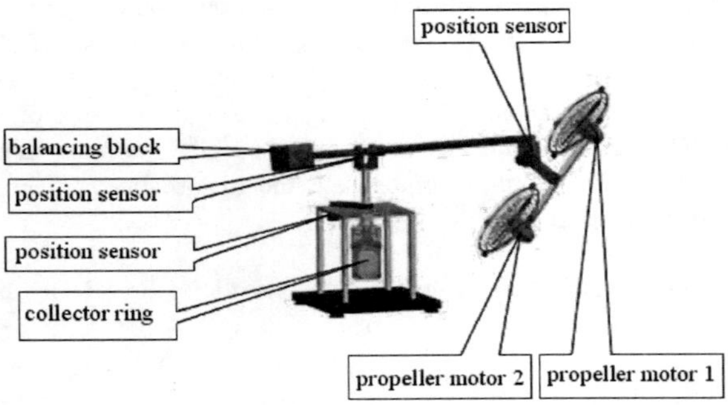


**Fig. 1.** Three degrees of freedom helicopter model [7]

The movement of three-DOF helicopter model system is subdivided by three degrees of freedom (including the pitch axis, the lateral axis and the rotation axis). Assuming the pitch angle is zero degree; the dynamic model of the helicopter is described by the motion equations [7] as follows

$$J_e\ddot{\varepsilon} = K_c l_1 (V_1 + V_2) - T_g = K_c l_1 V_s - T_g \tag{1}$$

$$J_p\ddot{p} = K_c l_p (V_1 - V_2) = K_c l_p V_d \tag{2}$$

$$J_t\ddot{r} = -G\sin(p) l_1 \tag{3}$$

where $V_1$ and $V_2$ are the voltage of two motors, $\varepsilon, p, r$ are the position of the pitch axis, lateral axis and the rotation axis. $T_g$ is the effective gravity moment generated by the pitch axis, $T_g = m_h g l_1 - m_b g l_2$. The other physical meanings and detail values of parameters are shown in Reference [7].

# 3 Helicopter Control System Design Based on Active Disturbance Rejection Controller

*Active disturbance rejection controller*

The ADRC controller consists of the tracking differentiator (TD), extended state observer (ESO) and nonlinear state error feedback (NLSEF). The TD is responsible for the arrangement of an appropriate transient process and provides proper differential signals of each order from the input signal. The core for an ADRC controller is ESO, which is capable of observing system uncertainties and providing feedback for compensation. The NLSEF determines the optimal combinations of states by the nonlinear algorithms [8]. A typical ADRC controller is shown in Figure 2.

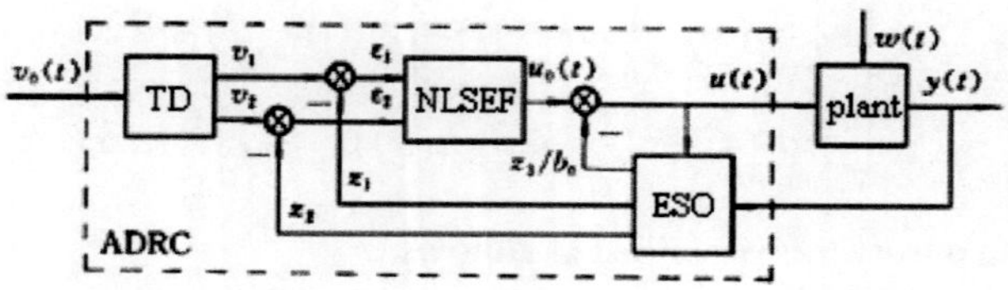


**Fig. 2.** Structure of active disturbances rejection controller [1]

The discrete equations of tracking differentiator, extended state observer and nonlinear state error feedback in two-order active disturbance rejection controller are shown respectively as follows[1]

The equation of tracking differentiator is

$$\begin{cases} v_1(k+1) = v_1(k) + hv_2(k) \\ v_2(k+1) = \\ v_2(k) + hfhan(v_1(k) - v_0(k), v_2(k), r, h) \end{cases} \tag{4}$$

where $v_0$ is the input signal of system, $v_1$ and $v_2$ are the output of tracking differentiator. $v_1$ tracks reference $v_0$ and $v_2$ approximates to the 'generalized differentiation' of $v_0$. $h$ is the step of the simulation, $r$ is the speed factor . The nonlinear function $fhan(x_1, x_2, r, h_0)$ is expressed as

$$\begin{cases} d = rh^2 \\ a_0 = hx_2 \\ y = x_1 + a_0 \\ a_1 = \sqrt{d(d+8|y|)} \\ a_2 = a_0 + sign(y)(a_1 - d)/2 \\ s_y = (sign(y+d) - sign(y-d))/2 \\ a = (a_0 + y - a_2)s_y + a_2 \\ s_a = (sign(a+d) - sign(a-d))/2 \\ fhan = -r(\frac{a}{d} - sign(a))s_a - rsign(a) \end{cases} \quad (5)$$

Extended state observer is described as follows

$$\begin{cases} e(k) = z_1(k) - y(k) \\ z_1(k+1) = z_1(k) + \\ h(z_2(k) - \beta_{01}e(k)) \\ z_2(k+1) = z_2(k) + \\ h(z_3(k) - \beta_{02} fal(e(k), 0.5, h) + b_0 u) \\ z_3(k+1) = z_3(k) - h\beta_{03} fal(e(k), 0.25, h) \end{cases} \quad (6)$$

Nonlinear state error feedback

$$\begin{cases} e_1(k) = v_1(k) - z_1(k) \\ e_2(k) = v_2(k) - z_2(k) \\ u_1(k) = k_1 fal(e_1(k), \alpha_1, \delta) + \\ k_2 fal(e_2(k), \alpha_2, \delta) \\ u(k) = u_1(k) - \frac{z_3(k)}{b_0} \end{cases} \quad (7)$$

where the $fal$ function is expressed as

$$fal(x, a, \delta) = \begin{cases} |x|^a sign(x), |x| \geq \delta \\ x/\delta^{a-1}, |x| < \delta \end{cases} \quad (8)$$

*Three-DOF helicopter control algorithm design based on the ADRC controller*

According to the motion equations of three-DOF helicopter, the position is changed in pitch axis, lateral axis and rotation axis. The rotation axis is controlled indirectly by angel $p$ in lateral axis. The active disturbance rejection controller is used to control three channels of helicopter respectively and ESO is used to observe the uncertainties of system, external disturbances and providing the compensation.

Solve the differential equations (2)

$$p = \frac{1}{J_p} K_c l_p (V_1 - V_2)^2 \tag{9}$$

(3) can then be written as

$$\ddot{r} = -G l_1 p = -\frac{G l_1 l_p K_c}{J_p J_r} (V_1 - V_2)^2 \tag{10}$$

where the lateral angel $p$ is set in a small range ($\pm$20 degrees), so that $\sin(p)$ is linearized to $p$.

Let $x_1 = \varepsilon, x_2 = p, x_3 = r$, $y$ is the output of system. Then the motion equations of the helicopter is

$$\begin{cases} \ddot{x}_1 = \dfrac{K_c l_1}{J_e} \cos(x_1) V_s - \dfrac{T_g}{J_e} \\ \ddot{x}_2 = \dfrac{K_c l_p}{J_p} V_d \\ \ddot{x}_3 = -\dfrac{G l_1 l_p K_c}{J_p J_r} V_d^2 \\ y_1 = x_1, y_2 = x_2, y_3 = x_3 \end{cases} \tag{11}$$

The expected pitch angle is set within $\pm$300 degrees and the position of the lateral axis influences the rotation axis indirectly. There are several uncertain factors in operation. The inertia of the degrees of freedom is time-varying and the motor force constant $K_c$ changes by input voltage. The frictions caused by the rotation axis and the gust also affect the precision of control [3]. The errors caused by non-determinacy, disturbances and simplification are compensated by ESO in each degree of freedom.

The control system is designed as the three two-order ADRC in parallel form, as shown in Figure 3. $\varepsilon_c, p_c, r_c$ are the set values of pitch axis, lateral axis and rotation axis respectively. $\varepsilon, p, r$ are the actual measured value and $\omega$ is the external interference.

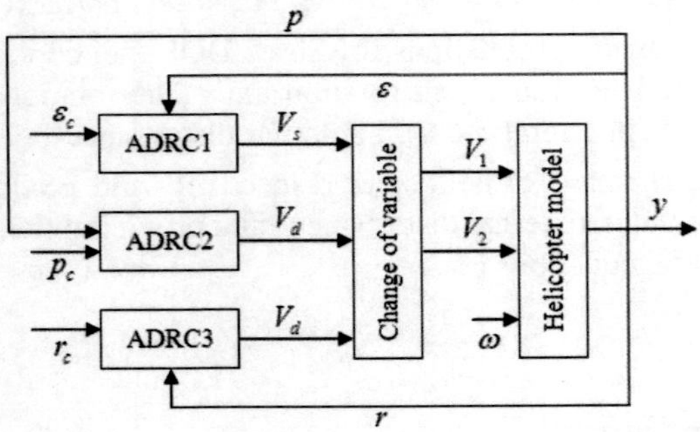


**Fig. 3.** Helicopter control system based on active disturbances rejection controller

## 4 Stability Analysis

Analyze the stability of ESO in pitch axis. (6) is written as [9]

$$\begin{cases} \dot{z}_1 = z_2 + L_1 g_1(e,\alpha,\delta) \\ \dot{z}_2 = z_3 + L_2 g_2(e,\alpha,\delta) + bu \\ \dot{z}_3 = L_3 g_3(e,\alpha,\delta) \end{cases} \tag{12}$$

where $x_m = [\varepsilon_{c1}, \varepsilon_{c2}, u_1]^T$ is the set value and $z = [z_1, z_2, z_3 / b_0]^T$ is the observed value, then $\hat{e} = [\hat{e}_1, \hat{e}_2, \hat{e}_3]^T = [\varepsilon_{c1} - z_1, \varepsilon_{c2} - z_2, u_1 - z_3 / b_0]^T$. The linear observer is designed to simplify the control algorithm as

$$\begin{bmatrix} \dot{\hat{e}}_1 \\ \dot{\hat{e}}_2 \\ \dot{\hat{e}}_3 \end{bmatrix} = \begin{bmatrix} -L_1 & 1 & 0 \\ -L_2 & 0 & 1 \\ -L_3 / b_0 & 0 & 0 \end{bmatrix} \begin{bmatrix} \hat{e}_1 \\ \hat{e}_2 \\ \hat{e}_3 \end{bmatrix} + \begin{bmatrix} 0 \\ -K_c l_1 / J_e \\ 0 \end{bmatrix} V_s \tag{13}$$

Simplify (13) as $\dot{\hat{e}} = \hat{A}\hat{e} + \hat{B}V_s$. It can be seen from the pole assignment that $\hat{A}$ is a stable matrix. Assume the Lyapunov function of observed error $\hat{e}$ as follows

$$V(\hat{e}) = \hat{e}^T P \hat{e} \tag{14}$$

if $P > 0$ and $A^T P + PA = -I$, then

$$\dot{V}(\hat{e}) = \hat{e}^T (A^T P + PA)\hat{e} + (B^T P\hat{e} + \hat{e}^T PB)V_s \\ = Q + M \tag{15}$$

where $Q = \hat{e}^T (A^T P + PA)\hat{e} \le 0$, $M = (B^T P\hat{e} + \hat{e}^T PB)V_s$. Let $M \le 0$ by choosing $V_s$, then $V(\hat{e}) \ge 0, \dot{V}(\hat{e}) < 0$, the system is asymptotically stable in large range. According to the stability analysis, the observed error of ESO $\hat{e}$ tends to zero in appropriate input. ESO can observe the internal state and weaken the external interference effectively.

## 5 Simulation Results

To test the performance of the helicopter control system based on the ADRC controller, the simulation was made in Matlab / Simulink. The initial positions of the helicopter are zero degrees. The objective is 30 degrees in pitch angel. Test the robust of the controller by disturbing the loop of pitch in time 10.

The parameter values of Auto Disturbance Rejection Controller are set as follows: $r = 30$, $h = 0.01$, $\beta_{01} = 100$, $\beta_{02} = 30$, $\beta_{03} = 15$, $b_0 = 1$, $k_1 = 100$, $k_2 = 10$, $\alpha_1 = 0.75$, $\alpha_2 = 1.25$, $\delta = 0$.

*Tracking the input set value*

Figure 4 (a) and (b) show the measure value of tracking input signal in pitch axis. The initial input position signal begins at time 0 and as the value of 30 degrees. Figure 4 (a) shows that the output of system can track the constant input value in time 8.4. Adding the step signal which value of 5 in time 10 to input, as shown in Figure 4 (b), the measure value of helicopter is still tracking input value quickly.

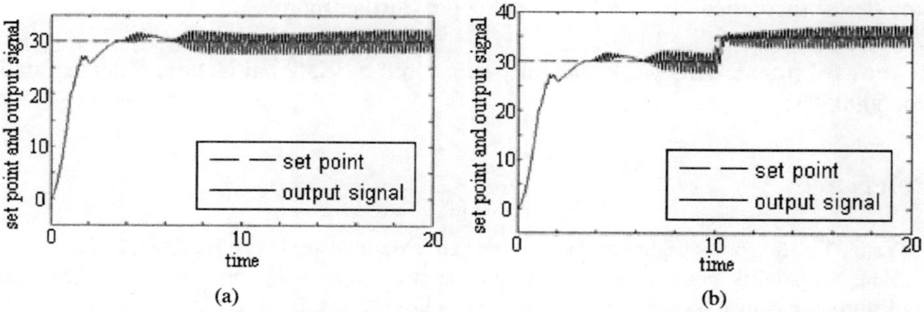


**Fig. 4.** Tracking the input value (a) tracking the constant input value(b) tracking the change input value

*External interference*

Figure 5 shows the measure value of tracking input signal in external interference in pitch axis. The interference in plant is joined in time 10 and as the value of 5. The result shows that there is no obvious change in measure value of output .It is proved that the ADRC controller has good robustness, in which the ESO can compensate the uncertainty, external disturbance and reduce the tracking error.

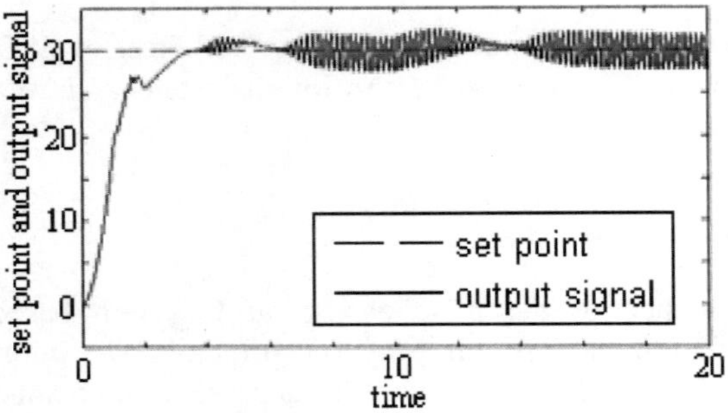


**Fig. 5.** Tracking the input value in external interference

# 6 Conclusion

The active disturbance rejection controller is used to tracking the input signal in degrees of freedom in three-DOF helicopter model. The algorithm is proposed to control the positions in each axis respectively and the stability of the system is analyzed by Lyapunov method. The simulation verifies the effect of the algorithm in the tracking control and external interference respectively. The results show that the controller can track the input reference signal, reject the disturbance, raise the expected stability and robust performance of the overall closed-loop system and provide the theoretical basis in actual helicopter further more.

**Acknowledgment.** This work was fully sponsored by National Natural Science Fund No. 60905051.

# References

1. Yang, Y.: Helicopter flight control. National Defense Industry Press, Beijing (2007)
2. Han, J.: ADRC control technology—the control technology of state estimation and disturbance compensation. National Defense Industry Press, Beijing (2008)
3. Qiao, J., Dai, Y., Liu, J.: Disturbance observer based on nonlinear inversion of helicopter Sliding Mode Control. Journal of Beijing Institute of Technology 29(3), 224–228 (2009)
4. Lopez-Martinez, M., Ortega, M.G., Vivas, C., Rubio, F.R.: Robust nonlinear control of a scaled helicopter with variable speed rotors. Revista Iberoamericana de Automatica e Informatica Industrial 4(3) (2007)

5. Kendoul, F., Fantoni, I., Lozano, R.: Modeling and control of a small autonomous aircraft having two tilting rotors. IEEE Transactions on Robotics 22(6), 1297–1302 (2006)
6. Lee, D., Kim, H.J., Sastry, S.: Feedback Linearization vs. Adaptive Sliding Mode Control for a Quadrotor Helicopter. International Journal of Control Automation and Systems 7(3), 419–428 (2009)
7. Three-DOF helicopter system experimental instructions. Googol technology (HK) Ltd. (2004)
8. Han, J.: The active disturbance rejection controller and its application. Control and Decision 13(1), 19–23 (1998)
9. Han, J.: Nonlinear Design methods for control system. In: The Proc of the 14th IFAC World Congress (1999)

# Study and Simulation of Crane Lifter Sensorless System Based on Sliding Mode Variable Structure

Feipeng Lou, Xinjin Wang, and Lingzhi Cao

Zhengzhou University of Light industry
School of Electrical and Electronic Engineering
Zhengzhou, China
feipeng1984@163.com

**Abstract.** The mathematical model of crane lifter system is established based three-phase permanent magnet synchronous motor as the power source. To achieve sensorless closed-loop control of crane lifter system, speed controller is designed using sliding mode variable structure and the system lifting speed is estimated applying model reference adaptive observer. Simulation studies have shown that adaptive observer can estimate the speed of movements accurately, sliding mode variable structure speed controller can ensure that the system has good static and dynamic speed performance and strong anti-disturbance capacity of the load.

**Keywords:** crane lifter system, sensorless, sliding mode variable structure, adaptive.

## 1 Introduction

In order to achieve high dynamic performance and closed-loop control of crane lifter system, the lifting speed is obtained by mechanical sensors generally. However, the use of mechanical sensors not only increase the cost of the system, but also have problems with the installation, it may affect the stability of the motor seriously if the improper installation. Non-sensor detection technology has gained widespread attention with modern control theory, digital signal processing technology and computer development. The literature [1] proposed a location algorithm based on flux, but the performance of the algorithm depends on the estimated flux and measured voltage and current accuracy, so the parameter changes in motor will affect the accuracy of position estimation. The literature [2] compared and analyzed of rotating high-frequency voltage injection method and pulsating high-frequency voltage injection method, but the algorithm needs motors have a higher rate of salient-pole as the external conditions. The literature [3] studied speed identification scheme based on model reference adaptive system for permanent magnet synchronous motor. The literature [4] analyzed the sliding mode variable structure predicted velocity estimation algorithm has a faster convergence and better anti-jamming capability, the system robustness had enhanced. In this paper, the speed controller is designed based on sliding mode variable structure in crane lifter system; the model reference adaptive method is applied to crane lifter system; through the design of adaptive observer to estimate the lifting speed, and ultimately achieve closed-loop control of the crane lifter system.

## 2 Mathematical Mode of Crane Lifter System

Crane lifting system mainly consists of three-phase permanent magnet synchronous motor, the reducer and transmission mechanism and the load. To facilitate the modeling and analysis, make the following assumptions: 1 excluding the quality of steel wire rope, 2 the flexibility of steel wire rope has nothing to do with the length, for a constant, 3 Brake is in good working condition, so the impact of brake can be neglected when modeling, 4 between deceleration and transmission the connection is rigid, there is no gap between the gears and so on.

### 2.1 Mathematical Modeling of Three-Phase PMSM

Under the premise of without affecting the control performance, ignoring motor core saturation, excluding the eddy current and hysteresis loss, three-phase windings are symmetrical, uniform, and windings of the induction electric potential waveform is a sine wave, in d-q coordinates the state equation of surface permanent magnet motor can be expressed as:

$$\begin{bmatrix} \dot{i}_d \\ \dot{i}_q \\ \dot{\omega} \end{bmatrix} = \begin{bmatrix} -\frac{R}{L} & p\omega & 0 \\ -p\omega & -\frac{R}{L} & -\frac{\psi_f p}{L} \\ 0 & \frac{3p\psi_f}{2J_1} & -\frac{B_1}{J_1} \end{bmatrix} \begin{bmatrix} i_d \\ i_q \\ \omega \end{bmatrix} + \begin{bmatrix} \frac{1}{L} & 0 & 0 \\ 0 & \frac{1}{L} & 0 \\ 0 & 0 & -\frac{T_m}{J_1} \end{bmatrix} \begin{bmatrix} u_d & u_q & 1 \end{bmatrix} \tag{1}$$

In the formula, $i_d$ and $i_q$ are respectively stator currents in $d$ axis and $q$ axis, $\omega$ is rotor angular velocity, $R$ is stator winding resistance, $L$ is axis inductance, $\psi_f$ is rotor flux, $p$ is motor pole pairs, $B_1$ is the damping coefficient with the speed of the rotor, $J_1$ is rotational inertia of the electric machine rotor, $T_m$ is load torque.

The torque equation of permanent magnet motor is:

$$T_e = \frac{3}{2} p\psi_f i_q \tag{2}$$

The mechanical speed equation of PMSM is:

$$T_e = J_1 \frac{d\omega}{dt} + B_1\omega + T_m \tag{3}$$

$T_m$ is motor shaft exporting mechanical torque, it can drive loads to move by driving mechanical transmission.

## 2.2 The Motion Equations of Transmission and Deceleration

Transmission mechanism and deceleration include the drive shaft, multi-stage reduction gear, and etc. $T_m$ is the load input of crane, namely overcomes the friction torque to drive load move. In order to facilitate research, inertia of each part of the moment, rotational friction coefficient and load torque are transformed the equivalent amount of the motor shaft: equivalent moment inertia $J_2$, equivalent rotational friction coefficient $B_2$, equivalent load torque $T_L$. Then the mathematical model of transmission and deceleration is such as:

$$T_m = T_L + J_2 \frac{d\omega}{dt} + B_2\omega \tag{4}$$

## 2.3 The Motion Equation of Load

Suppose $v$ is the load velocity, $F$ is the upward pull, $k_i$ is the ratio of load linear motion speed and motor speed, considering the dynamic load factor $\varphi_1$ when lifting and hoist load efficiency $\eta_L$, then:

$$k_i = v/\omega \tag{5}$$

$$T_L = \eta_L \varphi_1 F \tag{6}$$

It can be obtained movement equation of load lifting according to Newton's Second Law of Motion:

$$F = m\dot{v} + mg \tag{7}$$

Bringing formula (2)-(7) into formula (1) can obtain crane lifter system dynamic model as following:

$$\begin{bmatrix} \dot{i}_d \\ \dot{i}_q \\ \dot{v} \end{bmatrix} = \begin{bmatrix} -\frac{R}{L} & \frac{p}{k_i\eta_L\varphi_1}v & 0 \\ -\frac{p}{k_i\eta_L\varphi_1}v & -\frac{R}{L} & -\frac{\psi_f p}{L} \\ 0 & \frac{3p\psi_f}{2J1}a_1 & -a_1a_2 \end{bmatrix} \begin{bmatrix} i_d \\ i_q \\ v \end{bmatrix} + \frac{1}{L}\begin{bmatrix} u_d \\ u_q \\ 0 \end{bmatrix} + \begin{bmatrix} 0 \\ 0 \\ -\frac{k_i\eta_L\varphi_1 a_1 mg}{J1} \end{bmatrix} \tag{8}$$

There: $a_1 = \frac{\eta_L\varphi_1 k_i J_1}{k_i^2 m + J_1 + J_2}$, $a_2 = \frac{B_1 + B_2}{\eta_L\varphi_1 k_i J_1}$

## 3 Design of Speed Sliding Mode Controller

In the crane lifter system, the speed controller is designed as PID controller traditionally. The PID controller has simple structure, better stability and higher reliability, but it asks for higher precision object model. The controller can get good performance in the case of parameter matching, but it results in control performance degradation once the parameters changes or load torque disturbance occurs. The sliding mode variable structure control, not only has strong robustness, good stability, anti-jamming ability, but also has faster response characteristics and smaller overshoot.

It can be seen from the mathematical model of crane lifter system that the control of motor output torque is reduced to the control pairs of d-axis and q-axis current. Adopt $I_d$=0 vector control strategies, d-axis flux is produced by permanent magnets entirely. Then the mathematical model of crane lifter system can be translated into:

$$\begin{bmatrix} \dot{i}_q \\ \dot{v} \end{bmatrix} = \begin{bmatrix} -\frac{R}{L} & \frac{\psi_f p}{L} \\ \frac{3p\psi_f}{2J1}a_1 & -a_1a_2 \end{bmatrix} \begin{bmatrix} i_q \\ v \end{bmatrix} + \frac{1}{L}\begin{bmatrix} u_q \\ 0 \end{bmatrix} + \begin{bmatrix} d_1 \\ -\frac{\eta_L \varphi_1 k_i a_1 mg}{J1} + d_2 \end{bmatrix} \tag{9}$$

In the formula, other parameters remain unchanged, $d_1$ and $d_2$ are external disturbances of the system.

Let: $e = v - v^*$

Assume sliding surface as: $s = c_1 e + c_2 \int_0^t e dt + \dot{e}$

Then:

$$\begin{aligned} \dot{s} &= c_1\dot{e} + c_2 e + \ddot{e} \\ &= c_1\dot{e} + c_2 e + \frac{3p\psi_f a_1}{2J_1}\dot{i}_q - a_1a_2\dot{v} + \dot{d}_2 - \ddot{v}^* \\ &= c_1\dot{e} + c_2 e + \frac{3p\psi_f a_1}{2J_1}\left(-\frac{R}{L}i_q - \frac{\psi_f p}{L}v + \frac{1}{L}u_q + d_1\right) \\ &\quad - a_1a_2\dot{v} + \dot{d}_2 - \ddot{v}^* \\ &= f_1 + D_1 u_q + z_1 \end{aligned}$$

Where:

$$f_1 = c_1\dot{e} + c_2 e + \frac{3p\psi_f a_1}{2J_1}\left(-\frac{R}{L}i_q - \frac{\psi_f p}{L}v\right) - a_1a_2\dot{v} - \ddot{v}^*$$

$$D_1 = \frac{3p\psi_f a_1}{2J_1 L} \qquad z_1 = \frac{3p\psi_f a_1}{2J_1}d_1 + \dot{d}_2$$

Assume Lyapunov function is: $V=\frac{1}{2}s^2$

Assume sliding mode control rate is:

$$u_q=\frac{1}{D_1}[-k_1 s-\varepsilon sign(s)-f_1]$$

Where: $k_1>0\,,\varepsilon>|z_1|$

Then:

$$\dot{V}=s\dot{s}=s[-k_1 s-\varepsilon sign(s)+z_1]$$
$$=-k_1 s^2-\varepsilon s+z_1 s$$

$\dot{V}<0$ can be guaranteed as long as guarantee $\varepsilon>|z_1|$. The system has a wide range of asymptotically stable.

## 4 Design Model Reference Adaptive Observer

The main idea of model reference adaptive method is that let equation that has not included unknown parameter as a reference model, which let equation needs to be estimated as the adjustable model equation, both models have the same output of physical meaning. An appropriate adaptive law to regulate real-time adjustable model parameters can be obtained from output error of two models, so as to achieve the purpose that the output of adjustable model tracking the output of reference model.

The load speed can be estimated through the model reference adaptive observer and applied in crane lifter system to achieve sensorless vector control. In the d q coordinate, take the current model of mathematical model from the crane lifter system:

$$\begin{bmatrix}\dot{i}_d\\ \dot{i}_q\end{bmatrix}=\begin{bmatrix}-\frac{R}{L} & \frac{p}{\eta_L\varphi_1 k_i}v\\ -\frac{p}{\eta_L\varphi_1 k_i}v & -\frac{R}{L}\end{bmatrix}\begin{bmatrix}i_d\\ i_q\end{bmatrix}+\frac{1}{L}\begin{bmatrix}u_d\\ u_q\end{bmatrix} \tag{10}$$

It can obtain from formula (10) that current mode relates the speed of the load. So select crane lift system itself as a reference model and the current model as adjustable model, then use of parallel structure identification rate.

Let: $i'_d=i_d+\frac{p\psi_f}{\eta_L\varphi_1 k_i L}$

$$u'_d=u_d+\frac{Rp\psi_f}{\eta_L\varphi_1 k_i L}$$

$$i'_q=i_q,\quad u'_q=u_q$$

Then:

$$\begin{bmatrix} \dot{i}'_d \\ \dot{i}'_q \end{bmatrix} = \begin{bmatrix} -\dfrac{R}{L} & v \\ -v & -\dfrac{R}{L} \end{bmatrix} \begin{bmatrix} i'_d \\ i'_q \end{bmatrix} + \frac{1}{L} \begin{bmatrix} u'_d \\ u'_q \end{bmatrix} \tag{11}$$

Take parallel adjustable model as:

$$\begin{bmatrix} \hat{\dot{i}}'_d \\ \hat{\dot{i}}'_q \end{bmatrix} = \begin{bmatrix} -\dfrac{R}{L} & \hat{v} \\ -\hat{v} & -\dfrac{R}{L} \end{bmatrix} \begin{bmatrix} \hat{i}'_d \\ \hat{i}'_q \end{bmatrix} + \frac{1}{L} \begin{bmatrix} u'_d \\ u'_q \end{bmatrix} \tag{12}$$

$\hat{v}$ is the amount identified and required, other parameters remain unchanged. The operation process of the model reference adaptive observer is shown in Fig.1.

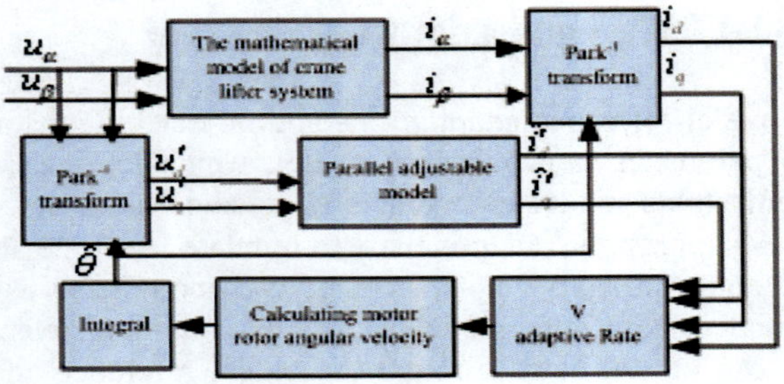


**Fig. 1.** The diagram of model reference adaptive algorithm

According to Popov stability theory and the super-positive real theorem [5], adopting proportional-integral adaptive rate, and using $i_d$ 、 $i_q$ instead of $i'_d$ 、 $i'_q$, the following equation can be gotten.

$$\begin{aligned} \hat{v} = & \int_0^t k_I [i_d \hat{i}'_q - i_q \hat{i}'_d - \frac{p\psi_f}{\eta_L \varphi_1 k_i L}(i_q - \hat{i}'_q)]dt \\ & + k_p [i_d \hat{i}'_q - i_q \hat{i}'_d - \frac{p\psi_f}{\eta_L \varphi_1 k_i L}(i_q - \hat{i}'_q)] + v(0) \end{aligned} \tag{13}$$

Where: $k_p$ is proportional rate, $k_I$ is integral rate, and both are greater than 0. $i_d$ 、 $i_q$ are obtained through the coordinate transformation. $\hat{i}'_d$ 、 $\hat{i}'_q$ are obtained by parallel adjustable model, as show in Fig.1. $v(0)$ is the initial value and can be determined according to the actual situation.

# 5 Simulation and Study

The diagram of crane lifter sensorless vector control system based adaptive sliding mode variable structure is shown in Fig.2. It mainly includes three-phase inverter, crane lifting module, coordinate transformation module, SPWM module, sliding-mode speed controller and model reference adaptive observer. Adopt $i_d = 0$ vector control strategies, *d*-axis flux produced by permanent magnets entirely. The lifting speed of load and the motor rotor position are obtained by the model reference adaptive observer.

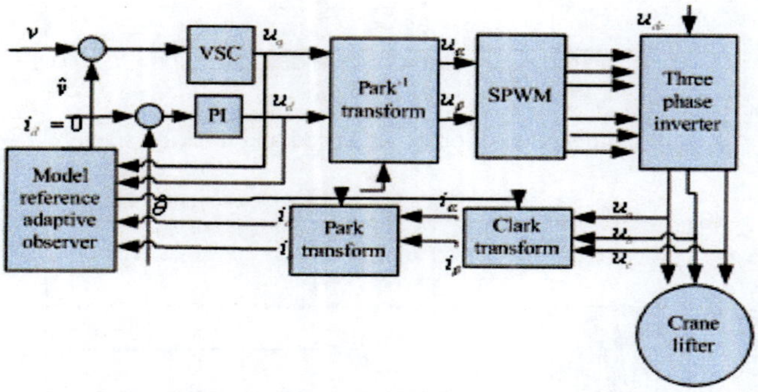


**Fig. 2.** The sliding mode adaptive control diagram

According to the establishment of the crane lifter system control diagram simulation is designed in the MATLAB software. Simulation parameters are used as follows: the motor pole pairs $P = 4$ , Stator resistance $R = 2.875\Omega$ , *d, q* axle inductance $L = 8.5mH$ , the rotational inertia of the electric machine rotor $J_1 = 0.0008kg.m^2$ , rotor flux $\psi_f = 0.175Wb$ , the damping coefficient $B_1 = 0$ , Equivalent rotational friction coefficient $B_2 = 0.01N.m.s$ ,the dynamic load factor when lifting $\varphi_1 = 1.5$ , hoist load efficiency $\eta_L = 0.75$ ,system disturbance $d_1$ is the movement of wire rope, such as the elastic deformation caused by a slight high-frequency oscillation disturbance, which are suppose as $d_1 = 2\sin(40\pi t)$ , $d_2$ mainly caused by fluctuation of the power supply voltage, which are supposed as $d_2 = 120 + 60pulse(100\pi t)$ , when light load $m = 50kg$ , when heavy load $m = 100kg$ .

The target speed of crane lifter system is given 2m/s, the load is given $m = 50kg$ . As shown in Fig.3, there is a certain error between the speed of model reference adaptive observer and the speed of sliding mode control when the system is starting. However, the observed speed can be tracked to the track speed precisely and fast. Taking the load into $m = 100kg$ when 0.05s, at this time speed tracking produced a small fluctuation, but can quickly keep stability of the track. Simulation results show that dynamic speed changes and the steady-state operation have a good dynamic characteristic by the model reference adaptive speed observer in the crane lifter system, as also as the speed estimation error is smaller.

As can be seen from the simulation diagram, the application of sliding mode variable structure to design speed controller can make the system have response time quickly and small overshoot and ensure system stability.

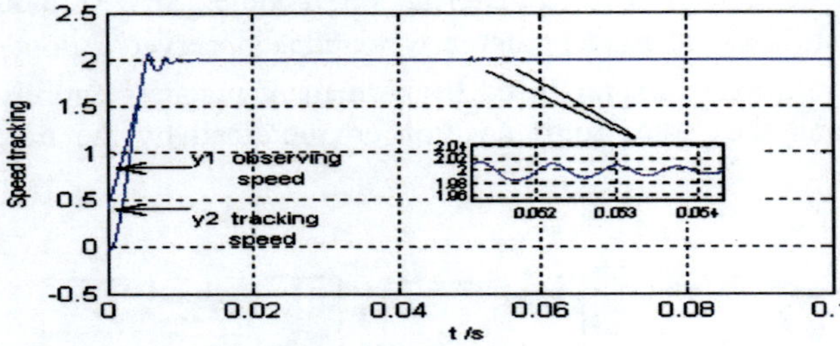


**Fig. 3.** Comparison of observed speed and tracking speed

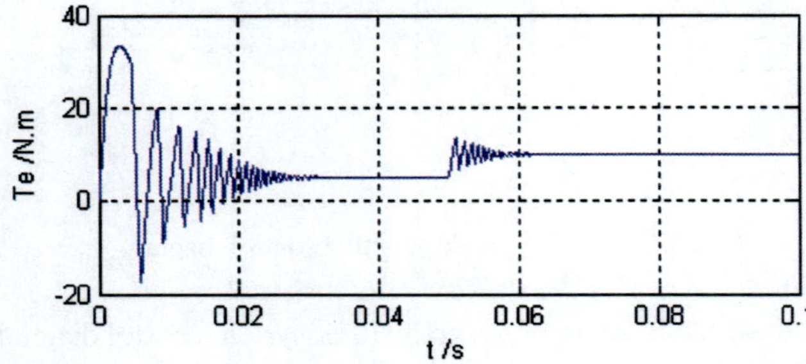


**Fig. 4.** The output torque waveform of motor

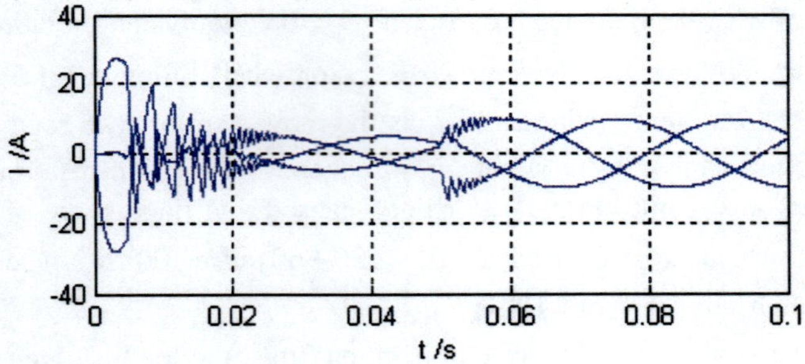


**Fig. 5.** The three-phase current waveform of motor

Fig.4 is the output torque waveform of three-phase permanent magnet motor. Fig.5 is the stator current waveforms. Motor output torque and current entered the stable state after a short period of oscillation, and the output speed changed little. These show that anti-disturbance ability of the designed system and the closed-loop load-adjustment ability is stronger.

# 6 Conclusion

This paper analyzed and established the mathematical model of the crane lifter system based on permanent magnet motor as power source. On the basis of this model, sliding mode speed controller and model reference adaptive observer are designed. At last the sensorless vector control system of crane lifter is studied through simulation experiments. Simulation results show that the adaptive observer can observe the lifting speed accurately, and the sliding mode speed controller can make the system have strong robustness and capacity against load disturbance. With the computer and digital signal processing development, it is especially suitable for industrial control DSP chip to achieve.

# References

1. Ertugru, N., Acarnley, P.P.: A New Algorithm For Sensorless Operation of Permanent Magnet Motors. IEEE Transactions on Industry Application 30(1), 126–133 (1992)
2. Feng-Qin, Yikang-He, Yi-Liu, Wei-Zhang: Comparative investigation of sensorless control with two high-frequency signal injection schemes. Proceeding of the CSEE 25(5), 116–121 (2005)
3. Wangpan-Zhang, Hong-Li, Licuo-Cui, Zengping-Qiao: Simulation of sensorless vector control for permanent magnet synchronous motor. Micro-motor 3, 51–53 (2009)
4. Jangru-Wan, Yonglin-Guo, Chuangeng-Cheng: Sliding mode variable structure control for sensorless PMSM. Electrical Applications 25(8), 75–78 (2006)
5. Chengyuan-Wang, Jiakuan-Xia: Motor modern control technology, p. 1. Machinery Industry Press (2006)
6. Chunwen-Li, Lingzhi-Cao, Weipu-Miao: Speed control system design of crane based on inverse system and sliding mode methods. Coal Mine Machinery 10, 26–30 (2008)

# Weigh in Motion Based on Dynamic Tire Force Separation

Zhou Zhi-feng

School of Mechanical Engineering, Shanghai University of Engineering Science
Shanghai, China
zhousjtu@126.com

**Abstract.** Dynamic tire forces are the main factor affecting the measurement accuracy of the axle weight of moving vehicles. This paper focuses on separating dynamic tire forces from axle-weight signal, and proposes a new method to measure the axle weigh of vehicle based on improved EMD (empirical mode decomposition). First, the characteristics of axle-weight signal and related concepts and process of EMD are briefly introduced. Then the causes and the influences on weighing accuracy of end effect and pseudo-IMF (intrinsic mode function) are analyzed in detail. The AR (auto-regressive) model prediction method and the correlation coefficient method are employed to extend the data points outside of signal end points and judge the pseudo-IMFs, respectively. The pseudo-IMFs and the residual are added to form a new residual, and the average value of new residual is regarded as the axle-weight estimation. Lastly, the simulation and real experiments are given to illustrate the proposed method. The max weighing error of axle-weight is 4.34% at the speed no higher than 20km/h.

**Keywords:** Weigh in motion, Dynamic tire force, EMD, Pseudo-IMF, End effect.

## 1 Introduction

Fast and accurate weighing the axle load of moving vehicle is significant for traffic management, maintenance and law enforcement. With the rapid development of traffic and transportation, the increasing phenomenon of overload in freight transportation has become a serious problem. The overload vehicles have caused various harms [1], such as shortening the lifespan of roads and bridges, causing the traffic accidents and resulting in the great loss of the state's regulation fees and tax income.

There exist various factors [2] that influence the weighing accuracy of the axle weight. The dynamic tire forces contained in the axle weight signal are the most main factor. System identification method [3] models the weighing system as an AR (auto-regressive) model with unit step input, but this method fails to eliminate the influence of the dynamic tire forces on the weighing accuracy. B-WIM[4-5] (bridge weigh in motion) is a good solution for eliminating the influence of dynamic tire forces, because B-WIM system uses a bridge as the weighing platform and has enough time to get the complete-cycle dynamic tire forces. But B-WIM system fails to accurately measure individual axle weights. If there are a lot of vehicles crossing the bridge simultaneously, the effect of a single axle will be very difficult to be distinguished from other vehicles' axles. MS-WIM [6-7] (multiple-sensor weigh in motion) is another method to reduce the effect of the dynamic tire forces in which a number of strip sensors are installed in the road at certain intervals. However, the cost of MS-WIM system is high. If the vehicle speed is 10 m/s and the lowest frequency of the dynamic tire forces is 2 Hz, 40 m instrumented road will be needed. If the strip sensors are placed at 0.4 m intervals along 40 m road, then 100 sensors will be needed. Furthermore, the design of sensor array and the calibration of sensors are difficult in practice.

In this paper, the method based on dynamic tire forces separation is presented to process the axle weight signal. In the following sections, the characteristics of the dynamic tire forces is analyzed; the concept and steps of EMD (empirical mode decomposition) are introduced; the causes of producing pseudo-IMF (intrinsic mode function) and end effect are analyzed, and their harms on weighing accuracy are also analyzed; the methods of restraining end effect and judging pseudo-IMF are employed; the simulation and field experiments are conducted to evaluate the performance of the proposed method.

## 2 Dynamic Tire Forces

When vehicle keeps static, the force of tires exerting on the ground is equal to the static axle weight (which is regarded as the real axle weight in practice). When vehicle moves, the tire force exerting on the ground contains the dynamic tire forces besides the static axle weight. The quarter-car model [8] is often used to study the dynamic tire forces.

Referencing the approach described in Appendix B of [8], we can simulate the vehicle response to road roughness. The dynamic tire forces mainly distribute in two frequency ranges [8]: (a) low frequencies, ranging from 1.5 to 4.5 Hz, and (b) high frequencies, ranging from 8 to 15 Hz. The Dynamic Load Coefficient (DLC) is usually used to characterize the magnitude of dynamic tire forces. Under normal operating conditions, the typical values of DLC are 0.1-0.3[8].

If we assume that the dynamic tire forces are the superposition of sine signals [9], then the force exerting on the weighing platform can be described as

$$f(t) = w + \sum_{i=1}^{n} A_i \sin(2\pi f_i t + \varphi_i), \tag{1}$$

where $t$ is the sampling time, $w$ is the static axle weight, $n$ is the number of the dynamic tire forces, $A_i$, $f_i$, and $\varphi_i$ are the amplitude, frequency and initial phase of the $i$-th dynamic tire force, respectively.

## 3 Empirical Mode Decomposition

EMD is the effective method for analyzing nonlinear and non-stationary signal. EMD can decompose a given signal into a collection of IMFs and a residue. An IMF is a function that satisfies two conditions [10]: (a) in the whole data set, the number of extrema and the number of zero crossing must either equal or differ at most by one; and (b) at any point, the mean value of the envelope defined by the local maxima and the envelope defined by the local minima is zero.

Given a signal $x(t)$, the effective algorithm of EMD can be summarized as follows[10]:

*1)* Identify all extrema of $x(t)$;

*2)* Interpolate between maxima (respectively minima), ending up with some envelope $e_{max}(t)$ (respectively $e_{min}(t)$);

*3)* Compute the average $m(t)=( e_{max}(t) + e_{min}(t))/2$;

*4)* Extract the detail $d(t)=x(t)-m(t)$;

*5)* Iterate on the residual $r(t)$.

In practice, the above procedure has to be refined by a sifting process which amounts to first iterating steps 1) - 4) upon the detail signal $d(t)$, until the latter can be considered as zero-mean according to the standard deviation[10]

$$SD = \sum_{t=0}^{T}\left[\frac{\left|d_{k-1}(t)-d_k(t)\right|^2}{d_{k-1}^2(t)}\right], \tag{2}$$

where $d_k$(t) denotes the $k$-th detail during the sifting process. The typical value for SD can be set from 0.2 to 0.3. Once SD reaches the typical value, the detail $d_k$(t) can be considered as the effective IMF, the corresponding residual is computed by

$$r(t) = x(t) - d_k(t)\,, \tag{3}$$

and step 5) applies. When the residual $r(t)$ becomes a monotonic function from which no more IMF can be extracted, we have

$$x(t) = \sum_{i=1}^{n} c_i(t) + r_n(t), \tag{4}$$

where $c_i(t)$ ($i$=1,2,…,$n$) denotes IMF and $r_n(t)$ denotes the residue, and $c_1$(t) has the locally highest frequency in $x(t)$, while $r_n(t)$ contains the slowest frequency, usually a trend of $x(t)$.

Comparing (1) and (4), we can see that the forms of two equations are same. We can decompose an axle weight signal into several IMFs and a residual with EMD method. The IMFs can be regarded as the dynamic tire forces contained in an axle weight signal. The residual can be regarded as the axle weight of moving vehicle. Computing the average value of the residual, we can get the estimation of axle weight.

## 4 Pseudo-IMF and End Effect

Let $x(t)$=2+sin(4$\pi t$)+sin(10$\pi t$+0.5$\pi$) be the axle weight signal, where 2 denotes the static axel weight, two sin functions denote the dynamic tire forces. The sampling interval is 0.01s. The sampling time is 2s. Fig.1 shows the decomposition of the signal $x(t)$. The signal has been decomposed into five IMFs and a residual. Imf1, imf2, imf3, imf4, imf5 and r5 denote IMFs and residual, respectively. The average value of r5 is 1.7968, which means the weighing error is 10.16%. Note that the inspected signal $x(t)$ only contains two frequency components. Therefore, only two among the five IMFs are the desirable IMFs, and the other three IMFs are the undesirable IMFs.

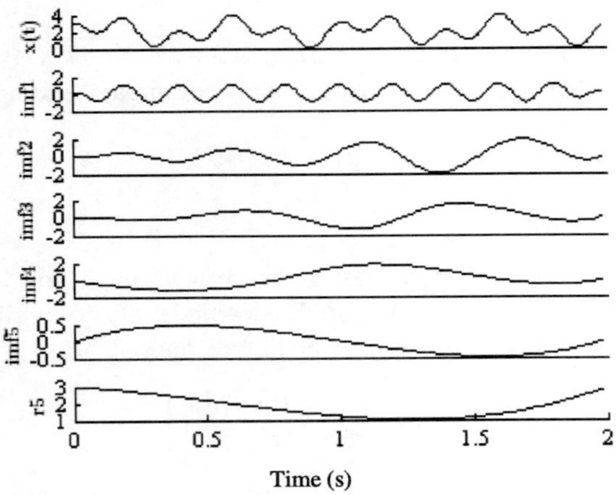


**Fig. 1.** EMD with end points

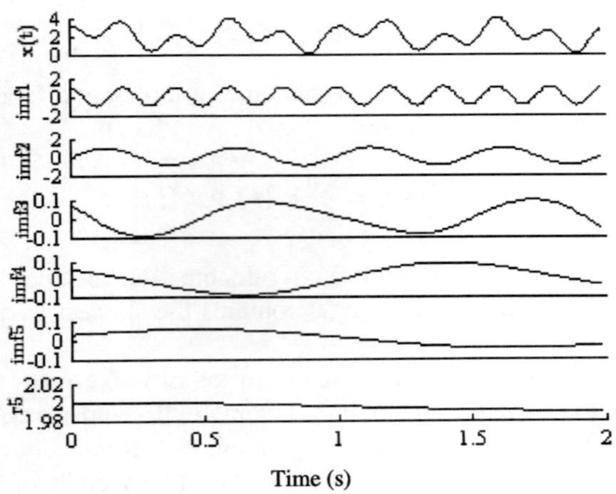


**Fig. 2.** EMD with AR model

In this paper, the undesirable IMF is called pseudo-IMF.

Since pseudo-IMF is not the real frequency component of the inspected signal, it should have weak correlation with the inspected signal. We can use the correlation coefficient of IMF with the inspected signal to judge whether an IMF is the pseudo-IMF. Here, the correlation coefficient can be described as [11]

$$\lambda_i = \frac{N\sum_{n=1}^{N} x(n)c_i(n) - \sum_{n=1}^{N} x(n)\sum_{n=1}^{N} c_i(n)}{\sqrt{\left[N\sum_{n=1}^{N} x^2(n) - (\sum_{n=1}^{N} x(n))^2\right]\left[N\sum_{n=1}^{N} c_i^2(n) - (\sum_{n=1}^{N} c_i(n))^2\right]}}, \tag{5}$$

where $x(n)$ denotes the inspected signal, $c_i(n)$ denotes the $i$-th IMF, $N$ denotes the samples number of the inspected signal. The correlation coefficients for imf1, imf2, imf3, imf4, imf5 with the inspected signal $x(t)$ are 0.6868, 0.5405, 0.0007, 0.0145 and 0.0081, respectively. In this study, $\lambda$=0.2 is used. Any IMF that has a correlation coefficient equal or higher than $\lambda$ will be retained; whilst, other IMFs will be regarded as pseudo-IMFs and added to the residual. Here, imf3, imf4 and imf5 are pseudo-IMF and added to r5 to form a new residual. The average value of the new residual is 1.9438 and the error is 2.81%. Compared with 10.16%, the weighing accuracy is improved.

The procedure of EMD is a kind of interpolating process based on signal extremum points. In practice, it is difficult to know whether two end points of signal are the extremum points. During the decomposition, two end points of signal are often simultaneously regarded as maximum point and minimum point. The envelope lines based extremum points will inevitably deform near the end points of signal, which results in deformation of IMFs derived from envelope lines. With the decomposition of signal, IMFs' deformation will extend from signal end to signal middle. When signal is shorter, IMFs' deformation is especially obvious. Inspecting imf1 and imf2 in fig.1, we can find that deformation of these two IMFs is obvious near the end points. The phenomenon of IMFs' deformation because of end points of signal is called end effect in this paper.

End effect is the main factor of producing pseudo-IMF and influencing the residual's accuracy. End points extending method is often employed to control end effect by predicting location and magnitude of data points outside of signal end points. Mirror method, statistical method, polynomial fitting method and neural network method are used to extend end points in [12]. In this paper, AR model extending method [13] is employed to estimate the data points outside of signal end points. AR model extending method can be summarized as follows:

1) Constructing 10th-order AR model;

2) Using the inspected signal and Burg algorithm to estimate the parameters of the constructed AR model;

3) Predicting the data points outside of end points of the inspected signal, and extending the inspected signal;

4) Decomposing the extended signal and getting IMFs with EMD method;

5) Subtracting the extended data points from the obtained IMFs and getting the true IMFs contained in the inspected signal.

Fig.2 shows the decomposition of the signal $x(t)$ using AR model extending method. Compared with fig.1, it is clear that the deformations of imf1and imf2 are obviously diminished near the end points, and also the amplitudes of imf3, imf4 and imf5 are diminished. The average value of r5 is 1.9946, which means the weighing error is 0.27%. Compared with 10.16%, the weighing accuracy is evidently improved. The correlation coefficients for imf1, imf2, imf3, imf4, imf5 with the $x(t)$ are 0.7104, 0.7040, 0.1513, 0.0736 and 0.0668, respectively. Imf3, imf4 and imf5 are pseudo-IMFs in fig.2. The average value of the new residual is 2.0018, which means the weighing error is 0.09%. Compared with 2.81%, the weighing accuracy is also improved.

# 5 Field Experiments

Fig. 3 shows the experiment site for this research. The weighing platform is 3.00 m in length and 0.75 m in width (direction of vehicle movement). The sampling frequency is 10 kHz. A two-axle truck runs through the weighing platform at 10 km/h, 15 km/h and 20km/h, respectively. Fig. 4 shows the sampled signal. The first maximum point of the ascent segment denotes tire has entered the weighing platform entirely and the first maximum point of the descent segment denotes tire is going to leave the weighing platform. The data segment between the two maximum points is selected as the valid signal. The dashed lines depict the selection for both front and rear axle signals in Fig. 4.

**Fig. 3.** Field experiment site

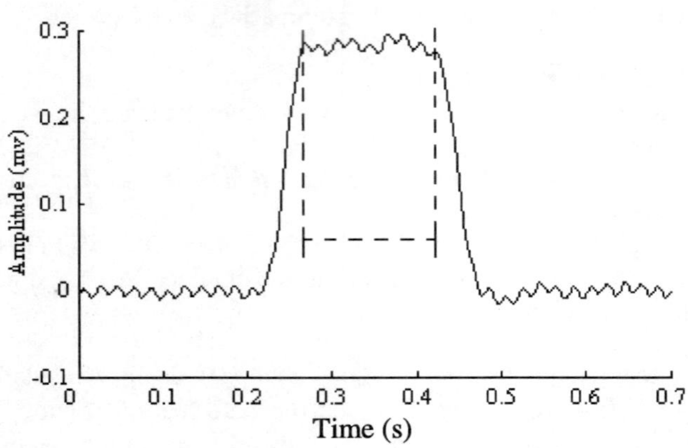


**Fig. 4.** Sampled signal at 10km/h

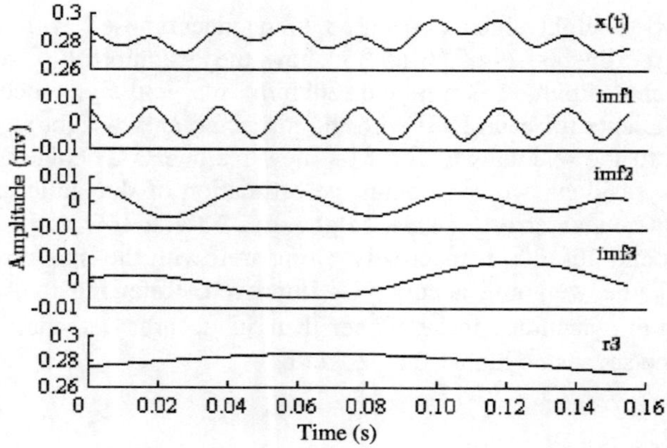


**Fig. 5.** Decomposition of axle weight signal at 10km/h

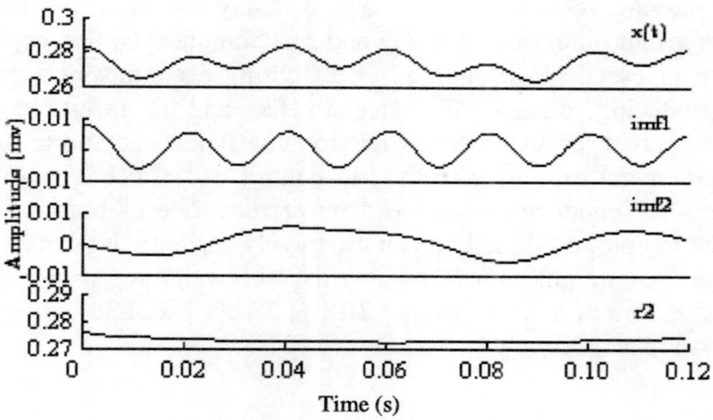


**Fig. 6.** Decomposition of axle weight signal at 15km/h

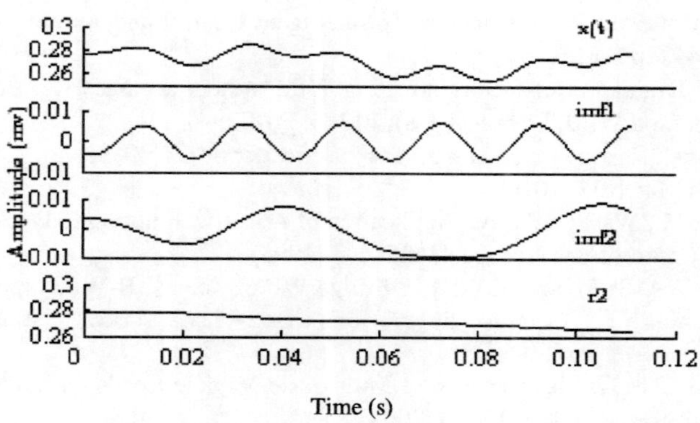


**Fig. 7.** Decomposition of axle weight signal at 20km/h

Forty-eight axle weight signals are used to be decomposed with the improved EMD in field experiments. Fig.5, 6 and 7 show the decomposition results of axle weight signals at 10km/h, 15km/h and 20km/h, respectively. According to the correlation coefficients for each IMF with the inspected axle weight signals, pseudo-IMFs are added to the residuals to form the new residuals. Calculating the average value of the new residual, we can obtain the estimation of the static axle weight of moving vehicle. The max errors of axle weight are 2.13%, 2.10% and 4.34% at speed 10km/h, 15km/h and 20km/h, respectively. Compared with the max errors of system identification [3], the weighing accuracy is improved. Referring to ASTM E1318-94[14], the weighing accuracy is far higher than III accuracy standard (axle weight error ±15%) at the speeds no higher than 20km/h.

## 6 Conclusions

The dynamic tire forces are the main factor that influences the weighing accuracy of axle weight of moving vehicles. On the basis of analyzing the characteristics of the dynamic tire forces and introducing EMD and related concepts, this paper points out the phenomenon of pseudo-IMF and end effect during decomposing signal, analyzes the causes of producing pseudo-IMF and end effect and the influence on weighing accuracy of axle weight, provides the correlation coefficient method to judge pseudo-IMF and the AR model extending method to control end effect. The simulation and field experiments are conducted to inspect the performance of improved EMD. The results show that the proposed method can effectively separate the dynamic tire forces contained in the axle weight signal and improve the weighing accuracy of the axle weight. The max errors of axle weight are 2.13%, 2.10% and 4.34% at speed 10km/h, 15km/h and 20km/h, respectively.

## References

1. Wang, J.F., Wu, M.G.: An overview of research on weigh-in-motion system. In: Proceeding of the 5th Word Congress on Intelligent Control and Automation, Hangzhou, pp. 5241–5244 (2004)
2. Scheuter, F.: Evaluation of factors affecting WIM system accuracy. In: Proc. of the 2nd European Conf. on WIM, Lisbon (1998); P1216, E0-E9, `http://www.haenni-scales.ch/e/produkte/pdf/P1216_E.pdf` (last accessed March 13, 2010)
3. Niedzwiecki, M., Wasilewski, A.: Application of Adaptive Filtering to Dynamic Weighing of Vehicles. Control Eng. Practice 4, 635–644 (1996)
4. O'Brien, E.: WAVE Report of WP1.2: Bridge WIM systems (B-WIM), pp. 1–99 (2001), `http://wim.zag.si/wave/download/wp12_final.pdf` (last accessed March 13, 2010)
5. Yu, L., Chan, T.H.T.: Identification of Multi-Axle Vehicle Loads on Bridges. Journal of Vibration and Acoustics 126, 17–26 (2004)
6. Cole, D.J., Cebon, D.: Performance and application of a capacitative strip sensor. In: Proc. 6th Int. Conf. on Road Traffic Monitoring, London, pp. 123–127 (1992)

7. Cebon, D.: WAVE Report of WP1.1: Multiple Sensor WIM, pp.1–73 (2001), http://wim.zag.si/wave/download/wp11_final.pdf (last accessed March 13, 2010)
8. Cebon, D.: Interaction between Heavy Vehicles and Roads. SAE International, 1–82 (1993)
9. Ono, T.: On Dynamic weighing of Highway Vehicles in Motion. In: SICE Annual Conference in Fukui, pp. 2108–2114 (2003)
10. Huang, N.E., Shen, Z., Long, S.R., Wu, M.C., Shih, H.H., Zheng, Q., et al.: The empirical mode decomposition and the Hilbert spectrum for nonlinear and non-stationary time series analysis. Proceedings of Royal Society, London, 903–995 (1998)
11. Peng, Z.K., Tse, P.W., Chu, F.L.: An improved Hilbert-Huang transform and its application in vibration signal analysis. Journal Sound and Vibration 286, 187–205 (2005)
12. Xu, B.-j., Zhang, J.-m., Xu, X.-l.: A Study on the Method of Restraining the Ending Effect of Empirical Mode Decomposition. Transactions of Beijing Institute of Technology 26, 197–200 (2006)
13. Yang, J.-w., Jia, M.-p.: Study on processing method and analysis of end problem of Hilbert-Huang spectrum. Journal of Vibration Engineering 19, 283–287 (2006)
14. ASTM Committee Standard specification for highway weigh-in-motion (WIM) systems with user requirements and test method. ASTM Designation E1318-94 (1994)

# The Improving of Initial Exchanges in IKEv2

Jieru Lu, Bing Su, Hongyuan Wang, and Yihan Zhang

School of Information Science & Engineering
Changzhou University
Changzhou, P.R. China
lujieru1980@sohu.com

**Abstract.** This paper proposed an approach to provide an efficient authentication and key agreement in IKEv2. We used Elliptic Curve Cryptography Diffie-Hellman (ECDH) to exchange keys between two communicating parties during IKE negotiation. Also we designed and implemented ECC X.509v3 certificate for identity authentication. The results of performance comparisons show that the proposed method provides greater security using fewer key sizes, and improves the execution efficiency of the initial exchanges in IKEv2.

**Keywords:** IKEv2, ECC, ECDH, X.509 digital certificate, mutual authentication, key agreement

## 1 Introduction

Internet Protocol Security (IPSec) is essentially as a set of security protocols and algorithms that ensure data security on the network layer. IPSec also includes protocols for establishing mutual authentication between agents at the beginning of the session and negotiation of cryptographic keys to be used during the session. IPSec is actually a framework that defines the external behavior of IPSec compliant nodes. It also defines associated protocols allowing IPSec compliant nodes to transfer data and control messages: Encapsulating Security Payload (ESP) [1] and optionally Authentication Header (AH) [2] for data transfer, as well as Internet Key Exchange version 2 (IKEv2) [3] for signaling purpose.

In IPSec, IKEv2 is used to provide a secure mechanism for establishing IPSec-protected connections. It is a set of protocols and mechanisms designed to perform two functions, creation of a protected environment which includes peers authentication that are unknown to each other in advance and to establish and manage Security Associations (SA) between the authenticated peers. Because of the number of exchange messages and the renewed choices of the encryption and authentication algorithms, IKEv2 is lesser complexity, greater simplicity and more security than IKEv1 [4].

The initial exchange protocol of IKEv2 consists of four messages and supports scenarios where the communication parties authenticate to each other using public key certificates or strong share secret keys. Digital certificated based authentication issue for remote access users is becoming increasingly popular for large IPSec VPN site-to-site deployments because it is more scalable than pre-shared keys [5]. In initial exchanges of IKEv2, the impact on efficiency and security of identifier authentication is different for each public-key algorithm. Although RSA is the most often used authentication method in IKEv2 [6], [7], it suffers serious performance impacts when key sizes increase much beyond those in common use day. ECC offers equivalent security with smaller key sizes, in less computation time and with less memory [8]. For this reason, in this paper we use ECC certificates for authentication instead of RSA and ECDH in IKEv2 to generate the shared key material for encryption.

Rest of paper is arranged as follows. First, in section 2, we give an overview of the IKEv2 protocol, especially some parts of the initial exchanges of necessary to understand key agreement and the certificate based authentication. Then in section 3, we explain the format of ECC X.509v3 certificates concerned about public key and signature information. Section 4 describes the design and implementation details. After performance comparisons in Section 5, Section 6 concludes the paper.

## 2 Overview of IKEv2 Initial Exchanges

The initial exchanges of IKEv2 have two main purposes: one to establish an IKE security association (IKE_SA) between itself and a given peer and another to establish one or more child security association (CHILD_SA). IKE_SA is an encrypted and integrity protected tunnel between two IKEv2 protocol implementations, which is used for both signaling and for carrying out processes like mutual authentication and CHILD_SA management. CHILD_SA is used for transferring data between peers and it is hence the central part of the whole IPSec architecture.

The initial exchanges of IKEv2 as given in [3], Section 1.2, are depicted in Fig. 1. These initial exchanges normally consist of four messages. The first pair of messages IKE_SA_INIT negotiates cryptographic algorithms, exchanges nonces and does a Diffie-Hellman exchange that will protect all subsequent exchanges including authentication. First the IKE initiator sends its proposed encryption, authentication, integrity algorithms and pseudo-random functions (SAi1). It furthermore sends the keying material for the Diffie-Hellman exchange (KEi) and some random value (Ni). Based on the received request, the responder replies with selected algorithm (SAr1), its own keying material (KEr), a random value (Nr) and supported certification authorities.

After IKE_SA_INIT exchanged, both peers now have the shared secret that allows them to encrypt and integrity protect all further exchanges. The second pair of messages IKE_AUTH authenticates the previous messages, exchanges identities and certificates and establishes the first CHILD_SA. During this exchange, the peers reveal their identities (IDi and IDr), and initiator presents its list of supported certification authorities (CERTREQ), then certificates (CERT) and the authentication data (AUTH) are exchanged. The remaining payloads, which are exchanged in order to create the first CHILD_SA include: proposals (SAi2), selected algorithms (SAr2), proposed traffic selectors (TSi, TSr) and narrowed traffic selectors (TSi, TSr flowing from responder to initiator). IKEv2 behaves as specified in [3]. With these messages first IKE_SA is created and traffic between initiator and responder is protected.

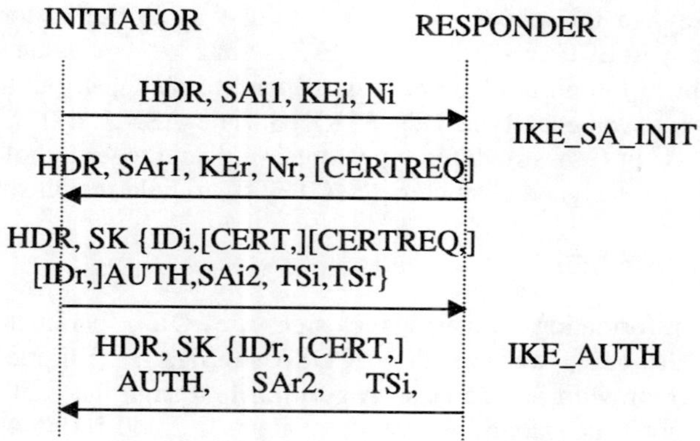


**Fig. 1.** Initial exchanges of IKEv2

# 3 ECC X.509v3 Certificates

Besides authentication based on pre-shared keys, RSA is the most often used authentication method in IKEv2. However, new concerns are rising about the security of 1024-bit RSA [9]. In this paper, we introduce the authentication of IKEv2 based on ECC Certificates. The ECC X.509v3 certificate format we discuss here is in [10]. The main difference between ECC and RSA certificates are subject public key information and Certification Authority (CA) signature information.

## 3.1 Subject Public Key Info.

Subject public key information contains the algorithm object identifier (OID), parameters and subject public key value. The OID of ECC public key is "1.2.840.10045.2.1" [11]. The ECC parameters object explicitly specifies the base field (incl. representation), the curve, the basis point, the generator, and some additional optional parameters (seed used for random generation of the curve, co-factor, etc.). The ASN.1 syntax for parameters in X.509 is [11]:

```
ECParameters ::= SEQUENCE {
          version    INTEGER { ecpVer1(1) (ecpVer1),
          fieldID    FieldID {{ FieldTypes }},
          curve      Curve,
          base       ECPoint,
          order      INTEGER,
          cofactor   INTEGER OPTIONAL,
      . . . . . . }
```

The parameters can be defined by three ways: (1) explicit specification, (2) choose an OID for a specific parameter set, and (3) inherited from the parameters associated with the CA key. In this paper, we use the third way to define the parameters of subject public key in ECC certificates. The ECC public key is a value of type ECPoint which is a point in the elliptic curve. It is mapped to a subject public key in X.509 format which is a value of type BIT STRING. The ASN.1 BIT STRING type is converted into a C or C++ structured type containing an integer to hold the number of bits and an array of unsigned characters ("OCTETs") to hold the bit string contents.

### 3.2 CA Signature Info.

CA signature information contains the signature OID, parameters and digital signature. In our case, we use the ECDSAwithSHA-1 (Elliptic Curve Digital Signature Algorithm with Secure Hash Algorithm 1) to sign the X.509v3 certificates. The signature shall be identified by the value "1.2.840.10045.4.1" and ASN.1 encoded using the following syntax [11]:

```
ECDSA-Sig-Value ::= SEQUENCE {
                r INTEGER,
                s INTEGER }
```

The parameters field here is inherited by CA key, because ECDSA parameters for signature verification must be obtained from subject public key information field of the issuer's certificate. The entire encoding of a value of ASN.1 type ECDSA-Sig-Value shall be the value of BIT STRING.

## 4 Design Details

In this section, we explain the general procedure to negotiate the shared key material by ECDH and authenticate by using ECC certificate in IKEv2 initial exchanges protocol. Then we implement these functions based on the open source Racoon2 [12]. After this exchange, IKE_SA and CHILD_SA are generated through four messages and an IPSec-protected tunnel is established.

There are two ways to trigger IKEv2 negotiation. One is requesting or updating SA from the kernel, another is requesting for negotiation from the other side of IKEv2 daemon thread. They are respectively corresponding to message 1 and 2 of IKE_SA_INIT interaction. SA payload states the cryptographic algorithms that initiator supports for the IKE_SA, including encryption algorithms, pseudo-random function (prf), integrity algorithms and Diffie-Hellman Group. KEi payload sends initiator's Diffie-Hellman value and Ni is initiator's nonce. Responder chooses a cryptographic suite form initiator's offered choices and expresses that choice in KEr payload, then sends its nonce in Nr payload. The optional CERTREQ field in the second message from responder to initiator is not used here.

The Diffie-Hellman exchange in the first pair of messages is used for two peers negotiating the key material which computes the shared secret key. This shared key is used to encrypt and decrypt the following data transformation. In this paper, we choose ECDH as the key agreement protocol [13]. Two peers negotiate an elliptic curve recommended in SEC2 (Standards for Efficient Cryptography) [14] and compute each public key $Qi=Ki\times G$ and $Qr=Kr\times G$ (Q is the peer's ECC public key, K is the corresponding private key, G is the base point). Then Qi is sent to responder as D-H value in KEi payload, while Qr is sent to initiator. Now, according to the ECC encryption algorithm, two peers generate the key material K=Qi×Kr=Qr ×Ki. The process of ECDH key exchange is finish by now. Then the shared keys are computed from the nonces in Ni and Nr, and the key material in g^ir. The shared key SKEYSEED is computed as: SKEYSEED=prf(Ni | Nr, g^ir). All messages following the initial exchanges are cryptographically protected using this shared secret key.

Messages in second phase are used for the authentication of the previous exchanged messages. In these encrypted messages initiator and responder introduce to each other by sending ECC certificates. In message 3 the initiator asserts its identity with IDi payload. Related information such as valid distinguished name, email, X.500 name can be used as identification. In this paper, we choose the subject unique ID in ID payload. ECC X.509 v3 certificate sent in CERT payload proves knowledge of the secret corresponding to IDi and integrity protects the contents of the first message using AUTH payload. The subject public key in this certificate is based on ECDSA, that is, the subject public key is specially used for computing digital signature. ECDSA signature value of the first message sent in AUTH payload is computed by the subject public key in ECC certificate. If initiator does not have a certificate of responder, a certificate request will be sent in CERTREQ payload.

In message 4, after receiving the ECDSA information from initiator, responder gets the subject public key Qi from initiator's ECC certificate to verify digital signature in AUTH field. If it is validated, initiator does have the private key Ki corresponding to the public key Qi, and its identity is authenticated. Then responder completes negotiation of a CHILD_SA with the additional fields. After that, it creates IDr payload by the subject information in its own ECC certificate and uses the corresponding private key Kr to compute ECDSA of message 2 sent in AUTH payload. If it receives the CERTREQ payload from message 3, it will pass its own ECC certificate in CERT payload to initiator for identity authentication and CHILD_SA negotiation. These messages are encrypted and integrity verified by the shared secret key generated through message 1 and 2. So the transmission is under security protection.

In order to use ECC certificates for authentication in IPSec VPN system, we design the interface and logic processing after we learned on the open-source Racoon2 [12].

- vchar_t * get_ecc_cert(const char * path)

This function writes the ECC certificate in PEM format [15] using vchar_t structure defined in Racoon2. The argument specifies the path of local ECC certificate. Vchar_t is a structure consists of a buffer value and its length. Under X386 system, the type of size_t is unsigned integer and caddr_t is unsigned character.

```
typedef struct vchar_tag {
size_t l;        /*length of the value*/
caddr_t v;  /*place holder to the pointer to the value*/
} vchar_t;
```

- int check_ecc_cert(vchar_t *ecc_cert, char *CApath, struct timeval *due_time)

This function attempts to validate an ECC certificate based on the root certificate of CA. The argument ecc_cert is the ECC certificate to be verified in vchar_t structure, while CApath specifies the path where the root certificate stored. The argument due_time gives current time for determining whether the certificate is expired. We get the subject public key information from the root certificate including CA's public key and corresponding OID of elliptic curve. Then we use ECDSA algorithm to verify the signature in ECC certificate by CA's public key Qca to determine if it is issued by CA.

- vchar_t *get_X509_ecc_pubkey(const char *ecc_path)

This function writes the ECC public key which is encoded using a SubjectPublicKeyInfo structure [16] in vchar_t structure from ECC certificate. Although the SubjectPublicKeyInfo object for ECC keys is an ASN.1 object, the value of the BIT STRING subjectPublicKey is the value of the ECC public key encoded as an octet string (sequence of bytes). So we write this function based on i2o_ECPublicKey from OpenSSL [16] which is used to change ECC public key from internal to octet string. The argument specifies the path for storing the certificate.

- vchar_t *get_ecc_privkey(const char *prikeypath)

This function writes the PKCS#8 (Private-Key Information Syntax Standard) [17] format private key in vchar_t structure. The argument gives the path of local private key.

- vchar_t *ecdsa_sign(const char *hash_type, vchar_t *octets, vchar_t*privkey)

This function computes the ECDSA signature of the message encoded as an octet string by the given hash function algorithm and supplied private key, and returns the created signature value in vchar_t structure. We implement it based on function ecdsa_do_sign in OpenSSL.

- int ecdsa_verify(const char *hash_type, vchar_t *octets, vchar_t* sig, vchar_t *pubkey)

This function, based on function ecdsa_do_verify in OpenSSL, verifies the signature "sig" is a valid ECDSA signature of the value "octets" by using the public key "pubkey".

## 5 Performance Advantages Comparisons

Currently, RSA certificates are the most widely used in the initial exchange of IKEv2. Under the same circumstances, we calculate the time taken to authenticate in IKEv2 using ECC and RSA certificates at equivalent security. The calculation begins with message 1 sent, ends with message 4 received by initiator including connecting with the kernel. Table 1 presents the testing result under two different equivalent security levels. The numbers show a clear performance advantage of ECC over RSA. We notice that it is more efficient to use ECC certificate for authentication in IKEv2. With the security level raises, the performance improves significantly. Take ECC-256 certificate for example, it offers a speed up to 18% over RSA-3072. These numbers not only show a significant performance advantage for ECC, they further illustrate that the ratio of the computation times for RSA and ECC is going to dramatically increase as higher security is needed.

Besides, we calculate the time taken to create the first CHILD_SA using two different D-H groups at approximately equivalent security. One is MODP1024 defined in traditional D-H group and another is EC2N163 defined in [13]. In this test, we choose elliptic curve sect163k1 defined in SEC2 [14] for negotiation, AES128-CBC for encryption, HMAC-SHA1 for integrity and ECC-233-bit certificate for authentication. Table 2 gives the comparison result. It shows the average time of creating the first CHILD_SA is 633.279ms by using EC2N163, decreasing almost 10% than using MODP1024.

**Table 1.** ECC and RSA based certificates performance

| Certificates Type | | ECC-233 | RSA-2048 | ECC-256 | RSA-3072 |
|---|---|---|---|---|---|
| parameter | Encryption Algorithm | AES128-CBC | | | |
| | Integrity Algorithm | HMAC-SHA1 | | | |
| | DH Group | MODP1024 | | | |
| Test 1 (ms) | | 695.283 | 720.429 | 700.029 | 859.944 |
| Test 2 (ms) | | 697.873 | 723.076 | 700.694 | 855.890 |
| Test 3 (ms) | | 695.225 | 728.206 | 703.896 | 859.824 |
| Test 4 (ms) | | 699.454 | 726.054 | 703.658 | 852.375 |
| Test 5 (ms) | | 695.915 | 723.435 | 705.974 | 852.368 |
| Average time to generate CHILD_SA (ms) | | 696.750 | 724.240 | 702.850 | 856.080 |

**Table 2.** The negotiation efficency between modp1024 and ec2n163 in ikev2

| DH Group | MODP1024 | EC2N163 |
|---|---|---|
| Test 1 (ms) | 695.283 | 634.133 |
| Test 2 (ms) | 697.873 | 634.703 |
| Test 3 (ms) | 695.225 | 631.621 |
| Test 4 (ms) | 699.454 | 631.192 |
| Test 5 (ms) | 695.915 | 634.746 |
| Average time to generate CHILD_SA (ms) | 696.750 | 633.279 |

## 6 Conclusion

In this paper, we propose, design and implement an efficient approach applied to the initial exchanges of IKEv2. We explain the procedure to compute the key material for shared key by ECDH and interface functions used to decode, verify the ECC certificates. With the use of this type of application, we aim to help to provide more security and stability than original system based on RSA. The testing results show that high level strength of security is achieved as well as the performance of the whole system is improved by using ECC certificates for identity authentication and ECDH for key negotiation in the initial exchanges of IKEv2.

**Acknowledgment.** This study was supported by the Natural Science Foundation of Jiangsu Province under grants No.BK2004039 and Youth Foundation of Changzhou University under grants No.JQ200805.

## References

1. Kent, S.: IP Encapsulating Security Payload (ESP), RFC4303 (December 2005)
2. Kent, S.: IP Authentication Header, RFC4302 (December 2005)
3. Kaufman, C. (ed.): Internet Key Exchange (IKEv2) Protocol, RFC4306 (December 2005)
4. Perlman, R., Kaufman, C.: Key Exchange in IPSec: Analysis of IKE. IEEE Internet Computing 4(6), 50–56 (2000)
5. Bollapragada, V., Khalid, M., Wainner, S.: IPSec VPN Design, pp. 118–120. Cisco, Indianapolis (2005)
6. Kukec, A., Groš, S., Glavinić, V.: Implementation of certificate based authentication in IKEv2 protocol. In: Luzar-Stiffler, V., Dobric, V.H. (eds.) Proceedings of the ITI 2007, 29th International Conference on Information Technology Interfaces, Cavtat, Croatia, pp. 697–702 (June 2007)
7. McDonald, D.L.: Internet Key Exchange Version 2 (IKEv2) Design Document. Solaris Security (December 2009)
8. Eberle, H., et al.: A Public-Key Cryptographic Processor for RSA and ECC. In: Proc. IEEE 15th Int'l Conf. Application-Specific Systems, Architectures and Processors (ASAP 2004), pp. 98–110. IEEE Press (July 2004)
9. Vanstone, S.: Next generation security for wireless: elliptic curve cryptography. Computers & Security 22(5), 412–415 (2003)
10. Certicom Research: X.509 WG Working Document: ECC in X.509Woking Group, Standards for efficient cryptography, Released Working Group Draft Version 0.2 (August 1999)
11. ANSI X9.62-2005. Public Key Cryptography for the Financial Services Industry: The Elliptic curve Digital Signature Algorithm (ECDSA) (November 2005)
12. Racoon2 Project, `http://www.racoon2.wide.ad.jp/w/`
13. Brown, D.: Certicom, "Additional ECC Groups For IKE and IKEv2", Internet draft (January 2006)
14. Certicom Research: SEC 2: Recommended Elliptic Curve Domain Parameters, Standards for Efficient Cryptography, Version 1.0 (September 2000)
15. Housley, R.: Internet X.509 Public Key Infrastructure Certificate and Certificate Revocation List (CRL) Profile, RFC 3280 (April 2002)
16. OpenSSL library, `http://www.openssl.org`
17. Kaliski, B.: Public-Key Cryptography Standards (PKCS) #8:Private-Key Information Syntax Specification Version 1.2, RFC5208 (May 2008)

# Research on Lane Detection Based on Machine Vision

Xining Yang, Dezhi Gao, Jianmin Duan, and Lei Yang

Intelligent Measure & Control Laboratory
Beijing University of Technology
Beijing, China
yangxining2003@163.com

**Abstract.** In order to enhance the real-time and stability of lane detection based on machine vision, a method for line detection based on combined road model is proposed. After image classification according to illumination, the image is processed by different algorithms during image pretreatment. For straight line and curve line in the vicinity, improved Hough Transform (HT) is adopted for line detection and tracking. Both of the road boundaries are fitted using Catmull-Rom Splines based on control points search algorithm for curve line in the distance. For various kinds of lanes on most structural road, experiment results indicate that the method has good robustness and stability.

**Keywords:** Lane detection, Hough transform, Image classify, Splines.

## 1 Introduction

It is an important social and economic issue to reduce traffic accidents and ensure the life to be safe. Intelligent vehicle assistant systems (IVAS) are being investigated and developed to fulfill such demand. At present, many researchers and vehicle makers are developing a driving assistance system and improving the driving safety. Road boundary detection is one of the key components of the IVAS. Many researchers focus on this area now [1].

In current literature, the methods of road detection mainly use vision or radar, which are used for structured or unstructured roads respectively. In this paper, we take the structured road with obvious lane line (such as highway) as the object of study.

The challenges of road lane detection include the ability to deal with various road types and environment condition (such as shadows, variable illumination) and to achieve real-time requirement at the same time. So, it is difficult for single algorithm to satisfy different road type. In this paper, the lane detection process is divided into several parts. First, the lane illumination classifier proposed in this paper can adaptively classify an image into a corresponding sort. Then, the system can assign the method that is suitable for the current illumination condition to perform appropriate pre-processing. We present the classification preprocessing mechanism in Section 2. In section 3, an improved method based Hough Transition (HT) was presented to detect straight lane, and curve lines in the distance are generated using Catmull-Rom Splines based on control points search algorithm. The experimental results are showed in section 4. Finally, we make conclusions in Section 5.

## 2 Image Pre-processing

### 2.1 Image Classify

The overall gray-scale mean value of image is a kind of global statistical feature, which has a strong ability to filter local noise [2]. So, the image gray-scale mean value is taken as mathematical criterion to describe the actual illuminating condition. Set image grey level is within the range of [1, $L$ ], and the image gray-scale mean value $\mu_T$ is defined as:

$$\mu_T = \sum_{i=1}^{L} ip(i), \ p(i) = \frac{n_i}{N}, \ N = \sum_{i=1}^{L} n_i \tag{1}$$

Where $n_i$ is the number of pixels on gray level $i$, $N$ is the total number of image pixels, $p(i)$ is the probability of gray level $i$.

A large number of road images in different illuminations condition were collected respectively. Experimental results show that the light intensity of environment can be well reflected by overall mean gray-scale of road images.

### 2.2 Image Preprocessing on Various Illumination Conditions

The road images under weak illumination has the characteristics of less gray-scale mean value and smaller contrast between objective and background, which causes lane line not clear enough. So that, before lane line recognition, we must do a series of pre-processing to the night illumination image in order to enhance the contrast of pixels on lane line, such as median filter, gray stretch and so on. And for strong illumination images, pixels grey level of lane marking and pavement is so nearly that the lane marking is difficult to be detected from the road image. Therefore an algorithm of histogram equalization had been adopted to enlarge the dynamic range of grey scale of the strong illumination image. From the pre-processing results shown as Fig.1 and Fig.2, we can see that the contrast between lane marking and pavement had been enhanced, which was beneficial to binary processing.

The Sobel edge detector which is easy to calculate and has good ability to suppress noise uses a simple convolution kernel to calculate the gradient of the image intensity at each point. It gives the direction of the largest possible increase from light to dark and the rate of change in that direction. Canny edge detector was designed to be an optimal edge detector, which includes four stages of image smoothing, gradient calculation, non maximum suppression and thresholding [3]. The algorithm can extract lane marking with single-pixel wide, but it is quite complicated. After image Classification, we only used canny edge detector to process strong illumination images, and use the sobel algorithm for other illumination images.

We make use of sobel and canny respectively to process the images in the situations of night illumination of Fig. 1(b) and strong illumination of Fig. 2(b), the processing result was showed in Fig.3.

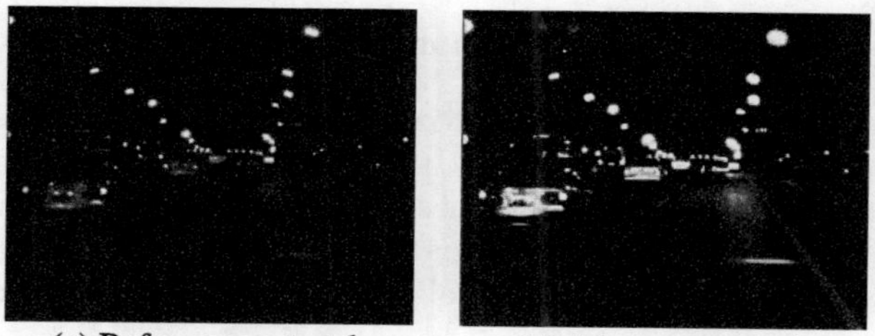

(a) Before grey stretch (b) After grey stretch

**Fig. 1.** Result of grey stretch for night image

(a) Before grey equalization (b) After grey equalization

**Fig. 2.** Result of grey equalization for sun image

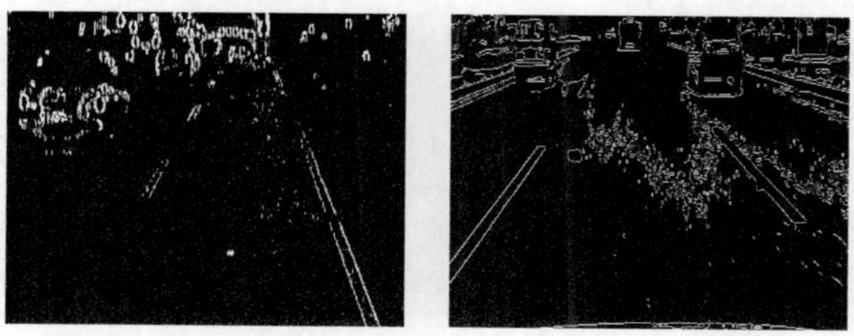

**Fig. 3.** Result after edge enhancement and two-value image

## 3 Lane Detection

We know that the high way is designed according to national standards. Generally speaking, the situations of the road mainly include four fundamental structure models, i.e. straight ahead road, left or right bound road, from straight ahead road to bound road and from bound road to straight ahead road. So, we used a method of improved HT to detect straight lane, and for branch roads we utilize the Catmull-Rom splines to fit the curve road in the distance.

### 3.1 Lane Line Detection Based on Improved HT

Lane line detection based on HT is a method used widely. Its basic idea is the point - line duality, that is, the collinear points of image space is corresponding to intersection of lines in parameter space. In turn, all lines intersected at the same point in parameter space have collinear points in the image space [4]. In order to solve the infinite slope of the vertical line, HT often carries on the following polar equation of the straight line.

$$\rho = x\cos\theta + y\sin\theta \tag{2}$$

$\rho$ presents the distance from line in image space to coordinate origin. $\theta$ is about the angle between x axis and line, $\rho \in (0, r)$, $\theta \in (0, 180^0)$ where $r$ is the diagonal length of the image.

Typically, a large number of target points are non-lane edge points (trees, sky, road signs and other interference points) in the binary image. In order to minimize these disturbance points and reduce the computational complexity of HT. HT has been improved through adjusting the value of $\rho, \theta$ dynamically in the paper.

*1) Create the ROI with polar angle constraints*

Generally, the left and right lane lines are distributed at the left and right side in the image. If the polar angles of left and right lane lines are $\theta_l$ and $\theta_r$ respectively, and the polar diameters are $\rho_l$ and $\rho_r$. According to a large number of calculations and analysis of left and right lane line images, we obtained the two lines' constraint domains of HT are showing as following which is called polar angle constraint region in the paper showed as Fig.4.

$$\theta_{l1} < \theta_l < \theta_{l2},\ \rho_{l1} < \rho_l < \rho_{l2}$$
$$\theta_{r1} < \theta_r < \theta_{r2},\ \rho_{r1} < \rho_r < \rho_{r2}$$

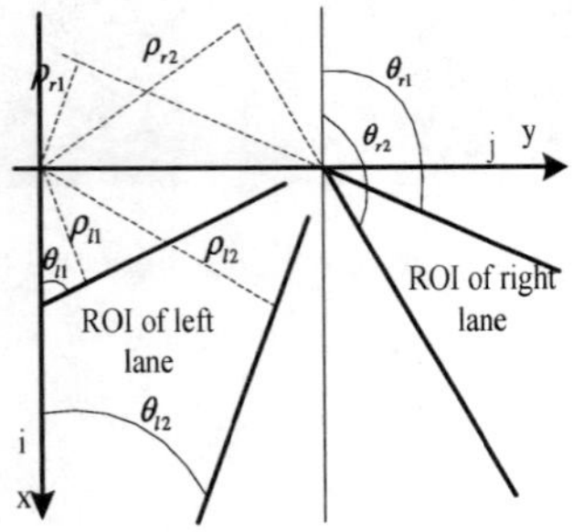


**Fig. 4.** ROI of polar angle constraints

In Fig.4, coordinate origin is defined at the top left corner of the image; x axis and y axis to the increase direction of the row and column in image. A large number of interference points are removed effectively through the establishment of the polar angle

(a) (b)

**Fig. 5.** ROI of polar angle constraints in binary image

constraints region. Fig. 5 (a) and (b) are the binary images of the road. The white lines present the constraint region, which shows that the interference beside lanes is filtered out.

*2) Lane track based on dynamic ROI-HT*

During the real-time process of image sequence, the position deviation of vehicle in adjacent images will be quite little due to the camera acquisition speed (such as at the speed of 30 frames/s), so the location of lane edge points has certain predictability. From the analysis of many road image frames, we find that the changes of lane line slope in the two continuous images are in the range of $\pm 3^\circ$, and the intercept are about $\pm 10$ pixels [5]. In this paper, the region of interest (ROI) is established based on the lane line parameters of last frame image dynamically. Dynamic ROI of both side lanes are showing as,

$$
\begin{aligned}
(kl_{up}, bl_{up}, kl_{down}, bl_{down}) &= (kl_0 + \Delta k, bl_0 + \Delta b, kl_0 - \Delta k, bl_0 - \Delta b) \\
(kr_{up}, br_{up}, kr_{down}, br_{down}) &= (kr_0 + \Delta k, br_0 + \Delta b, kr_0 - \Delta k, br_0 - \Delta b)
\end{aligned} \tag{3}
$$

Where $k_{up}, b_{up}$ and $k_{down}, b_{down}$ intercepts of ROI's upper and lower boundaries; $k_0, b_0$ is the slope and intercept of the last image; $\Delta k, \Delta b$ present the change of lane parameters in the two images.

The first step of lane detection and tracking was to transform parameters $(k, b)$ of ROI into $(\rho, \theta)$ parameters, and then, HT had been carried on in the dynamic ROI, which reduced the interference of the non-lane line edge points further and improved real-time ability of algorithm. During the experiment, we found that the detected lane lines often fluctuate between the left and right edges of the lane marking, because the lane marking have a certain width and the traditional HT considers the voting peak as the target line parameter. However, the HT based on dynamic ROI can reduce the fluctuation greatly. It also increases the stability of line detection, which makes the foundation for following work of the vehicle location.

### 3.2 Lane Detect Using Catmull-Rom Splines

*1) The Characteristic of Catmull-Rom splines*

Splines are smooth piecewise polynomial functions, which is proposed to describe curve or surface at first[6], and now they are widely used in pattern recognition, computer graphics process and image process and so on.

We make use of Catmull-Rom splines to represent curve lane lines because of because of its following characteristics:

- Catmull-Rom splines' control points are actually on the curve and it has local control, which means that modifying one control point only affects the part of the curve near that control point. These characteristics make it fit the curve fast.
- Compared with hyperbola and ellipse, Catmull-Rom splines can fit a curve of arbitrary shape, which is more appropriate for fitting a lane marking.

*2) The mathematical model of Catmull-Rom splines*

Given a series of points $P_0, P_1, ..., P_n$ on a plane, the Catmull-Rom spline is able to interpolate (pass through) the points from $P_1, P_2 ..., P_{n-1}$ only utilizing the tangent vector of $P_i$. In addition, the tangent vector at $P_i$ is parallel to the line connecting $P_{i-1}$ and $P_{i+1}$.

The formula is Catmull-Rom spline for one segment is presented as equation (4) [7]. Where $P_i$ is the control point which requires the Catmull-Rom spline to pass through, and $t \in [0,1]$.

$$P(t) = \frac{1}{2}[t^3 \ t^2 \ t \ 1] \begin{pmatrix} -1 & 3 & -3 & 1 \\ 2 & -5 & 4 & -1 \\ -1 & 0 & 1 & 0 \\ 0 & 2 & 0 & 0 \end{pmatrix} \begin{pmatrix} P_{i-1} \\ P_i \\ P_{i+1} \\ P_{i+2} \end{pmatrix} \tag{4}$$

*3) Curve lane fitting*

The key point of fitting curve lane is to find the control points that are needed to generate Catmull-Rom splines. We set the first and the last two control points equally for each spline since the Catmull-Rom spline interpolates all but the first and the last control points. Boundary is actually determined by four control points for left or right bound roads and three control points for other curve road, such as from straight ahead road to bound road, from bound road to straight ahead road.

For a curve lane load, a linear model is used to approximate the lane boundaries in near field, and each lane line denoted as $l_1, l_2$ respectively can be detected by HT, while in far field the lane boundaries are fitted with Catmull-Rom splines. The first control point is the intersection point of vertical edge of road image and $l_1$, $l_2$ respectively, which is denoted as $P_{l0}, P_{r0}$ showed in Fig. 6.

**Fig. 6.** Search of Splines control points

The intersection point of left and right lane is located on the horizon according to perspective projection principle. Here, we have $l_3$ instead of the horizon lane, where $l_3$ parallel to a lane which pass through the intersection point of $l_1$ and $l_2$. So, the last (farthest) control point is set to the end point of the farthest line segment, which is the intersection point of $l_3$ and the border of road image and denoted by $P_{ln}$, $P_{rn}$.

The second control point is locate between $P_{l0}$ and $P_{ln}$. In the paper, coordinate origin is defined at the top left corner of the image. $l_1, l_2$ can anastomose the curve lane marking very well in near field. But the deviation kept growing with the increase of distance. When the deviation is in excess of a threshold value, such as T, we set the current line number as $i$. The lane marking point on the line of $i$ will be regarded as the second control point.

Taking the right bound road image for example to show the searching of other control points. The steps are showing as following. Other control points on left lane marking all locate at the right side of $l_1$ and the right side of $l_2$ for right lane line.

The searching steps of control may be affected by some factors, such as discontinuous lane marking, disturbance point nearby lane and so on.

In order to reduce influence on the searching control points caused by the problem mentioned above, the following searching strategies are introduced:

1) When searching the i$^{th}$ row, take down the column m where the edge is if they is the first point. If there are no edge points in column of m+1, m+2,…, and m+j-1 except the m+j column, the point (i, j/2) is considered as a control point when j belongs to $h < j < H$ (where H is maximum width of the lane in near part, h is the minimum lane width). If not, stop search and turn to the next row. For the other rows, the points are searched from j/2 to left side and right side, which improved the searching efficiency and accuracy. Fig.7 shows the detail position information for control points.

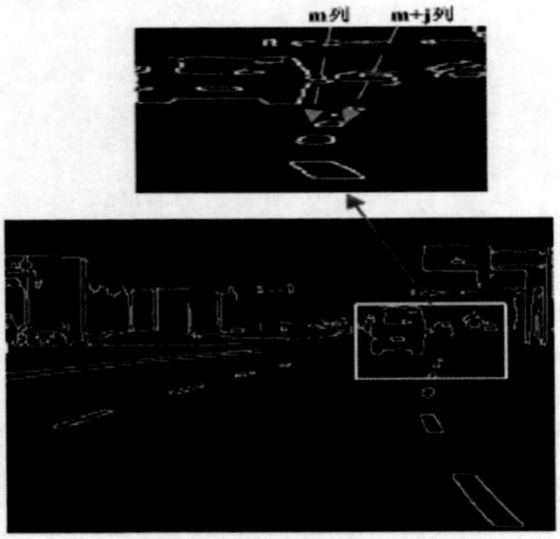


**Fig. 7.** Search of Splines control points for curved lines

2) According to the method above, other control points can be obtained with increasing the number of image's rows. If the column number of a possible control points is not monotonic, it is regarded as a sudden change interference point, which should be discarded.

After finishing searching control point, Spline function can be resolved, which is used to fit curve lane line.

## 4 Experimental Results

Algorithm is realized based on VC++ 6.0 and the OpenCV image processing library with PC configuration of CPU P4 2.0G and the memory 512M. The size of image is 640×480. Detection results of straight line based on improved HT are shown in Fig.8.

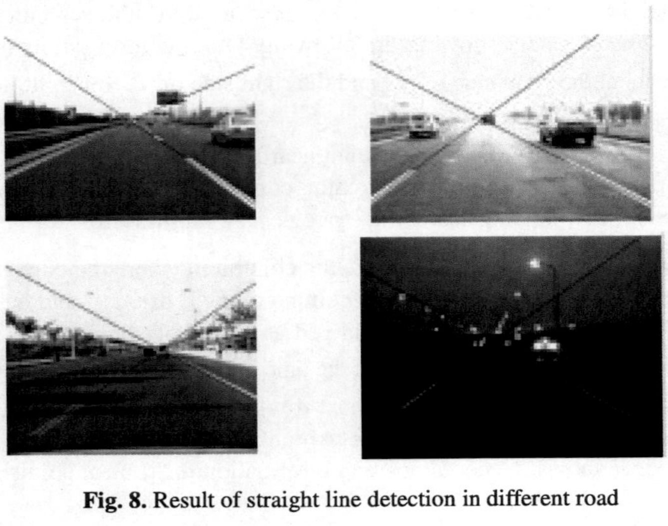

**Fig. 8.** Result of straight line detection in different road

Fig. 9. Result of curved lines detection

After detecting the near part of line with the improved HT and obtaining the control points, curve fitting results using the Spline function are shown in Fig. 9. Time spend on line detecting is about 4.5ms, and on curve fitting is about 7.5ms.

## 5 Conclusion

A method of lane detection based on combination model has been proposed in the paper. Improved HT is introduced to detect the straight line and near part of the curve, and Catmull-Rom Spline function is used to fit the far part of the curve. Experiments results show that the method has a good effect. The shortcoming of the method is that the abstraction of the control points for Catmull - Rom Spline could be improved.

## References

1. Lin, H., Ko, S., Kim, H., Kim, Y.: Road oundary detection with double filtering for intelligent vehicle. In: IEEE International Conference on Robotics and Biomimetics, ROBIO 2007, pp. 686–690 (2007)
2. Tianlei: Study on the Mark Recognition Algorithm Under Variable Illumination for CyberCar System. Jilin University (2005)
3. Hassan, M., Khalid, N.E.A., Ibrahim, A., Noor, N.M.: Evaluation of Sobel, Canny, Shen & Castan Using Sample Line Histogram Method. In: International Symposium on Information Technology, ITSim 2008, vol. 3, pp. 1–7 (2008)
4. Hough, P.V.C.: A method and means for recognizing complex patterns. U.S. Patent 3069654 (1962)
5. Zheng, B., Tian, B., Duan, J., Gao, D.: Automatic Detection Technique of Preceding Lane and Vehicle. In: 2008 IEEE International Conference on Automation and Logistics (ICAL 2008), Qingdao, China, September 1-3, pp. 1370–1375 (2008)
6. Guo, C., Mita, S., McAllester, D.: Stereovision-based road boundary detection for intelligent vehicles in challenging scenarios. In: The 2009 IEEE/RSJ International Conference on Intelligent Robots and Systems, October 11-15, pp. 1723–1728 (2009)
7. Foley, Van Dam, Feiner, Hughes: Computer Graphics: Principles and Practice, 2nd edn. Addision-Wesley Publishing Company (1990)

# Chaotic Dynamics and Chaos Control of Cournot Model with Heterogenous Players

Rong Hu and Qi Chen

School of Economics and Management
Nanjing University of Aeronautics and Astronautics
Nanjing, P.R. China

**Abstract.** We analyze a Cournot duopoly game, where players have heterogeneous expectations and nonlinear cost function. Two types of players are considered: bounded rational and naive expectation. By using the theory of bifurcations of dynamical systems, the existence and stability for the equilibrium points of this system are obtained. Numerical simulations used to show bifurcations diagrams for various parameters and sensitive dependence on initial conditions. We observe that an increase of the players' speed of adjustment may change the stability of Nash equilibrium point and cause bifurcation and chaos to occur, and the system can return to stable state from chaotic state by delayed feedback chaos control method while suitable controlling factor is chosen. The analysis and results in this paper are interesting in mathematics and economics.

**Keywords:** discrete dynamical system, Cournot model, chaos control, delayed feedback control, heterogenous expectations.

## 1 Introduction

Duopoly refers to a market situation in which the actions of several firms affect supply and price of same or homogeneous products. It is this characteristic of interdependence that takes oligopolists consider reaction of the other competitors. Under perfect competition all firms are output-takers or price-takers in oligopolistic competition. In the previous economics literatures, many models have been proposed to simulate behaviors of firms in duopoly markets. Cournot, in 1838 [1], introduced a famous formal theory of oligopoly. In Cournot model, each oligopolist assumes that other firms hold their outputs constant, and all oligopolist select a quantity to produce by other's outputs for maximizing profits in the model.

Expectations play an important role in modeling economic phenomena. A firm can choose his expectations rules of many available techniques to adjust his production outputs. There exist three different firms' expectations: naive, bounded rational and adaptive[2, 3]. In this paper we study the dynamics of Cournot model where each firm behaves with different expectations strategies, so we are going to apply this kind of beliefs to a duopoly game. This kind of beliefs is common in real world problems such as economic, biology and social sciences problems. Recently, a number of works of dynamic Cournot game model with different expectations have been done [4-7]. The conclusions in these articles have shown that the Cournot model may lead to complex behaviors such as cycle and chaotic. All these literatures had a hypothesis that firms had linear cost function. And there are no papers to discuss how to control the chaos in Cournot model while chaos control is an important work when firms' behaviors become chaotic.

The main aim of this work is to investigate the chaotic dynamics behaviors of a Cournot duopoly model representing two firms with heterogeneous expectations and nonlinear cost function, and to control the chaos by control method. The general scheme of this paper is as follows. In Section 2 we determine model of dynamical system of a Counot duopoly game with bounded rationality and naïve player. In Section 3 we study the model of duopoly game. Explicit parametric conditions of the existence and local stability of equilibrium points will be given. In Section 4 we simulate complex dynamic of this system via changing system parameters of the model. In Section 5 we use delayed feedback control method to control the chaos of Cournot duopoly game. Section 6 closes the paper with conclusions finally.

## 2 Model

We consider that two firms which choose quantities for their products in an oligopoly market. Let $q_i(t)$, $i=1,2$ represent the quantity of $i$th firm during period $t = 0,1,2\ldots$.The price each firms chooses $p_i$, a linear inverse demand function, is determined by the following equation[8]:

$$p_i = a - b(q_i + q_j) \quad i, j = 1,2. \quad i \neq j \tag{1}$$

where $a > 0$, $b > 0$. The cost function has the form

$$\begin{aligned} C_1 &= c_1 q_1^2 \\ C_2 &= c_2 q_2 \end{aligned} \tag{2}$$

where $c_i$, the positive parameters, are marginal costs of the firm $i$, $i = 1,2$, respectively, and $a > c_i$.

With above assumptions the profit of the $i$th firm in the single period is give by

$$\Pi_i(q_1, q_2) = p_i q_i - C_i \qquad i = 1,2. \tag{3}$$

Hence, $i$th firm's quantity for period $(t + 1)$ is decided by solving the optimization problem

$$\begin{aligned} q_1(t+1) &= \arg\max \Pi_1(q_1(t), q_2^*(t+1)) \\ q_2(t+1) &= \arg\max \Pi_2(q_1^*(t+1), q_2(t)) \end{aligned} \tag{4}$$

where $q_i^*(t+1)$ represents the expectation of $i$th firm about the production decision of $j$th firm during period $(t + 1)$ $(i = 1,2, i \neq j)$.

The marginal profit of firms in period $t$ is

$$\frac{\partial \Pi_1(q_1,q_2)}{\partial q_1} = a - 2(b + c_1)q_1 - bq_2$$
$$\frac{\partial \Pi_2(q_1,q_2)}{\partial q_2} = a - c_2 - bq_1 - 2bq_2 \tag{5}$$

This optimization problem has unique solution:

$$q_1 = \frac{a - bq_2}{2(b + c_1)}$$
$$q_2 = \frac{a - c_2 - bq_1}{2b} \tag{6}$$

If the two firms are naive players, then the duopoly game is described by using (6). While information in the market is incomplete usually, players can use more complicated expectations such as bounded rationality. The bounded rational player has no complete knowledge of the market; hence they try to use local information based on the marginal profit $\partial \Pi_i / \partial p_i$. It decides to increase (decrease) its quantity if it has a positive (negative) marginal profit. This adjustment mechanism has been called myopic by Dixit [9]. Thus, the dynamic adjustment mechanism can be modeled as

$$q_i(t+1) = q_i(t) + \alpha_i q_i \frac{\partial \Pi_i(q_1,q_2)}{\partial q_i} \quad t = 0, 1, 2\ldots \tag{7}$$

where $\alpha_i$ is a positive parameter which represents the speed of adjustment of $i$th firm.

In this study, we take firm 1 to have bounded rational expectation while firm 2 represents a naive player. Therefore, the dynamic equation of the bounded rational firm 1 is obtained from inserting (5) in (7) which has the form

$$q_1(t+1) = q_1(t) + \alpha q_1(t)(a - 2(b + c_1)q_1(t) - bq_2(t)) \tag{8}$$

firm 2 uses first equation of (6) to updates his outcome according to the dynamic equation

$$q_2(t+1) = \frac{a - c_2 - bq_1(t)}{2b} \tag{9}$$

Hence, from (8) and (9), the dynamical Cournot duopoly game with heterogeneous players whose cost function is nonlinear can be established as follows

$$\begin{cases} q_1(t+1) = q_1(t) + \alpha q_1(t)(a - 2(b + c_1)q_1(t) - bq_2(t)) \\ q_2(t+1) = \dfrac{a - c_2 - bq_1(t)}{2b} \end{cases} \tag{10}$$

# 3 Equilibrium Points and Local Stability

Because Cournot model is an economic model, we define the equilibrium points of the dynamic duopoly game as a nonnegative fixed point of the system (10). The fixed points of the system (10) are obtained as nonnegative solution of the nonlinear algebraic system

$$\begin{cases} \alpha q_1(a-2(b+c_1)q_1-bq_2)=0 \\ \dfrac{a-c_2-bq_1}{2b}-q_2=0 \end{cases} \tag{11}$$

which is obtained by setting $q_i(t+1)=q_i(t)=q_i$, $i=1,2$ in system (10). We have two fixed points of system (10)

$E_0=(0,\dfrac{a-c_2}{2b})$, $E^*=(q_1^*,q_2^*)$, where

$$q_1^*=\frac{a+c_2}{3b+4c_1}, q_2^*=\frac{ab+2ac_1-2bc_2-2c_1c_2}{b(3b+4c_1)} \tag{12}$$

Obviously, $E_0$ is boundary equilibria. The fixed point $E^*$ is a Nash equilibrium and has economic meaning when $a>\dfrac{2bc_2+2c_1c_2}{b+2c_1}$. For studying the local stability of equilibrium, the eigenvalues of the Jacobian matrix of system (10) on the complex plane must be considered. The Jacobian matrix of system (10) at the point ($q_1$, $q_2$) has the form

$$J(q_1,q_2)=\begin{bmatrix} 1+\alpha[a-4(b+c_1)q_1-bq_2] & -\alpha bq_1 \\ -1/2 & 0 \end{bmatrix} \tag{13}$$

**Theorem 1.** *The boundary equilibria $E_0$ of system* (10) *is unstable equilibrium points.*

**Proof.** In order to prove this result, we consider the eigenvalues of Jacobian matrix $J$ at $E_0$ which takes the form

$$J(E_0)=\begin{bmatrix} 1+\alpha(a+c_2)/2 & 0 \\ -1/2 & 0 \end{bmatrix} \tag{14}$$

whose eigenvalues are $\lambda_1=1+\alpha(a+c_2)/2$, $\lambda_2=0$.

From the condition that $a$, $c_i$, $\alpha$ are positive parameters, we have that $|\lambda_1|>1$. Then the equilibrium point $E_0$ is unstable fixed point (saddle point) for the system (10) and this completes the proof. □

At $E^*$, the Jacobian matrix is

$$J(E^*) = \begin{bmatrix} 1+\alpha[a-4(b+c_1)q_1^* - bq_2^*] & -\alpha b q_1^* \\ -1/2 & 0 \end{bmatrix} \tag{15}$$

The characteristic equations of the $J(E^*)$ has the form $f(\lambda) = \lambda^2 - Tr(J)\lambda + Det(J) = 0$, where $Tr(J)$ is the trace and $Det(J)$ is the determinant of the Jacobian matrix which are given by

$$Tr(J) = 1+\alpha[a-4(b+c_1)q_1^* - bq_2^*]$$

$$Det(J) = -\alpha b q_1^* / 2$$

Where $q_1^*, q_2^*$ are given by (12).

Since $Tr^2(J) - 4Det(J) \geq 0$, so the eigenvalues of Nash equilibrium are real. The local stability of Nash equilibrium is given by using Jury's conditions [10]. which are

(1). $1 - Tr(J) + Det(J) > 0$,

(2). $1 + Tr(J) + Det(J) > 0$, and

(3). $Det(J) < 1$

Then the necessary and sufficient of conditions of the stability region is

$$\begin{cases} 2a - (7b+8c_1)q_1^* - 2bq_2^* < 0 \\ \alpha < \dfrac{4}{(9b+8c_1)q_1^* + 2bq_2^* - 2a} \end{cases} \tag{16}$$

Obviously, the Nash equilibrium point $E^*$ is stable in the region defined by (16) and loses its stability through the boundary of this region. Set of inequalities (16) defines a region of stability of Nash equilibrium point $E^*$. So we have Theorem 2 about local stability of Nash equilibrium point $E^*$.

**Theorem 2.** *The Nash equilibrium $E^*$ of system* (10) *is stable in the region defined by* (16).

## 4 Numerical Simulations

In this section, we will use the bifurcation diagrams to illustrate the above results and find new dynamics of system (10) as the parameters varying. The complicated dynamic features of Cournot duopoly game with heterogeneous players will be shown.

To provide some numerical evidence for the chaotic behavior of system (10), we present various numerical results here to show the complex dynamics, including its bifurcations diagrams, Laypunov exponents, sensitive dependence on initial conditions. In order to study the local stability properties of the equilibrium points, it is convenient to take the parameters values as follows: $a = 6$, $b = 0.5$, $c_1 = 1$ and $c_2 = 4$.

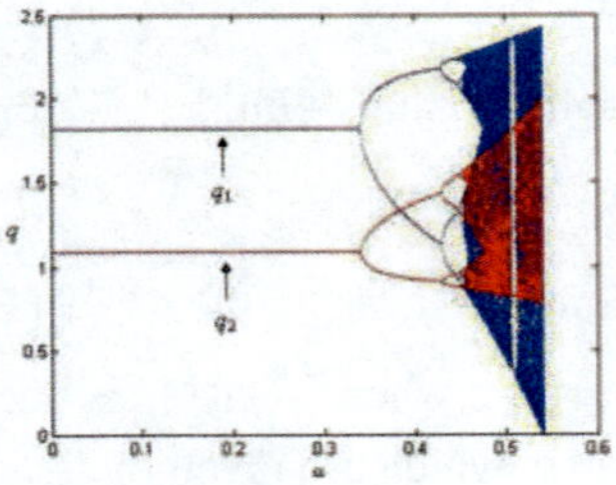

**Fig. 1.** Bifurcation diagram with respect to the parameter $\alpha$ speed of Adjustment of bounded rational player

Fig.1 shows the bifurcation diagram with respect to the parameter $\alpha$ (speed of adjustment of bounded rational firm 1), while the other parameters are fixed. From Fig.1, we can see that the orbit with initial values (2, 2) approaches to the stable fixed point $E^*(1.8182, 1.0909)$. If $\alpha$ increasing, the Nash equilibrium point $E^*$ becomes unstable and the bifurcation scenario is occurred. As $\alpha$ increases, infinitely many period-doubling bifurcation of quantity behavior become chaotic. It means for a large values of speed of adjustment of bounded rational firm 1, the system converge always to complex dynamics.

In order to analyze the parameter sets for which aperiodic behavior occurs, we can compute the maximal Lyapunov exponent depend on $\alpha$. For example, if the maximal Lyapunov exponent is positive, we can deduce that the phenomenon of chaos appears. Moreover, by comparing the standard bifurcation diagram in $\alpha$, we obtain a better understanding of the particular properties of the system. From the maximal Lyapunov exponent diagram it is easy to determine the parameter sets for which the system converges to cycles, aperiodic, chaotic behavior. The maximal Lyapunov exponent which corresponds to Fig.1 is plotted in Fig.2.

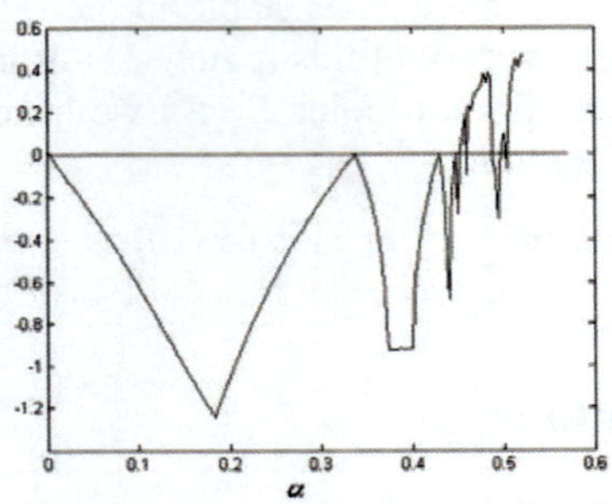


**Fig. 2.** Maximal Lyapunov exponent as a function of $\alpha$

To demonstrate the sensitivity to initial conditions of system (10), we compute two orbits with initial points $(q_{10}, q_{20})$ and $(q_{10}+0.00001, q_{20})$ at the parameter values $a = 6$, $b = 0.5$, $c_1 = 1$ and $c_2 = 4$ and $\alpha = 0.5$, respectively. We can get two firms' game results which are shown in Fig.3. At the beginning the results are indistinguishable, but after a number of games, the difference between them builds up rapidly.

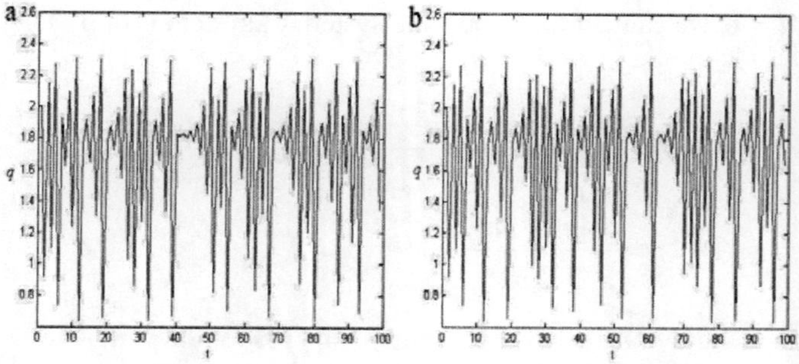


**Fig. 3.** Shows sensitive dependence on initial conditions, the two orbits of $q_1$-coordinates. (a) $(q_{10}, q_{20}) = (2, 2)$ and (b) $(q_{10}, q_{20}) = (2.00001, 2)$.

## 5 Chaos Control

From Fig.1, we can see that the firms are hard to forecast the quantity while chaos is occurred, and the firms' profits also become unstable. Chaos is harmful to firms, so it is necessary to control the chaos of Cournot duopoly game.

There are many methods to control chaos[11]. Delay feedback control (DFC) method is an important and widely-used method to control chaos which was proposed by Pyragas [12]. The basic idea of DFC method is to realize an active continuous control of a dynamic system by applying a feedback signal which is proportional to the difference between the dynamical variable and its delayed value. DFC is based on a self-controlling delayed feedback. Consider the form of delay feedback control system as

$$q(t+1) = f(q(t), u(t)) \tag{17}$$

Input signal u(·) is part of system output signals after a time delay. Pyragas proposed the following feedback in order to stabilize a $\tau$-periodic orbit

$$u(t) = k(q(t+1-\tau) - q(t+1)),\ t > \tau \tag{18}$$

where $\tau$ is a delay time and $k$ is a controlling factor.

As $\alpha$ increases, the bifurcation and chaotic scenario of system (10) is occurred. So we assume the bounded rational firm 1 takes measures to control the chaos of system (10) in this paper. Then we get the controlled system as follows

$$\begin{cases} q_1(t+1) = q_1(t) + \alpha q_1(t)(a - 2(b+c_1)q_1(t) - bq_2(t)) \\ \qquad + k(q_1(t+1-\tau) - q_1(t+1)) \\ q_2(t+1) = \dfrac{a - c_2 - bq_1(t)}{2b} \end{cases} \tag{19}$$

By choosing $\tau = 1$, we can get the controlled system of system (10) by DFC method as follows

$$\begin{cases} q_1(t+1) = q_1(t) + \dfrac{\alpha q_1(t)(a - 2(b + c_1)q_1(t) - bq_2(t))}{1+k} \\ q_2(t+1) = \dfrac{a - c_2 - bq_1(t)}{2b} \end{cases} \tag{20}$$

From Fig.1, we know that chaos exists in system(10) for the parameter valus $a = 6$, $b = 0.5$, $c_1 = 1$ and $c_2 = 4$ and $\alpha = 0.5$. Substituing Nash equilibrium point $E^*(1.8182, 1.0909)$ and the parameter values into Jacobian matrix of (20). By applying July conditions on Jacobian matrix of (20), we obtain that the absolute values of eigenvalues are less than 1 while $k > 0.5019$.

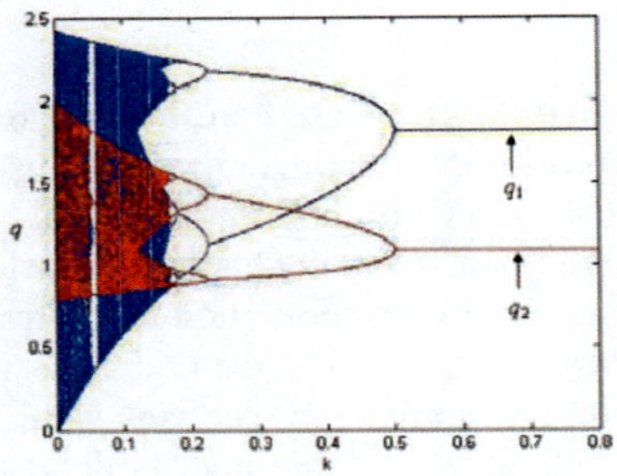

**Fig. 4.** Bifurcation diagram with respect to controlling factor $k$

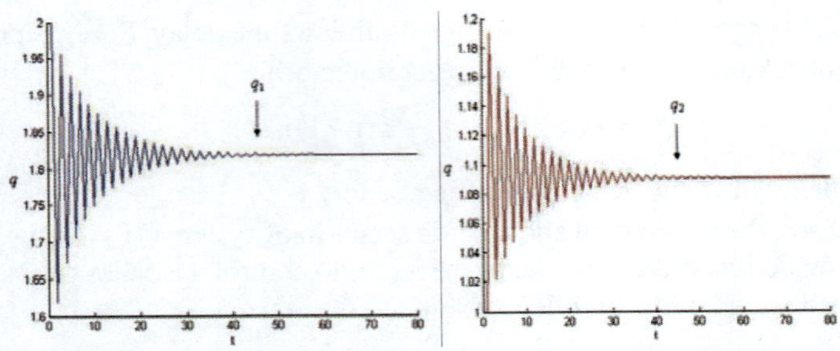

**Fig. 5.** Stabilization of fixed point for $k = 0.55$ (a) $q_1$-coordinates and (b) $q_2$-coordinates

From Fig.4 we can observe that the system is controlled from chaotic state to stable state when the controlling factor $k$ is greater than 0. 5019. Fig.5 shows firms' qualities from initial values (2, 2) converge to the fixed point when the controlling factor $k$ takes value 0.55.

## 6 Conclution

In this paper, we have analyzed the complex dynamics of a repeated oligopoly Cournot game model with heterogeneous players. The stability of equilibrium points is investigated in this game. We found the parameter $\alpha$ (speeds of adjustment of bounded rational player) may change the stability of Nash equilibrium point and cause bifurcation and chaos to occur. The system is sensitive dependence on initial condition when the system stays in chaotic state. By using the delay feedback control method, we have made the system from chaotic state to stable state while suitable controlling factor is chosen.

## References


1. Cournot, A.: Recherches sur les principes mathematics de la theorie de Richesse, Hachette (1838)
2. Agiza, H.N., Elsadany, A.A.: Chasotic dynamics in nonlinear duopoly game with hegerogeneous players. Appl. Math. Comput. 149, 843–860 (2004)
3. Agiza, H.N., Elsadany, A.A.: Nonlinear dynamics in the cournot duopoly game with heterogeneous players. Physica A 320, 512–524 (2003)
4. Kopel, M.: Simple and complex adjustment dynamics in cournot duopoly models. Chaos, Solitons & Fractals 12, 2031–2048 (1996)
5. Agiza, H.N.: On the stability, bifurcations, chaos and chaos control of Kopel map. Chaos, Solitons & Fractals 11, 1909–1906 (1998)
6. Elabbasy, E.M., Agiza, H.N., Elsadany, A.A.: Analysis of triopoly game with heterogeneous players. Computers & Mathematics with Applications 57(3), 488–499 (2009)
7. Elabbasy, E.M., Agiza, H.N., Elsadany, A.A., EL-Metwally, H.: The dynamics of triopoly game with heterogeneous players. International Journal of Nonlinear Science 3(2), 83–90 (2007)
8. Agiza, H.N.: On the stability, bifurcations, chaos and chaos control of Kopel map. Chaos, Solitons & Fractals 11, 1909–1906 (1998)
9. Dixit, A.: Comparative statics for oligopoly. International Economic Review 27, 107–122 (1986)
10. Puu, T.: Attractors, Bifurcations and Chaos: Nonlinear Phenomena in Economics. Springer, Berlin (2000)
11. Chen, L., Chen, G.: Controlling chaos in an economic model. Physica A 374(1), 349–358 (2007)
12. Pyragas, K.: Continous control of chaos by self-controlling feedback. Physics Letters A 170(6), 421–428 (1992)

# Study on SCM of Construction Enterprise Based on IT

Wang Lian-yue and Dong Lan-guo

College of Management, Qindao Technological University, Qingdao, China
wlybb001@163.com, qddlg@163.com

**Abstract.** The supply chain management idea originally from the manufacture industry provides a new management mode for the construction industry, and thus it has become an effective channel for creating and promoting the competitiveness to suit the increasingly intense market competition. This paper first discussed the concept and model of the supply chain management of a construction enterprise and analyzed the problems of construction supply chain Management; then on this basis, extends the construction enterprise's supply chain, including internal supply chain, external chain and customer supply chain based on the information technology; uses the appropriate information technology to build the integration, flexible and intelligent supply chain to enhance competitiveness.

**Keywords:** Information technology, construction enterprise, supply chain management, CSCM.

## 1 Introduction

Economic globalization, the integration process has led to increasingly fierce competition among the construction enterprises. However, the problems of the low centralization, rough technology, overheating competition, and poor economic benefit are still serious. Modern information technology in engineering management application can directly change the project strategies, implementation of project life-wide management, improve income of the project is to reduce mistakes and waste. Supply chain management rooted from the manufacture industry provides a new view angle to the managerial innovation of the construction industry.

## 2 Construction Supply Chain Management

### 2.1 Connotation of the Construction Supply Chain Management

CSC is focused on a general contracting enterprise, linking the material supplier, subcontractor, constructor (owner), consultant, designer, and etc. as a whole functional network through controlling the information flow, goods flow, and fund flow from material procurement, production and completion of partial or sectional works until the delivery of the works (as shown in Figure 1).

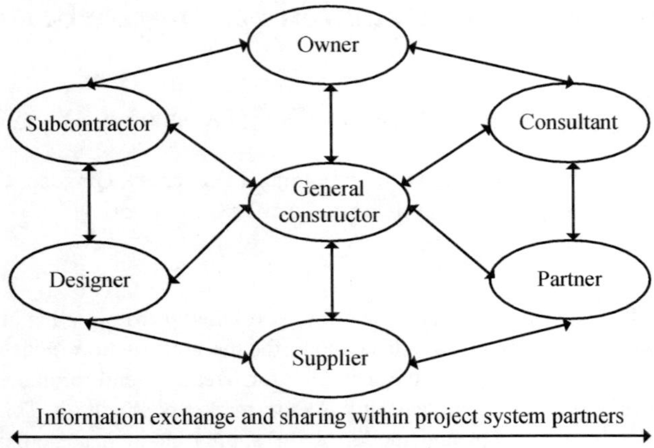


Management model of CSCM

## 2.2 Characteristic of CSCM in the Construction Industry

- Construction enterprise supply chain is the user-driven; each user has its own unique requirements, the needs of the owners is the driving force of information flow, goods flow, funds flow in the construction enterprises supply chain.
- All parties of the supply chain are connected through a contract. However, only the contract is not enough to complete projects in the agreed period for the least cost, best quality, you also have to rely on the parties to the trust, cooperation and information sharing.
- Construction enterprise supply chain is usually the project as the core, the cooperation between the parties on the supply chain is relative to the length of the project cycle, and namely, the construction supply chain has a significant dynamic in nature.
- Construction procurement basically bidding approach, the contractor often cannot intervene in the early stage of the project, which is not implementing the supply chain management of construction enterprises. In order to obtain a contract, contractor often using low-cost strategy, it is easy to form a win / lose business strategy
- The key berries of the supply chain are performance of independent information in all business systems (information silos phenomenon). As competition, corporate interests in order to carry out their own resources (material resources and information resources), self-enclosed, cooperation between parties is only a short-term projects, which increase the barriers of information and communication.

## 2.3 Problems of the Construction Supply Chain Management

For the construction enterprises, supply chains contain many processes, from production to undertake the task to completion and acceptance of construction to maintenance services and other aspects of logistics required for the organic integration of all the organizations. In this process, problems of the construction supply chain management as follows:

- Many large enterprises are still adopting a "vertically integrated" management model, the core enterprises and other enterprises is a kind of ownership relationship, and some leaders continue to pursue the "large", small and complete way of thinking.
- The design and operation of production and management system of construction enterprise are usually less to consider the impact of the supply chain, ignoring the external impact on the enterprise network. Within the enterprise system, supply, production, maintenance, service and other sectors did not form a coordinated "chain", their own way out of line with each other, so that enterprises cannot achieve the overall optimum.
- Between owners and contractors, contractors and suppliers is a hostile relationship. The supply chain did not form long-term close relationship that respond to slower, low operation efficiency.
- Enterprises lack the overall state of the supply chain operation evaluation and incentive mechanism which are confined within each department incentive system. The supply chain relationship between the various departments based on the "win / lose" basis, rather than based on win-win basis.
- Backward Logistics management technology, and unable to meet the requirements of dynamic response, information lags, and is not an effective use of supply chain resources to enable the stock to increase or interruption of information, and rarely peer enterprise's logistics supply points into them, which cannot form trade logistics alliance.
- Information systems are not against the uncertainty of changes in the supply chain to effectively track and manage.

## 3 Construction Enterprise SCM and IT

The characteristics of the construction industry determines that enterprise engaged in the production, to satisfy the requirement of the owner enterprises needs material suppliers, equipment suppliers, subcontractors, construction units, advisory units, design units effective cooperation. Construction enterprise itself is a complex system; in order to survival and development need constantly draw from the resources in the external environment, Construction companies to obtain and maintain a competitive advantage, not only depends on the knowledge and organization of its internal supply chain, and but also need understanding and adaptation of the entire external supply chain. In addition to the activities of internal and external, but also concern with the owner (customer) activities, a construction enterprise must keep a good relationship with the owner of a project to make the project completed smoothly and get trust of the owner, and to make him become a loyal customer, laying a foundation for the next project.

The problems of the CSCM, which divided into the construction enterprises supply chain into internal supply chain, external supply chain and customer supply chain, through integrated, intelligent and flexible the each supply chain to achieve a win-win objective.

In order to gain a competitive edge, companies must be the establishment of a dynamic, more responsive to internal and external supply chain. Now ERP, E-business, CRM and other information technology development, the construction enterprise itself and outside the supply chain had a tremendous impact. Construction of information technology should be used to develop a more sophisticated internal structure, external cooperation to promote business interaction and the combination of internal and external processes to reduce the time to communicate with customers, reduce logistics cost, benefits (as shown in Figure 2). Brought about by information technology, supply chain, information sharing is one of the main has strong vitality. Construction enterprises through the use of information technology to build a dynamic response to strong supply chain.

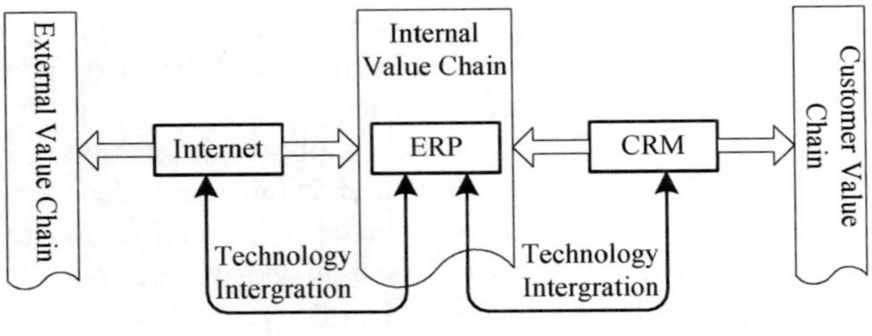


**Fig. 1.** SCM of construction enterprise based on IT

# 4 Role of IT in the CSCM

With the advent of the Internet era, fast, agile and value creation have become key factors in modern competition, the extend of use information become a judge sign of the competitiveness .,through real-time communications and transaction modes, which a positive impact and change connections of the internal supply chain, external supply chain and customer supply chain in the construction enterprises.

## 4.1 The Application of ERP Technology Internal Supply Chain

Construction enterprises should proceed from the whole supply chain, take horizontal integration strategies, enhance their core competitiveness, consider of the reorganization and optimization of enterprise supply chain management to achieve internal goals. Construction industry the use of ERP technology can be realized internal supply chain process reengineering, information sharing and improve supply chain management purposes. Construction companies through the use of ERP technology can be cross-sectoral the integration of real-time information, which are optimize the enterprise resources.

(1) Internal supply chain process reengineering. The heart of internal supply chain is optimize the business processes, use ERP technology for process re-engineering, examining all the activities in the enterprise, optimizing core activities.

(2) The achievement of the departmental information sharing. Construction enterprises through information real-time exchange in the upper and lower levels to strengthen the coordination of various departments to play the capacity and effectiveness of the various departments to achieve smooth running of the information flow, fund flow and goods flow.

(3) Improving supply chain management efficiency. ERP technology in the construction industry can be applied internal supply chain can change the department for their own array of difficulties, not the pursuit of maximizing the interests of this sector, but on the basis of mutual cooperation in order to maximize the overall interests of the enterprise.

## 4.2 Application of IT and E-Business the External Supply Chain

In order to improve the overall core competitive advantage, all members need to have good capacity for cooperation and close working relationship, can access effectively sharing information. Internet technology and E-business techniques adapt to their enterprises for agile, flexible requirements, to achieve a high degree of integration outside the enterprise of building a high degree of agile supply chain (as shown in Figure 3).

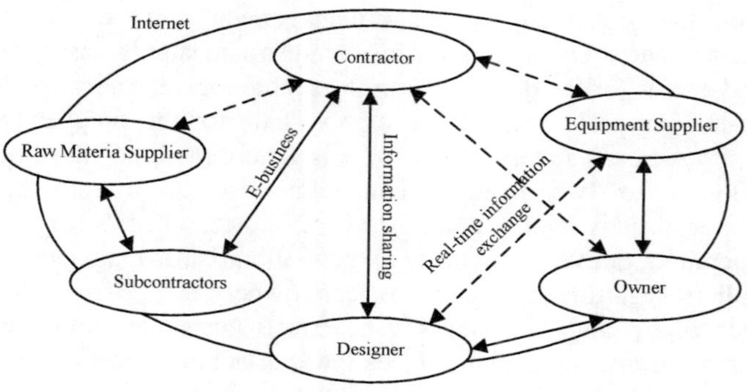


**Fig. 2.** External supply chain based on IT and E-business

## 4.3 Major Existing Problems

With the expansion of the construction project, more participants join the project. As the different roles, connection and strength in the design units, consultants, suppliers, subcontractors and contractors, can not be accurate and timely access to effective communication, the sharing of information, which is typical of the contractors and supply business relationship. All kinds of different of types and large quantities building materials used in building construction , even some need to be imported from abroad, so the supply and buy the construction materials become the key links of the supply chain management. Some uncertain factors, material suppliers are not able to provide materials on time. Mechanical failures, the upstream supplier delayed supply, suppliers and procurement side of the dispute, which may lead to delay in supply.

### 4.4 Strategy to Deal with the External Supply Chain

Internet Technology and E-business improve information-sharing level. Adoption of Internet technology and E-business applications can be an effective link suppliers, contractors, subcontractors, the strengthening of relations between information sharing, the supplier can access demand information through the web site, in accordance with the project plans to provide products, suppliers can also provide information to the contractor in time is formed the new requirements. In the supply chain members between different levels of direct communication can be achieved suppliers and contractors' interconnection.

Internet Technology and E-business to build the logistics network system. Third-party logistics system is effective methods and strategies of logistics supply chain integration, through the coordination of the transport system between enterprises, the enterprise's logistics business are outsourcing to a specialized department to undertake. Through outsourcing to a third-party logistics contractor, construction companies can put time and energy on their core business and improve efficiency. Third-party logistics system provides an integrated transport model, which achieves the Just in Time; zero inventories to make the inventory management of construction enterprises have become more economical. Second is a design logistics network. Optimization of logistics is not considered a node, but from an overall point of view, especially from the global supply chain management point of view. The main problems are select and identify supplier for the upstream supply chain, determine the distributors and agents for the downstream. The global logistics network design of the main decision-making problems and the supply chain for the supplier selection and identified on downstream supply chain for s is determined, so the global logistics network design make two markets constraints are taken into account. The construction enterprises is mainly consider of transportation costs, the advantages of technical cooperation, delivery reliability and collaborative management costs for the upstream suppliers, regardless of the downstream owners.

Internet technology and E-business can reduce costs and improve efficiency. Through electronic transactions and reduces the transaction process, greatly reducing the cycle from order to delivery to achieve a low-cost and improve efficiency. The main functions are on-line electronic document transmission, quickening the document transmission, shortening the disposal time of various engineering procedures; submission and checking of the subcontractor's progress and quality, the in-time online updating of the information of the fund usage; online organization and coordination and disposing the uncertain events during the project execution, such as design variation, project disputes, various burst-out events, and etc.

### 4.5 Application of CRM in Customer Supply Chain

For a construction company, the owner is customer. The owner plays a leading role in construction. This role not only dominates the selection of contractor, but also has a considerable effect on selection of subcontractor, material and equipment supplier. The owner's measure payment is directly reflected on the fund flow of the supply chain. Besides, the owner's supervision on the engineering quality and schedule is related to the project's smooth operation.

Contractor to use CRM (Customer Relationship Management), namely, customer relationship management, which is an effective way to share information resources between the core enterprise and customers. CRM system is a new business model based on "one to one customer relations theory", which integration customer information resource and sharing resources in order to respond rapidly to market needs and enhance their competitiveness in the market. Because of the contractor's number of customers is usually less compared with the manufacturer, and the owners usually take the initiative place in the construction market, therefore, customer relationship management, especially for contractors. Contractor establishing a CRM system that can conduct real-time exchange of information with the owners, can quickly grasp the needs of the owners, moreover owners can immediately grasp the progress of the project, quality, and other information. All participants can benefit from effective customer relationship management to achieve win-win purpose.

## 5 Conclusion

The implementation of construction supply chain management, through coordinate all aspects of enterprises and change concept of leader to build the internal supply chain; closely coordinate with suppliers, including upstream and downstream, understanding demands and goals, the fellowship cooperation relationship can really be set up based on mutual trust and sharing information to achieve a win-win goals; construction enterprise must keep a good relationship with the owner of a project to make the project completed smoothly and get trust of the owner, and to make him become a loyal customer, build the customer supply chain based on customer satisfaction.

## References

1. Song, Y., Jin, W., Zhang, W.: Analysis of the Core Competence of Construction Enterprises on the Basis of the Supply Chain Management. Architecture Economy (12), 17–19 (2004)
2. Su, H., Zhao, Y.: Heightening the Core Competence of General Contractors by Applying the Supply Chain Management. Value Engineering (3), 43–45 (2006)
3. Sunil, C., Peter, M.: Supply Chain Management – Strategy, Layout and Operation. Social Scientific Literature Publisher (2003)
4. Vrijhoef, R., Koskela, L.: The Four Roles of Supply Chain Management in Construction. European Journal of Purchasing & Supply Chain (6), 169–178 (2000)
5. Wang, Y.W., Xue, X.L.: Supply Chain Management Application Research in the Construction Industry. Journal of Civil Engineering 37(9), 86–89 (2007)

# A Comprehensive Approach to ERP Selection

Huo Lingyu[1], Liu Bingwu[1], Yan Fang[1], and Huo Lingyu[2]

[1] School of Information, Beijing Wuzi University, City Beijing, P.R. China
[2] School of Management and Economics, Beijing Institute of Technology, City Beijing, P.R. China
sboerp@163.com

**Abstract.** This Empirical research has shown that selecting an inappropriate system is a major reason for ERP (Enterprise Resource Planning) implementation failures. Review of literature has identified a need for approaches to assist organizations in the ERP system and vendor selection process. This paper therefore proposes an approach that differs from existing models in that it that has a broader, more comprehensive focus by simultaneously developing, defining and aligning the organization's strategies, infrastructure, processes and information with the ERP evaluation and selection process. The methodology consists of a conceptual framework, a risk model for ERP selection, and a process model for the ERP selection process. The final output for the organization is a choice of ERP system and vendor that ensures strategic fit and functional integration in the organization.

**Keywords:** ERP system selection, Alignment, Risk.

## 1 Introduction

An Enterprise Resource Planning (ERP) system is an integrated enterprise computing system to automate the flow of material, information, and financial resources among all functions within an enterprise on a common database [1]. Many companies have implemented ERP systems over the past decade. A wrong ERP project selection would either fail the project or weaken the system to an adverse impact on company performance [2,3]. Today there is therefore growing consensus among ERP implementers that selecting an inappropriate system is a major reason for ERP implementation failures [4].

Several methods have been proposed for selecting a suitable ERP project or management information system [5–8]. However, these mathematical programming methods can not contain sufficient detailed attributes, above all, which are not easy to quantify, so that the attributes were restricted to some financial factors, such as costs and benefits. Furthermore, many of them involved only the consideration of internal managers, but do not offer a comprehensive process for combining evaluations of the risk of the project, the functional fitness, and the process to select an ERP project objectively [9].

Research has indicated that about three quarters of attempted ERP projects are unsuccessful: a common problem encountered in adopting ERP software has been the issue of fit or alignment [10]. The misfit between ERP candidates and the enterprise's requirements is one of the main reasons of selection failure. An effective approach is needed which can make organizations more easily and systematically determine the locations of possible misfit and their degree of importance, thereby understanding the risk in their implementing an ERP.

ERP projects have often been found to be complex and risky to implement in enterprises. The organizational relevance and risk of ERP projects make it important for organizations to focus on ways to make ERP implementation successful. Controlling risk in software projects is considered to be a major contributor to project success [11]. Research focused on risk reducing in the ERP selection phrase is little.

It is revealed that little research has focused on the IT (or ERP) acquisition process (e.g., packaged software) itself. While many conceptual models of the buying behavior of organizations have been developed, there has been a lack of research to extend and test these models [12].Especially in the field of ERP selection, the lack of specificity of these models is one of the reasons why researchers have not designed follow-up research[13].

## 2 ERP Selection Approach

### 2.1 Conceptual Framework for ERP System Selection

A number of researchers have shown that normal accounting-oriented, cost–benefit analysis is unable to evaluate IT/IS effectively. There is now a belief that the financial perspective does not provide a sufficiently good evaluation of the potential and costs of an ERP system. The critical challenge of ERP implementation is mutual adaptation between the IT and the social organization It has been argued that the failure to realize the value of IT investments in part has been due to a lack of alignment between the business and IT strategies of organizations [14]. This assumption forms the backbone of the proposed ERP selection methodology. To succeed in exploiting IT's strategic role in supporting and shaping new business strategies, organizations must achieve:

- Strategic alignment in terms of the interrelationship between the business and IT.
- Process fit between the business process and the process that the ERP software supported.
- Architecture fit between the function area of the organization and the ERP software Architecture.
- Function fit between the functional integration and ERP software component.
- The fit between the information that the organization needed and the data that the ERP software can supported.
- Functional integration between the business and functional domains.
- The fit between the information that the organization needed and the data that the ERP software can supported.

The unfit between process, architecture, function and information decides the whether the enterprise need to BPR or not?

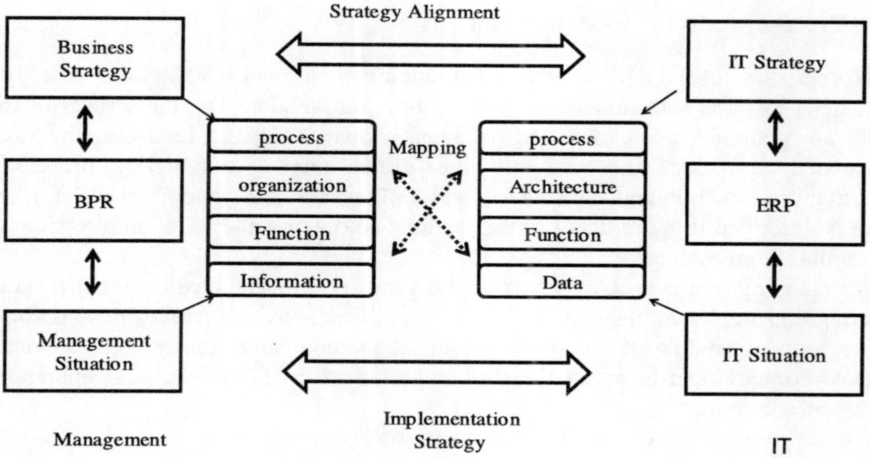


**Fig. 1.** Alignment Model for ERP selection

Because the components of the model are interrelated, the decisions made within one domain will affect the other domains. A fact that underlines the importance of achieving strategic fit, BPR or not decision, and functional integration between the four parts and the selected ERP solution, as shown in Fig.1.

## 2.2 The risk Model for ERP Selection

During the selection phase of ERP implementation, the main aim is to reducing the risk from outside to ensure the success of the project. The risk from software, consultant, methodology and institution are four key resources from outside, shown in Fig.2.

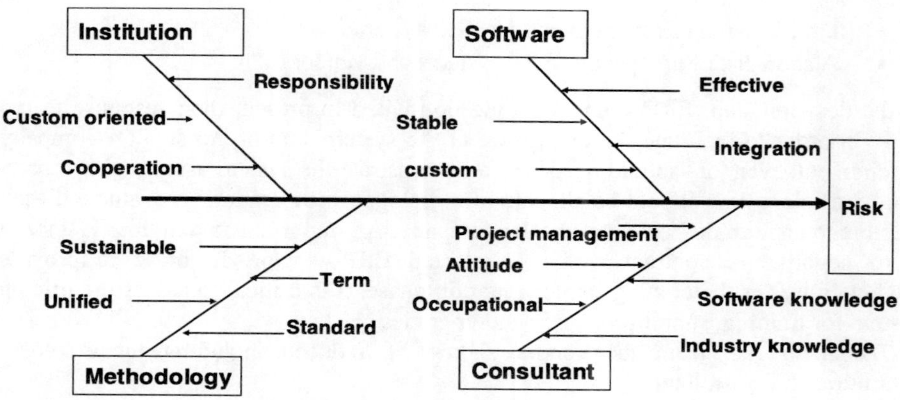


**Fig. 2.** Risk model for ERP selection

## 2.3 Process Model for ERP Selection

The corner stone of the ERP selection methodology is a process model that describes three steps towards selection of an ERP system and vendor. The three steps of the model are planning, evaluation and decision, shown in Fig.3. Each step includes interrelated activities that integrate the dimensions of strategies, processes, infrastructure, function and information and ERP system and vendor selection. Each phase is described in more detail in the sections below, together with an overview of key outputs from each step.

The planning step is used to prepare the organization for the challenging project of selecting and implementing a new ERP system. This activity involves developing a thorough understanding of the key strategies of the company in terms of the overall business strategy, and how functional strategies, such as IT, market and operations strategies, are defined.

The key outputs from the preparation phase will be:

- Business strategy and profile and characteristics
- IT strategy ( Including the planning how to developed ERP system and Long-list of potential ERP systems)

The evaluation step is used to identify potential misalignments between current strategies, and business and IT processes and infrastructure. This is done using established methods for mapping and analysis. This process will develop a deeper understanding of the future needs of the organization. In order to increase alignment of strategies, processes, infrastructure, function and information, a breakdown of the core business strategy is used to reduce the long-list into a shortlist of alternative ERP systems and vendors. As a supporting tool for the company, a structured set of general ERP system criteria should be produced [15].

The main risk from software should be reduced including the effective, stable, custom and the integration of the ERP system.

The key outputs from the analysis phase will be:

- Draft version of requirements for ERP system
- A short-list of prospective ERP systems and vendors

In the decision step, ERP vendors should be invited to present their response to the RFP. In order to facilitate the evaluation of the system's ability to support company requirements, vendors should be asked to demonstrate the system using real company data. Sufficient time should be allocated for dialogue with vendors to evaluate if they are able to capture the real needs of the company. An important step in this final stage is to negotiate a contract with the selected ERP vendor. In order to provide predictability for the coming implementation phase, it is critical to assess the pricing system for training, support, licenses and upgrades.

The main risk from the vendor should be reduced including methodology, institution and consultant.

The key outputs from the selection phase will be:

- Selected ERP system
- Contractual agreement with ERP vendor

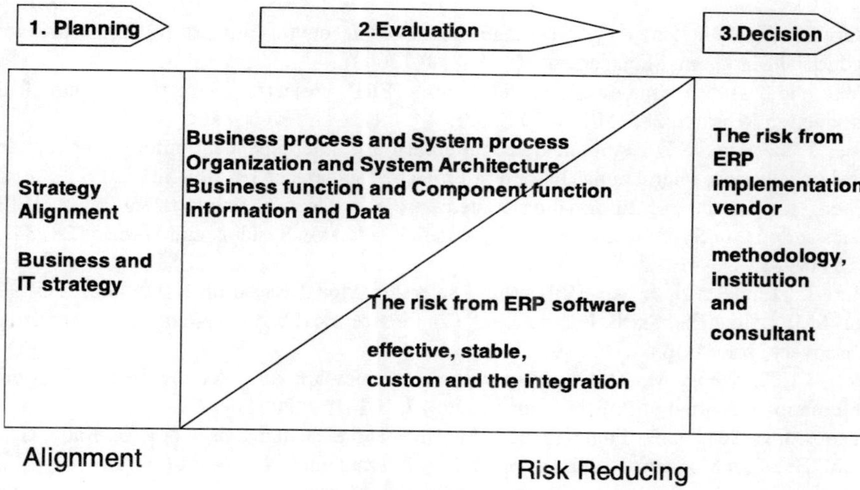


**Fig. 3.** Model for ERP selection

# 3 Conclusions

The main features of the proposed ERP selection approach have been described in terms of the underlying theoretical framework, a step-by-step process model and associated guidelines. An important characteristic of the approach is its emphasis on the alignment. The final output for the organization is a choice of ERP system and vendor that is appropriately aligned with clearly defined business and IT strategies, and defined and aligned business and IT processes, function, infrastructure and information. The proposed approach represents a useful tool for organizations with regards to the evaluation and selection of ERP systems and vendors.

**Acknowledgment.** This paper is supported by Funding Project for Academic Human Resources Development in Institutions of Higher Learning Under the Jurisdiction of Beijing Municipality (PHR200906210), Funding Project for Base Construction of Scientific Research of Beijing Municipal Commission of Education, Funding Project for Science and Technology Program of Beijing Municipal Commission of Education under the grant number KM200810037001.

# References

1. Kumar, V., Maheshwari, B., Kumar, U.: Enterprise resource planning systems adoption process: a survey of Canadian organizations. International Journal of Production Research 40, 509–523 (2002)
2. Wilson, F., Desmond, J., Roberts, H.: Success and failure of MRPII implementation. British Journal of Management 5, 221–240 (1994)
3. Hicks, D.A., Stecke, K.E.: The ERP maze: enterprise resource planning and other production and inventory control software. IIE Solutions 27(8), 12–16 (1995)

4. Verville, J., Halingten, A.: A six-stage model of the buying process for ERP software. Industrial Marketing Management 32, 585–594 (2003)
5. Teltumbde, A.: A framework of evaluating ERP projects. International Journal of Production Research 28, 4507–4520 (2000)
6. Lee, J.W., Kim, S.H.: An integrated approach for interdependent information system project selection. International Journal of Project Management 19, 509–523 (2001)
7. Chen, K., Gorla, N.: Information system project selection using fuzzy logic. IEEE Transactions on Systems, Man, and Cybernetics—Part A: Systems and Humans 28, 849–855 (1998)
8. Huo, L., Liu, B., Li, J.: An ERP System Selection Model Based on Fuzzy Grey TOPSIS for SMEs. In: The Sixth International Conference on Fuzzy System and Knowledge Discovery, vol. (8), pp. 244–248 (2009)
9. Wei, C.-C., Wang, M.-J.J.: A comprehensive framework for selecting an ERP system. International Journal of Project Management 22, 161–169 (2004)
10. Jacobs, I.S., Bean, C.P.: Fine particles, thin films and exchange anisotropy. In: Rado, G.T., Suhl, H. (eds.) Magnetism, vol. III, pp. 271–350. Academic, New York (1963)
11. Bannerman, P.L.: Risk and risk management in software projects: A reassessment. The Journal of Systems and Software 81, 2118–2133 (2008)
12. Ward, S., Webster Jr., F.E.: Organizational buying behavior. In: Robertson, T.S., Kassarjian, H.H. (eds.) Handbook of Consumer Behavior, pp. 419–458. Prentice-Hall, Englewood Cliffs (1991)
13. Verville, J., Halingten, A.: A six-stage model of the buying process for ERP software. Industrial Marketing Management 32, 585–594 (2003)
14. Henderson, J.C., Venkatraman, N.: Trategic alignment: Leveraging information technology for transforming (1999)
15. Wei, C.-C., Chien, C.-F., Wang, M.-J.J.: An AHP-based approach to ERP system selection. International Journal of Production Economics 96, 47–62 (2005)

# Research on High-Current Source Control and Protection Technology in the HVDC Synthetic Test System

Guang Ping Yao and Jian Dong Chen

Department of Information Science and Engineering, Jiangsu Polytechnic University, Changzhou, China
yaogp@263.net, cjd_820@139.com

**Abstract.** To evaluate the current bearing capacity of high voltage direct current transmission (HVDC) converter valve, "DC high-current source" is one of the important device used in the synthesis test system of converter valve, so it is very important to research the control and protection strategies of this DC high-current source. The paper introduces the steady flow control, protection measures, hardware features, software process, etc. The idea of DSP technology, CAN communication technology and embedded six pulse generator triggered technology integrated in the control and protection systems of three-phase full-controlled rectifier bridge is proposed, which forms a complete control and protection system of DC high-current source applied in the synthetic test for converter valve.

**Keywords:** High-current source, Protection and communication, Steady flow control, DSP Technology.

## 1 Introduction

High voltage direct current transmission (HVDC) is a high-tech of long-distance power transmission. DC converter valve is the core equipment to achieve HVDC, expensive, and its performance determines the high-voltage direct current transmission system reliability[1]. In the actual working conditions, the converter valve withstands high voltage and high current at the same time. In the manufacturing process of the converter valve, how to provide an equivalent environment of the actual condition for the examination and testing of converter valve's performance are the difficulties encountered by the people. If to provide high voltage and high current equivalent to the actual operating conditions at the same time, which means the need to build a huge power, huge test system, which is certainly not the economy but also difficult to achieve[2].

The synthetic test method is proposed in CIGRE, and the synthetic test system consists of a separate high-voltage source and DC high-current source [3-4]. In this paper, with regard to the characteristics of DC high-current source in the converter valve test system, to discuss its control and protection technology, this is undoubtedly necessary for the synthesis of converter valve test system in terms of reliable operation.

# 2 The Main Body of DC High-Current Source

## 2.1 Basic Working Principle of the First Loop

A team of three-phase full-controlled rectifier bridge composed of thyristors is the main part of the synthesis test of DC high-current source, Shown in Figure 1. VT1、VT3、VT5 are in the common cathode group, and VT4、VT6、VT2 are in the common anode group. At any one time there must be two thyristors conduction at the same time, and one in the common cathode group, the other in the common anode group to form a loop. Its turn-on order is +A, -C, +B, -A, +C, -B customarily. In conducting the synthetic tests, the bridge arm +A is replaced by test valves VT and isolation valves $V_{IS}$ which used as high-voltage isolation. When isolation valves $V_{IS}$ is stringed up might causes the bridge arm threshold voltage imbalance, resulting in DC transformer coils generate DC component. To use 1#, 2# diode string can set-off imbalance. The synthetic test control system provides switch signals between the bridge arm +A and test valves VT, DC-side of bridge by the resistance-inductance load to complete DC loop. By changing trigger angle $\alpha$ that is able to adjust current amplitude, so as to give practical working conditions for converter valve, then to have a series of type testing. According to three-phase full-controlled rectifier bridge theory (in the ideal case):

The resistive load output voltage should meet the following formula:

$$U_d = 2.34U_2COS\alpha \qquad \alpha \leq 60° \qquad (1)$$

or

$$U_d = 2.34U_2[1 + COS(\frac{\pi}{3} + \alpha)] \qquad \alpha > 60° \qquad (2)$$

In equation (1) and (2), $U_d$ is the average value of DC voltage across a resistive load, $U_2$ is the value of phase voltage of the rectifier bridge AC-side.

The inductive load output voltage should meet the following formula:

$$U_d = 2.34U_2COS\alpha \qquad \alpha \leq 90° \qquad (3)$$

In equation (3), $U_d$ is the average value of DC voltage across an inductive load, $U_2$ is the value of phase voltage of the rectifier bridge AC-side.

The output current shall be: $I_d = \frac{U_d}{R}$ , $R$ is DC-side load resistance.

## 2.2 Steady Flow Control

The bridge arm currents under test valves are the control target parameters for DC high-current source, its on-time is 120°electric angle, and its amplitude is equal to the DC side currents of the rectifier bridge. Therefore, as long as the DC-side current is stable so that the bridge arm +A current amplitude is stable too. To sample with the closed-loop current feedback control that can acquire a stable bridge output DC current, also enable the bridge arm current amplitude to stabilize. When the bridge

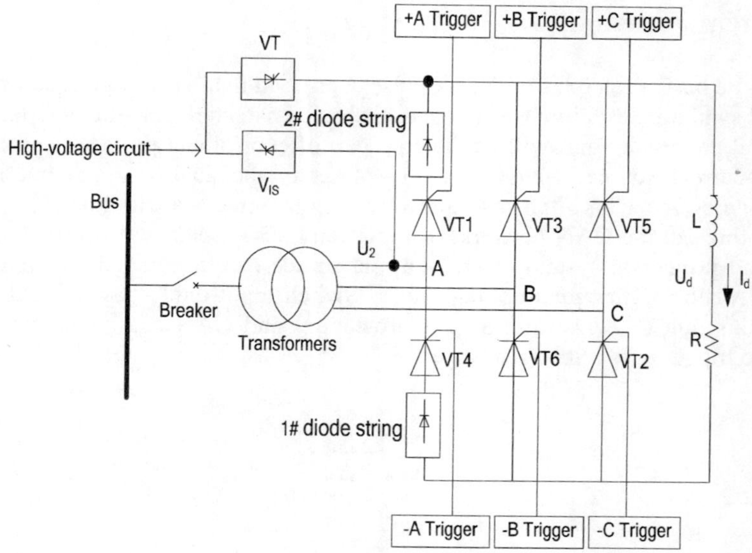


**Fig. 1.** Diagram of the main loop of DC high-current source

arm +A switches to test valves, there is the stable current amplitude via test valve. Shown in Figure 2, to compare the value of the DC current feedback $I_F$ with the default value $I_S$, by the PI regulator for conditioning (when the deviation between them is smaller, using PI control can improve the accuracy of steady flow; when the deviation between them is larger, using PI control can shorten the response time of steady flow[5]), then to convert obtained error amount to firing angle $\alpha$, through pulse triggering unit to trigger rectifier bridge, continuous amendment firing angle $\alpha$. Eventually to stabilize current on the setting value, for the completion of current negative feedback loop steady flow control.

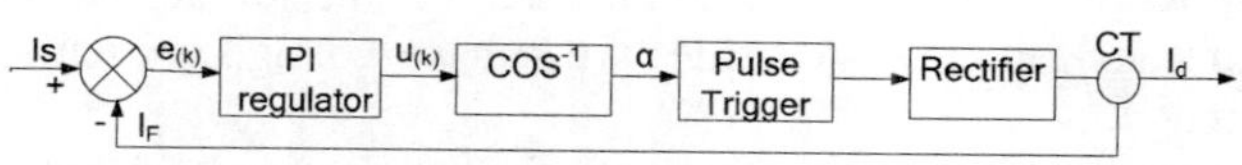


**Fig. 2.** Diagram of steady flow control

By the principle of three-phase full-controlled rectifier bridge:

$$\alpha = COS^{-1}\left(\frac{I_d * R}{2.34\,U_2}\right) \tag{4}$$

Equation (4) is the off-line calculations, to get the relational table about $\alpha$ and $I_d$. Look-up the table using computer technology, and obtain the corresponding firing angle $\alpha$.

# 3 System Function

Based on the basic principles of the six pulse controlled rectifier bridge used thyristor, combined synthetic test circuit works and the basic characteristics of converter valve. The central processing unit will be the core part of control and protection system for DC high-current source, which not only monitors the analog signal but also to monitor and drive each switch state instantly. It communicates with host computer in real-time through the CAN network, to report and accept data and instructions. The control and protection system of high-current source can be divided into five units, such as “Analog signal conditioning unit”, “Switch input unit”, “Switch output and synchronous unit”, “A/D, Control and protection, and CAN communication unit”, “Six pulse trigger unit”. Shown in Figure 3:

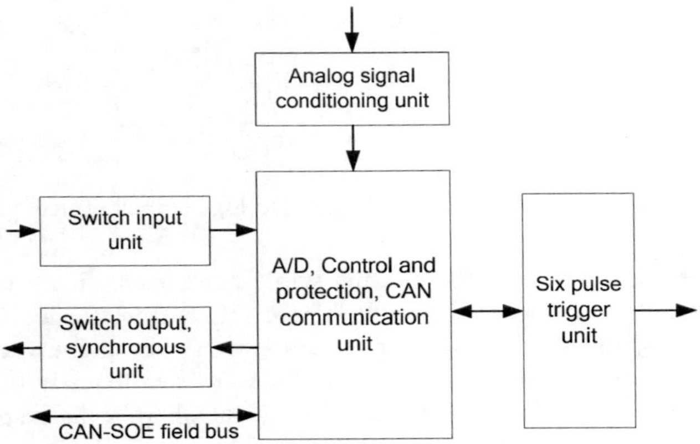


**Fig. 3.** Functional diagram of control and protection

Meanwhile, "Analog signal conditioning unit" is to complete the analog signal collected conditioning, in order to meet the A/D circuit for normal mode - digital conversion conditions. Such as: AC-side three-phase voltage and current, DC-side voltage and current.

"Switch input unit" will convert the state of all nodes in the external device into TTL logic levels through the optical isolation for the protection circuit monitoring, querying and conducting. Such as: breaker status, pressure switch status, temperature switch state of rectifier transformer, fast melt state, host computer switch.

"Switch output and synchronous unit" will isolate the internal logic signal through the optical coupler or relay, and generate enough power drive signals to drive external devices. Such as: to drive breakers, fiber optic firing head, and optical coupler, etc. "Synchronization Circuit" will be filtering and zero-crossing detection for synchronization signal (sinusoidal signal) of power, resulting in square-wave sync signal provided to the follow-up micro-controller unit for generating synchronized base of trigger pulses.

The core of "A/D, Control and protection, and CAN communication unit" is DSP microprocessors; its function is A/D conversion, switch Detection, as well as CAN communication. DSP real-time monitor each data of the analog signals and all status of the switch signals, and then report to the host computer processing through CAN-SOE network. Particularly when any one of these signals is unusual, DSP will be exit processing immediately. When working, the unit will receive the transmission command (data) of host computer through the CAN-SOE network, these commands include: running, quit, inquiry, inspection, timing.

"Trigger pulse unit" is base on the power synchronous signal, resulting in six-pulse trigger signal. In order to improve the transmission efficiency of pulse transformer, and reducing the power loss of drive circuit and thyristor gate, using pairs of narrow pulses and high-frequency modulation. If synchronization missing, fiber failure will be reported through the protection unit to the host computer processing, and then to protect them.

The above, the control and protection system of DC high-current source is to complete the analog signals control, the switch signals monitoring, all kinds of trouble reports, and six-phase trigger pulse driven.

## 4 Software Process

The main CPU of DC high-current source control and communications system use the TMS320 system's DSP chip TMS320LF2407of TI Company, this chip is built in 10-bit A/D converter and CAN2.0B Module, better to handle real-time tasks, but also simplify the design, easy to implement the required functionality. The main program flow is first to power-on initialization to detect the switching value, if not normal, through the SOE reporting failures, then sub-gate and re-testing until the normal, to handle the handing command parameters through the CAN-SOE, until normal to close breaker and delay. Followed by the detection of each analog, if not normal, then open breaker and re-submitted to SOE to wait for power-on, otherwise, to lift the pulse blockade, turn to monitor the status of inquiries to carry out the control and protection.

The whole control and protection and communication processing as shown in Figure 4:

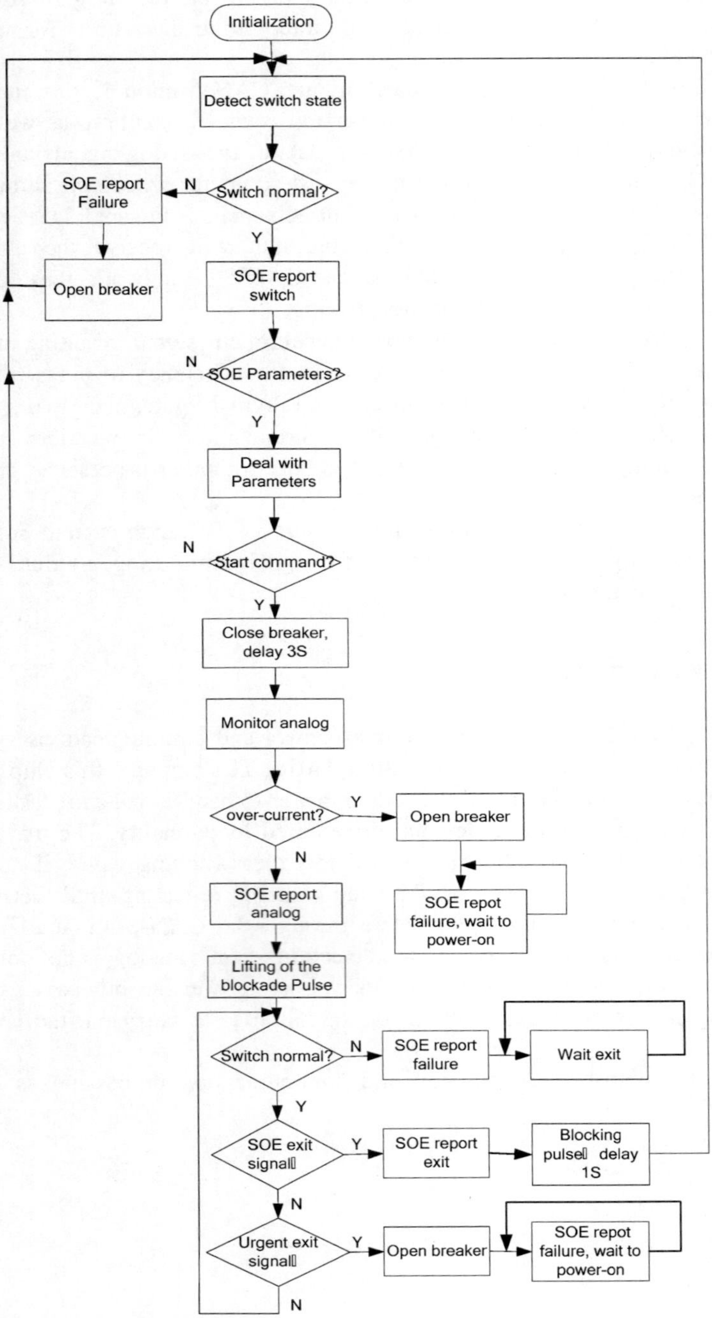


**Fig. 4.** Flow diagram of high-current source protection and communication

## 5 Conclusions

When the synthetic test system of HVDC converter valves is running, DC high-current source injects testing current for converter valves. Its control and protection technology directly affect the synthesis test results of converter valve. This paper introduces the technical route, and its legitimacy has been verified. The waveform of output DC is stable, non-super-harmonic and oscillations. The current accuracy is of 0.2%. Using DSP chip as the central controller is reasonable, and its resources and speed is very suitable for the system requirements. The functions of signal acquisition, operation, execution speed, protection response, and communications fully satisfy the requirements of the synthesis test. If the main circuit of the control and protection system is changed appropriately, that can be used in motors and electrical transmission areas. The application prospect is broad.

## References

1. Shu, Y.: Present status and prospect of HVDC transmission in China. High Voltage Engineering 30(11), 1–2 (2004)
2. Thyristor Valves for High Voltage Direct Current (HVDC) Power Transmission – Part 1: Electrical Testing, IEC 60700-1 (2003)
3. Task Force 03 of Working Group 14.01. Test Circuits for HVDC Thyristor Valves. CIGRE Technical Brochure 113 (1991)
4. Sheng, B.L., Jansson, E., Blomberg, A.: A new synthetic test circuit for the operational tests of HVDC thyristor modules. In: Proc. IEEE 16th APEC 2001 Conf., Anaheim, USA, pp. 1242–1246 (2001)
5. Wang, X., Lin, Y.: DSP-based digital synchronous motor excitation regulator. Electric Drive Automation 26(3), 59–62 (2004)

# Realization of Control Algorithm for Electro-Hydraulic Power Steering System Based on MC9S08AW32 Microcontroller

Xia Chang-gao, Zhu Zhong-ming, and Zheng Rong-liang

College of Automobile and Traffic Engineering, Jiangsu University, Zhenjiang, Jiangsu Province
xiacg@ujs.edu.cn

**Abstract.** Electro-hydraulic power steering system (EHPS) can get over the bug of assist power that can't be regulated in the traditional hydraulic power steering system (HPS). Because its assist power is greater than electronic power steering system (EPS), it is suitable for medium and heavy vehicles. In this paper, based on the EHPS system parameters, the control algorithm of the non-linear relations among steering wheel angular velocity, vehicle speed and motor speed of EHPS, which was designed to meet the steering portability at low speed and obvious steering feel and rapid response at high speed and save energy, was given. MC9S08AW32 was applied to design the control system of BLDCM and the scheme of hardware and software were given. The controller communicates with the computer through a special chip. Therefore the digital transmission between the controller and external device is realized. Finally the rationality and validity of the design is verified by the experimental results.

**Keywords:** Steering System, Electro-Hydraulic Assist Power, Controller.

## 1 Introduction

Vehicle steering system is the key assembly to vehicle active safety system and it affects the consumption of energy at the same time. Most modern vehicles assemble power steering system to meet people's increasing requirements of vehicle's handling comfort and portability. Today more than 90% of vehicle power steering systems assemble an engine-driven hydraulic power steering system (HPS). Due to the existence of HPS the phenomenon of decrease of driver's road feeling is likely to occur, when driving mileage is a straight road steering system do not need to work but the HPS system has always been in the engine-driven pump output of hydraulic energy, the energy consumes in vain; HPS is mainly designed by overcoming largest steering resistance at the low-speed parking condition to ensure the steering efforts. Engine speed is between 500 r/min and 800r/min at the low-speed parking condition while about 3 to 8 times the idle speed at medium-speed, high-speed driving at which time steering resistance needed decreases but engine-driven oil pump is in high-speed operation, the output flow is several times larger than that in the parking conditions and hydraulic energy output can be far beyond the needed so large quantities of energy is wasted. In order to overcome the shortcomings brought by engine-driven oil pump of traditional HPS system, some vehicle manufactures like TRW, ZF and Delphi research electro-hydraulic power steering system (EHPS) which use DC motors instead of engine to drive the steering pump in recent years.

EHPS sets the advantages of hydraulic power steering system and electric power steering system (EPS) in one. Compared to HPS, EHPS uses DC motor to drive oil pump so larger hydraulic flow is provided by the electric pump at engine idle while decreases at high speed according to the demands of steering gear. Variable assist power is gained according to the steering demands. Compared to EPS, EHPS is superior in the aspect of the comfort of steering feeling and the attenuation of the impact of the road due to the existence of hydraulic part. Therefore, it improves driving comfort and safety[1].

In this paper, based on the EHPS system parameters, the control algorithm of the non-linear relations among steering wheel angular velocity, vehicle speed and motor speed of EHPS which was designed to meet the steering portability at low speed and obvious steering feel and rapid response at high speed and energy-saving was given. MC9S08AW32 was applied to design the control system of a brushless DC motor (BLDCM).

## 2 Configuration and Working Principle of EHPS

The configuration and working principle of EHPS are shown in Figure 1[2][3]. System mainly consists of power steering gear, oil reservoir, electro-hydraulic pump, power steering ECU and related sensors so on.

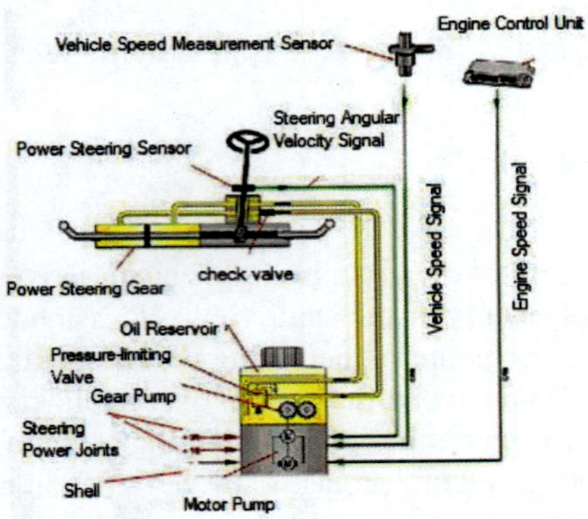


**Fig. 1.** Scheme of EHPS Operational Principle

First, power steering ECU calculates the appropriate motor speed according to the signals of the angular velocity sensor and vehicle speed sensor in real time and regulates the motor speed to reach the appropriate speed. Driven by the motor, hydraulic pump will pump high-pressure oil from a hole to diversion valve through a pipeline. High-pressure oil does not flow into the oil reservoir but flow back into the hydraulic pump directly when there is no steering operation. When there is a steering operation which makes valve spool and valve body move relative, causing high-pressure oil flowing into one of the oil cylinders to form pressure difference between the left and right cylinder, assist power was produced accordingly. Low-pressure oil in another cylinder is pressed back into the hydraulic pump.

## 3 The Control Algorithm of EHPS

Two situations must be considered about the control algorithm of EHPS: First, when there is no steering operation, namely the steering wheel is on-center (statistical data show most of the steering operation occurs in the region when the steering wheel is operated with small angle at high-speed driving). Because such cases account for the vast majority of vehicle mileage, motor speed should be as low as possible considering energy consumption. The relationship between motor idle speed ($n_0$) and vehicle speed ($V$) should be:

$$n_0=kV+b. \quad (1)$$

$k$, $b$ are constants. This design can meet the requirements that motor speed is at a higher idle when at the low-speed parking condition or low-speed driving and provides greater assist power and makes steering operation light and flexible. With the increase of the vehicle speed motor idle should be gradually reduced and the flow of the oil pump should be reduced accordingly . The assist power provided is small so the road feeling at the high-speed driving is increased.

The second situation occurs when there is steering operation at a constant vehicle speed. With the increase of the steering wheel angular velocity at a constant vehicle speed motor speed should increase rapidly to provide greater assist power to meet the requirements of enough assist power and rapid following. When the motor speed increases to a certain value it should be set not to be changed and into the assist power saturation region. To meet the above requirements motor speed can be expressed with the following:

$$n=\begin{cases} n_0 & (0<\omega<\omega_0) \\ f(\omega) & (\omega_0 \le \omega \le \omega_1) \\ n_{\max} & (\omega>\omega_1) \end{cases} \quad (2)$$

In the formula, $n_0$ is motor idle , motor idle speed is different at different vehicle speed; $n_{max}$ is the maximal speed which motor can achieve; $f(\omega)$ is a linear function.

Judging from real vehicle testing, the coefficients in the equations (1) and (2) are determined, the figure of the corresponding relationship among $V$, $\omega$ and $n$ is obtained, as shown in Figure 2(the corresponding curve is drawn by a large number of measured points).

Obtaining the characteristic curve as shown in Figure 2, it could be dispersed into several discrete curves of typical characteristic, and then control it by the characteristic curve which is closest to the typical curves. But as to the continuously variable vehicle speed, it would be difficult to get the exactly assist power at any vehicle speed, resulting in the control error, when the change of the speed is great, the phenomenon of non-uniform assist power and shake of steering wheel appears, therefore, the characteristic curve needs fitted.

As the neural network has better ability to approximate any complex non-linear curve, suitable for handling fuzzy and non-linear problems. It has not only strong robustness, but also a strong generalization[4][5]. So the neural network is used to fit the characteristic curves.

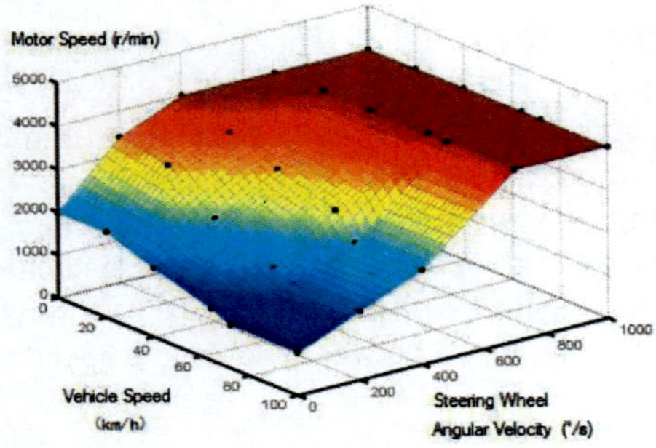


**Fig. 2.** Relation Curves Among Vehicle Speed, Steering Wheel Angular Velocity and Rotate Speed of BLDCM (Real Vehicle Testing)

Given any continuous function or mapping, a three-layer neural network can be used to realize it accurately according to Kolmogorov mapping neural network theory[6]. A three-layer BP neural network model is established for prediction. The BP structure model is shown in Figure 3, the input layer has two input neurons ($x_1$,$x_2$), the steering wheel angular velocity and vehicle speed respectively; output layer has one output neurons *n*, namely motor speed. Hidden layer nodes are determined by the performance of network influenced by the samples.

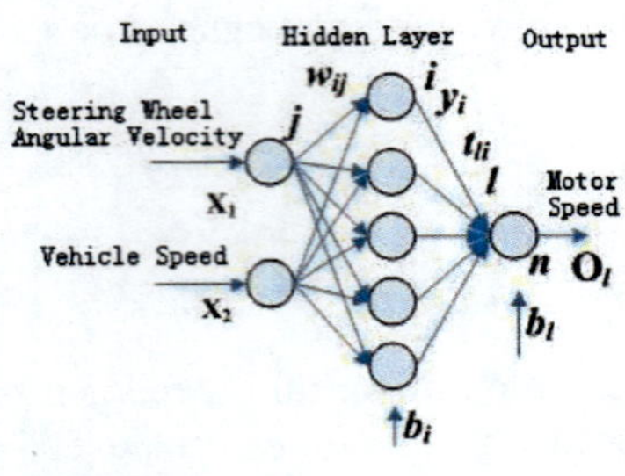


**Fig. 3.** BP Neural Network Chart

According to Figure 2 BP neural network is applied to fit the curves at different speed to realize the curves corresponding to motor speed at continuous angular velocity (as shown in Figure 4). There are two neurons in neural network model of the input layer. According to performance of the network influenced by the samples provided, the network converges fastest without over fitting by several attempts when the hidden layer node is 20. So the hidden layer nodes are determined to be twenty while the output layer one. Figure 4 shows the curve fitting result by the neural network while the vehicle speed is 0km/h, 20km/h, 40km/h, 60km/h, 80km/h, 100km/h, respectively.

In practical applications, according to Figure 4, a number of curves embedded into the controller for brushless DC motor control.

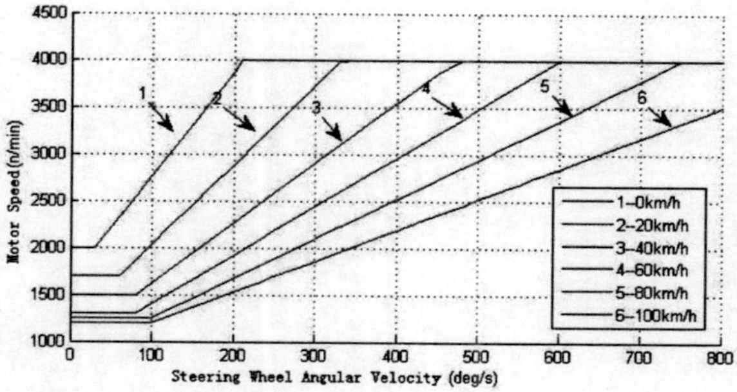


**Fig. 4.** BP Neural Network Fitting Curve

# 4 The Design of EHPS Controller

## 4.1 Hardware of EHPS Controller

The hardware of EHPS system controller includes several parts as follows: steering wheel angular velocity signal processing circuit, speed signal processing circuit, the direction of assist power circuit, H bridge driver circuit, controller power supply module, over-current protection and diagnosis and output of fault module and so on. Figure 5 shows the hardware structure of EHPS, single chip microcontroller (SCM) MC9S08AW32 is dedicated to the field of vehicles and can meet the vehicle bad working conditions.

Controller's power is supplied by the battery (12 volts). Current flows by two routes, one route 5-volt voltage which provides power to the SCM pin 22, 44, 54 is got by the power conversion circuit; another route the power is supplied to motor through the H-bridge driver circuit. Pin 21、45、55、59 grounded. Pin 3, 56 are connected to the reset pin of BDM and pin BKGD. The computer's USB port provides 5V power to simulator. Pin 4 is connected to over-current protection circuit and its level value reflects the over-current condition. Pin 11and 8 output two PWM signals and control the upper and lower tube of H-bridge by H-bridge driver circuit. Vehicle speed and steering wheel angular velocity signals are processed into pin 23 and 24. Pin 26 is connected to direction control circuit. Two motor current feedback signals are processed into pin 36 and 37. Pin 57 and 58 are connected to external crystal oscillator circuit. R1, R2, C1, C2, shown in the figure need the appropriate value, otherwise, crystal oscillator circuit does not start or stops vibration due to interference.

## 4.2 Control Software of EHPS

Figure 6 shows the flow chart of the main procedure of the EHPS control software.

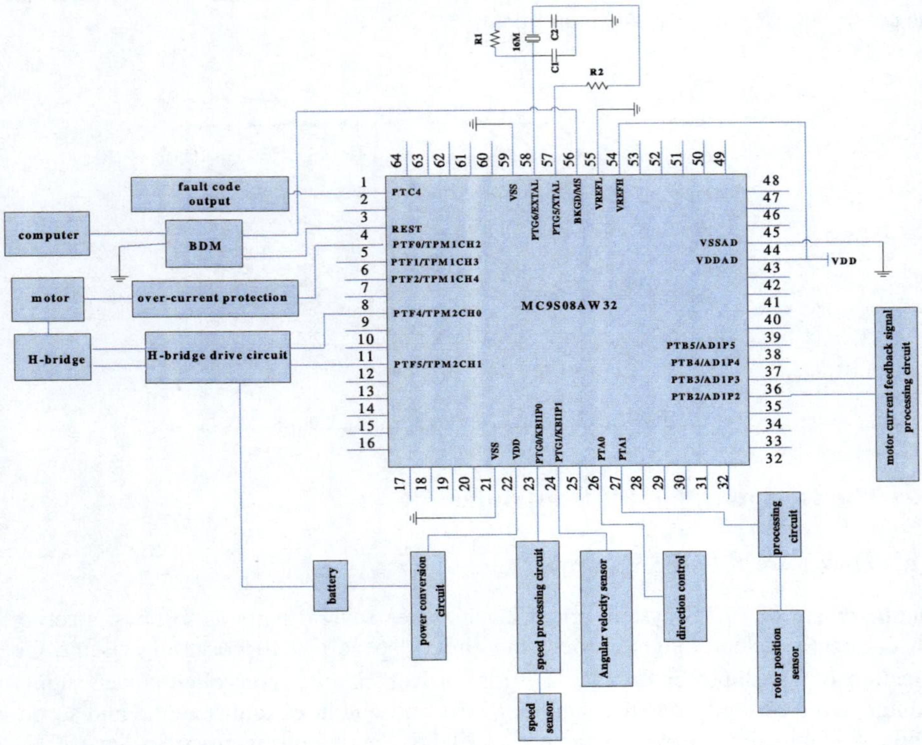


**Fig. 5.** Hardware of Controller

As shown in Figure 6 finishing the initialization of each function module of microcontroller the procedure detects the engine initial position by Hall position sensor so as to start motor normally. Then the procedure detects engine speed and judge whether it reaches the idle value. If the measured value is lower than the set idle speed value the procedure doesn't go into main loop in which the procedure detects whether the module is normal. If the fault is found the procedure goes into fault treatment loop and assist power is terminated. When the system is normal A/D sampling and conversion is started and the motor current is measured. Afterwards the level state of I/O ports, which reflects the signals of steering wheel angular velocity, is detected and judged to calculate the incremental PWM value. Then the procedure detects whether the new vehicle speed signals are detected according to which the incremental PWM value will be adjusted. Because vehicle speed signals are measured from external interrupt the last speed factor is adopted if there are no new speed signals. Afterwards the incremental PWM value is adjusted according to motor overheating and over-current conditions. Finally the PWM signals are outputted to control motor so as to provide assist power.

## 5 Test Results Analysis

After hydraulic pump is loaded to the motor and hydraulic oil is injected, according to the different speeds simulated by the procedure, the stable motor speed value without steering operation can be measured as shown in Table I . The maximal motor speed is 3600 r/min when duty cycle of PWM signal loaded in the power tube is 100%. According to Figure 4, the motor speed without steering operation tallies with the desired speed.

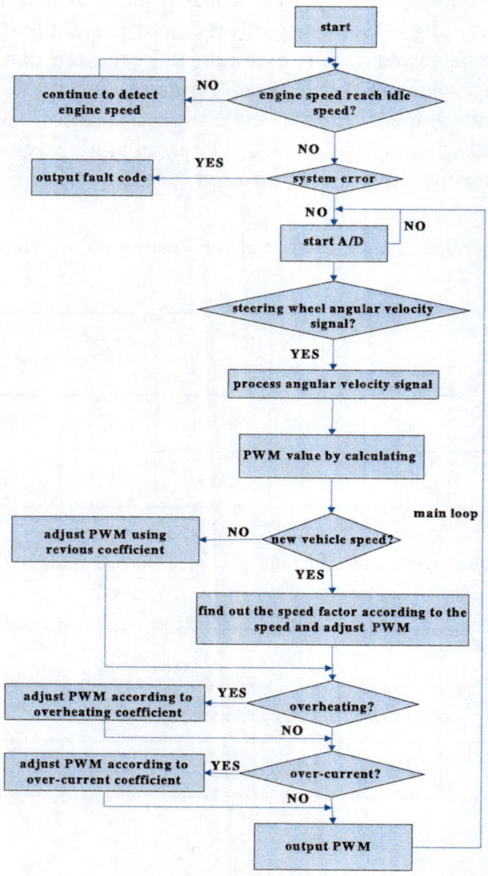


**Fig. 6.** Flow Chart of Main Procedure

Table 2 and Table 3, respectively, show the relationship between the steering wheel angular velocity and motor speed when operating the steering wheel at the vehicle speed of 20km/h and 100km/h.

**Table 1.** Function of Motor Speed and Vehicle Speed while loaded without Operating Steering Wheel

| Vehicle Speed(km/h) | 0 | 20 | 40 | 60 | 80 | 100 |
|---|---|---|---|---|---|---|
| Measured Motor Speed (r/min) | 1940 | 1667 | 1430 | 1279 | 1208 | 1245 |
| Desired Motor Speed (r/min) | 2000 | 1700 | 1500 | 1300 | 1250 | 1200 |

As shown from Table 2, the motor speed increases slowly and assist power lags behind a little when operating the steering wheel rapidly at low vehicle speed. The motor speed increases rapidly and steering efforts reduces and the steering operation is portable in other cases. The maximum hydraulic oil pressure can reach to 4.9 MPa when operating steering wheel according to oil pressure sensor.

As shown from Table 3, the change of motor speed is not obvious and only small assist power is provided when operating steering wheel at high vehicle speed. Steering feel is obvious thus steering sensitivity is ensured.

**Table 2.** Function of Motor Speed and Palstance of the Steering Wheel while Vehicle Speed is 20 km/h

| Palstance of the Steering Wheel (°/s) | 0 | 49 | 102 | 149 | 313 | 389 |
|---|---|---|---|---|---|---|
| Measured Motor Speed (r/min) | 1688 | 1702 | 2011 | 2399 | 3508 | 3564 |
| Desired Motor Speed (r/min) | 1700 | 1700 | 2057 | 2456 | 3850 | 4000 |

**Table 3.** Function of Motor Speed and Palstance of the Steering Wheel while Vehicle Speed is 100 km/h

| Palstance of the Steering Wheel (°/s) | 0 | 48 | 103 | 147 | 312 | 391 |
|---|---|---|---|---|---|---|
| Measured Motor Speed (r/min) | 1179 | 1187 | 1199 | 1334 | 1882 | 2142 |
| Desired Motor Speed (r/min) | 1200 | 1200 | 1210 | 1355 | 1900160icle speed is 12 | 2160 |

# 6 Conclusion

According to the analysis, the following conclusions can be drawn: the control algorithms of non-linear relationship of EHPS among steering wheel angular velocity, vehicle speed and motor speed , presented in this paper ,can meet the requirements that the steering operation is portable at low vehicle speed and steering feel is obvious and rapid response and energy-saving at high vehicle speed . Hardware and software system of EHPS developed by MC9S08AW32 chip can meet the requirements of assist power characteristic, steering feel and steering return-to-middle performance through tests.

## References

1. Shi, G.-B., Lin, Y., Zhang, X.: Power Steering of Car and Its Development. Transactions of the Chinese Society for Agricultural Machinery (10), 173–176 (2006)
2. Kato, H., Haga, K., Suzuki, M.: Energy-Saving Power Steering Pump. SAE paper 950582. Toyoda Machine Works, Ltd. (1995)
3. Inaguma, Y., Suzuki, K., Haga, K.: An Saving Technique in an Electro-Hydraulic Power Steering Energy (EHPS) System. SAE Paper 960934. Toyoda Machine Works Ltd. (1996)
4. Li, J.-Z.: A learning algorithm of artificial neural network based on genetic algorithm. Journal of Northwest Normal University (Natural Science) 38(2), 33–37 (2002)
5. Li, Z.-Y., Peng, L.-H.: Uncertain relationship between BP network learning and generalization ability. Science in China 10, 887–895 (2003)
6. Zhou, Z.-H., Cao-Cungen: Neural Network and Its Application. Tsinghua University Press, Beijing (2004)

# Research on the Improvement of Program Organization Structure Based on Information Management

Feixue Yang and Qianru Guo

School of Management, Tianjin University of Technology, Tianjin, China
yang_uqac@hotmail.com, loulou.80@163.com

**Abstract.** The existence of complexity of information and various interfaces within program leads to improvement on organizational structure. Based on literature review and in the perspective of information management, this paper summarizes essential characteristics of program organization structure, and finally makes three recommendations on optimization of organizational structure: establishing program management office, flattening organization structure and virtual organization.

**Keywords:** Organization structure, information management, structure improvement.

## 1 Introduction

Long-lasting and big size program becomes more common along with fast developing economy. Program management is a new issue in project management area, which has been paid close attention. In the fourth edition of PMBOK(2008), PMI' definite program as a series of interrelated projects that need be coordinated [1]. By the means of coordinate management of these projects, we can achieve strategy goals and benefits which we can not get from a single project. Along with informationization and globalization, people pay more attention to the influence of information management in program management.

## 2 Literature Review

Given the scope of this research, the literature review covered both the program organization structure and information management.

There are plenty results related to program management both in theory and practice. Most of researches focus on the flow and lifecycle of program, risk, conflict of program resources and so on. In general, there are few researches on program organizational structure.

In 1997, Peuegrinelli thought program is an organizational structure frame, which include present projects and future ones. It focuses on all activities needing attention, in order to obtain main benefits. Present researches pay more attention on the area of project-based organizations (PBO) and program management office (PMO). In 2003,

Anderson & Merna suggested the core of program management is program management office, which is a organizational entity, aiming at management all projects effectively in guide of program strategy [2]. In 2006, Thiry & Deguire discussed the landscape orientated integration of PBO in program management. They emphasized differences between organization management and project management, also the differences between organizational manager and project manager in work [3]. Peter Greiner invented "project control" method, which is used to support information management and decision making of big program. In 2007, Hu yi improved organization structure of Shanghai World Exposition program based on sustainable development strategy and project control method [4].

At the present time, most common approach to improve program management by means of information management is building information system platform.

In 2005, Guo changyou & Chen jun suggested the efficient and effective interaction between the manager of program and the managers of projects, which depends on the information framework and network to achieve, is the key of the program success [5]. Based on e-MIS, the Launch program management mode integrates project management P3 and different kinds of management theories and modules to establish the flat of program schedule and resources controlling, which carry out the benefit utmost of program management.

In 2006, Zhang lijun &Yang luyu suggested during construction period, the key for government investing program success is integrating information effectively. Public Building Service Project Information Portal (PBSPIP) is a good example [6]. In 2007, Wei faming, Liu zhongying & Zhang yanchen constructed an owner oriented program management information platform, based on analyzing of the characteristics of owner oriented progrma management [7].

At present, there are seldom researches on the relation between information management and organization structure of progrma. In 2001, Luo zuohan suggested, the implication of information technology changed enterprise environment and the foundation of information, promoting the reformation of organization system [8]. In 2005, Zhang lijun & Su ping suggested a good organization environment is an important guarantee to project information portal (PIP) [9].

In general, though some people did their researches on the relation between information management and organization structure, the researches are not enough and should be continue worked on.

## 3 The Limitations of Traditional Organization Structure

The characteristics of invalid organization are as following: first, slow or wrong decision making; second, no reaction to the changing environment; third, too many conflicts inside organization [10]. In the past time, when information technology were not so advanced, traditional organization structure ensure that the information collection and transmittal are effective, by means of strict bylaw system and clear accrual basis. Nowadays, as the development of information technology, the limitations are exposed.

Current project organization structure has many types, such as, matrix structure, line structure, project structure and ect. Generally speaking, these organization structures have too many levels, causing influent information transmission and difficulty of communication. Here are specific analysises.

First, there are problems in information communication. Take matrix structure for example, since information transmission has designated channel, maybe the communication is not so fluent. Nowadays, the transmit pattern becomes "one-to-more" instead of "one-to-one". Organizations based on new transmit technology have to coordinate activities among each department by ways of horizontal information transmission method.

Second point is decision-making problem. It is common to seen traditional organizations make decisions by supervisor. Today, supervisor needs to connect basic level since the exigency of information handling. Maybe certain extent divisionalization is good for organizations.

Last one is about organization coordination. Each department has it own roles, and members just have to finish their own work, lacking of coordination among departments. In program, information sources are complex, also there are amount of interfaces, so it is important to enhance coordination.

Therefore, current organization structure needs some changes, in order to improve information management and program quality.

# 4 Suggestions

## 4.1 Establishing PMO

PMO is an organization entity, which focuses on managing all projects of the program effectively and monitoring the whole program. In 2008, Jerzy Stawicki referred that "It is found PMO has five function types, that is, monitoring project performance, developing project management ability, multi-project management strategic management and learning organization [11]". Implement of PMO depends on the type project or program.

## 4.2 Flattening Organization Structure

*a)* Flattened organization structure caters to the improvement of information technology. The environment and information that organization facing is complex in program, in order to transmit information fast and fluent, basic level should have more autonomy rights [12]. As levels are less in flattened organization structure, the transmission will be fast. Also, it is good for unleashing people's ability. However, organization structure can't be too flattened. Certain flexibility is important [4].

## 4.3 Virtual Organization

Virtual organization is a networking organization made up by each independent organizations based on certain market opportunity. Virtual organization is a kind of dynamic alliance, composed by organizations providing complementary competition or resources. It works as a union providing products or service for the common interests. As for program, it is composed of many related projects. For each project, we can establish a virtual organization to support. What is more important is that we need coordinate all these related core virtual organization.

## 5 Summary

Through the analysis of disadvantages of traditional organizational structure, in the perspective of information management, this paper proposes some suggestions, which can be beneficial to improving information management within program. However, the suggestion mentioned above confines itself to impracticability and its application need to be tested in practice.

## References

1. PMI, A Guide to the project Management Body of Knowledge (PMBOK Guide), 4th edn. (2008)
2. Anderson, D., Merna, T.: Project Management Strategy–project management represented as a process based set of management domains and the consequences for project management strategy. International Journal of Project Management 21(6), 387–393 (2003)
3. Thiry, M., Deguire, M.: Meng meiyan. New Development of PBO. Project Management Technology 10, 24–28 (2006)
4. Hu, Y.: Programme Management of Shanghai World Exposition based on sustainable development strategy, pp. 60–61. Tongji University, Shanghai
5. Guo, C., Chen, J.: Study on program management mode based on information framework. Journal of Industrial Engineering and Engineering Management, 19–21 (2005)
6. Yang, L., Yang, L.: Feature and Inspiration of Public Building Service Project Information Portal. Construction Economy 9, 46–48 (2006)
7. Wei, F., Liu, Z., Zhang, C.: Implementation of owner oriented program management information platform. Economic & Trade Update 4, 75–76 (2007)
8. Luo, Z.: Study on the influence of information technology to enterprise organization structure. China Economist 1, 32–33 (2001)
9. Zhang, L., Su, P.: Analysis on organization environment leading to success of PIP. Modern Management Science 11, 40–42 (2005)
10. Zheng, B., Fu, Y.: Enterprise information technology and change of organization structure. Economic Management 15 (2002)
11. Jerzy, S., Li, Z.: PMO—means of transition to project organization. Project Management Technology 3, 68–72 (2008)
12. Chen, X.: Research on the adjustment of organization structure based on enterprise informarionization. Academic Research 2, 43–47 (2008)

# A Simulation Research of Nonlinear Control Method in Triple Inverted Pendulum

Zhang Ling, Qiao Wei, and Zheng Enrang

College of Electric and Communication Engineering, Shaanxi University of Science and Technology, Xi'an, Shaanxi, China
zhenger@sust.edu.cn

**Abstract.** To linearize the triple inverted pendulum model, which has the features of nonlinear, high-order,multivariable and strong-coupling, proposed a method of partial feedback leariation based on differential geometry theory. A Single neuron adaptive Proportion-Differential (SNPD) control was used to the system, compared with Linear Quadratic Regulator (LQR) strategy. The former improves the stability and enhances anti-jamming capability of the system. The results of simulation show that the proposed linearization and control method perform well.

**Keywords:** Triple inverted pendulum, differential geometry, partial feedback linearization, SNPD control, LQR control.

## 1 Introduction

The research about inverted pendulum was always a hotspot in the development of control theory, which is a typical, non-linear, high-level, multivariable, strong coupling and fast movement of natural unstable system. Hence, many scholars has proposed various control methods, such as neural network control[1]、H∞control[2]、fuzzy control[3-5]and so on. However, the model which used in these control methods often obtains through modeling or experimental modeling.The nonlinear model using Taylor expansion in the vicinity of equilibrium point, and then take the linear approximation. The main drawback of this linearization is that it has errors, and the error increases with the extending of region.

Based on the above issues, in this paper, used the theory of differential geometry to replace the method of linearization by Taylor series.The model after partial feedback linearization has a better effect than that only removes highend items,by which the error decreases, the model is closer to the original system model.And then used Proportion-Differential (PD) control compared with Linear Quadratic Regulator (LQR) control method to the linearized system model for simulation analysis. Simulation results show that, by the PD control can not only improve the system stability, but also enhance anti-jamming capability than by LQR control.

## 2 The Mathematical Model of Triple Inverted Pendulum System[6]

The principle structure of triple inverted pendulum system shown in Figure 1.

According to kinetic theory, the mathematical model of the system can be shown in the following:

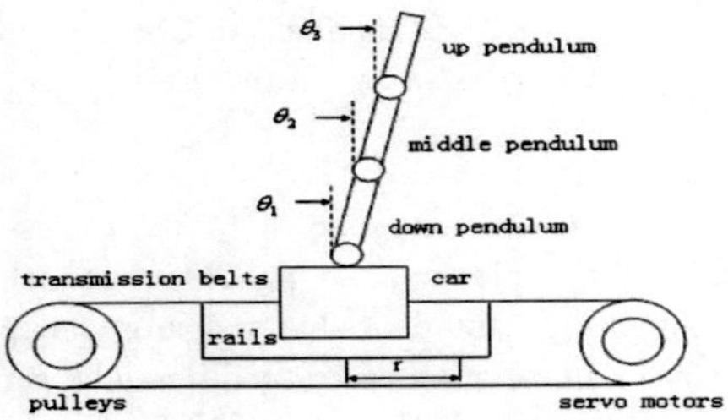


**Fig. 1.** Principle structure of triple inverted pendulum

$$\left[\ddot{r};\ddot{\theta}_1;\ddot{\theta}_2;\ddot{\theta}_3\right]^T = M(\theta_1,\theta_2,\theta_3)^{-1} F(\theta_1,\theta_2,\theta_3,\dot{\theta}_1,\dot{\theta}_2,\dot{\theta}_3)\left[\dot{r};\dot{\theta}_1;\dot{\theta}_2;\dot{\theta}_3\right]^T$$
$$+M(\theta_1,\theta_2,\theta_3)^{-1} N(\theta_1,\theta_2,\theta_3) + M(\theta_1,\theta_2,\theta_3)^{-1}\left[G_0;0;0;0\right]^T u$$

where:

$$M(\theta_1,\theta_2,\theta_3) = \begin{bmatrix} m_0+m_1+m_2+m_3 & (m_1d_1+m_2d_4+m_3d_4)\cos(\theta_1) & (m_2d_2+m_3d_5)\cos(\theta_2) & m_3d_3\cos(\theta_3) \\ (m_1d_1+m_2d_4+m_3d_4)\cos(\theta_1) & J_1+m_1d_1^2+m_2d_4^2 & (m_2d_2d_4+m_3d_4d_5)\cos(\theta_2-\theta_1) & m_3d_3d_4\cos(\theta_3-\theta_1) \\ (m_2d_2+m_3d_5)\cos(\theta_2) & (m_2d_2d_4+m_3d_4d_5)\cos(\theta_2-\theta_1) & J_2+m_2d_2^2+m_3d_5^2 & m_3d_3d_5\cos(\theta_3-\theta_2) \\ m_3d_3\cos(\theta_3) & m_3d_3d_4\cos(\theta_3-\theta_1) & m_3d_3d_5\cos(\theta_3-\theta_2) & J_3+m_3d_3^2 \end{bmatrix}$$

$$F(\theta_1,\theta_2,\theta_3,\dot{\theta}_1,\dot{\theta}_2,\dot{\theta}_3) = \begin{bmatrix} -f_0 & (m_1d_1+m_2d_4+m_3d_4)\sin(\theta_1)\dot{\theta}_1 & (m_2d_2+m_3d_5)\dot{\theta}_2\sin(\theta_2) & m_3d_3\dot{\theta}_3\sin(\theta_3) \\ 0 & -(f_1+f_2) & f_2+(m_2d_2d_4+m_3d_4d_5)\dot{\theta}_1\sin(\theta_2-\theta_1) & m_3d_3d_4\dot{\theta}_3\sin(\theta_3-\theta_1) \\ 0 & f_2-(m_2d_2d_4+m_3d_4d_5)\dot{\theta}_1\sin(\theta_2-\theta_1) & -(f_3+f_2) & f_3+m_3d_3d_5\dot{\theta}_2\sin(\theta_3-\theta_2) \\ 0 & -m_3d_3d_4\dot{\theta}_1\sin(\theta_3-\theta_1) & f_3-m_3d_3d_5\dot{\theta}_2\sin(\theta_3-\theta_2) & -f_3 \end{bmatrix}$$

$$N(\theta_1,\theta_2,\theta_3)=\begin{bmatrix} 0 \\ (m_1d_1+m_2d_4+m_3d_4)g\sin(\theta_1) \\ (m_2d_2+m_3d_5)g\sin(\theta_2) \\ m_3d_3g\sin(\theta_3) \end{bmatrix}$$

$r,\dot{r},\ddot{r},\theta_1,\dot{\theta}_1,\ddot{\theta}_1,\theta_2,\dot{\theta}_2,\ddot{\theta}_2,\theta_3,\dot{\theta}_3,\ddot{\theta}_3$ respectively for the car position, velocity, acceleration, hem, the pendulum and the pendulum and the vertical line, angle, angular velocity and angular acceleration; $u$ for the force imposed on car.

Where $m_0 \sim m_3$ respectively represents the mass of the car,the down pendulum, the middle pendulum and the up pendulum. $J_1 \sim J_3$ are moment of inertia of the down,the middle and the up pendulum. $d_1 \sim d_3$ respectively for the distance between the down pendulum to the shaft, the middle pendulum to the shaft center and the up pendulum to the shaft. $d_4$ and $d_5$ is the length of the down and the middle pendulum. $f_0 \sim f_3$ respectively for the coefficient of friction between the rail and Car, the shaft and the down pendulum, the shaft and the middle pendulum and the shaft and the up pendulum. $G_0$ is the ratio of force and control voltage. $g$ is acceleration gravity.

Take the state vector $x=[x_1,\cdots,x_8]=[r,\theta_1,\theta_2-\theta_1,\theta_3-\theta_2,\dot{r},\dot{\theta}_1,\dot{\theta}_2-\dot{\theta}_1,\dot{\theta}_3-\dot{\theta}_2]$, then, obtained the state equation of the system as follows:

$$\begin{cases} \dot{x}=f(x)+g(x)u \\ y=h(x) \end{cases} \tag{1}$$

## 3 Feedback Linearization

Triple inverted pendulum control for the role of the force on the movement can be regarded as a single-input system.

Based on mathematical model and the relevant theorems from differential geometry, at the $x=0$, the system's relative degree is $5<n$. Can be provened $g,ad_fg,...,ad_f^7g$ are linear dependence, and $ad_f^4g(x)\notin span\ \{g,\cdots,ad_f^7g\}$.

Therefore, triple inverted pendulum can not been completely feedback linearized.

After computing $ad_f^4g(x)\notin \bar{D}_3(x)=inv.cl.span$ $\{g,\cdots,ad_f^3g\}$, as the system model has more dimension, so by solving the partial differential equations $L_g\lambda(x)=L_{ad_fg}\lambda(x)$

$= \cdots = L_{ad_f^{n-2}} \lambda(x) = 0$ to get the output $\lambda(x)$ of the system at the relative equilibrium point of order $n$ is very difficult, then we can get a part of the state-feedback linearization of the system[7].

For the system (1), if the system relative degree $r < n$, through the state-feedback:

$$u = \frac{1}{a(x)}(-b(x) + v) \tag{2}$$

Can get one of the following systems:

$$\{\dot{z}_1 = z_2, \dot{z}_2 = z_3; \cdots, \dot{z}_{r-1} = z_r, \dot{z}_r = v, \dot{z}_{r+1} = q_{r+1}(z);$$

$$\dot{z}_{r+1} = q_{r+1}(z); \cdots \dot{z}_n = q_n(z)\} \tag{3}$$

It is clear that the system can be decomposed into a linear subsystem of dimension $r$ and one dimensional $n - r$ nonlinear subsystem, as shown in Figure 2:

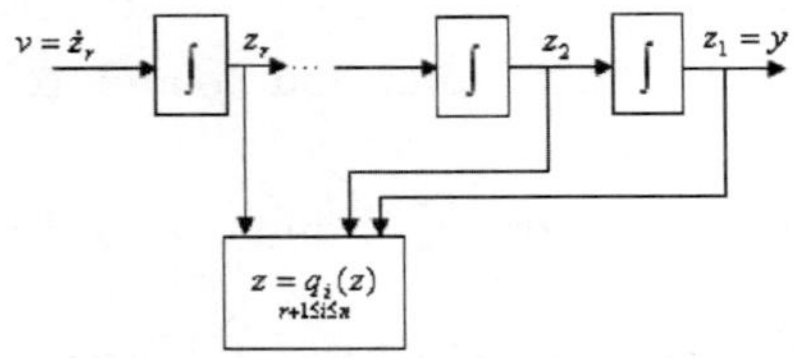


**Fig. 2.** Structure diagram of partial feedback linearization

Because the input force to control the car, so hem of the car selected angle output, $y = h(x) = x_2$, therefore, now the nonlinear triple inverted pendulum model by the method of state feedback control rule is :

$$u = -\frac{L_f^4(h(x))}{L_g L_f^3(h(x))} + \frac{1}{L_g L_f^4(h(x))} v = -\frac{f_5(x)}{g_5(x)} + \frac{1}{g_5(x)} v \tag{4}$$

The system model obtained by introducing feedback, that is take formula (4) into formula (1) as the partial feedback linearization model, then make the Taylor expansion at the equilibrium point of the obtained model, there is: $\sin x \approx x$, $\cos x \approx 1$, get the final linear model and the Matrix of the A,B:

$$\begin{cases} \dot{x} = Ax + Bu \\ y = Cx \end{cases} \tag{5}$$

where: $C = I_{8\times8}$

The open-loop pole of the triple inverted pendulum system $\{A,B,C\}$ is $(-13.2343,-7.1847,-3.1600,12.2089,\ 3.1483,6.8892,0,0)$, that is, the system is not stable. Where $rankQ_c=8$ , $rankQ_o=8$ , so, after partial feedback linearization of the triple inverted pendulum system is completely controllable and is complete observable.

## 4 Stabilization Control of Triple Inverted Pendulum

### 4.1 Linear Quadratic Regulator (LQR)

The optimal design of conventional LQR refers to design appropriately the state feedback controller K to minimized quadratic target function J, where the value of the K is only decided by the weight matrix Q and R. Control system structure shown in Figure3:

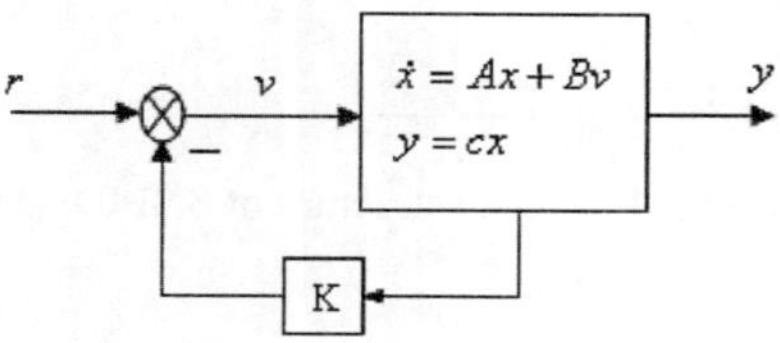


**Fig. 3.** Control structure of linear feedback

### 4.2 Single Neuron Adaptive PID (SNPID)

Single neuron adaptive PID (SNPID) inherited the all advantages of the neurons and the traditional PID, which has self-learning and adaptive ability and has strong robustness. SNPID is simple in structure, but can solve the traditional on-line real time PID tuning parameters of some system of lack of effective control, sush as, with a number of complex processes and parameters of slowly time-varying.

SNPID can realize the function of self-adaptive, self-organization by means of adjusting the weighting coefficients, which is based on supervised Hebb learning rules. The control and learning algorithm are as follows:

$$u(k)=u(k-1)+K\sum_{i=1}^{3}\overline{w_i(k)}x_i(k) \tag{6}$$

$$\overline{w_i(k)}=w_i(k)/\sum_{i=1}^{3}|w_i(k)| \tag{7}$$

$$\begin{cases} w_1(k)=w_1(k-1)+\eta_I z(k)u(k)x_1(k) \\ w_2(k)=w_2(k-1)+\eta_P z(k)u(k)x_2(k) \\ w_3(k)=w_3(k-1)+\eta_D z(k)u(k)x_3(k) \end{cases}$$

Where:

$$\begin{cases} x_1(k) = e(k) \\ x_2(k) = e(k) - e(k-1) \\ x_3(k) = \Delta^2 e(k) = e(k) - 2e(k-1) + e(k-2) \end{cases}$$

$\eta_I$, $\eta_P$, $\eta_D$ is integral, ratio, differential learning rate respectively, $K$ for the gain of neurons coefficient, notes that, $K > 0$.

Integral, proportionment and differential has adopted different learning rate $\eta_I$, $\eta_P$, $\eta_D$, in order to separate the different weights to adjust.

Control system structure shown in Figure 4:

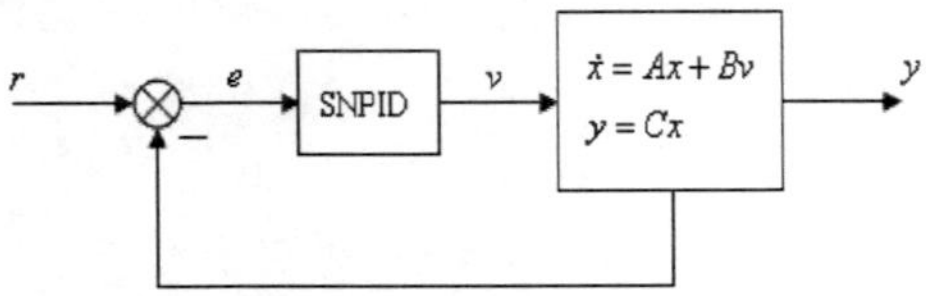


**Fig. 4.** Control structure of SNPID

## 5 Simulation and Analysis

First, designed the state feedback matrix $K$. According to the equation(5), after partial feedback linearization the system should be fully controlled. In accordance with literature[9], in the experience, selection $Q = diag(200, 500, 200, 200, 0, 0, 0, 0)$, $R = 1$. Use eight single-neurons as the input of PID controller. Simulation results are as follows.

Case 1, the initial position of the car at the right side of 0.2m in the rail, so in a very short period of time back to the central rail, and to maintain the non-inverted pendulum. Simulation of the system response curve shown in Figure 5, the black line for the PD control method, the green line for the LQR control methods. By comparised we can see, in the diagram a, the car changes in displacement, contrasted with LQR control, PD control can reduce the overshoot, but at the same time will increase the adjustment; in the diagram b, the angle of pendulum changes, PD control can significantly improve the down pendulum of the oscillation amplitude, regulating time is the same; in the diagram c, d plans were placed under the angle changes and the change in angle on the down-up pendulum, PD control is well placed on the middle-up pendulum of the amplitude of the shocks, and the adjustment time is slightly better than the LQR control.

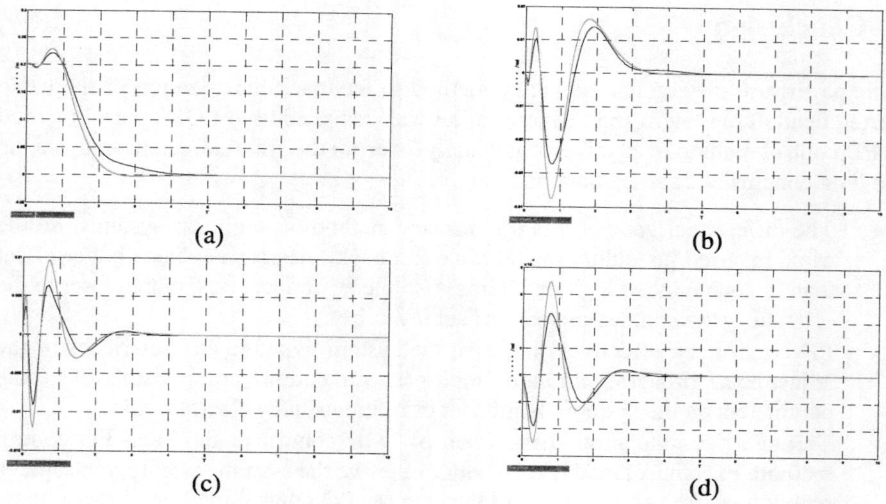


**Fig. 5.** Time response for case 1

Case 2, the initial condition is that the car is located in the central rail, and the triple inverted pendulum is in equilibrium, when the pendulum in the 2s to join 0.2rad disturbance, analysis of the inverted pendulum controller anti-jamming capability. Simulation of the system response curve shown in Figure 6, the comparison can be seen, figure a, in the PD control, car displacement slight increase in the magnitude of changes, and the adjustment time is increased; figue b, PD control of lower the amplitude of down pendulum, adjust time slightly increased; figue c and d, the angle range of the middle-down pendulum improved when use PD control method.And the regulation time of their both method is the same.

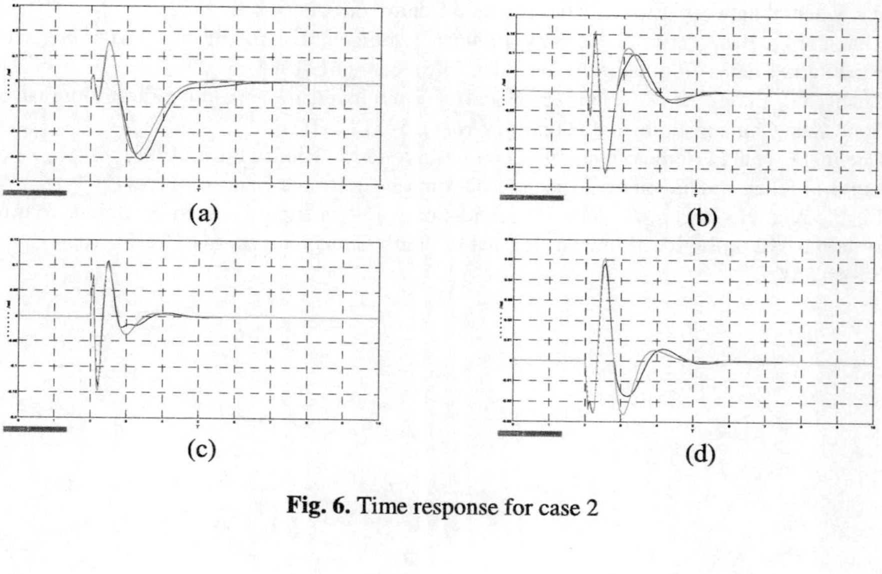


**Fig. 6.** Time response for case 2

## 6 Conclusion

In this paper, use differential geometry method to feedback linearization of the triple inverted pendulum system and then controller was designed used LQR control method and PD control method respectively, and analysis.From the simulation and analysis the following conclusions can be obtained:

- The differential geometry is the primary method of nonlinear systems, So the triple inverted pendulum model after feedback linearization has a better effect than the one remove high-end items.Also the error decreases, more closer to the original system and an increase of achievability.
- PD control can effectively improve the system dynamic characteristics of the adjustment process, through multiple simulation comparison of some parameters on the system, the impact of reference play a certain role.
- Through the simulation comparison of LQR control method and PD control method, PD control method can better improve the system stability, but same in regulating time; about the anti-interference, PD control method slightly better than LQR control method. In general, PD control can achieve better control effect.

## References

1. Yan, D., Saif, M.: Neural networks based controllers for non-linear systems. Control and Computers 23(3) (1995)
2. Van der Schaft, A.J.: L2-gain analysis of nonlinear systems and nonlinear H∞ control. IEEE Transaction on Automatic Control 37(6), 770–784 (1992)
3. Xing, J., Wang, L., Dai, G.: Accelerating fuzzy control process of inverted pendulum with BP Algorithm. Journal of Northwestern Polytechnical University 14(4), 539–544 (1996)
4. Yang, Z., Cheng, S., Pu, Y.: The control of a double inverted pendulum by using hierarchical fuzzy neural network. Electric Machines and Control 6(3), 246–248 (2002)
5. Zhang, M., Hao, J., et al.: Imitating-human intelligent control of triple link inverted-pendulum. Chinese Journal of Aeronautics 16(6), 654–661 (1995)
6. Zhang, G., Li, Z., Bi, X.: Nonliner model of triple inverted pendulum system. Journal of Soutnwest China Institute of Techmology 16(4), 5–15 (2001)
7. Cheng, D., et al.: Geometric theory of nonlinear systems. Science Press (1988)
8. Liu, J.: Matlab simulation of advanced PID control. Electronics Industry Press (2004)
9. Li, Y., Zhu, X., Zheng, X., Yin, L.: Simulation study on triple inverted-pendulum control system based on LQR Arithmetic. Journal of North China University of Technology 18(3), 35–40 (2006)

# Uncertainty Study of Measured Results for Electronic Thermometer Auto-verification System


Huang Yanyan and Xu Hongwei

Mechatronics Institute, China Jiliang University, Hangzhou, China
{hyy9791,xhw}@cjlu.edu.cn



**Abstract.** We have developed an electronic thermometer auto-verification system to improve efficiency of calibration. The uncertainty study of measured results is an important step to evaluate the performance of system. This paper gives the sources of uncertainty and their impact on the overall measurement. The standard uncertainty, composite uncertainty and extended uncertainty are calculated and discussed in detail. After the uncertainty is analyzed, it shows that the electronic thermometer auto-verification system is reliable.



**Keywords:** Uncertainty study, electronic thermometer, auto-verification.


## 1 Introduction

Mercury thermometer has been banned in many countries for its easily broken and mercury leak. Electronic thermometer is widely used in daily life for its intuitive readings, easy to carry and not easily damaged .At present, the electronic thermometer is calibrated by man. Human error and less efficient can not be eliminated. So we have developed an electronic thermometer auto-verification system. It is necessary to analysis the uncertainty of the system so that it can meet the testing requirements.

## 2 Electronic Thermometer Auto-verification System

To achieve auto-verification, it is the key to change the method of observing the readings by eyes. The thermometer is inserted vertically into a constant temperature slot. The image of the calibrated thermometer seized by the camera is input to the computer by the image acquisition card. Computer can control mechanical devices so six calibrated thermometers are moved to a suitable location by turns for a fixed camera to read. Use second-class standard platinum resistance thermometer for the measurement standard and its resistance values measured by the digital multimeter are acquired by computer automatically .Software system is modularized including data communications, image acquisition and processing, mechanical drive control, calibration judge, data processing and statement printing.

## 3 Mathematical Model

The indication error of electronic thermometer in a test point is calculated according to the formula (1).

$$\Delta t = \overline{t_i} - \overline{t_0} \tag{1}$$

In the formula:

$\Delta t$ ——indication error of calibrated electronic thermometer in each test point, °C.

$\overline{t_i}$ ——the average value of calibrated electronic thermometer in each test point, °C.

$\overline{t_0}$ ——the average value of standard thermometer in each test point, °C.

In each test point, $t_0$ is calculated according to the following steps.

$$W(t) = \frac{R_t}{R_{tp}} \tag{2}$$

$$\Delta W(t) = a_8 [W(t) - 1] + b_8 [W(t) - 1]^2 \tag{3}$$

$$W_r^*(t) = W(t) - \Delta W(t) \tag{4}$$

$R_t$ ——the measured value of second-class standard platinum resistance thermometer in each test point Ohm,

$R_{tp}$ ——the measured value of second-class standard platinum resistance thermometer in three-phase point of water Ohm,

$a_8$、$b_8$ ——the indexing constant of second-class standard platinum resistance thermometer which is available in its calibration certificate,

$W(t)$ ——the measured value of the resistance ratio for the second-rate standard platinum resistance thermometer in each test point,

$\Delta W(t)$ ——the margin function value of the resistance ratio for the second-rate standard platinum resistance thermometer in each test point,

$W_r^*(t)$ ——the reference function value of the resistance ratio for the second-rate standard platinum resistance thermometer in each test point.

After $W_r^*(t)$ is calculated, we can check the reference function table of standard platinum resistance thermometer to calculate $t_0$.

The uncertainty of $\overline{t_i}$ and $\overline{t_0}$ is analyzed respectively so that the uncertainty of measurement results can be acquired.

# 4 Uncertainty Assessment

## 4.1 Standard Uncertainty of Input Quantity

*1) uncertainty of* $\overline{t_i}$

a) uncertainty caused by measurement repeatability

In some test point, uncertainty caused by measurement repeatability:

$$u_{1.1} = \frac{\sqrt{\frac{1}{n-1}\sum_{i=1}^{n}(x_i - \overline{x})^2}}{\sqrt{n}} \ ^\circ\mathrm{C} \tag{5}$$

Degrees of freedom $v_{1.1}$ =n-1, normal distribution, type A.

b) uncertainty caused by little difference in temperature of slot

By the verification regulation, slot shall have a temperature gradient of not greater than ±0.01°C. The largest difference in temperature is considered.

$$u_{1.2} = \frac{0.01}{\sqrt{3}} = 0.00578\,^\circ\mathrm{C},$$

uniform distribution, estimates of related uncertainty is 20%,then $v_{1.2} = 12$, type B.

c) uncertainty caused by fluctuation of slot temperature

The slot temperature stability is not greater than ± 0.015°C /15min when thermometer is to be tested. The largest fluctuation of slot temperature is considered and it takes less than 15 minutes in some test point.

$$u_{1.3} = \frac{0.015}{\sqrt{2}} = 0.0106\,^\circ\mathrm{C},$$

inverse sine distribution, estimates of related uncertainty is 50%,then $v_{1.3} = 2$, type B.

d) uncertainty caused by quantized error

Quantized error of electronic thermometer with 0.1°Cresolving power is ± 0.05°C

$$u_{1.4} = \frac{0.05}{\sqrt{3}} = 0.0289\,^\circ\mathrm{C},$$

uniform distribution, estimates of related uncertainty is 20%,then $v_{1.4} = 12$, type B.

*2) uncertainty of* $\overline{t_0}$

By the formula (2), (3), (4), it is available that:

$$\begin{aligned} dW_r^*(t) &= dW(t) - a_8 dW(t) - 2b_8\left[W(t) - 1\right]dW(t) \\ &= [1 - a_8 + 2b_8 - 2b_8 W(t)]dW(t) \end{aligned} \tag{6}$$

$$U_{w_r^*}(t) = [1 - a_8 + 2b_8 - 2b_8 W(t)] U_{w(t)} \tag{7}$$

$$U_{\overline{t_0}} = \frac{U_{w_r^*}(t)}{dW_r / dt} = \frac{[1 - a_8 + 2b_8 - 2b_8 W(t)] U_{w(t)}}{dW_r / dt} \tag{8}$$

From above uncertainty of $\overline{t_0}$ can be calculated.

Uncertainty of $W(t)$

a) uncertainty caused by digital multimeter measurement

The Agillent Technologies 34401A which we use is a 6.5-digit digital multimeter. When its range of measurement is 100ohm, its testing current is 1mA. The multimeter's accuracy in one year is expressed in the form:

$$A = \pm(reading \times 100 \times 10^{-6} + range \times 40 \times 10^{-6}) \tag{9}$$

From formula (2) and (9) we can draw that:

$$dW(t) = \frac{1}{R_{tp}} dR_t - \frac{R_t}{R_{tp}^2} dR_{tp} \tag{10}$$

$$\begin{aligned} \delta_{2.1}' &= \frac{1}{R_{tp}} (R_t \times 100 \times 10^{-6} + 100 \times 40 \times 10^{-6}) \\ &\quad - \frac{W_t}{R_{tp}} (R_{tp} \times 100 \times 10^{-6} + 100 \times 40 \times 10^{-6}) \\ &= (1 - W_t) / R_{tp} \times 100 \times 40 \times 10^{-6} \end{aligned} \tag{11}$$

$u_{2.1}' = \delta_{2.1}' / \sqrt{3}$, uniform distribution, estimates of related uncertainty is 10%, then $v_{2.1}' = 50$, type B.

b) uncertainty caused by measurement repeatability

$u_{2.2}' = \dfrac{\sqrt{\dfrac{1}{n-1}\sum_{i=1}^{n}(R_{ti} - \overline{R_{ti}})^2}}{R_{tp}\sqrt{n}}$, Degrees of freedom $v_{2.2}' =$ n-1, normal distribution, type A

c) uncertainty caused by changes in the value of $R_{tp}$ for standard platinum resistance thermometer

The value of $R_{tp}$ for standard platinum resistance thermometer can be changed during the thermal process. Therefore, the changes will have a direct impact on the results. Changes of 5mK will be permitted. We use 5mK to calculate.It can be converted into the value of $W$ ($5\times10^{-3}\times3.9885407\times10^{-3}$ =$1.995\times10^{-5}$) .

$$u'_{2.3} = \frac{1.995\times10^{-5}}{2\times\sqrt{3}} = 5.7576\times10^{-6}$$

uniform distribution, estimates of related uncertainty is 20%,then $v'_{2.3} = 12$, type B.

d) uncertainty caused by calculating the temperature of standard platinum resistance thermometer

According to ITS-1990, the uncertainty of its interpolation formula is 1mK.Deuce distribution can be treated and it is very reliable.

$$u'_{2.4} = {dW}\Big/{dt} \times 0.5\times10^{-3}, v'_{2.4} = \infty, \text{ type B.}$$

e) The self-heating effects can be ignored since the standard platinum resistance thermometer is inserted directly into the media which is forced convection.

f) uncertainty caused by little difference and fluctuation of slot temperature has been included in the analysis of the calibrated thermometer, so it is not repeated here.

$$U_{\bar{t_0}} = \frac{[1-a_8+2b_8-2b_8W(t)]U_{w(t)}}{{dW_r}\Big/{dt}} = kU_{w(t)}, \tag{12}$$

## 4.2 Combined Standard Uncertainty, Effective Degrees of Freedom and Coverage Factor

The combined standard uncertainty in some test point:

$$u_c(t) = \sqrt{U^2_{\bar{t_i}} + U^2_{\bar{t_0}}}$$

$$= \sqrt{u^2_{1.1} + u^2_{1.2} + u^2_{1.3} + u^2_{1.4} + (ku'_{2.1})^2 + (ku'_{2.2})^2 + (ku'_{2.3})^2 + (ku'_{2.4})^2} \tag{13}$$

According to Welch-Satterthwaite formula, the DOF of $u_c(t)$ can be calculated

$$v_{eff} = \frac{u^4_c(t)}{\frac{u^4_{1.1}}{n-1} + \frac{u^4_{1.2}}{12} + \frac{u^4_{1.3}}{2} + \frac{u^4_{1.4}}{12} + \frac{(ku'_{2.1})^4}{50} + \frac{(ku'_{2.2})^4}{n-1} + \frac{(ku'_{2.3})4}{12}} \tag{14}$$

Coverage Factor, $t_{95}(v_{eff})$

## 4.3 Expanded Standard Uncertainty

$$u_{95} = t_{95}(v_{eff}) \times u_c(t), \tag{15}$$

Based on the above analysis, we measured four times at the test point of 37°C. The uncertainty of measurement results is as follows:

$$u_c(t) = 0.03\,^{\circ}\mathrm{C},$$

$$\upsilon_{eff} = 10\,,$$

$$u_{95} = t_{95}(v_{eff}) \times u_c(t) = 2.228 \times 0.043 = 0.068\,^{\circ}\mathrm{C}$$

## 5 Conclusion

After the uncertainty is analyzed, it shows that this auto-verification system is reliable. Software system which includes auto test, image acquisition and data processing can improve efficiency of calibration. LabVIEW is used for system development. It has friendly interface, rich feature and simple operation method. Therefore it can replace the verification system by manpower.

## References

1. JJF1059-1999 Evaluation and Expression of Uncertainty in Measurement
2. A Platinum Resistance Thermometer Calibration between 2°K and 273115. Journal of Scientific Instruments (Journal of Physics E) Series 2 1 (1968)
3. Vaughn, D., Strouse, G.F.: The Nist Industrial Thermometer Calibration Laboratory
4. Flores, N.C., Elizabeth, M.S., Boyle, A.E.: Thermometer Calibration Guide
5. JJG(Zhe)87-2006 Clinical Electronic Thermometer

# Design of a Novel Radix-2 Floating-Point FFT Processor Based on FPGA

Zhengyan Liu, Enrang Zheng, and Lingkun Ma

College of Electric and Communication Engineering, Shaanxi University of Science and Technology, Xi'an, Shaanxi, China
liuzhengyan1984@163.com

**Abstract.** Aiming at the speed and accuracy bottleneck of FFT processor in harmonic analyzer, proposed a method of using parallel processing, pipelinine structure floating-point operation, and an improved butterfly unit based on FPGA to design the processor. Used bottom-up method and VHDL language to program function modules, futher used the tool of quartus II to simulate the 1024 complex points. The results show that the processor improves the speed based on the high accuracy, could meet the real-time requirement of harmonic analyzer.

**Keywords:** FFT processor, parallel processing, pipeline structure, floating-point operation.

## 1 Introduction

A large number of non-linear components are used in power network with the development of power electronic technique, generate a lot of higher harmonic, it will affect the normal operation of power system, so the harmonic detection is necessary. The harmonic detection of power system usually via the Fast Fourier Transform(FFT),so designing of FFT processor is essential.

With the development of field-programmable gate arrays (FPGA)technology in recent years, design the FFT processor based on parallel and faster FPGA chip has become an inevitable trend. The key technology of FPGA is :the development cycle is short, the design cost is low, could process data parallel and easy to implement the pipeline structure, which are very suitable for realizing FFT algorithm[1].

The study on FFT algorithm always has the core to reduce the numbers of operation, however, the faster FFT has to pay more hardware, especially when the input data points larger. Simple control and regular achieved algorithm is better than the algorithm of only operation dominance in hardware system. There are many researches about FFT based on FPGA, but most are fixed-point format, has low accuracy[2].So a floating-point processor has designed facing with this trend in this paper[3].Used the pipeline technology and parallel processing with four data in the butterfly unit module to shorten the FFT operation cycles. Made the butterfly unit structure appropriate improvement with using 3-input adder to shorten the pipeline latency. The time of processing 1024 floating points is 67 $\mu$s when the external clock frequency is 100MHz. The processing time is far less than the sampling time of

1024-point sequence, so the sampling is continuous, could meet the real-time requirement of harmonic analyzer.

## 2 Design of the Overall Structure of FFT Processor

The overall structure of FFT processor is shown in Fig.1.

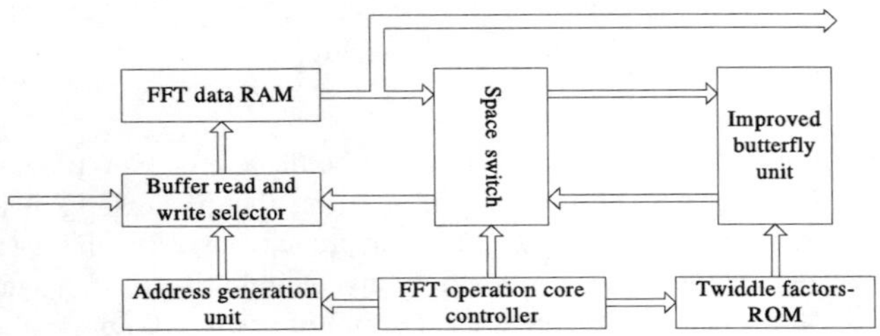


**Fig. 1.** Overall structure of FFT processor

This module is equivalent to a 4-addressable RAM when the buffer read and write selector selects the external circuit to operate the FFT data RAM to read and write, otherwise the module is on-going FFT operation.

On the one hand the FFT operation core controller provides the original addresses to address generation unit, on the other hand provides the read addresses to twiddle factors-ROM, enable it to output the twiddle factors. The address generation unit through operation to generate the new read and write addresses, then operates the FFT data RAM to ouput the data A and B for floating-point butterfly unit after throughing the buffer read and write selector.

Every step of FFT processor has regular addressing, however, the technology that once addresses four registers will be break the rule, especially in the first two steps of the processor. In order to solve this problem, the space switch is customized specially ,the output array in the wrong addressing is exchanged according to certain principle, the four arrays are got which are consistent with the butterfly operation. Therefore, these A and B are got from FFT data RAM which are adjusted by space switch.

When the operation of floating-point butterfly unit is completed, outputs the results to space switch, then the space switch adjusts these data again, outputs results to selector which let the FFT data RAM to read. The circuit could complete the complex FFT opertation under the FFT operation core controller.

The designing of an improved butterfly unit is mostly described in this paper.

## 3 Design of the Improved Butterfly Unit

### 3.1 Principle of the Improved Butterfly Unit

The diagram of radix-2 butterfly unit is shown in Fig.2[4].

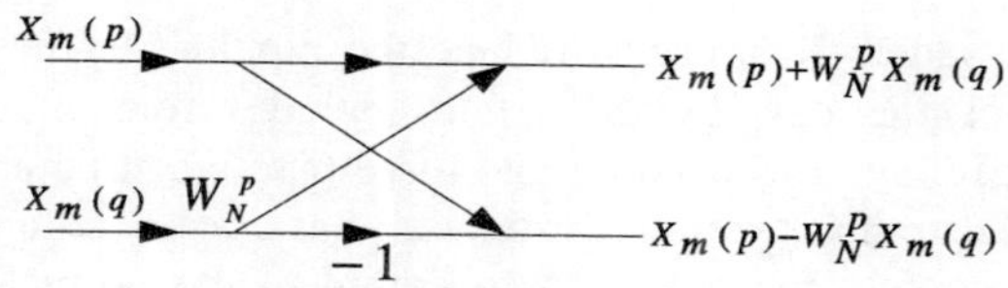


**Fig. 2.** Diagram of radix-2 butterfly unit

It will be complete the opertation of $\text{Re}\,sult = A \pm B \cdot J_x$, where $J_x$ is the twiddle factor.

Let $A = a + bi$, $B = c + di$, $J_x = \cos\theta + i\sin\theta$, where $a, b, c, d, \cos\theta, \sin\theta$ are all real floating-point numbers, and according to Fig.2 we get the (1):

$$A + B \cdot J_x = (a + c\cos\theta - d\sin\theta) + i(b + d\cos\theta + c\sin\theta)$$
$$A - B \cdot J_x = (a - c\cos\theta + d\sin\theta) + i(b - d\cos\theta - c\sin\theta) \quad (1)$$

As shown in (1), if we use 2-input adders, a butterfly unit includes four multipliers and six adders. It can be divided into five steps with using the floating-point operation, make these five steps into pipeline to improve the speed of the butterfly unit, as shown in Fig. 3(a). And the latency is the sum of one multiplier and two adder' latency. The floating-point addition can be divided into many operations: exponent shifting, addition, normalization, rounding and so on. So the latency is long.

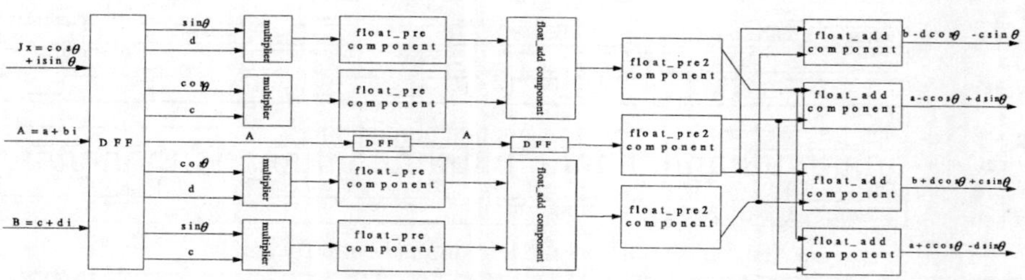


a. Five-stage pipeline flowchart of the traditional butterfly unit

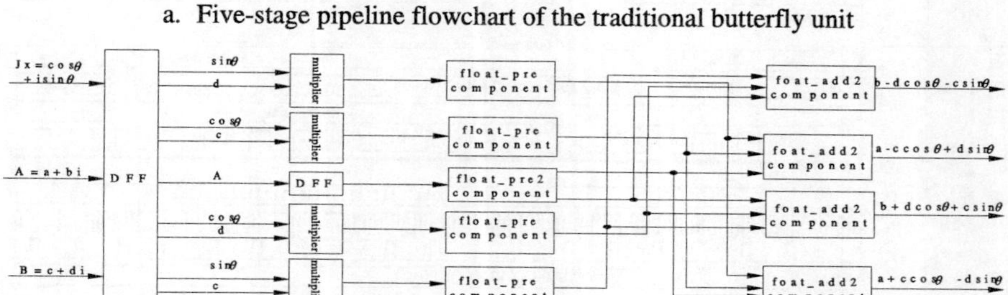


b. Three-stage pipeline flowchart of the improved butterfly unit

**Fig. 3.** Flowchart of the butterfly unit

Designed 3-input adders in order to reduce the pipeline latency of the butterfly unit, according to (1),the new butterfly unit includes four multipliers and four adders[5].The butterfly unit will become into three pipeline structure after improved, as shown in Fig.3(b),the pipeline structure has been simplified. In Fig.3, float_pre,float_pre2 components are used to complete the normalization, float_add component is used to complete the addition of 2-input, float_add2 component is used to complete the addition of 3-input.After analysised, the four 3-input adders use almost the same slices as six 2-input adders. So the new butterfly unit can reduce the pipeline latency as almost the same resources. And the latency is the sum of one multiplier and one adder' latency.

### 3.2 Flowchart of the Butterfly Unit Simulation

The floating-point notation uses two fixed set of data bits to represent the mantissa and exponent, the dynamic range is large, the overflow does not occur in the process of operating[6].A,B are the 36-bit binary numbers, D[35] is the sign of the real part, D[34..23] is the mantissa of the real part, D[22..18] is the exponent of the real part, D[17] is the sign of the imaginary part, D[16..5] is the mantissa of the imaginary part, D[4..0] is the exponent of the imaginary part. Expresses $J_x$ as the 34-bit binary numbers D [33 .. 0],where the D [33] is the sign of the real part, D [32] is the sign of imaginary part of the twiddle factor, D [31 .. 16] is the absolute value of the real part, D [15..0] is the absolute value of the imaginary part, let the 16-bit binary value divided by 65535,we can get the actual absolute value.

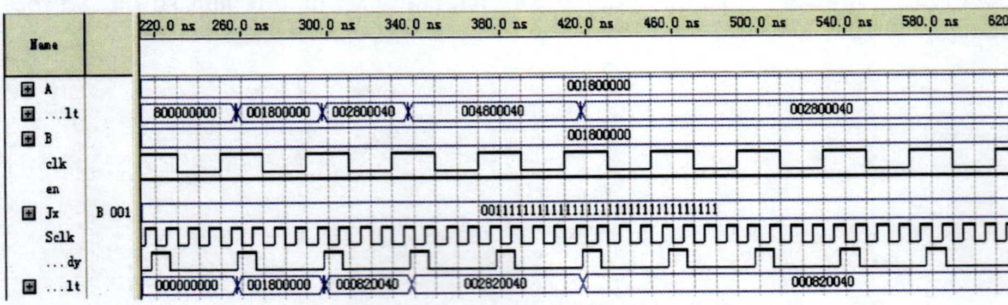


a. Timing simulation of the traditional butterfly unit

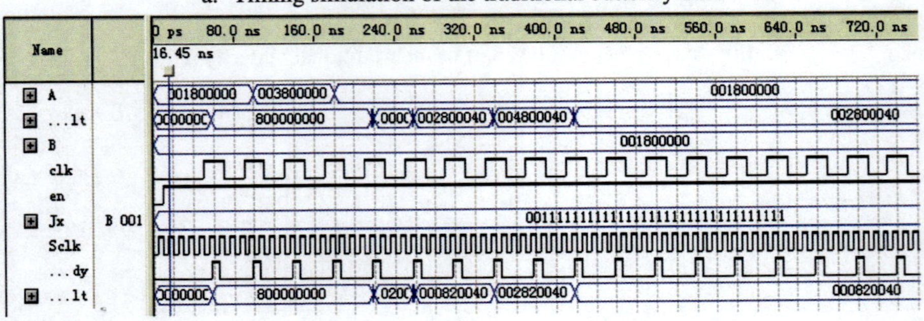


b. Timing simulation of the improved butterfly unit

**Fig. 4.** Timing simulation of the butterfly unit

The timing simulation of the traditional butterfly unit and the improved butterfly unit are shown in Fig.4.

We know the 3-input floating-point adder is designed correctly according to compare the two figures.

## 4 Simulation Results and Performance Analisis

Used the ALTERA company's Cyclone II EP2C35F672C6 chip to simulate. The functional simulation results of the FFT processor are shown in Fig.5.Set the system clock was 100MHz,Mode was low level, wrote the 16+18i complex sequence to FFT data RAM before FFT operating(ccorresponding value 008000240H).Mode became to high level at 2.72 $\mu$s, FFT processor began to operate.EOC became to high level at

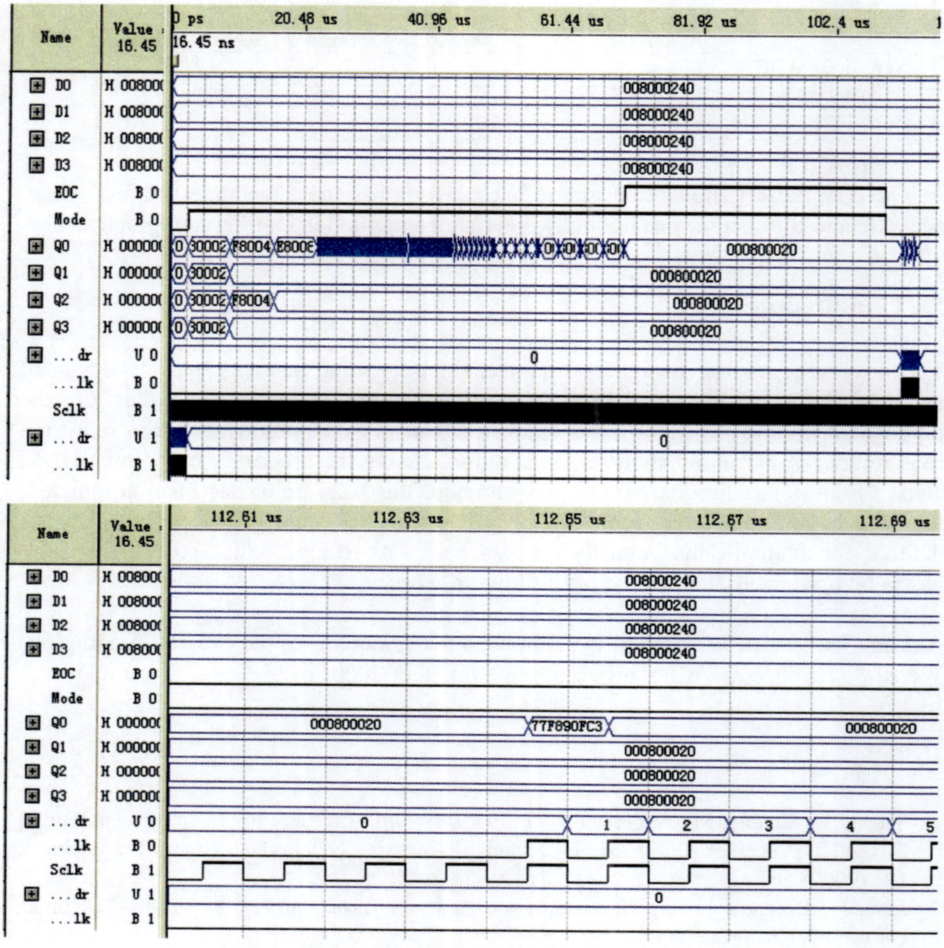


**Fig. 5.** The functional simulation results of FFT processor

69.965 $\mu$s, represented the FFT operation completed. As we can see, EOC always was 0 before Mode became to low level. The operating result of the original sequence were read by the external circuit after Mode became to low level. The result shows that the first point of output sequence is 77F890FC3, if it uses the decimal format ,the real part is 15356,the imaginary part is 17392.The theoretical output of the real part should be 16384,the imaginary part should be 18432,the error of the real part is 6.27%,the error of the imaginary part is 5.64%,this error we can accept.

The comparison of performance between the traditional FFT processor and our new processor is shown in Table 1.

**Table 1.** Comparison of Performance

| **FFT processor** | **Performance** | | | |
|---|---|---|---|---|
| | ***Processing time*** | ***Error of real part*** | ***Error of imaginary part*** | ***Latency of butterfly unit*** |
| Traditional FFT processor | 102.4 $\mu$ s | 11.26% | 10.71% | 300ns |
| Our FFT processor | 67 $\mu$ s | 6.27% | 5.64% | 260ns |

The comparison shows that our FFT processor has the advantages of fast speed, accurate and little latency of butterfly unit, can be applied in engineering practice.

## 5 Conclusion

The advantage of the design is that the floating-point butterfly unit is designed successfully, it can be complete the butterfly operation in less clock cycles, use the pipeline technology, whereby the FFT processor we designing meets the real-time requirement of harmonic analyzer. In addition, design the processor based on VHDL, since FPGA logic circuits can be programmed, the 1024-point has been completed, then the larger points sequence will become easily. Of course, the design also has defects, for example, the butterfly processor uses more logic units, so the algorithm has to be optimized, this will be my follow-up work.

**Acknowledgment.** We gratefully acknowledge the support of the Project Funding of Wenzhou Science & Techonolgy Bureau (NO.H20080001).

## References

1. Saeed, A., Elbably, M., Abdelfadeel, G., Eladawy, M.I.: Efficient FPGA implementation of FFT/IFFT Processor. International Jouranl of Circuits, Systems and Signal Processing 3, 103 (2009)
2. Juan, L.: Research on the New Harmonic Analyzer Based on FPGA, pp. 48–49. Hunan University, Changsha (2004)
3. Fan, Z., Liang, G., Liu, Y.: Design of a new floating-point FFT processor based on FPGA. Application of Electronic Technique 5, 23–26 (2008)

4. Scott Hemmert, K., Underwood, K.D.: An Analysis of the Double-Precision Floating-Point FFT on FPGAs. In: Proceedings of the 13th Annual IEEE Symposium on Field-Programmable Custom Computing Machines, pp. 2–3 (2005)
5. Mou, S., Yang, X.: Design of a High-speed FPGA-based 32-bit Floating-point FFT Processor. In: Eighth ACIS Internation Conference on Software Engineering, Artificial Intelligence, Networking, and Parallel/Distributed Computing, vol. 46, pp. 84–87. IEEE (2007)
6. Luo, X., Zhan, Y.: Using FPGA implement FFT algorithm. Electronic Products China 2, 12–13 (2002)

# Research on Fuzzy Clustering Algorithm WEKA-Based

Yang Jun-qiang[1], Du Jia[1], Zheng Shi-ming[2], Dong Lei[2], and Qiao Bing[3]

[1] Xi'an Communication Academy, Xi'an, China
[2] PLA University of Science and Technology, Nanjing, China
[3] Hebei Military Area of PLA, Shijiazhuang, China
zhengshimingnanjing@126.com, qiaobing@163.com

**Abstract.** The data in Internet has a large scale and dynamic peculiarity, the discovered knowledge or rules are likely to be imprecise or incomplete generally. Owing to the introduction of Fuzzy theory and information entropy into the clustering analysis, we ravel out the difficulties and achieve the best the results for clustering by looking for Fuzzy similarity upper approximation. Moreover, we improved on the algorithm by making use of building Fuzzy similarity relationship on the data sets. The process of embedding the Fuzzy approximation algorithm into the WEKA platform in which the classes and visualization functions of open source WEKA is fully utilized. The Fuzzy approximation algorithms extended the clustering algorithm in the WEKA system. The experiment proves that it has a higher accuracy for the nominal data.

**Keywords:** Fuzzy sets, Data mining, Fuzzy clustering, similarity upper approximation, WEKA.

## Introduction

The Fuzzy theory is one kind of research incomplete, the indefinite knowledge and the data expression, the induction method and the data analysis technology. The Fuzzy theory is a powerful tool that can deal with the uncertainty and the ambiquity hidden in the information, it has solved difficult problem for the general cluster algorithm in discovery knowledge which is not very complete and inaccurate [1].

## 1 Fuzzy Set

### 1.1 Concept of Fuzzy Set

The fuzzy set theory is one kind of new processing fuzzy and the uncertainty knowledge mathematical instrument. Besides the data itself, it does not need other prior background knowledge, to indefinite, incomplete and the fuzzy information can also give satisfaction the result.

**Definition 1:** Given a mapping in $U$ , $A:U \to [0,1]$ , $u \mapsto A(u)$ , then A is named fuzzy set, $A(u)$ is a subject function for A.

**Definition 2:** Given there are many fuzzy set in the given $U$ , $F(u)$ as the collectivity of fuzzy set, $F(U)=\{A|A:U\rightarrow[0,1]\}$.

**Definition 3:** Given $A\in F(u),\lambda\in[0,1]$, sign [2]

(1) $A_\lambda$ as a $\lambda$ cut-set of A, $\lambda$ is named level.

(2) $A_\lambda=\{u|u\in U,A(u)>\lambda\}$ $A_\lambda$ as a $\lambda$ strong cut-set of A.

**Definition 4:** upper approximation $\overline{R}(X)=\bigcup\{a\in U|[a]_R\cap X\neq\phi\}$

**Definition 5:** Tolerance-discrepancy relation [3]:

$$R(x)=\{y\in U|(x,y)\in R\}.$$

**Definition 6:** upper approximation based-Tolerance-discrepancy relation :

$$\overline{R}(X)=\bigcup\{X\in U|R(x)\cap X\neq\phi\}$$

### 1.2 Definition of Approximation Clustering

**Definition 7 :** similarity between the transaction M and N:

$$sim(M,N)=\frac{|syn(M_{C_i})\cap syn(N_{C_j})|}{|syn(M_{C_i})|+|syn(N_{C_j})|} \tag{1}$$

sim(M,N)∈[0,1].when transaction M and N is the same completely, sim(M,N) = 1 ; when transaction M and N is different completely, sim(M,N) = 0.

**Definition 8:** R(t) represent the same clustering about t, which is a set made up of all transactions that is similar to t. R(t)={s∈T, sRt}.

## 2 Clustering Algorithm Fuzzy-Based

### 2.1 Calculating for Subjection

$u_{ij}$ represents the extent that swatch j belong to sort, p is weight exponential [4].

$$u_{ij}=\left[\sum_{k=1}^{c}\left(\frac{\|x_j-Z_i\|^2}{\|x_j-Z_k\|^2}\right)^{1/p}\right]^{-1} \quad (1\le i\le num,1\le j\le n) \tag{2}$$

$$Z^{h+1}_{i} = \frac{\sum_{j=1}^{n} u_{ij} x_j}{\sum_{j=1}^{n} u_{ij}} \qquad (1 \le i \le num) \tag{3}$$

$WS_E$ is error square sum with weight, $c_j$ is the center $c_j = \sum_{i=1}^{m} w_{ij}^p X_i \Big/ \sum_{i=1}^{m} w_{ij}^p$

prove：

$$\frac{\partial}{\partial_{c_k}} WS_E = \frac{\partial}{\partial_{c_k}} \sum_{i=1}^{K} \sum_{x \in C_i} w_{ij}^p (c_i - x)^2 = \sum_{i=1}^{m} \sum_{x \in C_i} w_{ij}^p \frac{\partial}{\partial_{c_k}} (c_i - x)^2$$

$$= \sum_{x \in C_k} 2 w_{ij}^p (c_k - x_k) = 0$$

$$\sum_{x \in C_k} 2 w_{ij}^p (c_k - x_k) = 0 \Rightarrow \quad \sum_{i=1}^{m} (X_i - c_j) w_{ij}^p = 0 \Rightarrow \quad \sum_{i=1}^{m} X_i w_{ij}^p = \sum_{i=1}^{m} w_{ij}^p c_j \Rightarrow$$

$$c_j = \sum_{i=1}^{m} w_{ij}^p X_i \Big/ \sum_{i=1}^{m} w_{ij}^p$$

## 2.2 Calculating for Dissimilarity

Given data set include k kinds of different type variable, the dissimilarity $d(i, j)$ between the object i and j defined as follows:

$$d(i, j) = \frac{\sum_{t=1}^{k} \delta_{ij}^{(t)} d_{ij}^{(t)}}{\sum_{t=1}^{k} \delta_{ij}^{(t)}} \tag{4}$$

If $x_{it}$ or $x_{jt}$ lacked, $x_{it} = x_{jt} = 0$.the dissimilarity $d(i, j)$ has the relation to the type of the data:

- If t is binary variable or criterion variable: if $x_{it} = x_{jt}$, $d_{ij}^{(t)}$ **=0;otherwise** $d_{ij}^{(t)}$ **=1.**
- If t is interval-scaled variables: $d_{ij}^{(t)} = \frac{|x_{it} - x_{jt}|^2}{|\max_Q x_{Qt} - \min_Q x_{Qt}|^2}$.

## 2.3 Ascertain the Number of Clustering

Average information entropy defined as follows [5]:

$$H = -\sum_{j=1}^{n}\sum_{i=1}^{M}\left\{\left[u_{ij}\log_2 u_{ij} + (1-u_{ij})\log_2(1-u_{ij})\right]/M\right\} \tag{5}$$

$u_{ij}$ represents the extent that swatch j belong to sort , the minimum H corresponding to the M is as the best number for clustering.

### 2.4 Algorithm Description

Input : data set, scale proportion $N$, threshold $\varepsilon$, max iterative $L$, max num for clustering $Num$

Output : clustering C, num.

**Step 1.** ascertain the num of clustering randomly $num = \frac{n}{N} \le Num$ ($n$ is the total number of swatch), given $\{Z_1, Z_2, ..., Z_{num}\}$ as initial clustering center.

**Step 2.** calculating $d_{ij} = \|x_i - Z_j\| = \sqrt{\sum_{k=1}^{num}(x_{ik} - Z_{jk})^2}$

select $d_i = \min[d_{i1}, d_{i2}, ..., d_{im}]$, then calculating the center of all the clusters, as new centers $\{Z'_1, Z'_2, ..., Z'_{num}\}$.

**Step 3.** filtrate the center of clusters, looking for similarity upper approximation $S_i$ of $Z_i$, if $Z'_{ij} = \|Z'_i - Z'_j\| < \|x_i - Z'_j\|$, amalgamate $Z'_i$, $Z'_j$ .calculating $D_{Z'_i}, D_{Z'_j}$.

(1) if $D_{Z'_i} < D_{Z'_j}$, take $Z'_i$ as new center.

(2) if $D_{Z'_i} > D_{Z'_j}$, take $Z'_j$ as new center.

**Step 4.** according to expressions (2) and (3), chalk up $\{Z^{h+1}{}_1, Z^{h+1}{}_2, ..., Z^{h+1}{}_q\}$ ($q \le num$) clustering over again.

**Step 5.** if $\|Z^{h+1}{}_i - Z^h{}_i\| \le \varepsilon$ or $h > L$ stop, otherwise $h = h+1$, turn to step 3.

**Step 6.** according to expression (5) calculating $H^{(h)}$, if $H^{(h+1)} > H^{(h)}$ then the number of clustering $M = M + 1$ turn to step 2, otherwise the best number for clustering is M, ending.

### 2.5 Algorithm Improved

To improve the efficiency of the method farther, the following approach be improved for step 3 .

1) calculating the distance between the centers of the clustering $D_{ij} = \left\| Z_i - Z_j \right\|$, $i = 1, 2, ..., N_{num-1}$, $j = 1, 2, ..., N_{num}$

2) range the $D_{ij}$ that smaller than $K_c$ ascending. Given $D_{i_1 j_1} \le D_{i_2 j_2} \le ... \le D_{i_k j_k}$

3) according to l=1,2,...,L, unite $Z_{il}$ and $Z_{jl}$ that $D_{i_l j_l}$ corresponding to.

$$Z_l^* = \frac{1}{N_{i_l} + N_{j_l}} \left[ N_{i_l} Z_{il} + N_{j_l} Z_{jl} \right]$$

# 3 The Experiment Analyzes

## 3.1 WEKA Platform Introduction [6-11]

## 3.2 Empirical Datum

This experiment uses the famous iris flower (Iris) data set, this data set machine learning storehouse obtains from UC Ewen in the branch school (UCI). The iris flower data set contains 150 kind of iris flowers the information, every 50 kinds are from three iris flowers .Data pool detail: (a)Creator: R.A. Fisher (b)Donor: Michael Marshall RSHALL%PLU@io.arc.nasa.gov)(c) Date: July, 1988 [12].

## 3.3 Experimental Result Analysis

**Table 1.** Experiment data of Iris

| attribute | Min | Max | Mean | SD | Class Correlation |
|---|---|---|---|---|---|
| sepal length | 4.3 | 7.9 | 5.84 | 0.83 | 0.7826 |
| sepal width | 2.0 | 4.4 | 3.05 | 0.43 | -0.4194 |
| petal length | 1.0 | 6.9 | 3.76 | 1.76 | 0.9490 |
| petal width | 0.1 | 2.5 | 1.20 | 0.76 | 0.9565 |

Postil : SD= Standard Deviation

**Table 2.** Final Cluster Centers

| | Cluster 1 | Cluster 2 | Cluster 3 |
|---|---|---|---|
| Sepallength | 6.55 | 4.90 | 5.94 |
| Sepalwidth | 3.02 | 3.42 | 2.75 |
| Petallength | 5.56 | 1.46 | 4.31 |
| Petalwidth | 2.08 | .24 | 1.44 |

**Table 3.** Results of ANOVA

| | CMS | Cdf | EMS | F | Sig. |
|---|---|---|---|---|---|
| sepallength | 32.79 | 2 | .27 | 120.55 | .00 |
| sepalwidth | 6.22 | 2 | .11 | 58.70 | .00 |
| petallength | 218.71 | 2 | .18 | 1215.82 | .00 |
| petalwidth | 38.93 | 2 | .06 | 641.82 | .00 |

Postil: CMS =Cluster Mean Square Cdf =Cluster df EMS = Error Mean Square Sig.

**Table 4.** Results of Analysis of Variance

| Numbers | Rrse | Rmse | Mae | Ks | CC |
|---|---|---|---|---|---|
| 150 | 33.6% | 0.159 | 0.035 | 0.94 | 144(96%) |

postil：Rrse=Root relative squared error Rmse=Root mean squared error Mae=Mean absolute error Ks=Kappa statistic CC=Correctly Classified

**Table 5.** Detailed Accuracy By Class

| TP-Rate | FP-Rate | Precision | Recall | F-Measure | Class |
|---|---|---|---|---|---|
| 0.98 | 0 | 1 | 0.98 | 0.99 | setosa |
| 0.94 | 0.03 | 0.94 | 0.94 | 0.94 | versicolor |
| 0.96 | 0.03 | 0.941 | 0.96 | 0.95 | virginica |

Hathaway[13] gave this group in 1995 to test the data set the reality kind of central place respectively is: Z1=（6.58,2.97,5.55,2.02），Z2 = (5.00,3.42,1.46,0.24), Z3 = (5.93,2.77,4.26,1.32). The result which looking from Table 2 experimental results, this article proposed the fuzzy clustering algorithm which and Hathaway give to be already close, the cluster center was nearly close the actual kind of central place. Meanwhile from Table 1, tables 2 and Table 3 may see the flower petal length (petallength) and the flower petal width (petalwidth) these two attribute contribution is quite big, the sepal length (Sepallength) is next, sepal width (sepalwidth) the contribution is smallest[13].

**Table 6.** Confusion Matrix

| a | b | c |
|---|---|---|
| 49 | 1 | 0 |
| 0 | 47 | 3 |
| 0 | 2 | 48 |

postil：a = Iris-setosa b = Iris-versicolor c = Iris-virginica

Table 6 has given in the forecasting result each kind of example number. And the matrix line is the actual kind, the row of matrix is the kind which the forecast obtains, the matrix element is the corresponding test sample integer, what and in table 6 sees the cluster from Table 5 to be most perfect was Iris-setosa, the accuracy has achieved 100%.

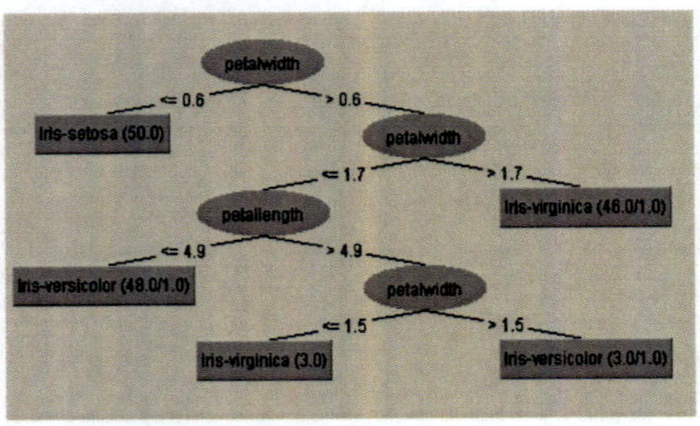


**Fig. 1.** Result of Classifying J48-DeTree-Based

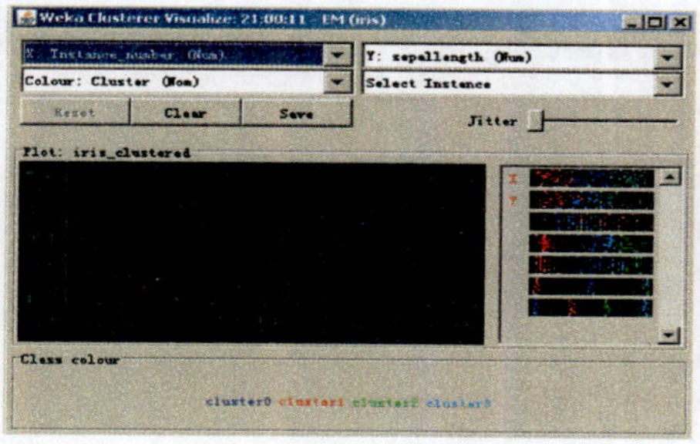


**Fig. 2.** Result of Clustering EM-Based

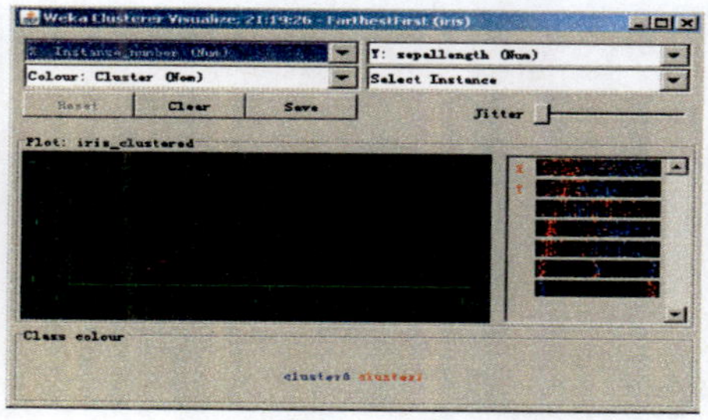


**Fig. 3.** Result of Clustering Farthestfirst-Based

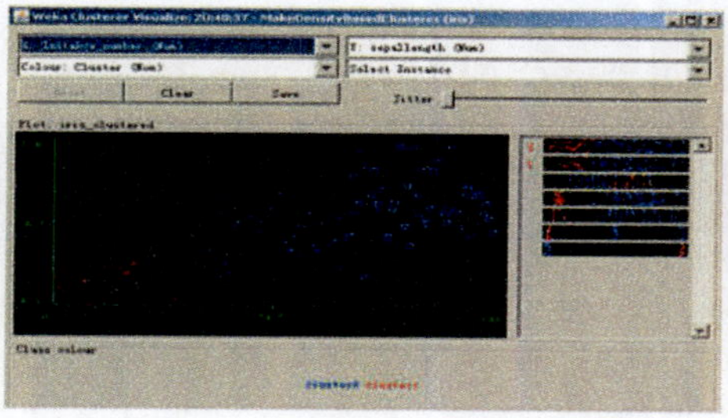

**Fig. 4.** Result of Clustering Density-based

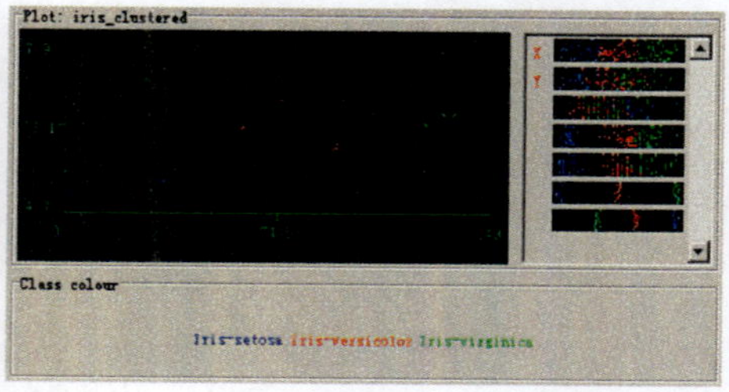


**Fig. 5.** Result of Clustering Fuzzy-based

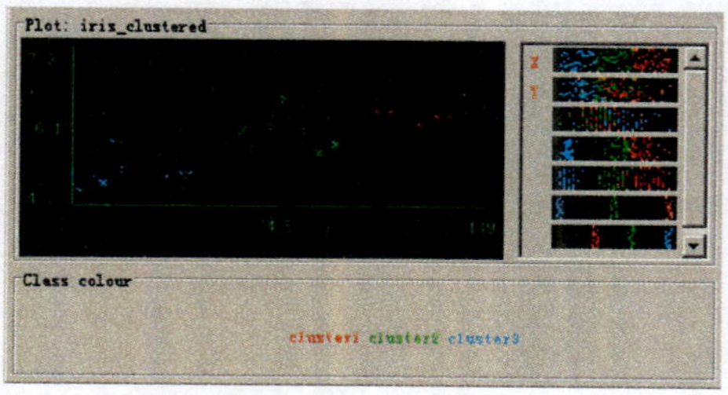


**Fig. 6.** Result of Clustering IM_Fuzzy-based

Moreover, to strengthen the comparison, but also has carried on the experiment to the Zoo data set, the Zoo sample is the zoo actual data set (the ARFF form, 14KB), and mixes the data set, contains 15 numerical attributes and 1 type attribute[14].

The method of computing average veracity as follows: suppose that there are k kinds of cluster in data set, $C_i$ represent the i$^{\text{th}}$ cluster , $n_i$ is the number of $C_i$ , $m_i$ is the correct number of $C_i$ ,Given average veracity P, as follows:

$$p=\begin{cases}\frac{1}{k}\sum_{i=1}^{k}(1-\frac{m_i-n_i}{m_i}) & m_i > n_i \\ \frac{1}{k}\sum_{i=1}^{k} m_i/n_i & m_i < n_i\end{cases} \tag{6}$$

**Table 7.** Compared Accurate with clustering algorithms

| Name | Iris（%） | Zoo（%） |
|---|---|---|
| EM | 75 | 73 |
| Farthestfirst | 35 | 80 |
| Density-based | 50 | 35 |
| K-mean | 80 | 40 |
| Cobweb | 83 | 10 |
| Fuzzy | 85 | 88 |
| IM_Fuzzy | 90 | 92 |

EM (Expectation Maximization) can process not only isomerism data, but also the multiple structure data, simultaneously produces the cluster result easy to explain. K-mean regarding the numerical attribute data, the cluster effect is obvious, but is not good regarding the category attribute data , and is more sensitive to the unusual data, in addition requests to assign the cluster number. The Farthestfirst algorithm in fact is one kind of fast K-mean, but when processes the non- numerical value data the effect surpasses K-mean. The Density-based algorithm has the good cluster effect for massive "the noise" data. The Cobweb algorithm is one general and simple increase-like cluster algorithm based on the model, the cluster effect and the probability distribution related, expandable is low, complexity of the time and the spatial rely on not only the attribute number, but also the numbers of each attribute value [15].

**Acknowledgment.** This work has been partially supported by National 863 plans projects (No.2007AA01z126). We are grateful to P. H. Miao Zhuang and Chen Li for their contributions to the paper. Thanks for Database Technology Research Lab and Software Engineering Lab. The authors are grateful for the anonymous reviewers who made constructive comments.

## References

1. Zhang, L., Li, R.H.: Designing of classifiers based on immune principles and fuzzy rules. Information Sciences 178, 1836–1847 (2008)
2. Sarkar, M.: Rough-fuzzy functions in classification. Fuzzy Sets and Systems 132, 353–369 (2002)
3. Li, J., Gao, X.-B., Jiao, L.-C.: A New Feature Weighted Fuzzy Clustering Algorithm. Acta Electronica Sinica (2006)
4. Meng, X.-M.: Text Mining Algorithm Based on Fuzzy Clustering. Computer Engineering and Design (2009)
5. Liu, Z.-Y., Geng, X.-Q.: Text Mining Algorithm Based on Fuzzy Clustering. Computer Engineering (2006)
6. Camastra, F., Verri, A.: A novel kernel method for clustering. IEEE Trans. Pattern Anal. Mach. Intell. 27(5), 801–805 (2005)
7. Fu, Y.-Q.: Research on Fuzzy Clustering and Clustering Ensemble in Data Mining. Southwest university (2008)
8. Qu, F.-H.: Research on Fuzzy Clustering Algorithm and Its Application. Jilin university (2009)

9. Kao, Y.-T., Zahara, E., Kao, I.-W.: A hybridized approach to data clustering. Expert Systems with Applications (2007)
10. Wang, B., Li, S.Y.: Simulation research on of fuzzy immune nonlinear PID control. Journal of Harbin University of Commerce 22(6), 72–75 (2006) (in Chinese)
11. Chen, H.-P.: Data mining platform-WEKA and secondary development on WEKA. Computer Engineering and Applications (2008)
12. Witten, I.H., Frank, E.: Data Mining practical Machine Learning Tools and Techniques. China Machine press, Beijing (2006)
13. Chen, N., Chen, A., Zhou, L.: Fuzzy K- prototypes algorithm for clustering mixed numeric and categorical valued data. Journal of Software (2001) (in English)
14. Zhang, Y.: A Fuzzy Clustering Algorithm Relief-Based. Application Technology (2009)
15. Ma, S., Wang, T.-J., Tang, S.-W., Yang, D.-Q., Gao, J.: A Fast Clustering Algorithm Based on Reference and Density. Journal of Software (2003)

# Research on Trustworthiness-Based Grid Resource Selection Optimized Algorithm

Xu Shun-fu[1], Wang Bing[1], Zhang Xue-jie[1], Zheng Shi-ming[2], and Shao Rong-ming[3]

[1] Nanjing Artillery Academy of PLA, Nanjing, China
[2] PLA University of Science and Technology, Nanjing, China
[3] Shenyang Artillery Academy Shenyang, China
{zhengshimingnanjing,shaorm}@126.com

**Abstract.** Resource selection is the central part and a prime concern of Grid research. This paper analyses the character of grid, taking into account the performance of algorithm and frequency used in grid domain. The reputation is calculated according to history experience and immediate situation. Our algorithm has dynamic and overall situation in resource selection, considering the trust for history accumulation, introducing into the function depend on time, It provides a method for selecting the value of recommend trust relationship in Grid resource selection, and suffices requirement of user and provider. Experimental results show that the model is exact and robust.

**Keywords:** Grid, Resource selection, Trustworthiness.

## 1 Introduction

Grid is a set of new technology and basic establishment which is built on Internet. It can connect all resources of Internet together. It aims to share the resources of dynamic and wide area distributing isomerous fictitious groups and solve the muti-domains of scientific and engineering problems. Under the condition of grid, the geography distributing and isomerous of resource affect the resource attemperation[1]. The distributing and dynamic nature of grid resource make it very hard to estimate accurately and control the resource estate. Resource selection is a process of selecting the resource from the candidate resource aggregate. The more perfect of resource finding mechanism and the resource, the more heavy task of selecting resource. At the present, resource finding has been resolved by the standard discreteness such as UDDI, WSDL and DAML-S in the OGSA and Web Service [2].

## 2 Reliability Summarization

### 2.1 Character of Credit Value[3]-[5]

Character of credit includes: ①Credit is a subjective concept; ②Credit is changing dynamically with time and crunode action change; ③Credit is attached to special crunode of filed, it need a carrier; ④Credit is dissymmetrical. That means the credit value from two crunode is equal to each other; ⑤Credit transfers with conditionally; ⑥Credit is context-sensitive.

## 2.2 The Definition of Reliability

**Definition 1:** Reliability expresses the credible level in the system. It use CF(T) as CF(T)∈[0,1]. Reliability is used to express the uncertainty. The rules express the resource selecting mode. Every resource, user and rule has a reliability.

**Definition 2:** Proof reliability CF(E) means the credible level of resource provider of estimator. CF(T)∈[0,1].The provider or estimator is called proof. When the proof is true at certain degree, CF(E)>0; When the proof is true , CF(E)=1; When the proof is fake , CF(E)=0.

**Definition 3:** If E THEN H(CF(H,E)), E means proof , H means assumption or conclusion. CF(H, E) is the reliability of this rule. CF(T)∈[0,1]. If CF(H,E)>0, it means the proof increase the true level of the conclusion. If the value of CF(H,E) is more larger, the conclusion H is more true. If CF(H,E)=1, it means the proof make the conclusion true. If CF(H,E)=0, it means the proof is fake.

**Definition 4:** $CF(H,E)=\sqrt{\dfrac{S_{HE}}{S_{HE}+F_{HE}}}$. This formula means the recommendation level of resource H from user E. $S_{HE}$ stands for the times of successful business among the user E and user H. $F_{HE}$ stands for the times of unsuccessful business among the user E and user H.

**Definition 5:** IF we know the rule that if E Then H(CF(H,E)), and the reliability of evidence E is CF(E),then reliability of conclusion H is CF(H)=CF(H,E)×CF(E).

# 3 The Reliability Model of Grid Resource Selection

*The Reliability Model Based on Web Service*

The reliability model of grid resource selection is based on the Web service frame. The model increases the reliability module.

The reliability of resource H is composed of history reliability $CF_A(H)$ and real time reliability $CF_B(H)$.

$$CF(H)=\alpha\, CF_A(H)+(1-\alpha)\, CF_B(H) \tag{1}$$

0≤α≤1. The value of α depends on the user's experience. On the condition of initialization, α=0.5. $CF_B(H)$ is calculated by resource value when the user link the resource[6]. The users get the resource value from resource provider, and compare with own needed value. According to the satisfaction level, the user get a value for $CF_B(E)$. The reliability of rule is:

$$CF_B(H)= CF_B(H,E)\times CF_B(E) \tag{2}$$

The register center calculates $CF_A(H)$ according to the recommendation level of all users. If there are two user $E_1$ and $E_2$:

$$CF_1(H)=CF(H,E_1)\times CF(E_1) \tag{3}$$

$$CF_2(H)=CF(H,E_2)\times CF(E_2) \tag{4}$$

Combine the (3) and (4):

$$CF_{1,2}(H)=CF_1(H)+CF_2(H)-CF_1(H)\times CF_2(H) \tag{5}$$

If resource H is recommended, then $CF_A(H)=CF_{1,2}(H)$, if it is recommended by more users, the register center calculates all recommended reliability in turn until all recommended reliability is combined and $CF_A(H)$ of provider is gained. The calculating mean of three or more users is in the same way.

$$\begin{aligned} CF_{1,2,3}(H) &= CF_1(H)+CF_2(H)+CF_3(H) \\ &-CF_1(H)\times CF_2(H)-CF_1(H)\times CF_3(H) \\ &-CF_2(H)\times CF_3(H)+CF_1(H)\times CF_2(H)\times CF_3(H) \end{aligned}$$

$$\begin{aligned} CF_{1,2,\ldots,n}(H) &= \sum_{i=1}^{n} CF_i(H)-\sum_{1\le i<j\le n} CF_i(H)CF_j(H) \\ &+\sum_{1\le i<j<k\le n} CF_i(H)CF_j(H)CF_k(H)-\ldots \\ &+(-1)^{n+1}CF_1(H)\times CF_2(H)\times\ldots\times CF_n(H), \\ &CF_1(H)\ge 0, CF_2(H)\ge 0,\ldots, CF_n(H)\ge 0 \end{aligned} \tag{6}$$

## 4 The Arithmetic of Grid Resource Selection

*The Arithmetic of Grid Resource Selection Based on Reliability*

Users find a set of resource which can match the demand according to the finding mechanism and calculate the resource reliability[7]. Users visit the register center and resource provider and calculate the reliability $CF(H_i)$ of resource $H_i$. Compared with all resource reliability, the most valuable resource serve for users. The arithmetic is as follows: Initialize the variable: T=0; M=0; α=0.5

For all the collected resource $H_i$ by finding mechanism

$Read(H_i, CF_A(H_i))$

Get(attribute variable table of resource, the successful times S, the unsuccessful F )

$$CF_B(H,E)=\sqrt{\frac{S}{S+F}};\ CF_B(H)=CF_B(H,E)\times CF_B(E);$$

$$CF(H)=\alpha\, CF_A(H)+(1-\alpha)\, CF_B(H);$$

IF $T<CF(H_i)$ Then Begin $T=CF(H_i)$ ;M=i; End;

END FOR

Select ( $H_M$ )

# 5 Model Improving

## 5.1 Model Improving 1

**Definition 6:** $CF(H,E)=\dfrac{S_{HE}-F_{HE}}{S_{HE}+F_{HE}}$ means the recommendation level of resource H from user E, $S_{HE}$ stands for the successful times of business among user E and H, $F_{HE}$ stands for the unsuccessful times of business among user E and H. We find that when the definition 4 is redefined, CF(H,E) is possible a negative. Therefore, the expression 5 can be transferred:

$$CF_{1,2}(H)=\begin{cases}CF_1(H)+CF_2(H)-CF_1(H)\times CF_2(H), CF_1(H)\ge 0, CF_2(H)\ge 0\\ CF_1(H)+CF_2(H)+CF_1(H)\times CF_2(H), CF_1(H)<0, CF_2(H)<0\\ \dfrac{CF_1(H)+CF_2(H)}{1-MIN\{|CF_1(H)|,|CF_2(H)|\}}, CF_1(H)\times CF_2(H)<0\end{cases}$$

the expression 6 can be transferred:

$$CF_{1,2,3}(H)=\begin{cases}CF_1(H)+CF_2(H)+CF_3(H)-CF_1(H)\times CF_2(H)-CF_1(H)\times CF_3(H)\\ -CF_2(H)\times CF_3(H)+CF_1(H)\times CF_2(H)\times CF_3(H), CF_1(H)\ge 0, CF_2(H)\ge 0, CF_3(H)\ge 0\\ CF_1(H)+CF_2(H)+CF_3(H)+CF_1(H)\times CF_2(H)\times CF_3(H), CF_1(H)<0, CF_2(H)<0, CF_3(H)<0\\ \dfrac{CF_1(H)+CF_2(H)+CF_3(H)}{1-MIN\{|CF_1(H)|,|CF_2(H)|,|CF_3(H)|\}}, CF_1(H)\times CF_2(H)\times CF_3(H)<0\end{cases}$$

if α=0.5, the reliability of resource $H_1$ is : $CF(H_1)=\alpha\times CF_A(H_1)+(1-\alpha)\times CF_B(H_1)=0.801$

the reliability of resource $H_2$ is :

$$CF(H_2)=\alpha\times CF_A(H_2)+(1-\alpha)\times CF_B(H_2)=0.82$$

The user can choose resource $H_2$ .the result is coincident with the former. But by compare, the model mending is more scientific on the description of reliability. CF(E)∈[-1,1]. When the evidence is true in some degree, CF(E)>0, when the evidence is fake in some degree [8], CF(E)<0, when the evidence is fake affirmatively, CF(E)=-1,when the evidence is unknown, CF(E)=0.

We can make some change on model and introduce the time depending function $f(t_i)=\frac{t_i-t_0}{\sum_i (t_i-t_0)}$, $t_0$ means the moment that the resource is firstly selected, $\sum_i (t_i-t_0)$ means the time summation of whole resource selection. The user's initial reliability $\mathrm{CF(E_i)}$ can be set again. The moment after set $\mathrm{CF(E_i)}$ is called $t_0^{'}$, $t_0^{'}$ is a new count moment. The new reliability $\mathrm{CF(E_i)}$ connects with $f^{'}(t_i)=\frac{t_i-t_0^{'}}{\sum_i (t_i-t_0^{'})}$ .the formula of CF(H)is changed to $\mathrm{CF(H)}=f(t_i)\cdot[\alpha \mathrm{CF_A(H)}+(1-\alpha)\,\mathrm{CF_B(H)}]$.

## 5.2 Model Improving 2

**Definition 7 (operator $\oplus$)** : suppose $X,Y$ and $Z$ are all row vector with n dimension, $X=(x_1,x_2,...,x_n)$, $Y=(y_1,y_2,...,y_n)$, $Z=(z_1,z_2,...,z_n)$ then

$$X\oplus Y\oplus Z=(x_1,x_2,...,x_n)\oplus(y_1,y_2,...,y_n)\oplus(z_1,z_2,...,z_n)$$
$$=\begin{bmatrix} x_1 & x_2 & \cdots & x_n \\ y_1 & y_2 & \cdots & y_n \\ z_1 & z_2 & \cdots & z_n \end{bmatrix}_{3\times n}$$

**Definition 8:** Recommendation trust for grid resource selection define as a 4-tuple $T$, $T=(DT_Q^A, RT_{Q(h)}^{A\leftarrow R_h}, RW_Q^{R_{i+1}\leftarrow R_i}, W_{dr})$, simplify sign $T=(DT,RT,RW,W)$, $DT$ denote direct trust vector, $RT$ denote indirect trust vector, $RW$ denote recommendation gene, $W$ denote recommendation weight vector.

**Definition 9:** $DT_Q^A$ ,represent direct trust from A to Q, $DT_Q^A=(b_Q^A(1),b_Q^A(2),...,b_Q^A(n))$ ; $RT_{Q(h)}^{A\leftarrow R_h}$ represent the recommendation information about Q that provided to A by $R_h$ , $RT_{Q(h)}^{A\leftarrow R_h}=(b_{Q(h)}^{A\leftarrow R_h}(1),b_{Q(h)}^{A\leftarrow R_h}(2),...,b_{Q(h)}^{A\leftarrow R_h}(n))$ , h represent $R_h$ is the $h_{th}$ recommender in all the recommendation path , reflecting the distance of recommendation trust a certain extent ; $RW_Q^{R_{i+1}\leftarrow R_i}$ represent the recommendation gene in the process of estimating about Q form $R_i$ to $R_{i+1}$, $R_i$ represent the reliability for provider; weight vector $W_{dr}=(w_d,w_r)$, $w_d$ represent the weight that grid entity endowed to direct trust, $w_r$ represent the weight that grid entity endowed to recommendation trust, and $w_d + w_r = 1$.

**Definition 10:** (combination information) $HT_Q^A$ represent A get hold of the information about Q get across combination; $W_{HT}$ represent the weight vectors in all the recommendation path for the recommendation information in the process of combining.

$$(RT_{Q(h_1)}^{A\leftarrow R_1} \oplus RT_{Q(h_2)}^{A\leftarrow R_2} \oplus \ldots \oplus RT_{Q(h_n)}^{A\leftarrow R_n})$$

$$= \begin{bmatrix} b_{Q(h_1)}^{A\leftarrow R_1}(1) & b_{Q(h_1)}^{A\leftarrow R_1}(2) & \ldots & b_{Q(h_1)}^{A\leftarrow R_1}(n) \\ b_{Q(h_2)}^{A\leftarrow R_2}(1) & b_{Q(h_2)}^{A\leftarrow R_2}(2) & \ldots & b_{Q(h_2)}^{A\leftarrow R_2}(n) \\ \ldots & \ldots & \ldots & \ldots \\ b_{Q(h_n)}^{A\leftarrow R_n}(1) & b_{Q(h_n)}^{A\leftarrow R_n}(2) & & b_{Q(h_n)}^{A\leftarrow R_n}(n) \end{bmatrix}$$

$$HT_Q^A = W_{HT} \bullet (RT_{Q(h_1)}^{A\leftarrow R_1} \oplus RT_{Q(h_2)}^{A\leftarrow R_2} \oplus \ldots \oplus RT_{Q(h_n)}^{A\leftarrow R_n})$$

$$= (HT_Q^A(1), HT_Q^A(2), \ldots, HT_Q^A(n))$$

where $$W_{HT} = \left(RW_Q^{A\leftarrow R_1} \Big/ \sum_{j=1}^{m} RW_Q^{A\leftarrow R_j}, \ldots, RW_Q^{A\leftarrow R_n} \Big/ \sum_{j=1}^{m} RW_Q^{A\leftarrow R_j}\right)$$

$$(RT_{Q(h_1)}^{A\leftarrow R_1} \oplus RT_{Q(h_2)}^{A\leftarrow R_2} \oplus \ldots \oplus RT_{Q(h_n)}^{A\leftarrow R_n})$$

$$= \begin{bmatrix} b_{Q(h_1)}^{A\leftarrow R_1}(1) & b_{Q(h_1)}^{A\leftarrow R_1}(2) & \ldots & b_{Q(h_1)}^{A\leftarrow R_1}(n) \\ b_{Q(h_2)}^{A\leftarrow R_2}(1) & b_{Q(h_2)}^{A\leftarrow R_2}(2) & \ldots & b_{Q(h_2)}^{A\leftarrow R_2}(n) \\ \ldots & \ldots & \ldots & \ldots \\ b_{Q(h_m)}^{A\leftarrow R_m}(1) & b_{Q(h_m)}^{A\leftarrow R_m}(2) & & b_{Q(h_m)}^{A\leftarrow R_m}(n) \end{bmatrix}_{m\times n}$$

$h_i (i = 1, 2, \ldots, m)$ represents the length in the $i_{th}$ recommendation path, increasing along with i, if $h_i < \varepsilon$, terminating combination. Figure 1 and 2 represent trust transfer model and trust combination model respectively.

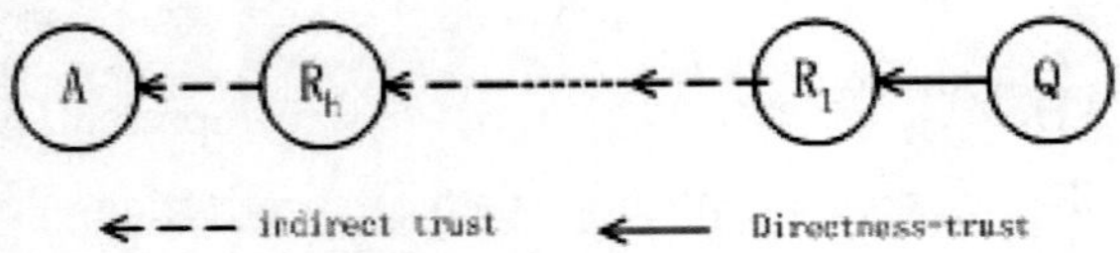


**Fig. 1.** Trust transfer model

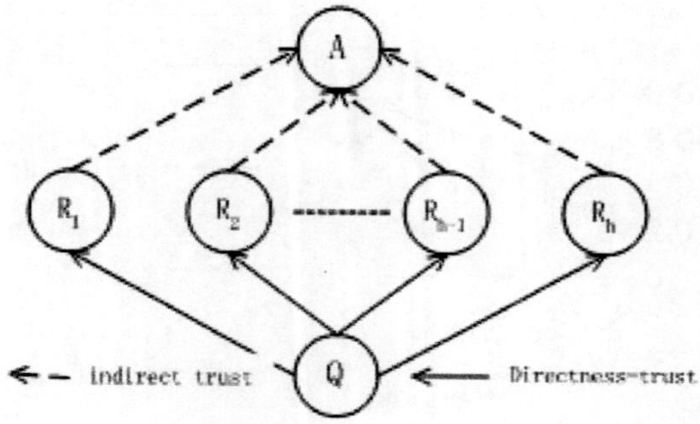


**Fig. 2.** Trust combination model

The final trust vector A to Q: $TRUST_Q^A = (w_{d(A)}, w_{r(A)}) \bullet (DT_Q^A \oplus HT_Q^A)$.

Where we improve on $RW_Q^{A \leftarrow R_i}$ :

$$RW_Q^{A \leftarrow R_i} = \frac{S_{HE} - F_{HE}}{S_{HE} + F_{HE}} * \beta^{\frac{1}{S_{HE}+F_{HE}}} * e^{-(f'(t_i)+\frac{h_i}{\tau_h})}, \text{ where } 0 < \beta < 1$$

$\beta^{\frac{1}{S_{HE}+F_{HE}}}$ represent the influence of alternation time on recommendation gene, the time more the reliability higher, $e^{-(f'(t_i)+\frac{h_i}{\tau_h})}$ reflect the alteration direction of recommendation gene along with the variation of time interval and length of path, the time and path increase more the reliability lower, $\tau_h$ is constant, $f'(t_i)$ and $h_i$ represent time attenuation gene and path attenuation gene respectively[9].

### 5.3 Experimental Result Analysis

This paper constructs a P2P share network for simulation. The nodes in the network can query files and respond to the query in the form of broadcasting, all parameters as follows, Table 1:

**Table 1.** List of parameters

| | |
|---|---|
| Category of file | 50 |
| Topology of network | P2P |
| TTL for querying | 5 |
| Number of node | 200 |
| Proportion of credible node | 80% |
| Proportion of spiteful node | 20% |
| Periodicity for simulating | 420 |

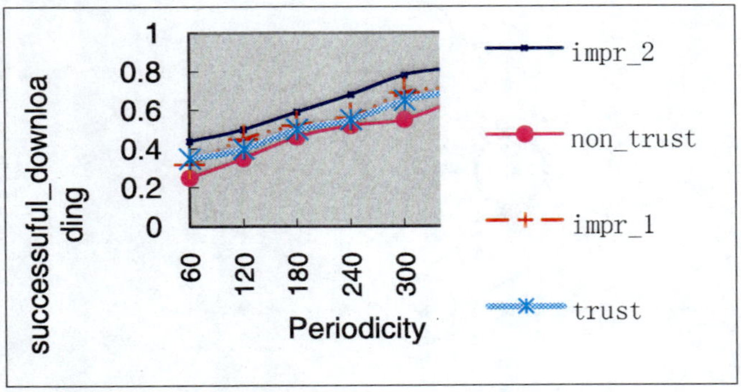


**Fig. 3.** Average successful ratios

For the Figure 3 above, non_trust represent that we don't adopt ant trust model in our experiment, trust represent make use of trust model presented in this paper at first, impr_1 and impr_2 represent Model improving 1, Model improving 2.From the figure 4 above we can see that the impr_2 has the best average ratios which introduced the improved trust model, along with the periodicity increasing, the average successful ratios get higher and higher. All the statistical results showed in figure 4.

## 6 Conclusion

The resource selection distribution under the grid circumstance has high dynamic. How to insure the QoS is the key problem. The paper introduced a kind of resource selection model. By comparing result under the old experience with the real time instance, it simplified the complexion of calculating, and overcame the unilateralism of evaluation. At the same time, we mended the model. It reflects the resource selection based on the grid circumstance impersonally, it can satisfy all performance requirement of users and resource providers. At last, the contrast of two model proved the validity of arithmetic[10].

**Acknowledgment.** This work has been partially supported by National 863 plans projects (No.2007AA01z126). We are grateful to P. H. Miao Zhuang and Chen Li for their contributions to the paper. Thanks for Database Technology Research Lab and Software Engineering Lab. The authors are grateful for the anonymous reviewers who made constructive comments.

## References

1. Foster, I.: The autonomy of the Grid, pp. 10–24 (2006)
2. Sreenath, R.M., Singh, M.P.: Agent-based service selection. In: Web Semantics: Science, Services and Agents on the World Wide Web, vol. 1, pp. 261–279 (2004)

3. Foster, I., et al.: Improving Data Availability through Dynamic Model-Driven Replication in Large Peer-to-Peer Communities. In: Global and Peer-to-Peer Computing on Large Scale Distributed Systems Workshop, Berlin (May 2002)
4. Li, W.-J.: Study on several trust models in grid environment. Journal of Fuzhou University (Natural Science Edition) 34(2) (2006)
5. Lu, Y.-B., Zhou, T.: Comparative Analysis Model Based-trust IN internet. Science and Technology Management Research (11) (2005)
6. Chen, J.: Study on grid trust models based- action. Northwest university (2007)
7. Berman, F., Fox, G., Hey, T.: Grid Computing Making the Global Infrastructure a Reality. Wiley Publishing Company (2003)
8. Duan, Z.-Y.: A Faul t-Tolerant Grid Resource Selection Algorithm. Computer Engineering & Science (4) (2005)
9. Qi, N.: QoS-based grid resource selection optimization model. Computer Engineering and Applications (8) (2008)
10. Zeug, L.Z., Benatatlah, B., Ngu, A.H.H.: QoS-aware Middleware for web services composition. IEEE Transactions on Software Engineering, 311–327 (2004)

# Research on the Evaluation of Contingency Plan System of Emergency Management in Supply Chain: Based on Two-Tuple Linguistic Information

Lu Shan

Hunan University of Commerce, Changsha, China
lushan20031020@163.com

**Abstract.** The emergency events on the supply chain may have adverse impact on the enterprises as well as the supply chain itself, which may also directly put threat on building up a harmonious society. As a result, a specific contingency plan for emergency management should be set up in order to deal with the emergency events. For the purpose of ensuring that the emergency events can be perfectly handled under the existing contingency plan, it is vital to make an accurate evaluation on the ability of the contingency plan to settle these events. This paper aims to set up the evaluation criteria for the contingency plan based on theoretical research and empirical study, and meanwhile, the contingency plan is assessed comprehensively by the two-tuple linguistic information model as well as IT-WAA operators. In addition, the effectiveness and practicability of the evaluation criteria and model are also verified through the empirical study.

**Keywords:** Supply Chain, Contingency Plan, Two-tuple Linguistic Information, Language Information.

## 1 Introduction

The emergency events on the supply chain are those happening abruptly with catastrophic results among the crisis enterprises should face. These events may have tremendously adverse impact on the enterprises as well as the supply chain itself and may cause other emergencies in public security, which may also directly put threat on the requirements of building up a harmonious society. Therefore, a specific contingency plan for emergency management should be set up in order to deal with the emergency events.

Recently, emergency management in the supply chain has caught widespread attention from both the business world and the academic circle. Yu G.（2006）did a research and proposed the strategy of production/inventory management under supply chain disruptions which is indicated as a constant or discrete Markov process and the demand is either defined or random [1]. Abboud（2001）focused on the model with a single supplier and illustrated separately the best strategy of production/inventory management under different circumstances [2]. Hayya（2004）did some further work on the model with two suppliers and proposed the optimal order quantity [3]. However, the above research is to evaluate the production-inventory management strategy under supply disruptions, while Tomlin (2005, 2006) further analyzed how to weigh and choose between preventive measures and contingency measures in emergency management [4-6]. This kind of research has just seen it start in China. Lei Zhen & Xu Jiuping (2004) introduced the emergency management of the unpredictable events in the supply chain and proposed the mechanism, organization and processes for the emergency management of the unpredictable events in the supply chain by defining and categorizing the events and analyzing the adverse impact of the events on the supply chain [7]. Yu Hui, Chen Jian & Yu Gang (2005) studied the impact of demand disruptions upon the supply chain coordinated by the classical quantity discount contract, gave the optimal response of the supply chain as a whole to the disruptions, thus coming to the suggestion of the new quantity discount contract with anti-disruption ability[8]. They also analyzed the impact on the supply chain coordinated by buyback contract by some emergent event which caused the retailer to face the different demand distribution and proposed an optimal strategy for the supply chain to the disruptions, and as optimal response for the supply chain to the disruptions, an adjusted buyback contract which has anti-disruption ability was also given in their paper[9]. Xu et al. (2003)did research and discussed the impact on the one-supplier-one-retailer supply chain when the demand suddenly changed, in which they take two occasions into consideration, namely, the decentralized decision making and the centralized decision making, so as to figure out how perfect coordination can be reached in the supply chain[10].

The above shows that currently lots of qualitative research has been done in terms of emergency management of sudden events on the supply chain, resulting in a series of preventive and risk-handling measures to set up a highly efficient and robust supply chain so as to better deal with the loss caused by the emergency. However, these measures take costs and should be given a tradeoff analysis on its effects and costs. Since the emergency events feature in their small probability and difficult forecasting but causing tremendous loss, it is important to set up a specific contingency plan for emergency management. Meanwhile, it is even more crucial and worth researching to choose a suitable tool to evaluate the plan. Therefore, the author gives his view that it is necessary to evaluate the contingency plan in a comprehensive way with scientific methods, so as to provide for references for setting up a real scientific and effective contingency plan for emergency management. For this reason, the method of two-tuple linguistic information is adopted in this thesis to give a systematic evaluation of the plan with examples and related calculations.

## 2 Evaluation Procedures of the Contingency Plan for Emergency Management in Supply Chain Based on Two-Tuple Linguistic Information

Suppose that there are m schemes to be evaluated, i.e. $A=\{a_1, a_2, \cdots, a_m\}$, s decision maker (DM), i.e. $DM=\{D_1, D_2, \cdots, D_s\}$, $q$ criteria, i.e. $C=\{C_1, C_2, \cdots, C_q\}$. The decision maker $D_p$ gives the criterion $C_l$ its weight of $\omega_l^p$, and the evaluation matrix for the plan is $C^p=(c_{il}^p)_{m\times q}$, in which $c_{il}^p$ indicates the value of $a_i$ under the criterion $C_l$ after the decision maker $D_p$ chooses the suitable standard language phrase for description from

$H$ according to his knowledge, experience and preference. The value may be some specified language evaluation grade, or between two sequential language evaluation grades, or the one grade that the decision maker could not define so that the importance of decision maker could not be defined. Then a rank should be determined among these schemes.

## 2.1 Two-Tuple Linguistic Information and Related Operators

In two-tuple linguistic information, $(H_i, \alpha_i)$ is used to indicate the language evaluation information by the decision maker, in which $H_i$ is a language evaluation grade, and $\alpha_i$ is the divergence between the language evaluation information given by the decision maker and the language evaluation grade $H_i$, $\alpha_i \in [-0.5, 0.5)$ [11].

**Definition 1[11]:** Suppose $H_i \in H$, then the corresponding two-tuple linguistic information can be given by the following transfer function $\phi$

$$\begin{cases} \phi : H \to H \times [-0.5, 0.5) \\ \phi(H_i) = (H_i, 0) \end{cases}$$

**Definition 2[11]:** Suppose there is a language evaluation set $H = \{H_0, H_1, H_2, \cdots, H_{2t}\}$ and $\theta \in [0, 2t]$ indicates the result of the operation of language symbols, the two-tuple linguistic information corresponding to $\theta$ can be given by the function $\Delta$

$$\begin{cases} \Delta : [0, 2t] \to H * [-0.5, 0.5) \\ \Delta(\theta) = (H_n, \alpha_n) = \begin{cases} H_n, n = Round(\theta) \\ \alpha_n = \theta - n \end{cases} \end{cases}$$

$Round$ means rounding operation.

**Definition 3[11]:** Suppose there is a language evaluation set $H = \{H_0, H_1, H_2, \cdots, H_{2t}\}$ and $(H_i, \alpha_i)$ is two-tuple linguistic information, then there should be an inverse function $\Delta^{-1}$, so that the two-tuple linguistic information could be transformed into the corresponding value $\theta \in [0, 2t]$, i.e.

$$\begin{cases} \Delta^{-1} : H \times [-0.5, 0.5) \to [0, 2t] \\ \Delta^{-1}(H_i, \alpha_i) = i + \alpha_i \end{cases}$$

**Definition 4:** Suppose $(H_i, \alpha_i)$ and $(H_j, \alpha_j)$ are two two-tuple linguistic information. If $i > j$, $(H_i, \alpha_i) > (H_j, \alpha_j)$, the sign "$>$" here means "superior to". If $i = j$, there are three cases as follows[11]:

(1) If $\alpha_i > \alpha_j$, then $(H_i, \alpha_i) > (H_j, \alpha_j)$;

(2) If $\alpha_i = \alpha_j$, then $(H_i, \alpha_i) = (H_j, \alpha_j)$, the symbol "$=$" here means "equal to";

(3) If $\alpha_i < \alpha_j$, then $(H_i, \alpha_i) < (H_j, \alpha_j)$, the symbol "$<$" here means "inferior to".

## 2.2 Language Evaluation Methods

For the condition in which the evaluation information given by the decision maker is between two language grades, $H_{n+\alpha}$ is used to indicate this case and $\alpha \in (0,1)$ which means that the evaluation information given by the decision maker is between $H_n$ and $H_{n+1}$, and the divergence is $\alpha$ relative to $H_n$ while it is $1-\alpha$ [12-13] relative to $H_{n+1}$.

It is easy to get, according to the properties of two-tuple linguistic information and the orderliness of $H$, that the incompleteness of the language phrase information can totally correspond to the function value of the weight language information $\Delta^{-1}$. If the decision maker's weight $\lambda_i$ is between $H_n$ and $H_{n+1}$, then the two-tuple linguistic information value of $\lambda_i$ is $[\Delta^{-1}((H_n,0)),\Delta^{-1}((H_{n+1},0))]$. If the decision maker's weight $\lambda_i$ cannot be defined, then the two-tuple linguistic information value of $\lambda_i$ is $[0,2t]$. Suppose that the two-tuple linguistic information, conversed from the language phrase of the degree of importance of the decision maker, corresponds to a set of information $\Omega$, and the two-tuple linguistic information, conversed from the language phrase of the degree of importance of the criteria given by the decision maker $D_p$, corresponds to a set of information $\Psi_p$.

The reliance degree of the evaluation information pertaining to the language evaluation grade can be deduced according to the language evaluation information given by the decision maker.

(1) If $D_p$ gives the value of $c_{ij}^p$ as $H_n$, then it can be defined that under criterion $C_j$ the reliance degree of scheme $a_i$ pertaining to $H_n$ is $T(p,i,j,n)=1$.

(2) If $D_p$ gives the value of $c_{ij}^p$ as $H_{n+\alpha}$, then it can be defined that under criterion $C_j$ the reliance degree of scheme $a_i$ pertaining to $H_n$ is $T(p,i,j,n)=1-\alpha$, while that pertaining to $H_{n+1}$ is $T(p,i,j,n+1)=\alpha$.

(3) If $D_p$ cannot give the value of $c_{ij}^p$, i.e. $D_p$ fails to decide whether scheme $a_i$ is superior or inferior under criterion $C_j$, then it can be defined that under criterion $C_j$ the reliance degree of scheme $a_i$ pertaining to any $H_n$ is $T(p,i,j,n)=0$ ($n=0,1,2,\cdots,2t$).

Under criterion $C_j$, the group reliance degree of scheme $a_i$ pertaining to $H_n$ is:

$$B(i,j,n)=\sum_{p=1}^{s}\lambda_p \times T(p,i,j,n)\ (n=0,1,2,\cdots,2t).$$

$\lambda_p$ is the weight for $D_p$.

# 3 Specific Calculation Examples

Suppose there are five contingency plans for emergency management in supply chain $a_1,a_2,a_3,a_4,a_5$, three evaluators $D_1,D_2,D_3$ and five criteria $C_1,C_2,C_3$, $C_4,C_5$, namely, the comprehensiveness, operability, quick execution, effectiveness and reasonability of the cost. The evaluation grade for each criterion is H={worst, worse, average, better, best}=$\{H_0, H_1, H_2, H_3, H_4\}$. The information of weights and evaluations given by the evaluators about every criterion are shown in Table 1, with which a rank can be done for the schemes.

**Table 1.** Evaluations of every plan by the evaluators

| $D_1$ | $C_1$ | $C_2$ | $C_3$ | $C_4$ | $C_5$ |
|---|---|---|---|---|---|
| $a_1$ | $[H_1,H_2]$ | $[H_3,H_4]$ | $[H_1,H_2]$ | $[H_3,H_4]$ | $[H_3,H_4]$ |
| $a_2$ | $[H_0,H_1]$ | $[H_1,H_2]$ | $[H_2,H_4]$ | $[H_1,H_2]$ | $[H_2,H_4]$ |
| $a_3$ | $[H_3,H_{3.5}]$ | $H_3$ | $H_{1.7}$ | $H_3$ | $[H_2,H_3]$ |
| $a_4$ | $[H_2,H_4]$ | $[H_3,H_4]$ | $H_1$ | $H_2$ | $[H_1,H_2]$ |
| $a_5$ | $H_0$ | $[H_2,H_4]$ | $[H_1,H_2]$ | $[H_2,H_4]$ | $H_2$ |

| $D_2$ | $C_1$ | $C_2$ | $C_3$ | $C_4$ | $C_5$ |
|---|---|---|---|---|---|
| $a_1$ | $[H_1,H_2]$ | $[H_4,H_5]$ | $[H_3,H_4]$ | $[H_3,H_4]$ | $H_4$ |
| $a_2$ | $[H_0,H_1]$ | $H_2$ | $[H_3,H_4]$ | $H_{3.5}$ | $[H_3,H_4]$ |
| $a_3$ | $H_0$ | $[H_0,H_1]$ | $H_{1.7}$ | $[H_2,H_3]$ | $H_2$ |
| $a_4$ | $[H_0,H_1]$ | $[H_0,H_1]$ | $H_{1.7}$ | $H_2$ | $H_3$ |
| $a_5$ | $[H_0,H_1]$ | $[H_2,H_3]$ | $[H_2,H_3]$ | $H_1$ | $H_2$ |

| $D_3$ | $C_1$ | $C_2$ | $C_3$ | $C_4$ | $C_5$ |
|---|---|---|---|---|---|
| $a_1$ | $H_2$ | $[H_1,H_2]$ | $[H_2,H_4]$ | $[H_2,H_4]$ | $H_4$ |
| $a_2$ | $[H_1,H_2]$ | $H_2$ | $H_2$ | $[H_3,H_4]$ | $[H_2,H_3]$ |
| $a_3$ | $[H_0,H_1]$ | $H_{2.3}$ | $H_2$ | $[H_2,H_3]$ | $[H_1,H_2]$ |
| $a_4$ | $H_{0.5}$ | $[H_0,H_1]$ | $[H_1,H_2]$ | $H_2$ | $[H_1,H_2]$ |
| $a_5$ | $H_1$ | $[H_1,H_2]$ | $[H_2,H_3]$ | $[H_2,H_3]$ | $H_2$ |

**Step 1:** Suppose the weights of the evaluators are (0.3,0.3,0.4) , then aggregate the interval two-tuple linguistic information decision-making matrix R1,R2,R3 given by the decision makers with the IT-WAA operator, thus gaining the group interval two-tuple linguistic information decision-making matrix:

**Table 2.** Group interval two-tuple linguistic information decision-making matrix

| | $C_1$ | $C_2$ | $C_3$ | $C_4$ | $C_5$ |
|---|---|---|---|---|---|
| $a_1$ | $[H_{1.4}, H_{1.4}]$ | $[H_{2.5}, H_{3.5}]$ | $[H_2, H_{3.4}]$ | $[H_{2.6}, H_4]$ | $[H_{3.7}, H_4]$ |
| $a_2$ | $[H_{0.4}, H_{1.4}]$ | $[H_{1.7}, H_2]$ | $[H_{2.3}, H_{3.2}]$ | $[H_{2.55}, H_{3.25}]$ | $[H_{2.3}, H_{3.6}]$ |
| $a_3$ | $[H_{0.9}, H_{1.45}]$ | $[H_{1.82}, H_{2.12}]$ | $[H_{1.82}, H_{1.82}]$ | $[H_{2.3}, H_3]$ | $[H_{1.6}, H_{2.3}]$ |
| $a_4$ | $[H_{0.8}, H_{1.7}]$ | $[H_{0.9}, H_{1.9}]$ | $[H_{1.21}, H_{1.61}]$ | $[H_2, H_2]$ | $[H_{1.6}, H_{2.3}]$ |
| $a_5$ | $[H_{0.4}, H_{0.7}]$ | $[H_{1.6}, H_{2.9}]$ | $[H_{1.7}, H_{2.7}]$ | $[H_{1.7}, H_{2.7}]$ | $[H_2, H_2]$ |

**Step 2:** Based on the research data, weights for every items can be reached by the method of factor analysis, i.e. transforming into the decision making under the defined weight. The information of the defined weight is: $\omega_i$ =(0.17 ,0.18 ,0.24 ,0.12 ,0.29). Then the group interval two-tuple linguistic information decision-making matrix is collected through weighted averaging, thus gaining $Z_1 = [H_{2.553}, H_{3.324}]$ , $Z_2 = [H_{1.899}, H_{2.8}]$, $Z_3 = [H_{1.6574}, H_{1.8889}]$, $Z_4 = [H_{1.2924}, H_{1.9244}]$, $Z_5 = [H_{1.548}, H_{2.193}]$.

**Step 3:** Possibility formula is adopted to compare the value of $Z_i$, which helps with ranking and selecting the best, so there comes j.

## 4 Conclusions

The emergency events, because of its abruptness, could cause shutdown or short of supplies in the enterprises with tremendous loss, and so establishing a perfect and effective contingency plan for emergency management in supply chain becomes the effective way to handle the emergency events and thus it is crucial to give accurate assessment of the ability of the contingency plan in handling the emergency events. At present, there are few of such research on the evaluation of the contingency plan for the supply chain. This paper, with several references as the basis, aims to set up the criteria for the contingency plan and the evaluation model based on the two-tuple linguistic information and related operators. With the model, the ranking of different schemes can be done, while the qualitative evaluation of the plan can be realized through the set of evaluation items. The specific calculation examples illustrates the effectiveness of the evaluation criteria and methods which are simple to use and easy to promote.

**Acknowledgement.** The work in the paper is supported by Hunan provincial commission of the scientific payoffs of philosophy and social sciences (0808041B), Outstanding Youth of Scientific Research Project of Hunan Provincial Education Department (09B056) and Hunan Social Sciences Fund item(09YBB228).

## References


1. Xiao, T., Yu, G.: Supply chain disruption management and evolutionarily stable strategies of retailers in the quantity-setting duopoly situation with homogeneous goods. European Journal of Operational Research 2, 648–668 (2006) (forthcoming)
2. Abboud, N.E.: A discrete-time Markov production-inventory model with machine breakdowns. Computers & Industrial Engineering, 95–107 (2001)
3. Li, Z.L., Xu, S., Hayya, J.: A periodic-review inventory system with supply interruptions. Probability in the Engineering and Informational Sciences 3, 33–53 (2004)
4. Tomlin, B.T.: Selecting a Disruption-Management Strategy for Short Life-cycle Products: Diversification, Contingent Sourcing, and Demand Management. In: Working Paper, Kenan-Flagler Business School. University of North Carolina (2005)
5. Tomlin, B., Wang, Y.: On the value of mix flexibility and dual sourcing in unreliable newsvendor networks. Manufacturing and Service Operations Management 1, 37–57 (2005)
6. Tomlin, B.: On the value of mitigation and contingency strategies for managing supply-chain disruption risks. Management Science 5, 639–657 (2006)
7. Lei, Z., Xu, J.: A Probe into the Emergency Management of Unpredictable Events in the Supply Chain. Project Management Technology 5, 26–29 (2004)
8. Yu, H., Chen, J., Yu, G.: How to Coordinate the Supply Chain under Disruptions. System Engineering-Theory & Practice 7, 9–16 (2005)
9. Yu, H., Chen, J., Yu, G.: Supply Chain Coordination under Disruptions with Buyback Contract. System Engineering-Theory & Practice 8, 38–43 (2005)
10. Xu, M., Qi, X., Yu, G., et al.: The demand disruption management problem for a supply chain system with nonlinear demand functions. Journal of Systems Science and Systems Engineering 1, 1–16 (2003)
11. Herrera, F., Martńez, L.: Fuzzy linguistic representation model for computing with words. IEEE Transactions on Fuzzy Systems 6, 746–757 (2000)
12. Wang, J.: Multi-criteria Group Decision-making Approach with Linguistic Assessment Information. Control and Decision 5, 545–548 (2007)
13. Wang, J.: Group Multi-criteria Linguistic Decision-making Method with Incomplete Certain Information. Control and Decision 4, 394–398 (2007)

# Research on Vendor Selection Based on Intuitionistic Fuzzy Sets

Lu Shan

Hunan University of Commerce, Changsha, China
lushan20031020@163.com

**Abstract.** Supply chain management is the advanced management mode of the present time for it can make the goal of win-win situation achievable for enterprises on the junctions of supply chain. And to evaluate and select vendors has become the key for enterprises to improve their competitive advantage because vendors are the very source of the whole supply chain. The objectives of the paper are twofold: One is to establish the criteria for evaluation of vendors on the basis of theoretical research and empirical investigation. The other is to make comprehensive evaluation of vendors, employing the TOPSIS approach which is based on intuitionistic fuzzy sets (IFS).Through empirical research, the validity and practicability of the evaluation criteria and method are verified.

**Keywords:** Supply chain, vendor selection, intuitionistic fuzzy sets.

## Introduction

As an important part of supply chain, vendors are not only the precondition for production organized by enterprises, but also a critical source for cost reduction and competitive advantage. Therefore, vendor management has become an important part of enterprise strategy. For this reason, the issue that what kind of method do we use to carry out evaluation of vendors and choose from a multitude of them the most suitable ones has received widespread attention by the academic circle. For example, Weber et al., after going through and analyzing 74 pieces of the literature concerning vendor selection. found that most of the articles mentioned price, delivery time, quality and ability criteria[1]. Through summarizing the purchasing practice of 170 purchasing agents and managers, Dickson developed 23 criteria for vendor selection[2]. In addition, Johnson adopted Enterprise Excellence Evaluation Method and pointed out that among the variety of factors which influence vendors, time, quality, costs and service are the key factors. Over the recent years, more and more literature have expanded the range of key factors to: quality, costs, delivery reliability, flexibility of quantity, information and service. Regarding to approaches and techniques adopted to evaluate and select vendors, a decision making technique initiated by T. L. Saaty.T.L in 1970s,namely Analytic Hierarchy Process (AHP),is prevalently used at the present time,next to this is Artificial Neural Network Approach. However, in this article, it is Intuitionistic Fuzzy Set Method that is introduced into.It is aimed to undertake systematical assessment of the vendors. Meanwhile, living examples are taken in this article to make calculation and proof the of this m validity of this method.

# 1 Intuitionistic Fuzzy Sets and Their Operation

Intuitionistic Fuzzy Sets,first suggested by Atanassov, are an extension and a development of the conventional fuzzy sets. A new attribute parameter-Degree of Non-membership Function is added to the former so they are able to describe and characterize the ambiguous nature of the objective world even more delicately, thus enabling a company of scholars to research and follow with interest.

Atanassov gives his definition of IFS as follows:

**Definition 1:** Let X be a given domain of discourse, then A, an intuitionistic fuzzy set in X, can be denoted as[3] :

$$A=\{<x,\mu_A(x),\nu_A(x)>| x\in X\}$$

In the above set, $\mu_A(x):X\to[0,1]$ and $\nu_A(x):X\to[0,1]$ represent A's membership function-- $\mu_A(x)$ and non-membership function-- $\nu_A(x)$ respectively. And all $x\in X, 0\le\mu_A(x)+\nu_A(x)\le 1$ on A hold.

Plainly, every conventional fuzzy subset is corresponding to the following intuitionistic fuzzy subset: $A=\{<x,\mu_A(x),1-\mu_A(x)>| x\in X\}$

As to each intuitionistic fuzzy subset in X, $\pi_A(x)=1-\mu_A(x)-\nu_A(x)$ is regarded as x 's intuitionistic index in A,and it is a hesitation degree of x corresponding to A.Apparently,for each $x\in X$ , $0\le\pi_A(x)\le 1$ .And for each conventional fuzzy subset A in X, $\forall x\in X, \pi_A(x)=1-\mu_A(x)-(1-\mu_A(x))=0$.

Intuitionistic fuzzy sets defined on domain of discourse X is written as IFS ( X).

**Definition 2:** Basic operations of intuitionistic fuzzy sets [4-7].

Let $A,B\in IFS(X)$, then:

(1) $A\subseteq B \Leftrightarrow \forall x\in X, \mu_A(x)\le\mu_B(x)$且$\nu_A(x)\ge\nu_B(x)$ ;

(2) $A\subset B \Leftrightarrow \forall x\in X, \mu_A(x)<\mu_B(x)$且$\nu_A(x)>\nu_B(x)$;

(3) $A=B \Leftrightarrow \forall x\in X, \mu_A(x)=\mu_B(x)$且$\nu_A(x)=\nu_B(x)$ ;

(4) $A\cap B=\{<x,\mu_A(x)\wedge\mu_B(x),\nu_A(x)\vee\nu_B(x)>| x\in X\}$ ;

(5) $A\cup B=\{<x,\mu_A(x)\vee\mu_B(x),\nu_A(x)\wedge\nu_B(x)>| x\in X\}$ ;

(6) $A^C=\overline{A}=\{<x,\nu_A(x),\mu_A(x)>| x\in X\}$ ;

(7) $A+B=\{<x,\mu_A(x)+\mu_B(x)-\mu_A(x)\mu_B(x),\nu_A(x)\nu_B(x)>| x\in X\}$ ;

(8) $A\cdot B=\{<x,\mu_A(x)\mu_B(x),\nu_A(x)+\nu_B(x)-\nu_A(x)\nu_B(x)>| x\in X\}$。

**Definition 3:** Let $A\in IFS(X)$, $<x,\mu_A(x),\nu_A(x)>\in A, <y,\mu_A(y),\nu_A(y)>\in A\}$, then:

(1) $x\ge_A y \Leftrightarrow \mu_A(x)\ge\mu_A(y)$且$\nu_A(x)\le\nu_A(y)$ ;

(2) $x>_A y \Leftrightarrow \mu_A(x)>\mu_A(y)$且$\nu_A(x)<\nu_A(y)$ ;

(3) If " $\mu_A(x)\geq\mu_A(y)$且$\nu_A(x)\leq\nu_A(y)$ " doesn't hold, and $\mu_A(x)-\nu_A(x)>\mu_A(y)-\nu_A(y)$, then $x\geq_A y$.

$x\geq_A y$ denotes that ,under corresponding fuzzy definition, $x$ is not inferior to $y$ , and $x>_A y$ means that, under corresponding fuzzy definition, $x$ is superior to $y$ .

**Definition 4:** Let X be limited domain of discourse which has n elements, $A=\{<x_j,\mu_A(x_j),\nu_A(x_j)>|x_j\in X\}$ , $B=\{<x_j,\mu_B(x_j),\nu_B(x_j)>|x_j\in X\}$ ,then the Hamming distance between the two IFS is defined as[6-9] :

$$d(A,B)=\frac{1}{2n}\sum_{j=1}^{n}([|\mu_A(x_j)-\mu_B(x_j)|+|\nu_A(x_j)-\nu_B(x_j)|+|\pi_A(x_j)-\pi_B(x_j)|)$$

If $A=\{<x,\mu_A,\nu_A>\}$ , then now, as a matter of convenience, it is denoted by $<\mu_A,\nu_A>$, such sets are denoted by IFS1.

**Definition 5:** Let $<\mu_A,\nu_A>\}$ and $<\mu_B,\nu_B>$ be two IFS, then [10]:

(1) $<\mu_A,\nu_A>+<\mu_B,\nu_B>=<\mu_A+\mu_B-\mu_A\mu_B,\nu_A\nu_B>$;

(2) $<\mu_A,\nu_A>\cdot<\mu_B,\nu_B>=<\mu_A\mu_B,\nu_A+\nu_B-\nu_{A_1}\nu_B)>$;

(3) $\lambda<\mu_A,\nu_A>=<1-(1-\mu_A)^{\lambda},\nu_A{}^{\lambda})>(\lambda\geq 0)$;

(4) $<\mu_A,\nu_A>^{\lambda}=<(\mu_A{}^{\lambda},1-(1-\nu_A)^{\lambda})>(\lambda\geq 0)$.

**Definition 6:** Let $f:IFS1^n\rightarrow IFS1$, and if

$$f_\omega(a_1,a_2,\cdots,a_n)=\sum_{j=1}^{n}\omega_j b_j\ ,$$

where $\omega=(\omega_1,\omega_2,\cdots,\omega_n)$ is a weighted vector correlated to funtion $f$ ,$\omega_i\in[0,1],\sum_{i=1}^{n}\omega_i=1$,and $b_j$ is the $j$ th major element(ranked according to definition 3 in the IFS $(a_1,a_2,\cdots,a_n)$,then we define function $f$ as the operator of intuitionistic fuzzy ordered weighted averaging(IFSOWA).

## 2 Vendor Selection Method Based on IFS

Let there be m vendors evaluated $A=\{a_1,\ a_2,\ \ \cdots,a_m\}$ , s decision makers $DM=\{D_1,D_2,\cdots,D_s\}$ and $q$ criteria denoted as $C=\{C_1,C_2,\cdots,C_q\}$ ,the weight of criterion $C_l$ given by decision maker $D_p$ is $\omega_l^p$ ,the value of vendor $a_i$ under criterion $C_j$ and relative to the fuzzy definition "suitable" is $<a_i,\mu_{ij}^P,\nu_{ij}^P>$,where $\mu_{ij}^P$ and $\nu_{ij}^P$ are the degree of membership and the degree of non-membership of $a_i$ under criterion $C_j$ and relative to the fuzzy definition respectively, and they satisfy $0\leq\mu_{ij}^P\leq 1$,

$0 \le v_{ij}^P \le 1$, $0 \le \mu_{ij}^P + v_{ij}^P \le 1$. $A_j = \{< a_i, \mu_{ij}^P, v_{ij}^P > | a_i \in A\}$ is an IFS. And $\pi_{ij}^P = 1 - \mu_{ij}^P - v_{ij}^P$ is an intuitionistic index for identifying the most excellent vendor.

## The TOPSIS Method Based on IFS

The steps of this method are demonstrated as below:

**Step 1:** Identify the ideal scheme and the negative ideal scheme.

All the values of ideal scheme $G^+$ under every criterion are $\{< g^+, 1, 0 >\}$.

All the values of negative ideal scheme $G^-$ under every criterion are $\{< g^-, 0, 1 >\}$.

**Step 2:** calculate the distance between the criterion value of every scheme and the ideal scheme and the negative ideal scheme.

The distance between $< a_i, \mu_{ij}^P, v_{ij}^P >$ and the ideal scheme's value--$\{< g^+, 1, 0 >\}$ under criterion $C_j$ is:

$$d_{ij}^{P+} = \frac{|\mu_{ij}^P - 1| + v_{ij}^+ + \pi_{ij}^P}{2}$$

The distance beween $< a_i, \mu_{ij}^P, v_{ij}^P >$ and the negative ideal scheme's value--$\{< g^-, 0, 1 >\}$ under criterion $C_j$ is:

$$d_{ij}^{P-} = \frac{\mu_{ij}^P + |v_{ij}^+ - 1| + \pi_{ij}^P}{2}$$

**Step 3:** Calculate the relative distance value of every scheme's criterion value. The relative distance value of scheme $a_i$'s criterion value is:

$$d_{ij}^P = \frac{d_{ij}^{P+}}{d_{ij}^{P+} + d_{ij}^{P-}},$$

Clearly, the smaller the value of $d_{ij}^P$ is, the better the schemes are.

**Step 4:** Calculate the schemes' relative distance value

The relative distance value of scheme $a_i$ is :

$$d_i^P = \sum_{j=1}^{q} \omega_j^P d_{ij}^P$$

**Step 5:** Group aggregation of the schemes' relative distance

After group aggregation of scheme $a_i$'s relative distance value--$d_i^P$,using OWA operator, comes out the value:

$d_i = f_W(d_i^1, d_i^2, \cdots, d_i^s) = \sum_{l=1}^{s} W_l d_i^l$ ,where $W_i$ is identified through the following steps:

Calculate using OWA and WC-OWA operator

**Definition 7:** Let $f: R^n \rightarrow R$, if $f_\omega(a_1,a_2,\cdots,a_n)=\sum_{j=1}^{n}\omega_j b_j$, Where $\omega=(\omega_1,\omega_2,\cdots,\omega_n)$ is the weighted vector correlated to function $f$, $\omega_i \in [0,1], \sum_{i=1}^{n}\omega_i = 1$, and $b_j$ is the $j$ th major element in the datum --$(a_1,a_2,\cdots,a_n)$, then we can define function $f$ as OWA operator[11].

The weighted vector correlated to function $f$ can be identified by the following formula: $\omega_i = Q(\frac{i}{n}) - Q(\frac{i-1}{n})$ Where the fuzzy semantic quantization function $Q$ is stated by the following formula[12]: $Q(r)=\begin{cases} 0, & r<\alpha \\ \frac{r-\alpha}{\beta-\alpha}, & \alpha \le r \le \beta \\ 1, & r>\beta \end{cases}$

Where, $\alpha,\beta,r \in [0,1]$. The parameters in function $Q$ and corresponding to the fuzzy semantic quantization criteria—"the majority", "at least half", "as many as possible" are $(\alpha,\beta)=(0.3,0.8)$, $(\alpha,\beta)=(0,0.5)$, $(\alpha,\beta)=(0.5,1.0)$.

**Step 6:** Identify the best scheme

We can conclude from the steps above that the smaller $d_i$ is, the better the schemes are. The order of the scheme sets is obtained through sorting $d_i$ from the largest to the smallest. The schemes which go down the hind part of the list are superior to those at the top and the rearest one is the best.

## 3 Calculations Based on Living Examples

For the actual data are very complicated, as a matter of convenience, we suppose that there are 4 criteria: $(C_1,C_2,C_3,C_4)$ which represent quality, price, delivery reliability and information respectively. The form of the values of the evaluation of the three vendors-- $(a_1,a_2,a_3)$ conducted by the three evaluators-- $(D_1,D_2,D_3)$ is $<a_i,\mu_{ij},v_{ij}>$, where $\mu_{ij}^P$ and $v_{ij}^P$ are $a_i$ 's degree of membership and degree of non-membership respectively under criterion $C_j$ and relative to the fuzzy definition, and they satisfy $0 \le \mu_{ij}^P \le 1$, $0 \le v_{ij}^P \le 1$, $0 \le \mu_{ij}^P + v_{ij}^P \le 1$. The data relevant to the calculation are indicated in table 1,2,3.To simplify these tables, the evaluation value $a_i$ is omitted.

**Table 1.** Schemes and respective weights given by decision maker $D_1$

| $D_1$ | $C_1$ | $C_2$ | $C_3$ | $C_4$ |
|---|---|---|---|---|
| $a_1$ | $<0.7,0.1>$ | $<0.6,0.4>$ | $<0.4,0.5>$ | $<0.4,0.5>$ |
| $a_2$ | $<0.35,0.5>$ | $<0.7,0.2>$ | $<0.7,0.1>$ | $<0.65,0.2>$ |
| $a_3$ | $<0.65,0.1>$ | $<0.6,0.3>$ | $<0.8,0.1>$ | $<0.5,0.36>$ |
| $\omega_l$ | 0.27 | 0.18 | 0.25 | 0.30 |

**Table 2.** Schemes and respective weights given by decision maker $D_2$

| $D_2$ | $C_1$ | $C_2$ | $C_3$ | $C_4$ |
|---|---|---|---|---|
| $a_1$ | $<0.8,0.2>$ | $<0.6,0.3>$ | $<0.4,0.4>$ | $<0.6,0.2>$ |
| $a_2$ | $<0.7,0.1>$ | $<0.65,0.25>$ | $<0.3,0.5>$ | $<0.6,0.2>$ |
| $a_3$ | $<0.5,0.3>$ | $<0.73,0.2>$ | $<0.6,0.2>$ | $<0.6,0.3>$ |
| $\omega_l$ | 0.2 | 0.35 | 0.21 | 0.24 |

**Table 3.** Schemes and respective weights given by decision maker $D_3$

| $D_3$ | $C_1$ | $C_2$ | $C_3$ | $C_4$ |
|---|---|---|---|---|
| $a_1$ | $<0.6,0.3>$ | $<0.6,0.2>$ | $<0.3,0.5>$ | $<0.5,0.5>$ |
| $a_2$ | $<0.6,0.32>$ | $<0.8,0.1>$ | $<0.7,0.3>$ | $<0.7,0.2>$ |
| $a_3$ | $<0.6,0.35>$ | $<0.8,0.2>$ | $<0.54,0.3>$ | $<0.8,0.2>$ |
| $\omega_l$ | 0.23 | 0.16 | 0.29 | 0.32 |

**Step 1:** Select the corresponding positive ideal solution and negative ideal solution under each criterion.

**Step 2:** By calculating the distance between the criterion value of every scheme and the positive ideal solutions and the negative ideal solutions, we get:

$d_{11}^{1+}=0.3$; $d_{11}^{1-}=0.9$; $d_{12}^{1+}=0.4$; $d_{12}^{1-}=0.6$; $d_{13}^{1+}=0.6$; $d_{13}^{1-}=0.5$; $d_{14}^{1+}=0.6$; $d_{14}^{1-}=0.5$; $d_{21}^{1+}=0.65$; $d_{21}^{1-}=0.5$; $d_{22}^{1+}=0.3$; $d_{22}^{1-}=0.8$; $d_{23}^{1+}=0.3$; $d_{23}^{1-}=0.9$; $d_{24}^{1+}=0.35$; $d_{24}^{1-}=0.8$; $d_{31}^{1+}=0.35$; $d_{31}^{1-}=0.9$; $d_{32}^{1+}=0.4$; $d_{32}^{1-}=0.7$; $d_{33}^{1+}=0.2$; $d_{33}^{1-}=0.9$; $d_{34}^{1+}=0.5$; $d_{34}^{1-}=0.64$; $d_{11}^{2+}=0.2$; $d_{11}^{2-}=0.8$; $d_{12}^{2+}=0.4$; $d_{12}^{2-}=0.7$; $d_{13}^{2+}=0.6$; $d_{13}^{2-}=0.6$; $d_{14}^{2+}=0.4$; $d_{14}^{2-}=0.8$; $d_{21}^{2+}=0.3$; $d_{21}^{2-}=0.9$; $d_{22}^{2+}=0.35$; $d_{22}^{2-}=0.75$; $d_{23}^{2+}=0.7$; $d_{23}^{2-}=0.5$; $d_{24}^{2+}=0.4$; $d_{24}^{2-}=0.8$; $d_{31}^{2+}=0.5$; $d_{31}^{2-}=0.7$; $d_{32}^{2+}=0.27$; $d_{32}^{2-}=0.8$; $d_{33}^{2+}=0.4$; $d_{33}^{2-}=0.8$; $d_{34}^{2+}=0.4$; $d_{34}^{2-}=0.7$; $d_{11}^{3+}=0.4$; $d_{11}^{3-}=0.7$; $d_{12}^{3+}=0.4$; $d_{12}^{3-}=0.8$; $d_{13}^{3+}=0.7$; $d_{13}^{3-}=0.5$; $d_{14}^{3+}=0.5$; $d_{14}^{3-}=0.5$; $d_{21}^{3+}=0.4$; $d_{21}^{3-}=0.68$; $d_{22}^{3+}=0.2$; $d_{22}^{3-}=0.9$; $d_{23}^{3+}=0.3$; $d_{23}^{3-}=0.7$; $d_{24}^{3+}=0.3$; $d_{24}^{3-}=0.8$; $d_{31}^{3+}=0.4$; $d_{31}^{3-}=0.65$; $d_{32}^{3+}=0.2$; $d_{32}^{3-}=0.8$; $d_{33}^{3+}=0.46$; $d_{33}^{3-}=0.7$; $d_{34}^{3+}=0.2$; $d_{34}^{3-}=0.8$.

**Step 3:** By calculating the relative distance value of every scheme's criterion values, we get:

$d_{11}^{1}=0.25$; $d_{12}^{1}=0.4$; $d_{13}^{1}=0.5455$; $d_{14}^{1}=0.5455$; $d_{21}^{1}=0.5652$; $d_{22}^{1}=0.2727$; $d_{23}^{1}=0.25$; $d_{24}^{1}=0.3043$; $d_{31}^{1}=0.28$; $d_{32}^{1}=0.3636$; $d_{33}^{1}=0.1818$; $d_{34}^{1}=0.4386$; $d_{11}^{2}=0.2$; $d_{12}^{2}=0.3636$; $d_{13}^{2}=0.5$; $d_{14}^{2}=0.3333$; $d_{21}^{2}=0.25$; $d_{22}^{2}=0.3182$; $d_{23}^{2}=0.5833$; $d_{24}^{2}=0.3333$; $d_{31}^{2}=0.4167$; $d_{32}^{2}=0.2523$; $d_{33}^{2}=0.3333$; $d_{34}^{2}=0.3636$; $d_{11}^{3}=0.3636$; $d_{12}^{3}=0.3333$; $d_{13}^{3}=0.5833$; $d_{14}^{3}=0.50$; $d_{21}^{3}=0.3704$; $d_{22}^{3}=0.1818$; $d_{23}^{3}=0.3$; $d_{24}^{3}=0.2727$; $d_{31}^{3}=0.3810$; $d_{32}^{3}=0.2$; $d_{33}^{3}=0.3966$; $d_{34}^{3}=0.2$

**Step 4:** By calculating the schemes' relative distance values, we get: $d_1^1=0.4395$; $d_2^1=0.3555$; $d_3^1=0.3181$; $d_1^2=0.3523$; $d_2^2=0.3639$; $d_3^2=0.3289$; $d_1^3=0.4661$; $d_2^3=0.2885$; $d_3^3=0.2986$.

**Step 5:** Group aggregation of the schemes' relative distances. Choose the fuzzy semantic quantization criteria—"the majority". Making use of definition 7,we can get the weighted vectors correlated to function f and they are $\omega_1=0.0667$; $\omega_2=0.6667$; $\omega_3=0.2666$.So the distance aggregation values are:

$$d_1=0.3511\ ;\ d_2=0.3538\ ;\ d_3=0.3030$$

**Step 6:** By sorting the three vendors, we can get $d_2>d_1>d_3$, so $d_3$ is the best vendor.

## 4 Conclusions

Vendor selection is an important element in purchasing decisions. For most enterprises, purchasing costs account for more than 70% of their gross costs. Therefore, reasonable vendor selection will exert direct influence on enterprises' strategy of costs reduction, flexibility enhancement and competitiveness improvement. with the globalization and intensity of market competition, lifecycle of products has become shorter and shorter. And emphasis on quality, delivery reliability, price and delivery advancement has made vendor selection more complicated and broadened its range, as well as made it become a question which preoccupies both the academic circle and the industrial circle. Relevant researching work has been and is still being done by many researchers. This article, on the basis of the relevant literature, establishes the vendor evaluation criteria and vendor evaluation model ground on IFS, and makes comprehensive evaluation of vendors. The establishment and application of the model can make an intuitive evaluation of vendors, however, it still has to experience repeated examinations in practice to amend and improve unceasingly.

**Acknowledgement.** The work in the paper is supported by Hunan provincial commission of the scientific payoffs of philosophy and social sciences (0808041B), Outstanding Youth of Scientific Research Project of Hunan Provincial Education Department (09B056) and Hunan Social Sciences Fund item(09YBB228).

## References

1. Weber, J.R., Current, W.C.: Benton Vendor selection criteria and methods. European Journal of Operational Research 50, 2–18 (1991)
2. Dickson, G.: An Analysis of Vendor Selection Systems and Decisions. Journal of Purchasing 2, 28–41 (1996)
3. Atanassov, K.T.: Intuitionistic fuzzy sets. Fuzzy Sets and Systems 20, 87–96 (1986)
4. Atanassov, K.T.: Intuitionistic fuzzy sets, pp. 123–154. Springer, Heidelberg (1999)
5. Atanassov, K.T.: More on intuitionistic fuzzy sets. Fuzzy Sets and Systems 1, 37–45 (1989)
6. Atanassov, K.T.: Remarks on the intuitionistic fuzzy sets. Fuzzy Sets and Systems 1, 117–118 (1992)
7. Atanassov, K.T.: New operations defined over the intuitionistic fuzzy sets. Fuzzy Sets and Systems 2, 137–142 (1994)
8. Atanassov, K.T., Gargov, G.: Interval valued intuitionistic fuzzy sets. Fuzzy Sets and Systems 3, 312–349 (1989)
9. Atanassov, K.T.: Operators over interval valued intuitionistic fuzzy sets. Fuzzy Sets and Systems 2, 159–174 (1994)
10. Bustince, H., Burillo, P.: Correlation of interval-valued intuitionistic fuzzy sets. Fuzzy Sets and Systems 2, 237–244 (1995)
11. Hong, D.H.: A note on correlation of interval-valued intuitionistic fuzzy sets. Fuzzy Sets and Systems 1, 113–118 (1998)
12. Xu, Z.: Methods for aggregating interval-valued intuitionistic fuzzy information and their application to decision making. Control and Decision 2, 215–219 (2007)

# An Empirical Study of the Influencing Factors of Relationship Capital in the Supply Chain

Lu Shan and Yu-lei Li

Hunan University of Commerce, Changsha, China
lushan20031020@163.com

**Abstract.** In recent years, more and more enterprises begin to realize that the relationship resources accumulated in the supply chain is one of the key parts of the social resources for a company. With the existing research done both at home and abroad as well as the current situation of Chinese enterprises, this paper proposes the structure and the influencing factors of relationship capital, and verifies the scientific attribute of the factors by reliability and validity analysis. It shows that the factors influencing relationship capital are ability, reputation and organizational adaptability of the partner enterprises, resources input and switching costs while the ability index can fall into two categorise: "managerial ability" and "production ability".

**Keywords:** Supply Chain, Relationship Capital, Reliability Analysis, Validity Analysis.

## 1 Introduction

At present, given the incompleteness of the contract and the uncertainty of the environment, many of the activities of the enterprises in the supply chain cannot be regulated by a formal contract but need to be coordinated through the relationship among those enterprises. The relationship management in the supply chain based on relationship capital focuses on social psychological factors among the enterprises, gives prominence to the action issue related to the establishment and maintenance of a cooperative relationship in the supply chain and points up one of the properties of relationship capital that relationship capital can bring about productivity ,that is to say the cooperative actions followed by the development of relationship capital is conducive to the transformation of potential value of supply chain into the real collaborative economic rent. Therefore, it is worth to give an in-depth study on the structure and influencing factors of the relationship capital. There are some scholars who have already done some research on the measurement indices of the relationship capital [1]. Sarkar holds the opinion that relationship capital consists of several dimensions, such as mutual trust, commitment on mutual benefit and cooperation and communication, all of which determine the strength and quality of the relationship [2]. Roy proposes that four indices, namely, strength, quality, tightness and quantity of relationship, can be used to measure the relationship capital between the enterprises [3]. Bao Gongmin, Wang Qingxi consider that relationship capital is mainly influenced bystructure of the alliance, the allies' act of showing credit and specific investment [4].

This paper, based on the existing research both at home and abroad, aims to propose the structure and influencing factors of the relationship capital and give the analysis of reliability and validity, by combining the research done by the author as well as the current situation of Chinese enterprises.

## 2 The System Design of the Structure and Influencing Factors of Relationship Capital in the Supply Chain

This thesis defines preliminarily that relationship capital in the supply chain mainly includes trust and commitment and that the influencing factors are ability, reputation and organizational adaptability of the partner enterprises, resources input and switching costs. The measurement indices of the above-mentioned variables are reviewed and modified based on the existing research both at home and abroad, the specific situation of Chinese enterprises as well as spot study and interviews.

## 3 Questionnaire Test

About 100 copies of the test are sent out and 84 of them are returned. After a test of their validity, 70 copies are left when deleting those invalid ones.

The research is carried out by adopting the method of SPSS15.0 to test the samples and analyze them in three steps:

**Step 1:** Cleaning up the measurement items of the latent variables and delete those with lower reliability. The method adopted is as follows: ①Selecting the measurement items by means of Corrected Item-Total Correction (CITC). For using CITC, two basic steps should be followed: first, selecting the items whose CITC index is less than 0.3; second, selecting the items whose CITC index is less than 0.3 and the deletion of which can increase the Cronbach'α index and this can help improve the reliability of the measuring scale as a whole [5]. ②Verifying the reliability of the measurement items by Cronbach's alpha. The value of Cronbach's alpha should not be lower than 0.5.

**Step 2:** Verifying the correlation among the measurement items and deciding whether factor analysis can be done according to KMO sample measurements. In the opinion of Ma Qingguo (2002) [6], those with KMO more than 0.7 are suitable for factor analysis, while those with KMO between 0.6 and 0.7 are not quite fit and those between 0.5 and 0.6 are barely proper. Based on this principle, those with KMO lower than 0.6 are not included in the analysis while those with KMO over 0.7 are going to be analyzed and for those with KMO between 0.6 and 0.7, they are used for theoretical study basically and whether they are analyzed or not depends on the fact later.

**Step 3:** Carrying out factor analysis for the measurement items with qualified KMO value. Variables with a single dimension are analyzed for aggregated validity while those with various dimensions are analyzed for validity in big samples by the method of confirmatory factor analysis, and the total variance explained of the common factors should be over 0.5.

## 4 Analysis of Reliability and Validity

The variables in the model are analyzed with data according to the above method, including mutual trust among the enterprises, mutual commitment among the enterprises, the ability, reputation and organizational adaptability of the partner enterprises and resources input.CITC and reliability analysis of the questionnaire indices system, as shown in Table 1.

**Table 1.** CITC and reliability analysis of the questionnaire indices system (N=70)

| Item | CITC at the beginning | CITC in the end | α ratio after deleting the item in the end | α ratio |
|---|---|---|---|---|
| TRU1 | .3574 | .3610 | .8939 | |
| TRU2 | .3988 | .4078 | .8931 | |
| TRU3 | .3563 | .3436 | .8941 | |
| TRU4 | .5257 | .5118 | .8913 | |
| TRU5 | .4394 | .4273 | .8927 | |
| TRU6 | .3802 | .3913 | .8934 | |
| COM1 | .4174 | .4088 | .8931 | |
| COM2 | .3151 | .3246 | .8946 | |
| COM3 | .5383 | .5364 | .8908 | Alpha at the begin-ning = .8956 |
| COM4 | .4156 | .4210 | .8929 | |
| COM5 | .4263 | .4212 | .8929 | |
| SC1 | .3994 | .3974 | .8932 | |
| SC2 | .3605 | .3663 | .8938 | |
| SC3 | .4146 | .4146 | .8930 | |
| CAP1 | .4772 | .4740 | .8918 | Alpha in the end = .8959 |
| CAP2 | .4886 | .4837 | .8917 | |
| CAP3 | .5295 | .5295 | .8908 | |
| CAP4 | .4657 | .4721 | .8919 | |
| CAP5 | .4429 | .4407 | .8925 | |
| SIN1 | .4987 | .4958 | .8917 | |
| SIN2 | .5687 | .5709 | .8900 | |
| SIN3 | .4977 | .5043 | .8912 | |
| SIN4 | .3850 | .3975 | .8934 | |
| REP1 | .4112 | .4065 | .8931 | |
| REP2 | .2754 | deleted | deleted | |
| REP3 | .4073 | .4078 | .8932 | |
| REP4 | .4122 | .4159 | .8929 | |
| TRC1 | .4377 | .4345 | .8926 | |
| TRC2 | .6076 | .6057 | .8891 | |
| TRC3 | .4469 | .4467 | .8924 | |
| CC1 | .5214 | .5213 | .8909 | |
| CC2 | .3111 | .3068 | .8948 | |

Table 1 shows that the CITC of the item corresponding to the reputation of the supplier, i.e. REP2, is 0.2754, which is lower than 0.3, and that when the item is deleted, the α ratio turns from 0.8956 to 0.8959. In this case, the item REP2 is finally deleted and other items are kept.

Next, the left items are analyzed through the method of exploratory factor analysis. Since there are more variables, the factor analysis is carried out separately on the variables with a single dimension.

1) Factor analysis of trust of the supplier

First, KMO and Bartlett's test is taken and verified on the items, the results of which are shown in Table 2. It is clear that KMO is 0.745 which is over 0.7 and that the statistical value of the Bartlett's test is lower than 0.001. The results show that there is strong relevance among the sub-indices of the trust of the supplier and that it is feasible to take the exploratory factor analysis on this index.

**Table 2.** KMO and Bartlett's Test

| Kaiser-Meyer-Olkin Measure of Sampling Adequacy. | | .745 |
|---|---|---|
| Bartlett's Test of Sphericity | Approx. Chi-Square | 26.776 |
| | Df | 15 |
| | Sig. | .000 |

Table 3 shows that there comes a new factor during the factor analysis of the trust index with the eigenvalue of 3.782, which can explain the total variance of 69.698%. This is a manifestation that the measurement structure is one-dimension and that the factor can be defined as "Trust".

**Table 3.** Dimensional analysis of the measurement of trust

| Component | Initial Eigenvalues | | | Extraction Sums of Squared Loadings | | |
|---|---|---|---|---|---|---|
| | Total | % of Variance | Cumulative % | Total | % of Variance | Cumulative % |
| 1 | 3.782 | 69.698 | 69.698 | 3.782 | 69.698 | 69.698 |
| 2 | .865 | 9.420 | 79.118 | | | |
| 3 | .668 | 7.141 | 86.259 | | | |
| 4 | .570 | 6.827 | 92.086 | | | |
| 5 | .490 | 4.502 | 96.588 | | | |
| 6 | .325 | 3.412 | 100.000 | | | |

Extraction Method: Principal Component Analysis.

2) Factor analysis of commitment index

According to calculations, KMO of the commitment index is 0.705 which is over 0.7 and the statistical value of the Bartlett's test is not significant. So it is feasible to take the exploratory factor analysis on this index. There comes a new factor during the factor analysis of the commitment index with the eigenvalue of 3.104, which can explain the total variance of 62.313%. This is a manifestation that the measurement structure is one-dimension and that the factor can be defined as "Commitment".

3) Factor analysis of relationship capital as a whole

The measurement items of Trust and Commitment are joined together for the factor analysis so as to verify whether the new factor formed therefrom is two-dimension or not.

In Table 4, KMO is 0.723 which is over 0.7 and the statistical value of the Bartlett's test is not significant. So it is feasible to take the exploratory factor analysis on this index.

**Table 4.** KMO and Bartlett's Test

| Kaiser-Meyer-Olkin Measure of Sampling Adequacy. | | .723 |
|---|---|---|
| Bartlett's Test of Sphericity | Approx. Chi-Square | 14.450 |
| | Df | 5 |
| | Sig. | .000 |

Table 5 shows the result of factor analysis, which says that there are only two factors gained from the analysis whose eigenvalue are 4.920 and 3.454 separately, with the cumulative explained variance reaching 59.764%.

From Table 6, it is clear that the factor analysis of relationship capital results in two factors, and that factor 1 contains the items of TR1, TR2, TR3, TR4, TR5 and TR6 which are all the variables of Trust as defined in this research while factor 2 contains the items of COM1, COM2, COM3, COM4 and COM5 which are all in conformity with the items of Commitment in this research.Therefore, , it is further verified that Trust and Commitment can be divided in the same space,and this bisection method is appropriate.

**Table 5.** Dimensional analysis of the measurements of relationship capital

| Component | Initial Eigenvalues | | | Extraction Sums of Squared Loadings | | |
|---|---|---|---|---|---|---|
| | Total | % of Variance | Cumulative % | Total | % of Variance | Cumulative % |
| 1 | 4.920 | 36.550 | 36.550 | 4.920 | 36.550 | 36.550 |
| 2 | 3.454 | 23.215 | 59.764 | 3.454 | 23.215 | 59.764 |
| 3 | .945 | 8.405 | 68.169 | | | |
| 4 | .865 | 7.685 | 75.854 | | | |
| 5 | .787 | 6.059 | 81.913 | | | |
| 6 | .644 | 5.764 | 86.677 | | | |
| 7 | .500 | 4.364 | 90.041 | | | |
| 8 | .438 | 3.798 | 93.839 | | | |
| 9 | .433 | 3.848 | 96.687 | | | |
| 10 | .378 | 2.343 | 98.031 | | | |
| 11 | .237 | 1.969 | 100.000 | | | |

Extraction Method: Principal Component Analysis.

4) Factory analysis of the organizational adaptability of the partner enterprises in the supply chain

When calculated based on the above means, KMO value is 0.715 which is over 0.7 and the statistical value of the Bartlett's test is not significant. So it is feasible to take the exploratory factor analysis on this index. There comes a new factor during the factor analysis of the index of organizational adaptability with the eigenvalue of 2.287, which can explain the total variance of 88.900%. This is a manifestation that the measurement structure is one-dimension and that the factor can be defined as "Organizational Adaptability".

**Table 6.** Factory analysis of relationship analysis

| | Component | |
|---|---|---|
| | 1 | 2 |
| TR1 | .722 | .207 |
| TR2 | .670 | .343 |
| TR3 | .630 | -.191 |
| TR4 | .633 | -.244 |
| TR5 | .595 | -.387 |
| TR6 | .641 | .458 |
| COM1 | .367 | .648 |
| COM2 | .356 | .750 |
| COM3 | .246 | .760 |
| COM4 | .398 | .615 |
| COM5 | .180 | .727 |

5) Factor analysis of the ability of the supplier

The KMO value is 0.758 which is over 0.7 and the statistical value of the Bartlett's test is not significant. So it is feasible to take the exploratory factor analysis on this index.

Table 7 shows that the factor analysis of the ability of the supplier results in two factors. Factor 1 contains the items of CAP 3andCAP 4, the sub-items of which are all related to the management aspect of the supplier and so factor 1 is defined as "Managerial Ability"; Factor 2 contains the items of CAP 1, CAP 2 and CAP 5, the sub-items of which are all related to the product aspect of the supplier and so factor 2 is defined as "Product Ability".

**Table 7.** Factor analysis of the ability of the supplier (Rotated)

| | Component | |
|---|---|---|
| | 1 | 2 |
| CAP1 | .337 | .653 |
| CAP2 | .063 | .659 |
| CAP3 | .802 | .203 |
| CAP4 | .860 | .022 |
| CAP5 | .009 | .765 |

Extraction Method: Principal Component Analysis.
Rotation Method: Varimax with Kaiser Normalization.

6) Factor analysis of specific resources input

When calculated based on the above means, KMO value is 0.754 which is over 0.7 and the statistical value of the Bartlett's test is not significant. So it is feasible to take the exploratory factor analysis on this index. There comes a new factor during the factor analysis of the index of organizational adaptability with the eigenvalue of 2.729, which can explain the total variance of 63.234%. This is a manifestation that the measurement structure is one-dimension and that the factor can be defined as "Specific Resources Input".

7) Factor analysis of reputation of the supplier

By the same method above, KMO value is 0.709 which is over 0.7 and the statistical value of the Bartlett's test is not significant. So it is feasible to take the exploratory factor analysis on this index. There comes a new factor during the factor analysis of the index of organizational adaptability with the eigenvalue of 2.281, which can explain the total variance of 62.684%. This is a manifestation that the measurement structure is one-dimension and that the factor can be defined as "Reputation of the Supplier".

8) Factor analysis of switching costs

The KMO value is 0.712 which is over 0.7 and the statistical value of the Bartlett's test is not significant. So it is feasible to take the exploratory factor analysis on this index. This analysis results in a factor with the eigenvalue of 2.416, which can explain the total variance of 67.210%. This is a manifestation that the measurement structure is one-dimension and that the factor can be defined as "Switching Costs".

## 5 Conclusion

With the above research, it is shown that the contents of relationship capital in the supply chain are mainly Trust and Commitment and that the factors influencing relationship capital are ability, reputation and organizational adaptability of the partner enterprises, resources input and switching costs while the ability index can fall into two categorise: "managerial ability" and "production ability". And more studies related to relationship capital can be further conducted on the basis of the conclusion in this paper.

**Acknowledgement.** The work in the paper is supported by Hunan provincial commission of the scientific payoffs of philosophy and social sciences (0808041B), Outstanding Youth of Scientific Research Project of Hunan Provincial Education Department (09B056) and Hunan Social Sciences Fund item(09YBB228).

## References

1. Daniel, H.Z., Hempel, D.J., Srinivasan, N.: A model of value assessment in collaborative R&D programs. Industrial Marketing Management 3, 653–664 (2002)
2. Sarkar, M.B., Echambadi, R., Cavusgil, S.T., Aulakh, P.S.: The influence of complementarity, compatibility, and relationship capital on alliance performance. Journal of the Academy of Marketing Science 4, 358–373 (2007)

3. Roy, J.P.: International joint venture partner selection and performance: The role of the host country legal environment. York University (2005)
4. Bao, G., Wang, Q.: On the Building and Maintenance of Strategic Alliance Relational Capital. R& D Management 3, 9–14 (2004)
5. Lu, W.: SPSS for Windows Statistical Analysis. Publishing House of Electronics Industry, Beijing (2002)
6. Ma, Q.: Management Statistics. Science Press, Beijing (2002)

# Low Complexity Synchronization Algorithms for HART C8PSK

Jin Zhenghua[1,2], Wang Hong[2], and Yang Zhijia[2]

[1] Graduate School, Chinese Academy of Sciences, Beijing 10039, China
[2] Shenyang Institute of Automation, Chinese Academy of Sciences, Liaoning Shenyang 110016, China
jinzh@sia.cn, wang.hong@microcyber.cn, yang@sia.cn

**Abstract.** This paper not only describes novel fast synchronization algorithms for HART C8PSK carrier recovery and symbol timing, but also provides a solution for HART C8PSK low power design. Synchronization is divided into two parts: initialization and tracking, the former part of which uses novel algorithms to attain equilibrium state quickly, and the later part of which is time-divided carried out. On basis of the proposed algorithms, high power component—equalizer and interpolator, can work at lowest rate, which indirectly reduces the system power. The implementation of those algorithms needs no additional hardware resource except a few lookup tables, and at the same time, they add negligible power consumption to the whole system. Simulations showed that the proposed synchronization algorithms can work very well in low SNR.

**Keywords:** Fast synchronization algorithms, low power design, HART C8PSK, Low Complexity.

## 1 Introduction

The HART (Highway Addressable Remote Transducer) Protocol is the global standard for sending and receiving digital information across analog wires between smart devices and control or monitoring system. The global installed base of HART-enabled devices is the largest of all communication protocols at more than 20 million. Because digital communication rate of the existing HART FSK protocol limits the use the HART, the HCF (HART communication Foundation) established the HART C8PSK (Coherent 8-way Phase Shift Keying) PHL (Physical Layer) Specification, which can provide 8-10 times higher digital rate [1]. In order to be intrinsic safe, HART has strict requirement on power dissipation, and HART C8PSK is significantly more complex signal than HART FSK, so the low power design of the HART C8PSK is very tough. An Austria company, called Volker Schindler, developed the first C8PSK HART modem chip by using lower power semiconductor technology in 2009, the current of which was 800uA, smaller than 1000uA requirement, but still larger than 750uA, the applicable level [2].

HART belongs to burst communication. HART C8PSK PHL specifies that receiver should adopt coherent demodulation, so rapid recovery of carrier phase and timing is essential. Short cyclic preamble is provided for channel estimation and synchronization. Digital phase-locked loop (PLL) with gear-shift algorithms can adjust the bandwidth of the PLL to solve both problems of fast acquisition and jitter reduction, so they are always solutions for this kind of problems [3, 4]. However, using the above technology, some industry giant, such as Rosemount, SMAR, and Siemens, organized by HCF, developed a HART C8PSK modem in 1998, only to find that the receiver need at least 58 preambles to stabilize and the power consumption of the loop was 569uA [5]. So techniques developed from DPLL, such as dual-loop DPLL [6], have difficulties in controlling the power dissipation in HART C8PSK synchronization. Huang Introduced a fast carrier synchronization algorithm for MPSK based on Mth power loop [7], but this kind of algorithm needs fine timing and high SNR, additionally it is hard to control power too. Based on a very short special carrier preamble modulated by a known pattern, Andronico introduced an algorithm to fast estimate synchronization parameters, but six complex multiples were involved [8]. Synchronization algorithms used in OFDM system always employ signal autocorrelation to get timing parameters fast, but they do not consider the carrier phase offset.

In this paper, we present novel algorithms with low complexity for HART C8PSK to fast estimate the synchronization parameters. The estimation is carried out before the equalizer and interpolator start working, and involves no multipliers. During the synchronization tracking, time-divided algorithms are used and they need low sampling rate which reduces the computation rate of high power components such as equalizer and interpolator.

In section 2 below, a brief description of the whole C8PSK receiver system is given. Section 3 dedicates to initialization algorithms for synchronization. Section 4 describes synchronization tracking. In section 5, simulation results of the proposed algorithms are presented. At last, we conclude our work in section 6.

## 2 Background

HART FSK underlies Bell-202 Telecomm standard, while HART C8PSK is similar to the signaling specified in the CCITT V.27 telecommunications standard, using a higher number of symbols per second and a wider bandwidth than V.27. HART C8PSK employs eight phases of a 3200 HZ carrier to transmit digital signals, and the preamble is 40 cyclic symbols of symbol 6(67.5°) and symbol 2(-67.5°). Symbol rate of HART C8PSK is 3200HZ. The receiver structure is shown as figure 1.

It should be noted that the equalizer is a preset CSD coded equalizer as shown in figure 2, which is made up of programmable shifters and adders. Those programmable shifters work according to a set of coefficients in the CSD coefficient ROM, which is decided by the result of channel estimation. When HART C8PSK signal is detected, channel estimation and synchronization initialization are carried out in parallel first. Equalizer and interpolator will not work until they finish. The adders of the equalizer are used for synchronization initialization as described in the next section. The sample rate of the ADC is 25600Hz, this is, 8 samples in a symbol time.

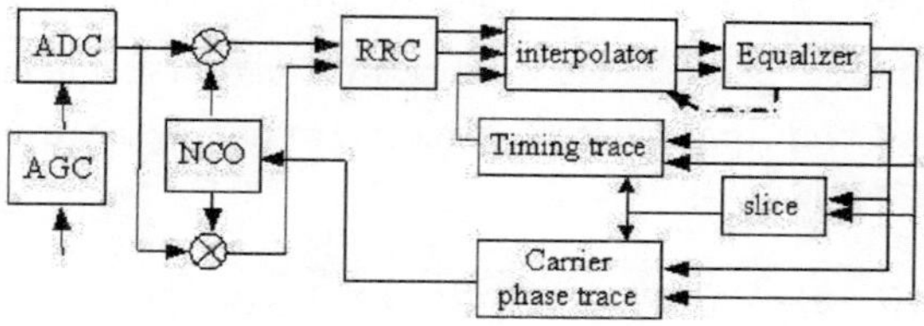


**Fig. 1.** Receiver structure

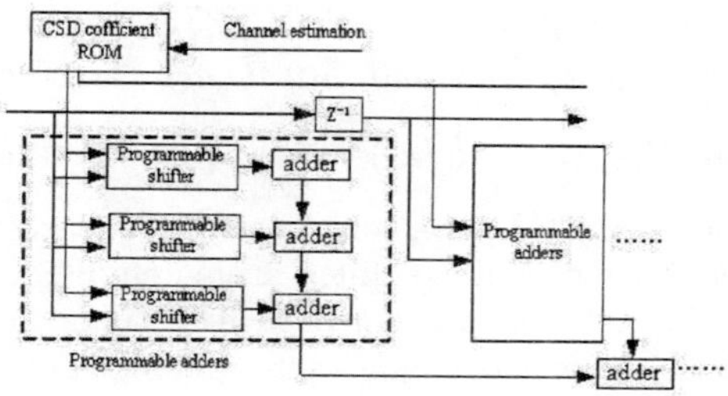


**Fig. 2.** CSD coded preset equalizer

# 3 Initialization of Synchronization

Synchronization initialization starts working after the AGC finish initialization, and uses the samples output from the RRC filters. Initialization algorithms reuse the adders and shift registers of equalizer for implementation.

Periodic preamble sent by HART C8PSK transmitter turns out periodic preamble signal to HART channel. The time for signal detection and AGC initialization is longer than combined impulse response of channel and shape filter, so the signal input to the initialization circuits can be seen as periodic signal. Digital signal of HART C8PSK output from the receiver shape filter (RRC) can be written as:

$$Q(n)=\sum_{i} g(nT_s-iT+\tau)\cos(\varphi_i+\Delta\theta)+N_Q(n) \tag{1}$$

$$I(n)=\sum_{i} g(nT_s-iT+\tau)\sin(\varphi_i+\Delta\theta)+N_I(n) \tag{2}$$

$$\Delta\theta=\theta_T-\theta_R \tag{3}$$

Where $\theta_T$ represents the phase of transmitter carrier and $\theta_R$ the phase of receiver carrier; $\varphi_i$ is the symbol phase for transmitted messages; $\tau<0.5*T_s$ is timing error; $T$ is symbol period; $T_s$=T/8 is sampling period; $g(n)$ represents the combined impulse response of shape filter, analog prefilter and channel; $N_Q(n)$ and $N_I(n)$ are noise. HART C8PSK PHL specifies the carrier frequency shall fluctuate within 3200±1 Hz at both transmitter and receiver end, so the frequency mismatch can be negligible here. Defining

$$SumQ(m)=\sum_{n=1}^{16}Q(n+m) \; ; \; SumI(m)=\sum_{n=1}^{16}I(n+m) \tag{4}$$

### 3.1 Proposition for Carrier Phase Estimation

**Proposition 1:** During the preamble of HART C8PSK (except the first and last 3 symbols of the preamble), $SumQ(m)$ and $SumI(m)$ can be written as:

$$SumQ(m)=A\cos(\Delta\theta) \;\; ; SumI(m)=A\sin(\Delta\theta)$$

**Prove:** The preamble of the C8PSK are cyclic symbol 6 and symbol 2 which can be viewed as a 1600 Hz periodic signal. Shape filter, analog prefilter and channel can filter out or distort some components but can not change the periodicity, so there are 16 samples in a period. AWGN noise is ignorable in the accumulation, so we can get that

$$SumQ(m)=\sum_{n=1}^{16}Q(n+m)=\sum_{n=1}^{16}Q(n)$$

$$=\sum_{n=1}^{16}\sum_{i}\{g[nT_s-(2i-1)T+\tau]\cos(67.5+\Delta\theta)$$

$$+g(nT_s-2iT+\tau)\cos(-67.5+\Delta\theta)\}$$

$$=\sum_{n=1}^{16}\sum_{i}\{\{g[nT_s-(2i-1)T+\tau]+g(nT_s-2iT+\tau)\}$$

$$*\cos 67.5*\cos\Delta\theta-\{g[nT_s-2(i-1)T+\tau]-g(nT_s-2iT+\tau)\}*\sin 67.5*\sin\Delta\theta \tag{5}$$

Defining

$$X(i,\tau)=\sum_{n=1}^{16}\sum_{i}g(nT_s-iT+\tau) \tag{6}$$

For $Ts=\mathrm{T}/8$, it is obviously that

$$X(i,\tau)=X(i+1,\tau) \tag{7}$$

So it is easy to get

$$X(2i,\tau)=X(2i-1,\tau) \tag{8}$$

Then the second half of (5) equals 0, and the first half is

$$SumQ(m)=2X(2i,\tau)*\cos 67.5*\cos\Delta\theta \tag{9}$$

For the same reason, we can get

$$SumI(m) = 2X(2i,\tau) * \cos 67.5 * \sin \Delta\theta \tag{10}$$

Proposition 1 is proved.

### 3.2 Proposition for Timing Phase Estiamtion

Defining

$$Ti(m) = \sum_{n=1}^{8} I(n+m) - \sum_{n=8}^{15} I(n+m) \tag{11}$$

**Proposition 2:** During the preamble of HART C8PSK (except the first and last 3 symbols of the preamble), if carrier phase is eliminated or small, and $Ti(q)$ is the nearest point to zero, then $Ti(q) \approx B\tau$

**Prove:** Substituting $\Delta\theta = 0$ into (2), we can see that $I(n)$ is the result of a cyclic sin67.5 and sin-67.5 series signal filtered by a low-pass filter (RRC), so it is easy to deduce that $I(n)$ is still a 1600HZ period signal. For the same reason, $Q(n)$ approximates a direct current signal.

Taking no account of ISI introduced by the channel, we can see that the best timing sample happens when $\tau = 0$ and $nT_s = iT$.

For periodic signal $I(n)$, Reference [10] has proved that $Ti(m)$ is a monotone function during in a cyclic of two symbols and equals 0 at the best sample time.

From reference [11], we can get that the coupling of carrier phase and timing phase is very small around the stable point $Ti(q) = 0$. Since the timing error $\tau$ is less than 1/16 of symbol time here, from the linearization of (4) we get our conclusion. Proposition 2 is proved.

### 3.3 Initialization Algorithm

Since synchronizer initialization algorithms reuse the C8PSK judgment module, we describe C8PSK judgment algorithm first. HART C8PSK judgment algorithm carries out in three steps: abstract values and signs of $I(nT)$ and $Q(Tn)$ output from the equalizer are computed first; then compare the abstract values; the result the comparing and the two signs are input to a 3-8 decoder to judge which symbol is received.

Based on proposition 1 and 2, synchronizer initialization algorithm of Multi speed HART is described as follow

(a) Sliding windows as (4) are calculated and the results go through mean filter;
(b) defining $\Delta\theta$ =45*k+m; the result $SumQ(m)$ and $SumI(m)$ go through the PSK judgment module, which judge the function parameter k; a little lookup table of abstract value $SumQ(m)$ is used to compute m;

(c) Feedback $\Delta\theta$ to oscillator to correct carrier phase error; wait three symbol time and begin timing phase estimation;
(d) Sliding windows as (11) are calculated; choose $Ti(q)$ that nearest to zero, and calculate timing offset $\tau$ ; Feedback $q$ and $\tau$ to Farrow interpolator.

## 4 Synchronization Tracking

Synchronization initialization provides coarse parameters estimation and only takes about 15 preambles to be done. It is necessary to use the equalized preambles to get better estimation of synchronization parameters. Additionally, the length of HART messages varies from several bytes to hundreds of bytes. In order to get better bit error rate, decision feedback synchronization are used for parameters tracking, employing digital PLL with gear-shift algorithms.

Equalizer and interpolator are high power component, and reduce their computing rate is effective way to lower power consumption. Fractionally spaced equalizer can work well with timing error. Therefore, to guarantee the performance of the system, T/2 spaced equalizer is adopted. We can get two equalized samples per symbol if the equalizer works at this lowest computation rate.

The timing offset is computed using Gardener algorithm:

$$eT(n) = I_p(n)*[I(n+1/2) - I(n-1/2) + Q_p(n)[Q(n+1/2) - Q(n-1/2)] \quad (12)$$

The carrier phase offset is computed as:

$$eC(n) = I_p(n)*Q(n) - I(n)*Q_p(n) \quad (13)$$

Where $I_p(n)$ and $Q_p(n)$ are known preamble during training sequence, and judgment feedbacks during the demodulation of the useful message; $I(n)$ and $Q(n)$ are samples from the equalizer. From (12) and (13), we can see that timing offset and carrier offset are calculated in different time of a symbol period and the multipliers can be implemented with time divided used lookup tables.

Both of the two offsets are filtered by one order lowpass filters before sending to oscillator or interpolator. The pole frequency of the lowpass filters decrease at the transition form preamble to useful data.

## 5 Analysis and Simulation

The proposed synchronization algorithms do not use multipliers, or any other complex computation resources. During the initialization phase, each sliding window of (4) and (11) needs only 2 additions for a sample output from the RRC filter, and they do not affect the maximum power of the system because equalizer and interpolator do not work at that time. During the tracking phase, only four additions and four table lookups are needed for the two offsets estimation in a symbol, and power consumption of those operations is negligible compared with that of equalizer. Further more, equalizer and interpolator can work at their lowest rate based on the proposed algorithms.

Computer simulations were performed to verify the performance of the proposed synchronization algorithms. Signed fixed-point arithmetic was used throughout simulation. The number precision of the receiver is as follow: 8 bits for A/D converter; 9 bits addresses and 10 bits precision lookup table for NCO; 18 central symmetric lookup tables for RRC, each table has 7 bits addresses and 8 bits precision; each of the equalizer adders has 10 bits precision and the output result precision reserves the high 8 bits. According to HCF, HART receiver should be able work under SNR=16 circumstance, while we add 13db noise to signals in simulation. Figure 3, 4, 5, and 6 are simulation results of the proposed algorithm.

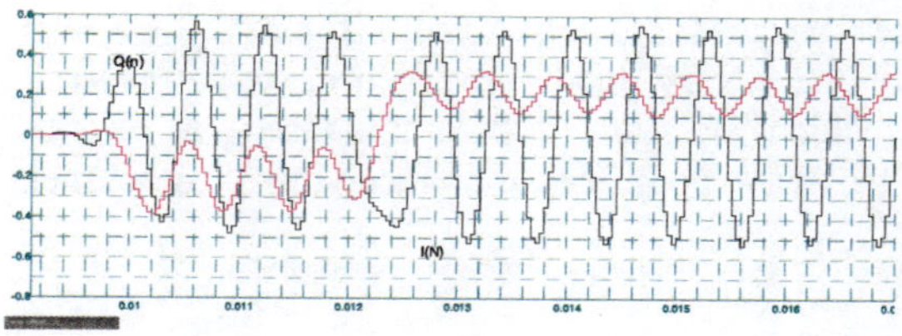


**Fig. 3.** Q(n) and I(n) from RRC (before equalizer)

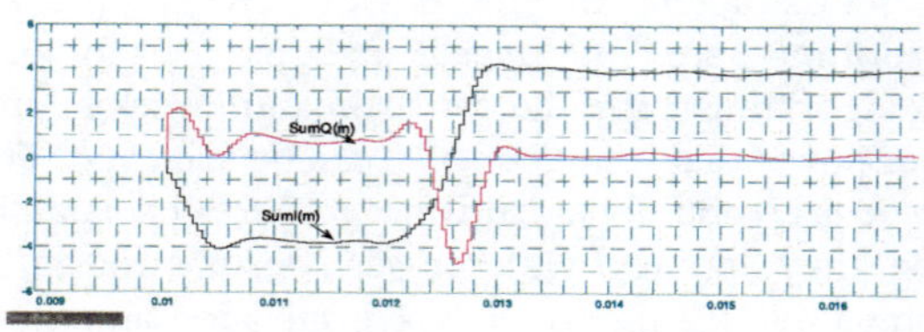


**Fig. 4.** SumQ(n) and SumI(n) output from equalizer

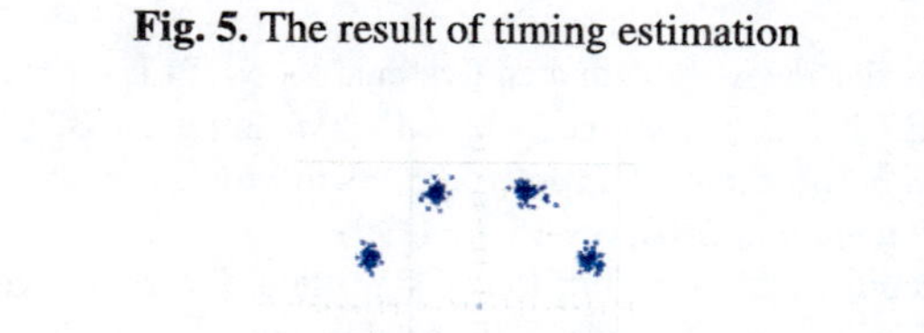


**Fig. 5.** The result of timing estimation

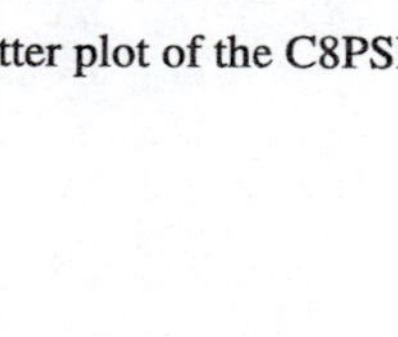

**Fig. 6.** scatter plot of the C8PSK receiver

Figure 3 shows the signal output from RRC filter, and we can see that there exists a great phase offset at the beginning of the simulation. As shown in figure 3, there still exists some phase offset when course carrier phase compensation carries out. Figure 4 shows the process and the result of course phase estimation in the sight of $SumQ(m)$ and $SumI(m)$, then we can see that $SumQ(m)$ and $SumI(m)$ fluctuate because of noise, so carrier recovery initialization algorithm just provides a course estimation of phase offset. Timing offset estimation is shown in figure 5 and we can see that initialization algorithm finds the q and $\tau$ for the interpolator first, so there is a big step at the beginning of the estimation. As shown in figure 5, timing tracking eliminates little offset and fluctuates because of noise. From figure 6, we can see that all the demodulated symbols have very good convergence, except several symbols of 6 and 2(67.5 and -67.5), which are part of preamble used to get fast initialization for synchronization, so they have no influence on receiving of useful message.

## 6 Conclusion

In this paper we have introduced novel synchronization algorithms for HART C8PSK. Based on the structure of the HART C8PSK preambles, novel fast synchronization algorithms are introduced and proved for course parameters estimation. Tracking algorithms make the two estimation work in different phases of a symbol time, so lower power and resource reuse can be achieved. The implementation of those algorithms needs no additional hardware resource except a few lookup tables, and at the same time, they add negligible power consumption to the whole system. Simulations have shown that the proposed synchronization algorithms can work very well in low SNR. At last, the proposed algorithms can use in similar case too.

## References


1. Pratt, W.: The HART PSK Physical Layer. A Status Report to HCF Members, 6–7 (2000)
2. Schindler, V., Voltopia: HCF Working Group Meeting Minutes Salzburg. In: HART® Communication Foundation Document Number, Austria, February 9-11, pp. 52–59 (2009)
3. Kim, B., Greco, J.D., Helman, D.N., Ngo, H.: A high speed data separator design using real time DSP for disk drive applications. In: IEEE ISCAS, San Diego, CA, pp. 625–628 (May 1992)
4. Kim, B.: Digital carrier recovery with adaptive dual loop DPLL for mobile communication applications. In: IEEE ICASSP, Minneapolis, MN, vol. 3, pp. 29–32 (April 1993)
5. Rosemount, F.: High Speed HART Prototype Modem Project. In: Prepared for The HART Communication Foundation, USA, pp. 35–60 (1999)
6. Kim, B.: Dual-Loop DPLL Gear-Shifting Algorithm for Fast Synchronization. IEEE Transactions on Circuits and Systems—II: Analog and Digital Signal Processing 44(7), 577–586 (1997)
7. Huang, L., Luo, M., Dong, X.: A Fast Carrier Synchronization Algorithm for Burst-Mode MPSK. In: Third International Conference on Natural Computation, ICNC, pp. 167–170 (2007)

8. Andronico, M., Casale, S., La Corte, A., Pinnisi, C.: A New Algorithm for Fast Synchronization in a Burst Mode PSK Demodulator. In: 1995 IEEE International Conference on Communications, ICC 1995, Seattle, vol. 3, pp. 1641–1646 (1995)
9. Goto, H., Hasegawa, Y., Tanaka, M.: Efficient Scheduling Focusing on the Duality of MPL Representatives. In: Proc. IEEE Symp. Computational Intelligence in Scheduling (SCIS 2007), pp. 57–64. IEEE Press (December 2007), doi:10.1109/SCIS.2007.357670
10. Feng, W.-J., Hou, J.-H.: A new algorithm for symbol timing and carrier phase offset acquisition in burst communication. Journal of Circuits and Systems 13, 119–122 (2008)
11. Meyr, H., Moeneclaey, M., Fechtel, S.A.: Digital communication receiver, pp. 271–315. John Wiley & Sons, Inc., NewYork (1998)

# Foreign Experts Foreign Languages Teaching Pattern Optimization Management under Current Chinese Educational Context

Dong Yang

Northern International College, Shandong Jiaotong University, Jinan, China
samyoung329@163.com

**Abstract.** Foreign experts who teach foreign languages in P.R. China, due to their group special characteristics, adopt and implement foreign languages teaching in their special own ways. Under current Chinese educational context, investigation and analysis on their special teaching patterns need to be conducted. Based on this, a realistic, practical and productive foreign experts foreign languages teaching pattern optimization plan is proposed combined with Chinese educational circumstances. The plan takes Northern International College of Shandong Jiaotong University as an example. English, Japanese and Russian are taken as the three main foreign languages at the school. With the development and implementation of the plan, foreign experts teaching are standardized and improved with perfect combination of western style under current Chinese educational context and reality.

**Keywords:** Foreign expert, foreign languages, teaching pattern, optimization management.

## 1 Introduction

With the ever-increasing of China's foreign exchange and co-operation, more and more foreign experts who work in the fields of culture and education have come to the People's Republic of China to conduct various teaching tasks and obligations. Their numbers and qualifications witness a steady increase in recent decades.

As a very special group, foreign experts who work in the fields of education and culture have diverse understandings and interpretations about everything in China. For the experts who teach foreign languages, they adopt and implement foreign languages teaching in their special own ways. Therefore, conducting investigation on foreign experts foreign languages teaching pattern optimization plan is of vital importance and of productive value, especially under Chinese current educational circumstances.

## 2 A Comparison of Foreign Languages Teaching Pattern between Foreign Experts and Their Chinese Counterparts

### 2.1 The Advantages of Foreign Experts

Firstly, foreign experts lay great emphasis on students-centeredness. They encourage students in class to try language practices and seldom correct students' utterance mid-way.

This helps students greatly to overcome their psychological problems of foreign language acquisition and foster them to express their thoughts and ideas clearly and smoothly.

Secondly, the teaching and learning environment is active and productive. Foreign experts, on the whole, are very easy-going, humorous, vigorous and flexiable. They love using real things to illustrate with the aid of pictures, photos, or models, etc. Students in the class are relaxed and eager to learn.

Thirdly, foreign experts encourage students' divergent thinking and their social practices. There is always no unified correct answer in foreign experts class. They encourage students' full participation and innovation.

Fourthly, but not lastly, the foreign language they teach is their mother-tongue. Their position is irreplaceable in terms of oral and aural language teaching, foreign culture, etc.

### 2.2 The Advantages of Their Chinese Counterparts

Firstly, foreign experts Chinese counterparts---the localized Chinese foreign languages instructors have systematic knowledge of different disciplines, such as education, linguistics, psychology, etc, which enable them to provide all-sided language assistance for their students.

Secondly, Chinese teachers understand Chinese teaching environment much better than their foreign counterparts. They also have reasonable and proper evaluation of the students due to their cultural backgrounds.

Thirdly, but not lastly, Chinese teachers lay great emphasis on students language input, which helps them significantly in their foreign languages reading and writing.

From the above-mentioned advantages of foreign experts and their Chinese counterparts, we can see that both of them have unique characteristics and are important parts in Chinese foreign languages teaching. Both of them need to learn from each other to enhance their own advantages and overcome the disadvantages. In order to bring foreign experts teaching effectiveness into full play, a realistic, practical and productive foreign experts foreign languages teaching pattern optimization plan is not only significant, but very necessary as well.

## 3 Foreign Experts Foreign Languages Teaching Pattern Optimization Plan under Current Chinese Educational Context—Take NIC as Example

### 3.1 Background Introduction

With the entry into WTO and the globalization of world economy, many universities in China believe that it is important for the undergraduates and future professionals to develop an international perspective. As a result, some universities start offering programs and encourage students to study abroad. They are exploring opportunities to collaborate with various foreign universities in relation to such kind of programs. Furthermore, the education authority in China encourages and supports this initiative. Policies regarding such foreign degree programs recognize these qualifications. Authorized by Shandong Education Department, Northern International College (NIC) of Shandong Jiaotong University (SJU) is a college affiliated directly to SJU. The college that is a platform for the university's international exchange and co-operation is within SJU. The college mainly provides program education. Its enrollment is in accordance with SJU's general enrollment plan. According to the enrollment plan issued by Shandong Education Department, the college recruits students from Shandong province. The qualified graduates from the college will receive a diploma and a degree certificate of SJU program education. While carrying out the program educational teaching plan, the college has mainly introduced part of the foreign universities' teaching plan, core subjects, original teaching materials and teaching quality supervision system, employed foreign teachers and implemented bilingual teaching. Three foreign languages are conducted at the college, namely English, Japanese and Russian. Every year the college employs 8-10 foreign teachers to conduct foreign languages teaching tasks.

### 3.2 Optimization Plan Proposed

As a product of Sino-foreign co-operation and exchange, NIC cannot reply on regular foreign languages teaching practices. It needs teaching reform and innovation. After taking NIC overall conditions into full consideration, we can easily see that a realistic, practical and productive foreign experts foreign languages teaching pattern optimization plan is of importance and necessity.

### 3.3 NIC of SJU Foreign Experts Foreign Languages Teaching Pattern Optimization Implementation Plan

*1) Foreign Experts are Divided into Three Categories:* English language teaching group, Japanese language teaching group and Russian language teaching group. The reason for this is that the current foreign languages offered at NIC of SJU are English, Japanese and Russian.

*2) Foreign Languages Course Packages as the Main Body of the Optimization Implementation Plan.*
Foreign languages course packages are mainly based on the core courses offered by SJU international co-operative partners. It absorbs the advantages of international partner universities current teaching theories and practices while taking Chinese learning context into full account.

*3) Foreign Experts Co-operate with Their Chinese Counterparts to Conduct Classes.*
Foreign experts and their Chinese counterparts work closely and interactively together to conduct classes. It is of great significance and necessity. It brings innovation and vitality into the foreign languages classroom. Students satisfaction rate tend to be on the high side compared with traditional teaching pattern---Chinese foreign language instructors' full and overall dominance of the foreign languages classroom. Chinese teachers traditional teaching goal is knowledge-and- result-oriented, while foreign experts teaching goal is process, experience and application-oriented. Thus by co-operating with each other, two goals can stand side by side in perfect harmony.
*4) Overcome the Traditional Foreign Languages Teaching Pattern.*
Since the students at NIC are from Sino-foreign co-operation projects, most of them will complete their university study outside of China after they meet the language requirement of each individual SJU international partner. Therefore, the teaching objective and motivation for both the teachers and the students is language level test before the students go broad to seek further study. There are three foreign languages offered at the college, namely, English, Japanese and Russian. Thus at the college,

IELTS for English, JLPT for Japanese and RLPT for Russian are melted into the college teaching curriculum. Traditional foreign language teaching pattern is not suitable and proper for these group students due to their lack of innovation and low rate of stimulating students' learning interest and motivation. A new binary teaching pattern is formed at NIC, namely, foreign language experts and their Chinese counterparts co-operate closely and interactively for one individual class.

*5) The Teaching Focus Is Not Only on Language Input, More Emphasis Should be Laid on Language Output.*

For the ideal foreign language learners, language input and output are of great importance. The co-operation between the foreign experts together with their Chinese counterparts is of perfect combination for the language input and output, especially for the latter. When the students go abroad to seek further study, language output is the one more needed and frequently used.

*6) Quality Courses Offered at SJU International Partners Should Be Melted with Chinese Characteristics.*

The quality courses offered at SJU international partners are melted with Chinese characteristics and flavors. The courses need to be localized and adjusted to Chinese learning and teaching context to make the maximum use of them.

As for the details of the core courses and optional courses offered by NIC of SJU international co-operative partners, Table 1 & Table 2 are listed as examples:

**Table 1.** Core courses example offered by NIC of sju international co-operation partners

| Module Number | Module Name | Nominal Hours |
|---|---|---|
| 9571A | Speaking1 | 50 |
| 9571B | Listening1 | 50 |
| 9571C | Pronunciation1 | 30 |
| 9571D | Reading1 | 40 |
| 9571E | Writing1 | 40 |
| 9571F | Study Skills1 | 40 |

**Table 2.** Optional courses example offered by NIC of sju international co-operation partners

| Module No. | Module Name | Nominal Hours |
|---|---|---|
| 3674A | Exploring the internet | 10 |
| 3677A | Computer essentials | 24 |
| 9572K | English for basic computing | 30 |
| 4953A | Technical instructions | 25 |
| 4953B | Technical descriptions | 25 |
| 4953C | Explanations of processes | 25 |
| 4953D | Technical recounts | 25 |
| 4953E | Combined texts 1 | 25 |
| 4953F | Numerical reports | 30 |
| 4953G | Cause and effect explanations | 30 |
| 4953H | Technical comparisons and contrasts | 30 |
| 4953J | Combined texts 2 | 35 |
| 4953K | Reports: proposals | 30 |
| 4953L | Historical explanations | 30 |
| 4953M | Comparisons& contrasts of proposals | 30 |
| 4953N | Combined texts 3 | 35 |

### 3.4 A Survey of Students Satisfaction of NIC Foreign Experts Foreign Language Teaching Pattern is Conducted.

We have conducted a survey of students' satisfaction of NIC foreign experts foreign language teaching pattern. Three languages are involved, namely English, Japanese and Russian. There are 58 subjects for the English language survey; 12 subjects for the Japanese language survey and 109 subjects for the Russian survey. 11 foreign language experts were evaluated by the subjects of the survey. As for details of the survey, we can see TABLE 3.

**Table 3.** A Survey of students' satisfaction of nic foreign experts foreign language teaching pattern

| Number | Languages | Subjects Number | Satisfaction Rate |
|---|---|---|---|
| 1 | English | 58(55) | 94.82% |
| 2 | Japanese | 12(11) | 91.67% |
| 3 | Russian | 109(106) | 97.25% |

For the English language foreign experts, 55 of 58 subjects express strong satisfaction with the satisfaction rate of 94.82%; for the Japanese language foreign experts, 11 of 12 subjects express strong satisfaction with the satisfaction rate of 91.67%; for the Russian language foreign experts, 106 of 109 subjects express strong satisfaction with the satisfaction rate of 97.25%.

From the survey, we can see that generally speaking, students of NIC tend to be very satisfied with foreign language experts teaching. To them, the work of the experts is valuable, effective and beneficial.

## 4 Conclusion

Under current Chinese educational context, investigation and analysis on foreign experts special teaching patterns need to be conducted. Based on this, a realistic, practical and productive foreign experts foreign languages teaching pattern optimization plan is proposed combined with Chinese educational circumstances. The plan takes Northern International College of Shandong Jiaotong University as an example. English, Japanese and Russian are taken as the three main foreign languages at the school. With the development and implementation of the plan, foreign experts teaching are standardized and improved with perfect combination of western style under current Chinese educational context and reality.

**Acknowledgment.** I would like to thank my family for the support they provided me through my entire life and in particular, I must acknowledge my wife and my son, without whose love, encouragement, I would not have completed this paper.

## References


1. Yang, D.: China-Australia Co-operative Program Teaching Management Pattern And Implementation Plan. In: 2009 ISECS International Colloquium on Computing, Communication, Control, and Management, vol. I, pp. 149–152 (2009)
2. Shang, Y., Yang, D.: Chinese Culture All-Dimensional Introduction for Foreign Experts of Higher Education in China. In: 2009 International Conference of Management Science and Information System, vol. 1-4, pp. 234–237 (2009)
3. Gen, D.: Study on Internationalized Co-operated Programs Pattern. Higher Education Development & Evaluation (1) (2007)
4. Tao, L., Shen, J.: A Comprehensive Study on Sino-Foreign Co-operated Programs in the Past Decade. Chinese Medicine Education (7) (2006)
5. Liu, M.: Ten Countermeasures to Improved Sino-Foreign Co-operated Programs Quality. China Higher Education Evaluation (2) (2006)
6. Shen, J., Gao, S.: Rational Thoughts on Sino-Foreign Co-operated Programs. Universities Educational Science (1) (2006)
7. Xu, S.: The Existing Problems, Supervision and Criterion of Sino-Foreign Co-operated Programs. Journal of Zhengzhou University (2) (2005)
8. Bai, L., Zhang, C.: Sino-Foreign Co-operated Programs in Higher Education Internationalization Context. Liao Ning Education Research (11) (2005)
9. Cong, X., Hua, Y.: Thoughts on Sino-Foreign Co-operated Programs. Journal of Tang Shan Institute (9) (2005)
10. Jiang, L.: Current Conditions and Problems Analysis on Sino-Foreign Co-operated Programs. Journal of Guangdong Industrial University (6) (2005)
11. Gong, S.: Current Analysis on Sino-Australia Univresities Co-operated Programs. Education Development Research (2) (2005)
12. Wang, F.: Culture and Language. Foreign Languages Teaching and Research Press, Beijing (1998)
13. Hu, W.: Foreign Language Teaching and Culture. Hunan Education Press (1997)
14. Deng, Y.: Language and Culture. Foreign Languages Teaching and Research Press, Beijing (1989)
15. Kramsch, C.: Language and Culture. Shanghai Foreign Language Education Press (2000)
16. Krasnick, H.: The culture puzzle: cross-cultural communication for English as a 2nd language. TESOL Quarterly (22) (1998)
17. Walton, S.J.: Stress management training for overseas effectiveness. International Journal of Intercultural Relations (14) (1990)
18. Eschbach, D.M., Parker, G.E., Stoeberl, P.A.: American repatriate employees' retrospective assessments of the effects of cross-cultural training on their adaptation to international assignments. International Journal of Human Resource Management (12) (2001)
19. Dornyei, Z., Canale, M., Swain, M.: Theoretical bases of communicative approaches to second language teaching and testing. Applied Linguistics 1(1), 1–47 (1980)
20. Tarone, E.: Some influences on the syllable structure on interlanguage phonology. International Review of Applied Linguistics 18, 139–152 (1980)
21. Ellis, R.: Understanding Second Language Acquisition. Oxford University Press (1985)
22. Domyei, Z., Thurrell, S.: Strategic competence and how to teach it. ELT Journal (45), 16–23 (1991)

23. Lan, W., Wong, J.: The effects of strategy training on developing discussion skills in an ESL classroom. ELT Journal 54(3), 245–253 (2000)
24. Cohen, A.: Strategies in Learning and Using a Second Language, pp. 65–106. Longman, Harlow (1998); Hability of communication strategies. TESOL Quarterly (29), 55–85 (1995)
25. Bialystok, E.: Communication Strategies, pp. 37–83. Blackwell, Oxford (1990)
26. Jiao, G.: The Review on the Sino-Foreign Co-operative Programs of Higher Learning Institutions in China. Higher Education in China (1998)

# The Application of Geomagnetic Spatial Frequency Spectrum in Geomagnetic Localization

Yunxiao Lv, Shitu Luo, and Meiping Wu

National University of Defense Technology, Changsha, China

**Abstract.** In order to reduce the bad impact of the geomagnetic signal's error on geomagnetic localization, this paper pay much attention on studying the character of geomagnetic signal and error using method of spatial frequency spectrum analysis. First, Fourier analysis method is used to study frequency feature of geomagnetic signal and error, an important result is get that in Fourier domain the character of geomagnetic signal is almost the same with signal which contains noise. Secondly, based on the above result, Time-Frequency Matching (TFM) method is used to carry out geomagnetic localization. Finally, this paper compares TFM with Contour Constraint Matching method in application of geomagnetic localization, and the results show that when the geomagnetic signal contains white noise, the localization error of TFM is smaller. So the analysis of Geomagnetic Spatial Frequency Spectrum has great importance in geomagnetic localization application.

**Keywords:** Amplitude spectrum, pending matching locus, matching criteria.

## 1 Introduction

Geomagnetism Aided Navigation(GAN) is a new branch of the integrated navigation techniques [1].This technology is based on the Geomagnetic field which is inherent resource of the earth. Felix Goldenberg [2] have pointed that geomagnetic field can be used as a reliable navigation reference on at least 98% of the Earth's surface including water covered areas.

At present, inertial navigation system(INS) is widely used. Although it can provide continuous, real-time and multidimensional navigation information, the position error of it accumulate over time. However, it can limits geomagnetism matching area, and combining with Geomagnetism navigation information, the precision of INS can be effectively readjusted, the system reliability and navigation precision is improved, and the localization error doesn't accumulate over time. But in fact, the mensurated geomagnetism is easily influenced by outside interferential magnetic field, and includes kinds of noise such as changing magnetism, magnetism of instrument etc. The noise has bad impact on the precision of matching localization.

Some researchers have put forward some methods to reduce the noises, such as adaptive calibration of the diversionary error [3], combination of the prepositive filter and digital lowpass,etc. But these methods have little contribution to geomagnetism matching. As well known that the spectrum energy of white noise is approaching zero in frequency domain. So this paper studies the feature of geomagnetic signal in frequency, and put forward a new matching method, which is named Time-Frequency Matching(TFM) method.

## 2 Spatial Frequency Spectrum of Geomagnetic Signal

### 2.1 Fourier Expansion of Discrete Geomagnetic Signal

By using fourier expansion, geomagnetic signal can be expressed as Harmonic function. Suppose the geomagnetic signal of the measuring line with the length of $2L$ is $S(x)$, then the periodic signal $S(x)$ can be denoted as (1),

$$S(x)=\sum_{k=0}^{\infty}A_k \sin(kwx+\phi_k),(k=0,1,2,\cdots\cdots) \tag{1}$$

Where $A_k$ is amplitude (AM), $\phi_k$ is phase, $w$ is angular wave number which is equal with pulsatance, $w=2\pi f$, $f$ is wave number, $f=\frac{1}{2L}$.

### 2.2 Description of Geomagnetic Noise

For mensurated geomagnetic signal $x(n)$, suppose that contains real geomagnetic signal $s(n)$ and noise $u(n)$ .Most of the time, the noise is additive white noise, then the mensurated geomagnetic signal $x(n)$ can be denoted as (2),

$$x(n)=s(n)+u(n) \tag{2}$$

Generally, $s(n)$ is steady signal. However, $u(n)$ is random signal, which is independent of steady signal. It can't be described whit mathematic formula, and can't be forecasted . Thereby, this paper will simulate $u(n)$ by white noise. Figure.1 shows that white noise has all frequency in frequency domain.

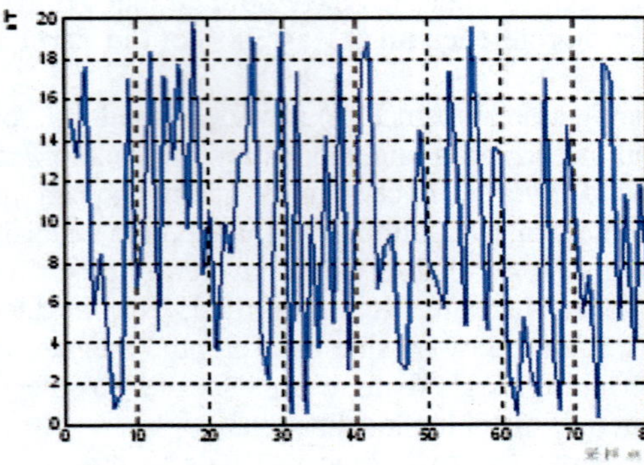


**Fig. 1.** Description of White noise in frequency domain

### 2.3 Frequency Feature of Geomagnetic Signal

If treat the period of geomagnetic signal is infinity in spatial, it can be analyzed as harmonic of different frequency. Its amplitude isn't changing with time, but with space.

Harmonics of different frequency have different amplitudes and phases. All the amplitudes and phases are the function of frequency, and we call them amplitude spectrum and phase spectrum.

For studying the feature of geomagnetic signal in frequency domain, this paper have got geomagnetic data at 80 sampling points (the sampling distance is 100m) in some experimental area. By simulation, we get amplitude spectrum and phase spectrum of real geomagnetic signal and geomagnetic signal which contains white noise of 20nT.

In "Fig.2", sub a presents amplitude spectrum and phase spectrum of geomagnetic signal, sub b presents amplitude spectrum and phase spectrum of geomagnetic signal which contains white noise of (0,20nT),while 0 is the mean of white noise, and 20nT is the mean square deviation. when the geomagnetic signal contains white noise, its amplitude spectrum of it is approximately the same as the real signal, except the first frequency point. However, the phase spectrum is effected by white noise. So we try to locate with amplitude spectrum in this paper.

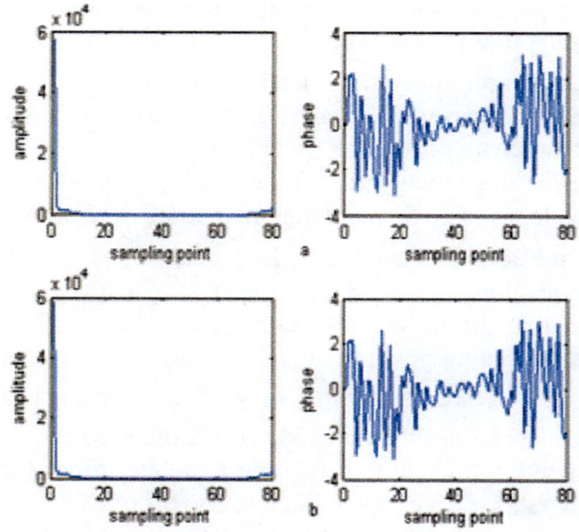


**Fig. 2.** Signal feature of frequency

## 3 Prepare Your Paper Before Styling

Recently, Phase Correlation algorithm and Fast Fourier Translation (FFT) have been wildly used in vision matching, and the precision is high. In this paper, we use amplitude spectrum in geomagnetism matching. But in frequency, amplitude has no space information. For solving this problem, we will appoint matching policy and find pending matching locus in spatial domain, and use information of frequency to match.

### 3.1 Matching Policy in Frequency

In frequency, we use related function to verify the comparability of the pending matching locus and real locus. When the comparability of them is the biggest, the real locus is got, and the INS error can be readjusted. Suppose $A_M^k$ is the amplitude of mensurated geomagnetic signal, and $A_S^k$ is the amplitude of pending matching locus. Mean square error(MSE) of the measured track and the pending matching locus is denoted as (3),

$$msd_{ij} = \frac{1}{N}\sum_{k=1}^{N}(A_{S(i,j)}^k - A_M^k)^2 \tag{3}$$

When MSE is minimum, the pending matching locus is the best matching locus.

### 3.2 Generation of Pending Matching Locus

Before matching in frequency, pending matching locus should be generated in spatial space. Summarizing the process as follows:

**Stepl:** make sure the length of pending matching locus;
**Step2:** obtain the output locus of INS based on its error;
**Step3:** obtain geomagnetism of the first point ;
**Step4:** generate the initial geomagnetic isoline of the first point;
**Stept5:** translate the locus of INS, and generate a series of pending matching locus.

During this process, the area of matching place determines the amount of pending matching locus and matching time. In this paper, we use the initial error of INS compress matching area. Suppose that east error and north error of INS is (600m,1000m), the initial output position of INS is (a,b), and the searching radii is r, so the matching area is $(a \pm r, b \pm r)$. Table 1 shows that searching radii effect matching time and the amount of pending matching locus. If increase the length of searching radii, the amount of pending matching locus and matching time will increase. When the length of searching radii is bigger than the initial error of INS, we can't get the right locus.The area of experimental place is $14km \times 14km$, and we have translated it to grid coordinate. Grid spacing is 100m. "Fig.3"shows some pending matching locus. In "Fig.3",red line is the real locus of vehicle, black dashed is output locus of INS, green star line is initial isoline, the blue point line is pending matching line, and the color bar represents magnitude of geomagnetism.

**Table 1.** Determination of matching area

| Searching radii $r$ (km) | Pending matching locus (piece) | Matching time (s) |
|---|---|---|
| 2.0 | 157 | 119.8 |
| 1.8 | 131 | 98.1 |
| 1.6 | 114 | 84.7 |
| 1.4 | 100 | 74.8 |
| 1.2 | 81 | 60.9 |
| 1.0 | 56 | 42.4 |
| 0.8 | 36 | 28.0 |
| 0.6 | 19 | NULL |

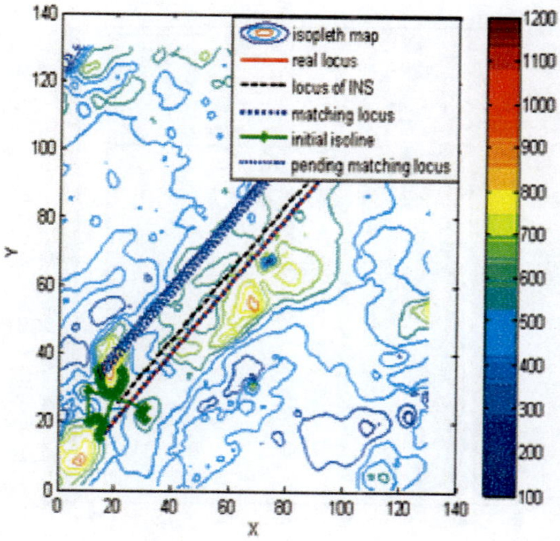


**Fig. 3.** Genermation of pending matching locus

# 4 Simulation of Time-Frequency Matching(TFM) Algorithm

Table 2 shows the main parameters of simulation. The length of pending matching locus is 8000m, and the initial error of INS is (600m, 1000m). The coordinate is the same with "Fig.3". "Fig.4" ,"Fig.5",and "Fig.6" show the matching results when geomagnetic signal contains different noise, and Table 3 shows the biggest matching error.

Contrast with Contour Constraint Matching(CCM)[4] which is a time domain algorithm, we find that the localization error of TFM is smaller than CCM, when geomagnetic signal contain white noise.

**Table 2.** Simulation parameters

| Parameter | value |
|---|---|
| Initial error of INS | (600 m, 1000 m) |
| Grid spacing | 100m |
| Sample interval of INS | 0.01s |
| Geomagnetic measure interval | 20s |
| Correlation principle | MSE |

**Table 3.** Matching error

| | White noise of (0, 20nT) | White noise of (10nT, 20nT) | System error of 10nT |
|---|---|---|---|
| **Matching error of FTM(m)** | 10.13m | 7.74m | 13.02m |
| **Matching error of CCM(m)** | 19.72m | 20.70m | 13.02m |

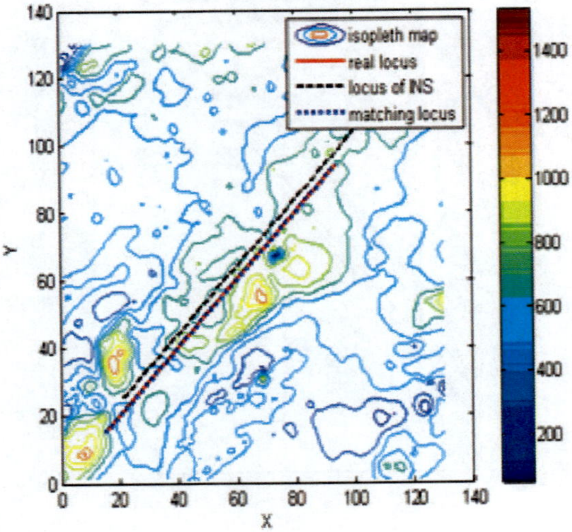


**Fig. 4.** Matching result (geomagnetic signal contain white noise of (0,20nT).)

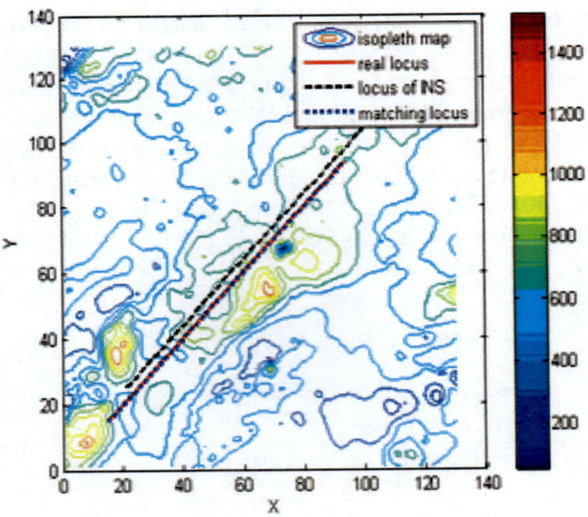


**Fig. 5.** Matching result (geomagnetic signal contain white noise of (10nT,20nT).)

## 5 Conclusions

Geomagnetic amplitude spectrum can be use in geomagnetic localization, and it can restrain white noise when matching. Compare with CCM, when geomagnetic signal contains white noise, the matching error of it is smaller, but for system error, FTM have the same effect with CCM. Thereby, we can use geomagnetic spatial spectrum to localize in navigation.

However, this method have to translate signal between time domain and frequency domain, and the consumption of time increases. This paper just study the geomagnetic spatial spectrum easily, and the further study should go on in future.

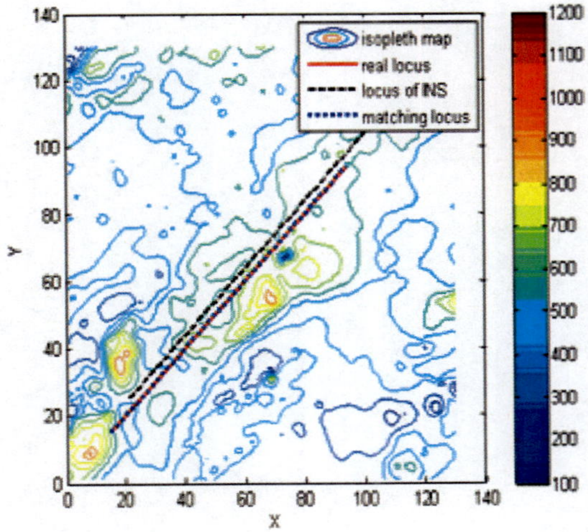


**Fig. 6.** Matching result (geomagnetic signal contain 10nT system error)

**Acknowledgment.** Thanks Professor Meiping Wu and associate professor Shitu Luo for their help during the complementation of the paper. This paper was funded by Inertial Navigation Laboratory, Mechatronics and automation college, National University of Defense Technology.

## References

1. Luo, S., Wang, Y., Liu, Y., Hu, X.: Research on Geomagnetic-matching Technology Based on Improved ICP Algorithm. In: The 2008 IEEE International Conference on Information and Automation (2008)
2. Goldenberg, F.: Geomagnetic Navigation beyond Magnetic Compass. Presented at PLAN, San Diego, California (2006)
3. Lin, C.-S., Xiang, Q., Gong, S.-G.: Ortho-error Analysis and Rectification of Three-axis Magnetometer. Journal of Detection and Control 27(2), 9–12 (2005)
4. Liu, Y., Wu, M., Hu, X., Xie, H.: Contour Constraint Based Geomagnetic Matching Method. Space Science 27, 505–510 (2007)

# A RF Card System for the Management of Cow Health

He Baoxiang, Lu Guirong, Chu Kaibin, and Zhu Zhengwei

Department of Computer Science and Engineering, Jiangsu Polytechnic University,
Changzhou, Jiangsu, China
{hbx,lgr}@em.jpu.edu.cn, ckb910@163.com, zzhuw@jpu.edu.cn

**Abstract.** RF reader mainly consists of transponder and reader. Coding on the transponder based on AT89C2051 can be realized. Coded signals are modulated through resistance attached coupling LC circuit. Decoding and displaying can be completed using SCM on reader. Carrier signals at 125KHz produced by oscillator and frequency divider composed of CD4060 are emitted by a push-pull circuit consisted of transistor. The design has a novel thinking in the advantage of small volume, low cost, low power dissipation, high reliability and self-test function.

**Keywords:** RF card, reader, wireless identification, circuit, cow, SCM.

## 1 Introduction

It's very important both to monitor the health of cows and timely to find the sick and estrus ones in the industry of feeding cow. In China, most of the monitoring of dairy cattle, which is very difficult to the breeder, is in charged by the breeder. With development of modern electronic technology, the computer management system of dairy farm has been widely adopted by a large number of foreign countries in the breeding industry, and this system can insure the cow's health in accordance with each cow's steps everyday. In Israel, each cow has a pedometer, which is used for statistical cows' steps. However, the cost of the introduction of this system is very high. The system with RF(radio frequency) card technology [1], which can carry out the activities of dairy cattle statistical management, is cost saving.

## 2 Relationship between Cows' Health and Activity

There are some links between cows' daily activity amount and the number of their walking steps. The more steps, the higher activity amount. Contrarily, the activity amount is lower. Each healthy cow has a daily average activity amount, namely, the walking steps is changed in a small scape. When one cow's activity has a sudden decrease, it indicates that it may be sick. When one cow has a sudden activity increase, it indicates that it may has an estrus. Based on this unique feature of the cows, the system in this thesis aims at the health management of dairy cattle with the calculation of each cow's daily activity level.

## 3 Main Structure of System and Working Principles

The RF Card system is a wireless identification devices, which can be divided into two parts, namely, the readers and the radio frequency card [2]. The system is mainly used in modern life, such as in the attendance and consumption field. It avoids the malfunction cause by the reading and writing of the osculatory IC card, oil or chip exposure, thus it has the advantages of free access, high reliability and wireless [3]. At present, the RF card system has gradually become the best choice, replacing the optical cards, barcode cards, and contactless IC card [4]. The RF card is a device installed on the cow's legs, whose requirements are Lower power consumption and small size. Figure 1 is the RF Card system block diagram.

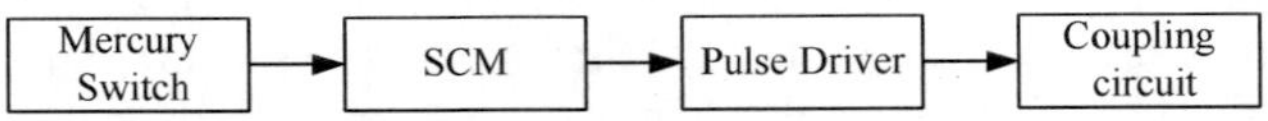


**Fig. 1.** The RF Card system block diagram

The working principle of RF card is SCM software in each card setting a unique encoding (card number, which is equivalent to our ID cards). Besides, the MCU is connected with a mercury switch. Therefore, whenever the cow takes one step, the mercury switches off for once. In this way, the cow's activity was recorded. When the RF card gets close to the reader, radio wave energy of a reader coupled circuit is received and supplied power for RF cards after being filtered by the rectifier. Meanwhile, the microcontroller fetches a interrupts instruction, encodes the card number and steps number for Manchester encoded data, then emits the codes by pulse drive circuit. According to the encoded signals sent by MCU, Pulse drive circuit controls the resistance at both ends of the RF card coupling circuit. When the level is high, the resistance is connected into the coupling circuit, otherwise it is not connected. According to PSA Device theory, the voltage and the current of the transformer all change when the load changes, thus the signals can be transferred to the reader [5].

The RF Card reader, made up of the AT89C51 microcontroller and peripheral circuits, is a system which reads the card number and count-steps date. Figure 2 is the reader block diagram.

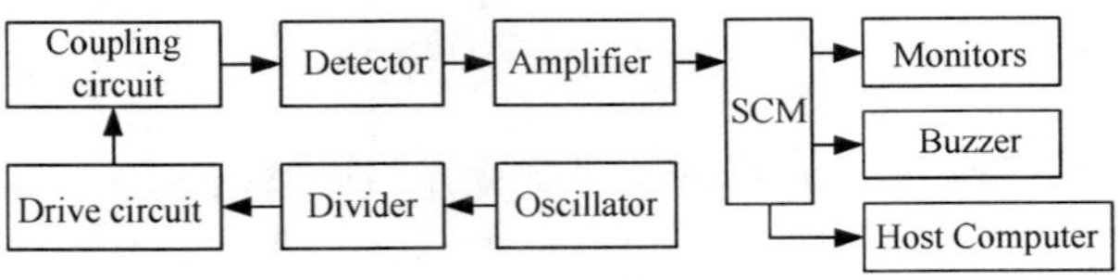


**Fig. 2.** The RF Card reader block diagram

The principle of reader is: 8MHz square wave signal is generated by oscillator and transformed into 125KHz square wave signal by frequency divider. The drive circuit drives the coupling circuit coils and emits 125KHz radio waves. When the coupling circuit coil is far away from radio card, a constant amplitude signals appear in the coupling circuit, so it shows no demodulated signal output in the detector circuit. In accordance with principles of transformers, when the above-mentioned RF card gets close to the coupling circuit coil, it will transfer energy to the RF card, and the change of load-resistance at both ends of the RF card coil is equivalent to the load change of reader[6]. So the voltage at both ends of coil changes along with the resistance of coupled circuit in RF card, and it can achieve signal loading and amplitude modulation. The signals are transferred from the AM signal to the encoding of radio frequency card by detector, amplified and shaped by the amplifier circuit, sent to the microcontroller for decoding, and changed into the relevant card number and count-steps of data.

# 4 System Hardware Design

## 4.1 Reader Hardware Circuit

The reader comprises the oscillator, frequency divider, driving circuit, coupled circuits, detection circuits, amplifiers, microcontroller circuits, buzzer and displayer and so on [7]. 125KHz square wave signal generate circuit is made up of the pulse oscillation circuit and frequency dividing circuit, such as Figure3.

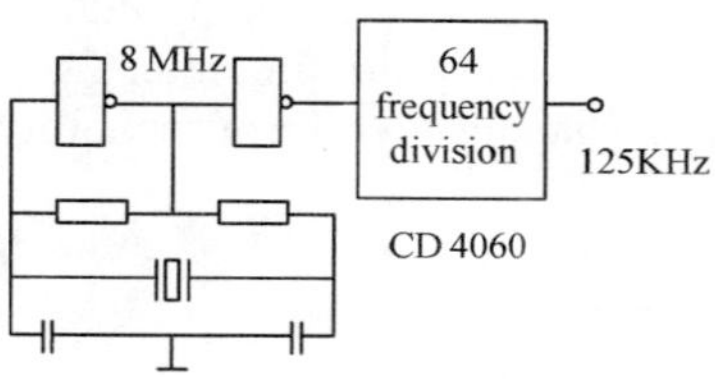


**Fig. 3.** 125KHz square wave signal generate circuit

The oscillator consists of quartz crystal, which generates 8MHz square wave signal, and CD4060, by which 8MHz signal is transformed into 125KHz square wave by using 64 frequency division.

Figure 4 is the signal transmitter driving circuit. Because the brushing card mode of cows is not as same as human's, which needs the card getting in the vicinity of the reader directly, it needs a far distance between the reader and the card. Thus, the reading can be achieved so as long as the cows go by a reader. In order to achieve long-distance reader, driver circuit is adopting a push-pull drive mode. It consists of transistors $T_1 \sim T_4$, Non-gate 74HC04 and coupling coils $L_1$ composition. Among them, $T_1 \sim T_4$ Composition of bridge push-pull circuit, intended for drive by the series resonant circuit which consists of $L_1$ and $C_1$ ($C_1$ is Using high-precision 3300pF capacitor). When $L_1$ and $C_1$ is in the series resonance,

$$f_0 = \frac{1}{2\pi\sqrt{L_1 C_1}} \qquad (1)$$

Because of $f_0$ is 125 kHz, According to formula (1), $L_1$ is about 492 pH.

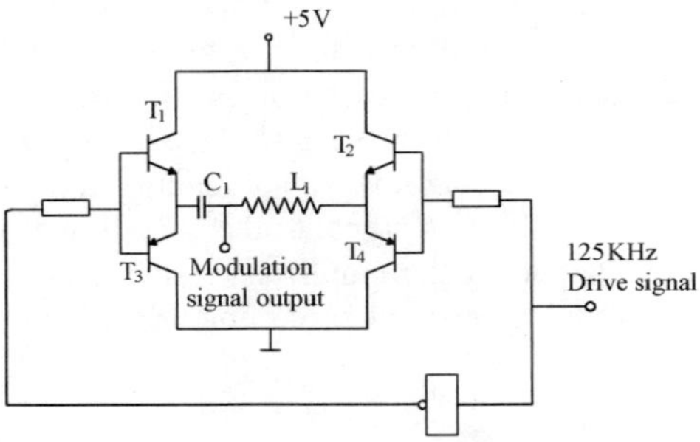


**Fig. 4.** Signal transmitter driving circuit

The signal transmitted from the $L_1$ is an amplitude modulation wave when reader is close to a radio frequency card, which can be achieved with ordinary detector circuit. The circuit as shown in Figure 5, which can complete low-pass filtering by using the IN4148 switching diode detector diode, 100kΩ resistor and the completion of 1000pF capacitor. The pulse signal from detector should be received by microcontroller must amplified by amplification circuit because its amplitude is small. Amplifier circuit consists of the CD4069 reverses, as shown in Figure 6. The last two amplifier circuits form a low-pass filter including shaping-waveform capability.

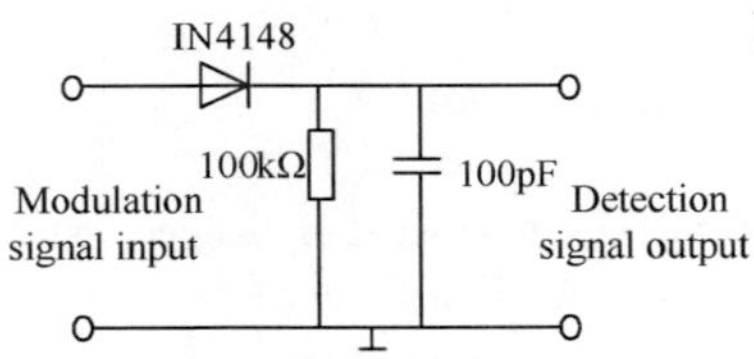


**Fig. 5.** Ordinary detector circuit

The microcontroller uses AT89C51 to decode encoded signals which received. After decoding, the signal is transmitted to the bit machine for processing via the RS232 interface, and the circuit as shown in Figure 7.

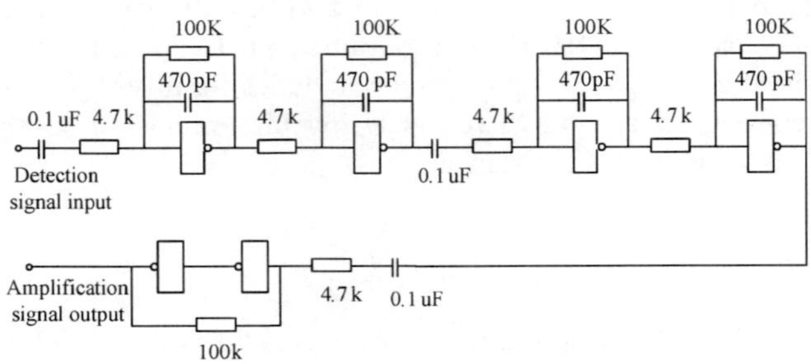


**Fig. 6.** Amplifier circuit consists of the CD4069 reverses

## 4.2 RF Card Hardware Circuit

The RF card circuit consists of counting pulse generator circuit, microcontroller circuits, pulse drive circuit and coupled circuits and so on[8]. Schematic diagram as shown in Figure 1.

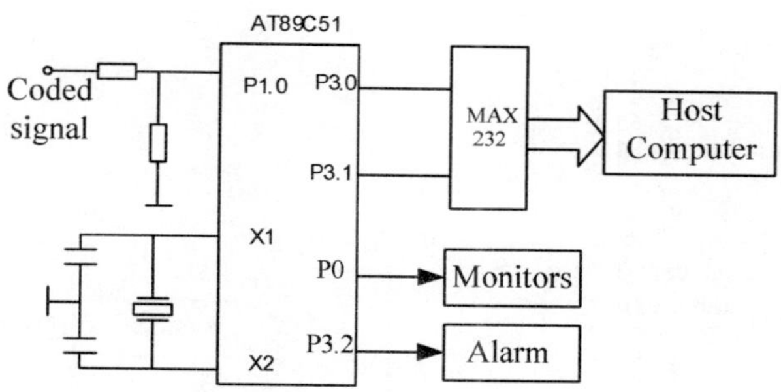


**Fig. 7.** The microcontroller circuit

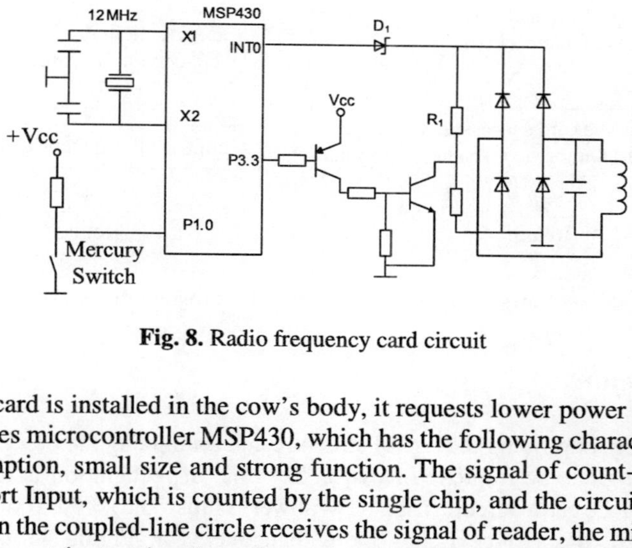


**Fig. 8.** Radio frequency card circuit

As the RF card is installed in the cow's body, it requests lower power consumption. The system uses microcontroller MSP430, which has the following characteristics: low power consumption, small size and strong function. The signal of count-steps is input by the P1.0 port Input, which is counted by the single chip, and the circuit as shown in Figure 8. When the coupled-line circle receives the signal of reader, the microcontroller fetches an interrupt instruction from $D_1$, and outputs the encoding signals of number and steps-count data by the P3.3 port. Two transistors are adopting the pulse drive mode, which is used to check whether the $R_1$ (510Ω) is connected effectively. When P3.3 is low level, $R_1$ is connected into coupling circuit and the circuit load becomes heavy. When the card gets close to the reader, due to the load of reader changes, it drops the voltage at both ends of coupling circuit in reader, thus generates the amplitude-modulation effect. When P3.3 is high, the transistor is cut off and resistors connection is invalid, it has no effect on the coupling circuit. Thus, it can not only achieve the normal transmission of encoding, but also improve its efficiency.

## 5 System Software Design

The system software design includes RF card and reader. Due to the single function of each module, software design is relatively simple. The program flows of the reader and the radio-frequency card are respectively as shown in Figure 9 and Figure 10.

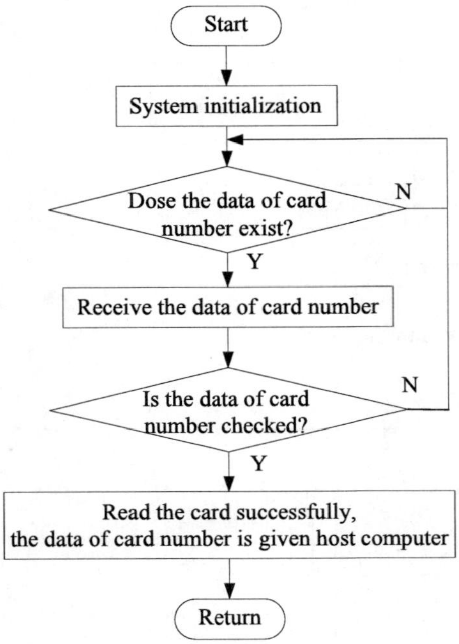


**Fig. 9.** Program flows of the reader

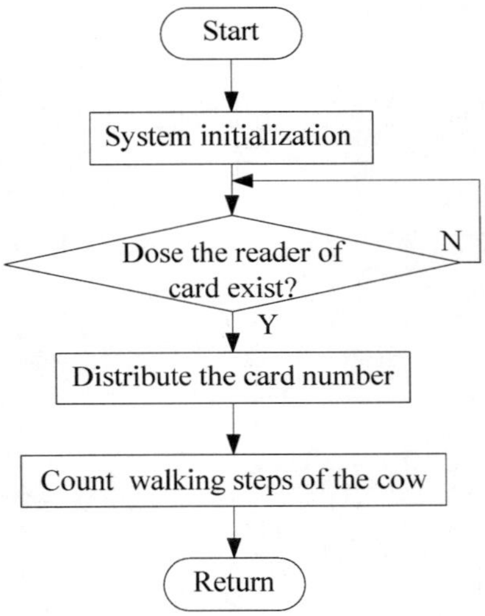


**Fig. 10.** Program flows of the RF card

## 6 Conclusion

After debugging, improving the circuit structure and optimizing the design parameters, the final reading can be improved to 98% (the time of recognition is less than 1 s). Because the RF card circuit uses low-power chips and 3V, 1000mA lithium battery-powered, the service life is more than one year. Figure 11 is the voltage waveforms of the actual measurement of both ends of reader coils. Figure 11(a) is the waveform of reader coils far from the RF Card. Figure 11(b) is the waveform of reader coils near the RF card (the square wave signal is added into the resonant circuit, so it is not an ordinary sine wave form). Obviously, the waveform amplitude of coil near a RF card is lower than one far from the RF card, and the lagging edge of the former becomes flatter than one of the latter. The differences between two waveforms are related to the distance of the coupling coils in the reader and the RF card. The waveforms of the amplified modulation-output signal are shown as Figure 12. Figure 12(a) is the waveform when reader is far from the RF card. Figure 12(b) is the waveform when the

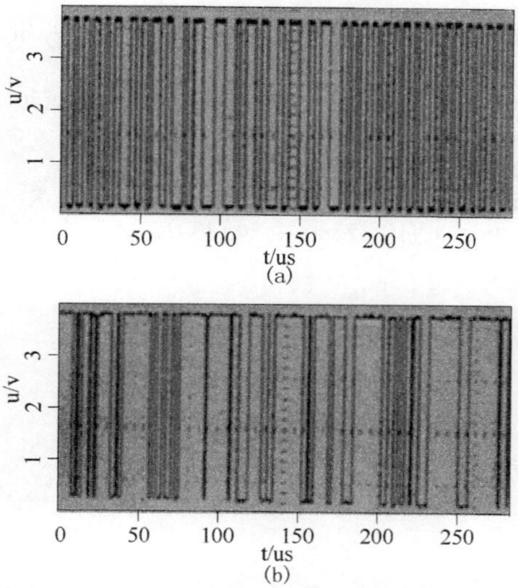


**Fig. 11.** The voltage waveforms of the actual measurement of both ends of reader coils:(a) reader coils far from the RF card, (b) reader coils near the RF card

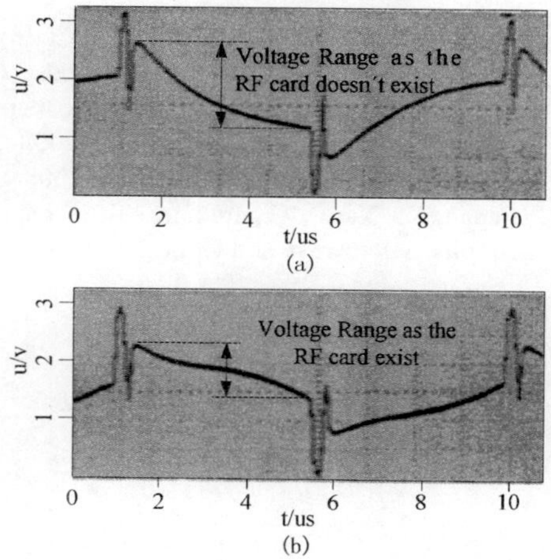


**Fig. 12.** The waveforms of the amplified modulation-output signal: (a)the RF card doesn´t exist, (b) the RF card exist

reader is near the card. By all appearance, the output waveform is unchanged in the absence of the RF card, the output waveform in the duty cycle appears large variation when coil is near the card, which shows that coded signal input.

Due to a special signaling mode, the system can read faster than normal magnetic cards and realize high-speed read. In addition, the use of standard manchester encoding make the accuracy and resolution of the reader raise significantly. If the card can be used in RF low-power chips and laid by the charging circuit, its market prospects will be even more significant.

## References

1. Griffin Joshua, D., Durgin Gregory, D., Andreas, H., Bernard, K.: How to construct a test bed for RFID antenna measurements. In: IEEE Antennas and Propagation Society International Symposium, APS 2006 Digest, pp. 457–460 (2006)
2. Murrill, J.: Reconfigurable modular RF interface. In: Autotestcon (Proceedings), pp. 267–271 (2007); Proceedings of 2006 IEEE Autotestcon - IEEE Systems Readiness Technology Conference (2006)
3. Valeri, A., Alexander, B., Stefan, C., Gevorg, P., Kay, R., Simrock Stefan, N., Petr, V.: Digital RF control system for the DESY FLASH linear accelerator. In: EUROCON 2007 - The International Conference on Computer as a Tool, pp. 1178–1185 (2007)
4. Canan, K., Yasar, G., Ibrahim, T.: A coplanar waveguide on-chip RF choke for wlan RF circuits. Microwave and Optical Technology Letters 49(10), 2530–2534 (2007)
5. Cao, N., Sun, Y.: Principle and Application of Radio Frequency Identification Card. Electronic Technology 46(1) (January 2009)

6. Shao, R., Yang, C.: Application of Wireless Radio Frequency Identification Technology in Mine Management. Afety in Coal Mines 37(11) (November 2009)
7. Abhilash, G., Madhavan, S.: A low-cost approach for testing embedded RF Passive circuits based on oscillation principle. In: 2008 Electrical Design of Advanced Packaging and Systems Symposium, IEEE EDAPS 2008 - Proceedings, pp. 194–197 (2008)
8. Canan, K., Yasar, G., Ibrahim, T.: A coplanar waveguide on-chip RF choke for wlan RF circuits. Microwave and Optical Technology Letters 49(10), 2530–2534 (2007)

# Theoretical Derivation of Terahertz Birefringence in Liquid Crystal

Xiaoguang Zhang[1], Guangda Yang[2], You Zhou[3], Xiusheng Yan[2], and Huai Yang[4]

[1] School of Materials Science and Engineering, University of Science and Technology Beijing / National Laboratory of Electro-, ptics System Technology, Beijing/Sanhe, China
[2] National Laboratory of Electro-Optics System Technology, Sanhe, China
[3] School of Automation, Beijing Institute of Technology, Beijing, China
[4] School of Materials Science and Engineering, University of Science and Technology Beijing, Beijing, China
copa9999@hotmail.com, copa9999@yahoo.com.cn, xzy21221@sina.com, yanxs686@163.com, anghuai@mater.ustb.edu.cn

**Abstract.** We discussed the use and the meaning of the study in Terahertz phase shifter field and presented the argumentation on birefringence in liquid crystals.

**Keywords:** terahertz, liquid crystal, phase shifter.

## 1 Introduction

The characters in terahertz wave band (0.1- 10 THz, 3000- 30 $\mu m$) was studied rarely until 1980s. With the development of ultrashort laser technology, the mechanism, testing technology and the applications of terahertz has undergone remarkable rapid growth.

From the view of long wave, the study on terahertz mainly depends on electronics, while from the view of short wave, the study on it bases on photon electronics. Consequently, a gap which is called "terahertz gap" appeared. The terahertz gap exhibits particularity which is of great difference with other electromagnetism wave lengths. Accordingly, the technology of terahertz imaging and time- domain spectroscopy have been playing key roles in fields of biomedicine, safety detection, industry lossless surveillance, space physics, environment detection, chemical analysis and network communication.[1]

Phase shifter is the hard core of the phased-array radar and the satellite communication. Most of the phased array for the moment is controlled by ferrite or diode phase shifter. With the manifold of radiation units, the phase shifter can hardly satisfy the development of the phased-array antenna: the lumpishness and the unresponsive of ferrite phase shifter caused difficulty of its adaptation, while diode phase shifter is of poor performance high cost. In order to meet the needs of the development, it is desiderate in studying high performance low cost phase shifter. For actual and further applications in terahertz, several groups have demonstrated devices based on optically or electrically controlled carrier concentration in quantum well structures [2-4]. However, the quantum well based terahertz phase shifters have not yet demonstrated 2-pi phase shift and need to be operated at temperatures far below room temperature.

Some groups have managed in the development of phase shifter with the application of liquid crystal birefringence, and they have accomplished phase control in microwave, millimeter wave and terahertz wave in room temperature [5.6]. Through the phase modulation of terahertz, the inmechanical control on the direction of terahertz wave transmission can be achieved. It is meaningful in the use of terahertz imaging and terahertz transmission. In this paper, we will discuss the mechanism of the application of liquid crystal birefringence in terahertz wave length.

## 2 Discussion of the Terahertz Birefringence in Liquid Crystal

A refracted light divides into two beams when a beam of parallel natural light incidents the surface of an anisotropic medium. One beam of light is called ordinary light (o light), the other beam is called extraordinary light (e light).

Terahertz wave has the same character when it incidents into anisotropic medium such as liquid crystal [7].

When a beam of terahertz wave incidents into the surface of liquid crystal, the relationship of electric displacement vector $\mathbf{D}$ and electric field intensity $\mathbf{E}$ is:

$$\mathbf{D} = \varepsilon\varepsilon_0 \mathbf{E} \tag{1}$$

In space right angle reference frame (in space rectangular coordinate system), (1) can be written as

$$\begin{bmatrix} D_x \\ D_y \\ D_z \end{bmatrix} = \varepsilon\varepsilon_0 \begin{bmatrix} E_x \\ E_y \\ E_z \end{bmatrix} \tag{2}$$

In equation (2), $\varepsilon = \begin{bmatrix} \varepsilon_{xx} & \varepsilon_{xy} & \varepsilon_{xz} \\ \varepsilon_{yx} & \varepsilon_{yy} & \varepsilon_{yz} \\ \varepsilon_{zx} & \varepsilon_{zy} & \varepsilon_{zz} \end{bmatrix}$ is a second-order covariant tensor.

If $0 < \varepsilon_{xx} < \varepsilon_{yy} < \varepsilon_{zz}$, (2) is diagonalized as

$$\begin{cases} D_x = \varepsilon_{xx}\varepsilon_0 E_x \\ D_y = \varepsilon_{yy}\varepsilon_0 E_y \\ D_z = \varepsilon_{zz}\varepsilon_0 E_z \end{cases} \tag{3}$$

The Maxwell equations is

$$\nabla \times (\nabla \times \mathbf{E}) + \frac{1}{C^2}\frac{\partial^2 \mathbf{E}}{\partial t^2} = -\frac{1}{C^2}\varepsilon\varepsilon_0 \frac{\partial^2 \mathbf{E}}{\partial t^2} \tag{4}$$

If $\frac{\partial}{\partial t} \rightarrow -i\omega, \nabla \rightarrow i\mathbf{K}$, then (4) is changed as

$$\mathbf{K}\times(\mathbf{K}\times\mathbf{E})+\frac{\omega^2}{\mathbf{C}^2}\mathbf{E}=-\frac{\omega^2}{\mathbf{C}^2}\varepsilon\varepsilon_0\mathbf{E} \tag{5}$$

(5) can be written as

$$\begin{cases} (-K_y^2-K_z^2+\frac{\omega^2}{C^2})E_x \\ +K_xK_y\mathbf{E}_y+K_xK_zE_z \\ =-\frac{\omega^2}{C^2}\varepsilon_{xx}\varepsilon_0 E_x \\ K_yK_xE_x+ \\ (-K_x^2-K_z^2+\frac{\omega^2}{C^2})E_y \\ +K_yK_zE_z \\ =-\frac{\omega^2}{C^2}\varepsilon_{yy}\varepsilon_0 E_y \\ K_zK_xE_x+K_zK_yE_y \\ +(-K_x^2-K_y^2+\frac{\omega^2}{C^2})E_z \\ =-\frac{\omega^2}{C^2}\varepsilon_{zz}\varepsilon_0 E_z \end{cases} \tag{6}$$

If the waves can be transmitted along any directions, $E_x$, $E_y$, $E_z$ have solutions. The determinant of the coefficients in (6)

$$\begin{vmatrix} -K_y^2-K_z^2+(\frac{\omega}{C}\sqrt{1+\varepsilon_{xx}\varepsilon_0})^2 & K_xK_y & K_xK_z \\ K_yK_x & -K_x^2-K_z^2+(\frac{\omega}{C}\sqrt{1+\varepsilon_{yy}\varepsilon_0})^2 & K_yK_z \\ K_zK_x & K_zK_y & -K_x^2-K_y^2+(\frac{\omega}{C}\sqrt{1+\varepsilon_{zz}\varepsilon_0})^2 \end{vmatrix}=0$$

In the xoy plane of coordinates, $K_z = 0$. The determinant is simplified as

$$\begin{aligned}&[-K_x^2 - K_y^2 + (\frac{\omega}{C}\sqrt{1+\varepsilon_{zz}\varepsilon_0})]\\&\{[-K_y^2 + (\frac{\omega}{C}\sqrt{1+\varepsilon_{xx}\varepsilon_0})^2]\\&[-K_x^2 + (\frac{\omega}{C}\sqrt{1+\varepsilon_{yy}\varepsilon_0})^2]\\&-K_x^2 K_y^2\} = 0\end{aligned} \tag{7}$$

If $\frac{\omega}{C}\sqrt{1+\varepsilon_{xx}\varepsilon_0} = K_1$, $\frac{\omega}{C}\sqrt{1+\varepsilon_{yy}\varepsilon_0} = K_2$, $\frac{\omega}{C}\sqrt{1+\varepsilon_{zz}\varepsilon_0} = K_3$, (7) can be written as

$$\begin{cases} K_x^2 + K_y^2 = K_3^2 \\ \dfrac{K_x^2}{K_2^2} + \dfrac{K_y^2}{K_1^2} = 1 \end{cases} \tag{8}$$

In a similar way, the project in yoz plane of coordinates is the circle as the radius of $K_1$ and the ellipse as the major semi-axis of $K_3$ and the minor semi-axis of $K_2$. The project in xoz plane of coordinates is the circle as the radius of $K_2$ and the ellipse as the major semi-axis of $K_3$ and the minor semi-axis of $K_1$. Figure 1 shows that the projects of waves in three planes of coordinates are circles and the ellipses. Choose the x, y, z coordinate axes according to $\varepsilon_{xx} < \varepsilon_{yy} < \varepsilon_{zz}$. In the xoz plane of coordinates, the two waves intersect at the point M in each quadrant. The common

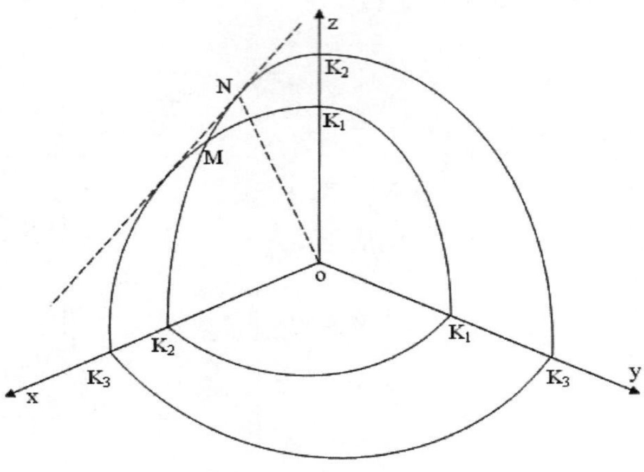


**Fig. 1.** Wave front of terahertz wave in liquid crystal ( $\varepsilon_{xx} < \varepsilon_{yy} < \varepsilon_{zz}$ )

tangent of circle and ellipse in xoz plant is showed in Figure 1 and the tangent point N is on the circle. ON is a light ax of liquid crystal. The wave has unique phase velocity when transmitted in this direction.

As was shown in figure 2, when $\varepsilon_{xx} = \varepsilon_{yy} \neq \varepsilon_{zz}$ , in xoz plant, the two wave front meets at a point M, there are two cross points. Terahertz wave has unique wave number and unique phase velocity in this direction, while there have two wave numbers and two phase velocity in any other direction. The two phase velocity response to the states of orthogonal polarization. So that, terahertz wave was formed by two separated waves that the orthogonal polarization and in different phase velocity. We can say that, terahertz wave has the character of birefringence when it incidents into liquid crystal.

In terahertz phase shifter, the effect of phase modulation depends on the scale of the birefringence of liquid crystal. So it is needed in choosing liquid crystals which have bigger birefringence. The effective depth of liquid is also considerable in phase modulation. the deeper the liquid crystal, the better the modulation shift. But the depth should be limited because that the depth increase results in the shorter of responding time.

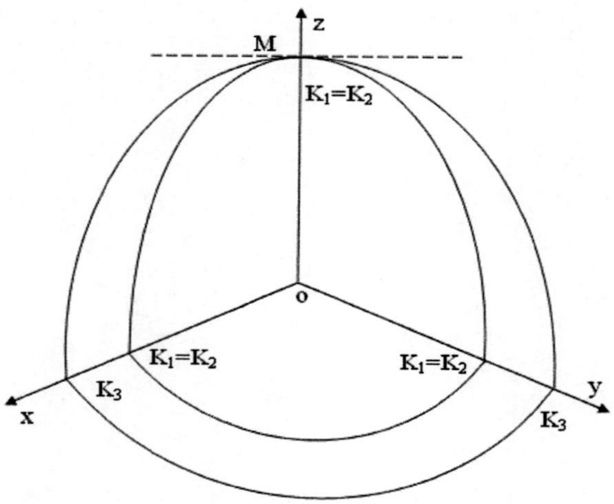


**Fig. 2.** Wave front of terahertz wave in liquid crystal ( $\varepsilon_{xx} < \varepsilon_{yy} \neq \varepsilon_{zz}$ )

## 3 Conclusions

In this paper, we proved terahertz birefringence in liquid crystal by theory equations. For better effect of phase modulation, it is needed to choose liquid crystals which have bigger birefringence, design better electrode structure, matching phase shifter and driving voltage. The issues studied above need further discussion.

## References

1. Zhang, X.C., Jin, Y., Ma, X.F.: Coherent measurement of THz optical rectification from electro-optic crystals. Appl. Phys. Lett. 61(23), 2764–2766 (1992)
2. Libon, I.H., Baumgärtner, S., Hempel, M., Hecker, N.E., Feldmann, J., Koch, M., Dawson, P.: An optically controllable terahertz filter. Appl. Phys. Lett. 76, 2821–2823 (2000)
3. Kersting, R., Strasser, G., Unterrainer, K.: Terahertz phase modulator. Electron. Lett. 36, 1156–1158 (2000)
4. Kleine Ostmann, T., Koch, M., Dawson, P.: Modulation THz radiation by semiconductor nanostructures. Microwave Opt. Tech. Lett. 35, 343–345 (2002)
5. Lim, K.C., Margerum, J.D., Lackner, A.M.: Liquid crystal millimeter wave electronic phase shifter. Appl. Phys. Lett. 62, 1065–1067 (1993)
6. Dolfi, D., Labeyrie, M., Joffre, P., Huignard, J.P.: Liquid crystal microwave phase shifter. Electron. Lett. 29, 926–928 (1993)
7. Chen, C.Y., Tsai, T.R., Pan, C.L., Pan, R.P.: Room temperature terahertz phase shifter based on magnetically controlled birefringence in liquid crystals. Applied Physics Letters 83(22), 4497–4499 (2003)

# Cognitive Radio Decision Engine Based on CMOPSO

Zhijun Teng[1,2], Zhiqian Li[1], Na Wang[1], Xiaoxia Li[1], and Lijuan Zhao[1]

[1] School of Information Engineering, Northeast Dianli University, Jilin City, Jilin Province, China
[2] Department of Information & Communication Engineering Harbin Engineering University, Harbin City, Heilongjiang Province, China
tengzhijun@163.com

**Abstract.** Cognitive radio decision engine based on CMOPSO is proposed. Through the introduction of external species, adaptive mutation, adaptive grid method, multi-objective particle swarm optimization is improved. Multi-objective functions to be optimized are derived. CMOPSO is used to adjust and optimize the radio parameters. MATLAB-Simulink simulation is used to analyze the performance of algorithm. The correctness and effectiveness of the module of multi-objective particle swarm optimization is shown.

**Keywords:** cognitive radio, CMOPSO, performance, optimization.

## 1 Introduction

Cognitive radio can optimize performance of radio according to the perceived environmental parameters by artificial intelligence technology. Optimization minimizes consumption of resources, such as minimum bandwidth, power consumption r while meeting users' performance. Parameters to be adjusted in physical layer (PHY) and media control layer (MAC) may include the center frequency, modulation scheme, symbol rate; transmit power, coding strategies and so on. This is a multi-parameter non-linear multi-objective optimization process.

Cognitive engine based on genetic algorithm (GA) is used to achieve basic functions of cognitive radio (CR) (see [1]). Multi-objective GA is used to adjust the radio parameters, verifying the effectiveness of GA (see [2], [3], [4]). Multi-objective optimization is converted to single-objective by a simple weighted method, using a simple GA to adjust parameters (see [5], [6]). This article uses the concept of multi-objective particle swarm optimization presented by Coello Coello (Coello Multi-Objective Particle Swarm Optimization, CMOPSO), proposes cognitive radio performance based on CMOPSO, introduces external species, adaptive mutation, adaptive grid, thereby improving the performance of the algorithm.

## 2 Particle Swarm Optimization

### 2.1 Standard Particle Swarm Optimization

In standard Particle Swarm Optimization (PSO), the speed and position of the particles in the search space are determined as follows:

$$\mathbf{v}_{t+1} = w\mathbf{v}_t + r_1 \cdot rand() + (\mathbf{P}_t - x_t) + r_2 \cdot rand() \cdot (\mathbf{G}_t - x_t) \quad (1)$$

$$\mathbf{x}_{t+1} = \mathbf{x}_t + \mathbf{v}_t \quad (2)$$

Where w is inertia weight; r1and r2 are speed constant; rand () are random number uniformly distributed in the interval [0, 1]; Pt and Gt are the best position of particle's own pbest and global best position gbest at the time of t respectively. pbest is particle's own best position, while the gbest particle corresponds to the global best position, which is the best position of the entire group. xt = (xt1, xt2, ..., xtm) and vt = (vt1, vt2, ..., vtm) are the position and speed at the time of t respectively.

## 2.2 CMOPSO

In Multi-Objective Particle Swarm Optimization presented by Coello Coello (MOPSO) [7], the external particle swarm is used to guide the flight of other particles out of the group. The adaptive grid method is used to maintain the external particle swarm. In order to distinguish the shorthand of multi-objective particle swarm optimization algorithm--MOPSO, the algorithm is abbreviated as CMOPSO.

CMOPSO steps are as follows:

- (1) Initialize the particle swarm P, determine its initial position and initial velocity for each particle, calculate target vector for each particle in the particle swarm P;
- (2) Store some particles in the particle swarm P into the external particle swarm NP. The position of these particles is non-inferior solution;
- (3) Determine initial their own best position for each particle, namely initial position for each particle;
- (4) Divide target space into many lattices (hypercube), and determine which each particle belongs to according to target vector;
- (5) Define the fitness value of the lattice which contains at least one external particle for each particle, and select a grid plate according to wheel gamble method for each particle, and randomly select a individual from external particle swarm as an particles' gbest;
- (6) Update all particles' position and velocity according to the basic formula of PSO;
- (7) Use the following measures to avoid particles flying out of the search space: Once the particle flies out of the boundaries of decision variable, the particle remains on the border, while changing the flight direction;
- (8) Calculate target vector for each particle of in the particle swarm P;
- (9) Update and safeguard external particle swarm NP using adaptive grid method;
- (10) Update particles' pbest. Comparing a new solution with its own best position pbest, if the new solution dominates pbest, then the new solution is the new pbest; otherwise, pbest remain unchanged;
- (11) If the termination conditions are true, then the search stops; otherwise, go to 6.

# 3 Cognitive Radio Optimization Based on Cmopso

The position information of m particles indicates the solution of optimization problem, namely each adjustable transmission parameters. With the additional flight speed for each particle, the position of particle can be changed. We can judge the pros and cons of performance of each particle through the priority and inferior of objective functions. Global optimal solution can be obtained by iteration of m particles. A set of objective functions represent multiple objectives to be optimized, each objective function representing a target. These targets change with the channel conditions, so CMOPSO need to dynamically select objective functions according to the channel conditions. Process is as follows:

- The word "data" is plural, not singular.
- Step 1 Sense the spectral environment, including the channel conditions, the user's service requirements and spectral regulations. If there are changes, go to step 2;
- Step 2 If the channel conditions change, re-select a group of fitness functions, otherwise, find out the Pareto-optimal set corresponding to the current channel conditions from a previously saved results, and go to step 4;
- Step 3 Add the spectral rules limit, implement multi-objective particle swarm optimization and obtain a Pareto-optimal set;
- Step 4 Choose a most satisfactory solution from this Pareto-optimal set according to users' service demand, and notify the cognitive radio to update their transmission parameters. Then turn to Step 1.

# 4 Simulation and Results

## 4.1 Experimental Framework

Simulation of this article refers to Bio-CR model simulation experiments [1], based on Matlab-Simulink in the IEEE802. 11a. CR performance optimization engine replaces the traditional adaptive controller, so CR can adjust transmission parameters according to wireless channel and users' service demand.

## 4.2 Experimental Steps

Headings, or heads, are organizational devices that guide the reader through your paper. There are two types: component heads and text heads.

- Step 1 Transmitter and receiver transmit the initial transmission parameters in a wireless channel;
- Step 2 Transmitter and receiver sense wireless channel to determine the channel type;
- Step 3 If the channel type has changed, select a learning device used to calculate BER according to the channel type, run multi-objective particle swarm optimization module to obtain a Pareto-optimal set, or else turn to step 2;

- Step 4 Select the most satisfactory solution from this Pareto-optimal set and use this solution to update the transmission parameters of transmitter and receiver. Go to step 2.

### 4.3 Experimental Analysis

Take dispersive fading channel as an example:

- (1) If service type is voice, we concern about network latency rather than the BER, so data rate is improved by multi-objective particle swarm optimization.
- (2) If service type is video, video is similar to voice, but more concerns about network latency than voice, so data rate is further enhanced, although the BER has also increased a bit.
- (3) Next, when the service type is e-mail, we concern about BER rather than the data rate, so data rate is reduced, while BER greatly reduced.
- (4) Finally, when the service type is a download, download is similar to Email, but Email more concerns about data rate, so data rate is enhanced while BER is very low.

## 5 Conclusions

This article regards cognitive radio performance optimization as a dynamic multi-objective optimization problem. Through the introduction of external species, adaptive mutation, adaptive grid method, multi-objective particle swarm optimization is improved. CMOPSO determines a set of fitness functions according to the channel conditions, runs the multi-objective particle swarm optimization algorithm to obtain a Pareto-optimal set, selects the most satisfactory solution according to users' service demand and notify the cognitive radio to update their transmission parameters. Matlab simulation proves the correctness and effectiveness of CMOPSO. The next work is: cross-level co-simulation of cognitive radio engine using OPNET network simulation.

**Acknowledgment.** The authors thank Wei Wang for supporting comments.

## References

1. Rieser, C.J.: Biologically Inspired Cognitive Radio Engine Model utilizing distributed genetic algorithms for secure and robust wire-less communications and networking. Dept. of Electrical Engineering in Virginia Tech., Blacksburg (2004)
2. Rondeau, T.W., Rieser, C.J., Bostian, C.W.: In: SDR Forum Technical Conference (2004)
3. Zhao, Z.J., Zheng, S.L., Xing, G.J., Shang, J.N.: Piezoelectrics and Acoustooptics 2990 (2007) (in Chinese); Jing, Z.Z., Chain, Z.: Xing International, Hisatoshi Na piezoelectric and acousto-optic 2990 (2007)
4. Rondeau, T.W., Le, B., Maldonado, D., Scaperoth, D., Bostian, C.W.: In: The First International Conference on Cognitive Radio Oriented Wireless Networks and Communication (2006)

5. Hauris, J.F.: In: IEEE International Symposium on Computational Intelligence in Robotics and Automation, vol. 427 (2007)
6. Newman, T.R., Barker, B.A., Wyglinski, A.M., Agah, A., Evans, J.B., Minden, G.J.: Wiley Wireless Communications and Mobile Computing 71129 (2007)
7. Coello Coello, C.A., Pulido, G.T., Lechuga, M.S.: Handling multiple objectives with particle swarm optimization. IEEE Transactions on Evolutionary Computation 8(3), 256–279 (2004)

# Combination Method of Rules and Statistics for Abbreviation and Its Full Name Recognition

Yang Hua[1], Hong Yu[1], Hua Zhenwei[2], Yao Jianmin[1], Zhao Mingming[1], and Feng Yanhui[1]

[1] Provincial Key Laboratory of Computer Information Processing Technology, Soochow University, Suzhou 215006, China, Suzhou
[2] Suzhou Municipal Science and Technology Bureau, Suzhou 215002, China, Suzhou
{20084227065085,20084227065101,jyao}@suda.edu.cn, tianxianer@gmail.com, nckj@szkj.gov.cn

**Abstract.** This paper proposes a Combination of rules and statistics algorithm. Firstly, this research uses the rule-based approach to identify the abbreviation. Secondly, the full name candidates of the abbreviation are recognized based on n-gram feature. Thirdly, we use the rule and statistic based algorithm to identify the best candidate for the abbreviation. The method of abbreviation recognition has achieved a high accuracy rate, and it is independent, portable and efficient. The method of the full name recognition is superior to the approach introduced in related work on the basis of analyzing and comparing experiment results.

**Keywords:** Biomedical information extraction, Abbreviation recognition, Abbreviation full name recognition, Text mining, Combining method of rules and statistics.

## 1 Introduction

Abbreviation of bio-medicine is the language structure which is shorted from a phrase or nouns of full name. It is a relative term to non-shorten language structure and the meaning of it is equal to the full name. They are two forms of abbreviation, acronym-type abbreviations and non-acronym-type abbreviations. For instance, acronym-type abbreviation: the full name of "AMI" is "acute myocardial infarction", non-acronym-type abbreviation: the full name of "AW" is "water activity", and so on. Our task is to identify <abbreviation, full name>[1] from the bio-medicine text.

## 2 Related Work

AFP (Acronym Finding Program)[2] is the earlier research on the recognition of abbreviation and its full name. In the AFP, they set the length of the full name from 3 to 10. However, the AFP can not recognize the abbreviations which lengths are 2, such as AI (Artificial Intelligence), DB (Database) and so on. Recently there are a variety of text mining methods which are used in the recognition of abbreviation and its full name.

Pustejovsky J, Castano J, Cochran B [1] recognized the abbreviation and its full name by hand-built regular expressions and syntactic information. Ao and Takagi [3] build a ALICE system, their algorithm is based on heuristic pattern-match rules. Larkey LS, Ogilvie P, Price MA, Tamilio B [4], Yu H, Hripcsak G, Friedman C [5], Schwartz As and Hearst MA[6] respectively put forward their own pattern-matching rules. The method of Schwartz As and Hearst MA [6] is simple, easy to implement and has a relatively high accuracy rate of up to 95.8%. However, their approach has some drawbacks. It cannot recognize the mirror abbreviations and non-acronym-type abbreviations such as CNSI (cyclophilin seven suppressor).

Wei Zhou, Vetle I. Torvik, Neil R. Smalheiser [7] analysis the form of the pair of <abbreviation, full name> in MEDLINE, and build a abbreviation database which is called ADAM. Okazaki N and Ananiadou S [8] build a abbreviation dictionary on the basis of MEDLINE. Statistically based approach can recognize the abbreviation both acronym-type and non-acronym-type. However, they are affected by the capacity of MEDLINE. If we can not find the abbreviation in the database, we can not use this statistically based approach. In the experiment we find that the statistically based approach can not recognize the abbreviations which have many full names, and the frequencies of these full names are similar.

Chang JT, Schutze H, Altman RB [9] proposed a supervised machine learning approach to recognize the abbreviation and its full name, the precision rate is 76.5%. It is well-known the supervised machine learning method requires a lot of labor and time, and the method have a lower precision rate. Therefore, avoiding machine learning approach, this paper addresses some of the existing problems and proposes a method which based on the rules and statistics to recognize the abbreviation and its full name.

The contents of this paper are as follows: in the 3th section abbreviation recognition and the candidates of full name will be discussed; in section 4 the recognition of the best full name; in section 5, experimental design will be presented in detail and the outcomes will be delivered and analyzed; at last, there is a conclusion of this article.

# 3 The Recognition of Abbreviation and the Candidates of Its Full Name

In the process of recognition, we use a hypothesis proposed by Gale W.A., Church K.W., Yarowsky D [11] that an abbreviation only has one sense in a discourse. This hypothesis made the process of recognition become simple and effective. In this section, we mainly solve the problem of the abbreviation recognition and the extraction of candidates. The first step of the recognition is to indentify abbreviation from text.

## 3.1 The Abbreviation Recognition Based on Rule

In this paper, a more simple and complete method is proposed. We improve the rule-based recognition algorithm which was proposed by Park and Byrd [10]. Their rule set was faultiness. Some words they recognized were not the abbreviation, and some abbreviations were not been recognized. We increase the rule set, and improve the accuracy rate. The set is showed in the table below:

**Table 1.** The rule set about the English abbreviation

| Rules Type | The contents and description |
|---|---|
| First Character Type | the numbers or letters |
| Length | between 2 to 10 |
| Match Case | at least one uppercase letter |
| Dictionary | Abbreviation doesn't appear in the dictionary |
| Separating Character | Abbreviation doesn't contain (, ), [, and so on |
| Unit of Measure | Abbreviation is not the number and quantity unit. Such as12Kg. |
| Special Character | Abbreviation is not the number and quantity unit. Such as+,=. |

In the table 1, the first four categories of the rule are the rules in Park and Byrd [10], the last three are the increased rules. In this paper, we improve the Park's algorithm and increase the rule set and improve the correct rate greatly.

## 3.2 The Extraction of the Full Name Candidate

Firstly, we define the search space. The sentence where the abbreviations and their full names appear is divided into two fragments as the search space by the abbreviation.

Secondly, we define the length of the full name of the abbreviation. We set the maximum length of the full name of abbreviation is as follows:

$$Max.|DEF| = \min(|ABB|+5, |ABB|*2) \tag{1}$$

In the equation (1), $Max.|DEF|$ is the maximum length of full name of the abbreviation, $|ABB|$ is the length of abbreviation.

Thirdly, we define the size of n (n-gram). The size of n is the maximum length of search space. The maximum length of the search space is $Max.|DEF|$, if the length of search space is greater than $Max.|DEF|$, else the maximum length of the search space is the search space.

Take the sentence of "we investigate the effect of thyroid transcription factor 1" as an example, the search space is "we investigate the effect of thyroid transcription factor 1", the size of n is 8(8-gram). We extract candidates from the search space on the basis of n-gram (n=8). On the basis of 1-gram, the candidates contain "we", "investigate", "the" and so on. On the basis of 2-gram, the candidates were "we investigate", "investigate the", "the effect" and so on.

## 3.3 The Candidates Filtering of the Full Name of Abbreviation

In Abbreviation, the first and last word can not be prepositions, auxiliaries, conjunctions, modal verbs or pronouns and so on[12]. Therefore, we filter these candidates whose first word and last word is in the list of stop words.

As narrated above, using the "TTF-1" as an example, we will filter these candidates which their first or last word is "we", "of", "the".

## 4 The Recognition of the Best Full Name

In order to identify the best candidates of full name of abbreviation, this paper propose a comprehensive and effective method, as the equation (2).

$$S(A,F_C)_i = \alpha * apr_i + (1-\alpha)sr_i \tag{2}$$

$apr_i$ is the value of the $i^{th}$ candidate by statistic-based method, $sr_i$ is the weight of the $i^{th}$ candidate on the basis of rule set, $\alpha$ is a variable parameter, $S(A,F_C)_i$ is the weight of the $i^{th}$ candidate. In the experiment, we adjust the variable parameter $\alpha$ to gain the best result. At last, we set the parameter value 0.35. We chose the best full name on the basis of $S(A,F_C)_i$.

### 4.1 Using the Rrule-Based Method to Realize the Value of $sr_i$

In this part, we use our own rule-based method to gain the value of $sr_i$:

$$sr_i = \frac{\mathrm{F}_{\text{Augment}}}{N_{LD} + N_{SW}} \tag{3}$$

Words Filter: we set the augmentable value of $\mathrm{F}_{\text{Augment}}$ by computing the number of the characters in the abbreviation which are matched with the character in the candidate.

$$\mathrm{F}_{\text{Augment}} = \frac{\lambda * N_F + (1-\lambda) * N_{NF}}{|A|} \tag{4}$$

In this formula, $N_F$ to show that the number of characters in the abbreviation which can be matched to the first character of the words in the candidates, $N_{NF}$ is the number of characters in the abbreviation which can not be matched to the first character of the word in the candidates but can be matched to the internal characters, $|A|$ is the length of abbreviation, and the value of $\lambda$ is set as 0.8 [12].

Character length difference: As been shown in formula (5), $N_{LD}$ is the Subtraction between the number of characters in abbreviation and the number of characters in the candidate. $|A|$ is the same as the above, $|F_C|$ is the number of characters in the candidate of full name.

$$N_{LD} = \left| |A| - |F_C| \right| \quad (5)$$

Number of stop words: $N_{SW}$ is the number of stop words in the candidate. In fact, the more stop words in the abbreviation, the less possibility the candidate is the true full name. So we put this rule to reduce weight.

### 4.2 Using the Statistic-Based Method to Realize the Value of $apr_i$

In this part we use the statistic-based method to compute the value of $apr_i$. The method is similar to the algorithm proposed by Wei Zhou, Vetle I. Torvik, Neil R. Smalheiser [7].

Take "TTF-1" as an example, Schwartz AS, Hearst MA [6] and Wen-Hsiang Lu, Jiun-Hung Lin, and Yao-Sheng Chang [12] could not recognize its full name. We send the word of TTF-1 to PubMed. We will gain 687 abstracts. The number of abstracts which contain the fragment "1(TTF-1)" is 523, the number of abstracts which contain the fragment "factor 1(TTF-1)" is 519. It is an apparent exception that the number drops significantly from "thyroid transcription factor 1(TTF-1)" (the number is 517) to "of thyroid transcription factor 1(TTF-1)" (the number is 18), the stupendous difference is our breakthrough point. The whole process is formalized as follows:

Firstly, we send the abbreviation to the PubMed, extract all of the abstracts, and computer the number of the abstracts which contain the candidate.

Secondly, we calculate the $apr_i$, $apr_i$ is optimized from $pr_i$ to recognize the abbreviation full name as Wei Zhou, Vetle I. Torvik, Neil R. Smalheiser [7]:

$$apr_i = pr_i - 2 * \sqrt{\frac{pr_i * (1 - pr_i)}{count\left["w_{i-1}...w_2 w_1" and "abb"\right]}}, \quad i \geq 1 \quad (6)$$

$$pr_i = \frac{count\left["w_i w_{i-1}...w_2 w_1" and "abb"\right] - 1}{count\left["w_{i-1}...w_2 w_1" and "abb"\right]}, \quad i \geq 1 \quad (7)$$

Where $count\left["w_i w_{i-1}...w_2 w_1" and "abb"\right]$ represent the number of the fragment $w_i w_{i-1}...w_2 w_1$ occurs in the abstracts.

Thirdly, we reset the value of $apr_i$. In the formula (8), if $count\left["w_1" and "abb"\right]$ is less than 10, this represents that there are no sufficient data to be used for counting abstracts, we can not gain accurate result, so the value of $apr_i$ is 0. If the database has enough abstracts, when the $apr_i \geq 0.05$ and $apr_{i+1} < 0.05$, we remain the value of $apr_i$ unchanged, in the other cases, the value of $apr_i$ is 0.

$$apr_i = \begin{cases} \begin{cases} apr_i & \begin{cases} apr_i \geq 0.05 \\ apr_{i+1} < 0.05 \end{cases} \\ 0 & else \end{cases} & count["w_1"and"abb"] \geq 10 \\ 0 & count["w_1"and"abb"] < 10 \end{cases} \quad (8)$$

In the experiment, by statistic-based method $apr_4$ is 0.998 and $apr_5$ is 0.0172. So we obtain that the value of $apr_4$ is 0.988, the other $apr$ is 0 in the end. Then combining the rule-based method, we can recognize the full name of "TTF-1", the full name of "TTF-1" is "thyroid transcription factor 1".

## 5 Experiments and Results

The experiment consists of two parts: the abbreviation recognition and the full name recognition, and the full name of abbreviation recognition.

Corpus: the corpus of abbreviation recognition is the 100 abstracts downloaded from MedlinePlus. MedlinePlus is a database of authoritative, up-to-date, and extensive consumer health information compiled by the National Library of Medicine. The 100 abstracts contain 21577 words and 1383 abbreviations. The corpus of the full name recognition is the Medstract Gold Standard Corpus which is publicly available tagged corpus. It is composed of the development corpus with 126 tagged pairs of <abbreviation, long form> and the evaluation corpus with 168 tagged pairs. The development corpus has been used for determining the value of $\alpha$, and the evaluation corpus is used for comparing with related works.

Evaluation: we use precision and recall that are commonly used in the field to evaluate the abbreviation and full name recognition, but in the full name recognition we also use the harmonic mean (F-measure).

### 5.1 The Abbreviation Recognition Experiment

Firstly, we use our corpus to test the system proposed by Park. The result has been shown in Table 2. We can see from the result, they also regard the chemical formula and quantifier, some phrase which contains separating character and so on as the abbreviation, such as "NH(4)(+)", "Ar-O-CH(2)", "12kg", "P<0"and so on. At the same time there are 44 abbreviations have not been recognized, and most of them can be find in the dictionary, they are ambiguous abbreviations. By the above analysis, this research increases the chemical formula, quantifier and separating character rules into the rule set. To recognize the ambiguous abbreviations, we remove the words in the dictionary. On the basis of our method this is very effective. The result is showed as follows:

**Table 2.** The result of abbreviation recognition

| Method | Precision | Recall |
|---|---|---|
| Park et al. | 98.2% | 95.1% |
| Our method | 99.9% | 99.6% |

By analyzing the result, out method regard “Fat” (it occurs one time), “All” (it occurs two times), “Rib” (it occurs one time) as abbreviation. It is because they are all ambiguous abbreviations and all appear at the first of the sentence, and are to comply with the rule standard. We also can not recognize the “IV”, the word can be find from dictionary because it is four in Latin.

## 5.2 The Full Name Recognition Experiment

This paper make the rule-based method as baseline to compare with the rule-base method proposed by Schwartz AS, Hearst MA [6] and ALICE algorithm proposed by Ao and Takagi[3], a supervised machine learning approach proposed by Chang JT, Schutze H, Altman RB [9] using the evaluation corpus. The result is showed as follows:

**Table 3.** The result of full name recognition

| Method | Precision | Recall | F-measure |
|---|---|---|---|
| Chang(score=0.03) | 76.5% | 80.1% | 78.3% |
| S&H | 95.8% | 81.5% | 88.1% |
| ALICE | 82.7% | 76.5% | 79.5% |
| Rule-base method | 88.7% | 89.2% | 88.9% |

From the table we can see that the S&H algorithm has a high precision compare with our method. It is because they have a low recall. They identify 143 abbreviations but 137 abbreviations are correct. In our experiment we recognize 169 abbreviations, and one of them is not the abbreviation. The reason is that we regard the phrase “Fas” as abbreviation. In fact, it is the full name of “CD95”. In our result, there are 150 abbreviations have the correct full name. This research improves the recall, because we improve the abbreviation recognition method.

By analyzing the result, the abbreviations which have the wrong full name contain the non-acronym-type abbreviations and partial matching abbreviations, Such as <bHLH, basic helix-loop-helix>, <5-HT, serotonin> and so on. The statistic-based method is very effective for recognizing the non-acronym-type abbreviations. When there are no enough data for statistic-based method, the rule-based method will come into play. Then we combine the rule-based method and statistic-based method. The result of combination method can be seen as follows:

**Table 4.** The result of combination method

| | |
|---|---|
| Precision | 97.0% |
| Recall | 97.6% |
| F-measure | 97.3% |

In the Table 4, we can see that our method play a role. Only four full names can not be recognized, the abbreviations are “Pax6”, “Ci”, “Fu”, “O-glycans”. Firstly, they are all non-acronym-type abbreviations, the rule-based method failures for them. Secondly, they all have low frequency in the database, so all of the value of $apr$ is zero. This problem may be solved by increasing the database.

## 6 Conclusions

In this paper, a combination of statistical and rule algorithm has been proposed to recognize abbreviations and their full name. The approach significantly improves accuracy. This research is recognizing the full name of abbreviation in text. Sometimes there is only the abbreviation in the text, because the full name and abbreviations are not bound to appear simultaneously in literature. When we identify the full name, there is no corresponding text, so the recognition is wrong. In the future work, recognition of the abbreviation whose full name is not in the text can be considered in order to enhance accuracy.

**Acknowledgment.** The work is supported by the National Natural Science Foundation of China under Grant No 60970057.

## References

1. Pustejovsky, J., Castano, J., Cochran, B.: Automatic extraction of acronym-meaning pairs from medline databases. Stud. Health Technol. Inform. 10(1), 371–375 (2001)
2. Taghva, K., Gilbreth, J.: Recognizing Acronym and their Definitions, Technical Report 95-03, Information Science Research Institute, University of Nevada, Las Vegas (June 1995)
3. Ao, H., Takagi, T.: Alice: An Algorithm to Extract Abbreviations from MEDLINE. J. AM. Med. Inform. Assoc. 12, 576–586 (2005)
4. Larkey, L.S., Ogilvie, P., Price, M.A., Tamilio, B.: Acrophile: an automated acronym extractor and server. In: Nurnberg, P.J., Hicks, D.L., Furuta, R. (eds.) Proceedings of the 5th ACM International Conference on Digital Libraries, SanAntonio, June 02-07, pp. 205–214. ACM Press (2000)
5. Yu, H., Hripcsak, G., Friedman, C.: Mapping abbreviations to full forms in biomedical articles. J. AM. Med. Inform. Assoc. 9, 262–272 (2002)
6. Schwartz, A.S., Hearst, M.A.: A simple algorithm for identifying abbreviation definitions in biomedical text. In: Pacific Symposium on Biocomputing, pp. 451–462 (2003)
7. Zhou, W., Torvik, V.I., Smalheiser, N.R.: ADAM: another database of abbreviations in MEDLINE. Bioinformatics, 2813–2818 (October 2006)
8. Okazaki, N., Ananiadou, S.: A general method applicable to the search for simi-larities in the amino acid sequence of two proteins. J. Mol. Biol. 48, 443–453 (1970)
9. Chang, J.T., Schutze, H., Altman, R.B.: Creating an Online Dictionary of Abbreviations from MEDLINE. J. AM. Med. Inform. Assoc. 9, 612–620 (2002)
10. Park, Y., Byrd, R.J.: Hybrid Text Mining for Finding Abbreviations and Their Definitions. In: Lee, L., Harman, D. (eds.) Proceedings of the 6th Conference on Empirical Methods in Natural Language Processing, Pittsburgh, June 03-04, pp. 126–133. Association for Computational Linguistics Press (2001)
11. Gale, W.A., Church, K.W., Yarowsky, D.: One Sense Per Discourse. In: Proceedings of the ARPA Workshop on Speech and Natural Language Processing, pp. 233–237 (1992)
12. Lu, W.-H., Lin, J.-H., Chang, Y.-S.: Improving Translation of Queries with Infrequent Unknown Abbreviations and Proper Names. In: The Association for Computational Linguistics and Chinese Language Processing, pp. 7–9 (2008)

# Author Index